Lecture Notes in Computer Science 4619

Commenced Publication in 1973
Founding and Former Series Editors:
Gerhard Goos, Juris Hartmanis, and Jan van Leeuwen

Frank Dehne Jörg-Rüdiger Sack
Norbert Zeh (Eds.)

Algorithms and Data Structures

10th International Workshop, WADS 2007
Halifax, Canada, August 15-17, 2007
Proceedings

Volume Editors

Frank Dehne
Carleton University
Ottawa, Canada
E-mail: frank@dehne.net

Jörg-Rüdiger Sack
Carleton University
Ottawa, Canada
E-mail: sack@scs.carleton.ca

Norbert Zeh
Dalhousie University
Halifax, Canada
E-mail: nzeh@cs.dal.ca

Library of Congress Control Number: 2007931625

CR Subject Classification (1998): F.2, E.1, G.2, I.3.5, G.1

LNCS Sublibrary: SL 1 – Theoretical Computer Science and General Issues

ISSN 0302-9743
ISBN-10 3-540-73948-3 Springer Berlin Heidelberg New York
ISBN-13 978-3-540-73948-7 Springer Berlin Heidelberg New York

Springer is a part of Springer Science+Business Media

springer.com

Printed in Germany

Typesetting: Camera-ready by author, data conversion by Scientific Publishing Services, Chennai, India
Printed on acid-free paper SPIN: 12100453 06/3180 5 4 3 2 1 0

Preface

The papers in this volume were presented at the 10th Workshop on Algorithms and Data Structures (WADS 2005). The workshop took place August 15 - 17, 2007, at Dalhousie University, Halifax, Canada. The workshop alternates with the Scandinavian Workshop on Algorithm Theory (SWAT), continuing the tradition of SWAT and WADS starting with SWAT 1988 and WADS 1989. From 142 submissions, the Program Committee selected 54 papers for presentation at the workshop. In addition, invited lectures were given by the following distinguished researchers: Jeff Erickson (University of Illinois at Urbana-Champaign) and Mike Langston (University of Tennessee).

On behalf of the Program Committee, we would like to express our sincere appreciation to the many persons whose effort contributed to making WADS 2007 a success. These include the invited speakers, members of the Steering and Program Committees, the authors who submitted papers, and the many referees who assisted the Program Committee. We are indebted to Gerardo Reynaga for installing and modifying the submission software, maintaining the submission server and interacting with authors as well as for helping with the preparation of the program.

August 2007

Frank Dehne
Jörg-Rüdiger Sack
Norbert Zeh

WADS Organization

Organizing Institutions

Steering Committee

Frank Dehne	Carleton University, Canada
Ian Munro	University of Waterloo, Canada
Jörg-Rüdiger Sack	Carleton University, Canada
Roberto Tamassia	Brown University, Canada

Program Co-Chairs

Frank Dehne	Carleton University, Canada
Jörg-Rüdiger Sack	Carleton University, Canada
Norbert Zeh	Dalhousie University, Canada

Conference Chair

Norbert Zeh	Dalhousie University, Canada

Sponsor

Program Committee

Susanne Albers	University of Freiburg
Alberto Apostolico	University of Padova
Tetsuo Asano	JAIST
Mike Atallah	Purdue University
Mark de Berg	University of Eindhoven
Gerth Brodal	University of Aarhus
Timothy Chan	University of Waterloo

Danny Ziyi Chen	University of Notre Dame
Tom Cormen	Dartmouth College
Frank Dehne	Carleton University
Camil Demetrescu	University of Rome "La Sapienza"
David Eppstein	University of California at Irvine
Susanne Hambrusch	Purdue University
Rolf Klein	University of Bonn
Mike Langston	University of Tennessee
Andrzej Lingas	Lund University
Joe Mitchell	SUNY Stonybrook
David Mount	University of Maryland
Andrew Rau-Chaplin	Dalhousie University
Arnold Rosenberg	University of Massachusetts
Jörg-Rüdiger Sack	Carleton University
Roberto Tamassia	Brown University
Norbert Zeh	Dalhousie University

Referees

Mohammad Abam
Luca Allulli
Spyros Angelopoulos
Yossi Azar
Florian Berger
Daniel Bienstock
Henrik Blunck
Vincenzo Bonifaci
Prosenjit Bose
Sergio Cabello
Amitabh Chaudhary
Minkyoung Cho
Artur Czumaj
Emilio Di Giacomo
Florian Diedrich
Martin Dietzfelbinger
John Eblen
Leah Epstein
Jeff Erickson
Tomas Feder
Uri Feige
Jeremy Fineman
Guilherme Fonseca
Sorelle Friedler
Leszek Gasieniec
Chris Gray
Allan Grønlund Jørgensen
Ansgar Grüne
Joachim Gudmundsson
Xin Han
Jason Hartline
Refael Hassin
Mathias Hauptmann
Herman Haverkort
Gorel Hedin
Thore Husfeldt
Piotr Indyk
Riko Jacob
Tom Kamphans
Haim Kaplan
Menelaos Karavelas
Juha Kärkkäinen
Matthew Katz
Iordanis Kerenidis
Martin Köhler
Darek Kowalski
Miroslaw Kowaluk
Elmar Langetepe
Luigi Laura
Christos Levcopoulos
David Manlove
Conrado Martinez

Nargess Memarsadeghi
Matus Mihalak
Gabriel Moruz
Elena Mumford
Seffi Naor
Zeev Netov
Bengt Nilsson
Shawn T. O'Neil
Eli Packer
Rasmus Pagh
Christian N.S. Pedersen
Mia Persson
Andrea Ribichini
Gary Rogers
Valentin Polishchuk
Adi Rosen
Piotr Sankowski
Ulrich M. Schwarz
Michiel Smid
Shakhar Smorodinsky
Daniel Spielman
Tami Tamir
Ralf Thöle
Kasturi Varadarajan
Gert Vegter
Martin Wahlen
Haitao Wang
Barry Wittman
Neal Young
Hamid Zarrabi-Zadeh
Yun Zhang

Table of Contents

Session 1

Session 2A

Session 2B

Session 3A

Session 3B

Session 4A

Session 4B

Session 5

Session 6A

Session 6B

Session 7A

Session 7B

Session 8A

Session 8B

Session 9A

Session 9B

Session 10A

Session 10B

Session 11A

Session 11B

Finding Small Holes
A Brief Foray into Computational Topology

Jeff Erickson

University of Illinois, Urbana-Champaign
jeffe@cs.uiuc.edu
http://www.cs.uiuc.edu/~jeffe

Numerous applications call for the detection of small topological features in various spaces; examples include simplification of surfaces reconstructed from point clouds, efficient algorithms for graphs embedded on surfaces, coverage analysis for ad-hoc/sensor networks, and topological analysis of high-dimensional data. This talk is a survey algorithms for one of the simplest problems of this type: finding the shortest cycle in a given topological space that cannot be continuously contracted to a point. Spaces of interest include polygons with holes, combinatorial surfaces, piecewise-linear 2-manifolds, Rips-Vietoris complexes, and general simplicial complexes. Almost no optimal algorithms are known, even in settings where the problem has a straightforward polynomial-time solution; consequently, the talk will include several open problems. No prior knowledge of topology will be assumed.

F. Dehne, J.-R. Sack, and N. Zeh (Eds.): WADS 2007, LNCS 4619, p. 1, 2007.

Approximate Range Searching: The Absolute Model*

Guilherme D. da Fonseca

Department of Computer Science
University of Maryland
College Park, Maryland 20742
fonseca@cs.umd.edu

Abstract. Range searching is a well known problem in the area of geometric data structures. We consider this problem in the context of approximation, where an approximation parameter $\varepsilon > 0$ is provided. Most prior work on this problem has focused on the case of relative errors, where each range shape R is bounded, and points within distance $\varepsilon \cdot diam(R)$ of the range's boundary may or may not be included. We consider a different approximation model, called the *absolute model*, in which points within distance ε of the range's boundary may or may not be included, regardless of the diameter of the range. We consider range spaces consisting of halfspaces, Euclidean balls, simplices, axis-aligned rectangles, and general convex bodies. We consider a variety of problem formulations, including range searching under general commutative semigroups, idempotent semigroups, groups, and range emptiness. We show how idempotence can be used to improve not only approximate, but also exact halfspace range searching. Our data structures are much simpler than both their exact and relative model counterparts, and so are amenable to efficient implementation.

1 Introduction

The *range searching problem* involves preprocessing a set P of n points in $\mathbb{R}^d$ so that given a region R, that is drawn from a predefined set of shapes $\mathcal{R}$, a predefined function $f(P \cap R)$ can be computed efficiently. The set of possible shapes $\mathcal{R}$ characterizes the *range space*, and the elements of $\mathcal{R}$ are called *ranges*. Examples of range spaces include halfspaces, axis-aligned rectangles, and simplices. We let $q(R) = f(P \cap R)$ denote the result of the *query*. The points $p \in P$ are called *data points*.

Range searching is a well-studied problem in computational geometry. Excellent surveys have been written by Matoušek [17] and Agarwal and Erickson [1]. The two most common examples include *range counting*, where $f(Q) = |Q|$ and *range reporting*, where $f(Q) = Q$.

* The work of the author has been supported in part by CAPES, Brazil, under grant BEX-1319027. The author would like to thank his advisor David M. Mount.

F. Dehne, J.-R. Sack, and N. Zeh (Eds.): WADS 2007, LNCS 4619, pp. 2–14, 2007.

More generally, we have a commutative semigroup $(\mathbf{S}, +)$, a *weight* function $w : P \rightarrow \mathbf{S}$, and want to compute $f(Q) = \sum_{p \in Q} w(p)$. This general version of the problem is called the *semigroup version*. In the *group* version of the problem, $(\mathbf{S}, +)$ is a commutative (Abelian) group. The group version may admit more efficient solutions, because both addition and subtraction may be used to compute the answer to the query. In the *idempotent* version of the problem, $(\mathbf{S}, +)$ is an idempotent group, i.e. $x + x = x$ for all $x \in \mathbf{S}$. One example of idempotent semigroup is $(\mathbb{R}, \max)$. The *emptiness version* can be modeled by the idempotent semigroup $(\{0, 1\}, \vee)$ and assigning $w(p) = 1$ for all $p \in P$.

The relatively high complexity of exact range searching has led researchers to consider the problem in the context of approximation. A natural way to do this is to consider the range shape to be "fuzzy," and allow points that are close to the range boundary to either be counted or not. There are two natural ways to define approximation. In both cases a user-supplied approximation parameter $\varepsilon > 0$ is given, either at preprocessing time or at query time. In the *relative error model* (or simply *relative model* for short) it is assumed that the range shape R is bounded, and points lying within distance $\varepsilon \cdot diam(R)$ of the boundary of the range may or may not be included. In contrast, in the *absolute error model* (or simply *absolute model*) points lying within distance ε of the range's boundary may or may not be included.

Note that, in the absolute model, some type of scaling is needed, for otherwise it would be possible to answer queries with arbitrarily high precision by applying some high scale factor to the point and range coordinates, while keeping the error parameter fixed. Without loss of generality, we assume throughout that the point set P has been transformed (through a uniform scaling and translation) to lie within the unit hypercube $[0, 1]^d$. We assume that the ranges have been similarly transformed, and the parameter ε has been scaled correspondingly.

Approximate range searching in the relative model is studied in [6, 3, 4, 5]. Chazelle, Liu and Magen [12] studied approximate range searching in the absolute model, but considered the problem in spaces of high dimension. They presented a data structure which answers halfspace queries in $O((d/\varepsilon)^2 \log^{O(1)}(d/\varepsilon))$ time with $dn^{O(1/\varepsilon^2)}$ storage. Throughout this paper, we assume that the dimension d is constant.

There are a number of reasons for studying approximate range searching in the absolute model. First, the absolute model admits much simpler solutions. While the most efficient data structures for answering range queries both in the relative model and exact case tend to be quite complex, our techniques are extremely simple (involving simple structures such as grids and quadtrees) and so are amenable to efficient implementation. In the absolute model, the data structures are not sensitive to the point distribution. Therefore, the absolute model allows us to reason about the range searching problem (for example, the best size and shape of the generators) in a simpler context, and may lead to more efficient and simpler structures for exact range searching.

Second, the absolute model is better suited for several applications, when compared to the relative model. If the coordinates of a point represent an object that exists within some extent in space, or data that is subject to measurement errors or noise, then the approximation quality should be based on the expected error of the point locations, not on the diameter of the query range. A shortcoming of the relative model is that it cannot directly handle unbounded ranges, such as halfspaces and unbounded polyhedra.

Finally, the storage space and query time of the absolute model data structures is independent of n, and hence our results can be adapted to work in the *data stream model* [18, 7]. In the data stream model, the data set is too large to fit in memory, therefore the storage space should be independent of n (sometimes polylogarithmic functions of n are acceptable). Also, the data points are examined one at a time, in a single pass, while queries regarding the points that have already been seen, or that have been seen recently, need to be answered efficiently. Exact range searching clearly requires $\Omega(n)$ storage for all reasonable sets of ranges, and approximate range searching in the relative model requires $\Omega(n)$ storage when the query ranges can be scaled arbitrarily. Suri, Tóth, and Zhou [19] consider approximate range counting in the data stream model, approximating the number of points inside the query region.

We assume a model of computation that supports integer division. We use $\widetilde{O}(x)$ to denote $O(x \log^{O(1)} x)$, $\widetilde{\Omega}(x)$ to denote $\Omega(x/\log^{O(1)} x)$, and $\alpha(m, n)$ to denote the inverse Ackermann function [20].

The main results of this paper include the introduction of the absolute model, and approximate data structures for the most common range spaces, including a data structure that benefits from idempotence, as well as the *halfbox quadtree* data structure. Several data structures provide a space-time tradeoff, where the query time can be made arbitrarily low, at the cost of extra storage space. The complexities of our approximate data structures are summarized in Table 1.

We also relate approximate data structures in the absolute model to exact data structures. Assuming uniformly distributed data points, we provide an exact data structure for halfspace range searching, which makes use of idempotence to improve over the most efficient exact data structure previously known. This exact data structure is defined in the semigroup arithmetic model [11, 9, 5], and has $O(n^{1-2/(d+1)})$ expected query time with $O(n)$ space, matching the lower bound proved in [9] up to logarithmic factors. The theoretical importance of the

Table 1. Complexities of several approximate range searching data structures

Range	Version	Storage space	Query time	Preprocessing	Section
Halfspace	semigroup	$O(1/\varepsilon^d)$	$O(1)$	$\widetilde{O}(n+1/\varepsilon^d)$	3.1
	idempotent	$m \geq 1/\varepsilon^{(d+1)/2}$	$O(1/m\varepsilon^d)$	$\widetilde{O}(n+1/\varepsilon^d)$	3.2
	emptiness	$m \geq 1/\varepsilon^{(d-1)/2}$	$O(1/m\varepsilon^{d-1})$	$O(n+1/\varepsilon^d)$	3.2
Spherical	semigroup	$\widetilde{O}(1/\varepsilon^d)$	$O(1/\varepsilon^{(d-1)/2})$	$\widetilde{O}(n+1/\varepsilon^d)$	4.1
	semigroup	$m \geq 1/\varepsilon^{d+1}$	$O(1/m^{(1/2)-O(1/d)}\varepsilon^{d-O(1)})$	$O(n+m/\varepsilon^{(d-1)/2})$	4.1
Simplex	group	$\widetilde{O}(1/\varepsilon^d)$	$O(1/\varepsilon^{d-2}+\log(1/\varepsilon))$	$\widetilde{O}(n+1/\varepsilon^d)$	4.2
Orthogonal	semigroup	$m \geq 1/\varepsilon^d$	$O(\alpha(m, 1/\varepsilon^d))$	$O(n+m)$	5
	group	$O(1/\varepsilon^d)$	$O(1)$	$O(n+1/\varepsilon^d)$	5
Convex	semigroup	$O(1/\varepsilon^d)$	$O(1/\varepsilon^{d-1})$	$O(n+1/\varepsilon^d)$	5

data structure relies on the fact that uniform distribution and the semigroup arithmetic model are also assumed in the lower bound proved in [9]. Therefore, we open some important theoretical and practical questions: Is the average case complexity for uniformly distributed data strictly lower than the worst case complexity? Does the semigroup arithmetic model allow more efficient idempotent halfspace range searching data structures than the real RAM model? An improved data structure that worked in the real RAM model would be of practical interest, even if it relied on the uniform distribution of the points.

In Section 2, we formalize some definitions. In Section 3, we introduce halfspace range searching data structures for different versions of the problem. In Section 4, we introduce the halfbox quadtree, which answers spherical and simplex queries. In Section 5, we briefly mention approximate data structures for orthogonal and general convex ranges.

Some details and proofs are omitted from this conference version due to space limitations. Complete details are presented in the full paper [14].

2 Preliminaries

In this section, we provide basic definitions and discuss some results that are used throughout the paper. We start by formally defining the absolute model.

Given a range $R \in \mathcal{R}$ and an approximation error $\varepsilon > 0$, we define R^+ as the locus of points x such that $dist(x, R) \leq \varepsilon$. We define R^- as the locus of points x such that $dist(x, \overline{R}) \geq \varepsilon$, where $\overline{R}$ is the complement of R. We say that R_ε *ε-approximates* R *within* B if $R^- \cap B \subseteq R_\varepsilon \cap B \subseteq R^+ \cap B$. We say that R_ε *ε-approximates* R if R_ε ε-approximates R within $[0,1]^d$. We say $q_\varepsilon(R)$ is an *ε-approximation* of $q(R)$ if there is R_ε such that $q_\varepsilon(R) = q(R_\varepsilon)$ and R_ε approximates R.

We define a computational model, called the *Approximate Semigroup Arithmetic model* (*ASA model* for short), which makes it easier to describe our data structures. The ASA model is similar to the semigroup arithmetic model [11,9,5]. We explain how to convert our approximate data structures from the ASA model to the real RAM model (with integer division), preserving the same query time and storage space.

Given a collection of sets $\mathcal{S}$, let $\bigcup(\mathcal{S}) = \bigcup_{S \in \mathcal{S}} S$. In the semigroup version, we say that a set of regions $\mathcal{G}$ and a function $g : \mathcal{R} \to 2^{\mathcal{G}}$ *ε-generates* $\mathcal{R}$ if, for all $R \in \mathcal{R}$, the sets in $g(R)$ are pairwise disjoint, and $\bigcup(g(R))$ ε-approximates R. The elements of $\mathcal{G}$ are called *generators.* In the idempotent version the elements of $g(R)$ do not need to be pairwise disjoint, because $x + x = x$ for all $x \in \mathbf{S}$. In the group version, as the elements of the semigroup $(\mathbf{S}, +)$ have an inverse, the generators can be summed and subtracted in a multiset fashion, as long as the final result is a set, that is, no point is counted more than once, or counted a negative number of times. For simplicity, we use $\bigcup(g(R))$ to refer to the sums and subtractions of generators in the group version.

In the ASA model, a set $\mathcal{G}$ and a function g that ε-generates $\mathcal{R}$ is called a *data structure* for $\mathcal{R}$. The *storage space* of the data structure is defined as $|\mathcal{G}|$, and the *query time* is defined as $T(\mathcal{G},\mathcal{R}) = \max_{R\in\mathcal{R}} |g(R)|$. We say that a data structure provides *internal approximation* if $\bigcup(g(R)) \subseteq R$, for all $R \in \mathcal{R}$, and provides *external approximation* if $R \subseteq \bigcup(g(R))$, for all $R \in \mathcal{R}$. Modifying our data structures to provide either internal approximation or external approximation is straightforward.

There would be little use to develop upper bounds using the ASA model, if we could not convert the data structures from the ASA model to a more standard model of computation such as the real RAM model. The ASA model (as well as the semigroup arithmetic model) considers neither preprocessing time nor the time to identify the proper generators for a given range. Identifying the proper generators consists of computing $g(R)$ efficiently, and is a simple task for the data structures we present, with the exception of the exact halfspace data structure from Section 3.3. *Preprocessing* consists of computing $w(G) = \sum_{p\in P\cap G} w(p)$, for all $G \in \mathcal{G}$. We may modify our set of generators $\mathcal{G}$ in order to obtain a set that is faster to preprocess, by using an approximation of each $G \in \mathcal{G}$, instead of G itself, when computing $w(G)$. We provide details on how to preprocess our data structures efficiently throughout the text. Therefore, our data structures work in the real RAM model, with the exception of the exact halfspace data structure from Section 3.3, which works in the semigroup arithmetic model.

3 Halfspace Range Searching

In the halfspace range searching problem, the range space is the set of all halfspaces. Brönnimann, Chazelle, and Pach [9] showed that a data structure with $m \geq n$ storage space takes $\widetilde{\Omega}(n^{1-\frac{d-1}{d(d+1)}}/m^{\frac{1}{d}})$ query time. The lower bound uses the semigroup arithmetic model, and holds on the expected case when the data points are uniformly distributed in the unit hypercube.

The most efficient exact data structure known for the semigroup version is due to Matoušek [16] and has $\widetilde{O}(n/m^{1/d})$ query time with m storage space. For small d, the gap between the best general lower bound and the best upper bound is significant. For example, when $m = O(n)$ and $d = 2$, there is a $\widetilde{\Omega}(n^{1/3})$ lower bound, and a $O(n^{1/2})$ upper bound.

Arya, Malamatos, and Mount [5] studied the importance of the semigroup being idempotent. A semigroup $(\mathbf{S}, +)$ is *idempotent* if $x + x = x$ for all $x \in \mathbf{S}$, and is *integral* if $kx \neq x$ for all $x \in \mathbf{S}\setminus\{0\}$, and $k \in \mathbb{N}^+$. They showed that, when the semigroup is integral, the lower bound for exact halfspace range searching can be improved to $\widetilde{\Omega}(n/m^{(d+1)/(d^2+1)})$. They also used idempotence to develop more efficient spherical range searching data structures, in the relative model. We show that idempotence can be used not only to improve halfspace range searching in the absolute model, but also in the exact version assuming uniform distribution.

In Section 3.1, we show that the approximate version can be solved in $O(1)$ query time, $O(1/\varepsilon^d)$ space, and $\widetilde{O}(n + 1/\varepsilon^d)$ preprocessing time. This is noteworthy, given the high complexity of the exact version. In Section 3.2, we make use of idempotence to achieve a space-time tradeoff, building a data structure with $m \geq 1/\varepsilon^{(d+1)/2}$ storage space and $O(1/m\varepsilon^d)$ query time. We also mention how to improve the idempotent data structure for the case of emptiness queries. In Section 3.3, we use the approximate idempotent data structure to improve the best known bounds of exact range searching, in the semigroup arithmetic model, when the points are uniformly distributed in the unit cube.

Without loss of generality, we consider the range space $\mathcal{R}$ to be the set of halfspaces of the form $x_d \leq b + a_1x_1 + \cdots + a_{d-1}x_{d-1}$ with $-1 \leq a_1, \ldots, a_{d-1} \leq 1$. We call the terms $a_1, \ldots, a_{d-1}$ *slopes* and b the x_d*-intercept*. An arbitrary halfspace can be converted into this form through an appropriate rotation. As only $2d$ different rotations are necessary, one data structure can be kept for each rotated set of points, without changing our asymptotic results.

3.1 Approximate Semigroup Version

In this section, we provide a data structure to solve approximate halfspace range searching with $O(1)$ query time, $O(1/\varepsilon^d)$ storage space, and $\widetilde{O}(n+1/\varepsilon^d)$ preprocessing time. The general idea is to define a sufficiently large set $\mathcal{G}$ of halfspaces, so that any query halfspace is approximated by some halfspace in $\mathcal{G}$. As no two halfspaces in $\mathcal{G}$ are too similar to each other, efficient preprocessing requires building approximate data structures for subdivisions of the unit cube.

We define the set of generators $\mathcal{G}$ as the set that contains $\emptyset$, $[0,1]^d$, and the halfspaces whose boundary intersect the unit hypercube and have the slopes and the x_d-intercept as multiples of a parameter ε' to be specified later. Therefore, $\mathcal{G}$ contains $O(1/\varepsilon'^d)$ halfspaces. Given a halfspace $R \in \mathcal{R}$, the function $g(R)$ is the set containing the single halfspace obtained by rounding all slopes and the x_d-intercept of R to the closest multiple of ε'. If the boundary of R does not intersect the unit hypercube, then $g(R)$ is defined as $\{[0,1]^d\}$ if $[0,1]^d \subseteq R$, and $\{\emptyset\}$ if $R \cap [0,1]^d = \emptyset$. It is straightforward to prove the following lemma.

Lemma 1. *$g(R)$ $(d\varepsilon'/2)$-approximates R.*

To build an ε-approximate data structure, we set $\varepsilon' = 2\varepsilon/d$. Using Lemma 1, we have:

Theorem 1. *$(\mathcal{G}, g)$ is an ε-approximate halfspace range searching data structure with $O(1/\varepsilon^d)$ storage space, $O(1)$ query time, and $\widetilde{O}(n+1/\varepsilon^d)$ preprocessing time.*

It is easy to implement the query algorithm in the real RAM model, without changing the storage space or the query time, as long as integer division is available. Preprocessing the data structure efficiently is not trivial, though. Consider the following preprocessing algorithm.

Divide the unit hypercube in 2^d identical hypercubes. Recursively compute an approximate halfspace range searching data structure for each subdivision, using the case when the hypercube has diameter ε as a base case. For each generator $G \in \mathcal{G}$ ($|\mathcal{G}| = 1/\varepsilon^d$), and each subdivision s (out of 2^d subdivisions), perform a query for the intersection of G and s using the data structure for the subdivision s. Make $w(G)$ the sum of the results of all queries from generator G.

Disregarding the additive term of $O(n)$, the preprocessing time $T(\delta)$ for a data structure of diameter δ satisfies $T(\varepsilon) = O(1)$, and $T(\delta) = O(\delta/\varepsilon^d) + 2^d T(\delta/2) = O(\delta \log(\delta/\varepsilon)/\varepsilon^d)$.

We should note that the error accumulates through $O(\log(1/\varepsilon))$ levels. Consequently, the data structure is $O(\varepsilon \log(1/\varepsilon))$-approximate. We can set $\varepsilon = \varepsilon'/\log(1/\varepsilon')$, in order to obtain an $O(\varepsilon')$-approximate data structure in $O(n + \log^{d+1}(1/\varepsilon')/\varepsilon'^d)$ preprocessing time.

3.2 Approximate Idempotent Version

In this section, we make use of idempotence to achieve a space-time tradeoff, building a data structure with $m \geq 1/\varepsilon^{(d+1)/2}$ storage space and $O(1/m\varepsilon^d)$ query time. The idea is to use a set of properly placed large balls as generators. There are two sources of approximation error: one comes from the fact that we are approximating flat surfaces with balls, and the second one comes from the fact that we may use balls that are not exactly tangent to the surface being approximated. First, we present a scheme involving an infinite number of generators, which adressess the first issue. Then, we reduce this to a finite set of generators.

Let $r > \sqrt{d}+1$ be a constant, and let $\varepsilon < 1/2$ be an approximation parameter. We define $\mathcal{G}'$ to be the set of balls B of radius r such that B is centered at $(x_1, \ldots, x_d)$ where $x_1, \ldots, x_{d-1}$ are multiples of $\sqrt{r\varepsilon}/d$. Note that $|\mathcal{G}'|$ is infinite. Given a halfspace $R \in \mathcal{R}$, let $g'(R)$ be the subset of balls from $\mathcal{G}'$ that are tangent to the boundary of R and are contained in R. The proof of the following lemma is ommited due to space limitations.

Lemma 2. $(\mathcal{G}', g')$ *$(\varepsilon/2)$-generates* $\mathcal{R}$.

We now define $\mathcal{G} \subset \mathcal{G}'$ as the set of balls B such that: (i) B has radius r; (ii) there is a hyperplane h with slopes between -1 and 1, that is tangent to B at point p, with $p \in [-\sqrt{r\varepsilon}/d, 1 + \sqrt{r\varepsilon}/d]^{d-1} \times [0, 1 + \sqrt{d}]$; and (iii) B is centered at $(x_1, \ldots, x_d)$ where $x_1, \ldots, x_{d-1}$ are multiples of $\sqrt{r\varepsilon}/d$, and x_d is a multiple of $\varepsilon/2$.

The function $g(R)$ is obtained by replacing each ball B' from $g'(R)$ that approximates the boundary of R within $[0,1]^d$, with the closest ball $B \in \mathcal{G}$ such that $B \subset R$. The preprocessing uses Theorem 5.

Theorem 2. *$(\mathcal{G}, g)$ is an ε-approximate halfspace range searching data structure, for the idempotent version, with internal approximation, $m \geq 1/\varepsilon^{(d+1)/2}$ storage space, $O(1/m\varepsilon^d)$ query time, and $O(n + \log^{d+1}(1/\varepsilon)/\varepsilon^d)$ preprocessing time.*

For the emptiness version, the generators are bullet-shaped objects, formed by each ball and a semifinite cylinder extending downwards, and we use a simple compression argument to reduce the storage space by a factor of ε. Details are ommited due to space limitations.

Theorem 3. *There is an ε-approximate halfspace range searching data structure for the emptiness version with $m \geq 1/\varepsilon^{(d-1)/2}$ storage space, $O(1/m\varepsilon^{d-1})$ query time, and $O(n + 1/\varepsilon^d)$ preprocessing time.*

Another approach for the emptiness version consists of computing an ε-kernel [2, 10] containing $O(1/\varepsilon^{\frac{d-1}{2}})$ points, and then using an exact halfspace emptiness data structure with the ε-kernel as the set of points. The latter approach attains lower query times for the case of $O(1/\varepsilon^{(d-1)/2})$ space, but involves complex data structures from [15] for $d > 3$.

3.3 Exact Idempotent Version

In this section, we show how to use the approximate idempotent data structure to build an exact halfspace range searching data structure in the semigroup arithmetic model. To our knowledge, this is the first data structure to match the $\widetilde{\Omega}(n^{1-2/(d+1)})$ lower bound [9] when the storage space is linear, and the data points are uniformly distributed within the unit cube. We do not provide an efficient way to determine the set of generators for a given query. Consequently, the results hold only for the semigroup arithmetic model.

The general idea of the data structure is to properly set the parameter ε used in the data structure from Section 3.2, in order to make the expected number of points in the fuzzy boundary equal to the query time of the approximate data structure. Generators for individual points can be used to count the points in the fuzzy boundary.

In this section, we assume that the n data points are uniformly distributed in the unit hypercube $[0,1]^d$. Let $m \geq n$ denote the storage space, and $\varepsilon = O((nm)^{-1/(d+1)})$. We apply Theorem 2 to build an ε-approximate data structure $(\mathcal{G}_1, g_1)$ with $O(m)$ storage space and query time

$$O\left(\frac{1}{m\varepsilon^d}\right) = O\left(\frac{n^{1-1/(d+1)}}{m^{1/(d+1)}}\right) = O(n\varepsilon).$$

As the data structure (G_1, g_1) provides internal approximation, no data points outside R are counted, but some data points inside R and within distance ε from the boundary may not be counted. To answer the query exactly, we set $\mathcal{G}_2 = \{\{p\} : p \in P\}$ and define $g_2(R)$ as the set of generators $\{p\} \in \mathcal{G}_2$ such that $p \in R$ and p is within distance ε from the boundary of R. As the data points are uniformly distributed within the unit hypercube, for any fixed $R \in \mathcal{R}$, the expected number of generators in $g_2(R)$ is $E(|g_2(R)|) = O(n\varepsilon)$.

Let $\mathcal{G}' = \mathcal{G}_1 \cup \mathcal{G}_2$, and $g'(R) = g_1(R) \cup g_2(R)$. To use the standard semigroup arithmetic model definition of generators, let $\mathcal{G}$ denote the set of linear forms $\sum_{p \in P \cap G'} w(p)$ for all $G' \in \mathcal{G}'$.

Theorem 4. *$\mathcal{G}$ is an exact halfspace range searching data structure in the semigroup arithmetic model with $m \geq n$ storage space, and $O(n^{1-1/(d+1)}/m^{1/(d+1)})$ expected query time assuming that the points are uniformly distributed in a hypercube. When $m = n$, the query time is $O(n^{1-2/(d+1)})$.*

Theorem 4 matches the lower bound of $\widetilde{\Omega}(n^{1-(d-1)/d(d+1)}/m^{1/d})$ [9] for the case of $m = n$ (up to logarithmic factors). The lower bound is also in the semigroup arithmetic model, and also assumes that the points are uniformly distributed inside the unit hypercube. Therefore, improving the lower bound for the case of $m = n$, if at all possible, would require either a different model of computation or a different set of data points.

4 Halfbox Quadtree

In this section, we introduce a data structure called the *halfbox quadtree*, and present some applications. In Section 4.1, we analyze the query time of the halfbox quadtree for spherical ranges. We also show how an additional set of generators can be used to provide a space-time tradeoff. In Section 4.2, we analyze the query time of the halfbox quadtree for simplex ranges, in the group version.

A *quadtree box* is a box that can be obtained by recursively dividing the unit hypercube into 2^d identical hypercubes. We define a *half quadtree box* (*halfbox* for short) as the intersection of a quadtree box and a halfspace. The *size* of a halfbox is the diameter of the corresponding quadtree box. A *halfbox quadtree* is formed by associating each quadtree box Q of diameter at least ε with an ε-approximate halfspace range searching data structure for the bounding box Q. If we use the halfspace data structure from Theorem 1, the resulting halfbox quadtree has $O(\log(1/\varepsilon)/\varepsilon^d)$ storage space, $O(n+\log^{d+1}(1/\varepsilon)/\varepsilon^d)$ preprocessing time, $O(1)$ query time for halfbox ranges.

4.1 Approximate Spherical Range Searching

In spherical range searching, the ranges are Euclidean balls. The exact version of the problem can be reduced to halfspace range searching by projecting the points onto an appropriate $(d+1)$-dimensional paraboloid [8]. Spherical range queries in P are equivalent to halfspace range queries in P'. Arya, Malamatos and Mount [3,5] present approximate data structures, in the relative model, with $m \geq n\log(1/\varepsilon)$ space and $\widetilde{O}(n^{1-1/d}/m^{1/d}\varepsilon^{d-1})$ query time, for general semigroups, and $\widetilde{O}(n^{1/2-1/2d}/m^{1/2d}\varepsilon^{(d-1)/2})$ query time for idempotent semigroups.

In this section, we show that the halfbox quadtree answers spherical range queries in $O(1/\varepsilon^{\frac{d-1}{2}})$ time. The data structure works for general semigroups. We also show how to reduce the query time by increasing the space, adding some extra generators to the data structure. Lemma 3 shows that a small number of halfboxes can approximate a ball significantly better than a larger number of quadtree boxes.

Lemma 3. *A ball B of radius r can be ε-approximated by a set of disjoint half quadtree boxes, where each half quadtree box has size $\Omega(\sqrt{r\varepsilon})$.*

The following packing lemma allows us to bound the query time as a function of the size of the quadtree boxes (this follows from Lemma 3 in [6]).

Lemma 4. *If $\mathcal{Q}$ is a set of pairwise disjoint quadtree boxes, each of diameter at least δ, that intersect the boundary of a convex range of diameter $\Delta \geq \delta$, then $|\mathcal{Q}| = O((\Delta/\delta)^{d-1})$.*

It follows that:

Theorem 5. *The halfbox quadtree is an ε-approximate range searching data structure for spherical ranges with $O(\log(1/\varepsilon)/\varepsilon^d)$ storage space, $O(1/\varepsilon^{(d-1)/2})$ query time, and $O(n + \log^{d+1}(1/\varepsilon)/\varepsilon^d)$ preprocessing time. If the query ball B has radius r, then the query time is $O((\min(r,1)/(r+1)\varepsilon)^{(d-1)/2})$.*

The spherical range searching data structure from Theorem 5 has minimum storage space, except for a logarithmic factor (because we could place $1/\varepsilon^d$ points in a grid, and retrieve each individual weight using ranges of radius ε). It is natural to ask how we could improve the query time by increasing the storage space. Let $r > 1$ be a parameter, and consider the set of generators $\mathcal{G}$ formed by $\emptyset$, $[0,1]^d$, and the balls B such that (i) B has radius at most r, (ii) the radius of B is a multiple of $\varepsilon/2$, (iii) the boundary of B intersects $[0,1]^d$, and (iv) B is centered at $(x_1, \ldots, x_d)$ where $x_1, \ldots, x_d$ are multiples of $\varepsilon/2\sqrt{d}$.

Theorem 6. *The halfbox quadtree, together with $(\mathcal{G}, g)$, is an ε-approximate range searching data structure for spherical ranges with $m = r^d/\varepsilon^{d+1} > 1/\varepsilon^{d+1}$ storage space, $O(1/m^{\frac{1}{2}-\frac{1}{2d}}\varepsilon^{d-\frac{1}{2}-\frac{1}{2d}})$ query time, and $O(n + m/\varepsilon^{(d-1)/2})$ preprocessing time.*

4.2 Approximate Simplex Range Searching

In simplex range searching, the ranges are d-dimensional simplices. Chazelle [11] proved that, if m units of storage are allowed, then the query time is $\Omega(n/\sqrt{m})$ in the plane, and $\Omega((n/\log n)/m^{1/d})$ in d-dimensional space. In the exact version, the most efficient linear size data structure is due to Matoušek [16] and has $O(n^{1-1/d})$ query time.

We show how to answer ε-approximate simplex queries in $O(\log(1/\varepsilon))$ time for $d = 2$ and $O(1/\varepsilon^{d-2})$ time for $d \geq 3$, using the halfbox quadtree. Our query algorithm requires the use of a subtraction operation, and so applies only in the group setting.

We recursively answer a query $q(Q, R)$ with a simplex range R in the quadtree box Q, starting with Q as the unit hypercube. If $Q \cap R = \emptyset$, then we return 0. If $Q \cap R = B$, then we return the precomputed $w(Q) = \sum_{p \in P \cap Q} w(p)$. If Q does not contain any $(d-2)$-faces of R, then there is a set H of at most $d+1$ halfspaces that form the complement of R, and the halfspaces in H are pairwise disjoint. Then, we can return $w(Q) - \sum_{h \in H} q_\varepsilon(h \cap Q)$, where $h \cap B$ is a halfbox, and $q_\varepsilon(\cdot)$ is the result of the approximate query using the halfbox quadtree. If the diameter of Q is less than ε, then we verify whether R contains the center of Q and answer accordingly. Otherwise, we return the sum of $q(Q', R)$ for all 2^d subdivisions Q' of Q.

We can use the following packing lemma to bound the query time.

Lemma 5. *If C is a set of pairwise disjoint quadtree boxes, each of diameter at least δ, that intersect a $(d-2)$-dimensional convex polytope of constant diameter, then $|C| = O(1 + 1/\delta^{d-2})$.*

To analyze the query time, we look at the recursion tree of the query algorithm. The diameter of the quadtree boxes at level ℓ is $\Theta(2^{-\ell})$. The query algorithm only makes a recursive call when Q intersects the a $(d-2)$-face of R. As R is the intersection of a constant number of halfspaces, the number of $(d-2)$-faces of R is also constant. As R is convex, each $(d-2)$-face of R inside the unit box is a convex polytope of constant diameter. It follows from Lemma 5 that the number of recursive calls at level ℓ is $\Theta(2^{\ell(d-2)})$. Summing the number of recursive calls for all $\log(\Theta(1/\varepsilon))$ levels we conclude:

Theorem 7. *The halfbox quadtree is an ε-approximate range searching data structure for simplex ranges, in the group version, with $O(1/\varepsilon^d \log(1/\varepsilon))$ storage space, $O(\log(1/\varepsilon))$ query time for $d = 2$, $O(1/\varepsilon^{d-2})$ query time for $d \geq 3$, and $O(n + \log^{d+1}(1/\varepsilon)/\varepsilon^d)$ preprocessing time.*

5 Other Results

An important aspect of this work is the observation that the absolute model allows efficient approximate range searching with very simple methods. As further evidence of this, in this section, we briefly examine approximate range searching data structures for two additional problems: orthogonal ranges and convex ranges. The set of *orthogonal ranges* is the set of all axis-aligned hyper-rectangles. The set of *convex ranges* is the set of all d-dimensional convex shapes.

For the case of orthogonal ranges, we create a set P' of $O(1/\varepsilon^d)$ grid aligned points where, for any point $p \in P$, there is a point $a(p) \in P'$ with $dist(p, a(p)) \leq \varepsilon$. We define the weight $w(p')$, for $p' \in P'$, as the sum of $w(p)$ for the points p with $p' = a(p)$. Then, we build a d-dimensional array A, ranging from 1 to $O(1/\varepsilon)$ in each dimension, using the weights of the points in P'. A partial sum query in A is equivalent to an ε-approximate orthogonal range query for P. If subtraction is allowed, partial sum queries can be answered in $O(1)$ time with $O(1/\varepsilon^d)$ storage space. Without subtraction, partial sum queries take $O(\alpha(m, 1/\varepsilon^d))$ time with $m \geq 1/\varepsilon^d$ storage space [21, 13].

For the case of convex ranges, we define $\mathcal{G}$ as the set of quadtree boxes of diameter at least 2ε. Note that $|\mathcal{G}| = O(1/\varepsilon^d)$. The function $g(R)$ is defined as the set of quadtree boxes from $\mathcal{G}$ whose centers are contained in $\mathcal{R}$. The data structure $(\mathcal{G}, g)$ answers convex range queries in $O(1/\varepsilon^{d-1})$ time. We can efficiently compute $g(R)$ in the real RAM model if we assume that we can determine whether a range intersects a quadtree box in $O(1)$ time (as in [6]). A variation of this data structure uses a compressed quadtree to store arbitrarily small quadtree boxes. In this variation, the parameter ε only needs to be provided at query time, and the storage space is $O(n)$.

References

1. Agarwal, P.K., Erickson, J.: Geometric range searching and its relatives. In: Chazelle, B., Goodman, J., Pollack, R. (eds.) Advances in Discrete and Computational Geometry, pp. 1–56. American Mathematical Society, Providence, RI (1998)
2. Agarwal, P.K., Har-Peled, S., Varadarajan, K.R.: Geometric approximation via coresets. In: Goodman, J.E., Pach, J., Welzl, E. (eds.) Combinatorial and Computational Geometry, MSRI Publications/Cambridge Univ. Press, Cambridge (2005)
3. Arya, S., Malamatos, T., Mount, D.M.: Space-time tradeoffs for approximate spherical range counting. In: SODA'05. 16th Ann. ACM-SIAM Symposium on Discrete Algorithms, pp. 535–544. ACM Press, New York (2005)
4. Arya, S., Malamatos, T., Mount, D.M.: The effect of corners on the complexity of approximate range searching. In: SoCG'06. Proceedings of the 22nd ACM Symp. on Computational Geometry, pp. 11–20. ACM Press, New York (2006)
5. Arya, S., Malamatos, T., Mount, D.M.: On the importance of idempotence. In: STOC'06. Proc. 38th ACM Symp. on Theory of Computing, pp. 564–573. ACM Press, New York (2006)
6. Arya, S., Mount, D.M.: Approximate range searching. Comput. Geom. Theory Appl. 17(3-4), 135–152 (2000)
7. Babcock, B., Babu, S., Data, M., Motwani, R., Widom, J.: Models and issues in data stream systems. In: Proc. ACM Princ. Database Systems, pp. 1–16. ACM Press, New York (2002)
8. de Berg, M., van Kreveld, M., Overmars, M., Schwartzkopf, O.: Computational Geometry: Algorithms and Applications. Springer, Heidelberg (2000)
9. Brönnimann, H., Chazelle, B., Pach, J.: How hard is half-space range searching. Discrete & Computational Geometry 10, 143–155 (1993)
10. Chan, T.M.: Faster core-set constructions and data stream algorithms in fixed dimensions. In: SoCG'04. Proceedings of the 20th ACM Symp. on Computational Geometry, pp. 152–159. ACM Press, New York (2004)
11. Chazelle, B.: Lower bounds on the complexity of polytope range searching. J. Amer. Math. Soc. 2, 637–666 (1989)
12. Chazelle, B., Liu, D., Magen, A.: Approximate range searching in higher dimension. In: CCCG'04. Proceedings of the 16th Canadian Conference on Computational Geometry, pp. 154–157 (2004)
13. Chazelle, B., Rosenberg, B.: Computing partial sums in multidimensional arrays. In: SoCG'89. Proc. 5th ACM Symp. Comput. Geom., pp. 131–139. ACM Press, New York (1989)
14. da Fonseca, G.D., Mount, D.M.: Approximate range searching: The absolute model. Technical Report CS-TR-4873, Computer Science Department, University of Maryland (2007)
15. Matoušek, J.: Reporting points in halfspaces. Comput. Geom. Theory Appl. 2(3), 169–186 (1992)
16. Matoušek, J.: Range searching with efficient hiearchical cuttings. Discrete & Computational Geometry 10, 157–182 (1993)
17. Matoušek, J.: Geometric range searching. ACM Computing Surveys 26(4), 421–461 (1994)

18. Muthukrishnan, S.: Data streams: Algorithms and applications. Now Publishers (2005)
19. Suri, S., Tóth, C.D., Zhou, Y.: Range counting over multidimensional data streams. Discrete & Computational Geometry 36(4), 633–655 (2006)
20. Tarjan, R.E.: Efficiency of a good but not linear set union algorithm. J. ACM 22(2), 215–225 (1975)
21. Yao, A.C.: Space-time tradeoff for answering range queries. In: STOC'82. Proc. 14th ACM Symp. on Theory of Computing, pp. 128–136. ACM Press, New York (1982)

Orthogonal Range Searching in Linear and Almost-Linear Space

Yakov Nekrich

Department of Computer Science, University of Bonn
yasha@cs.uni-bonn.de

Abstract. In this paper we describe space-efficient data structures for two-dimensional range searching problem.

We present a dynamic linear space data structure that supports orthogonal range reporting queries in $O(\log n + k \log^{\varepsilon} n)$ time, where k is the size of the answer. Our data structure also supports emptiness and one-reporting queries in $O(\log n)$ time and thus achieves optimal time and space for this type of queries. In the case of integer point coordinates, we describe a static linear space data structure that supports range reporting queries in $O(\log n/\log\log n + k\log^{\varepsilon} n)$ time and emptiness and one-reporting queries in $O(\log n/\log\log n)$ time. This is the first linear space data structure for these problems that achieves sub-logarithmic query time.

We also present a dynamic linear space data structure for range counting queries with $O((\log n/\log\log n)^2)$ time and a dynamic $O(n\log n/\log\log n)$ space data structure for semi-group range sum queries with query time $O((\log n/\log\log n)^2)$.

1 Introduction

In the orthogonal range searching problem we store the set of d-dimensional points P in a data structure so that for an arbitrary d-dimensional query rectangle Q information about the points in $Q \cap P$ can be provided efficiently. In the case of the orthogonal range reporting problem, all points in $Q \cap P$ must be reported. In the case of the orthogonal semi-group range sum problem, we associate an element $g(p)$ of a commutative semi-group G with each point p of P, and for a query rectangle Q, $\sum_{p\in Q\cap P} g(p)$ can be computed. In the case of orthogonal range counting problem, $|Q \cap P|$ must be reported. In the case of emptiness queries, we must determine if $P \cap Q = \emptyset$. One-reporting queries are a special case of range reporting queries: for a query-rectangle Q we report an arbitrary point $p \in P \cap Q$, if $P \cap Q \neq \emptyset$. In this paper we describe linear space dynamic data structures for two-dimensional range reporting and counting. We also improve the space usage and query time of the dynamic data structure for the two-dimensional semi-group range sum problem.

Two-dimensional range reporting queries can be answered in $O(\log n+k)$ time by a data structure that uses $O(n\log^{\varepsilon} n)$ space both in static [6] and dynamic [14] scenarios. Here and further k denotes the size of the answer and ε is an arbitrary

F. Dehne, J.-R. Sack, and N. Zeh (Eds.): WADS 2007, LNCS 4619, pp. 15–26, 2007.

positive constant. The fastest linear space data structure [11] supports queries in $O(\sqrt{n \log n} + k)$ and updates in $O(\log n)$ time. If we allow penalties for each point in the answer, the best previously known linear space static data structure of Chazelle [6] supports queries in $O(\log n + k \log^{\varepsilon} n)$ time. The dynamic data structure of [15] supports queries in $O(\log n \log\log n + k \log^{\varepsilon} n)$ time and $O(n)$ space. Two previous results imply that emptiness and one-reporting queries can be answered in $O(\log n)$ time in the static case and $O(\log n \log\log n)$ time in the dynamic case.

In this paper, we present a dynamic linear space data structure that supports orthogonal range reporting queries in $O(\log n + k \log^{\varepsilon} n)$ time and updates in $O(\log^{7/2} n)$ time. Our data structure supports emptiness and one-reporting queries in $O(\log n)$ time thus matching the static data structure of Chazelle [6]. We also present a linear space static data structure with $O(\log n/\log\log n + k \log^{\varepsilon} n)$ time (respectively $O(\log n/\log\log n)$ for the emptiness and one-reporting queries) in the case when point coordinates are integers. This is the first data structure with linear space and sublogarithmic query time for these problems. We describe a linear space dynamic data structure for range counting queries with $O((\log n/\log\log n)^2)$ query time. This is an $O((\log\log n)^2)$ factor improvement over the fastest previously known linear space data structure of Chazelle [6]. Our dynamic data structure for two-dimensional semigroup range sum queries supports queries in $O((\log n/\log\log n)^2)$ time and uses $O(n \log n/\log\log n)$ space. As follows from the lower bound recently proven by Pătraşcu [17], this query time is optimal for any data structure with polylogarithmic update time.

Our approach is based the construction of efficient data structures for the narrow grid: the y-coordinates of points are (different) integers in the interval $[1, O(n)]$ and the x-coordinates are integers in the interval $[1, \log^{c} n]$ for some constant $c < 1$. Using a modification of the standard range tree technique [4] a general two-dimensional range searching query can be reduced to $O(\log n/\log\log n)$ queries to data structures on a narrow grid. In sections 2 and 3 we describe data structures for range reporting and range counting on a narrow grid. Results for two-dimensional orthogonal range reporting and orthogonal range counting are given in section 4. In section 5 we present a data structure for orthogonal semigroup range sum queries. The model of computation used in this paper is the unit cost word RAM: we assume that point coordinates of an arbitrary point fit into one machine word and all operations, including multiplication and bitwise operations, can be performed in $O(1)$ time.

2 Range Reporting on the Narrow Grid

In this section we describe a data structure for range reporting queries on a $W \times H$ grid, so that $W = O(\log^{1/4} n)$, $H = O(n)$, and all points have different y-coordinates. We divide the grid into $O(n/\sqrt{\log n})$ rows $R_i = [1, W] \times [r_{i-1}, r_i)$, $i = 1, 2, \ldots, O(n/\sqrt{\log n})$ for $r_i = i\sqrt{\log n}$. Thus, the grid is divided into $O(n/\sqrt{\log n})$ rows and every row contains $O(\sqrt{\log n})$ elements. Our data structure for the range reporting queries consists of the following components: For every pair i, j, $1 \le i \le j \le W$, there is a data structure C_{ij} for one-dimensional range reporting queries. If there is at least one point $p = (x, y)$ with $r_{s-1} \le y \le r_s$ and $i \le x \le j$, then we store an element s in C_{ij}. For every row R_i, we also store the coordinates of points that belong to R_i. We will show below that range reporting queries to a single row R_i can be supported in $O(k)$ time, and the total space necessary to store all points in all rows R_i is $O(n \log\log n)$ bits. Then, we will show that each data structure C_{ij} can be implemented with $O((n/\sqrt{\log n}))$ bits, so that one-dimensional range reporting queries are supported efficiently. Finally, we will describe how the queries and updates on the narrow grid are supported.

First, we describe a data structure for range reporting queries on R_i. For every point p in R_i we store the difference d_p between the y-coordinate of p and r_{i-1}. Since $d_p - r_{i-1} \le \sqrt{\log n}$ and the total number of points in R_i does not exceed $\sqrt{\log n}$, the values of d_p for all $p \in R_i$ can be stored in $O(\sqrt{\log n}\log\log n)$ bits. Since the x-coordinates of all points belong to $[1, O(\log^{1/4} n)]$, x-coordinates of all points can be also stored in $O(\sqrt{\log n}\log\log n)$ bits. All points in R_i are stored in a fixed order (e.g. sorted by their y-coordinates). Let d_j be the value of d_p for the j-th point p. Let v be the number of points in R_i. We store the values of d_j in a word D_i; D_i consists of v components of $s = \log\log n$ bits each; components are separated by special bits called flag bits. In the same way we store the x-coordinates of points in a word X_i. Since each point needs $O(\log\log n)$ bits, we can store words X_i and D_i for all i in a list L, so that L uses $O(n \log\log n)$ bits.

Since both x- and y-coordinates of all points in R_i can be stored in one machine word, we can answer two-dimensional range-reporting queries for R_i in constant time by exploiting the bit parallelism. Here and further we denote by `AND` and `OR` the bitwise AND and OR operations; x^v denotes the string x repeated v times. Consider a two-dimensional range reporting query $[a, b] \times [c, d]$, so that $1 \le a \le b \le W$, $r_{i-1} \le c \le d < r_i$. Let $c' = c - r_{i-1} - 1$, $d' = d - r_{i-1} + 1$. Using standard techniques we can construct words M_c and M_d that contain v copies of c' and d' respectively. We multiply c' with $1(0^s1)^v$ and obtain the word M_c that contains v copies of c' separated by zeros. We set all flag bits of D_i to 1 by `AND` with $(10^s)^v$. Let $M_1 = D_i - M_c$. The sj-th bit of M_1 equals to 1 iff the j-th component stored in D_i is greater than c'. We set $M_1 = (M_1 \texttt{ AND } (10^s)^v)$. Analogously, we multiply d' with $1(0^s1)^v$ and obtain the word M_d that contains v copies of d' separated by zeros. We set the flag bits of M_d to 1 by `OR` with $(10^s)^v$. Let $M_2 = ((M_d - D_i) \texttt{ AND } (10^s)^v)$. The j-th flag bit of M_2 is 1, iff the j-th component of D_i is smaller than d'. Let $M = M_1 \texttt{ AND } M_2$. The sj-th bit of the mask M equals to 1, iff $c' < d_j < d'$. But $c' < d_j < d'$ iff $c \le y_j \le d$, where y_j is the y-coordinate of the j-th point. In the same way, we can construct the mask M', such that the sj-th bit of M' equals to 1, iff $a \le x_j \le b$. Hence, the sj-th bit of $M'' = M \texttt{ AND } M'$ is 1 iff the j-th point in R_i is contained in $[a, b] \times [c, d]$. Using a look-up table of size $o(n)$ we can identify the positions of all non-zero bits in M'' and output the coordinates of corresponding points using X_i and D_i.

A data structure C_{ij} contains elements from $[1, m]$ for $m = O(n/\sqrt{\log n})$. This fact can be used to implement C_{ij} with $O(m) = O(n/\sqrt{\log n})$ bits. With a slight misuse of notation we will also denote by C_{ij} the set of elements stored in C_{ij}. We store all elements of C_{ij} in the compact list $\mathcal{L}$ organized in the same way as the data structure of [5] (a similar approach is also used in [15]). All elements of $\mathcal{L}$ are divided into groups, and the elements of each group except of the first element are difference encoded: if the i-th group G_i consists of elements $e_{i1}, e_{i2}, \ldots e_{in_i}$, then we store $e_{i1}, \delta_{i2} = e_{i2} - e_{i1}, \ldots \delta_{in_i} = e_{in_i} - e_{in_i - 1}$ in G_i. Each difference δ_{ij} is gamma coded[7], so that δ_{ij} is stored with $O(\log \delta_{ij})$ bits. We choose the size of each group G_i in such a way that all encoded elements in G_i require at most $8 \log m - 4$ bits and at least $2 \log m - 1$ bits. It can be shown (s. [5]) that all groups G_i require $O(m)$ bits; hence, there are $O(m/\log m)$ groups. In [5] it is also shown that since G_i is stored in $O(1)$ words, we can insert and delete elements into G_i and search in G_i in $O(1)$ time using table look-up. We store the first elements of each group in a data structure C'_{ij}. If we implement C'_{ij} using a linear space data structure, then C'_{ij} uses $O((m/\log m) \log \frac{m}{\log m}) = O(m)$ bits. We can implement C'_{ij} as a van Emde Boas data structure with $O(\log \log n + k)$ query time and $O(\log \log n)$ update time. Alternatively, C'_{ij} can be implemented as a static data structure with $O(k)$ query time, or as a randomized data structure of [16] with $O(\log \log \log n + k)$ query time and $O(\log \log n)$ update time. We also store an array V with $m/\log m$ entries: $V[s]$ stores a pointer to the smallest element e_s in C_{ij} that belongs to the interval $[(s-1)\log m, s \log m]$. $V[s]$ consists of the pointer to a group G_i in which e_s is stored and the index of e_s in G_i; hence $V[s]$ can be stored with $O(\log m)$ bits and V requires $O(m)$ bits. Due to the concavity of the logarithm function, the total number of bits to encode all elements $e \in [(i-1)\log m, i \log m]$ is $O(\nu \log \frac{\log m}{\nu})$ for $\nu = |\{e | e \in [(i-1)\log m, i \log m] \cap C_{ij}\}|$. Since $\nu \log \frac{\log m}{\nu} \leq \log m$ for $\nu \leq \log m$, all difference coded values can be stored in $O(1)$ words. Hence, all elements of C_{ij} that belong to an interval $[(s-1)\log m, s \log m]$ for some s are stored in $O(1)$ consecutive groups G_i.

A one-dimensional query $[c, d]$ to C_{ij} is answered as follows. We compute $c' = c/\log n$ and $d' = d/\log n$. If $V[c']$ is empty, we set $c'' = \lceil c/\log n \rceil$, otherwise we find $succ(c, C_{ij})$. Here and further the predecessor of an integer e in a set S is $pred(e, S) = \max\{x \in S \cup \{-\infty\} | x \leq e\}$, and the successor of e in a S is $succ(e, S) = \min\{x \in S \cup \{+\infty\} | x \geq e\}$. Observe that since at least one element of C_{ij} belongs to $[c' \log n, (c'+1)\log n]$, either a predecessor or a successor of c is contained in $C_{ij} \cap [c' \log n, (c'+1)\log n]$. Since all elements from $[c' \log n, (c'+1)\log n]$ belong to a constant number of groups G_i that follow the group to which $V[c']$ points, we can find $succ(c, C_{ij})$ in $O(1)$ time. Let G_j be the group to which $succ(a, C_{ij})$ belongs. We can find all elements in G_j that belong to $[c, d]$ in $O(1)$ time per element and set c'' to be the first element in the group G_{j+1} that follows G_j. If $V[d']$ is empty, we set $d'' = d' \log n$, otherwise we find a group G_l that contains the predecessor of d in G_l. We report all elements of G_l that belong to $[c, d]$ in $O(1)$ time per element, and set d'' to be the first element in G_l. We find all groups G_m, such that all elements of G_m belong to $[c'', d'']$ using C'_{ij}.

For every element e of C'_{ij} that belongs to $[c'', d'']$ we output all elements of the corresponding group G_e. Thus a one-dimensional query to C_{ij} can be answered in $O(f(n) + k)$ time, where $f(n)$ is the query time of the one-dimensional linear space data structure C'_{ij} and k is the size of the answer.

To answer a two-dimensional range reporting query $[a, b] \times [c, d]$ on a narrow grid, we find $f = \lceil c/\sqrt{\log n}\rceil$ and $l = \lceil d/\sqrt{\log n}\rceil$. If $f = l$, $[a, b] \times [c, d]$ is contained in a row R_f, and we can report all points in $[a, b] \times [c, d]$ using X_f and D_f as described above. If $f < l$, $[a, b] \times [c, d] = [a, b] \times [r_f, r_{l-1}] \cup [a, b] \times (r_{l-1}, d] \cup [a, b] \times [c, r_f)$. Thus a range reporting query is reduced to a one-dimensional range reporting query to C_{ab} and two three-sided queries to rows R_f and R_l. If $f < l$ we find all relevant rows R_s that contain at least one element with x-coordinate in $[a, b]$ by a one-dimensional query $[f+1, l-1]$ to C_{ab}. For every found element s we report all points in R_s whose x-coordinates belong to $[a, b]$ using X_s. Finally we answer queries $([a, b] \times (r_l, d]) \cap R_l$ and $([a, b] \times [c, r_f)) \cap R_f$. The search time is dominated by the search time of the one-dimensional query to C_{ab}.

When a new point $p = (x, y)$ is inserted, we find $s = \lceil y/\log n\rceil$. For every pair a, b such that $1 \le a \le x \le b \le W$, we insert s into C_{ab} if necessary. If the row R_s is empty, we construct the words X_s and D_s and insert them into the list L. Deletions are processed in a similar way.

We can sum up the results of this section in the following:

Lemma 1. *There exists a data structure for an $O(\log^{1/4} n) \times O(n)$ grid that uses $O(n \log\log n)$ bits and supports range reporting queries in $O(\log\log n + k)$ time and updates in $O(\log^{1/2} n \log\log n)$ time.*

There exists a static data structure for an $O(\log^{1/4} n) \times O(n)$ grid that uses $O(n \log\log n)$ bits and supports range reporting queries in $O(k)$ time.

There exists a randomized data structure for a $O(\log^{1/4} n) \times O(n)$ grid that uses $O(n \log\log n)$ bits and supports range reporting queries in $O(\log\log\log n + k)$ time and updates in $O(\log^{1/2} n \log\log n)$ time.

3 Range Counting Queries on the Narrow Grid

In this section we describe a data structure for orthogonal range counting queries on a $W \times H$ grid for $W = O(\log^{1/4} n)$ and $H = O(n)$. We can use an approach similar to the approach of section 2. In the same way as in the previous section, the grid is divided into $n/\sqrt{\log n}$ rows $R_i = [1, O(\log^{1/4} n)] \times [r_{i-1}, r_i)$, so that $r_i - r_{i-1} = \sqrt{\log n}$ for $i = 1, 2, \ldots, O(n/\sqrt{\log n})$.

For every row R_i we store words X_i and D_i defined in section 2. The range counting queries for each row R_i can be answered in $O(1)$ time: As described in the previous section, for an arbitrary $[a, b] \times [c, d]$ contained in a row R_i, we can construct the word M of $o(\log n)$ bits such that the number of 1's in M equals to the number of points in $[a, b] \times [c, d] \cap R_i$. We can count the number of 1's in M in $O(1)$ time with a look-up table of size $o(n/\log n)$. We also store for each pair i, j, $1 \le i \le j \le W$, a one-dimensional data structure C_{ij} for the special case of range sum queries: we associate an integer value $w(e)$, $w(e) \le \sqrt{\log n}$, with each element e of C_{ij}, so that for an arbitrary interval $[c,d]$, the sum $\sum_{e\in[c,d]} w(e)$ can be computed. Observe that this problem is easier than the semi-group range sum problem, because subtractions are allowed and bit-packing techniques can be applied. We associate with each element s stored in C_{ij} the number of points $(x,y) \in R_s$ with $i \le x \le j$. We will show later in this section that each C_{ij} uses $O(n \log\log n/\sqrt{\log n})$ bits.

Analogously to the range reporting query processing, a range counting query $[a,b] \times [c,d]$ is answered by answering at most three queries: if $l > f$ for $f = \lceil c/\sqrt{\log n}\rceil$ and $l = \lceil d/\sqrt{\log n}\rceil$, $[a,b]\times[c,d] = [a,b]\times[r_f, r_{l-1}] \cup [a,b]\times(r_{l-1}, d] \cup [a,b]\times[c, r_f)$; if $l = f$, it suffices to answer a query $[a,b]\times[c,d]$ for the row R_f. Range counting queries to a single row R_i can be answered as described above.

Now we turn to the description of a data structure C_{ij}. Elements of C_{ij} belong to the interval $[1,U]$ for $U = O(n/\sqrt{\log n})$. We divide $[1,U]$ into $O(n\log\log n/\log n)$ intervals of size $\log n/\log\log n$ each. For the m-th interval $U_m = [(m-1)\log n/\log\log n, m\log n/\log\log n]$, $m = 1, 2, \ldots, O(n\log\log n/\log^{3/2} n)$, we store the prefix sums sum_e of all elements of the interval U_m in a word W_m. Here for an element $e \in U_m$, $sum_e = \sum_{e'\in U_m, e'\le e} w(e)$. Observe that each value associated with an element of C_{ij} does not exceed $\sqrt{\log n}$. Therefore prefix sums stored in W_i do not exceed $\log^{3/2} n$ and each prefix sum can be stored in $O(\log\log n)$ bits. When a value associated with some element of U_m is incremented or decremented, W_m can be updated in $O(1)$ time using the bit parallelism. We associate with an element m of the data structure C'_{ij} the sum of values associated with all elements of C_{ij} that belong to U_m. We will show below that C'_{ij} uses linear space in the number of its elements. Since C'_{ij} contains $O(n\log\log n/\log^{3/2} n)$ elements, C'_{ij} uses $O(n\log\log n/\sqrt{\log n})$ bits. The sum of values associated with elements of C_{ij} in the interval $[c,d]$ for $c \in U_m$ and $d \in U_m$, can be computed as $sum_d - sum_{c-1}$, where sum_d and sum_c are the prefix sums of d and c respectively. Both sum_d and sum_c can be computed in $O(1)$ time with help of W_m. Let $v = \log n/\log\log n$. We can answer a general range sum query $[c,d]$ to C_{ij} by answering a range sum query $[\lceil c/v\rceil, \lfloor d/v\rfloor]$ to C'_{ij} and two range sum queries $[c, \lceil c/v\rceil v]$ and $[\lfloor d/v\rfloor v, d]$ to $W_{\lceil c/v\rceil}$ and $W_{\lfloor d/v\rfloor+1}$ respectively. It remains to show how a linear space data structure C'_{ij} for range sum queries can be constructed.

We start with a description of the dynamic variant of the data structure C'_{ij}. Later in this section we will describe a static data structure. Dynamic data structure C'_{ij} is implemented as a B-tree T with node degree $\log^{1/4} n$. The values associated with elements of C'_{ij} are stored in the leaves of T. In every internal node v of T we store the data structure S_v that supports range sum queries in the interval $[1, 2\log^{1/4} n]$. A query (i,j) to S_v returns the total number of leaves in children $v_i, v_{i+1}, \ldots, v_j$ of the node v.

We say that a node v is on level l if the path from v to a leaf node consists of l edges. In every node v on level $l > 1$ the data structure S_v consists of two parts: array A_v with $\log^{1/2} n$ entries and a word B_v with $\log^{1/2} n$ components of $\log\log n$ bits. Following the idea of [18], we store the information about $\log^{1/2} n$ most recent updates in the word B_v and the information about the older updates in array A_v. Components in B_v and entries in A_v correspond to pairs r, s, $1 \le r \le s \le \log^{1/4} n$. In the entry of A_v that corresponds to a pair r, s we store the total number of elements in children $v_r, v_{r+1}, \ldots, v_s$ of v at some previous time. We keep track of the recent changes with help of the word B_v. When the i-th child of v is updated, i.e. when the number of elements in v_i is incremented or decremented, we update all components r, s of B_v with $r \le i \le s$. Since all components of B_v fit into one machine word, we can increment or decrement all relevant components in $O(1)$ time. During the t-th update operation we set $A_v[r, s] = A_v[r, s] + B_v[r, s]$ and $B_v[r, s] = 0$ for $r = (t \mod \log^{1/2} n)/\log^{1/4} n$ and $s = (t \mod \log^{1/2} n) \mod \log^{1/4} n$. Hence, the components of B_v keep track of the changes in the sums during the last $\log^{1/2} n$ updates. Therefore the values of components do not exceed $\log^{1/2} n$ and can be stored in less than $\log\log n$ bits each. Thus each data structure S_v for a node v on level $l \ge 2$ uses $\log^{1/2} n$ words of memory and supports updates in $O(1)$ time. Data structures S_v for internal nodes on level $l = 1$ consist only of a word B_v. Since the children of B_v are leaves, all components of B_v fit into one word. If a new leaf child of v is inserted (an old child is deleted), we insert or remove one component of B_v and update all relevant components of B_v in $O(1)$ time. Since the total number of nodes on levels $l \ge 2$ is $O(n/\log^{1/2} n)$, all data structures S_v use $O(n)$ words. Suppose that the total number of points in an interval $[c, d]$ must be found. Using a standard searching data structure, we identify the predecessor t_d of d and the successor t_c of c among the leaves of T. Let q be the lowest common ancestor of t_c and t_d. Then a range sum query can be answered by answering at most two queries to internal nodes of T on each level l, $1 \le l \le l_q$, where l_q is the level of the node q. Hence, a range sum query can be answered in $O(\log n/\log\log n)$ time.

When a new element is inserted into C'_{ij} we insert the new leaf u into T. For the parent v of u we update S_v by inserting new components into B_v in $O(1)$ time. For every node w on the path from v to the root, we update S_w in $O(1)$ time as described above. If the number of children of some node x exceeds $2\log^{1/4} n$, we split x and insert the new element into the data structure S_y for the parent y of x and rebuild S_y into $O(\log^{1/4} n)$ time. Since a node on level l is split after $O(B^l)$ updates the amortized cost of rebuilding data structures S_y is $O(\log n/\log\log n)$. Deletions can be processed in a similar way. We can update the value associated with some element e of C'_{ij} by removing e from C'_{ij} and re-inserting e with the new associated value. Thus C'_{ij} (and hence, C_{ij}) supports update operations in $O(\log n/\log\log n)$ time. When a new point is inserted or removed, $O(\log^{1/2} n)$ elements in data structures C_{ij} may be updated; hence, the total update time is $O(\log^{3/2} n/\log\log n)$.

In the static case we can easily construct a data structure C'_{ij} that uses linear space in the number of its elements and supports range sum queries in $O(1)$ time: with every element e we store the sum S_e of values associated with elements that precede e, $S_e = \sum_{e' \le e} w(e')$, and the sum of values in the range $[c, d]$ equals to $S_d - S_{c-1}$. We sum up our results in the following:

Lemma 2. *There exists a dynamic data structure for range counting queries on an* $O(\log^{1/4} n) \times O(n)$ *grid with* $O(\log n/\log\log n)$ *query time and* $O(\log^{3/2} n/\log\log n)$ *update time that uses* $O(n \log\log n)$ *bits.*

There exists a static data structure for range counting queries on an $O(\log^{1/4} n) \times O(n)$ *grid with* $O(1)$ *query time that uses* $O(n \log\log n)$ *bits.*

4 Two-Dimensional Range Searching

Let P be the set of points stored in the data structure; let P_x and P_y be the sets of x- and y-coordinates of points in P. We assume w.l.o.g. that all points have different y-coordinates.

We can transform a data structure for a narrow grid into a two-dimensional range searching data structure using the standard range trees technique [4]. We build a tree T_x over the set of x-coordinates P_x. In order to support re-balancing operations efficiently, we implement T_x as a WBB tree [2] with branching parameter $\log^{1/4} n$ and leaf parameter 1. Every internal node has at most $4\log^{1/4} n$ children [2]. With every leaf x we associate a range $[x, x')$, where x' is the successor of x in P_x; a range associated with an internal node is a union of ranges associated with its children.

In every internal node v we store a data structure D_v. For every point (x, y) such that x belongs to the range of v, we store a point $(i, l_v(y))$ in D_v. We choose i so that (x, y) belongs to the range of the i-th child of v, and $l_v(y)$ is the *v-label* assigned to the point (x, y) in the node v. The labels are assigned in such way that for any points (x_1, y_1), (x_2, y_2) whose labels are stored in D_v, $y_1 < y_2 \Rightarrow l_v(y_1) < l_v(y_2)$. Besides that, for all (x, y) that belong to the range of v, $l_v(y) \in [1, O(m)]$, where m is the total number of points that belong to the range of v. Since all elements $(i, l_v(y))$ stored in D_v are points on a $O(\log^{1/4} n) \times O(m)$ grid, we can implement D_v as one of the data structures described in sections 2 and 3.

To answer a query $[a, b] \times [c, d]$ we find $a_x = succ(a, P_x)$ and $b_x = pred(b, P_x)$. For every node v on the path from the root of T_x to a_x and for every node v on the path from the root to b_x, we answer a two-dimensional range searching query $[i, j] \times [c_v, d_v]$, such that the ranges of the children $v_i, v_{i+1}, \ldots, v_j$ of v are contained in $[a, b]$. The interval $[c_v, d_v]$ contains a label $l_v(y)$, iff $y \in [c, d]$. We will show later how intervals $[c_v, d_v]$ can be found. Since at most two queries to data structures D_v are answered on each level of T_x, the total number of queries on a narrow grid is $O(\log n/\log\log n)$. Hence, a two-dimensional range counting query can be answered in $O((\log n/\log\log n)q(n))$ time, where $q(n)$ is the query time for a data structure on a $O(\log^{1/4} n) \times O(n)$ grid. A two-dimensional range reporting query can also be answered in $O((\log n/\log\log n)q(n))$ time, if we ignore the time to output the points in the answer.

The labeling l_v can be constructed and maintained in the same way as in [15]. Let Y_u be the set of u-labels of the elements that belong to node u; let $Y_{v,u}$ be the set of v-labels of the elements that belong to u, where v is the parent node of u. For every internal node v and every child u of v we store a mapping

$f_u : Y_{v,u} \rightarrow Y_u$. Thus f_u maps the v-labels of points that belong to u into their u-labels. In the root r of T_x we store an additional mapping f_r that assigns a label l_r to every element of P_y. The mappings f_u can be maintained with the sparse table technique [10], [19]. As shown in [3], to maintain the mapping f_u it suffices to store $|Y_u|/\log(|Y_u|)$ auxiliary records of $\log(|Y_u|)$ bits each. When the set Y_u is updated, we may change the values of $f_u(e_1), f_u(e_2), \ldots, f_u(e_s)$ for some elements $e_1, e_2, \ldots, e_s \in Y_{v,u}$, so that the order-preserving property of f_u is maintained. In this case we say that elements $e_1, \ldots, e_s$ are f_u-*moved*. Using e.g. the algorithm of [3], we f_u-move $O(\log^2 |Y_{v,u}|)$ elements of $Y_{v,u}$ in the case of an update operation, so that the properties of the mapping f_u are preserved. It was shown in [15] that given a mapping $f : S \rightarrow V$, such that $S \subset U$ and $V \subset U$ for $|U| < n$, we can store a data structure of $O(|S| \log \frac{|U|}{|S|})$ bits, so that for every element $e \in S$ $f(e)$ can be computed in $O(\log \log n)$ time and an element e with the corresponding $f(e)$ can be inserted into (removed from) S in $O(\log \log n)$ time. We can also construct a static data structure that uses $O(|S| \log \frac{|U|}{|S|})$ bits, such that for every $e \in S$ $f(e)$ can be computed in $O(1)$ time. Besides that, we store for every child v_s of each node v a data structure that supports predecessor and successor queries in $O(\log \log n)$ time and uses $O(|Y_{v_s}| \log \log n)$ bits. Alternatively, we can store a static data structure that supports predecessor and successor queries in $O(1)$ time and also uses $O(|Y_{v_s}| \log \log n)$ bits. A detailed description of these data structures will be given in the full version of this paper.

To answer a two-dimensional range reporting query $[a, b] \times [c, d]$, we compute $c_r = f_r(succ(c, P_y))$ and $d_r = f_r(pred(d, P_y))$. We visit all nodes on the paths from r to a_x and from r to b_x as described above. In every visited node v we proceed as follows: If the ranges of children $v_i, v_{i+1}, \ldots, v_j$ of v are contained in $[a, b]$, we answer a query $[i, j] \times [c_v, d_v]$. By Lemma 1 such a query can be answered in $O(\log \log n + k)$ time in the dynamic case or in $O(k)$ time in the static case. If the child v_s of v must also be visited, we set $c_{v_s} = f_{v_s}(succ(c_v, Y_{v,v_s}))$ and $d_{v_s} = f_{v_s}(pred(d_v, Y_{v,v_s}))$. Since the total number of visited nodes is $O(\log n / \log \log n)$, we can find the labels of all points in $[a, b] \times [c, d]$ in $O(\log n + k)$ time. If point coordinates are integers, a static data structure can answer queries in $O(\log n / \log \log n + k)$ time. For every found label $e_v \in Y_v$, such that $c_v \leq e_v \leq d_v$, we identify the corresponding point $p = (x, y)$, such that e_v is a v-label of p. This can be done by computing $e_{u_1} = f_v^{-1}(e_v)$, $e_{u_2} = f_{u_1}^{-1}(e_{u_1})$, ..., $e_r = f_{u_s}^{-1}(e_{u_s})$, where $u_1, u_2, \ldots, u_s$ are nodes of T_x on the path from v to the root r. Since all points have different y-coordinates, we can identify a point by its r-label using a look-up table of linear size. This procedure incurs a $O(\log n)$ penalty for each point in the answer. The penalty can be reduced to $O(\log^{\varepsilon} n)$ using the method described in [15], section 4.

When a new point (x, y) is inserted into the dynamic data structure, we insert the v-labels of p into data structures D_v for $O(\log n / \log \log n)$ nodes $v_1, v_2, \ldots, v_t$ (i.e. nodes whose ranges contain x). When a new v-label is inserted into the set of v-labels Y_v, we may have to change the values of $O(\log^2 n)$ other v-labels. This means that $O(\log^2 n)$ elements must be removed from and re-inserted into D_v. Hence the cost of an insertion is $O(\log^2 n \log^{1/2} n \log \log n(\log n / \log \log n)) =$

$O(\log^{7/2} n)$. When a new element is inserted into a WBB tree T_x, some internal nodes can be split and associated data structures can be re-built. It follows from [2] that if a node v on level l is split, then at least $\log^{l/4} n/2$ insertions below v must be performed before it splits again. We can show that the amortized cost incurred by re-building data structures associated with split nodes is $O(\log^{5/2} n/\log\log n)$. Details will be given in the full version. Hence the total insertion time is $O(\log^{7/2} n)$. Deletions are processed in the same way as insertions.

Results for two-dimensional range reporting queries are summed up in the following theorem:

Theorem 1. *There is a linear space data structure A for two-dimensional orthogonal range reporting queries with $O(\log n + k\log^{\varepsilon} n)$ query time and $O(\log^{7/2} n)$ update time. A supports emptiness and one-reporting queries in $O(\log n)$ time.*

If all point coordinates are integers, there exists a linear space static data structure B for two-dimensional orthogonal range reporting queries with $O(\log n/\log\log n + k\log^{\varepsilon} n)$ query time. B supports emptiness and one-reporting queries in $O(\log n/\log\log n)$ time.

The data structure for the two-dimensional dynamic range counting problem is almost identical with the data structure of Theorem 1. The only difference is that we store in every node v a data structure S_v of Lemma 2 that supports range counting queries on a narrow grid in $O(\log n/\log\log n)$ time.

Theorem 2. *There is a linear space data structure C for orthogonal range counting queries with $O((\log n/\log\log n)^2)$ query time and $O(\log^{9/2} n/(\log\log n)^2)$ update time.*

5 Two-Dimensional Semi-group Range Counting

In this section we assume that an element $g(p)$ of a commutative semi-group G is associated with each point p. The goal of the semi-group range sum query is to find for a query rectangle $[a,b]\times[c,d]$ the sum of values associated with points p that belong to $[a,b]\times[c,d]$. We assume that an arbitrary element of G fits into one word of memory. Observe that unlike in the previous sections it is not possible to pack more than one element of G into a machine word.

In the following Lemma we show how semi-group range counting queries on the narrow grid can be processed.

Lemma 3. *There exists a linear space data structure S for semi-group range sum queries on $O(\log^{1/4} n)\times O(n)$ grid that supports queries in $O(\log n/\log\log n)$ time and updates in $O(\log^{3/2+\varepsilon} n)$ time.*

Proof. Suppose that the x-coordinates of all points belong to the range $[1,W]$ for $W = O(\log^{1/4} n)$. Data structure S consists of the same components as data structures in Lemmas 1 and 2: the grid is divided into rows $R_i = [1, O(\log^{1/4} n)]\times$

$[r_{i-1}, r_i)$, where $r_t = t\sqrt{\log n}$, $t = 0, 1, \ldots, O(n/\sqrt{\log n})$. Elements of each row R_i are stored in a list L_i. For every pair i, j, such that $1 \le i \le j \le W$, we store a data structure C_{ij} that supports one-dimensional semi-group range sum queries and contains $O(n/\sqrt{\log n})$ elements. We associate a value $v = \sum_{p \in [r_{s-1}, r_s) \times [i,j]} g(p)$ with en element s stored in C_{ij}. All data structures C_{ij}, $1 \le i \le j \le W$, contain $O(n)$ elements and use $O(n)$ space.

To answer a query $[a, b] \times [c, d]$ we find $r_f = \lceil c/\sqrt{\log n} \rceil$ and $r_l = \lceil d/\sqrt{\log n} \rceil$. If $r_f = r_l$, the query rectangle is contained in row R_f. Hence, we can examine all $O(\sqrt{\log n})$ points in L_f and compute the sum of values associated with points $p \in [a, b] \times [c, d]$ in $O(\sqrt{\log n})$ time. If $r_f < r_l$, we find $q_1 = \sum_{p \in ([a,b] \times (r_{l-1}, d])} g(p)$, $q_2 = \sum_{p \in ([a,b] \times [c, r_f))} g(p)$, and $q_3 = \sum_{p \in ([a,b] \times [r_f, r_{l-1}])} g(p)$. We can compute q_1 and q_2 in $O(\sqrt{\log n})$ time; q_3 is computed with a range sum query $[f, l]$ to C_{ab}.

C_{ij} can be implemented as a B-tree T_B with node degree $O(\log^{\varepsilon} n)$. Elements and their values are stored in the leaves of T_B. In every internal node v we store a data structure S_v with $O(\log^{\varepsilon} n)$ elements. Let v_i be the i-th child of v and $set(i)$ be the set of all leaf descendants of v_i. We associate with an element i of S_v the value $w(i)$, such that $w(i)$ equals to the sum of values associated with leaf descendants of v_i, $w(i) = \sum_{e \in set(i)} g(e)$. For each node v on level $l \ge 2$, we additionally store in S_v the sum $\sum_{a=i}^{j} w(a)$ for all pairs i, j. Every data structure S_v for a node v on level $l \ge 2$ uses $O(\log^{2\varepsilon} n)$ space and supports updates in $O(\log^{2\varepsilon} n)$ time and semi-group range sum queries in $O(1)$ time. Every data structure S_v for a node v on level $l = 1$ supports range-sum queries in $O(\log^{\varepsilon} n)$ time and updates in $O(1)$ time. To answer a one-dimensional semi-group range sum query, we visit all nodes ν on the path from l_c to q and on the path from l_d to q, where l_c and l_d are the leaves of T_B corresponding to c and d, and q is the least common ancestor of l_c and l_d. In every visited node ν we answer a range sum query to S_ν. There are at most two visited nodes on level $l = 1$, and the total number of visited nodes is $O(\log n/ \log\log n)$. Hence, the total time to answer a query is $O(\log n/ \log\log n)$. Thus C_{ij} uses $O(n/\sqrt{\log n})$ space, semi-group range sum queries are supported in $O(\log n/ \log\log n)$ time, and updates are supported in $O(\log^{1+2\varepsilon} n)$ time.

When a new point (x, y) is inserted into (removed from) the data structure, we update the value associated with $s = \lceil y/\sqrt{\log n} \rceil$ in all C_{ij} for $1 \le i \le x \le j \le W$. We also add (x, y) into (remove (x, y) from) the list L_s. Hence, insertions and deletions are supported in $O(\log^{3/2+2\varepsilon} n)$ time. To obtain the result of the Lemma, it remains to substitute $\varepsilon' = \varepsilon/2$ into the above proof.

Theorem 3. *There exists a data structure for two-dimensional orthogonal semi-group range sum queries that uses* $O(n \log n/ \log\log n)$ *space and supports queries in* $O((\log n/ \log\log n)^2))$ *time and updates in* $O(\log^{9/2+\varepsilon} n)$ *time.*

Proof. The data structure is organized in the same way as the data structures of Theorems 1 and 2, but in every node v of the range tree T_x a data structure F_v of Lemma 3 is stored. A two-dimensional range sum query can be answered by answering $O(\log n/ \log\log n)$ queries to data structures F_v. An update operation leads to updating $O(\log n/ \log\log n)$ data structures F_v in $O(\log^{5/2} n/ \log\log n)$.

Besides that we must update the labeling l_v in $O(\log n/\log\log n)$ nodes v. In the same way as in the proof of Theorem 1, inserting or deleting a v-label can lead to v-moving $O(\log^2 n)$ elements. Hence, $O(\log^2 n)$ elements must be removed and re-inserted into F_v for $O(\log n/\log\log n)$ nodes v, and the total cost of an update operation is $O(\log^2 n \log^{3/2+\varepsilon} n(\log n/\log\log n)) = O(\log^{9/2+\varepsilon} n)$.

References

1. Alstrup, S., Brodal, G.S., Rauhe, T.: New Data Structures for Orthogonal Range Searching. In: Proc. 41st FOCS 2000, pp. 198–207 (2000)
2. Arge, L., Vitter, J.S.: Optimal External Memory Interval Management. SIAM J. on Computing 32(6), 1488–1508 (2003)
3. Bender, M.A., Cole, R., Demaine, E.D., Farach-Colton, M., Zito, J.: Two Simplified Algorithms for Maintaining Order in a List. In: Möhring, R.H., Raman, R. (eds.) Proc. 10th ESA 2002. LNCS, vol. 2461, pp. 152–164. Springer, Heidelberg (2002)
4. Bentley, J.L.: Multidimensional Divide-and-Conquer. Commun. ACM 23, 214–229 (1980)
5. Blandford, D.K., Blelloch, G.E.: Compact Representations of Ordered Sets. In: Proc. 15th SODA 2004, pp. 11–19 (2004)
6. Chazelle, B.: A Functional Approach to Data Structures and its Use in Multidimensional Searching. SIAM J. on Computing 17, 427–462 (1988)
7. Elias, P.: Universal Codeword Sets and Representations of the Integers. IEEE Transactions on Information Theory 21, 194–203 (1975)
8. van Emde Boas, P.: Preserving Order in a Forest in Less Than Logarithmic Time and Linear Space. Inf. Process. Lett. 6(3), 80–82 (1977)
9. van Emde Boas, P., Kaas, R., Zijlstra, E.: Design and Implementation of an Efficient Priority Queue. Mathematical Systems Theory 10, 99–127 (1977)
10. Itai, A., Konheim, A.G., Rodeh, M.: A Sparse Table Implementation of Priority Queues. In: Even, S., Kariv, O. (eds.) Proc. 8th ICALP 1981. LNCS, vol. 115, pp. 417–431. Springer, Heidelberg (1981)
11. Van Kreveld, M., Overmars, M.H.: Divided K-d Trees. Algorithmica 6(6), 840–858 (1991)
12. Mehlhorn, K.: Data Structures and Algorithms 1: Sorting and Searching. Springer, Heidelberg (1984)
13. Mortensen, C.W.: Fully Dynamic Orthogonal Range Reporting on RAM. SIAM J. on Computing 35, 1494–1525 (2006)
14. Nekrich, Y.: Space Efficient Dynamic Orthogonal Range Reporting. In: Proc. 21st SoCG 2005, pp. 306–313 (2005)
15. Nekrich, Y.: A Linear Space Data Structure for Orthogonal Range Reporting and Emptiness Queries. In: Proc. 18th CCCG 2006, pp. 159–163 (2006)
16. Mortensen, C.W., Pagh, R., Pătraşcu, M.: On Dynamic Range Reporting in One Dimension. In: Proc. 37th STOC 2005, pp. 104–111 (2005)
17. Pătraşcu, M.: Lower Bounds for 2-Dimensional Range Counting. In: Proc. 39th STOC 2007 (to appear)
18. Pătraşcu, M., Demaine, E.D.: Tight Bounds for the Partial-Sums Problem. In: Proc. 15th SODA 2004, pp. 20–29 (2004)
19. Willard, D.E.: A Density Control Algorithm for Doing Insertions and Deletions in a Sequentially Ordered File in Good Worst-Case Time. Information and Computation 97, 150–204 (1992)

Spherical LSH for Approximate Nearest Neighbor Search on Unit Hypersphere

Kengo Terasawa and Yuzuru Tanaka

Meme Media Laboratory, Hokkaido University,
N-13, W-8, Sapporo, 060–8628, Japan
{terasawa,tanaka}@meme.hokudai.ac.jp

Abstract. LSH (Locality Sensitive Hashing) is one of the best known methods for solving the c-approximate nearest neighbor problem in high dimensional spaces. This paper presents a variant of the LSH algorithm, focusing on the special case of where all points in the dataset lie on the surface of the unit hypersphere in a d-dimensional Euclidean space. The LSH scheme is based on a family of hash functions that preserves locality of points. This paper points out that when all points are constrained to lie on the surface of the unit hypersphere, there exist hash functions that partition the space more efficiently than the previously proposed methods. The design of these hash functions uses randomly rotated regular polytopes and it partitions the surface of the unit hypersphere like a Voronoi diagram. Our new scheme improves the exponent ρ, the main indicator of the performance of the LSH algorithm.

1 Introduction

Nearest Neighbor Search is one of the fundamental problems in computer science with applications in information retrieval, pattern recognition, clustering, machine learning, data mining, and so forth. If we have a set of n data points in d-dimensional space, a brute-force search can find the nearest neighbor in $O(dn)$ time. Nowadays, since the size of the data we should treat tends to become larger and larger, the linearity of the complexity to the size of the population has posed a problem. Therefore, we need some new data structure that returns the nearest neighbor of an arbitrarily given point faster than the brute-force method. The main objective of the algorithms for a nearest neighbor search is to build a data structure which, given any query point q, quickly reports the data point that is closest to q.

As a result of an intensive research efforts, this problem was well solved, particularly for low dimensional spaces. However, all of them tend to fail when the dimensionality goes higher: sometimes the time complexity asymptotically tends to be $O(dn)$, which means no improvement over the brute-force method, and sometimes they need memory space exponential to d, which is of course infeasible when d is large. In this way the nearest neighbor search in high dimensions is still a difficult problem. This difficulty is known as "the curse of dimensionality."

F. Dehne, J.-R. Sack, and N. Zeh (Eds.): WADS 2007, LNCS 4619, pp. 27–38, 2007.

In recent years, approximation methods have been proposed to overcome the curse of dimensionality. The c-approximate nearest neighbor problem is the relaxed problem that allows an output point to be at most c times distant than the exact nearest neighbor is. A randomized (or probabilistic) algorithm is also often employed to overcome the curse of dimensionality. With a fixed parameter δ, the randomized algorithm should return requested point with a probability of no less than $1 - \delta$. Using these approximations, many algorithms have been proposed. Among them, one of the best known algorithms is Locality Sensitive Hashing[1,2,3], which is also called LSH.

LSH [1,2,3] is a randomized algorithm for the approximate nearest neighbor problem that runs significantly faster than other existing methods, especially in high dimensional spaces. The basic idea of LSH is to hash each data point into hash tables using a hash function randomly chosen from the locality sensitive hash function family. In finding the nearest neighbor, LSH scans only the points that have the same hash index as the query point. It runs in $\tilde{O}(n^{\rho})$ time, where ρ is the main indicator of the performance, and it satisfies $\rho < 1$. The value of ρ has been improved several times during this decade. It has been reported that state-of-the-art algorithm [3] for high dimensional spaces has reduced the value of ρ to 0.5563 for $c = 1.5$, and to 0.3641 for $c = 2.0$.

The main indicator of performance, noted as ρ, was determined by the design of the hash functions. The firstly proposed LSH [1] used random bit extraction from unary expressions, and showed the performance to be $\rho = 1/c$. In a later improvement of LSH [2], a random projection based on p-stable distributions was employed and ρ was slightly improved from earlier versions. The most recent improvement of LSH is found in [3], which proposed novel hash functions and ρ was significantly improved, especially for $d \leq 24$.

As seen before, much effort has been made to solve the approximate nearest neighbor problem in $\mathbb{R}^d$ space. In practical applications of pattern recognition, however, it is often the case that to find the nearest neighbor on the hypersphere is more important than that in the entire $\mathbb{R}^d$ space. In such cases, descriptor-vectors are sometimes normalized to unit length in preprocessing because their magnitudes are less important than their directions. For example, the SIFT (Scale Invariant Feature Transform) descriptor [4,5], which is one of the most famous descriptors in computer vision, uses a 128-dimensional descriptor-vector normalized to unit length. In text processing, the cosine similarity [6] is widely used. Note that the nearest neighbor search with a cosine similarity measure is equivalent to the nearest neighbor search with a Euclidean distance measure after normalizing all the vectors to unit length. Other than mentioned here, we have many examples of pattern recognition algorithms that use vectors normalized to unit length.

In this paper, we focus on the special case of the approximate nearest neighbor problem — all points in the dataset are constrained to lie on the surface of the unit hypersphere — and propose a variant of LSH that efficiently solves this special problem. Since our algorithm works for datasets of points on a hypersphere, we named our algorithm as Spherical LSH (SLSH).

2 Locality Sensitive Hashing (LSH)

Since our proposal is a variant of the LSH algorithm, we must describe LSH. The following is a brief introduction to the LSH algorithm. A more detailed description will be found in [1].

LSH is a randomized algorithm for solving the (R, c)-NN problem. The (R, c)-NN problem is a decision version of the approximate nearest neighbor problem. It is known that the c-approximate nearest neighbor problem can be reduced to (R, c)-NN problem with complexity $O(\log(n/\varepsilon))$. In the following, c is an approximation factor, and let $c = 1 + \varepsilon$.

Definition 1. The c-approximate nearest neighbor problem *is defined as follows: Given a set P of points in a d-dimensional space $\mathbb{R}^d$, devise a data structure which for any query point $q \in \mathbb{R}^d$ find a point $p \in P$ that satisfies for all $p' \in P$, $d(p, q) \leq c \cdot d(p', q)$.*

Definition 2. The (R, c)-NN problem *is defined as follows: Given a set P of points in a d-dimensional space $\mathbb{R}^d$, and a parameter $R > 0$, devise a data structure which for any query point $q \in \mathbb{R}^d$ does the following:*

- *if there exists a point $p \in P$ s.t. $d(p, q) \leq R$ then return YES and a point $p' \in P$ s.t. $d(p', q) \leq cR$,*
- *if $d(p, q) > cR$ for all $p \in P$ then return NO.*

In [7], Har-Peled showed the following theorem.

Theorem 1. *The c-approximate nearest neighbor problem can be reduced to a (R, c)-NN problem with complexity $O(\log(n/\varepsilon))$.*

LSH can solve the (R, c)-NN problem significantly faster than other existing methods, especially in high dimensional spaces. The basic idea of LSH is to hash every point in the dataset into hash tables using a hash function randomly chosen from the locality sensitive hash function family. Finding the nearest neighbor of a query point involves applying the hash functions to the query point and accumulating the points in the dataset that appear in the corresponding buckets.

The locality sensitive hash function family is an important constituent of the LSH algorithm. For a domain S of the point set, an LSH family is defined as follows:

Definition 3. *A family $\mathcal{H} = \{h : S \to U\}$ is called (r_1, r_2, p_1, p_2)-sensitive if for any $u, v \in S$,*

- *if $d(u, v) \leq r_1$ then $Pr_{\mathcal{H}}[h(u) = h(v)] \geq p_1$,*
- *if $d(u, v) > r_2$ then $Pr_{\mathcal{H}}[h(u) = h(v)] \leq p_2$,*

where $d(u, v)$ is the distance between u and v.

In order for an LSH family to be useful, it has to satisfy the inequalities $p_1 > p_2$ and $r_1 < r_2$.

For solving the (R, c)-NN problem, LSH sets $r_1 = R$ and $r_2 = cR$, and then amplifies the difference of collision probabilities by taking direct product of hash functions, i.e.,

$$g(p) = \{ h_1(p), h_2(p), \ldots, h_k(p) \}, \tag{1}$$

where h_i is a (r_1, r_2, p_1, p_2)-sensitive hash function randomly chosen from the family $\mathcal{H}$. For a query point q, LSH scans only the points that stay in the same bucket as $g(q)$. Since the process is probabilistic, it could occur that the query point and the nearest point stay away from each other. In order to reduce such false negatives, the LSH algorithm makes L hash tables, and scans the points in the union of the buckets corresponding to each of $g_1(p), g_2(p), \ldots, g_L(p)$.

From the settings above, we can obtain the following theorem:

Theorem 2. *LSH can solve the (R, c)-NN problem with $O(dn+n^{1+\rho})$ space and $\tilde{O}(n^\rho)$ time, where $\rho = \frac{\log 1/p_1}{\log 1/p_2}$*

Now, the remaining problem is to design the locality sensitive hash functions. The firstly proposed LSH [1], which worked for the Hamming metric space, used a random bit extraction from unary expressions. It showed the performance to be $\rho = 1/c$. The later improvement of LSH [2] extended the target metric space to arbitrary l_p-norm space with $p \in (0, 2]$, and it also improved the index ρ by using a random projection based on p-stable distributions. The most recent improvement of LSH [3] employed "ball partitioning" instead of the former "grid partitioning" to partition the space and bounded the complexity by

$$\rho = \frac{1}{c^2} + O\left(\frac{\log \log n}{\log^{1/3} n}\right). \tag{2}$$

For the practical variant, the paper [3] also proposed to use the Leech Lattice-based partitioning, which is likely to perform better than the aforementioned "ball partitioning" due to much lower "big-Oh" constants. It uses a lattice called Leech Lattice [8], which is a very symmetric lattice embedded in a 24-dimensional space. When the dimensionality is larger than 24, dimensionality reduction is needed before using Leech Lattice.

These improvements are all concerned with the problem of: "How to partition the space well?" Leech Lattice-based partitioning, which is round and symmetric, is quite a nice partitioning method except that it can be applied only to a 24-dimensional space.

Now recall that the purpose of this paper is to solve the nearest neighbor problem on a unit hypersphere. While the examples presented above are all considering the general problem to partition the entire $\mathbb{R}^d$ space, we have to solve the special problem of partitioning the unit hypersphere embedded in $\mathbb{R}^d$. Is there any partitioning that works nicely especially for the hypersphere? Our answer will be described in the next section.

3 Spherical LSH (SLSH)

Here we propose a novel locality sensitive hash function that performs better than the previously proposed ones. While earlier LSH families are considering arbitrary points in $\mathbb{R}^d$ space as in [1,2,3], we are considering arbitrary points on the unit $(d-1)$-sphere embedded in $\mathbb{R}^d$ space with center at the origin. In other words, all we have to do is to partition the surface of the unit hypersphere in $\mathbb{R}^d$, in contrast to the fact that the earlier LSH algorithm [1,2,3] had to partition the entire $\mathbb{R}^d$ space. This section describes the locality sensitive hash functions for partitioning the surface of the unit hypersphere in high dimensions. We named this process as SLSH (Spherical LSH).

3.1 Problem Description

The problem of interest in this paper is defined as follows.

Definition 4. The (R, c)-NN problem on the Unit Hypersphere: *Given a set P of points in a d-dimensional space $\mathbb{R}^d$, and all points $p \in P$ satisfies that $||p|| = 1$. Given a parameter $R > 0$, devise a data structure which for any query point $q \in \mathbb{R}^d$ satisfying $||q|| = 1$ does the following:*

- *if there exists a point p s.t. $d(p, q) \leq R$ then return YES and a point p' s.t. $d(p', q) \leq cR$,*
- *if $d(p, q) > cR$ for all $p \in P$ then return NO.*

This is a special case of the (R, c)-NN problem (Def. 2), but with a wide application area as described in Section 1.

3.2 Locality Sensitive Hash Functions Using a Regular Polytope

SLSH uses a randomly rotated regular polytope for partitioning the surface of the unit hypersphere. After rotating the polytope at random, the hash function $h(p)$ is then defined as the number assigned to the vertex which is nearest to p. In other words, our hash function partitions the surface of the unit hypersphere like a Voronoi diagram.

First let us go through some notations.

Hypersphere: A hypersphere is the generalization of the sphere to higher dimensions. Often the symbol $\mathbb{S}^n$ is used to represent the n-sphere that has an n surface dimension and is embedded in an $(n+1)$-dimensional space.
The unit hypersphere is the hypersphere whose radius is unity. From now on we will consider the unit $(d-1)$-sphere, whose center is located at the origin. Note that the $(d-1)$-sphere is embedded in $\mathbb{R}^d$.

Regular polytope: A regular polytope is the generalization of the regular polygon (in two-dimensional space) and the regular polyhedron (in three-dimensional space) to higher dimensions. It has a high degree of symmetry such as the following:

- All edges have an equal length, which means that the distance between the adjacent vertices are always the same.
- All faces are congruent.

It is known that there exist only three kinds of regular polytopes in higher ($d \geq 5$) dimensions, namely,

Simplex, having $d+1$ vertices, is analogous to the tetrahedron.
Orthoplex (Cross polytope), having $2d$ vertices, is analogous to the octahedron.
Hypercube (Measure polytope), having 2^d vertices, is analogous to the cube.

Suppose that we randomly rotate the regular polytope inscribed in a unit $(d-1)$-sphere. We can partition the $(d-1)$-sphere so that all points belong to the nearest vertex of the rotated regular polytope.

Definition 5. (The Key idea of our algorithm): *Let* $\{\tilde{v}_1, \tilde{v}_2, \ldots, \tilde{v}_N\}$ $(||\tilde{v}_i||^2 = 1)$ *be a set of vertices that forms a regular polytope in* $\mathbb{R}^d$ *where* N *represents the number of vertices of the employed polytope, and let* A *be a rotation matrix. For an arbitrary unit vector* p*, a hash function* $h_A(p)$ *is defined as the following:*

$$h_A(p) = \text{argmin}_i \ ||A\tilde{v}_i - p||^2. \tag{3}$$

Note that for a given p, we can obtain $h_A(p)$ in $O(d^2)$ time for every type of regular polytope (it will be discussed later).

By considering A as an arbitrary rotation matrix in $\mathbb{R}^d$ space, $\mathcal{H} = \{h_A\}$ satisfies the definition of the locality sensitive hash function family. SLSH uses this LSH family for hashing.

3.3 The Algorithm

Here we will describe the details of the algorithm.

The coordinates of the vertices of the regular polytope in d-dimensional space are given by the following:

Simplex:

$$[\tilde{\boldsymbol{v}}_i]_j = \delta_{ij} - \frac{d+1-\sqrt{d+1}}{d(d+1)} \qquad (i = 1, 2, \ldots, N) \tag{4}$$

$$[\tilde{\boldsymbol{v}}_{N+1}]_j = \frac{1-\sqrt{d+1}}{d} - \frac{d+1-\sqrt{d+1}}{d(d+1)} \tag{5}$$

where $[\tilde{\boldsymbol{v}}_i]_j$ represents the j-th coordinate of the i-th vertex $\tilde{\boldsymbol{v}}_i$, and δ_{ij} is 1 for $i = j$ and 0 otherwise.

Orthoplex: All permutations of $(\pm 1, 0, 0, \cdots, 0)$ give the coordinates of the vertices. It follows that orthoplex has $2d$ vertices.

Hypercube: $\frac{1}{\sqrt{d}} \times (\pm 1, \pm 1, \cdots, \pm 1)$ give the coordinates of the vertices. It follows that the hypercube has 2^d vertices.

Let us consider how to obtain the nearest vertex efficiently. Instead of directly solving Eq.(3), it is computationally easier to solve

$$h_A(p) = \mathrm{argmax}_i(A\tilde{v}_i \cdot p). \tag{6}$$

If we calculate $\{v_i = A\tilde{v}_i \mid i = 1, \cdots, N\}$ in advance, a $d+1$ dot-product calculation would suffice to return $h_A(p)$ for the simplex. For the orthoplex, doing a $2d$ dot-product calculation is also possible; however, a more efficient way exists. If v is a vertex of the orthoplex, then $-v$ is also a vertex of the orthoplex. The dot-product of $-v$ and p is just $-(v \cdot p)$. Therefore, we do not have to calculate the dot-product $2d$ times — a calculation of only d times will suffice. For hypercube, naively solving Eq.(6) needs a 2^d times dot-product calculation that is of course infeasible. However, a way exists to avoid such prohibitive calculations. The same partitioning can be obtained by a d times bisection using orthonormal basis vectors $\{e_1, \cdots, e_d\}$. We can bisect the hypersphere using each one of the basis vectors, i.e., $Bisec_{e_i}(p) = 1$ if $(e_i \cdot p) \geq 0$ and $Bisec_{e_i}(p) = 0$ otherwise. Then we could construct the map $p \in \mathbb{S}^{d-1} \mapsto \{0,1\}^d \mapsto \{0, 1, \cdots, 2^d - 1\}$. This partitioning is equivalent to a partitioning based on the nearest vertices of a hypercube. Thus, we can conclude the following: for every type of regular polytope, $h_A(p)$ can be calculated in $O(d^2)$ time.

Let us discuss the preprocessing cost. Preprocessing of SLSH costs $O(d^3 + d^2 n)$ time for one hash table. The former d^3 is the cost to make random rotation matrix. The latter $d^2 n$ is to hash all points in the dataset. The memory space overhead is $O(d^2 L)$ to store the rotated vertices, and $O(nL)$ to store the hash index of all points in the datasets. Note that the number of the hash tables L can be bounded by n^ρ.

The algorithm is summerized in Fig. 3 and Fig. 4.

4 Performance Evaluation

As mentioned before, the performance of LSH is evaluated by the index $\rho = \frac{\log 1/p_1}{\log 1/p_2}$. In this section we describe the evaluation of the ρ.

Again, p_1 is the probability of collision of two points with a distance of R, and p_2 is the probability of collision of two points with a distance of cR. In the original LSH they can change the scale of the coordinate; thus, they can assume $R = 1$ without loss of generality. On the other hand, we cannot scale the coordinate because SLSH works only to $\mathbb{S}^{d-1}$. Therefore the performance index ρ of SLSH must be evaluated for several Rs.

For ease of comparison, we evaluated several types of locality sensitive hash functions. The candidates were as follows:

SLSH (Our proposal): Partitioning based on the rotated regular polytope.
Leech Lattice: Proposed in [3], with the dimensionality reduction of d to 24.
Spherical Bisection: Another candidate for partitioning the hypersphere. The Sperical Bisection bisects the hypersphere with a *random* set of vectors, in contrast to our "hypercube-partitioning" which bisects the hypersphere with an *orthonormal* set of vectors.

Table 1. Probabilities of collision for two points with the distance of r. Except for the value of the Leech Lattice, which is cited from [3], the values displayed below are obtained through our Monte-Carlo simulation for 10^6 trials (We omitted the value of spherical bisection. It is just $1 - \cos^{-1}(1 - r^2/2)/\pi$).

Distance r	LeechLattice ($d > 24$)	16-dim simplex	16-dim orthoplex	16-dim hypercube	64-dim simplex	64-dim orthoplex	64-dim hypercube
0.10		0.90133	0.88612	0.59084	0.87169	0.85846	0.12152
0.20		0.80746	0.77939	0.33587	0.75356	0.73061	0.01271
0.30		0.71800	0.67894	0.18092	0.64590	0.61412	0.00116
0.40		0.63309	0.58535	0.09186	0.54531	0.50879	0.00008
0.50		0.55276	0.49754	0.04315	0.45407	0.41365	0.00000
0.60		0.47649	0.41595	0.01836	0.37078	0.32937	0.00000
0.70	0.08535	0.40459	0.34066	0.00676	0.29652	0.25503	0.00000
0.80	0.05259	0.33750	0.27211	0.00212	0.23071	0.19144	0.00000
0.90	0.03117	0.27473	0.21051	0.00050	0.17326	0.13748	0.00000
1.00	0.01779	0.21676	0.15533	0.00006	0.12449	0.09314	0.00000
1.10	0.00975	0.16419	0.10797	0.00000	0.08456	0.05854	0.00000
1.20	0.00515	0.11785	0.06906	0.00000	0.05260	0.03326	0.00000
1.30	0.00266	0.07826	0.03872	0.00000	0.02921	0.01656	0.00000
1.40	0.00133	0.04622	0.01789	0.00000	0.01378	0.00644	0.00000
1.50	0.00067	0.02253	0.00587	0.00000	0.00504	0.00181	0.00000
1.60	0.00033	0.00775	0.00108	0.00000	0.00117	0.00028	0.00000
1.70	0.00016	0.00124	0.00006	0.00000	0.00010	0.00001	0.00000
1.80	0.00008	0.00003	0.00000	0.00000	0.00000	0.00000	0.00000
1.90	0.00004	0.00000	0.00000	0.00000	0.00000	0.00000	0.00000
2.00	0.00002	0.00000	0.00000	0.00000	0.00000	0.00000	0.00000

Let $p(r)$ represent the collision probability of a single hash function with respect to the distance r, i.e., $p(r) = \Pr_{\mathcal{H}}[h(u) = h(v)]$ for the two point u and v that satisfy $||u - v|| = r$.

In Table 1, the values of $p(r)$ are displayed. The values of the Leech Lattice were cited from [3]. The values of the Spherical Bisection were analytically obtained by the equation $p(r) = 1 - \theta/\pi$, where θ represents the angle between two vectors measured in radians. The cosine of θ and the Euclidean distance r has the relationship as $r^2 = 2(1 - \cos\theta)$. The values $p(r)$ of the SLSH were computed by the Monte-Carlo simulation for 10^6 trials. Note that the value $p(r)$ does not depend on input sets, distributions, or anything like that: it relies on the probability of two points being hashed to the same value, which can be computed using Monte-Carlo methods fairly accurately.

Figures 1 and 2 plot the value of ρ vs. $R = r_1$. As mentioned before, the performance of the Leech Lattice method does not depend on R because the coordinate scale is changeable in the Leech Lattice method. Therefore, only the best ρ over R is plotted for the Leech Lattice. For the SLSH and the Spherical Bisection, the value of $\rho = \frac{\log 1/p_1}{\log 1/p_2}$ was calculated from $p(r)$ displayed in Table 1.

Table 2. The values of ρ

R	c	Leech Lattice ($d > 24$)	SLSH for 64-dim orthoplex
0.64	1.5	0.5563	0.5471
0.72	1.5	0.5563	0.5189
0.80	1.5	0.5563	0.4858
0.56	2.0	0.3641	0.3456
0.64	2.0	0.3641	0.3063

In the calculation we did not use the value $p(r)$ less than 0.00001, since such values are less reliable and inappropriate for SLSH implementation.

From these figures, we can observe the following: For any dimensionalities and for any polytopes, SLSH performs better than the Spherical Bisection method. Comparing ρ on the same polytopes, larger dimensionality implies better ρ. It means that our method could avoid the curse of dimensionality. Comparing ρ on various types of polytopes, simplex and orthoplex tend to show a similar result except that the orthoplex shows slightly better result when R becomes larger. The hypercube shows quite different behavior from the others. The collision probability $p(r)$ of the hypercube rapidly drops to near zero, especially in high dimensional spaces. Although the index ρ is lower than the others are, its use for a practical application is difficult because too small $p(r)$ prevent the proper adjustment of k and L, that are restricted to positive integer. Comparing SLSH with the Leech Lattice-based method, in the case of $c = 1.5$, for almost all dimensionalities, SLSH performs better than the Leech Lattice-based method when R is larger than 0.60–0.64. In the case of $c = 2.0$, for almost all dimensionalities, SLSH performs better than the Leech Lattice-based method when R is larger than 0.48–0.52. We displayed some of the representative values in Table 2.

Here we note that $R = 0.65$ and $c = 1.5$ are a practical values found in the literature. The paper [9] says that all but 3% of the data points have their nearest neighbor within a distance of $R = 0.65$ in their experiment. Furthermore, if we want to implement the c-approximate nearest neighbor, its reduction to (R, c)-NN needs the hierarchical implementation of LSH such as $R = C_0 c^\lambda$ for $\lambda = 1, 2, \cdots, \Lambda$. SLSH helps substitute for the LSH-based c-approximate nearest neighbor in some hierarchies of the implementation.

5 Discussion

Why could SLSH outperform the original LSH? One of the reasons is that SLSH could work on the original dimension. It does not need the dimensionality reduction. In dimensionality reduction, one cannot avoid the chance of far away points colliding to the near point. It is the disadvantage of dimensionality reduction. We can avoid such disadvantages by not using the dimensionality reduction. We have developed a good partitioning method that can be applied to any dimensional hypersphre. In our partitioning, the point $\boldsymbol{p}_1$ and $\boldsymbol{p}_2 = -\boldsymbol{p}_1 + \boldsymbol{\varepsilon}$, where $\boldsymbol{\varepsilon}$

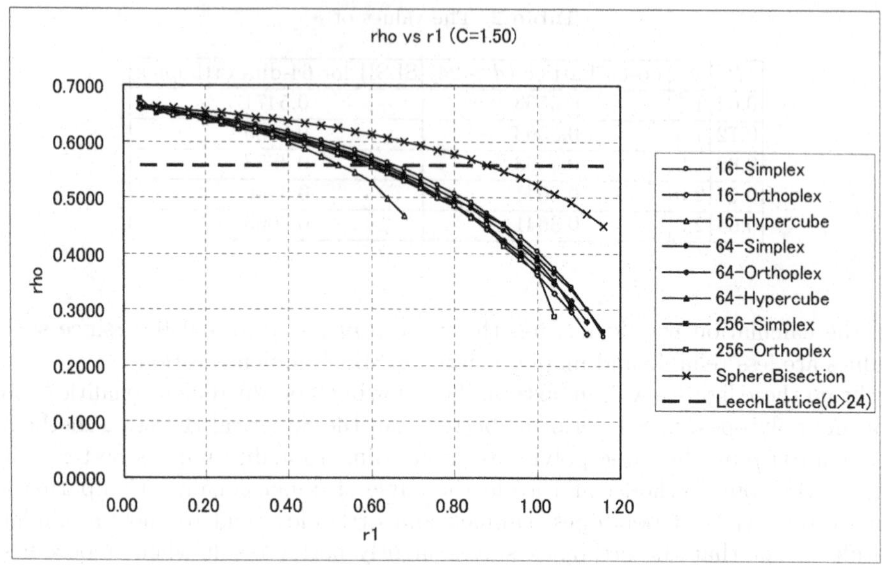

Fig. 1. The values of $\rho = \frac{\log p_1}{\log p_2}$ in case $c = 1.5$

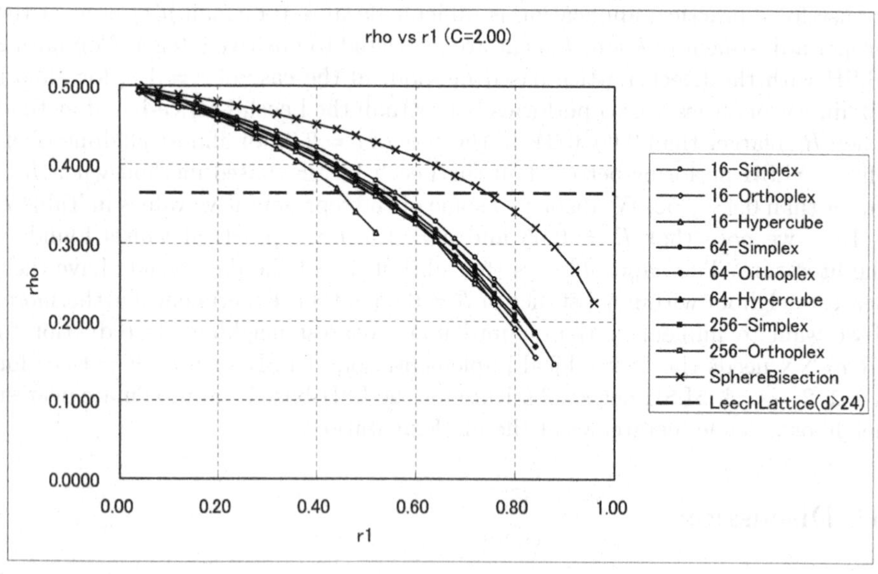

Fig. 2. The values of $\rho = \frac{\log p_1}{\log p_2}$ in case $c = 2.0$

is an arbitrary small vector, will *never* collide for any type of polytope, say, simplex, orthoplex, or hypercube. If we try to guarantee that $\boldsymbol{p_1}$ and $\boldsymbol{p_2}$ will never collide by the Spherical Bisection method, we need at least d times partitioning

simplex — The range of the function is: $h_A(p) \in \{1, \ldots, d+1\}$
Preprocessing:
Let $\{\tilde{v}_i \mid i = 1, \ldots, d+1\}$ be a set of vectors described in Eq. (4) and (5).
Let A be a random rotation matrix.
For $i \leftarrow 1$ **to** $d+1$ **do**
$v_i \leftarrow A\tilde{v}_i$
(Then $\{v_1, v_2, \cdots, v_d\}$ forms a randomly rotated simplex.)
Calculation of $h_A(p)$ **for input** p**:**
$h_A(p) \leftarrow \text{argmax}_i(v_i \cdot p)$
Return $h_A(p)$

orthoplex — The range of the function is: $h_A(p) \in \{1, \ldots, 2d\}$
Preprocessing:
Let v_i be the i-th column of a random rotation matrix A.
(Then $\{v_1, v_2, \cdots, v_d, -v_1, -v_2, \cdots, -v_d\}$ forms a randomly rotated orthoplex.)
Calculation of $h_A(p)$ **for input** p**:**
$h_A(p) \leftarrow \text{argmax}_i |v_i \cdot p|$.
if $(v_{h_A(p)} \cdot p) < 0$ **then** $h_A(p) \leftarrow h_A(p) + d$
Return $h_A(p)$

hypercube — The range of the function is: $h_A(p) \in \{0, \ldots, 2^d - 1\}$
Preprocessing:
Let v_i be the i-th column of a random rotation matrix A.
(Then $\{u_{s_1 s_2 \cdots s_d} = \sum s_i v_i \mid s_i = \pm 1\}$ forms a randomly rotated hypercube.)
Calculation of $h_A(p)$ **for input** p**:**
$h_A(p) \leftarrow 0$
For $i \leftarrow 1$ **to** d **do**
if $(v_i \cdot p) \geq 0$ **then** $h_A(p) \leftarrow h_A(p) + 2^{i-1}$
Return $h_A(p)$

Fig. 3. The implementation of $h_A(p)$ for each type of regular polytope

How to make a random rotation matrix:

For $i \leftarrow 1$ **to** d **do**
Let v_i be a random vector from the d-dimensional Gaussian distribution.
For $j \leftarrow 1$ **to** $i - 1$ **do**
$v_i \leftarrow v_i - (v_i \cdot v_j) v_j / |v_j|$ *(Gram-Schmidt orthogonalization)*
$v_i \leftarrow v_i / |v_i|$ *(normalize to unit length)*
Return $(v_1 \; v_2 \; \cdots \; v_d)$

Fig. 4. Algorithm for making a random rotation matrix

(It is similar to our hypercube method). However, like our hypercube method, it tends to partition the space too thin. It may be understood by the fact that it partitions the space into 2^d fragments. On the other hand, our simplex method and orthoplex method partition the space into $d+1$ or $2d$ fragments. It is a milder partitioning than that of the partitioning into 2^d fragments, but points far away will collide with very little probability. The aforementioned supports the efficacy of our algorithm.

6 Conclusion

In this paper we have proposed an algorithm to solve the approximate nearest neighbor problem when all points are constrained to lie on the surface of the unit hypersphere. Our algorithm, named SLSH, is based on the LSH scheme, and outperforms state-of-the-art LSH variants.

References

1. Gionis, A., Indyk, P., Motwani, R.: Similarity Search in High Dimensions via Hashing. In: VLDB 1999. Proc. 25th International Conference on Very Large Data Bases, pp. 518–529 (1999)
2. Datar, M., Indyk, P., Immorlica, N., Mirrokni, V.: Locality-Sensitive Hashing Scheme Based on p-Stable Distributions. In: Proc. Symposium on Computational Geometry 2004, pp. 253–262 (2004)
3. Andoni, A., Indyk, P.: Near-Optimal Hashing Algorithms for Approximate Nearest Neighbor in High Dimensions. In: FOCS'06. Proc. 47th Annual IEEE Symposium on Foundations of Computer Science, pp. 459–468 (2006)
4. Lowe, D.G.: Object recognition from local scale-invariant features. In: ICCV'99. Proc. 7th International Conference on Computer Vision, vol. 2, pp. 1150–1157 (1999)
5. Lowe, D.G.: Distinctive image features from scale-invariant keypoints. International Journal of Computer Vision 60(2), 91–110 (2004)
6. Salton, G., McGill, M.J.: Introduction to Modern Information Retrieval. McGraw-Hill, New York (1983)
7. Har-Peled, S.: A Replacement for Voronoi Diagrams of Near Linear Size. In: FOCS'01. Proc. 42nd Annual Symposium on Foundations of Computer Science, pp. 94–103 (2001)
8. Leech, J.: Notes on sphere packings. Canadian Journal of Mathematics, 251–267 (1967)
9. Darrell, T., Indyk, P., Shakhnarovich, G. (eds.): Nearest Neighbor Methods in Learning and Vision: Theory and Practice. MIT Press, Cambridge (2006)

A 4/3-Approximation Algorithm for Minimum 3-Edge-Connectivity

Prabhakar Gubbala[1] and Balaji Raghavachari[2]

[1] Microsoft Corporation, Redmond, WA 98052
[2] Computer Science Department, University of Texas at Dallas, Richardson, TX 75080, USA
{prabha,rbk}@utdallas.edu

Abstract. The minimum cardinality 3-edge-connected spanning subgraph problem is considered. An approximation algorithm with a performance ratio of $4/3 \approx 1.33$ is presented. This improves the previous best ratio of 3/2 for the problem. The algorithm also works on multigraphs and guarantees the same approximation ratio.

Keywords: Approximation algorithms, Graph and network algorithms, Connectivity, Combinatorial Optimization.

1 Introduction

We study the minimum-cardinality 3-edge-connected spanning subgraph (3-ECSS) problem. The corresponding vertex-connectivity problem is 3-VCSS. A graph is *k-edge-connected* if the deletion of up to $k-1$ edges leaves a connected graph. It can also be expressed as a graph that has k edge-disjoint paths between every pair of vertices. The 3-ECSS problem is a fundamentally important problem. We demonstrate using examples that the previous approaches can not guarantee a ratio of 4/3. We prove a new lower bound for 3-ECSS and use it to derive a 4/3-approximation algorithm.

It is known that an inclusion-wise minimal subgraph that is 3-connected is a 2-approximation. In a landmark paper, Cheriyan and Thurimella [1] presented an elegant algorithm (that we will call the CT algorithm) that achieves a performance ratio of $1 + \frac{2}{k+1}$ for k-ECSS. For 3-ECSS, the ratio is 3/2 for simple graphs. Gabow [2] has given a 3/2 approximation algorithm for 3-ECSS in multigraphs. Gabow et al [3] have analyzed algorithms for k-ECSS by using a linear programming approach. Interestingly, the ratio for 3-VCSS obtained by the CT algorithm is 4/3, which is smaller than the 3/2 ratio for 3-ECSS. In fact, if the given graph is 3-vertex-connected, then their 3-VCSS algorithm finds a 4/3-approximate solution for the 3-ECSS problem! The gap between the performance ratios of 3-VCSS and 3-ECSS is only in graphs that are 3-edge-connected, but are not 3-vertex-connected. We close this gap by presenting an algorithm for 3-ECSS with a performance ratio of 4/3.

Khuller and Vishkin [6] obtained approximation algorithms for 2-ECSS and 2-VCSS with ratios 3/2 and 5/3 respectively. Vempala and Vetta [9] improved

F. Dehne, J.-R. Sack, and N. Zeh (Eds.): WADS 2007, LNCS 4619, pp. 39–51, 2007.

the ratios for both problems to 4/3. The ratio for 2-ECSS was subsequently improved to 5/4 by Jothi et al [5], and the ratio for 2-VCSS was improved to 9/7 by Gubbala and Raghavachari [4]. There are two surveys that highlight the work in this area by Khuller [7], and Kortsarz and Nutov [8].

2 Definitions

Let $G = (V, E)$ be the given graph, with $|V| = n$. Let *Opt* be an optimal 3-ECSS of G. We will also use *Opt* to denote the cardinality of an optimal 3-ECSS of G, and this should cause no confusion. The term "connectivity" in this paper refers to edge-connectivity, unless explicitly stated as "vertex-connectivity".

A subgraph $H = (V, E')$ with $E' \subseteq E$ is called a *2-edge cover* if the degree of each node is at least 2. A *2-matching* is a subgraph, all of whose vertices have degree 2 or less. The minimal 2-edge cover and the maximal 2-matching problems are closely related, and a solution for one can be easily converted into a solution for the other. Both problems are solved using algorithms for matching.

Given a cycle C, we say that two chords (a, b) and (x, y) *cross* each other if the vertices appear in the order a, x, b, y or a, y, b, x when we go around the cycle. We call this two edge set a crossing. We define two different kinds of edge crossings in a specific setting. A crossing can be a useful crossing (UFC) or a useless crossing (ULC) with respect to a pair of vertices in a cycle. Let u and v be any two vertices on C. We call a crossing $ULC(u, v)$ if one edge in the crossing is incident on u and the other edge is incident on v. We call all other kinds of crossings $UFC(u, v)$.

3 Locally Connected Cycles and a New Lower Bound

We call a cycle of length k a *locally connected cycle* (lcc) if the following conditions are met: (a) $k \geq 6$, (b) $k-2$ vertices of the cycle have no outside neighbors, and (c) the cycle has no useful crossings with respect to the vertices that have outside neighbors. Figure 12 shows an lcc in which all nodes outside the cycle have been grouped into a set called the core. It can be shown that $2k - 3$ edges are incident on these $k - 2$ vertices even in an optimal solution, and all of these edges are contained within the nodes of the cycle. We now prove a new lower bound for minimum cardinality 3-ECSS problem. We use it in the analysis of our algorithm to prove that the approximation ratio is 4/3.

Lemma 1. *Consider $G = (V, E)$, a 3-edge-connected graph on n vertices that has no cut vertices. Let p be the number of paths in a maximum cardinality 2-matching. Let L be a set of locally connected cycles, all of whose vertices are disjoint. Let $|L| = lc$. Then, $Opt \geq 3n/2 + 3p/4 + 3lc/4$.*

Proof. Due to lack of space, we provide just a sketch of the proof. Suppose we shrink the vertices of an lcc of k vertices into a single vertex. This operation decreases the number of vertices in the graph by $k - 1$, but *Opt* decreases by

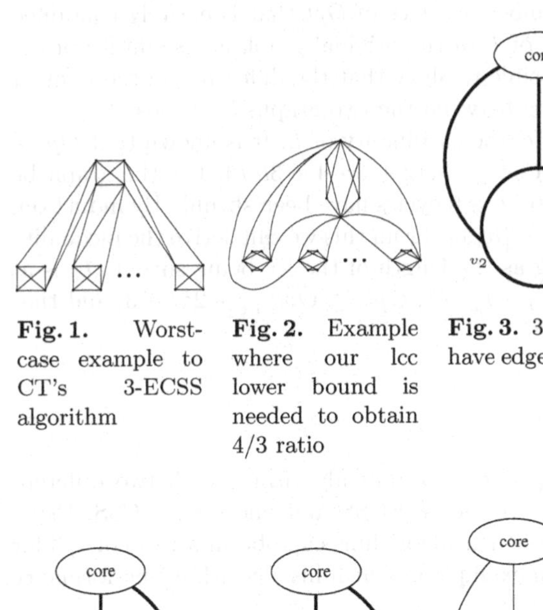

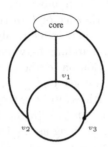

Fig. 1. Worst-case example to CT's 3-ECSS algorithm

Fig. 2. Example where our lcc lower bound is needed to obtain 4/3 ratio

Fig. 3. 3 vertices have edges to core

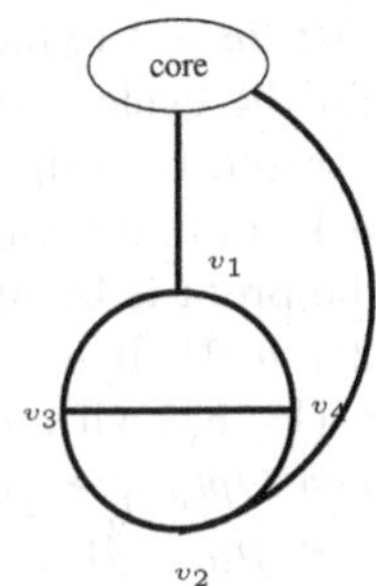

Fig. 4. (v_3, v_4) crosses (v_1, v_2)

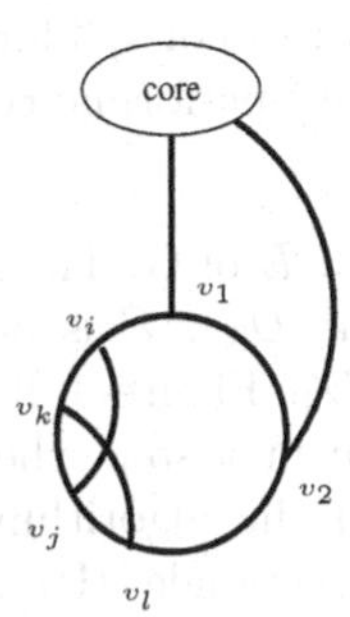

Fig. 5. (v_i, v_j) crosses (v_k, v_l)

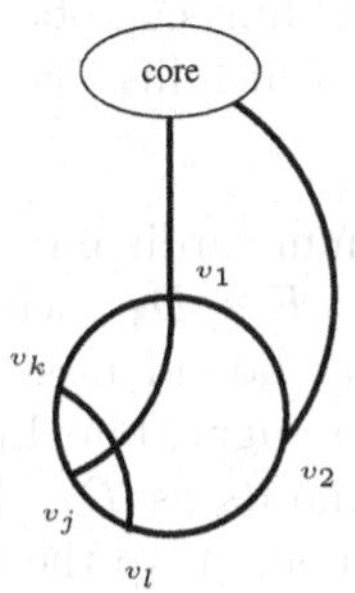

Fig. 6. (v_1, v_j) crosses (v_k, v_l)

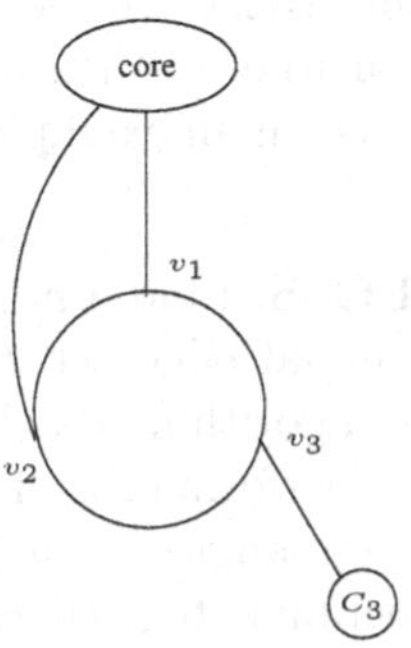

Fig. 7. v_3 has descendant c_3

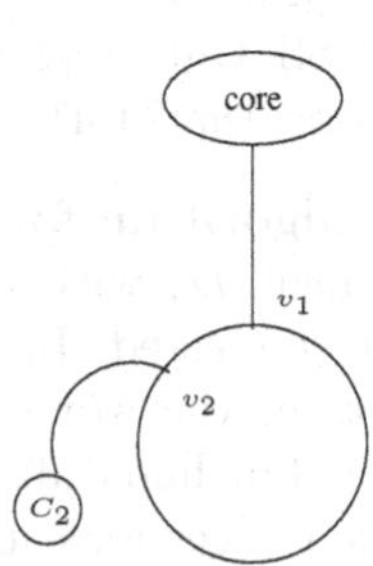

Fig. 8. v_2 has a descendant

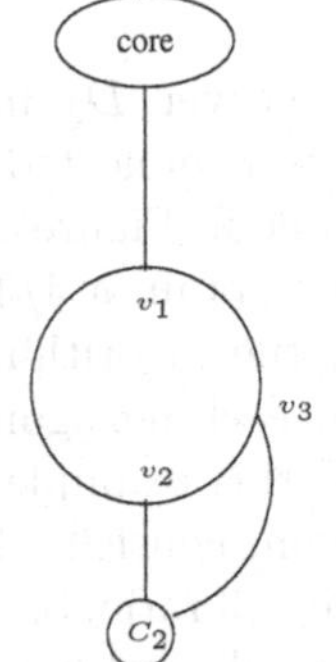

Fig. 9. desc. of v_2 has edge to v_3

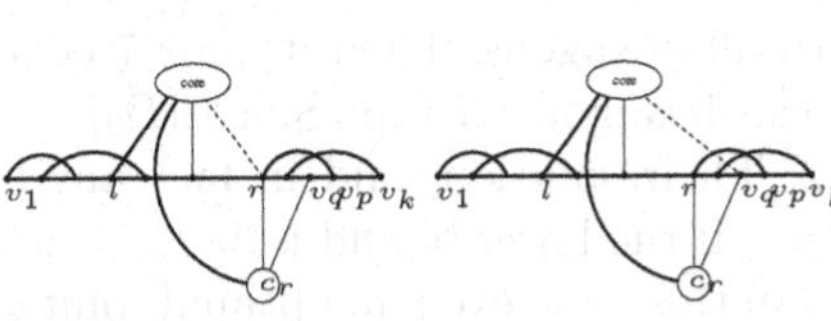

Fig. 10. Closed ear generated at r

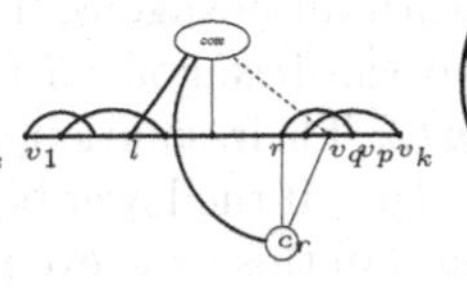

Fig. 11. Closed ear becomes open on changing UTV

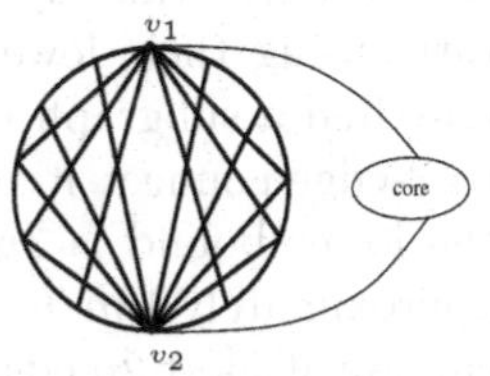

Fig. 12. A locally connected cycle

at least $2k-3$ (which is the number of edges of *Opt* that is entirely contained within the cycle). Any other lcc of L in the original graph is also an lcc of the new graph. In addition, it is possible to show that the difference in the number of paths in a maximal 2-matching between the two graphs is at most 1.

The proof is by induction on lc, the cardinality of L. It is known that $Opt \geq 3n/2+p$ [1]. If $lc=0$, we get $Opt \geq 3n/2+3p/4+3lc/4$. Let the graph be $G_c=(V_c,E_c)$ with n_c vertices after $c \geq 0$ cycles have been shrunk. By induction, we have $Opt_{c+1} \geq \frac{3}{2}n_{c+1}+\frac{3}{4}p_{c+1}+\frac{3}{4}lc_{c+1}$. From this we can derive the inequality $Opt_c \geq \frac{3}{2}n_c+\frac{3}{4}p_c+\frac{3}{4}lc_c$ as long as the length of the lcc being shrunk, k_c is at least 6, using the facts $n_c=n_{c+1}+k_c-1$, $Opt_c \geq Opt_{c+1}+2k_c-3$, and that $|p_c-p_{c+1}| \leq 1$ and $|lc_c-lc_{c+1}| \leq 1$.

4 Examples

Cheriyan and Thurimella [1] gave two elegant algorithms with two different lower bounds for k-connectivity, one for k-ECSS and one for k-VCSS. Using examples, we show that neither of their algorithms can obtain a ratio of 4/3 for 3-ECSS. Gabow [2] has shown an example in which his algorithm's performance is worse than 1.41.

CT algorithm for k-ECSS: Find a minimum cardinality $D_3 \subseteq E$ of G. Then augment D_3 with a minimal edge set $F \subseteq E-D_3$ such that $D_3 \cup F$ is 3-edge-connected. In this algorithm, the lower bound used is D_3. Figure 1 illustrates the worst case performance of this algorithm. In the first step, the algorithm finds all the rectangles with 2 chords as D_3. Then the algorithm adds 3 edges for each rectangle to 3-edge-connect it to the top rectangle. Here, $|D_3|=3n/2$ and the output has approximately $9n/4$ edges, a 3/2 approximation. Note that in reality the CT algorithm's performance in this example is better than 3/2, but the lower bounds used by them are unable to establish a better ratio. Our algorithm finds a solution with about $8n/4=2n$ edges.

CT algorithm for k-VCSS: Find a minimum cardinality 2-edge cover, D_2, in G and add to it a minimal edge set $F \subseteq E-D_2$ such that $D_2 \cup F$ is 3-connected. In this algorithm, the lower bound used is $|D_2|+n/2$. The example in Figure 2 shows that if this algorithm is used for 3-ECSS, then it does not obtain a 4/3 ratio using their lower bound of $|D_2|+n/2$. In the first step, the algorithm may find a subgraph with all hexagons. Then it adds 7 edges for each hexagon to 3-edge-connect it to the hexagon on top. Since $|D_2|=n$ in this example, the lower bound is $3n/2$. Their solution, and in fact any solution, consists of approximately $13n/6$ edges. If the lower bound is $3n/2$, to achieve 4/3 ratio, one can use at most $2n$ edges. In this case, even an optimal solution uses about $13n/6$ edges. So, with their lower bound, it is not possible to achieve a ratio of 4/3. Our new lower bound is useful here, since all the 6-cycles are locally connected cycles.

5 Overview of the Algorithm

It is known that the input graph can be assumed to be 2-vertex-connected when we are solving edge connectivity problems because graphs with cut vertices can be split into multiple graphs that are all 2-vertex-connected, and solutions to these graphs can be combined into a solution for the input graph (in an approximation-preserving fashion).

Our algorithm proceeds as follows. We find a maximum cardinality 2-matching (subgraph in which each vertex is incident to at most two edges). We process it to ensure that end vertices of paths are not adjacent to its vertices. One of its cycles is selected as the "core". We 2-connect all the paths and cycles one at a time to the core. We call this the *2-connect step.* In the final step, we add more edges from G to this subgraph to make it 3-connected. We call this the *3-connect step.*

In a 2-connected graph, a set of vertices that are 3-connected among themselves is called a *cluster.* For example, a Hamiltonian cycle of a graph with n vertices has n clusters, each with a singleton vertex. Since edge connectivity forms an equivalence relation among vertices, when a vertex in cluster A is 3-edge-connected to a vertex in cluster B, then all the vertices in both clusters A and B are 3-edge-connected, thus forming a single cluster. When we add an edge that collapses clusters together, we call it a *merger.* Each merger decreases the number of clusters in the graph by one. Sometimes, adding two edges in a 2-connected subgraph may cause three mergers. Identifying and adding these kind of useful edges helps us in finding a 3-connected subgraph with fewer edges. When we 2-connect the graph, we simultaneously try to minimize the number of edges while maximizing the number of mergers. The more the number of mergers in the 2-connect step, the fewer the number of edges that are added in the 3-connect step.

We proved in Lemma 1 that $Opt \geq 3n/2 + 3p/4 + 3lc/4$. Since our goal is a 4/3-approximation, we can use up to $4/3(3n/2 + 3p/4 + 3lc/4) = 2n + p + lc$ edges. We allocate a charge of 2 to each vertex, 1 to each path and 1 to each lcc, and pay for the edges selected using a combination of these charges.

5.1 Virtual Edges

We process the cycles and paths of the 2-matching in a depth-first order (explained in more detail later). An edge (u, v) is called a *virtual* edge if there is a vertex u, whose proper descendant (in the DFS tree) is adjacent to v. We use the term component to refer to either a path or a cycle. While running DFS, the vertex from which we enter a component is called an upper tree vertex (UTV). When we 2-connect a component x (a path or a cycle), we always make sure that there is an edge incident on UTV and x. This is important because it makes the concept of virtual edges work in the algorithm. When we consider selecting a virtual edge (u, v) in the solution from the perspective of vertex u, we actually select the back edge from u's descendant component c_u. The connectivity from u to that descendant component c_u is ensured during the progress of the algorithm because we always make sure that there is an edge incident on UTV and the component that we 2-connect. We also ensure that the edges from a component fall on two different vertices in the core.

5.2 Algorithm 3EdgeConnect(G)

1. Find a maximum 2-matching (maximum cardinality subgraph in which each vertex is incident to at most two edges) in the graph. Let the 2-matching that we obtain be a collection of cycles C and a set of paths P. If an end vertex of a path is adjacent to a vertex on the same path, we can split the path into a cycle and a path. After doing so, the number of paths and the number of edges in the 2-matching are unchanged, but the number of cycles increases by 1. We repeat this transformation until no end vertex of a path is adjacent to a vertex on the same path. After this step, edges from end vertices of a path always fall on other components. This property is useful when we 3-connect paths.
2. Select a cycle as core. Let its length be k. Assign it a credit of k edges for the 3-connect step.
3. Repeat until there is only one component: Select a target component A. If A is a cycle, call 2connectCycle(A) else call 2connectPath(A).
4. Add a minimal set of edges from G that 3-connects the currently 2-connected subgraph.

5.3 Selecting a Target Component

With C and P (cycles and paths in a maximum 2-matching) on hand, we select a cycle and call it "core". We call paths and cycles as components. We build a depth-first (DFS) tree on a graph induced by the components, starting from the core. The set of edges we add in this process forms a spanning tree on the graph induced by the components. We call the vertices of this tree as "nodes" to distinguish them from the vertices of the given graph. We select the target node in DFS order. Because each node is a component, we need a priority scheme within the vertices of a node which determines the order in which edges from each node are considered. Vertices of a path are given decreasing priority from one end to the other, with one end chosen arbitrarily as the "left" end. When a path component is encountered by DFS, we give more priority to edges on vertices that are more to the left. The left-most vertex has highest priority and the priority decreases as we go from left to right. The only exception to this priority scheme is when we give least priority to the vertex through which DFS enters a path. When a cycle is encountered by DFS, we give least priority to the vertex we entered. The adjacent vertex that is left (one of the two directions is arbitrarily chosen as "left") of the vertex we entered has highest priority. As we go further, the priority decreases and we finally hit the vertex we entered that has the least priority.

5.4 2-Connecting a Cycle

In general, a cycle of length k has a charge of $2k$ edges because it has k vertices and each vertex has a charge of 2 edges. So, between the 2-connecting and 3-connecting steps, we are allowed to spend $2k$ edges. But when we encounter an lcc, we can use $2k + 1$ edges because each vertex has a charge of 2 edges and an lcc has a charge of 1 edge. After finishing the 2-connecting step, we make sure that we have enough charge left for the 3-connecting step. In addition, when 2-connecting a cycle, we make sure that there is at least one edge that is incident on UTV and any vertex in the cycle. We also ensure that the cycle is connected to at least two different vertices in the core.

Lemma 2. *Consider a graph $G(V, E)$ with no cut vertices. If c is a cycle of length k, then one of the following is true:*

1. *It can be 3-connected to the core with a charge of at most $2k$ edges, or,*
2. *It is an lcc and it can be 3-connected to the core with a charge of at most $2k + 1$ edges.*

Proof. There has to be at least one edge (that is not a virtual edge) incident on UTV in core and a vertex v_1 in the cycle because we enter the cycle using that particular DFS edge. We will refer to that vertex as v_1, and the corresponding edge as the upper tree edge. We do not need to keep the upper tree edge in the solution. All we need to ensure is there is an edge incident on UTV and one of the vertices of the cycle. We also make sure that whenever we 3-connect a cycle, edges from the cycle are incident on at least two different vertices in the core.

Case 1: There exists at least three vertices, namely v_1, v_2, and v_3, in c that have edges (possibly, virtual edges) to the core: We select the edges of the cycle c and these three edges from three different vertices to the core (see Fig. 3). The 3-connected components in the graph processed so far are $\{core, v_1, v_2, v_3\}$ and $k - 3$ singleton components. We need at most $k - 3$ edges to 3-connect these 3-connected components. We have used $k + 3$ edges in the 2-connect step and leave a charge of $k - 3$ edges for the 3-connect step.

The following idea is used in all of the cases, and we will discuss it just once here. We make sure that all the three edges from the cycle do not fall on the same vertex in the core, which is possible because G does not have cut vertices. The 2-connect algorithm needs to make sure that there is at least one edge incident on UTV. If none of these three edges selected falls on UTV, we add an edge incident on UTV and delete one of the selected three edges.

Case 2: All edges to the core from c come from just two vertices v_1 and v_2:

1. If there exists an edge (v_3, v_4) that crosses (v_1, v_2) in c, we do the following (see Fig. 4). We select all the k edges of the cycle, $(v_1, core)$, $(v_2, core)$ and (v_3, v_4). When selecting edges incident on core, we make sure that both edges do not fall on the same vertex. Now we have spent $k + 3$ edges to 2-connect this cycle of length k. The 3-connected components in the graph are $\{v_1, v_2, v_3, v_4\}$, core and $k - 4$ singleton vertices. So, we need at most $k - 3$

edges to 3-connect this cycle to the core. As we have spent $k + 3$ edges in the 2-connect step, we have enough charge available to finish the 3-connect step.

2. If there exists a $UFC(v_1, v_2)$ in the cycle, we do the following. Let the crossing be $\{(v_i, v_j), (v_k, v_l)\}$.
   (a) If none of the edges that cross each other falls on v_1 or v_2 (see Fig. 5), we select the cycle c with the edges (v_i, v_j), (v_k, v_l), $(v_1, core)$, and $(v_2, core)$. Now we have selected $k + 4$ edges. The 3-connected components in the subgraph we selected are $\{core\}$, $\{v_1, v_2\}$, $\{v_i, v_j, v_k, v_l\}$, and $k - 6$ singleton vertices. We need at most $k - 4$ edges to 3-connect this cycle to the core. As we have spent $k + 4$ edges in the 2-connect step, we have enough charge to finish 3-connect step.
   (b) If one of the edges that cross each other falls on v_1 or v_2 (see Fig. 6), we do the following. Without loss of generality, let $v_i = v_1$. We select edges of the cycle c, (v_1, v_j), (v_k, v_l), $(v_1, core)$ and $(v_2, core)$. Now we have selected $k + 4$ edges. The 3-connected components in the subgraph we selected are $\{core\}$, $\{v_1, v_2, v_j, v_k, v_l\}$, and $k - 5$ singleton vertices. We need at most $k - 4$ edges to 3-connect this cycle to the core. Because have spent $k + 4$ edges in the 2-connect step, we have enough charge to finish the 3-connect step.
3. If one of the $k - 2$ vertices in the cycle other than v_1 and v_2, say v_3, has a descendant component c_3 (see Fig. 7), we do the following. We know that after c_3 is 3-connected, at least one edge falls on v_3 because v_3 is UTV of c_3. In addition, edges from c_3 should fall on at least two different vertices on "core". From the perspective of the descendant component, "core" means the cycle c and its core. So, after c_3 is 3-connected, v_3 is 3-connected to one of the vertices in c or the core. So, we select $k + 2$ edges: the cycle c, $(v_1, core)$ and $(v_2, core)$. The 3-connected components in the subgraph we selected are $\{core\}$, $\{v_1, v_2\}$, and $k - 2$ singleton vertices. But we know that v_3 is 3-connected to one of the vertices in c or core after c_3 is 3-connected. So, we only need $k - 2$ edges to 3-connect this current cycle to the core.
4. If v_1 and v_2 are adjacent vertices in the cycle and if v_2 has a descendant component c_2 (see Fig. 8), we do the following. As in the previous case, when c_2 is 3-connected, we add an edge incident to v_2 and an edge incident to one of the vertices in the cycle or core other than v_2 (say v_a). So, we select $k + 2$ edges: the cycle c, $(v_1, core)$ and $(v_2, core)$. The 3-connected components in the subgraph we selected are $\{core\}$, $\{v_1, v_2\}$, and $k - 2$ singleton vertices. In all cases except when $v_a = v_1$, we save an edge because v_2 is 3-connected to some vertex in cycle or core other than v_1 due to 3-connecting c_2. So, we only need $k - 2$ edges to 3-connect this current cycle to the core. As we spent $k + 2$ edges, we have the required charge of $k - 2$ edges to 3-connect this current cycle. In the case when $v_a = v_1$, we can delete edge (v_1, v_2) from the solution because v_1 and v_2 are 4-connected. So, we have selected only $k + 1$ edges and we have the required charge of $k - 1$ edges to 3-connect this current cycle to the core.

5. If v_1 and v_2 are non adjacent vertices in the cycle and edges from v_2's descendant falls on at least one vertex v_3 other than v_1 or v_2, we do the following. Let v_1 be the vertex through which DFS entered the cycle. So, v_1 has a direct edge (not virtual) to the core (see Fig. 9). Let that descendant component of v_2 that has edges incident on v_3 be c_2. If v_2 has an edge to the core that is independent of c_2, we change UTV of c_2 to v_3. Now this case is exactly same as Case 3. If all edges from v_2 to the core are virtual edges through c_2, we do the following. Consider the vertices on left hand side of $\{v_1, v_2\}$ in the cycle. Let us call them LHS. Note that c_2 can not have edges incident on LHS because of our priority scheme. So, v_3 is on the right hand side of $\{v_1, v_2\}$. We already considered the case when at least one vertex other than v_1 or v_2 have a descendant component. So, when the algorithm enters this step, we know that LHS does not have any descendant components. Also, LHS can not have edges to the core because only v_1 or v_2 have edges incident to the core. In a scenario such as this, in simple graphs, there must be a crossing on the LHS. Because we considered all $UFC(v_1, v_2)$'s, that crossing must be a $ULC(v_1, v_2)$. Now, we change the UTV of c_2 from v_2 to v_3. After this change, only v_1 and v_3 have edges to the core. Observe that a $ULC(v_1, v_2)$ is a $UFC(v_1, v_3)$. So, this case is same as Case 2.
6. When algorithm enters this step, we know that the cycle does not have any useful crossings with respect to vertices v_1 and v_2. Also, none of the vertices other than v_1 and v_2 has outside neighbors. So, the cycle we are considering is an lcc (see Fig. 12). We can spend a charge of $2k + 1$ edges to 3-connect this cycle. We select the edges of the cycle, (v_1, core), (v_2, core). We spent $k + 2$ edges to 2-connect this cycle. In the process, we 3-connected v_1 and v_2. That leaves us with the following 3-connected components in the graph: $\{\mathit{core}\}$, $\{v_1, v_2\}$ and $k - 2$ singleton vertices in the cycle. So, we need $k - 1$ edges to 3-connect these components. As we spent $k + 2$ edges out of $2k + 1$ edges, we have the required charge of $k - 1$ edges to 3-connect this cycle.

5.5 2-Connecting a Path

A path with k vertices has a charge of $2k + 1$ edges, because each vertex has a charge of 2 and each path has a charge of 1. We will show that this charge is sufficient to 3-connect a path in this section. After the 2-connecting step, we make sure that we have enough charge left for the 3-connecting step. When we 2-connect a path, we also ensure that there is at least one edge that is incident on UTV and any vertex in the path, and that the path is connected to at least two different vertices in the core.

Algorithm for 2-connecting a path: Let k be the number of vertices in the path under consideration. One end of the path is arbitrarily chosen as the "left" end. We start from leftmost vertex in the path and try to add an edge that is incident on the right most vertex in the path. But there are some exceptions. We try to generate open ears wherever possible. Let l be the left-most vertex of the path that is connected to the core, and let r be the right-most vertex in it connected to the core. A special ear is an ear that has two edges: one from l to the core, and the other from r to the core. When we add a special ear, we try to add it to two different vertices in the core whenever possible. In the following discussion, assume that the vertices of the path are $v_1 \dots v_k$, where v_1 is the left-most vertex. During the DFS on components, we give more priority to edges on vertices that are further to the left. So, v_1's child can have an edge to v_2, but not the other way around (except when DFS enters the path through v_1). When v_1 is the vertex from which DFS enters the path, v_2's descendant can have edges on v_1 because when we do DFS, we give least priority to v_1's neighbors. In the general case, v_2's descendant c_2 can not have edges incident on v_1 because otherwise c_2 would have been v_1's descendant component. But, while 2-connecting a path we may change the UTV of the descendants. That means we can make v_1's child as v_2's child if we do not use those virtual edges from v_1's child. Note that we can not change UTV of current component (i.e., the path that is being currently processed). But, we can change UTV of a descendant component (future component).

Rule-1: Whenever possible, we try to add open ears. We add a closed ear only when we are unable to find an open ear. Rule-2: In general, we prefer ears that cover more vertices. But, special ears have higher priority than other ears even if it covers fewer vertices. Notice that rule-1 has maximum priority.

Lemma 3. *Consider a graph $G(V, E)$ with no cut vertices. If p is a path of length k, p can be 3-connected to the core with a charge of at most $2k + 1$ edges.*

Proof. Let the number of ears we added be e_1, and we know that it causes either $e_1 - 1$ or $e_1 - 2$ mergers. In the following discussion, Case 1 deals with the situation when there are $e_1 - 1$ mergers. If only $e_1 - 2$ mergers are caused and it generates a closed ear, we handle it in Case 2a. Only possibility left now is that there are only $e_1 - 2$ mergers and two ears cause the same merger. This case is further divided in two subcases. First subcase (Case 2(b)i) is when leftmost vertex in the path has an edge incident to the core. Second subcase (Case 2(b)ii) is when leftmost vertex in the path does not have an edge to the core.

1. If there are $e_1 - 1$ mergers we do the following. If both edges $(l, core)$ and $(r, core)$ fall of the same vertex in the core, we see whether that vertex is UTV. If it is UTV, we add another edge that is incident on any vertex other than UTV in core and some vertex in path. This edge causes a merger. If it is not UTV, we add an edge incident on UTV and any vertex in path. This also causes a merger. If both edges $(l, core)$ and $(r, core)$ fall on two different vertices in the core, we take a look at the two vertices. If one of them is UTV, we do not add the third edge. If neither of them is UTV, we add an edge incident on UTV and any vertex in path. This also causes a merger. Every time we add a third edge, we caused a merger. So, we can delete that third edge and analyze the performance of the algorithm.

   Note that in all the cases, whenever we 3-connect a path, we need to make sure that at least one edge falls on UTV and edges from the path are incident on at least two different vertices in the core. We can ensure these two things using the above steps. So, here onwards we will not repeat this part. We have selected $k + e_1$ edges. If the path is just 2-connected to the core, there are $k + 1$ 3-connected components including core and the singleton vertices. But we have $e_1 - 1$ mergers. So, number of 3-connected components in the subgraph selected is $k - e_1 + 2$. Therefore, we need $k - e_1 + 1$ edges to 3-connect the path to the core. As we have spent only $k + e_1$ edges, we have enough charge to finish the 3-connecting step.

2. If there are $e_1 - 2$ mergers in the path, we do the following.
   (a) If we get a closed ear (see Fig 10), we do the following. A closed ear can only start at r because if it starts at any other vertex, we have a cut vertex there and we know that input graph is processed for cut vertices. When we get a closed ear at r, to avoid a cut vertex at r, a descendant component of r must exist such that it has at least one edge incident to the core and at least one edge incident to one of the vertices that are right of r in the path. Let the descendant component of r be c_r and the ear that starts at r be $e_r = (r, v_p)$. Let v_q be the right-most vertex that has an edge incident to c_r. We know that $r < q \leq p$, because we considered ears that cover more vertices.

      In this case, we delete $(r, core)$ from the solution, change the UTV of c_r from r to v_q and add the edge $(v_q, core)$ to the solution. We know that edge $(v_q, core)$ exists because the descendant component c_r has at least an edge incident to the core. If the ear added just before special ear uses a virtual edge through c_r, that edge will now fall on v_q and that is fine. If $q = p$, we delete the edge (r, v_p). Otherwise we will keep that edge. Now, e_1 ears cause $(e_1 - 1)$ mergers (see Fig 11). This case falls under Case 1.
   (b) If two ears cause the same merger, we do the following.
      i. If leftmost vertex v_1 has an edge incident to the core, we do the following. In this case, the first ear selected by the algorithm is the special ear. After selecting that first ear, every open ear causes a merger. We already considered the case when we have a closed ear. So, we assume that all ears are open. As a result of that we have $e_1 - 1$ mergers and this case falls under Case 1.
      ii. If leftmost vertex v_1 does not have an edge incident to the core, it has a descendant component due to the following reason. We know that the end vertices of the path can have edges only to other components. Leftmost vertex is an end vertex in a path. So, it can have edges incident only to other components. Also, the leftmost vertex is generally given the highest priority when finding the target components. Only case when leftmost vertex is not given highest priority is when DFS enters the path through its leftmost vertex. In this case, it has an edge to the core. So, leftmost vertex of a path either has a descendant component or it has an edge incident to the core. But, in this case, leftmost vertex v_1 does not have any edges incident to the core. So, it has a descendant component.

The only possibility left is that the special ear and the next ear we add 3-connects same two vertices l and r.Note that, the algorithm selects only two edges incident on v_1. That is because v_1 does not have any edges to the core and we first add an ear that starts at v_1 and covers maximum number of vertices. We will not add any edges incident to v_1 later. So, v_1 just has two edges incident on it and that means it is not 3-connected to any other vertex. But we know that after the descendant component c_1 of v_1 is 3-connected, v_1 will be 3-connected to at least one vertex in the path or core because v_1 is the UTV of c_1. So, we save an edge.

We have selected $k + e_1$ edges. If path is just 2-connected to the core, there are $k + 1$ 3-connected components including the core and the singleton vertices. But we have $e_1 - 2$ mergers. So, the number of 3-connected components in the subgraph selected is $k - e_1 + 3$. Therefore, we need $k - e_1 + 2$ edges to 3-connect the path to the core. Because we saved an edge due to the existence of c_1, we have enough charge to finish the 3-connecting step.

6 Summary

In summary, we preprocess the graph for cut vertices then apply our algorithm. We find a maximum 2-matching in the given graph. First we reserve enough edges for a cycle to 3-edge-connect itself and call it core. Then we 2-connect one path or cycle at a time and always save enough charge to 3-connect that particular path or cycle. In the end, we have a 2-connected subgraph of the given graph with enough charge available to 3-connect it by adding a minimal set of edges to it.

References

1. Cheriyan, J., Thurimella, R.: Approximating minimum-size k-connected spanning subgraphs via matching. SIAM J. Comput. 30(2), 528–560 (2000)
2. Gabow, H.N.: An ear decomposition approach to approximating the smallest 3-edge connected spanning subgraph of a multigraph. SIAM J. Discrete Math. 18(1), 41–70 (2004)
3. Gabow, H.N., Goemans, M.X., Tardos, É., Williamson, D.P.: Approximating the smallest k-edge connected spanning subgraph by LP-rounding. In: SODA, pp. 562–571. SIAM (2005)
4. Gubbala, P., Raghavachari, B.: Approximation algorithms for the minimum cardinality two-connected spanning subgraph problem. In: Jünger, M., Kaibel, V. (eds.) IPCO. LNCS, vol. 3509, pp. 422–436. Springer, Heidelberg (2005)
5. Jothi, R., Raghavachari, B., Varadarajan, S.: A 5/4-approximation algorithm for minimum 2-edge-connectivity. In: SODA, pp. 725–734 (2003)
6. Khuller, S., Vishkin, U.: Biconnectivity approximations and graph carvings. J. ACM 41(2), 214–235 (1994)

7. Khuller, S.: Approximation algorithms for finding highly connected subgraphs. In: Approximation algorithms for NP-hard problems, ch. 6, pp. 236–265. PWS Publishing Co., Boston, MA (1997)
8. Kortsarz, G., Nutov, Z.: Approximating minimum cost connectivity problems. In: Handbook of Approximation Algorithms and Metaheuristics. Computer & Information Science Series, vol. 10, ch. 58, Chapman & Hall/CRC, Sydney, Australia (2007)
9. Vempala, S., Vetta, A.: Factor 4/3 approximations for minimum 2-connected subgraphs. In: Jansen, K., Khuller, S. (eds.) APPROX 2000. LNCS, vol. 1913, pp. 262–273. Springer, Heidelberg (2000)

Approximating the Maximum Sharing Problem

Amitabh Chaudhary[1,*], Danny Z. Chen[1,**], Rudolf Fleischer[2,***], Xiaobo S. Hu[1], Jian Li[2], Michael T. Niemier[1,3,†], Zhiyi Xie[2], and Hong Zhu[2,‡]

[1] Department of Computer Science and Engineering, University of Notre Dame, Notre Dame, IN 46556, USA
{achaudha, dchen, shu, mniemier}@cse.nd.edu

[2] Department of Computer Science and Engineering, Shanghai Key Laboratory of Intelligent Information Processing, Fudan University, Shanghai 200433, China
{lijian83, rudolf, xie_zhiyi, hzhu}@fudan.edu.cn

[3] College of Computing, Georgia Institute of Technology, Atlanta, GA 30332, USA
mniemier@cc.gatech.edu

Abstract. In the *maximum sharing problem (*MS*)*, we want to compute a set of (non-simple) paths in an undirected bipartite graph covering as many nodes as possible of the first node layer of the graph, with the constraint that all paths have both endpoints in the second node layer and no node in that layer is covered more than once. MS is equivalent to the *node-duplication based crossing elimination problem (*NDCE*)* that arises in the design of molecular quantum-dot cellular automata (QCA) circuits and the physical synthesis of BDD based regular circuit structures in VLSI design. We show that MS is NP-hard, present a polynomial-time 1.5-approximation algorithm, and show that MS cannot be approximated with a factor better than $\frac{740}{739}$ unless $P = NP$.

1 Introduction

Let $G = (U, V; E)$ be an undirected bipartite graph in which U is the *upper* node layer and V is the *lower* node layer. Let $m = |E|$ and $n = |U| + |V|$. In the *maximum sharing (*MS*) problem* we want to find a set of (non-simple) paths

* The order of authors follows the international standard of alphabetic order of the last name. In China, where first-authorship is a particularly important aspect of a publication, the order of authors should be J. Li, A. Chaudhary, D.Z. Chen, R. Fleischer, X.S. Hu, M.T. Niemier, Z. Xie, and H. Zhu.

** The research of this author was supported in part by the US National Science Foundation under Grant CCF-0515203. This work was partially done while the author was visiting and supported in part by a grant from the Shanghai Key Laboratory of Intelligent Information Processing at Fudan University, China.

*** We gratefully acknowledge support from the National Natural Science Fund China (grant no. 60573025).

† The research of this author was supported in part by the US National Science Foundation under Grant CCF-0404011.

‡ This work is supported by the China National Natural Science Fund (grant no. 60496321).

F. Dehne, J.-R. Sack, and N. Zeh (Eds.): WADS 2007, LNCS 4619, pp. 52–63, 2007.

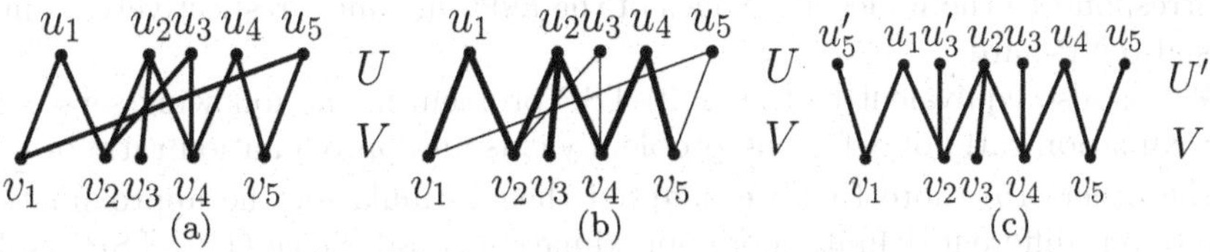

Fig. 1. MS and NDCE: (a) A layout of a two-layer graph G with ten edge crossings; (b) an MS solution with four sharings; (c) the corresponding NDCE solution with two node duplications

with both endpoints in V maximizing the total length of the paths. Every edge and every node of V can appear at most once in all the paths, except for edges to nodes in V of degree one which may appear twice consecutively in the same path. Note that a node in U can occur multiple times in the same path, and in multiple paths as well. See Fig. 1(b) for an example. Intuitively, a path in an MS solution can be viewed as a concatenated sequence of sharings. A *sharing* is a path (x, u, y) with $u \in U$ and $x, y \in V$, $x \neq y$.

MS has important applications in circuit design. Consider, for example, the (two-layer) *crossing minimization problem (*CM*)*. The two layers of G are laid out on parallel straight lines and each edge is drawn as a straight line segment between an upper node and a lower node. The objective is to minimize the number of edge crossings by reordering the nodes in the two layers. In some applications it is not enough to minimize the number of edge crossings — we must have no crossings at all. This can be achieved by duplicating nodes of the upper layer [3,4]. When we *duplicate* a node u we create a new node u' and partition the edges incident to u into a set of edges that remain incident to u and a set of edges that become incident to u'. A node may be duplicated multiple times; at each instance some of the currently incident edges are transferred to the new copy. In the *node-duplication based crossing elimination problem (*NDCE*)* we want to minimize the number of upper nodes to be duplicated such that, after a suitable reordering of the nodes, all edges can be drawn crossing-free as straight lines. See Fig. 1(c) for an example.

An MS solution with s sharings corresponds to an NDCE solution with $m-n-s$ node duplications [4]. Thus, minimizing node duplications in NDCE is equivalent to maximizing sharings in MS.

Good approximations for MS do not necessarily translate into good approximations for NDCE (just consider the case when NDCE has a solution with zero duplications), but a good MS-approximation can serve as a good heuristic for solving NDCE. For example, if the number of sharings is at most a fraction $k/(k+1)$ of the number of upper vertices, for some $k \geq 1$, then NDCE can be approximated to a $(1 + k/3)$ factor in polynomial time, using our 1.5-approximation for MS.

Our results. MS generalizes the NP-hard *maximum weight node-disjoint path cover problem (*MWPC*)*, where we want to find a set of node-disjoint paths in an undirected graph maximizing the number (or total weight) of the edges used by the paths. MWPC is equivalent to MS when all nodes in U have degree two (V and

U correspond to the nodes and edges of the MWPC instance, respectively). Thus, MS is also NP-hard.

MWPC is also equivalent to the (1, 2)-TSP problem in the following sense. An approximation ratio of γ for one problem yields an approximation ratio of $\frac{1}{2-\gamma}$ for the other [15] (note that we adapted their formula for the approximation ratio to our different definition of approximation ratio). Since (1, 2)-TSP can be approximated with a factor of $\frac{8}{7}$ [2], MWPC, and thus the case of MS where all nodes in U have degree two can be approximated with a factor of $\frac{7}{6}$. On the other hand, it is NP-hard to approximate (1, 2)-TSP better than with a factor of $\frac{741}{740}$ [10]. Thus, MS cannot be approximated with a factor better than $\frac{740}{739}$ unless $P = NP$. This lower bound could also be derived directly from an approximation preserving reduction from (1, 2)-TSP to MWPC.

The main result of this paper is a polynomial-time 1.5-approximation algorithm for MS by a reduction to the *color path packing problem (*CPP*)*. A maximum matching in the graph of the CPP instance would give us a 2-approximation for MS. We can solve a relaxation of CPP, the *color path-cycle packing problem (*CPCP*)*, optimally in polynomal time by computing a maximum matching in a related graph. In a non-trivial step we can then transform an optimal CPCP solution back to a 1.5-approximation solution for MS. The 2-matching algorithm would also give us a 1.5-approximation for MWPC.

Related work. CM has been studied in the context of graph drawing [7], visualization [6], DNA mapping [17], and optimization of circuit layouts in terms of wiring congestion, total wire length, and layout area (e.g., see [3,14]). It is NP-hard [13], even when one layer of nodes is already in a fixed order [8] (so-called *fixed-layer* CM). No approximability results are known for CM.

A related problem is to determine whether a given bipartite graph has a (not necessarily induced) planar subgraph with at least k edges, for a given k. This problem, too, was shown to be NP-complete by Eades and Whitesides [7]. It remains NP-complete even for the fixed-layer case. If both layers have a fixed ordering, then there is a polynomial-time algorithm. Another related problem is when only a given set of edge crossings is considered as restricted, and the objective is to minimize the restricted crossings; Finocchi [11] gave a 2-approximation solution for this problem.

NDCE was introduced (and solved by an integer linear program formulation) by Chaudhary et al. [4] to solve layout problems in the design of molecular quantum-dot cellular automata (QCA) circuits [1,16]. QCA circuits are currently the focus of increasingly intense research efforts aimed at building logic gates at the nanoscale. A major obstacle to building QCA circuits is that chemists are finding that it is considerably difficult to fabricate wire crossings in molecular QCA circuit layouts. Thus, some of the current research efforts are focused on building QCA circuits with no crossings in their layouts. Hence the need for solving NDCE.

Another application for crossing elimination can be found in the physical synthesis of Binary Decision Diagram (BDD) based regular circuit structures [3]. In contrast to CM, fixed-layer NDCE (where the order of the nodes in V is fixed) can be solved in linear time [4]. For NDCE on general non-bipartite graphs, a heuristic method was proposed by Cao and Koh [3] but they did not give any guarantees on the quality of the solution.

Sharings were introduced by Chaudhary et al. [4]. In an earlier paper [5], the authors studied the *maximum simple sharing problem (*MSS*)*, where also the upper nodes can only be visited at most once by the paths. By relaxing the path constraint to also allow cycles, they were able to obtain a $\frac{5}{3}$-approximation for MSS. Although the two problems seem to be closely related, the techniques used for solving MSS are not helpful for solving MS.

2 Approximating MS: A Special Case

In this section we present a polynomial-time 1.5-approximation algorithm for MS under the assumption that all lower nodes in V have degree at least two. In the next section we show how to extend it to also handle degree-one nodes.

Let $G = (U, V; E)$ be an undirected bipartite graph such that the nodes in V have degree at least two. We will transform the MS problem on G into a color path-cycle packing problem (CPCP) on a graph G_V generated from G. CPCP captures the key structure of MS, and we can solve it optimally in polynomial time by reduction to a maximum matching problem. From an optimal CPCP solution on G_V we can then obtain a set of feasible paths forming a 1.5-approximate solution for MS on G.

2.1 The Color Path-Cycle Packing Problem (CPCP)

Consider the following undirected graph $G_V = (V, E_V)$, which intuitively collapses every sharing in G into a single edge between two lower nodes. The node set of G_V is just the lower node set V. For any two distinct nodes v, w in V, if v and w can form a sharing in G, then we put the edge (v, w) in E_V. We associate with each edge $e = (v, w)$ in E_V a set C_e of upper nodes of G, called *colors*, such that if v and w can form a sharing in G through an upper node u, then u is included as a color in C_e.

MS is equivalent to the *color path packing problem (*CPP*)* on G_V, where we want to find a subgraph H of G_V with the maximum number of edges consisting of a set of node-disjoint simple paths in G_V such that we can color each edge e of H by a color in C_e without coloring any two consecutive edges in any path of H by the same color. Since MS is NP-hard, CPP is also NP-hard.

To approximate MS, we actually need to relax CPP by allowing simple cycles as well as simple paths in the sought subgraph of G_V. We also relax the color constraints. We call this relaxed version the *color path-cycle packing problem (*CPCP*)*. Formally, CPCP is the problem to find a subgraph H of G_V with the maximum number of edges such that every node in H has degree at most two and we can color both endnodes of every edge e of H by a color in C_e such that the two colors corresponding to the two incident edges of a node are different. Note that an edge may assign different colors to its two endpoints. In other words, an edge may or may not change its color halfway between the endpoints.

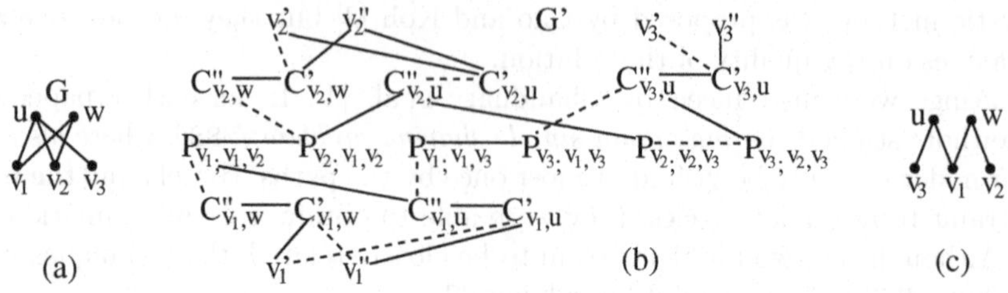

Fig. 2. Transforming an MS problem on G into a maximum matching problem on G'. (a) A simple bipartite graph G; (b) the corresponding graph G' with a maximum matching (the dashed edges); (c) the corresponding sharing path (v_3, u, v_1, w, v_2) in G.

Clearly, the optimal objective function value of CPCP is at least as large as that of CPP. We show below that we can obtain a CPP solution SOL_{CPP} from any CPCP solution SOL_{CPCP} such that $|SOL_{CPP}| \geq \frac{2}{3}|SOL_{CPCP}|$. Since we can solve CPCP optimally in polynomial time by reduction to a maximum matching problem, we obtain a $\frac{2}{3}$-approximation for CPP.

2.2 Solving CPCP

We construct an undirected graph $G' = (V', E')$ from G and G_V as follows (see Fig. 2). For every node $v \in V$ we put two nodes v' and v'' in V', called *V-type nodes*. For every edge (v, u) in E with $v \in V$ and $u \in U$ we add to V' two nodes $C'_{v,u}$ and $C''_{v,u}$, called *C-type nodes*, and to E' three edges $(v', C'_{v,u})$, $(v'', C'_{v,u})$, and $(C'_{v,u}, C''_{v,u})$. For any two nodes v_1, v_2 in G that can form a sharing through an upper node $u \in U$ we add to V' two nodes $P_{v_1;v_1,v_2}$ and $P_{v_2;v_1,v_2}$, called *P-type nodes*, and to E' an edge $(P_{v_1;v_1,v_2}, P_{v_2;v_1,v_2})$; moreover, we add edges $(P_{v_1;v_1,v_2}, C''_{v_1,u})$ and $(P_{v_2;v_1,v_2}, C''_{v_2,u})$ to E'. These are the nodes and edges of G'. Note that V' contains exactly $2 \cdot |V|$ V-type nodes, $2 \cdot |E|$ C-type nodes, and $2 \cdot |E_V|$ P-type nodes.

This construction transforms the MS problem on G to a maximum matching problem on G' (by relaxing some constraints of the MS problem). Fig. 2(c) gives an example showing a sharing path in G corresponding to a matching in G'.

Theorem 1. *Suppose G_V has an optimal* CPCP *solution SOL whose value is $|SOL|$, and G' has a maximum matching M. Then, $|M| = |E_V| + |E| + |SOL|$.*

Proof. First we prove $|M| \geq |E_V| + |E| + |SOL|$. Given an optimal CPCP solution SOL on G_V, we construct a matching M' of size $|E_V| + |E| + |SOL|$ in G' as follows. For every edge $e = (v, w) \in SOL$ in G_V whose endpoints are colored by $c_{e,v}$ and $c_{e,w}$ (possibly $c_{e,v} = c_{e,w}$), we add the edges $(v', C'_{v,c_{e,v}})$, $(C''_{v,c_{e,v}}, P_{v;v,w})$, $(w', C'_{w,c_{e,w}})$, and $(C''_{w,c_{e,w}}, P_{w;v,w})$ of G' to M'. Note that v' (or w') should be changed to v'' (or w'') if v' (or w') is already matched by an edge of M'. Clearly, these edges are part of a matching in G'.

After adding these edges to M', each unsaturated pair of P-type nodes $P_{v_1;v_1,v_2}$ and $P_{v_2;v_1,v_2}$ can be matched by adding the edge $(P_{v_1;v_1,v_2}, P_{v_2;v_1,v_2})$

to M'. Each pair of unsaturated nodes $C'_{v,u}$ and $C''_{v,u}$ can be matched by adding the edge $(C'_{v,u}, C''_{v,u})$ to M'. Note that for every edge in SOL, our construction of M' saturates two V-type nodes in G'; further, all C-type nodes and P-type nodes are saturated. Therefore, the number of saturated nodes in G' is exactly $2|SOL| + 2|E| + 2|E_V|$, and $|M'| = |SOL| + |E| + |E_V|$.

Now we prove $|M| \leq |E_V| + |E| + |SOL|$. Given a maximum matching M in G', it is sufficient to construct a CPCP solution SOL' of size $|SOL'| = |M| - |E_V| - |E|$ in G_V. Note that there exists a maximum matching in G' such that all the P-type nodes and C-type nodes are saturated (the set of all edges of the types of $(C'_{v,u}, C''_{v,u})$ and $(P_{v_1;v_1,v_2}, P_{v_2;v_1,v_2})$ forms a matching in G' saturating all P-type and C-type nodes; then start Edmonds' algorithm [9]). Suppose w.l.o.g. that M has this property. Let n_s denote the number of saturated V-type nodes in M.

Note that if any two nodes $P_{v_1;v_1,v_2}$ and $P_{v_2;v_1,v_2}$ are not matched by the edge $(P_{v_1;v_1,v_2}, P_{v_2;v_1,v_2})$ in M, then both nodes must each be matched with some C-type nodes. Let the corresponding edges in M be $(P_{v_1;v_1,v_2}, C''_{v_1,c_1})$ and $(P_{v_2;v_1,v_2}, C''_{v_2,c_2})$. Then, SOL' contains the edge $e = (v_1, v_2)$, and the colors assigned to (v_1, v_2) in SOL' are $c_{e,v_1} = c_1$ and $c_{e,v_2} = c_2$.

We first argue that $n_s = 2 \cdot |SOL'|$. Since all P-type and all C-type nodes are saturated, for each pair of P-type nodes, say $P_{v_1;v_1,v_2}$ and $P_{v_2;v_1,v_2}$, that are not matched by the edge $(P_{v_1;v_1,v_2}, P_{v_2;v_1,v_2})$ in M, there is a one-to-one correspondence with a pair of saturated V-type nodes: one of v'_1 or v''_1 and one of v'_2 or v''_2.

Next, we prove that SOL' is indeed a CPCP solution on G_V. It is easy to see that the degree of every node v in SOL' is at most two, because each of $v', v'' \in V'$ can contribute to at most one edge adjacent to v in SOL'. Suppose $e_1 = (u, v), e_2 = (v, w) \in SOL'$ are two adjacent edges, which means that in G', $P_{v;u,v}$ is not matched with $(P_{v;u,v}, P_{u;u,v})$ and $P_{v;v,w}$ is not matched with $(P_{v;v,w}, P_{w;v,w})$. W.l.o.g. assume that $P_{v;u,v}$ is matched with $(P_{v;u,v}, C''_{v,c_1})$ and $P_{v;v,w}$ is matched with $(P_{v;v,w}, C''_{v,c_2})$. Then we have $c_1 \neq c_2$, and the label colors $c_{e_1,v} = c_1 \neq c_2 = c_{e_2,v}$ (otherwise, $C''_{v,c_1} = C''_{v,c_2}$ would be adjacent to two different edges in the matching M, a contradiction). Therefore, SOL' is a feasible CPCP solution. Since $2|M| = n_s + 2|E_V| + 2|E|$ and $n_s = 2|SOL'|$, we conclude $|SOL'| = |M| - |E_V| - |E|$. □

Corollary 1. *There is a polynomial-time algorithm for computing an optimal* CPCP *solution on G_V.*

Proof. The maximum matching problem on G' can be solved in $O(\sqrt{|V'|} \cdot |E'|)$ time [12]. The proof of Theorem 1 shows how to obtain in polynomial time an optimal CPCP solution SOL' in G_V from a maximum matching M in G'. □

2.3 A 1.5-Approximation for CPP and MS

In this subsection, we show how to obtain a 1.5-approximate CPP solution SOL on G_V from an optimal CPCP solution SOL' on G_V. This immediately gives a 1.5-approximation for MS on G. First, we illustrate on an example why a CPCP solution SOL' may fail to be a feasible CPP solution, and how we can make it feasible by removing some edges from SOL'. Let $P = (v_1, v_2, v_3, v_4)$ be a path in G_V with $C_{(v_1,v_2)} = \{c_1\}$, $C_{(v_2,v_3)} = \{c_1, c_2\}$, and $C_{(v_3,v_4)} = \{c_2\}$. A CPCP solution could contain all three edges with the following color labeling: $c_{(v_1,v_2),v_1} = c_{(v_1,v_2),v_2} = c_1$, $c_{(v_2,v_3),v_2} = c_2$, $c_{(v_2,v_3),v_3} = c_1$, and $c_{(v_3,v_4),v_3} = c_{(v_3,v_4),v_4} = c_2$. This is not a feasible CPP solution. But we can remove the middle edge (v_2, v_3) to make it feasible.

Below we show how we can remove some edges from SOL' to obtain a CPP solution. Remember that we want to remove at most one third of all edges from SOL'.

Let $P = (v_1, \ldots, v_\ell)$ be a path in SOL' with label colors such that $c_{(v_{i-1},v_i),v_i} \neq c_{(v_i,v_{i+1}),v_i}$ for any i with $1 < i < \ell$. We say P can be *feasibly colored* if the coloring of P can be converted to a feasible CPP coloring by carefully choosing color labels for its *edges* (not for its *nodes*) from the color labels of its nodes. To be more precise, we can label each edge (v_{i-1}, v_i) of P by a color $c_{(v_{i-1},v_i)} \in \{c_{(v_{i-1},v_i),v_{i-1}}, c_{(v_{i-1},v_i),v_i}\}$ for $1 < i \leq \ell$, such that $c_{(v_{i-1},v_i)} \neq c_{(v_i,v_{i+1})}$, for $1 < i < \ell$.

Lemma 1. *Paths in SOL' of length at most 2 are feasibly colorable.* □

Note that in G two different nodes $v_1, v_2 \in V$ cannot simultaneously have sharings with two distinct upper nodes $u, w \in U$ in a feasible MS solution. Such short cycles cannot occur in SOL' (for example, in Fig. 2(a), $v_1, v_2 \in V$ cannot simultaneously have sharings with $u, w \in U$). It might appear that such infeasible simultaneous sharings might correspond to a "degenerate" cycle (v_1, v_2, v_1) in SOL'. The next lemma shows that there is no such "degenerate" cycle in SOL'.

Lemma 2. *Every cycle in SOL' has length at least three.* □

SOL' may contain simple paths and simple cycles. We first deal with the paths in SOL'. Let $P = (v_0, \ldots, v_{t-1})$ be a path in SOL'. We remove the edges (v_{3k-1}, v_{3k}) from P, for $k = 1, \ldots, \lfloor \frac{t}{3} \rfloor$. The remaining parts of P are a set of paths of length at most two. By Lemma 1, these paths can be feasibly colored. Note that we deleted no more than one third of the edges of SOL'.

Handling cycles in SOL' is more complicated. By Lemma 2, the length of each cycle in SOL' is at least three. We distinguish three cases based on the cycle length. An edge $e = (u, v)$ in SOL' with label colors $c_{e,u} = c_{e,v}$ is called a *1-color edge*, otherwise a *2-color edge*.

Lemma 3. *(a) A simple path consisting of successive* 1-color edges *and at most two* 2-color edges, *one at each end of the path, that does not form a cycle can be feasibly colored.*
(b) A simple path consisting of successive 2-color edges *and at most one* 1-color edge *at one end of the path that does not form a cycle can be feasibly colored.*

Proof. We only show part (b). Consider a simple path $P = (v_1, \ldots, v_\ell)$, in which (v_1, v_2) is a 1-color edge and the other edges are 2-color edges. First, we label (v_1, v_2) with color $c_{(v_1,v_2),v_1}$. Since $c_{(v_2,v_3),v_2} \neq c_{(v_2,v_3),v_3}$, at least one of them is not equal to $c_{(v_1,v_2),v_1}$. Thus, we can label (v_2, v_3) with this color. Similarly, we can feasibly color the other 2-color edges of P. □

Lemma 4. *For any cycle $C \in SOL'$ of length at least four, there exists a subpath of length three in C that can be feasibly colored.*

Proof. If C contains only 2-color edges or only 1-color edges (but not both types), then, by Lemma 3, any three consecutive edges of C can be feasibly colored. If C includes edges of both types, then consider a maximal subpath P of C consisting of only 1-color edges. P is certainly not a cycle. If $|P| \geq 3$, then the lemma holds for P. If $|P|$ is 1 (or 2), then we take two (or one) of the edges adjacent to the endnodes of P (P together with these edges does not form a cycle, because $|C| > 3$). The subpath of C formed by P and these edges can be feasibly colored by Lemma 3(a). □

Lemma 5. *For any cycle $C \in SOL'$ of length at least five in which the 1-color edges and 2-color edges do not appear alternatingly, there exists a subpath of length four in C that can be feasibly colored.* □

Case 1: $C = (v_0, v_1, \ldots, v_{3t-1}, v_0)$ is a cycle of length $3t$. We remove the edges (v_{3k}, v_{3k+1}) from C, for $k = 0, \ldots, t-1$. The remaining parts of C are a set of paths of length exactly two which can be feasibly colored by Lemma 1.

Case 2: C is a cycle of length $3t + 1$. We want to remove t edges, resulting in one path of length three and $t - 1$ paths of length two which can all be feasibly colored. By Lemma 4, we can find three successive edges of C that can be feasibly colored (and thus be kept in the CPP solution SOL). Next, we remove the two edges of C adjacent to this length-three subpath (if $t = 1$, then there is only one such adjacent edge). If $t \geq 2$, what is left from C at this point is a path P of length $3t - 4$. By using the same scheme as for handling the path case, we remove $t - 2$ edges from P and obtain a set of paths of length at most two. Note that we remove a total of t edges from C, which is less than a third of all edges.

Case 3: C is a cycle of length $3t + 2$. Similarly to Case 2, if we can find four successive edges in C that can be colored feasibly by Lemma 5, then we are done. Suppose we cannot; then by Lemma 5, the 1-color edges and 2-color edges in C must appear alternatingly, and thus t must be an even integer. Let $C = (v_0, \ldots, v_{t'-1})$, where $t' = 3t + 2$, be a cycle in which the 1-color edges and 2-color edges appear alternatingly and (v_0, v_1) is a 1-color edge. We remove the edges (v_{4k}, v_{4k+1}), for $k = 0, \ldots, \lfloor \frac{t'-1}{4} \rfloor$. The remaining parts of C are all paths of length at most three which can be feasibly colored by Lemma 3(a). Hence, in this case we remove a total of $\lfloor \frac{t'-1}{4} \rfloor + 1 \leq t$ edges, which is less than a third of all edges.

Theorem 2. MS *can be approximated within a factor of* 1.5 *in polynomial time if there are no degree-one lower nodes.*

Proof. The claim follows from Corollary 1. The running time of our MS approximation algorithm is dominated by the step of computing a maximum matching in the graph G', whose numbers of nodes and edges are a low degree polynomial in the numbers of nodes and edges of the input graph G. □

2.4 A More Practical 1.5-Approximation Algorithm

The technique described in the previous subsection gives the currently best approximation ratio for MS. In practice, however, we can do better, since many edges removed are unnecessary. Here, we propose a more practical algorithm to compute a CPP solution by removing edges from a CPCP solution. Note that, in the worst case, the scheme in this subsection also gives a 1.5-approximation.

Given a CPCP solution SOL_{CPCP} that consists of some cycles and paths, we first analyze the structures that make SOL_{CPCP} not a feasible CPP solution. One reason could be the existence of cycles, and we need to remove at least one edge on each cycle.

Another problem might be a subpath formed by 2-color edges with two 1-color edges at each end. For example, consider a simple path $P = (v_1, \ldots, v_\ell)$ in which ℓ is even, (v_1, v_2) and $(v_{\ell-1}, v_\ell)$ are both 1-color edges, and the other edges are 2-color edges. Moreover, colors are assigned as follows: $c_{(v_1,v_2),v_1} = c_{(v_2,v_3),v_3} = \cdots = c_{(v_{\ell-2},v_{\ell-1}),v_{\ell-1}} = c_1$, and $c_{(v_2,v_3),v_2} = c_{(v_3,v_4),v_3} = \cdots = c_{(v_{\ell-2},v_{\ell-1}),v_{\ell-2}} = c_{(v_{\ell-1},v_\ell),v_\ell} = c_2$. Clearly, this subpath cannot be feasibly colored without edge removals. To see why this is the only case when we cannot feasibly color all edges, consider an optimal deletion of a 1-color edge (v_1, v_2). It should be a 1-color edge because we cannot feasibly color (v_1, v_2) with its color $c_{(v_1,v_2),v_1} = c_1$. This is because one of its adjacent 2-color edges, say (v_2, v_3), would be forced to be colored with $c_1 = c_{(v_2,v_3),v_3}$ (note that $c_{(v_2,v_3),v_2} \neq c_1$). If there is another choice to color (v_2, v_3) (i.e., using color $c_{(v_2,v_3),v_2}$) without changing other edges' colors, we could add (v_1, v_2) to the solution SOL_{CPCP}, contradicting the optimality. But why is (v_2, v_3) forced to be colored with $c_1 = c_{(v_2,v_3),v_3}$? That is because its adjacent 2-color edge (v_3, v_4) is forced to be colored with $c_1 = c_{(v_2,v_3),v_2}$. We can continue this argument until we encounter a 1-color edge $(v_{\ell-1}, v_\ell)$ which must be colored with $c_{(v_{\ell-1},v_\ell),v_\ell}$.

If a 2-color edge is deleted in an optimal deletion, the argument is similar and proceeds until meeting the 1-color edges at the two ends of the path. If such a structure occurs, then we cannot feasibly color all edges in the structure, and at least one edge must be deleted. At the same time, it is easy to see that it is sufficient to delete only one 1-color edge in such a subpath. Thus, we can w.l.o.g. assume that an optimal method always removes 1-color edges. Moreover, such subpaths are mutually disjoint except that some pairs of them may share a common 1-color edge (at the ends of such a pair of subpaths).

We construct the following graph D for the edge removal. The nodes of D are all 1-color edges of the CPCP solution. Two nodes of D are linked by an edge if the two corresponding 1-color edges are at the two ends of one subpath described above. Thus any edge removal that makes the CPCP solution a feasible CPP solution corresponds to a vertex cover in D. Since D is a graph of maximum degree two, we can compute a minimum vertex cover in D in polynomial time.

3 Approximating MS: The General Case

In this section we show how to modify the approximation algorithm of Section 2 to accommodate lower nodes of degree one. We will use the fact that we may assume w.l.o.g. that each upper node is connected to at most one degree-one lower node (if we can connect crossing-free to one such node, we can connect crossing-free to many such nodes).

Given G, we construct G_V exactly as before. We must slightly relax the color constraint in the definition of CPP. If a path visits a degree-one node in V, it may have two consecutive occurrences of the same edge leading to that node, which of course must have the same color. Now, MS on G is again equivalent to CPP on G_V. For CPCP, we need a similar relaxation of the color constraint at degree-one nodes of V. The optimal objective value of CPCP on G_V is then at least as large as that of CPP.

Now we construct the graph G' from G. Let V_S be the set of degree-one lower nodes in G. For each $v \in V$ we add two V-type nodes v' and v'', for a total of $2 \cdot |V|$ such nodes. For each edge (v, u) in G with $v \in V$ and $u \in U$, if $v \notin V_S$, we add two C-type nodes $C'_{v,u}$ and $C''_{v,u}$, and we add edges as described before. If, however, $v \in V_S$, we add four C-type nodes: $C'_{v_a,u}$ and $C''_{v_a,u}$, as well as $C'_{v_b,u}$ and $C''_{v_b,u}$. Further, we add the following six edges: $(v', C'_{v_a,u})$, $(v'', C'_{v_a,u})$, and $(C'_{v_a,u}, C''_{v_a,u})$, and $(v', C'_{v_b,u})$, $(v'', C'_{v_b,u})$, and $(C'_{v_b,u}, C''_{v_b,u})$. Thus the total number of C-type nodes is $2 \cdot (|E + |V_S|)$. Finally, we add $2 \cdot |E_V|$ P-type nodes exactly as before. The edges between P-type and C-type nodes are added just as before, except when for an edge $(v_1, v_2) \in E_V$ in which one (and there can be only one) of the nodes, say v_2, is in V_S. In that case, let v_1 and v_2 have a common upper node neighbor u (again, only one common neighbor is possible). We add the following four edges to E': $(P_{v1;v_1,v_2}, P_{v2;v_1,v_2})$, $(P_{v1;v_1,v_2}, C''_{v1,u})$, $(P_{v_2;v_1,v_2}, C''_{v_{2a},u})$, and $(P_{v_2;v_1,v_2}, C''_{v_{2b},u})$.

Theorem 3. *Suppose G_V has an optimal* CPCP *solution SOL whose value is $|SOL|$, and G' has a maximum matching M. Then, $|M| = |E_V| + |E| + |V_S| + |SOL|$.* □

Informally, the only change in G' is that there are now more C-type nodes — $2 \cdot |V_S|$ more nodes. See Fig. 3 for an illustration.

The final step is to show that given any CPCP solution on G_V, say SOL', we can obtain a solution SOL for CPP on G_V such that $|SOL| \geq \frac{2}{3}|SOL'|$. Let $v \in V_S$ be a degree-one lower node that is a neighbor of the upper node $u \in U$. Suppose SOL' consists of a path or cycle with the three nodes $\sigma_v = (v_1, v, v_2)$ in order, and with corresponding colors $c_{(v_1,v),v_1}$, $c_{(v_1,v),v}$, $c_{(v,v_2),v}$, and $c_{(v,v_2),v_2}$. Now $c_{(v_1,v),v} = c_{(v,v_2),v} = u$ (this is a legal coloring). Observe that if there is no such sequence σ_v in SOL' for any $v \in V_S$, then SOL' is also a solution to CPCP on G_V, and we can directly use the algorithm in Section 2.3 to obtain a solution SOL for CPP on G_V.

So suppose such a sequence σ_v exists. There can be at most one sequence for each $v \in V_S$. Take the path (or cycle) P' containing v and break it into two paths (or one path, respectively) at the node v. In other words, the sequence (v_1, v) is

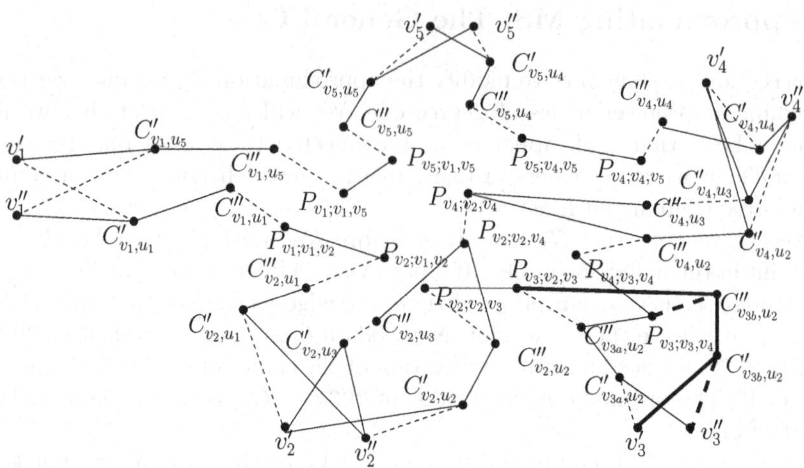

Fig. 3. The graph G' corresponding to G in Fig. 1(a). The thick edges are the extra edges required due to degree-one lower nodes. The dashed edges are the matching corresponding to the CPCP solution.

separated from (v, v_2). Let the resulting path(s) be denoted by P. Remove the third edge for each path in P, as in Section 2.3, followed by an application of Lemma 1, to obtain paths in P in which exactly one color from C_e is assigned to each edge e, and no two consecutive edges have the same color. P may, however, have two occurrences of v and thus cannot be part of a solution for CPP. We remedy this by combining back the two separate ends (if they both exist) to form the original sequence (v_1, v, v_2). This will lead to two consecutive edges having the same color, but that is a legal coloring.

Does this recombination create a cycle? Observe that if the original sequence σ_v is part of a cycle, the cycle has at least three edges. This follows from reasoning very similar to the proof of Lemma 2. If P consists of a single path, it has at least three edges, and thus at least one edge is removed. Thus, recombinations do not form a cycle. The process described above is performed on every sequence σ_v, for each $v \in V_S$. For paths or cycles in SOL' not containing such sequences, the process is exactly as described in Section 2.3.

Theorem 4. MS *can be approximated within a factor of* 1.5 *in polynomial time.*

□

References

1. Antonelli, D.A., Chen, D.Z., Dysart, T.J., Hu, X.S., Khang, A.B., Kogge, P.M., Murphy, R.C., Niemier, M.T.: Quantum-dot cellular automata (QCA) circuit partitioning: problem modeling and solutions. In: DAC'04. Proc. 41st ACM/IEEE Design Automation Conf., pp. 363–368. IEEE Computer Society Press, Los Alamitos (2004)

2. Berman, P., Karpinski, M.: $\frac{8}{7}$-approximation algorithm for (1, 2)-TSP. In: SODA'06. Proc. 17th Annual ACM-SIAM Symp. on Discrete Algorithms, pp. 641–648. ACM Press, New York (2006)
3. Cao, A., Koh, C.-K.: Non-crossing ordered BDD for physical synthesis of regular circuit structure. In: Proc. International Workshop on Logic and Synthesis, pp. 200–206 (2003)
4. Chaudhary, A., Chen, D.Z., Hu, X.S., Niemier, M.T., Ravinchandran, R., Whitton, K.M.: Eliminating wire crossings for molecular quantum-dot cellular automata implementation. In: Proc. of IEEE/ACM International Conference on Computer-Aided Design, pp. 565–571. ACM Press, New York (2005)
5. Chen, D.Z., Fleischer, R., Li, J., Xie, Z., Zhu, H.: On approximating the maximum simple sharing problem. In: Asano, T. (ed.) ISAAC 2006. LNCS, vol. 4288, pp. 547–556. Springer, Heidelberg (2006)
6. Di Battista, G., Eades, P., Tamassia, R., Tollis, I.: Graph Drawing: Algorithms for the Visualization of Graphs. Prentice-Hall, Englewood Cliffs (1998)
7. Eades, P., Whitesides, S.: Drawing graphs in two layers. Theor. Comput. Sci. 131, 361–374 (1994)
8. Eades, P., Wormald, N.C.: Edge crossings in drawings of bipartite graphs. Algorithmica 11(4), 379–403 (1994)
9. Edmonds, J.: Paths, trees, and flowers. Canadian Journal of Mathematics 17, 449–467 (1965)
10. Engebretsen, L., Karpinski, M.: TSP with bounded metrics. Journal of Computer and System Sciences 72(4), 509–546 (2006)
11. Finocchi, I.: Layered Drawings of Graphs with Crossing Constraints. In: Proc. 9th Annual International Computing and Combinatorics Conference, pp. 357–367 (2001)
12. Gabow, H.N.: Data Structures for Weighted Matching and Nearest Common Ancestors with Linking. In: SODA'90. Proc. 7th Ann. ACM-SIAM Symp. on Discrete Algorithms, pp. 434–443. ACM Press, New York (1990)
13. Garey, M.R., Johnson, D.S.: Crossing number is NP-complete. SIAM Journal on Algebraic and Discrete Methods 4(3), 312–316 (1983)
14. Lengauer, T.: Combinatorial Algorithms for Integrated Circuit Layout. Wiley, Chichester (1990)
15. Papadimitriou, C.H., Yannakakis, M.: The traveling salesman problem with distances one and two. Mathematics of Operations Research 18(1), 1–11 (1993)
16. Tougaw, P.D., Lent, C.S.: Logical devices implemented using quantum cellular automata. J. of App. Phys. 75, 1818 (1994)
17. Waterman, M.S., Griggs, J.R.: Interval graphs and maps of DNA. Bull. Math. Biol. 48(2), 189–195 (1986)

The Stackelberg Minimum Spanning Tree Game*

Jean Cardinal[1], Erik D. Demaine[2], Samuel Fiorini[3],
Gwenaël Joret[1,**], Stefan Langerman[1,***],
Ilan Newman[4], and Oren Weimann[2]

[1] Computer Science Department, Université Libre de Bruxelles,
B-1050 Brussels, Belgium
{jcardin,gjoret,slanger}@ulb.ac.be
[2] MIT Computer Science and Artificial Intelligence Laboratory,
Cambridge, MA 02139, USA
{edemaine,oweimann}@mit.edu
[3] Department of Mathematics, Université Libre de Bruxelles,
B-1050 Brussels, Belgium
sfiorini@ulb.ac.be
[4] Department of Computer Science, University of Haifa,
Haifa 31905, Israel
ilan@cs.haifa.ac.il

Abstract. We consider a one-round two-player network pricing game, the *Stackelberg Minimum Spanning Tree* game or StackMST. The game is played on a graph (representing a network), whose edges are colored either red or blue, and where the red edges have a given fixed cost (representing the competitor's prices). The first player chooses an assignment of prices to the blue edges, and the second player then buys the cheapest possible minimum spanning tree, using any combination of red and blue edges. The goal of the first player is to maximize the total price of purchased blue edges. This game is the minimum spanning tree analog of the well-studied Stackelberg shortest-path game.

We analyze the complexity and approximability of the first player's best strategy in StackMST. In particular, we prove that the problem is APX-hard even if there are only two different red costs, and give an approximation algorithm whose approximation ratio is at most $\min\{k, 3 + 2\ln b, 1 + \ln W\}$, where k is the number of distinct red costs, b is the number of blue edges, and W is the maximum ratio between red costs. We also give a natural integer linear programming formulation of the problem, and show that the integrality gap of the fractional relaxation asymptotically matches the approximation guarantee of our algorithm.

* This work was partially supported by the *Actions de Recherche Concertées (ARC)* fund of the *Communauté française de Belgique*.

** Aspirant F.R.S. – FNRS.

*** Chercheur qualifié F.R.S. – FNRS.

F. Dehne, J.-R. Sack, and N. Zeh (Eds.): WADS 2007, LNCS 4619, pp. 64–76, 2007.

1 Introduction

Suppose that you work for a networking company that owns many point-to-point connections between several locations, and your job is to sell these connections. A customer wants to construct a network connecting all pairs of locations in the form of a spanning tree. The customer can buy connections that you are selling, but can also buy connections offered by your competitors. The customer will always buy the cheapest possible spanning tree. Your company has researched the price of each connection offered by the competitors. The problem considered in this paper is how to set the price of each of your connections in order to maximize your revenue, that is, the sum of the prices of the connections that the customer buys from you.

This problem can be cast as a *Stackelberg game*, a type of two-player game named introduced by the German economist Heinrich Freiherr von Stackelberg [18]. In a Stackelberg game, there are two players: the *leader* moves first, then the *follower* moves, and then the game is over. The follower thus optimizes its own objective function, knowing the leader's move. The leader has to optimize its own objective function by anticipating the optimal response of the follower. In the scenario depicted in the preceding paragraph, you were the leader and the customer was the follower: you decided how to set the prices for the connections that you own, and then the customer selected a minimum spanning tree. In this situation, there is an obvious tradeoff: the leader should not put too a high price on the connections—otherwise the customer will not buy them—but on the other hand the leader needs to put sufficiently high prices to optimize revenue.

Formally, the problem we consider is defined as follows. We are given an undirected graph $G = (V, E)$ whose edge set is partitioned into a *red edge set* R and a *blue edge set* B. Each red edge $e \in R$ has a nonnegative fixed *cost* $c(e)$ (the best competitor's price). The leader owns every blue edge $e \in B$ and has to set a *price* $p(e)$ for each of these edges. The cost function c and price function p together define a *weight* function w on the whole edge set. By "weight of edge e" we mean either "cost of edge e" if e is red or "price of edge e" if e is blue. A spanning tree T is a *minimum spanning tree* (MST) if its *total weight*

$$\sum_{e \in E(T)} w(e) = \sum_{e \in E(T) \cap R} c(e) + \sum_{e \in E(T) \cap B} p(e) \tag{1}$$

is minimum. The *revenue* of T is then

$$\sum_{e \in E(T) \cap B} p(e). \tag{2}$$

The Stackelberg Minimum Spanning Tree problem, StackMST, asks for a price function p that maximizes the revenue of an MST. Throughout, we assume that the graph contains a spanning tree whose edges are all red; otherwise, there is a cut consisting only of blue edges and the optimum value is unbounded. Moreover, to avoid being distracted by epsilons, we assume that among all edges of the same weight, blue edges are always preferred to red edges; this is a standard assumption. As a consequence, all minimum spanning trees for a given price function p have the same revenue; see Section 2 for details.

Related work. A similar pricing problem, where the customer wants to construct a shortest path between two vertices instead of a spanning tree, has been studied in the literature; see van Hoesel [17] for a survey. Complexity and approximability results have recently been obtained by Roch, Savard and Marcotte [14], and by Grigoriev, van Hoesel, Kraaij, Uetz, and Bouhtou [9]: the problem is strongly NP-hard and $O(\log |B|)$-approximable. A generalization of the problem to more than one customer has been tackled using mathematical programming tools, in particular bilevel programming; see Labbé, Marcotte, and Savard [12]. This generalization was motivated by the problem of setting tolls on highway networks. Cardinal, Labbé, Langerman, and Palop [3] give a geometric version of the problem.

Sometimes the goal of the leader is not to invite the followers to use his/her part of the network and maximize his/her own revenue but to encourage socially acceptable or optimal behaviors among the followers (the users of the network) so as to maximize some global objective. These kinds of Stackelberg games have been studied recently, e.g., by Cole, Dodis and Roughgarden [4] and Swamy [16]. An extensive bibliography on similar networking games has been compiled recently by Altman et al. [2]. In another Stackelberg game studied by Roughgarden [15], the leader is a job scheduler whose goal is to compute a scheduling strategy for the jobs he/she controls such that total latency in the system is minimized after the followers have selfishly scheduled their jobs.

Hartline and Koltun [10] propose approximation algorithms for several APX-hard pricing problems, where the goal is to find the best prices for a set of items, given knowledge of the consumer's behavior in the form of a combinatorial preference structure.

Finally, our problem should not be confused with other spanning tree games found in cooperative games and mechanism design theory [8,11], with parametric spanning tree problems [7,6], or with two-stage stochastic minimum spanning tree problems [5].

Our results. We analyze the complexity and approximability of the StackMST problem. Specifically, we prove the following:

1. StackMST is APX-hard, even if there are only two red costs, 1 and 2 (Section 3). This result is also the first NP-hardness proof for this problem. The reduction is from SetCover.
2. StackMST is $O(\log n)$-approximable, and is $O(1)$-approximable when the red costs either fall in a constant-size range or have a constant number of distinct values (Section 4). More precisely, we analyze the following simple approximation algorithm, called *Best-out-of-k*: for all i between 1 and k, consider the price function for which all blue edges have price c_i, and output the best of these k price functions. Here, and throughout the paper, c_i denotes the ith smallest cost of a red edge and k the number of distinct red costs.

We prove that the approximation ratio of this algorithm is bounded above by $\min\{k, 3 + 2\ln b, 1 + \ln(c_k/c_1)\}$, where b is the number of blue edges.
3. The integrality gap of a natural integer linear programming formulation asymptotically matches the approximation guarantee of Best-out-of-k (Section 5). Thus, effectively, any approximation algorithm based on the linear programming relaxation of our integer program (or any weaker relaxation) cannot do better than Best-out-of-k. Of course, this result does not imply that Best-out-of-k is optimal. In fact, a central open question about StackMST is to determine if it admits a constant factor approximation algorithm.

Some of the proofs are omitted and will appear in the full version of this paper.

2 Basic Results

Before we proceed to our main results, we prove a few basic lemmas about StackMST.

We claimed in the introduction that the revenue of the leader depends on the price function p only, and not on the particular MST picked by the follower. To see this, let $w_1 < w_2 < \cdots < w_\ell$ denote the different edge weights. The greedy algorithm (a.k.a. Kruskal's algorithm) will work in ℓ phases: in its ith phase, it will consider all blue edges of price w_i (if any) and then all red edges of cost w_i (if any). The number of blue edges selected in the ith phase will not depend on the order in which blue or red edges of weight w_i are considered. This shows the claim. Moreover, if there is no red edge of cost w_i then p is not an optimal price function because the leader can raise the price of every blue edge of price w_i to the next weight w_{i+1} and thus increase his/her revenue. This implies the following lemma.

Lemma 1. *In every optimal price function, the prices assigned to the blue edges appearing in some MST belong to the set* $\{c(e) : e \in R\}$. □

Notice that the prices given to the blue edges that are in no MST do not really matter (as long as they are high enough). We find it convenient to see them as equaling ∞. This has the same effect as deleting those blue edges. A direct consequence of Lemma 1 is that the decision version of StackMST belongs to NP, using some price function p with $p(e) \in \{c(e) : e \in R\} \cup \{\infty\}$ for all $e \in B$ as a certificate. Another possibility for a certificate is an acyclic set of blue edges F, interpreted as the set of blue edges in any MST. Given F, we can easily compute an optimal price function such that F is the set of blue edges in any MST, with the help of Lemma 2 below. In the lemma, we denote by $\mathcal{C}_e$ the set of cycles of $G = (V, E)$ that include some edge e. (Notice that $\mathcal{C}_e$ is nonempty whenever e is blue because $G_r = (V, R)$ is connected.)

Lemma 2. *Consider a price function p, a corresponding minimum spanning tree T, and let $F = E(T) \cap B$. Then for every $e \in F$, we have*

$$p(e) \leq \min_{C \in \mathcal{C}_e} \max_{e' \in E(C) \cap R} c(e'). \tag{3}$$

Moreover, whenever F is any acyclic set of blue edges and we set $p(e)$ equal to the right hand side of (3) *for $e \in F$ and $p(e) = \infty$ for $e \in B - F$, we have $E(T') \cap B = F$ for any corresponding MST T'.*

It follows from the above lemma that StackMST is fixed parameter tractable with respect to the number of blue edges. Indeed, to solve the problem, one could try all acyclic subsets F of B, and for each of them put the prices as above (this can easily be done in polynomial time), and finally take the solution yielding the highest revenue. We conclude this section by stating a useful property satisfied by all optimal solutions of StackMST.

Lemma 3. *Let p be an optimal price function and T be a corresponding MST. Suppose that there exists a red edge e in T and a blue edge f not in T such that e belongs to the unique cycle C in $T+f$. Then there exists a blue edge f' distinct from f in C such that $c(e) < p(f') \leq p(f)$.*

3 Complexity and Inapproximability

By Lemma 1, StackMST is trivially solved when the cost of every red edge is exactly 1, i.e., when $c(e) = 1$ for all $e \in R$. In this section, we show that the problem is APX-hard even when the costs of the red edges are only 1 and 2, i.e., when $c(e) \in \{1, 2\}$ for all $e \in R$. We start with NP-hardness:

Theorem 1. StackMST *is NP-hard even when $c(e) \in \{1, 2\}$ for all $e \in R$.*

Proof. We present a reduction from SetCover (in its decision version). Let $(\mathcal{U}, \mathcal{S})$ and the integer t be an instance of SetCover, where $\mathcal{U} = \{u_1, u_2, \ldots, u_n\}$, and $\mathcal{S} = \{S_1, S_2, \ldots, S_m\}$. Without loss of generality, we assume that $u_n \in S_i$ for every $i = 1, 2, \ldots, m$ (we can always add one element to $\mathcal{U}$ and to every S_i to make sure this holds).

We construct a graph $G = (V, E)$ with edge set $E = R \cup B$ and a cost function $c : R \to \{1, 2\}$ as follows. The vertex set of G is $\mathcal{U} \cup \mathcal{S} = \{u_1, u_2, \ldots, u_n\} \cup \{S_1, S_2, \ldots, S_m\}$. The edge set of G and cost function c are defined as follows:

- there is a red edge of cost 1 linking u_i and u_{i+1} for every $1 \leq i < n$;
- there is a red edge of cost 2 linking u_n and S_1, and linking S_j and S_{j+1} for every $1 \leq j < m$;
- whenever $u_i \in S_j$ we link u_i and S_j by a blue edge.

We illustrate such a construction in Fig. 1. We claim that $(\mathcal{U}, \mathcal{S})$ has a set cover of size t if and only if there exists a price function $p : B \to \{1, 2, \infty\}$ for the blue edges of G whose revenue is $n + 2m - t - 1$.

($\Rightarrow$) Suppose $(\mathcal{U}, \mathcal{S})$ has a set cover of size t. We construct p as follows: for every blue edge $e = u_i S_j$, we set $p(e)$ to be 1 if S_j is in the set cover, and 2 otherwise. We show that the revenue of p equals $n + 2m - t - 1$ by running Kruskal's MST algorithm starting with an empty tree, T. Because the blue edges of weight 1 are the lightest, we start with adding them one by one to T such

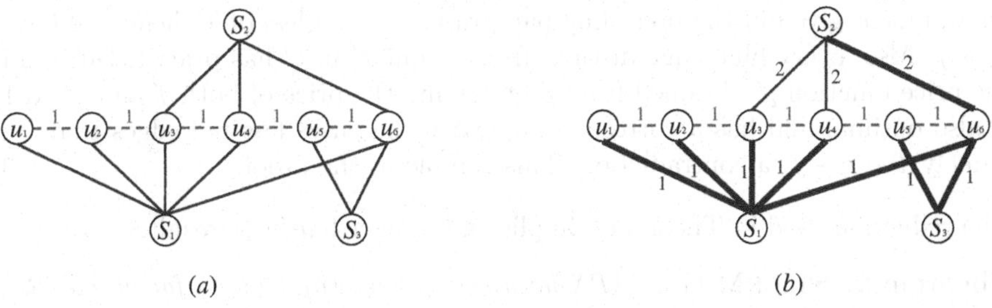

Fig. 1. (a) The graph G constructed for $n = 6$, $m = 3$ with $S_1 = \{u_1, u_2, u_3, u_4, u_6\}$, $S_2 = \{u_3, u_4, u_6\}$ and $S_3 = \{u_5, u_6\}$. The red edges of cost 2 are omitted for clarity. The red edges of cost 1 are dashed, and the blue edges are solid. (b) An optimal price function p on the blue edges that yields a revenue of 9, an example MST is depicted in bold.

that we add an edge only if it doesn't close a cycle in T. After going over all blue edges of weight 1, we are guaranteed that T is a tree that spans all the vertices u_i for every $i = 1, \ldots, n$, and every vertex S_j such that S_j is in the set cover. This is because these vertices are connected through u_n with only blue edges of weight 1. So the current weight of T is $|T| - 1 = n + t - 1$. We next try to add the red edges of weight 1, but every such edge connects two vertices, u_i and u_{i+1}, already spanned by T and therefore closes a cycle, so we add none of them. Next we add the blue edges of weight 2. For every S_j not in the set cover, we connect S_j to T with one blue edge of weight 2 (the second one will close a cycle). Therefore, after going over all the blue edges of weight 2, we added a weight of $2(m - t)$ to T. Furthermore, T spans the entire graph so there is no need to add any red edges of weight 2. All the edges in T are blue and the revenue of T is $(n + t - 1) + 2(m - t) = n + 2m - t - 1$.

($\Leftarrow$) Suppose that there exists a price function $p : B \to \{1, 2, \infty\}$ for the blue edges of G whose revenue is $n + 2m - t - 1$ for some t. By Lemma 1, there exists such a function p that is optimal. Choose then $p : B \to \{1, 2, \infty\}$ as an optimal price function that minimizes the number of red edges in an MST T.

Assume first that T contains only blue edges. Then every vertex u_i is incident to some blue edge in T with price 1. Thus the set $\mathcal{S}'$ of those S_j's that are linked to some blue edge in T with price 1 is a set cover of $(\mathcal{U}, \mathcal{S})$. On the other hand, notice that any $S_j \in \mathcal{S} \setminus \mathcal{S}'$ is a leaf of T, because if there were two blue edges $u_i S_j, u_{i+\ell} S_j$ in T then none of them could have a price of 2 because of the cycle $S_j u_i u_{i+1} \ldots u_{i+\ell} S_j$. Therefore, the revenue of p equals $(n+|\mathcal{S}'|-1)+2(m-|\mathcal{S}'|) = n + 2m - |\mathcal{S}'| - 1$. As by hypothesis this is at least $n + 2m - t - 1$, we deduce that the set cover $\mathcal{S}'$ has size at most t.

Suppose now that T contains some red edge e and denote by X_1 and X_2 the two components of $T - e$. There exists some blue edge $f = u_i S_j$ in G that connects X_1 and X_2 because the graph (V, B) induced by the blue edges is connected (because u_n is linked with blue edges to every S_j). By Lemma 3, there exists a blue edge $f' = u_{i'} S_{j'}$ distinct from f in the unique cycle C in $T + f$ such that $c(e) < p(f') \leq p(f)$. In particular, we have $c(e) = 1$ and $p(f') = 2$. By an argument given in the preceding paragraph, $S_{j'}$ is a leaf of T, hence we have $j' = j$. Also, every blue edge distinct from f and f' in C has price 1. But then the price function p' obtained from p by setting the price of both f and f' to 1 is also optimal and has a corresponding MST that uses less red edges than T, namely $T - e + f$, a contradiction. This completes the proof. □

The reduction used in Theorem 1 implies a stronger hardness result.

Theorem 2. StackMST *is APX-hard even when* $c(e) \in \{1, 2\}$ *for all* $e \in R$.

The above theorem is proved by showing that, for any $\varepsilon > 0$, a $(1 - \varepsilon)$-approximation for StackMST implies a $(1 + 8\varepsilon)$-approximation for Vertex-Cover in graphs of maximum degree at most 3. The theorem then follows from the APX-hardness of the latter problem [1,13]. The formal details will appear in the full version of this paper.

4 The Best-Out-Of-k Algorithm

As before, let k denote the number of distinct red costs, and let $c_1 < c_2 < \cdots < c_k$ denote those costs. Without loss of generality, we assume that all weights are positive (otherwise we contract all red edges of cost 0). Recall that the Best-out-of-k algorithm is as follows. For each j between 1 and k, set $p(e) = c_j$ for all blue edges $e \in B$ and compute an MST T_j. Then pick j such that the revenue of T_j is maximum and output the corresponding feasible solution. We analyze the approximation ratio ensured by this algorithm.

Theorem 3. *Best-out-of-k is a* ρ*-approximation, where*

$$\rho = 1 + \sum_{i=2}^{k} \frac{c_i - c_{i-1}}{c_i} = 1 + \frac{c_2 - c_1}{c_2} + \cdots + \frac{c_k - c_{k-1}}{c_k}.$$

Proof. Consider a minimum cost red spanning tree, that is, an MST obtained after setting the prices on all blue edges to ∞. For $j = 1, \ldots, k$, let m_j denote the number of red edges of cost c_j in that tree. Similarly, denote by m'_j the number of red edges of cost c_j in any MST the follower can select when all blue edges are for free (i.e., have a price of 0). Then we can bound the optimal revenue as follows:

$$\mathrm{OPT} \leq \sum_{i=1}^{k} c_i m_i - \sum_{i=1}^{k} c_i m'_i = \sum_{i=1}^{k} c_i (m_i - m'_i). \tag{4}$$

Let b_j be the number of blue edges in T_j, the MST computed by Best-out-of-k at step j, and set $b_{k+1} = 0$. Let also R_j denote the red edges of cost at most c_j. For $E' \subseteq E$, we denote by G/E' the graph obtained from G after contraction of every edge in E'. Because the number of edges in a set $F \subseteq E$ inducing a forest in G equals $|G| - |G/F|$ (where $|H|$ denotes the number of vertices of graph H), we obtain

$$\begin{aligned} m_j &= |G/R_{j-1}| - |G/R_j|, \\ m'_j &= |G/(R_{j-1} \cup B)| - |G/(R_j \cup B)|, \\ b_j &= |G/R_{j-1}| - |G/(R_{j-1} \cup B)|, \end{aligned}$$

and so we deduce

$$m_j - m'_j = b_j - b_{j+1}. \tag{5}$$

Thus the revenue given by T_j, which we denote q_j, is exactly

$$q_j = c_j b_j = \sum_{i=j}^{k} c_j (b_i - b_{i+1}) = \sum_{i=j}^{k} c_j (m_i - m'_i).$$

Then (5) implies:

$$m_j - m'_j = \frac{q_j}{c_j} - \frac{q_{j+1}}{c_{j+1}} \quad \text{for } 1 \le j \le k-1 \quad \text{and} \quad m_k - m'_k = \frac{q_k}{c_k}.$$

Therefore we can rewrite our upper bound on OPT given in (4) in terms of the q_j's:

$$\begin{aligned} \sum_{i=1}^{k} c_i (m_i - m'_i) &= c_1 \left(\frac{q_1}{c_1} - \frac{q_2}{c_2} \right) + c_2 \left(\frac{q_2}{c_2} - \frac{q_3}{c_3} \right) + \cdots + c_k \frac{q_k}{c_k} \\ &= q_1 + \frac{c_2 - c_1}{c_2} q_2 + \cdots + \frac{c_k - c_{k-1}}{c_k} q_k. \end{aligned}$$

Now consider the index j such that q_j is maximum. The above equation and (4) together imply:

$$\frac{\text{OPT}}{q_j} \le \frac{\sum_{i=1}^{k} c_i (m_i - m'_i)}{q_j} = \frac{q_1}{q_j} + \frac{c_2 - c_1}{c_2} \cdot \frac{q_2}{q_j} + \cdots + \frac{c_k - c_{k-1}}{c_k} \cdot \frac{q_k}{q_j} \tag{6}$$

$$\le 1 + \frac{c_2 - c_1}{c_2} + \cdots + \frac{c_k - c_{k-1}}{c_k} = \rho.$$

So Best-out-of-k is a ρ-approximation algorithm. (The last inequality follows from the maximality of q_j.) □

Observe that the above proof shows that Best-out-of-k can be implemented to run within the same time complexity as an MST algorithm. Indeed, when the price of all blue edges is set to c_j, the resulting revenue is $\sum_{i=j}^{k} c_j (m_i - m'_i)$. Thus we can find which c_j would maximize the revenue simply by computing the m_i's and m'_i's, which can be done by computing an MST of (V, R) and $(V, R \cup B)$, respectively (where the edges in B have price 0).

Note also that if the costs are exactly 1, 2, ..., k, then ρ equals the kth harmonic number $H_k = 1 + 1/2 + \cdots + 1/k$. In general, $\rho - 1$ can be regarded as a Riemann (under-)approximation of the integral $\int_{c_1}^{c_k} \frac{1}{t}\, dt$. So we have

$$\rho \le 1 + \int_{c_1}^{c_k} \frac{1}{t}\, dt = 1 + \ln c_k - \ln c_1 = 1 + \ln W,$$

where $W = c_k/c_1$. We also have $\rho \leq k$, because ρ is the sum of k terms, all not exceeding 1. Moreover, as we now prove, Equation (6) implies the following result leading to the conclusion that Best-out-of-k is, in particular, a $\min\{k, 3 + 2\ln|B|), 1 + \ln(\frac{c_k}{c_1})\}$ - approximation.

Proposition 1. *Best-out-of-k is a* $(3 + 2\ln b)$*-approximation, where* b *is the number of blue edges.*

Proof. Recall q_i is the revenue of the ith tree computed by Best-out-of-k, and q_j is their maximum. Let $c_0 = 0$ and ℓ be the index where $c_{\ell-1} < \frac{c_k}{b^2}$ and $c_\ell \geq \frac{c_k}{b^2}$. Without loss of generality, we assume $q_k \neq 0$ (otherwise we focus on the last q_i that is non-zero) so we have $c_k \leq q_k \leq q_j$. We then deduce

$$q_i \leq c_i \cdot m < \frac{c_k}{b^2} \cdot b \leq \frac{q_j}{b} \text{ for every } 1 \leq i \leq \ell - 1. \tag{7}$$

Equation (6) in the proof of Theorem 3 and Equation (7) together imply:

$$\begin{aligned}
\frac{\text{OPT}}{q_j} &\leq \sum_{i=1}^{\ell-1} \frac{c_i - c_{i-1}}{c_i} \frac{q_i}{q_j} + \sum_{i=\ell}^{k} \frac{c_i - c_{i-1}}{c_i} \frac{q_i}{q_j} \\
&\leq \sum_{i=1}^{\ell-1} \frac{q_i}{q_j} + \sum_{i=\ell}^{k} \frac{c_i - c_{i-1}}{c_i} \\
&< \sum_{i=1}^{\ell-1} \frac{1}{b} + 1 + \int_{c_\ell}^{c_k} \frac{1}{t} \mathrm{d}t \\
&< \ell/b + 1 + \ln \frac{c_k}{c_\ell} \leq \ell/b + 1 + \ln(b^2) = \ell/b + 1 + 2\ln b.
\end{aligned}$$

To complete the proof we describe a procedure to simplify the StackMST instance in order to ensure $\ell/b \leq 2$. First, as long as some vertex v of the graph has no blue edge incident to it, contract the cheapest edge in $\delta(v) = \{e \in E : v \in e\}$. Next, remove the most expensive edge in every red cycle in the graph, until the red edges form a spanning tree. As is easily verified, the resulting StackMST instance is equivalent to the original. That is, the set of blue edges does not change and the revenue of every price function is the same for both instances. So for the analysis we can assume that every vertex has some blue edge incident to it and the red edges form a spanning tree. Therefore, we have $b \geq n/2 \geq (k+1)/2 \geq \ell/2$ and Best-out-of-k is a $(3 + 2\ln b)$-approximation. □

A natural generalization of StackMST to matroids is as follows. Given a matroid $(S, \mathcal{I})$ with $\mathcal{I}$ partitioned into two sets $\mathcal{R}$ and $\mathcal{B}$, and nonnegative costs on the elements of $\mathcal{R}$, assign prices on the elements of $\mathcal{B}$ in such a way that the revenue given by a minimum weight basis of $(S, \mathcal{I})$ is maximized. We mention that the analysis of Best-out-of-k given in the proof of Theorem 3 extends swiftly to the case of matroids, yielding a ρ-approximation algorithm in this more general setting.

5 Linear Programming Relaxation

In this section, we give an integer programming formulation for the problem and study its linear programming relaxation. In this section, all costs c_i are assumed to be positive. For each $j = 1, \ldots, k$, and each blue edge $e \in B$ we define a variable $x_{j,e}$. The interpretation of these variables is as follows: think of a feasible solution $p : B \to \{c_1, c_2, \ldots, c_k\}$ and a minimum spanning tree T with respect to p. Then $x_{j,e} = 1$ means that the blue edge e appears in T, with a price $p(e)$ of at least c_j.

We let $c_0 = 0$ and denote by R_j the set of red edges of cost at most c_j. For t pairwise disjoint sets of vertices $C_1, \ldots, C_t$, we denote by $\delta_B(C_1 : C_2 : \cdots : C_t)$ the set of blue edges that are in the cut defined by these sets. The integer programming formulation then reads:

$$
\text{(IP)} \quad \max \sum_{\substack{e \in B \\ 1 \le j \le k}} (c_j - c_{j-1}) x_{j,e}
$$

$$
\begin{aligned}
\text{s.t.} \quad & \sum_{e \in \delta_B(C_1:C_2:\cdots:C_t)} x_{j,e} \le t - 1 && \forall j \ge 1, && (8)\\
& && \forall C_1, ..., C_t \text{ components of } (V, R_{j-1}); &&\\
& \sum_{e \in P \cap B} x_{1,e} + x_{j,f} \le |P \cap B| && \forall f = ab \in B, \forall j \ge 2, && (9)\\
& && \forall P \ ab\text{-path in } (B \cup R_{j-1}) - f; &&\\
& x_{1,e} \ge x_{2,e} \ge \cdots \ge x_{k,e} \ge 0 && \forall e \in B; && (10)\\
& x_{j,e} \in \{0, 1\} && \forall j, \forall e \in B. && (11)
\end{aligned}
$$

Proposition 2. *The integer program above is a formulation of* StackMST.

As already noted, $\sum_{j=1}^{k} c_j(m_j - m'_j)$ is an upper bound on OPT (see Section 4). The rest of this section is devoted to the LP relaxation of the above IP, obtained by dropping constraint (11). We will show that the LP is tractable and that it provides an upper bound on OPT at least as good as $\sum_{j=1}^{k} c_j(m_j - m'_j)$. On the other hand, its integrality gap turns out to be k on instances with k distinct costs, thus matching the guarantee given by the Best-out-of-k algorithm. (Let us recall that the integrality gap of the LP on a specified set of instances is defined as the supremum of the ratio (LP)/(IP) over these instances.)

Proposition 3. *The LP can be separated in polynomial time.*

Proposition 4. *We have* (IP) $\le$ (LP) $\le \sum_{j=1}^{k} c_j(m_j - m'_j)$.

Proposition 5. *The integrality gap of the LP is k on instances with k distinct costs.*

Proof. We already know from Proposition 4 that the integrality gap is at most k on instances with k distinct costs. In order to show that it is also at least k, we define an instance of StackMST as follows:

- choose an integer $a \geq 2$;
- the graph has $a^{k-1}+1$ vertices, the set of whose is denoted $V = \{v_0, v_1, \ldots, v_{a^{k-1}}\}$;
- the set of blue edges is a spanning star with v_0 as center, i.e. $B = \{v_0 v_i | 1 \leq i \leq a^{k-1}\}$;
- the ith red cost is $c_i = a^{i-1}$, for $1 \leq i \leq k$;
- the components of the graph (V, R_i), where R_i is the set of red edges of cost at most c_i (and $R_0 = \emptyset$), are $\{v_1, \ldots, v_{a^i}\}, \{v_{a^i+1}, \ldots, v_{2a^i}\}, \ldots, \{v_{(a^{k-i}-1)a^i+1}, \ldots, v_{a^{k-i}a^i}\}$, for $1 \leq i \leq k-1$;
- the unique component of (V, R_k) is V.

We didn't define explicitly the set of red edges in the above description. This is because, as shown by the IP formulation, it is sufficient to give the components of the graph (V, R_i) for $i = 1, 2, \ldots, k$. (Notice for instance that we may always 'realize' these components with a set of red edges inducing a path.)

Consider an optimal solution of the StackMST problem for the instance defined above, and let T be a corresponding MST. Look at any blue edge e in T, of price c_i, and let C_e be the unique component of $(V - v_0, R_{i-1})$ that contains an endpoint of e. No other blue edge of T has an endpoint in C_e, because otherwise T has not minimum weight. Thus, if e and f are two distinct blue edges of T, then $C_e \cap C_f = \emptyset$. Noticing that the price given to e is $c_i = a^{i-1} = |C_e|$, we deduce that the revenue given by T is

$$\sum_{e \in B \cap T} |C_e| \leq a^{k-1}.$$

Moreover, a revenue of a^{k-1} is easily achieved, set for instance all blue edges to the same price c_i for some $i \in \{1, \ldots, k\}$. Hence, (IP) $= a^{k-1}$.

We now define a feasible solution x^* for the LP. The point x^* will have the property that $x^*_{i,e} = x^*_{i,f}$ for $1 \leq i \leq k$ and all $e, f \in B$. We thus let $y_i = x^*_{i,e}$ for $e \in B$. The constraints on the y_i's imposed by the LP are then:

$$\begin{aligned} a^{i-1} y_i &\leq 1 && \text{for } 1 \leq i \leq k; \\ y_1 + y_i &\leq 1 && \text{for } 2 \leq i \leq k; \\ y_1 \geq y_2 \geq \cdots &\geq y_k \geq 0. \end{aligned}$$

Set $y_1 = (a-1)/a$ and $y_i = 1/a^{i-1}$ for $2 \leq i \leq k$, which satisfies the above constraints. The value of the objective function of the LP for the point x^* is

$$\begin{aligned} \mathrm{LP}(x^*) &= \sum_{\substack{e \in B \\ 1 \leq i \leq k}} (c_i - c_{i-1}) x^*_{i,e} \\ &= a^{k-1} \left(\frac{a-1}{a} + \sum_{2 \leq i \leq k} (a_{i-1} - a_{i-2}) \frac{1}{a^{i-1}} \right) = k a^{k-1} - k a^{k-2}. \end{aligned}$$

Therefore, the ratio $\mathrm{LP}(x^*)/(\mathrm{IP})$ tends to k as $a \to \infty$, implying the claim. □

To conclude this section, let us mention that we know of additional families of valid inequalities that cut the fractional point used in the above proof. We leave the study of those for future research.

Acknowledgments

We thank Martine Labbé and Gilles Savard for preliminary discussions concerning the Stackelberg minimum spanning tree problem. We also thank Martin Hoefer for his comments which led us to prove Proposition 1.

References

1. Alimonti, P., Kann, V.: Some APX-completeness results for cubic graphs. Theoret. Comput. Sci. 237(1-2), 123–134 (2000)
2. Altman, E., Boulogne, T., El-Azouzi, R., Jiménez, T., Wynter, L.: A survey on networking games in telecommunications. Computers and Operations Research 33(2), 286–311 (2006)
3. Cardinal, J., Labbé, M., Langerman, S., Palop, B.: Pricing of geometric transportation networks. In: CCCG. Proc. Canadian Conference on Computational Geometry, pp. 92–96 (2005)
4. Cole, R., Dodis, Y., Roughgarden, T.: Pricing network edges for heterogeneous selfish users. In: STOC. Proc. Symp. Theory of Computing, pp. 521–530 (2003)
5. Dhamdhere, K., Ravi, R., Singh, M.: On two-stage stochastic minimum spanning trees. In: Jünger, M., Kaibel, V. (eds.) IPCO. LNCS, vol. 3509, pp. 321–334. Springer, Heidelberg (2005)
6. Eppstein, D.: Setting parameters by example. SIAM Journal on Computing 32(3), 643–653 (2003)
7. Fernández-Baca, D., Slutzki, G., Eppstein, D.: Using sparsification for parametric minimum spanning tree problems. Nordic J. Computing 3(4), 352–366 (1996)
8. Granot, D., Huberman, G.: Minimum cost spanning tree games. Mathematical Programming 21(1), 1–18 (1981)
9. Grigoriev, A., van Hoesel, S., van der Kraaij, A., Uetz, M., Bouhtou, M.: Pricing network edges to cross a river. In: Persiano, G., Solis-Oba, R. (eds.) WAOA 2004. LNCS, vol. 3351, pp. 140–153. Springer, Heidelberg (2005)
10. Hartline, J.D., Koltun, V.: Near-optimal pricing in near-linear time. In: Dehne, F., López-Ortiz, A., Sack, J.-R. (eds.) WADS 2005. LNCS, vol. 3608, pp. 422–431. Springer, Heidelberg (2005)
11. Karlin, A., Kempe, D., Tamir, T.: Beyond VCG: Frugality of truthful mechanisms. In: FOCS. Proc. 46th Annual IEEE Symposium on Foundations of Computer Science, pp. 615–626. IEEE Computer Society Press, Los Alamitos (2005)
12. Labbé, M., Marcotte, P., Savard, G.: A bilevel model of taxation and its application to optimal highway pricing. Management Science 44(12), 1608–1622 (1998)
13. Papadimitriou, C.H., Yannakakis, M.: Optimization, approximation, and complexity classes. J. Comput. System Sci. 43(3), 425–440 (1991)
14. Roch, S., Savard, G., Marcotte, P.: An approximation algorithm for Stackelberg network pricing. Networks 46(1), 57–67 (2005)
15. Roughgarden, T.: Stackelberg scheduling strategies. SIAM Journal on Computing 33(2), 332–350 (2004)

16. Swamy, C.: The effectiveness of Stackelberg strategies and tolls for network congestion games. In: SODA. Proc. Symp. on Discrete Algorithms (to appear)
17. van Hoesel, S.: An overview of Stackelberg pricing in networks. Research Memoranda 042, Maastricht : METEOR, Maastricht Research School of Economics of Technology and Organization (2006)
18. von Stackelberg, H.: Marktform und Gleichgewicht (Market and Equilibrium). Verlag von Julius Springer, Vienna (1934)

Edges and Switches, Tunnels and Bridges

D. Eppstein[1], M. van Kreveld[2], E. Mumford[3], and B. Speckmann[3]

[1] Department of Computer Science, University of California, Irvine
eppstein@ics.uci.edu
[2] Department of Information and Computing Sciences, Utrecht University
marc@cs.uu.nl
[3] Department of Mathematics and Computer Science, TU Eindhoven
e.mumford@tue.nl and speckman@win.tue.nl

Abstract. Edge casing is a well-known method to improve the readability of drawings of non-planar graphs. A cased drawing orders the edges of each edge crossing and interrupts the lower edge in an appropriate neighborhood of the crossing. Certain orders will lead to a more readable drawing than others. We formulate several optimization criteria that try to capture the concept of a "good" cased drawing. Further, we address the algorithmic question of how to turn a given drawing into an optimal cased drawing. For many of the resulting optimization problems, we either find polynomial time algorithms or NP-hardness results.

1 Introduction

Drawings of non-planar graphs necessarily contain edge crossings. The vertices of a drawing are commonly marked with a disk, but it can still be difficult to detect a vertex within a dense cluster of edge crossings. *Edge casing* is a well-known method—used, for example, in electrical drawings and, more generally, in information visualization—to alleviate this problem and to improve the readability of a drawing. A *cased drawing* orders the edges of each crossing and interrupts the lower edge in an appropriate neighborhood of the crossing. One can also envision that every edge is encased in a strip of the background color and that the casing of the upper edge covers the lower edge at the crossing. See Fig. 1 for an example.

If there are no application specific restrictions that dictate the order of the edges at each crossing, then we can in principle choose freely how to arrange

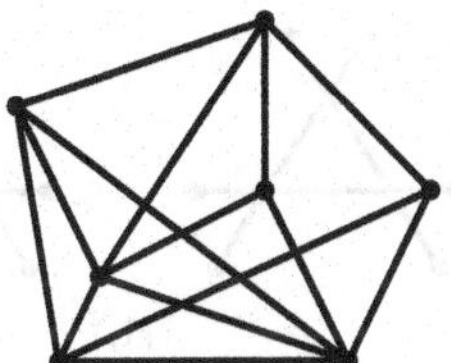
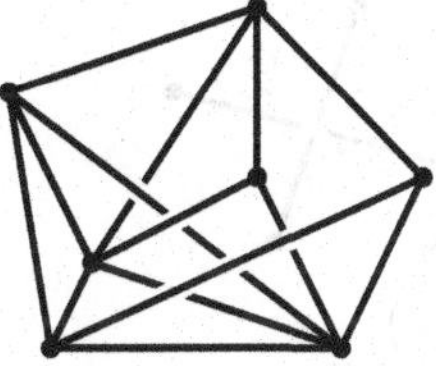

Fig. 1. Normal and cased drawing of a graph

F. Dehne, J.-R. Sack, and N. Zeh (Eds.): WADS 2007, LNCS 4619, pp. 77–88, 2007.

them. Certain orders will lead to a more readable drawing than others. In this paper we formulate several optimization criteria that try to capture the concept of a "good" cased drawing. Further, we address the algorithmic question of how to turn a given drawing into an optimal cased drawing.

Definitions. Let G be a graph with n vertices and m edges and let D be a drawing of G with k crossings. We want to turn D into a cased drawing where the width of the casing is given in the variable *casingwidth*. To avoid that the casing of an edge covers a vertex we assume that no vertex v of D lies on (or very close to) an edge e of D unless v is an endpoint of e. Further, no more than two edges of D cross in one point and any two crossings are far enough apart so that the casings of the edges involved do not interfere. With these assumptions we can consider crossings independently. Without these restrictions the problem changes significantly—optimization problems that are solvable in polynomial time can become NP-hard. Additional details can be found in the full paper.

We define the *edge crossing graph* G_{DC} for D as follows. G_{DC} contains a vertex for every edge of D and an edge for any two edges of D that cross. Let C be a crossing between two edges e_1 and e_2. In a cased drawing either e_1 is drawn on top of e_2 or vice versa. If e_1 is drawn on top of e_2 then we say that C is a *bridge* for e_1 and a *tunnel* for e_2. In Fig. 2, C_1 is a bridge for e_1 and a tunnel for e_2. The *length* of a tunnel is *casingwidth*$/ \sin \alpha$, where $\alpha \leq \pi/2$ is the angle of the edges at the crossing. A pair of consecutive crossings C_1 and C_2 along an edge e is called a *switch* if C_1 is a bridge for e and C_2 is a tunnel for e, or vice versa. In Fig. 2(a), (C_1, C_2) is a switch.

Stacking and weaving. When we turn a given drawing into a cased drawing, we need to define a drawing order for every edge crossing. We can choose to establish a global top-to-bottom order on the edges, or to treat each edge crossing individually. We call the first option the *stacking model* and the second the *weaving model*, since cyclic overlap of three or more edges can occur (see Fig. 2(b)).

Quality of a drawing. Globally speaking, two factors may influence the readability of a cased drawing in a negative way. Firstly, if there are many switches along an edge then it might become difficult to follow that edge. Drawings that have many switches can appear somewhat chaotic. Secondly, if an edge is frequently below other edges, then it might become hardly visible. These two considerations lead to the following optimization problems for a drawing D.

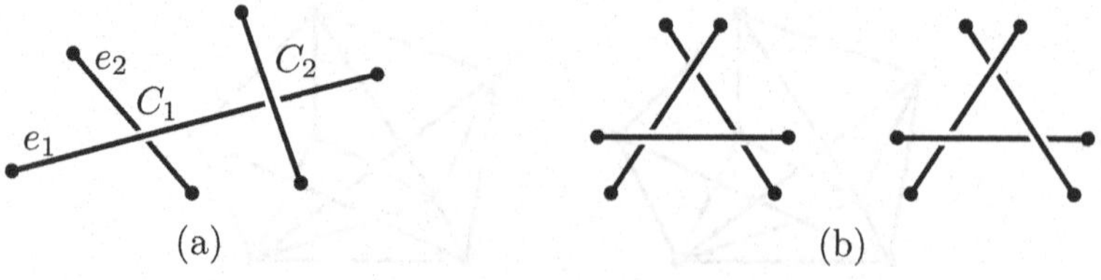

Fig. 2. (a) Tunnels and bridges. (b) Stacking and weaving.

MinTotalSwitches Minimize the total number of switches.
MinMaxSwitches Minimize the maximum number of switches for any edge.
MinMaxTunnels Minimize the maximum number of tunnels for any edge.
MinMaxTunnelLength Minimize the maximum total length of tunnels for any edge.
MaxMinTunnelDistance Maximize the minimum distance between any two consecutive tunnels.

Fig. 3 illustrates that the weaving model is stronger than the stacking model for MinTotalSwitches—no cased drawing of this graph in the stacking model can reach the optimum of four switches. For, the thickly drawn bundles of $c > 4$ parallel edges must be cased as shown (or its mirror image) else there would be at least c switches in a bundle, the four vertical and horizontal segments must cross the bundles consistently with the casing of the bundles, and this already leads to the four switches that occur as drawn near the midpoint of each vertical or horizontal segment. Thus, any deviation from the drawing in the casing of the four crossings between vertical and horizontal segments would create additional switches. However, the drawing shown is not a stacked drawing.

Related work. If we consider only simple arrangements of line segments in the plane as our initial drawing, then there is a third model to consider, an intermediate between stacking and weaving: drawings which are plane projections of line segments in three dimensions. We call this model the *realizable model*. Clearly every cased drawing in the stacking model is also a drawing in the realizable model, but not every cased drawing in the weaving model can be realized (see [8]). The optimal drawing in Fig. 3 can be realized, hence the realizable model is stronger than the stacking model. In the full paper we show that the weaving model is stronger than the realizable model.

Results. For many of the problems described above, we either find polynomial time algorithms or NP-hardness results in both the stacking and weaving models. We summarize our results in Table 1. In this paper we assume that our input drawing is a straight line drawing, but several of our results also generalize to curved drawings. Section 2 presents the results concerning the optimization problems that seek to minimize the number of switches and Section 3 discusses our solutions to the optimization problems that concern the tunnels. In the full paper we show that MinTotalSwitches becomes NP-hard in both the weaving

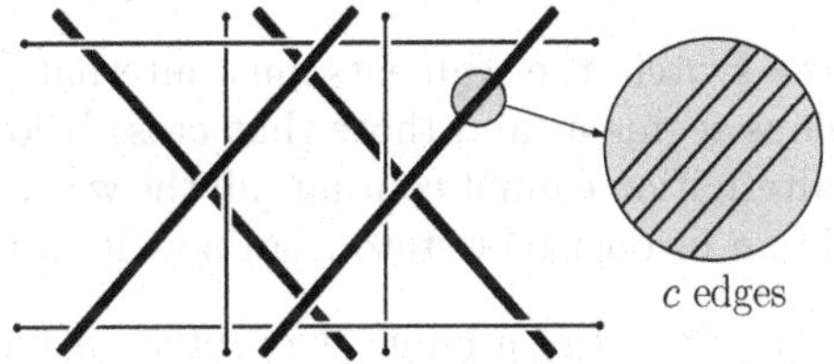

Fig. 3. Optimal drawing in the weaving model for MinTotalSwitches

Table 1. Table of results: n is the number of vertices, $m = \Omega(n)$ is the number of edges, $K = O(m^3)$ is the total number of pairs of crossings on the same edge, $k = O(m^2)$ is the number of crossings of the input drawing, and $q = O(k)$ is the number of its *odd face polygons*

Model	Stacking	Weaving
MinTotalSwitches	*open*	$O(qk + q^{5/2} \log^{3/2} k)$
MinMaxSwitches	*open*	*open*
MinMaxTunnels	$O(m \log m + k)$ *exp.*	$O(m^4)$
MinMaxTunnelLength	$O(m \log m + k)$ *exp.*	NP-hard
MaxMinTunnelDistance	$O(m \log m + k \log m)$ *exp.*	$O((m + K) \log m)$ *exp.*

and the stacking model if we allow more than three edges to cross in one point. We conclude with some open problems.

2 Minimizing Switches

In this section we discuss results related to the MinTotalSwitches and MinMaxSwitches problems. We first discuss some non-algorithmic results giving simple bounds on the number of switches needed, and recognition algorithms for graphs needing no switches. As we know little about these problems for the stacking model, all results stated in this section will be for the weaving model.

Lemma 1. *Given a drawing D of a graph we can turn D into a cased drawing without any switches if and only if the edge crossing graph G_{DC} is bipartite.*

Corollary 1. *Given a drawing D of a graph we can decide in $O((n+m)\log(n+m))$ time if D can be turned into a cased drawing without any switches.*

Proof. We apply the bipartiteness algorithm of [3]. Note that this does not construct the arrangement, so there is no term with k in the runtime. □

Define a *vertex-free cycle* in a drawing of a graph G to be a face f formed by the arrangement of the edges in the drawing, such that there are no vertices of G on the boundary of f. An *odd vertex-free cycle* is a vertex-free cycle composed of an odd number of segments of the arrangement.

Lemma 2. *Let f be an odd vertex-free cycle in a drawing D. Then in any casing of D, there must be a switch on one of the segments of f.*

Proof. Unless there is a switch, the segments must alternate between those that cross above the previous segment, and those that cross below the previous segment. However, this alternation cannot continue all the way around an odd cycle, for it would end up in an inconsistent state from how it started. □

Lemma 3. *Given a drawing D of a graph the minimum number of switches of any cased drawing obtained from D is at least half of the number of odd vertex-free cycles in D.*

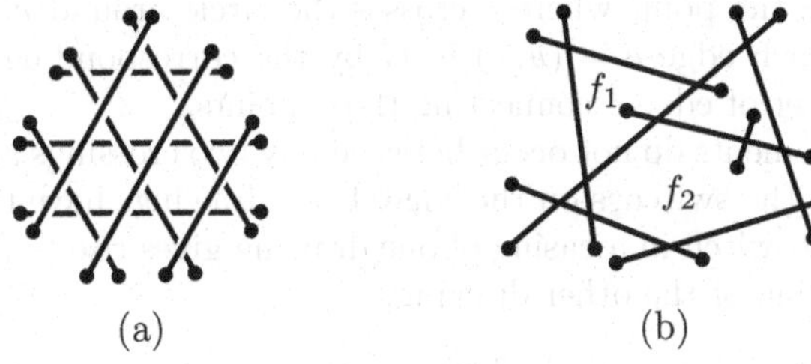

Fig. 4. (a) A construction with $O(n)$ edges and $\Omega(n^2)$ triangles. (b) A degree-one graph, f_1 is an odd polygon and f_2 is an even polygon.

Proof. Let o be the number of odd vertex-free cycles in D. By Lemma 2, each odd vertex-free cycle must have a switch on one of its segments. Choose one such switch for each cycle; then each segment belongs to at most two vertex-free cycles, so these choices group the odd cycles into pairs of cycles sharing a common switch, together with possibly some unpaired cycles. The number of pairs and unpaired cycles must be at least $o/2$, so the number of switches must also be this large.

Lemma 4. *For any n large enough, a drawing of a graph G with n vertices and $O(n)$ edges exists for which any crossing choice gives rise to $\Omega(n^2)$ switches.*

Proof. A construction with three sets of parallel lines, each of linear size, gives $\Omega(n^2)$ vertex-free triangles, and each triangle gives at least one switch (see Fig. 4(a)). □

Lemma 5. *For any n large enough, a drawing of a graph G with n vertices and $O(n^2)$ edges exists for which any crossing choice gives rise to $\Omega(n^4)$ switches.*

Proof. We build our graph as follows: make a very elongated rectangle, place $n/6$ vertices equally spaced on each short edge, and draw the complete bipartite graph. This graph has $(n/6)^2$ edges. One can prove that there is a strip parallel to the short side of the rectangle, such that the parts of the edges inside the strip behave in the same way as parallel ones do with respect to creating triangles when overlapped the way it is described in the previous lemma. This gives us the desired graph with $\Omega(n^4)$ triangles, and hence with $\Omega(n^4)$ switches. □

We define a *degree-one graph* to be a graph in which every vertex is incident to exactly one edge; that is, it must consist of a collection of disconnected edges.

Lemma 6. *Let D be a drawing of a graph G. Then there exists a drawing D' of a degree-one graph G', such that the edges of D correspond one-for-one with the edges of D', casings of D correspond one-for-one to casings of D', and switches of D correspond one-for-one with switches of D'.*

Proof. Form G' by placing a small circle around each vertex of G. Given an edge $e = (u, v)$ in G, let u_e be the point where e crosses the circle around u and similarly let v_e be the point where e crosses the circle around v. Form D' and G' by replacing each edge $e = (u, v)$ in G by the corresponding edge (u_e, v_e), drawn as the subset of edge e connecting those points.

As these replacements do not occur between any two crossings along any edge, they do not affect the switches on the edge. Both drawings have the same set of crossings, and any switch in a casing of one drawing gives rise to a switch in the corresponding casing of the other drawing. □

In a drawing of a degree-one graph, define a *polygon* to be a sequence of segments of the arrangement formed by the drawing edges that forms the boundary of a simple polygon in the plane. Define a *face polygon* to be a polygon that forms the boundary of the closure of a face of the arrangement; note that there may be edges drawn in the interior of this polygon, as long as they do not separate it into multiple components.

Lemma 7. *In a drawing of a degree-one graph, there can be no vertex on any segment of a polygon.*

Proof. We have already required that no vertex can lie on an edge unless it is the endpoint of an edge. And, if a segment contains the endpoint of an edge, it cannot continue past the endpoint to form the boundary of a polygon. □

Note, however, that a polygon can contain vertices in its interior. Define the *complexity* of a polygon to be the number of segments forming it, plus the number of graph vertices interior to the polygon. We say that a polygon is *odd* if its complexity is an odd number, and *even* if its complexity is an even number (see Fig. 4(b)).

Lemma 8. *Let p be a polygon in a drawing of a degree-one graph. Then, modulo two, the complexity of p is equal to the sum of the complexities of the face polygons of faces within p.*

Proof. Each segment of p contributes one to the complexity of p and one to the complexity of some face polygon. Each vertex within p contributes one to the complexity of p and one to the complexity of the face that contains it. Each segment within the interior of p either separates two faces, and contributes two to the total complexity of faces within p, or does not separate any face and contributes nothing to the complexity. Thus in each case the contribution to p and to the sum of its faces is the same modulo two. □

Lemma 9. *Let p be an odd polygon in a drawing of a degree-one graph. Then there exists an odd face polygon in the same drawing.*

Proof. By Lemma 8, the complexity of p has the same parity as the sum of the complexities of its faces. Therefore, if p is odd, it has an odd number of odd faces, and in particular there must be a nonzero number of odd faces. □

Lemma 10. *Let D be a drawing of a degree-one graph. Then D has a casing with no switches if and only if it has no odd face polygon.*

Proof. As we have seen, D has a casing with no switches if and only if the edge crossing graph is bipartite. This graph is bipartite if and only if it has no odd cycles, and an odd cycle in the edge crossing graph corresponds to an odd polygon in D. For, if C is an odd cycle in the edge crossing graph, it must lie on a polygon p of D. Each crossing in C contributes one to the complexity of this polygon. Each edge of D that crosses p without belonging to C either crosses it an even number of times (contributing that number of additional segments to the complexity of p) and has both endpoints inside p or both outside p, or it crosses an odd number of times and has one endpoint inside p; thus, it contributes an even amount to the complexity of p. Thus, p must be an odd polygon. By Lemma 9, there is an odd face polygon in D. Conversely, any odd face polygon in D can be shown to form an odd cycle in the edge crossing graph. □

Theorem 1. MinTotalSwitches *in the weaving model can be solved in time* $O(qk + q^{5/2}\log^{3/2} k)$*, where* k *denotes the number of crossings in the input drawing and* q *denotes the number of its odd face polygons.*

Proof. Let D be the drawing which we wish to case for the minimum number of switches. By Lemma 6, we may assume without loss of generality that each vertex of D has degree one.

We apply a solution technique related to the Chinese Postman problem, and also to the problem of via minimization in VLSI design [2]: form an auxiliary graph G^o, and include in G^o a single vertex for each odd face polygon in D. Also include in G^o an edge connecting each pair of vertices, and label this edge by the number of segments of the drawing that are crossed in a path connecting the corresponding two faces in D that crosses as few segments as possible. We claim that the minimum weight of a perfect matching in G^o equals the minimum total number of switches in any casing of D.

In one direction, we can case D with a number of switches equal to or better than the weight of the matching, as follows: for each edge of the matching, insert a small break into each of the segments in the path corresponding to the edge. The resulting broken arrangement has no odd face cycles, for the breaks connect pairs of odd face cycles in D to form larger even cycles. Therefore, by Lemma 10, we can case the drawing with the breaks, without any switches. Forming a drawing of D by reconnecting all the break points adds at most one switch per break point, so the total number of switches equals at most the weight of the perfect matching.

In the other direction, suppose that we have a casing of D with a minimum number of switches; we must show that there exists an equally good matching in G^o. To show this, consider the drawing formed by inserting a small break in each segment of D having a switch. This eliminates all switches in the drawing, so by Lemma 10, the modified drawing has no odd face polygons. Consider any face polygon in the modified drawing; by Lemma 9 it must include an even number of odd faces in the original drawing. Thus, the odd faces of D are connected in groups of evenly many faces in the modified drawing, and within each such group we can connect the odd faces in pairs by paths of breaks in the drawing, giving a matching in G^o with total weight at most equal to the number of switches in D.

The number of vertices of the graph G^o is $O(q)$, where q is the number of odd face polygons in D. We can construct G^o in time $O(qk)$ where k is the number of crossings in D by using breadth-first search in the arrangement dual to D to find the distances from each vertex to all other vertices. A minimum weight perfect matching in a complete weighted graph with integer weights bounded by k can be found in time $O(q^{5/2} \log^{3/2} k)$ using the algorithm of Gabow and Tarjan [5]. Therefore the time for this algorithm is $O(qk + q^{5/2} \log^{3/2} k)$. □

3 Minimizing Tunnels

In this section we present three algorithms that solve MinMaxTunnels, MinMaxTunnelLength, and MaxMinTunnelDistance in the stacking model. We also present algorithms for MinMaxTunnels and MaxMinTunnelDistance in the weaving model. MinMaxTunnelLength is NP-hard in the weaving model.

3.1 Stacking Model

In the stacking model, some edge e has to be bottommost. This immediately gives the number of tunnels of e, the total length of tunnels of e, and the shortest distance between two tunnels of e. The idea of the algorithm is to determine for each edge what its value would be if it were bottommost, and then choose the edge that is best for the optimization to be bottommost (smallest value for MinMaxTunnels and MinMaxTunnelLength, and largest value for MaxMinTunnelDistance). The other $m - 1$ edges are stacked iteratively above this edge. It is easy to see that such an approach indeed maximizes the minimum, or minimizes the maximum. We next give an efficient implementation of the approach. The idea is to maintain the values of all not yet selected edges under consecutive selections of bottommost edges instead of recomputing it.

We start by computing the arrangement of edges in $O(m \log m + k)$ expected time, for instance using Mulmuley's algorithm [7]. This allows us to determine the value for all edges in $O(k)$ additional time.

For MinMaxTunnels and MinMaxTunnelLength, we keep all edges in a Fibonnacci heap on this value. One selection involves an extract-min, giving an edge e, and traversing e in the arrangement to find all edges it crosses. For these edges we update the value and perform a decrease-key operation on the Fibonnacci heap. For MinMaxTunnels we decrease the value by one and for MinMaxTunnelLength we decrease by the length of the crossing, which is $\mathit{casingwidth}/ \sin\alpha$, where α is the angle the crossing edges make. For MinMaxTunnels and MinMaxTunnelLength this is all that we need. We perform m extract-min and k decrease-key operations. The total traversal time along the edges throughout the whole algorithm is $O(k)$. Thus, the algorithm runs in $O(m \log m + k)$ expected time.

For MaxMinTunnelDistance we use a Fibonnacci heap that allows extract-max and increase-key. For the selected edge we again traverse the arrangement to update the values of the crossing edges. However, we cannot update the value of an edge in constant time for this optimization. We maintain a data structure for each edge that maintains the minimum tunnel distance in $O(\log m)$ time under updates. The structure is an augmented balanced binary search tree that stores the edge parts in between consecutive crossings in its leaves. Each leaf stores the distance between these crossings. Each internal node is augmented such that it stores the minimum distance for the subtree in a variable. The root stores the minimum distance of the edge if it were the bottommost one of the remaining edges. An update involves merging two adjacent leaves of the tree and computing the distance between two crossings. Augmentation allows us to have the new minimum in the root of the tree in $O(\log m)$ time per update. In total this takes $O(m \log m + k \log m)$ expected time.

Theorem 2. *Given a straight-line drawing of a graph with n vertices, $m = \Omega(n)$ edges, and k edge crossings, we can solve* MINMAXTUNNELS *and* MINMAXTUNNELLENGTH *in $O(m \log m + k)$ expected time and* MAXMINTUNNELDISTANCE *in $O(m \log m + k \log m)$ expected time in the stacking model.*

3.2 Weaving Model

In the weaving model, the polynomial time algorithm for MINMAXTUNNELS comes from the fact that the problem of directing an undirected graph, and minimizing the maximum indegree, can be solved in time quadratic in the number of edges [9]. We apply this on the edge crossing graph of the drawing, and hence we get $O(m^4)$ time. For minimizing tunnel length per edge, we can show:

Theorem 3. MINMAXTUNNELLENGTH *is NP-hard in the weaving model.*

Proof. The reduction is from PLANAR 3-SAT, shown NP-hard by Lichtenstein [6]. The reduction is similar to the one for maximizing minimum visible perimeter length in sets of opaque disks of unit size [1]. Note that the proof implies that no PTAS exists. The reduction only uses edges that intersect two or three other edges, so restricting the number of intersections per edge to be constant leaves the problem NP-hard. Also, the number of orientations of edges is constant.

A cased drawing of a set of line segments has property (A) if every line segment has at most two tunnels at crossings with a perpendicular segment, or one tunnel at a crossing with a non-perpendicular segment. Our reduction is such that a PLANAR 3-SAT instance is satisfiable if and only if a set of line segments has a cased drawing with property (A).

We arrange a set of line segments of equal length, using only four orientations. The slopes are -4, $-\frac{1}{4}$, $+\frac{1}{4}$, and $+4$. If two perpendicular line segments cross, then one has tunnel length equal to the width w of the casing at the crossing. If two other line segments cross, then one edge has tunnel length $w/\sin(\gamma) = 2,125 \cdot w$ at the crossing, where $\gamma = 2 \cdot \arctan(\frac{1}{4})$ is the (acute) angle between the line segments. Therefore, a cased drawing with property (A) has tunnel length at most $2,125 \cdot w$, whereas a cased drawing that does not satisfy property (A) has an edge that has tunnel length at least $3 \cdot w$. This shows the direct relation between property (A) and MINMAXTUNNELLENGTH, and provides the gap that shows that no PTAS exists.

A Boolean variable x_i is modeled by a cycle of crossing line segments as in Fig. 5. Along the cycle, crossings alternate between perpendicular and non-perpendicular, and hence it has even length. The variable satisfies property (A) iff the cycle has cyclic overlap, which can be clockwise or counterclockwise. One state is associated with $x_i = \text{TRUE}$, the other is associated with $x_i = \text{FALSE}$. In each state, the line segments of the cycle alternate in allowing an additional, perpendicular line segment to have a bridge over the line segment of the cycle. In the figure, where the cycle is in the TRUE-state, the line segments with slope $+\frac{1}{4}$ and $+4$ allow such an extra tunnel under a line segment that is not from the cycle. If the cycle is in the FALSE-state, the line segments with slope -4 and $-\frac{1}{4}$ allow the extra tunnel. We use the line segments of slope $-\frac{1}{4}$ to make connections and channels to clauses where $\overline{x_i}$ occurs, and the line segments with slope $+\frac{1}{4}$ for clauses where x_i occurs. Note that the variable can be made larger easily to allow more connections, in case the variable occurs in many clauses.

Channels are formed by line segments that do not cross perpendicularly. So any line segment of the channel can have a tunnel at at most one of its two crossings, or else property (A) is violated. Note that a sequence of crossing line segments with slopes such as -4, $+4$, $+\frac{1}{4}$, $-\frac{1}{4}$ gives a turn in the channel. The exact position of the crossing is not essential and hence we can easily reach any part of the plane with a channel, and ending with a line segment of any orientation. A 3-SAT clause is formed by a single line segment that is crossed perpendicularly by three other line segments, see Fig. 6. Property (A) holds if the clause line segment has at most two tunnels. This corresponds directly to satisfiability of the clause.

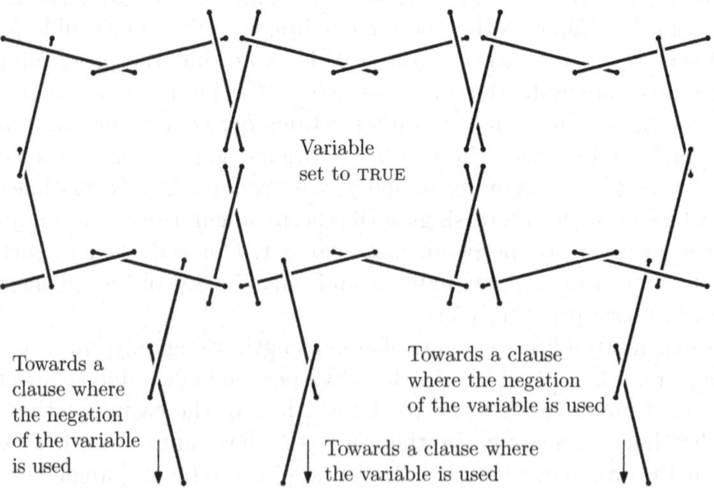

Fig. 5. Boolean variable and the connection of channels

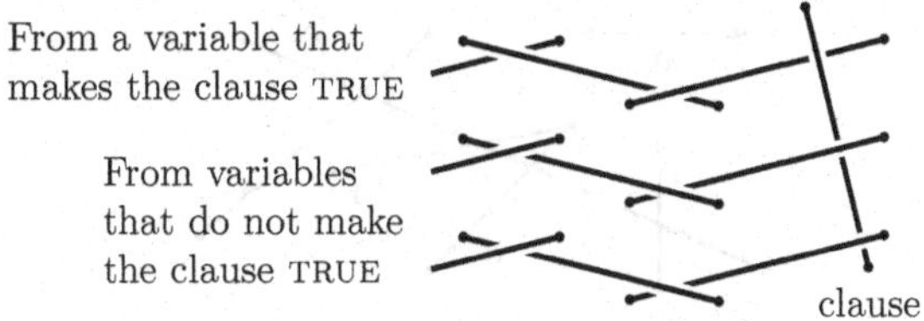

Fig. 6. A clause construction

With this reduction, testing if property (A) holds is equivalent to testing if the PLANAR 3-SAT instance is satisfiable, and NP-hardness follows. □

In the remainder of this section we show how to solve MAXMINTUNNELDISTANCE. We observe that there are polynomially many possible values for the smallest tunnel distance, and perform a binary search on these, using 2-SAT instances as the decision tool.

We first compute the arrangement of the m edges to determine all crossings. Only distances between two—not necessarily consecutive—crossings along any edge can give the minimum tunnel distance. One edge crosses at most $m-1$ other edges, and hence the number of candidate distances, K, is $O(m^3)$. Obviously, K is also $O(k^2)$. From the arrangement of edges we can determine all of these distances in $O(m \log m + K)$ time. We sort them in $O(K \log K)$ time to set up a binary search. We will show that the decision step takes $O(m + K)$ time, and hence the whole algorithm takes $O(m \log m + K \log K) = O((m + K) \log m)$ time.

Let δ be a value and we wish to decide if we can set the crossings of edges such that all distances between two tunnels along any edge is at least δ. For every two edges e_i and e_j that cross and $i < j$, we have a Boolean variable x_{ij}. We associate x_{ij} with TRUE if e_i has a bridge at its crossing with e_j, and with FALSE otherwise. Now we traverse the arrangement of edges and construct a 2-SAT formula. Let e_i, e_j, and e_h be three edges such that the latter two cross e_i. If the distance between the crossings is less than δ, then e_i should not have the crossings with e_j and e_h as tunnels. Hence, we make a clause for the 2-SAT formula as follows (Fig. 7): if $i < j$ and $i < h$, then the clause is $(x_{ij} \vee x_{ih})$; the other three cases ($i > j$ and/or $i > h$) are similar. The conjunction of all clauses gives a 2-SAT formula that is satisfiable if and only if we can set the crossings such that the minimum tunnel distance is at least δ. We can construct the whole 2-SAT instance in $O(m+K)$ time since we have the arrangement, and satisfiability of 2-SAT can be determined in linear time [4].

Theorem 4. *Given a straight-line drawing of a graph with n vertices and $m = \Omega(n)$ edges, we can solve* MAXMINTUNNELDISTANCE *in $O((m + K) \log m)$ expected time in the weaving model, where $K = O(m^3)$ is the total number of pairs of crossings on the same edge.*

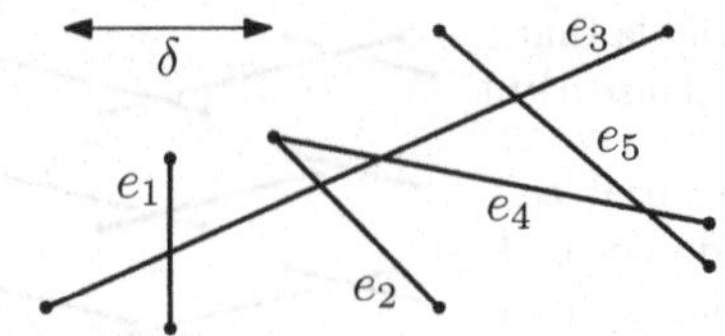

Fig. 7. The 2-SAT formula $(\overline{x}_{13} \vee \overline{x}_{23}) \wedge (\overline{x}_{23} \vee x_{34}) \wedge (\overline{x}_{23} \vee x_{35}) \wedge (x_{34} \vee x_{35})$

4 Conclusions and Open Problems

We presented polynomial time algorithms or NP-hardness results for a number of optimization problems that are motivated by cased drawings. Naturally, we would like to establish the difficulty of the MINMAXSWITCHES problem. We would also like to implement our algorithms to visually evaluate the quality of the resulting drawings.

References

1. Cabello, S., Haverkort, H., van Kreveld, M., Speckmann, B.: Algorithmic aspects of proportional symbol maps. In: Azar, Y., Erlebach, T. (eds.) ESA 2006. LNCS, vol. 4168, pp. 720–731. Springer, Heidelberg (2006)
2. Chen, R.-W., Kajitani, Y., Chan, S.-P.: A graph-theoretic via minimization algorithm for two-layer printed circuit boards. IEEE Transactions on Circuits and Systems 30(5), 284–299 (1983)
3. Eppstein, D.: Testing bipartiteness of geometric intersection graphs. In: Proc. 15th ACM-SIAM Symposium on Discrete Algorithms, pp. 853–861. ACM Press, New York (2004)
4. Even, S., Itai, A., Shamir, A.: On the complexity of timetable and multicommodity flow problems. SIAM Journal on Computing 5(4), 691–703 (1976)
5. Gabow, H.N., Tarjan, R.E.: Faster scaling algorithms for general graph-matching problems. Journal of the ACM 38(4), 815–853 (1991)
6. Lichtenstein, D.: Planar formulae and their uses. SIAM Journal on Computing 11(2), 329–343 (1982)
7. Mulmuley, K.: Computational Geometry: An Introduction through Randomized Algorithms. Prentice-Hall, Englewood Cliffs (1994)
8. Pach, J., Pollack, R., Welzl, E.: Weaving patterns of lines and line segments in space. Algorithmica 9(6), 561–571 (1993)
9. Venkateswaran, V.: Minimizing maximum indegree. Discrete Applied Mathematics 143, 374–378 (2004)

How to Draw a Clustered Tree*

Giuseppe Di Battista, Guido Drovandi, and Fabrizio Frati

Dipartimento di Informatica e Automazione – Università di Roma Tre

Abstract. The visualization of clustered graphs is a classical algorithmic topic that has several practical applications and is attracting increasing research interest. In this paper we deal with the visualization of clustered trees, a problem that is somehow foundational with respect to the one of visualizing a general clustered graph. We show many, in our opinion, surprising results that put in evidence how drawing clustered trees has many sharp differences with respect to drawing "plain" trees. We study a wide class of drawing standards, giving both negative and positive results. Namely, we show that there are clustered trees that do not have any drawing in certain standards and others that require exponential area. On the contrary, for many drawing conventions there are efficient algorithms that allow to draw clustered trees with polynomial asymptotic optimal area.

1 Introduction and Overview

The problem of drawing trees is a classical topic of investigation in algorithmics. Contributions on that field span almost three decades, from the groundbreaking work of Valiant [13] to the recent papers of Garg and Rusu that investigate how to obtain optimal area drawings with prescribed aspect ratio [11]. Algorithms for drawing trees have been proposed within many drawing conventions. To give a few examples, upward drawings have been studied in [9], straight-line upward drawings in [12,2], and straight-line orthogonal drawings in [3,1].

Despite such a large amount of investigation on algorithms for drawing trees no contribution has been presented in the literature on how to draw *clustered trees*. A *clustered graph* (c-graph) is a pair $C = (G, T)$, where G is a graph and T is a rooted tree such that the leaves of T are the vertices of G. Graph G and tree T are called *underlying graph* and *inclusion tree*, respectively. A *clustered tree* (c-tree), is a c-graph whose underlying graph is a tree. Each internal node ν of T corresponds to the subset $V(\nu)$ of the vertices of G (called *cluster*) that are the leaves of the subtree rooted at ν. The subgraph of G induced by $V(\nu)$ is denoted by $G(\nu)$, where ν is a cluster of T. If each cluster induces a connected subgraph of G, then C is *c-connected.*

A drawing of a c-graph $C = (G, T)$ consists of a drawing of G and of a representation of each node ν of T as a simple closed region $R(\nu)$ such that: (i)

* Work partially supported by MUR under Project MAINSTREAM Algorithms for Massive Information Structures and Data Streams.

F. Dehne, J.-R. Sack, and N. Zeh (Eds.): WADS 2007, LNCS 4619, pp. 89–101, 2007.

$R(\nu)$ contains the drawing of $G(\nu)$; (ii) $R(\nu)$ contains a region $R(\mu)$ iff μ is a descendant of ν in T; and (iii) the borders of any two regions don't intersect. Consider an edge e and a node ν of T. If e crosses the boundary of $R(\nu)$ more than once, we say that e and $R(\nu)$ have an *edge-region crossing.* A drawing of a *c*-graph is *c-planar* if it doesn't have edge crossings or edge-region crossings and a graph is *c*-planar if it has a *c*-planar drawing.

Many papers have been presented for constructing *c*-planar drawings within many drawing conventions. Namely, in [7] it is shown how to construct $O(n^2)$ area *c*-planar orthogonal drawings of *c*-graphs with max degree 4. Eades et al. [6] present an algorithm for constructing *c*-planar straight-line (SL) drawings of *c*-graphs, where clusters are drawn as convex regions. Such an algorithm requires, in general, exponential area. However, in [8] it is shown that such a bound is asymptotically optimal in the worst case.

In this paper we look for algorithms for constructing *c*-planar drawings of *c*-trees in efficient area, by considering the most investigated drawing standards for the underlying tree (see e.g. [1,2,9,10]). We deal both with *c*-connected and non-*c*-connected *c*-trees and consider drawings in which the clusters are represented by rectangles (R-drawings), by convex polygons (C-drawings), and also by eventually non-convex polygons (NC-drawings). In most cases we are able to find asymptotically optimal area bounds. After preliminaries (Section 2), in Section 3 we deal with *c*-connected *c*-trees and show that quadratic area is achievable for many drawing styles, namely strictly upward order-preserving poly-line R-drawings, strictly upward non-order-preserving SL R-drawings, and upward orthogonal order-preserving R-drawings (if the underlying graph is a binary tree). Such results are interesting to compare with the results in the above mentioned [8], where it is shown that for general *c*-connected *c*-graphs exponential area can be needed. Furthermore, such bounds are asymptotically optimal in the worst-case. On the other hand, we show that orthogonal SL R-drawings are generally not realizable. In Section 4 we deal with non-*c*-connected *c*-trees: we show that SL C-drawings generally require exponential area, and that poly-line order-preserving drawings can be realized in optimal quadratic area. Moreover, we show that upward drawings of non-*c*-connected *c*-trees aren't generally feasible. In Section 5 we show that if the clusters can be represented by non-convex regions, then polynomial area is achievable in many cases. Some proofs are omitted, for space limitations. Details can be found in [4].

A summary of the results presented in this paper is given in Tables 1 and 2, where "UB" and "LB" stand for *Upper Bound* and *Lower Bound*, respectively. "Upward" means *upward* when referred to orthogonal drawings and means *strictly upward* otherwise. If the straight-line column doesn't have a "✓", then the drawing is poly-line. Orthogonal drawings are referred to binary trees. An "X" means that in general a drawing with the corresponding features does not exist. Observe that an area upper bound obtained within a certain drawing convention (say, upward straight-line) for R-drawings is also an upper bound for C-drawings and for NC-drawings. On the contrary, a lower bound for NC-drawings implies a lower bound for C-drawings and for R-drawings.

Tables 1 and 2 are also a reference point for classifying open problems. They correspond to question marks, to cells where upper and lower bounds do not match, and to cells where a drawing is in general not feasible. Such latter cells open the problem of recognizing the c-trees that have one. Among the open problems, we underline the one of determining the area requirement of SL (strictly upward) order-preserving C-drawings of c-connected trees. In fact, an $O(n^2)$ area bound on this problem would imply most of our positive results on c-connected trees. Further, it is interesting in our opinion to state whether SL R-drawings of non-c-connected c-trees always exist. As far as we know the same problem is open also for c-connected c-graphs. About (upward) SL NC-drawings of non-c-connected c-trees we conjecture that polynomial area is always achievable. However, it is not clear which would be the appropriate bound.

Table 1. Summary of the results on minimum area drawings of c-connected c-trees

				R-Drawings				C-Drawings				NC-Drawings			
upward	straight-line	ordered	orthogonal	UB	ref.	LB	ref.	UB	ref.	LB	ref.	UB	ref.	LB	ref.
✓	✓			$O(n^2)$	Th. 2	$\Omega(n^2)$	Le. 1	$O(n^2)$	Th. 2	$\Omega(n^2)$	Le. 1	$O(n^2)$	Th. 2	$\Omega(n^2)$	Le. 1
✓	✓	✓		?	-	$\Omega(n^2)$	Le. 1	?	-	$\Omega(n^2)$	Le. 1	$O(n^4)$	Th. 8	$\Omega(n^2)$	Le. 1
✓		✓		$O(n^2)$	Th. 1	$\Omega(n^2)$	Le. 1	$O(n^2)$	Th. 1	$\Omega(n^2)$	Le. 1	$O(n^2)$	Th. 1	$\Omega(n^2)$	Le. 1
✓		✓	✓	$O(n^2)$	Th. 3	$\Omega(n^2)$	Le. 1	$O(n^2)$	Th. 3	$\Omega(n^2)$	Le. 1	$O(n^2)$	Th. 3	$\Omega(n^2)$	Le. 1
	✓		✓	**X**		Th. 4		?	-	$\Omega(n^2)$	Le. 1	$O(n^3 \log n)$	Th. 9	$\Omega(n^2)$	Le. 1
		✓	✓	$O(n^2)$	[7]	$\Omega(n^2)$	Le. 1	$O(n^2)$	[7]	$\Omega(n^2)$	Le. 1	$O(n^2)$	[7]	$\Omega(n^2)$	Le. 1

Table 2. Summary of the results on min area drawings of non-c-connected c-trees

				R-Drawings				C-Drawings				NC-Drawings			
upward	straight-line	ordered	orthogonal	UB	ref.	LB	ref.	UB	ref.	LB	ref.	UB	ref.	LB	ref.
✓				**X**		Th. 5		**X**		Th. 5		?	-	$\Omega(n^2)$	Le. 1
	✓			?	-	$\Omega(2^n)$	Th. 6	$O(2^n)$	[6]	$\Omega(2^n)$	Th. 6	?	-	$\Omega(n^2)$	Le. 1
		✓	✓	$O(n^2)$	[7]	$\Omega(n^2)$	Le. 1	$O(n^2)$	[7]	$\Omega(n^2)$	Le. 1	$O(n^2)$	[7]	$\Omega(n^2)$	Le. 1
		✓		$O(n^2)$	Th. 7	$\Omega(n^2)$	Le. 1	$O(n^2)$	Th. 7	$\Omega(n^2)$	Le. 1	$O(n^2)$	Th. 7	$\Omega(n^2)$	Le. 1

2 Preliminaries

A *grid drawing* of a graph is a mapping of each vertex v to a point $(x(v), y(v))$ in the plane, where $x(v)$ and $y(v)$ are integers, and of each edge to a Jordan curve between the endpoints of the edge. A *planar drawing* is such that no two edges intersect. A *planar graph* is a graph that admits a planar drawing. A *poly-line* (*PL*) drawing is such that the edges are sequences of rectilinear segments. An *orthogonal drawing* is such that the edges are sequences of axis-parallel rectilinear segments. A *straight-line* (*SL*) drawing is such that all edges are rectilinear segments. The smallest rectangle with sides parallel to the axes that covers a drawing completely is called *bounding box* of the drawing. The *height* (*width*) of a drawing is one plus the height (resp. width) of its bounding box. The *area* a drawing is the product of its height by its width. A *rooted tree* is a tree with one distinguished node called *root*. A *binary tree* is a rooted tree such that each node has at most 2 children. For an underlying tree G (for an inclusion tree T) the subtree of G (the subtree of T) rooted at a vertex v is denoted by $G(v)$ (by $T(v)$). A drawing of a tree is *upward* (*strictly upward*) if every node is placed not below (above) its children and each edge is represented by a curve non-increasing (monotonically decreasing) in the vertical direction. A drawing of a tree is *order-preserving* if the order of the edges incident on each node is the same of one specified in advance.

We define the following drawing conventions for *c*-graphs. A polygon with vertices having integer coordinates is a *lattice polygon*. A drawing of a *c*-tree $C = (G, T)$ is an *NC*-drawing (*Non-Convex*-drawing) if it is *c*-planar, the vertices of G and the bends on the edges of G (if any) have integer coordinates, and the border of each cluster is a lattice polygon. An *NC*-drawing is a *C*-drawing (*Convex*-drawing) if the border of each cluster is a convex lattice polygon. A *C*-drawing is an *R*-drawing (*Rectangle*-drawing) if the border of each cluster is an axis-parallel rectangle with corners having integer coordinates.

Lemma 1. *There exist n-vertex c-trees requiring $\Omega(n^2)$ area in any NC-drawing.*

3 R-Drawings of C-Connected C-Trees

We show that quadratic area is sufficient (and necessary) to construct *R*-drawings of *c*-connected *c*-trees in which the underlying tree is represented within several drawing standards. Namely, we present an algorithm for constructing $\Theta(n^2)$ area strictly upward order-preserving PL *R*-drawings of n-vertex *c*-connected *c*-trees. Then, we slightly modify such an algorithm to obtain different drawings.

Let $C = (G, T)$ be a *c*-connected *c*-tree and suppose G is rooted at any vertex r. For each cluster of T we add dummy vertices and edges to G as follows. Consider any arbitrary order of the clusters. For each cluster, vertices and edges are added to the *c*-tree obtained from the augmentations performed when considering the previous clusters. For each cluster μ inducing a subtree $G(\mu)$ of G, consider the root r_μ of $G(\mu)$ and the parent p_μ of r_μ (if it exists) in G (see Fig. 1.a). Split edge (p_μ, r_μ) by inserting a dummy vertex s_μ into edges (p_μ, s_μ) and (s_μ, r_μ). For the clusters μ such that the root of $G(\mu)$ is r, insert a dummy vertex s_μ and an edge (s_μ, r_μ). In any case, add also two dummy vertices c^1_μ and c^2_μ as children of s_μ and a dummy vertex c^3_μ as child of c^2_μ. The counter-clockwise order of the children of s_μ is c^1_μ, r_μ, and c^2_μ (Fig. 1.b). Vertices s_μ, c^1_μ, c^2_μ, and c^3_μ belong to cluster μ. After having performed the described augmentation on each cluster, we obtain a *c*-tree $C' = (G', T')$. We call r' the root of G'.

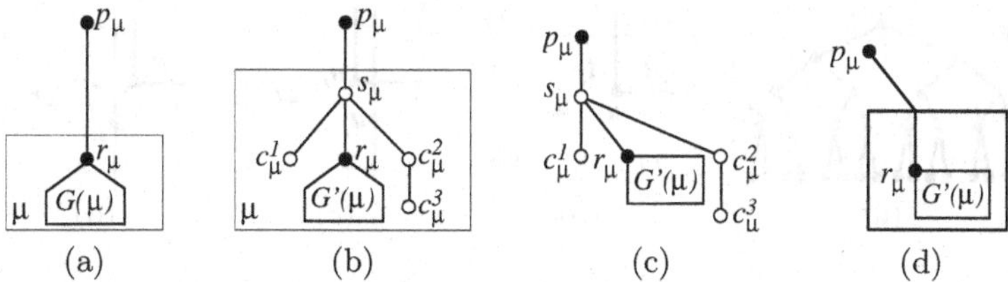

Fig. 1. (a) Edge (r_μ, p_μ). (b) Dummy vertices and edges for μ. (c) Position of the dummy vertices of μ. (d) Replacing dummy vertices with a rectangle representing μ.

We now construct a strictly upward drawing of G'. Denote by $p(v)$ the parent of v. First, assign an x-coordinate to each vertex in G' with a depth-first traversal of G'. Set $x(r') = 1$. Then, suppose that the x-coordinate has been already assigned to a vertex v. Let $v_1, v_2, \ldots, v_m$ be the children of v in counter-clockwise order. Set $x(v_1) = x(v)$; for each child v_i of v, $i = 2, \ldots, m$, set $x(v_i) = 1 + \max_{u \in G'(v_{i-1})}\{x(u)\}$. Concerning the y-coordinates, we perform a traversal of G' with the following properties: When a vertex in $V'(\mu)$ is encountered, every vertex in $V'(\mu)$ will be visited before each not yet visited vertex $w \notin V'(\mu)$. Observe that this implies that when you are visiting μ and you encounter a vertex that belongs to $V'(\mu)$ and to $V'(\nu)$, with ν descendant of μ in T', then every vertex in $V'(\nu)$ will be visited before each not yet visited vertex $w \in V'(\mu)$ such that $w \notin V'(\nu)$. Set $y(r') = 1$. Let μ_r be the smallest cluster containing r'. Now suppose that you are currently analyzing a cluster μ; if there is more than one vertex in $V'(\mu)$ that isn't yet visited, then consider the first not yet visited vertex $v \in V'(\mu)$ that is encountered in a depth-first traversal of $G'(\mu)$; if the smallest cluster ν containing v is the same cluster or is a descendant of the smallest cluster containing $p(v)$, then set $y(v) = y(p(v)) - 1$, otherwise set $y(v)$ equal to the minimum y-coordinate of a vertex in the biggest cluster containing $p(v)$ and not containing v minus one. In every case cluster ν will be the new current cluster. If there is exactly one vertex in $V'(\mu)$ that isn't yet visited, then such vertex is c^3_μ; set $y(c^3_\mu)$ equal to the minimum y-coordinate of a vertex in $V'(\mu)$ minus one. If all the vertices in $V'(\mu)$ have been already visited, then let the cluster parent of μ be the new current cluster.

For each cluster μ remove vertices s_μ, c^1_μ, c^2_μ, and c^3_μ and their incident edges, and insert a rectangle R_μ:$[x(s_\mu), x(c^3_\mu)] \times [y(c^3_\mu), y(s_\mu)]$ representing μ (Fig. 1.c-d). Draw the edges of G: for each edge $(p(v), v)$ in G, if $y(v) < y(p(v) - 1)$, then draw a polygonal line composed of two segments, the first between $p(v)$ and point $(x(v), y(p(v)) - 1)$, and the second between point $(x(v), y(p(v)) - 1)$ and v. Otherwise $(y(v) = y(p(v)) - 1)$ draw a SL segment between $p(v)$ and v.

Theorem 1. *For every n-vertex c-connected c-tree $C = (G, T)$ a $\Theta(n^2)$ area strictly upward order-preserving PL R-drawing can be constructed in $O(n^2)$ time.*

The above described algorithm can be slightly modified in order to produce R-drawings within different drawing conventions for the underlying tree.

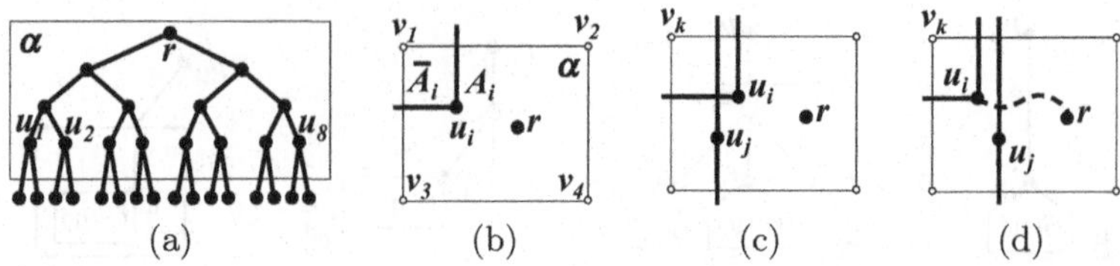

Fig. 2. (a) A *c*-planar drawing of C. (b) Regions A_i and $\overline{A}_i$. (c) The edges connecting u_i to its children and the edges connecting u_j to its children cross. (d) $\overline{A}_i$ is inside $\overline{A}_j$.

Theorem 2. *For every n-vertex c-connected c-tree* $C = (G, T)$ *a* $\Theta(n^2)$ *area strictly upward non-order-preserving SL R-drawing can be constructed in* $O(n^2)$ *time.*

Theorem 3. *For every n-vertex c-connected binary c-tree a* $\Theta(n^2)$ *area upward orthogonal order-preserving R-drawing can be constructed in* $O(n^2)$ *time.*

Contrasting with the above positive results, we prove Theorem 4, also contrasting with the fact that each binary tree has an orthogonal SL drawing.

Theorem 4. *There exists a c-connected c-planar binary c-tree that doesn't admit any orthogonal SL R-drawing.*

Proof. Consider the *c*-tree $C = (G, T)$ defined as follows: G is a complete rooted binary tree with 31 vertices; all the non-leaf vertices of G belong to the same cluster α that is the only non-root cluster, and all the leaves of G don't belong to α. It's easy to see that C is *c*-planar (see Fig. 2.a).

Consider the rectangle A representing α in any orthogonal SL R-drawing Γ of C. Let r be the root of G, let $u_1, \ldots, u_8$ be the 8 vertices that are leaves in $G(\alpha)$, and let $v_1, \ldots, v_4$ be the corners of A. Consider any placement of r inside A. The two edges connecting a vertex u_i and its children divide A in two regions A_i and $\overline{A}_i$ the first containing r and the second not. Since Γ is a SL orthogonal drawing, then both A_i and $\overline{A}_i$ contain at least one corner v_k (see Fig. 2.b). If two regions $\overline{A}_i$ and $\overline{A}_j$, with $i \neq j$, contain the same corner v_k, then either the edges connecting u_i to its children and the edges connecting u_j to its children cross (see Fig. 2.c), or $\overline{A}_i$ ($\overline{A}_j$) is enclosed inside $\overline{A}_j$ (resp. $\overline{A}_i$), and so the path connecting r and u_i (resp. connecting r and u_j) crosses one of the edges connecting u_j and its children (resp. connecting u_i and its children) (see Fig. 2.d). Since (i) each region $\overline{A}_i$ contains at least one corner v_k of A, (ii) any two regions $\overline{A}_i$ and $\overline{A}_j$, with $i \neq j$, cannot contain the same corner v_k, and (iii) there are 4 corners v_k and 8 regions $\overline{A}_i$, then Γ cannot be an orthogonal SL R-drawing of C. □

4 R-Drawings and C-Drawings of Non-C-Connected C-Trees

We consider R-drawings and C-drawings of non-*c*-connected *c*-trees. Most of the positive results presented for *c*-connected trees are not achievable for non-*c*-connected trees, that seem to have the same area requirement of general *c*-graphs.

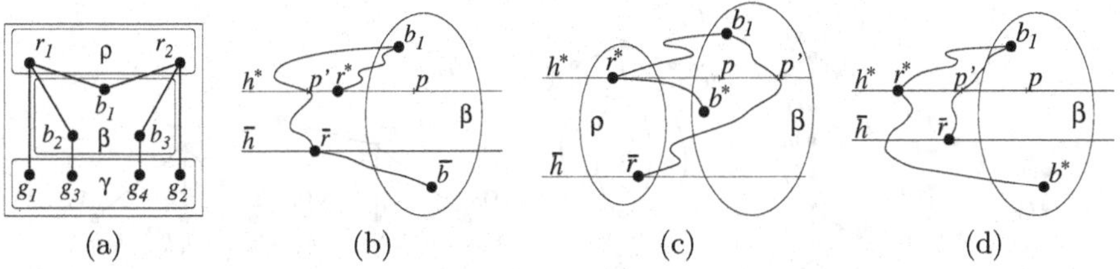

Fig. 3. (a) A c-planar drawing of C; (b) p' is outside β and r^* is closer than p' to p; (c) p' is inside β or r^* is farther than p' to p, and (r^*, b^*) has no intersection with $\overline{h}$; (d) p' is inside β or r^* is farther than p' to p, and (r^*, b^*) has any intersection with $\overline{h}$

Theorem 5. *There exists a non-c-connected c-planar c-tree that doesn't admit any upward C-drawing.*

Proof. Consider the c-tree $C = (G, T)$ defined as follows: G has root b_1, that has two children r_1 and r_2. Vertex r_1 (vertex r_2) has two children b_2 and g_1 (b_3 and g_2). Vertex b_2 (vertex b_3) has a child g_3 (resp. g_4). Vertices b_i, with $i \in \{1, 2, 3\}$, belong to cluster β, vertices r_i, with $i \in \{1, 2\}$ belong to cluster ρ, and vertices g_i, with $i \in \{1, 2, 3, 4\}$ belong to cluster γ. The inclusion tree T has root α that has three children β, ρ, and γ. It's easy to see that C is c-planar (see Fig. 3.a). Suppose that an upward C-drawing Γ of C exists. Let l be the line through b_1 and through the one of b_2 and b_3 that has minimum y-coordinate. The upwardness of Γ implies that $\min_{i \in \{2,3\}}(y(b_i)) \leq y(r_1), y(r_2) \leq y(b_1)$. It follows that vertices r_1 and r_2 are both on the same of the half-planes induced by l, since otherwise β would cross ρ. We claim that either there exists a vertex r_i, with $i \in \{1, 2\}$, that is enclosed inside a region R delimited by cluster β and by edges (b_j, r_k), with $j \in \{1, 2, 3\}$, $k \in \{1, 2\}$, and $k \neq i$, or there exists a vertex b_i, with $i \in \{2, 3\}$ that is enclosed inside a region R delimited by cluster ρ and by edges (b_j, r_k), with $j \in \{1, 2, 3\}$, $k \in \{1, 2\}$, and $j \neq i$.

Consider the horizontal line h^* through the one between r_1 and r_2 that has greater y-coordinate. Consider any intersection point p between h^* and β. Let r^* ($\overline{r}$) be the one between r_1 and r_2 that has greater (resp. smaller) y-coordinate. If $y(r_1) = y(r_2)$, then let r^* ($\overline{r}$) be the one between r_1 and r_2 that is closer (resp. farther) to p. Let b^* ($\overline{b}$) be the only child of r^* ($\overline{r}$) in β. Consider any intersection point p' between h^* and edge $(b_1, \overline{r})$. If p' is outside β and r^* is closer than p' to p, then by simple topologic arguments r^* is closed inside the region R delimited by cluster β, by edge $(b_1, \overline{r})$, and by edge $(\overline{r}, \overline{b})$ (Fig. 3.b). If p' is inside β or if r^* is farther than p' to p, let $\overline{h}$ be the horizontal line through $\overline{r}$. If edge (r^*, b^*) has no intersection with $\overline{h}$, then we have that b^* is closed inside the region R delimited by cluster ρ, by edge (b_1, r^*), and by edge $(b_1, \overline{r})$ (see Fig. 3.c). If edge (r^*, b^*) has intersection with $\overline{h}$, then vertex $\overline{r}$ is closed inside the region R delimited by cluster β, by edge (b_1, r^*), and by edge (r^*, b^*) (Fig. 3.d). Observe that every vertex b_i or r_j, with $i \in \{2, 3\}$ and $j \in \{1, 2\}$, has a child g_k, with $k \in \{1, 2, 3, 4\}$, belonging to cluster γ. Hence, the child g_k of the vertex b_i or r_j that is closed inside R must lie inside R, as well, since placing g_k outside R would imply an edge crossing or an edge-region crossing. Moreover, the child g'_k

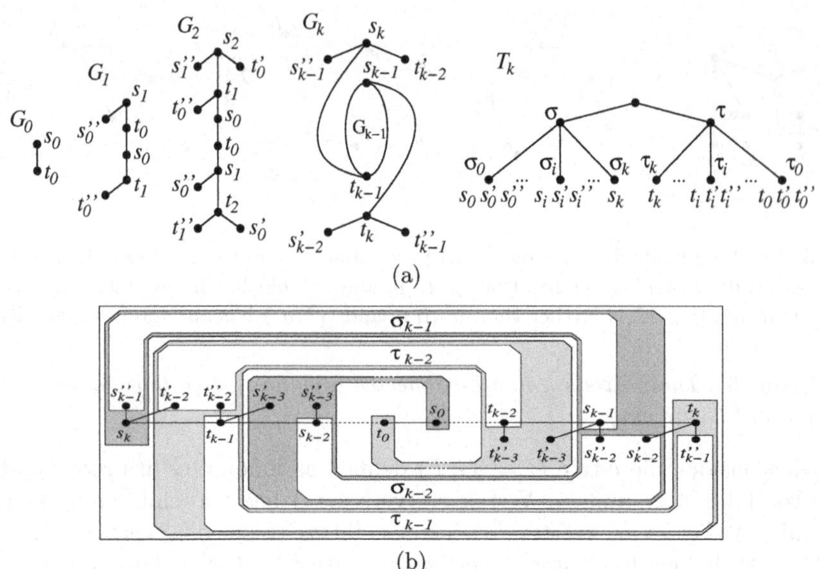

Fig. 4. (a) Inductive construction of the c-tree C_k. (b) c-planar drawing of C_k.

of the vertex b_i that has minimum y-coordinate among the vertices of cluster β lies outside R, with $k \in \{3, 4\}$ and $i \in \{2, 3\}$. It follows that γ crosses region R, implying an edge-region crossing or a region-region crossing. □

Now we show that SL drawings of non-c-connected c-trees may require exponential area. Let $C_k = (G_k, T_k)$ be the family of non-c-connected c-planar c-trees described below (Fig. 4.a). G_0 has vertices s_0 and t_0 and edge (s_0, t_0). T_0 has a root node with two children σ and τ. Node σ (node τ) has one child σ_0 (τ_0), where $s_0 \in V(\sigma_0)$ ($t_0 \in V(\tau_0)$). G_1 is obtained from G_0 by adding vertices s_1, t_1, s''_0, and t''_0 and edges (s_1, t_0), (s_1, s''_0), (t_1, s_0) and (t_1, t''_0). T_1 is obtained from T_0 by adding σ_1 to the children of σ and τ_1 to the children of τ, where $s''_0 \in V(\sigma_0)$, $s_1 \in V(\sigma_1)$, $t''_0 \in V(\tau_0)$, and $t_1 \in V(\tau_1)$. In general, C_k ($k > 1$) is defined as follows. G_k is obtained from G_{k-1} by adding vertices s_k, t_k, s''_{k-1}, t''_{k-1}, s'_{k-2}, and t'_{k-2}, and edges (s_k, t_{k-1}), (s_k, s''_{k-1}), (t_k, s_{k-1}), (t_k, t''_{k-1}), (s_k, t'_{k-2}), and (t_k, s'_{k-2}). T_k is obtained from T_{k-1} by adding σ_k to the children of σ and τ_k to the children of τ, where $s'_{k-2} \in V(\sigma_{k-2})$, $s''_{k-1} \in V(\sigma_{k-1})$, $s_k \in V(\sigma_k)$, $t'_{k-2} \in V(\tau_{k-2})$, $t''_{k-1} \in V(\tau_{k-1})$, and $t_k \in V(\tau_k)$. It is easy to see (Fig. 4.b) that C_k is c-planar. Also, $G(\sigma)$, $G(\tau)$, $G(\sigma_i)$, and $G(\tau_i)$ ($i = 0, \ldots, k-1$) are not connected. For simplifying the notation, in the following we assume k is odd.

Lemma 2. *In any c-planar drawing of C_k we can find polygonal lines $l(s_0, s_1)$ connecting s_0 to s_1, $l(t_0, t_1)$ connecting t_0 to t_1 and, for $i = 2, \ldots, k$, $l(s_{i-1}, s_i)$ connecting s_{i-1} to s_i, $l(t_{i-1}, t_i)$ connecting t_{i-1} to t_i, $l(t_{i-2}, s_i)$ connecting t_{i-2} to s_i, and $l(s_{i-2}, t_i)$ connecting s_{i-2} to t_i such that those lines do not cross between them, don't cross any edge of G_k, and: (1) $l(s_0, s_1)$ crosses only the border of σ_0 and σ_1; (2) $l(t_0, t_1)$ crosses only the border of τ_0 and τ_1; (3) $l(s_{i-1}, s_i)$ crosses*

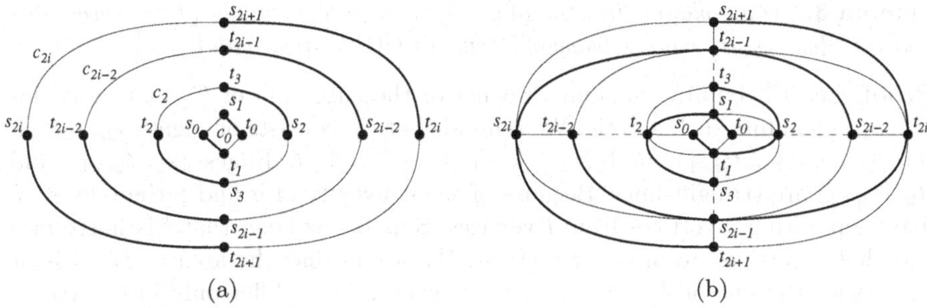

Fig. 5. Graph G'_k. (a) Cycles c_{2i} and (b) their interconnections.

only the border of σ_{i-1} *and* σ_i*; (4)* $l(t_{i-1}, t_i)$ *crosses only the border of* τ_{i-1} *and* τ_i*; (5)* $l(t_{i-2}, s_i)$ *crosses only the border of* τ_{i-2}*,* τ*,* σ_i*, and* σ*; and (6)* $l(s_{i-2}, t_i)$ *crosses only the border of* σ_{i-2}*,* σ*,* τ_i*, and* τ*.*

To study the c-planar drawings of C_k we study those of c-graphs $C'_k = (G'_k, T'_k)$. See [8]. Roughly, the vertices of G'_k are the vertices of G_k but for those vertices with apex "prime" or "double prime". The edges are those induced by such vertices plus the edges corresponding to the polygonal lines of Lemma 2.

More formally, graph G'_k is defined as follows (Fig. 5). Let c_{2i} $(i = 0, \ldots, \frac{k-1}{2})$ be the cycle composed by (s_{2i}, s_{2i+1}), (t_{2i}, t_{2i+1}), (s_{2i}, t_{2i+1}), and (s_{2i+1}, t_{2i}). Each c_{2i} is connected to c_{2i+2} by (s_{2i}, t_{2i+2}), (t_{2i}, s_{2i+2}), (s_{2i+1}, t_{2i+2}), (t_{2i+1}, s_{2i+2}), (s_{2i+1}, s_{2i+2}), (t_{2i+1}, t_{2i+2}), (s_{2i+1}, t_{2i+3}), and (t_{2i+1}, s_{2i+3}). The graph resulting from the connection of all the c_{2i} is G'_k. The inclusion tree T'_k is the subtree of T_k restricted to the vertices of G'_k. Since a c-planar drawing of C'_k can be obtained from a c-planar drawing of C_k by inserting the polygonal lines of Lemma 2, and by removing the vertices with apex "prime" or "double prime" and their incident edges, then C'_k is c-planar. Also, it is easy to see that G'_k is triconnected.

Since G'_k is triconnected the embeddings G'_k differ only in the external face. Consider any face f of G'_k as external. Three cases are possible: (i) $f = c_0$; (ii) $f = c_{k-1}$; (iii) c_{2h} and c_{2h+2} are the cycles that contain the vertices of f. Selecting f as external face induces a nesting of the cycles c_{2i} of G'_k. In case (i) c_{2i+2} is contained into c_{2i} $(i = 0, \ldots, \frac{k-1}{2})$. In case (ii) c_{2i} is contained into c_{2i+2} $(i = 0, \ldots, \frac{k-1}{2})$. In case (iii) c_{2i} is contained into c_{2i+2} for $i = 0, \ldots, h$ and c_{2i+2} is contained into c_{2i} for $i = h+1, \ldots, \frac{k-1}{2}$. In all cases there is a nesting of the cycles with depth greater or equal than $(k-1)/4$. Consider only such a nesting. Also, for sake of simplicity, renumber the vertices of G'_k (and of G_k) according to such a nesting. Namely, call c_0 the most nested cycle, s_0, t_0, s_1, and t_1 its vertices. Call c_2 the cycle surrounding c_0 etc. Denote the external cycle by c_{2d}.

The following lemma generalizes Theorem 4 of [8]. In [8] the drawings of C'_k are studied where all the edges are straight-lines, while in the following lemma only the edges of G'_k that are also edges of G_k are required to be straight.

Lemma 3. *Any c-planar drawing of C'_k where the edges of G_k are SL segments and the clusters are convex polygons requires $\Omega(b^n)$ area, with $b > 1$.*

Proof. Let Γ'_{2i} be any c-planar drawing of the subgraph of C'_k embedded inside c_{2i}, including such a cycle. We remind that c_{2i} consists of edges (s_{2i}, s_{2i+1}), (t_{2i}, t_{2i+1}), (s_{2i}, t_{2i+1}), and (t_{2i}, s_{2i+1}), $0 \leq i \leq d$. Edges (s_{2i}, t_{2i+1}) and (t_{2i}, s_{2i+1}) are straight-lines. Because of the convexity of σ and τ there exists a line l separating s-vertices from t-vertices. Suppose w.l.o.g. that l is horizontal and that s-vertices are above t-vertices. We argue that the area of Γ'_{2i+2} is at least twice the one of Γ'_{2i}, $0 \leq i \leq d-1$, even if they differ only in 6 vertices. The thesis follows from this argument. First, we show that $y(s_j) < y(s_{j+1})$, $0 \leq j \leq 2d-2$. Suppose $y(s_j) \geq y(s_{j+1})$. Consider any placement of t_{j+2} and any drawing of its incident edges. It is easy to see that s_{j+2} is forced inside a face that is internal to the cycle composed by (s_{j+1}, t_j), (t_{j+1}, t_j), (t_{j+1}, t_{j+2}), and (t_{j+2}, s_{j+1}). Hence, if j is even (if j is odd) c_{j+2} cannot be external to c_j (resp. c_{j+1} cannot be external to c_{j-1}), contradicting the assumption to draw c_{2i+2} externally with respect to c_{2i}. Analogously, we have that $y(t_j) > y(t_{j+1})$, $0 \leq j \leq 2d-2$. Since it has to be connected to vertex s_{2i+1} with a SL segment, vertex t_{2i+2} can lie in three possible regions. (A) the half-plane cut by (t_{2i}, s_{2i+1}) and not including s_{2i}, (B) the half-plane cut by (s_{2i}, s_{2i+1}), not including t_{2i} and above the horizontal line through t_{2i+1}, and (C) the half-plane cut by (s_{2i}, s_{2i+1}), not including t_{2i} and below the horizontal line through t_{2i+1}. Region (A) is discarded because placing t_{2i+2} in such a region has the effect of forcing s_{2i+2} inside a face that is internal to the cycle composed by edges (s_{2i+1}, t_{2i}), (t_{2i+1}, t_{2i}), (t_{2i+1}, t_{2i+2}), and (t_{2i+2}, s_{2i+1}). This contradicts the assumption to draw cycle c_{2i+2} externally with respect to cycle c_{2i}. Placing t_{2i+2} inside Region (B) implies that $y(t_{2i+2}) \leq y(t_{2i+1})$, hence such a region is also discarded. Hence, we have that Region (C) is the only possible placement of t_{2i+2}. This implies the duplication of the area, as it can be shown with the same technique applied in [5]. □

Lemma 4. *If it exists a SL C-drawing of C_k with area a, then it exists a c-planar drawing of C'_k such that the edges of G_k are SL segments, the clusters are represented by convex polygons, and the area is less or equal than a.*

Proof. Consider any C-drawing of C_k with area a. It can be augmented without increasing the area with the polygonal lines of Lemma 2, still remaining c-planar. The vertices that are not of G'_k and their incident edges can be removed obtaining a c-planar drawing of C'_k with area less or equal than a. □

From the above lemmas we have:

Theorem 6. *There exist n-vertex non-c-connected c-planar c-trees requiring $\Omega(b^n)$ area in any SL C-drawing, with $b > 1$.*

The above lower bound can be matched by an exponential upper bound. Namely, one can augment the non-c-connected c-planar c-tree in a c-connected c-planar c-graph, that admits an exponential area C-drawing, by the results in [6]. If we relax the SL constraints, then better results can be obtained:

Theorem 7. *There exists an algorithm for computing a $\Theta(n^2)$ area order-preserving 2-bends PL R-drawing of every non-c-connected c-planar c-tree.*

Proof. The proof is based on the results in [7] that show how to construct a $\Theta(n^2)$ area visibility representation of a *c*-graph; such a representation can be easily turned in an order-preserving 2-bends PL *R*-drawing. □

5 NC-Drawings of C-Trees

We show that polynomial area is sufficient for strictly upward order-preserving *SL NC*-drawings of *c*-connected *c*-trees. Note that in the same drawing convention, if *R*- and *C*-drawings require polynomial or exponential area is open.

We show an inductive algorithm to construct a strictly upward order-preserving *SL NC*-drawing of a *c*-connected *c*-tree $C = (G, T)$. Let r be the root of G and let $G(r_1)$, $G(r_2)$, ..., $G(r_k)$ be the subtrees of G rooted at the children r_1, r_2, ..., r_k of r. Suppose that, for each $C_i = (G(r_i), T_i)$, $(1 \leq i \leq k)$, where T_i is the subtree of T induced by the clusters that contain at least one vertex of $G(r_i)$, it can be constructed a strictly upward *NC*-drawing Γ_i. Suppose also that each cluster μ in T_i is represented by a polygonal line composed by four parts: an horizontal segment $T(\mu)$ delimiting the top side of the cluster and lying on the line $y = y_T(\mu)$, two vertical segments $L(\mu)$ and $R(\mu)$ delimiting the left and right sides of the cluster and lying on the lines $x = x_L(\mu)$ and $x = x_R(\mu)$, respectively, and one polygonal line $B(\mu)$ monotonically increasing in the x-direction delimiting the bottom side of the cluster. Notice that the above inductive hypothesis is easily verified in the base case. Namely, if $G(r_i)$ has only one vertex v, it is drawn on a grid point. The clusters containing v are drawn as squares enclosing each other. Now, suppose to have a drawing Γ_i of each C_i. For each i such that $1 \leq i \leq k$, consider the set V_i of vertices of $G(r_i)$ and the set S_i of clusters belonging to T_i that don't contain r. Let $x_L(\Gamma_i) = \min_{v \in V_i, \mu \in S_i}\{x(v), x_L(\mu)\}$, $x_R(\Gamma_i) = \max_{v \in V_i, \mu \in S_i}\{x(v), x_R(\mu)\}$, and $y_T(\Gamma_i) = \max_{\mu \in S_i}\{y(r_i), y_T(\mu)\}$. For each i such that $1 \leq i \leq k$, remove the part of Γ_i that is inside one of the three half-planes $x < x_L(\Gamma_i)$, $x > x_R(\Gamma_i)$, and $y > y_T(\Gamma_i)$. This gives us partial drawings Γ'_i of all the C_i's, where the notations $x_L(\Gamma)$, $x_R(\Gamma)$, and $y_T(\Gamma)$ are extended to $x_L(\Gamma')$, $x_R(\Gamma')$, and $y_T(\Gamma')$, respectively, in the obvious way. Place the Γ'_i's one beside the other, with $x_L(\Gamma'_{i+1}) = x_R(\Gamma'_i) + 1$, and so that all the r_i's lie on the same horizontal line h. Place r $2n^2$ units above and on the same vertical line of r_1. Draw *SL* edges between r and its children. Consider the clusters $\mu_1, \mu_2, \ldots, \mu_l$ containing r ordered so that μ_j is a sub-cluster of μ_{j+1}, for $1 \leq j < l$. For each μ_j, draw $T(\mu_j)$ as an horizontal segment between points $(x_L(\Gamma'_1) - j, y(r) + j)$ and $(x_R(\Gamma'_k) + j, y(r) + j)$; draw $L(\mu_j)$ as a vertical segment between points $(x_L(\Gamma'_1) - j, y(r) + j)$ and $(x_L(\Gamma'_1) - j, y_T(\Gamma'_1) + l - j + 1)$; draw $R(\mu_j)$ as a vertical segment between endpoints $(x_R(\Gamma'_k) + j, y(r) + j)$ and $(x_R(\Gamma'_k) + j, y_T(\Gamma'_k) + l - j + 1)$. Now we draw each $B(\mu_j)$. For each Γ'_i and each μ_j such that T_i doesn't contain μ_j, with $1 \leq i \leq k$ and $1 \leq j \leq l$, draw an horizontal segment between points $(x_L(\Gamma'_i), y_T(\Gamma'_i) + l - j + 1)$ and

$(x_R(\Gamma'_i), y_T(\Gamma'_i) + l - j + 1)$. Notice that now for each Γ'_i and each μ_j the part $B(\Gamma'_i, \mu_j)$ of $B(\mu_j)$ between x-coordinates $x_L(\Gamma'_i)$ and $x_R(\Gamma'_i)$ has been drawn. For each pair $(\Gamma'_i, \Gamma'_{i+1})$ and each μ_j, with $1 \leq i < k$ and $1 \leq j \leq l$, connect $B(\Gamma'_i, \mu_j)$ and $B(\Gamma'_{i+1}, \mu_j)$ by a segment between the rightmost point of $B(\Gamma'_i, \mu_j)$ and the leftmost point of $B(\Gamma'_{i+1}, \mu_j)$. Close the polygon representing μ_j with a segment between $(x_L(\Gamma'_1) - j, y_T(\Gamma'_1) + l - j + 1)$ and the leftmost point of $B(\Gamma'_1, \mu_j)$ and a segment between $(x_R(\Gamma'_k) + j, y_T(\Gamma'_k) + l - j + 1)$ and the rightmost point of $B(\Gamma'_k, \mu_j)$. It's easy to see that in the obtained drawing Γ each cluster is drawn as a polygon with the properties described in the inductive hypothesis of the construction. Hence we obtain:

Theorem 8. *For every c-connected c-tree there exists a strictly upward order-preserving SL NC-drawing in $O(n^4)$ area.*

Theorem 9. *For every c-connected binary c-tree $C = (G, T)$ there exists a SL orthogonal upward NC-drawing with $O(n^3 \log n)$ area.*

Proof. We construct an *hv*-drawing [3] of G with $O(n)$ height and $O(\log n)$ width, with the algorithm in [3]. Then, for each cluster μ we augment the grid inserting $O(n)$ horizontal lines and $O(\log n)$ vertical lines. Such lines are used to insert new tracks between vertices. Such tracks are exploited to route the edges of the boundary of μ. Since we have $O(n)$ clusters the thesis follows. □

References

1. Chan, T., Goodrich, M., Kosaraju, S.R., Tamassia, R.: Optimizing area and aspect ratio in straight-line orthogonal tree drawings. Comput. Geom. 23(2) (2002)
2. Chan, T.M.: A near-linear area bound for drawing binary trees. Algorithmica 34(1) (2002)
3. Crescenzi, P., Di Battista, G., Piperno, A.: A note on optimal area algorithms for upward drawings of binary trees. Comput. Geom. 2, 187–200 (1992)
4. Di Battista, G., Drovandi, G., Frati, F.: How to draw a clustered tree. Tech. Rep. RT-DIA-115-2007, Dip. Informatica e Auto., Univ. Roma Tre (2007)
5. Di Battista, G., Tamassia, R., Tollis, I.G.: Area requirement and symmetry display of planar upward drawings. Discrete & Comp. Geometry 7, 381–401 (1992)
6. Eades, P., Feng, Q., Lin, X., Nagamochi, H.: Straight-line drawing algorithms for hierarchical graphs and clustered graphs. Algorithmica 44(1), 1–32 (2006)
7. Eades, P., Feng, Q., Nagamochi, H.: Drawing clustered graphs on an orthogonal grid. J. Graph Algorithms Appl. 3(4), 3–29 (1999)
8. Feng, Q., Cohen, R.F., Eades, P.: How to draw a planar clustered graph. In: Li, M., Du, D.-Z. (eds.) COCOON 1995. LNCS, vol. 959, pp. 21–30. Springer, Heidelberg (1995)
9. Garg, A., Goodrich, M.T., Tamassia, R.: Planar upward tree drawings with optimal area. Int. J. Comput. Geometry Appl. 6(3) (1996)
10. Garg, A., Rusu, A.: Straight-line drawings of binary trees with linear area and arbitrary aspect ratio. In: Goodrich, M.T., Kobourov, S.G. (eds.) GD 2002. LNCS, vol. 2528, pp. 320–331. Springer, Heidelberg (2002)

11. Garg, A., Rusu, A.: Straight-line drawings of binary trees with linear area and arbitrary aspect ratio. J. Graph Algorithms Appl. 8(2), 135–160 (2004)
12. Trevisan, L.: A note on minimum-area upward drawing of complete and Fibonacci trees. Information Processing Letters 57(5), 231–236 (1996)
13. Valiant, L.G.: Universality considerations in VLSI circuits. IEEE Trans. Comp. 30 (1981)

Drawing Colored Graphs on Colored Points*

Melanie Badent[1], Emilio Di Giacomo[2], and Giuseppe Liotta[2]

[1] Department of Computer and Information Science, University of Konstanz
melanie.badent@uni-konstanz.de

[2] Dip. di Ingegneria Elettronica e dell'Informazione, Università degli Studi di Perugia
{digiacomo, liotta}@diei.unipg.it

Abstract. Let G be a planar graph with n vertices whose vertex set is partitioned into subsets $V_0, \ldots, V_{k-1}$ for a positive integer $1 \leq k \leq n$ and let S be a set of n distinct points in the plane partitioned into subsets $S_0, \ldots, S_{k-1}$ with $|V_i| = |S_i|$ $(0 \leq i \leq k-1)$. This paper studies the problem of computing a crossing-free drawing of G such that each vertex of V_i is mapped to a distinct point of S_i. Lower and upper bounds on the number of bends per edge are proved for any $3 \leq k \leq n$. As a special case, we improve the upper and lower bounds presented in a paper by Pach and Wenger for $k = n$ [*Graphs and Combinatorics* (2001), 17:717–728].

1 Introduction and Overview

Let G be a planar graph with n vertices whose vertex set is partitioned into subsets $V_0, \ldots, V_{k-1}$ for some positive integer $1 \leq k \leq n$ and let S be a set of n distinct points in the plane partitioned into subsets $S_0, \ldots, S_{k-1}$ with $|V_i| = |S_i|$ $(0 \leq i \leq k-1)$. Each index i is a *color*, G is a *k-colored planar graph*, and S is a *k-colored set of points compatible with G*. This paper studies the problem of computing a *k-colored point-set embedding of G on S*, i.e. a crossing-free drawing of G such that each vertex of V_i is mapped to a distinct point of S_i.

Computing k-colored point-set embeddings of k-colored planar graphs has applications in graph drawing, where the *semantic constraints* for the vertices of a graph G define the placement that these vertices must have in a readable visualization of G (see, e.g., [7]). For example, in the context of data base systems design some particularly relevant entities of an ER schema may be required to be drawn in the center and/or along the boundary of the diagram (see, e.g., [18]); in social network analysis, a typical technique to visualize and navigate large networks is to group the vertices into clusters and to draw the vertices of the same cluster close to each other and relatively far from those of other clusters (see, e.g., [6]). A natural way of modelling these types of semantic constraints is to color a (sub)set of the vertices of the input graph and to specify a set of locations having the same color for their placement in the drawing.

The problem of computing k-colored point-set embeddings of k-colored planar graphs has therefore attracted considerable interest in the graph drawing

* This work is partially supported by the MIUR Project "MAINSTREAM: Algorithms for massive information structures and data streams".

F. Dehne, J.-R. Sack, and N. Zeh (Eds.): WADS 2007, LNCS 4619, pp. 102–113, 2007.

and computational geometry communities, where particular attention has been devoted to the *curve complexity* of the computed drawings, i.e. the maximum number of bends along each edge. Namely, reducing the number of bends along the edges is a fundamental optimization goal when computing aesthetically pleasing drawings of graphs (see, e.g., [7]). Before presenting our results, we briefly review the literature on the subject. Since there is not a unified terminology, we slightly rephrase some of the known results; in what follows, n denotes both the number of vertices of a k-colored planar graph and the number of points of a k-colored set of points compatible with the graph.

Kaufmann and Wiese [16] study the "mono-chromatic version" of the problem, that is they focus on 1-colored point-set embeddings. Given a 1-colored planar graph G (i.e. a planar graph G) and a (1-colored) set S of points in the plane they show how to compute a 1-colored point-set embedding of G on S such that the curve complexity is at most two, which is proved to be worst case optimal. Further studies on 1-chromatic point-set embeddings can be found in [4,5,11]; these papers are devoted to characterizing which 1-colored planar graphs with n vertices admit 1-colored point-set embeddings of curve complexity zero on any set of n points and to presenting efficient algorithms for the computation of such drawings.

2-colored point-set embeddings are studied in [10] where it is proved that subclasses of outerplanar graphs, including paths, cycles, caterpillars, and wreaths all admit a 2-colored point-set embedding on any 2-colored set of points such that the resulting drawing has constant curve complexity. It is also shown in [10] that there exists a 3-connected 2-colored planar graph G and a 2-colored set of points S such that every 2-colored point-set embedding of G on S has at least one edge requiring $\Omega(n)$ bends. These results are extended in [8], where an $O(n \log n)$-time algorithm is described to compute a 2-colored point-set embedding with constant curve complexity for every 2-colored outerplanar graph and it is proved that for any positive integer h there exists a 3-colored outerplanar graph G and a 3-colored set of points such that any 3-colored point-set embedding of G on S has at least one edge having more than h bends. Characterizations of families of 2-colored planar graphs which admit a 2-colored point-set embedding having curve complexity zero on any compatible 2-colored set of points can be found in [1,2,13,14,15].

Key references for the "n-chromatic version" of the problem are the works by Halton [12] and by Pach and Wenger [17]. Halton [12] proves that an n-colored planar graph always admits an n-colored point-set embedding on any n-colored set of points; however, he does not address the problem of optimizing the curve complexity of the computed drawing. About ten years later, Pach and Wenger [17] re-visit the question and show that an n-colored planar graph G always has an n-colored point-set embedding on any n-colored set of points such that each edge of the drawing has at most $120n$ bends; they also give a probabilistic argument to prove that, asymptotically, the upper bound on the curve complexity is tight for a linear number of edges. More precisely, let G be an n-colored planar graph with m independent edges and let S be a set of n

points in convex position such that each point is colored at random with one of n distinct colors. Pach and Wenger prove that, almost surely, at least $\frac{m}{20}$ edges of G have at least $\frac{m}{40^3}$ bends on any n-colored point-set embedding of G on S.

The present paper describes a unified approach to the problem of computing k-colored point-set embeddings for $3 \leq k \leq n$. The research is motivated by the following observations: (i) The literature has either focused on very few colors or on the n colors case; in spite of the practical relevance of the problem, little seems to be known about how to draw graphs where the vertices are grouped into $3 \leq k \leq n$ clusters and there are semantic constraints for the placement of these vertices. (ii) The $\Omega(n)$ lower bound on the curve complexity for 2-colored point-set embeddings described in [10] implies that for any $2 \leq k \leq n$ there can be k-colored point-set embeddings which require a linear number of bends per edge. This could lead to the conclusion that in order to compute k-colored point-set embeddings that are optimal in terms of curve complexity one can arbitrarily n-color the input graph, consistently color the input set of points, and then use the drawing algorithm by Pach and Wenger [17]. However, the lower bound of [10] shows $\Omega(n)$ curve complexity for a *constant number* of edges, whereas the drawing technique of Pach and Wenger gives rise to a *linear number* of edges each having a linear number of bends. Hence, the total number of bends in a drawing obtained by the technique of [17] is $O(n^2)$ and it is not known whether there are small values of k for which $o(n^2)$ bends would always be possible. (iii) There is a large gap between the multiplicative constant factors that define the upper and the lower bound of the curve complexity of n-colored point-set embeddings [17]. Since the readability of a drawing of a graph is strongly affected by the number of bends along the edges, it is natural to study whether there exists an algorithm that guarantees curve complexity less than $120n$. Our main results are as follows.

- A lower bound on the curve complexity of k-colored point-set embeddings is presented which establishes that $\Omega(n^2)$ bends may be necessary even for small values of k. Namely, it is shown that for any k such that $3 \leq k \leq n$ there exists a k-colored planar graph G and a k-colored set of points S compatible with G such that any k-colored point-set embedding of G on S has at least $\frac{n}{6} - 1$ edges each having at least $\frac{n}{6} - 1$ bends. This lower bound generalizes and improves the one in [17] for $k = n$.
- An $O(n^2 \log n)$-time algorithm is described that receives as input a k-colored planar graph G ($3 \leq k \leq n$), a k-colored set of points S compatible with G, and computes a k-colored point-set embedding of G on S with curve complexity at most $3n + 2$. This reduces by about forty times the previously known upper bound for $k = n$ [17].
- Motivated by the previously described lower bound, special colorings of the input graph are studied which can guarantee a curve complexity that does not depend on n. Namely, it is shown that if the k-colored planar graph G has $k - 1$ vertices each having a distinct color and $n - k + 1$ vertices of the same color, it is always possible to compute a k-colored point-set embedding whose curve complexity is at most $9k - 1$.

For proofs omitted in this abstract refer to the full version of this paper [3].

2 Preliminaries

A *drawing* of a graph G is a geometric representation of G such that each vertex is a distinct point of the Euclidean plane and each edge is a simple Jordan curve connecting the points which represent its end-vertices. A drawing is *planar* if any two edges can only share the points that represent common end-vertices. A graph is *planar* if it admits a planar drawing.

Let $G = (V, E)$ be a graph. A *k-coloring* of G is a partition $\{V_0, V_1, \ldots, V_{k-1}\}$ of V where the integers $0, 1, \ldots, k-1$ are called *colors*. In the rest of this section the index i is $0 \leq i \leq k-1$ if not differently specified. For each vertex $v \in V_i$ we denote by $col(v)$ the color i of v. A graph G with a k-coloring is called a *k-colored graph*. Let S be a set of distinct points in the plane. We always assume that the points of S have distinct x-coordinates (this condition can always be satisfied by means of a suitable rotation of the plane). For any point $p \in S$ we denote by $x(p)$ and $y(p)$ the x- and y-coordinates of p, respectively. A *k-coloring* of S is a partition $\{S_0, S_1, \ldots, S_{k-1}\}$ of S. A set S of distinct points in the plane with a k-coloring is called a *k-colored set of points*. For each point $p \in S_i$ $col(p)$ denotes the color i of p. A k-colored set of points S is *compatible with* a k-colored graph G if $|V_i| = |S_i|$ for every i; if G is planar, we say that G has a *k-colored point-set embedding* on S if there exists a planar drawing of G such that: (i) every vertex v is mapped to a distinct point p of S with $col(p) = col(v)$, (ii) each edge e of G is drawn as a polyline λ; a point shared by any two consecutive segments of λ is called a *bend* of e. The *curve complexity* of a drawing is the maximum number of bends per edge. Throughout the paper n denotes the number of vertices of graph and m the number of its edges.

3 Lower Bound on the Curve Complexity

A *diamond graph* is a 3-colored planar graph as the one depicted in Figure 1(a). More formally, let $n \geq 12$, let $n'' = (n \bmod 12)$ and let $n' = n - n'' = 12h$ for some $h > 0$; a *diamond graph* $G_n = (V, E)$ is defined as follows: $V = V_0 \cup V_1 \cup V_2$; $V_0 = \{v_i \mid 0 \leq i \leq \frac{n'}{3} + \lceil\frac{n''}{2}\rceil\}$; $V_1 = \{u_i \mid 0 \leq i \leq \frac{n'}{3} + \lfloor\frac{n''}{2}\rfloor\}$; $V_2 = \{w_i \mid 0 \leq i \leq \frac{n'}{3}\}$; $E = E_0 \cup E_1 \cup E_2 \cup E_3 \cup E_4$; $E_0 = \{(v_i, v_{i+1}) \mid 0 \leq i \leq \frac{n'}{3} + \lceil\frac{n''}{2}\rceil - 1\}$; $E_1 = \{(u_i, u_{i+1}) \mid 0 \leq i \leq \frac{n'}{3} + \lfloor\frac{n''}{2}\rfloor - 1\}$; $E_2 = \{(w_i, w_{i+1}), (w_{i+1}, w_{i+2}), (w_{i+2}, w_{i+3}), (w_{i+3}, w_i) \mid 0 \leq i \leq 4h-1, i \bmod 4 = 0\}$; $E_3 = \{(w_{i+1}, w_{i+4}), (w_{i+3}, w_{i+4}), (w_{i+1}, w_{i+6}), (w_{i+3}, w_{i+6}) \mid 0 \leq i \leq 4h-5, i \bmod 4 = 0\}$; $E_4 = \{(w_{4h-1}, v_{\frac{n'}{3}+\lceil\frac{n''}{2}\rceil}), (w_{4h-3}, v_0), (w_0, u_0), (w_2, u_{\frac{n'}{3}+\lfloor\frac{n''}{2}\rfloor})\}$.

Let $S' = S_0 \cup S_1$ be a 2-colored set of points all belonging to a horizontal straight line ℓ; S' is a *bi-colored sequence* if $|S_0| = |S_1|$ or $|S_0| = |S_1|+1$ and given two points p and q of S' such that there is no point r with $x(p) < x(r) < x(q)$, then $col(p) \neq col(q)$. A *3-colored set of points with an alternating bi-colored sequence* is a 3-colored set of points $S = S_0 \cup S_1 \cup S_2$ such that $S' = S_0 \cup S_1$ is an alternating bi-colored sequence and no point of S_2 is on ℓ. A 3-colored set of points with an alternating bi-colored sequence is shown in Figure 1(b).

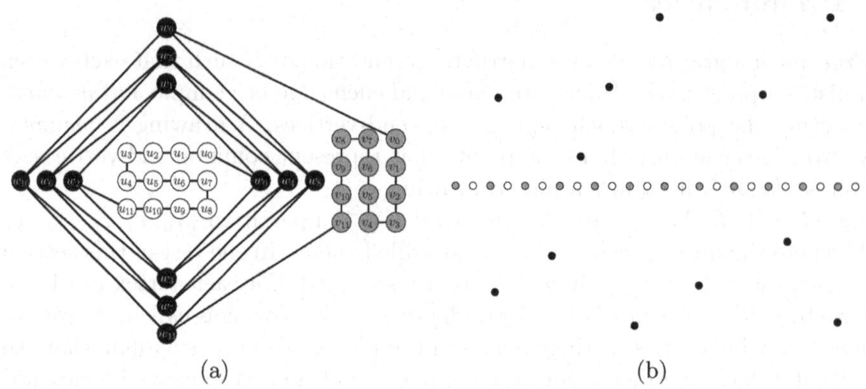

Fig. 1. (a) A diamond graph for $h = 3$. (b) A 3-colored set of points with an alternating bi-colored sequence.

Let G_n ($n \geq 12$) be the diamond graph with n vertices and let S be a 3-colored set of points with an alternating bi-colored sequence and compatible with G_n. Let Γ_n be a 3-colored point-set embedding of G_n on S. Let $p_0, p_1, \ldots, p_{8h+n''-1}$ be the points of the bi-colored sequence of S ordered according to their x-coordinates. Denote with z_i the vertex of G_n which is mapped to p_i. Notice that z_i and z_{i+1} are not adjacent in Γ_n because one of them belongs to V_0 and the other one belongs to V_1 in G_n. Connect in Γ_n z_i and z_{i+1} with a straight-line segment ($i = 0, \ldots, 8h+n''-2$); the obtained path is called *bi-colored path on* Γ_n.

Lemma 1. *Let G_n ($n \geq 12$) be a diamond graph and let S be a 3-colored set of points with an alternating bi-colored sequence such that S is compatible with G_n. Let Γ_n be a 3-colored point-set embedding of G_n on S, let e be an edge of Γ_n, and let Π be the bi-colored path on Γ_n. If Π crosses e b times, then e has at least $b-1$ bends.*

Lemma 2. *Let G_n ($n \geq 12$) be a diamond graph and let S be a 3-colored set of points with an alternating bi-colored sequence such that S is compatible with G_n. Let Γ_n be a 3-colored point-set embedding of G_n on S and let Π be the bi-colored path on Γ_n. Π crosses at least $\frac{n'}{6} - 1$ edges of Γ_n, where $n' = n - (n \mod 12)$; also, Π crosses each of these edges at least $\frac{n'}{6}$ times.*

Proof. For a planar drawing of G_n and a cycle $C \in G_n$ we say that C *separates* a subset $V' \subset V$ from a subset $V'' \subset V$ if all vertices of V' lie in the interior of the region bounded by C and all vertices of V'' are in the exterior of this region. In every planar drawing of G_n each of the h cycles defined by the edges in the set E_2 separates all vertices in V_0 from all vertices in V_1. Thus, every edge of Π must cross these h cycles. Analogously, in every planar drawing of G_n, each of the $h-1$ cycles defined by the edges in the set E_3 separates all vertices in V_0 from all vertices in V_1. Therefore, every edge of Π must also cross these $h-1$ cycles. The number of edges in Π is $\frac{2n'}{3} + n'' - 1$, where $n'' = n - n' = n \mod 12$, and hence each cycle is crossed $\frac{2n'}{3} + n'' - 1$ times. Since each cycle has four edges, we have that at least $2h - 1 = \frac{n'}{6} - 1$ edges (one per cycle) are crossed at least $\lceil \frac{n'}{6} + \frac{n''}{4} - \frac{1}{4} \rceil \geq \lceil \frac{12h}{6} - \frac{1}{4} \rceil = \lceil 2h - \frac{1}{4} \rceil = 2h = \frac{n'}{6}$ times. □

Theorem 1. *For every $n \geq 12$ and for every $3 \leq k \leq n$ there exists a k-colored planar graph G with n vertices and a k-colored set of points S compatible with G such that any k-colored point-set embedding of G on S has at least $\frac{n'}{6} - 1$ edges each having at least $\frac{n'}{6} - 1$ bends, where $n' = n - (n \mod 12)$.*

We conclude this section comparing the result of Theorem 1 with the known lower bound for $k = n$ [17]. Let G be an n-colored graph with m independent edges and let S be a set of n points in convex position such that each point is colored at random with one of n distinct colors. In [17] it is proved that, almost surely, at least $\frac{m}{20}$ edges of G have at least $\frac{m}{40^3}$ bends on any possible n-colored point-set embedding of G on S. A comparison with the result in Theorem 1 can be easily done by observing that the maximum number of independent edges in a graph with n vertices is at most $n/2$.

4 Upper Bound on the Curve Complexity

Theorem 1 shows that in terms of curve complexity the problem of computing a k-colored point-set embedding for any $k \geq 3$ is as difficult as computing an n-colored point-set embedding. Therefore, a drawing algorithm that is asymptotically optimal in terms of curve complexity for all values of k such that $1 \leq k \leq n$ could be designed as follows: (1) Randomly assign each vertex of color i of the input graph to a distinct point of color i of the input set of points. (2) Apply the drawing algorithm of Pach and Wenger [17], which constructs an n-colored point-set embedding whose curve complexity is at most $120n$. However, since optimizing the number of bends per edge is an important requirement that guarantees the readability of a drawing of a graph [7], we present in this section a new approach to the computation of n-colored point-set embeddings which reduces the maximum number of bends per edge from at most $120n$ to at most $3n + 2$.

The key idea is to translate the geometric problem into an equivalent topological problem, namely that of suitably augmenting a planar graph by adding dummy edges that do not cross the real edges too many times. The main ingredients for this approach are: (i) The notion of *augmenting k-colored Hamiltonian path* for a k-colored planar graph G. (ii) A theorem that proves that the number of crossings between the edges of an augmenting k-colored Hamiltonian path and the edges of a k-colored planar graph give an upper bound on the curve complexity of a k-colored point-set embedding of G. (iii) An augmentation algorithm that, for any linear ordering of the vertices of G, computes an augmenting k-colored Hamiltonian path which visits the vertices according to this ordering and that crosses each edge of G at most $3n - 1$ times.

A *k-colored sequence* σ is a linear sequence of (possibly repeated) colors $c_0, c_1, \ldots, c_{n-1}$ such that $0 \leq c_j \leq k - 1$ $(0 \leq j \leq n - 1)$. We say that σ is *compatible* *with* a k-colored graph G if, for every $0 \leq i \leq k-1$, color i occurs $|V_i|$ times in σ. Let S be a k-colored set of points and let $p_0, p_1, \ldots, p_{n-1}$ be the points of S ordered according to their x-coordinates. We say that S *induces* the k-colored sequence $\sigma = col(p_0), col(p_1), \ldots, col(p_{n-1})$.

A graph G has a *Hamiltonian path* if it has a simple path that contains all the vertices of G. If G is a k-colored graph and $\sigma = c_0, c_1, \ldots, c_{n-1}$ is a k-colored sequence compatible with G, a *k-colored Hamiltonian path of G consistent with σ* is a Hamiltonian path $v_0, v_1, \ldots, v_{n-1}$ such that $col(v_i) = c_i$ $(0 \leq i \leq n-1)$. Suppose that G is a k-colored planar graph and that G does not have a k-colored Hamiltonian path consistent with σ. One can augment G to a (not necessarily planar) k-colored graph G' by adding to G a suitable number of dummy edges and such that G' has a k-colored Hamiltonian path $\mathcal{H}'$ consistent with σ and that includes all dummy edges.

If G' is not planar, we can apply a planarization algorithm (see, e.g., [7]) to G' with the constraint that only crossings between dummy edges and edges of $G - \mathcal{H}'$ are allowed. Such a planarization algorithm constructs an embedded planar graph G'' where each edge crossing is replaced with a dummy vertex, called *division vertex*. By this procedure an edge e of $\mathcal{H}'$ can be transformed into a path whose internal vertices are division vertices. The subdivision of $\mathcal{H}'$ obtained this way is called an *augmenting k-colored Hamiltonian path of G consistent with σ* and is denoted as $\mathcal{H}''$. If every edge e of G is crossed at most d times in G' (i.e. e is split by at most d division vertices in G''), $\mathcal{H}''$ is said to be an *augmenting k-colored Hamiltonian path of G consistent with σ and inducing at most d division vertices per edge.* If G' is planar, then $\mathcal{H}''$ coincides with $\mathcal{H}'$. If both end-vertices of $\mathcal{H}''$ are on the external face of the augmented Hamiltonian form of G, then $\mathcal{H}''$ is said to be *external.*

Let v_d be a division vertex for an edge e of G. Since a division vertex corresponds to a crossing between e and an edge of $\mathcal{H}'$, there are four edges incident on v_d in G''; two of them are dummy edges that belong to $\mathcal{H}''$, the other two are two "pieces" of edge e obtained by splitting e with v_d. Let (u, v_d) and (v, v_d) be the latter two edges. We say that v_d is a *flat division vertex* if it is encountered after u and before v while walking along $\mathcal{H}''$; v_d is a *pointy division vertex* otherwise. The following theorem refines and improves a result presented in [8].

Theorem 2. *Let G be a k-colored planar graph, let σ be a k-colored sequence compatible with G, and let $\mathcal{H}$ be an augmenting k-colored Hamiltonian path of G consistent with σ having at most d_f flat and d_p pointy division vertices per edge. If $\mathcal{H}$ is external then G admits a k-colored point-set embedding on any set of points that induces σ such that the maximum number of bends along each edge is $d_f + 2d_p + 1$.*

Based on Theorem 2, we show our upper bound by proving that for *any* n-colored sequence σ an n-colored planar graph G always admits an augmenting k-colored Hamiltonian path of G consistent with σ such that $d_f \leq 3n - 3$ and $d_p \leq 2$, which implies a curve complexity of $3n + 2$. The algorithm to compute an augmenting k-colored Hamiltonian path of G consistent with σ relies on a morphing technique that starts with a special type of planar drawing where all vertices are aligned and transforms it into a drawing with aligned vertices that respects the given linear ordering.

Let $G = (V, E)$ be a planar graph. A *topological book embedding* of G is a planar drawing such that all vertices of G are represented as points of a horizontal straight line ℓ called *spine* and each edge intersects the spine a finite number of times. The straight line ℓ defines two half-planes one above and one below ℓ which are called the *top page* and the *bottom page*, respectively. In a topological book embedding each edge can be either completely contained in the top page, or completely contained in the bottom page, or can cross the spine. A crossing between an edge and the spine is called a *spine crossing*. In order to simplify the description of our results, we assume that a topological book embedding is such that every edge is a sequence of circular arcs; each circular arc of an edge e is called an *arc* of e. It is also assumed that if an edge e crosses the spine at a point p, the two arcs of e sharing p belong to opposite pages.

A *monotone topological book embedding* is a topological book embedding such that each edge crosses the spine at most once. Also, let $e = (u, v)$ be an edge of a monotone topological book embedding that crosses the spine at a point p; e is such that if u precedes v in the left-to-right order along the spine then p is between u and v, the arc with endpoints u and p is in the bottom page, and the arc with endpoints u and v is in the top page.

Theorem 3. [9] *Every planar graph admits a monotone topological book embedding. Also, a monotone topological book embedding can be computed in $O(n)$ time, where n is the number of the vertices in the graph.*

Given a monotone topological book embedding Γ of a planar graph G, we transform Γ into a new topological book embedding Γ' such that the linear ordering of the vertices along the spine coincides with an arbitrary given linear ordering λ of the vertices of G. Every vertex v of G has a *source position* $s(v)$ defined by the point representing v in Γ and a target position $t(v)$ in Γ' defined by the point representing v in Γ'. The linear ordering of the target positions of the vertices of G in Γ' coincides with λ. The transformation from Γ to Γ' moves each vertex of G from its source to its target position by processing the vertices in Γ from left to right. The *trajectory* of vertex v is the straight-line segment $\overline{s(v)t(v)}$. When v is moved to its target position the shape of those edges that are incident to v and of those edges that are intersected by the trajectory of v is changed in order to guarantee the planarity of the drawing.

To better explain the various steps of this morphing technique from Γ to Γ', we introduce the notion of 2-*spine drawing* of a planar graph G which generalizes the definition of topological book embedding. A 2-spine drawing Γ^* of G is a planar drawing such that each vertex is represented as a point of one among two parallel horizontal lines called *spines* of Γ^*. Each edge $e = (u, v)$ of G can have both end-vertices represented in Γ^* as points both in the same spine or in different spines. If both u and v are in the same spine, edge e is drawn in Γ^* as a sequence of arcs; if u is in the upper spine and v is in the lower spine, then when going from u to v along e in Γ^* we find a (possibly empty) sequence of arcs whose endpoints are in the upper spine, a straight-line segment between the two spines, and a (possibly empty) sequence of arcs whose endpoints are in the lower spine. A sequence of arcs of an edge e whose endpoints are in the upper (respectively, lower) spine is called an *upper sequence* of e (respectively, a *lower sequence* of e). The straight-line segment of an edge e between the two spines is called the *inter-spine segment* of e. Note that a 2-spine drawing such that all vertices are points of one of the spines is a topological book embedding.

Lemma 3. *Let G be a planar graph and let λ be a given linear ordering of the vertices of G. G admits a topological book embedding such that the left-to-right order of the vertices along the spine is λ.*

Sketch of Proof. Based on Theorem 3, G has a monotone topological book embedding that we call Γ. Let ℓ be the spine of Γ and let $v_0, \dots, v_{n-1}$ be the vertices of G in the left-to-right order they have along ℓ. Let ℓ' be a horizontal line below ℓ. For each vertex v of G we define a target position on ℓ' such that the left-to-right order of these target positions corresponds to λ.

We process each vertex of Γ in the left-to-right order along ℓ. At each step a vertex is moved to its target position on ℓ' and a 2-spine drawing with spines ℓ and ℓ' of G is computed. Indeed, in order to compute a topological book embedding Γ' of G such that Γ' satisfies the statement, we compute a sequence $\Gamma_0, \dots, \Gamma_n$ of 2-spine drawings with spines ℓ and ℓ' such that Γ_0 coincides with Γ and Γ_n coincides with Γ'. At Step i ($0 \leq i \leq n-1$) the 2-spine drawing Γ_i is transformed into Γ_{i+1} by moving v_i to its target position on ℓ' and by changing the shape of the edges accordingly.

When vertex v_i is moved to its target position, we maintain the planar embedding and only change the shape of the edges incident on v_i and the shape of any edge that is intersected by the trajectory of v_i. In the remainder, we assume that the target positions along ℓ' are such that a trajectory of a vertex intersects an arc with end-points p and q only if one of the end-points of the trajectory is in the closed interval defined by p and q. (This assumption can be satisfied by suitably choosing the radii of the arcs of the edges and the distance between spines ℓ and ℓ'.)

Transformation of the shape of the edges intersected by the trajectory of v_i: The trajectory τ of v_i can intersect both inter-spine segments of some edges or arcs belonging to the the lower sequence of some edges. Notice that if τ intersects both inter-spine segments and arcs, then the inter-spine segments are encountered before the arcs when going from $s(v_i)$ to $t(v_i)$; see also Figure 2. Let $s_0, s_1, \dots, s_{h-1}$ be the segments crossed by τ in the order they are encountered when going from $s(v_i)$ to $t(v_i)$ along τ; denote by x_j the endpoint of s_j that is on ℓ and by x'_j the endpoint of s_j that is on ℓ' ($0 \leq j \leq h-1$). Two cases are possible: **Case a:** x'_j is to the left of x'_{j+1} along ℓ', and therefore x_j is to the left of x_{j+1} along ℓ (see Figure 2(a)); **Case b:** x'_j is to the right of x'_{j+1} along ℓ', and therefore x_j is to the right of x_{j+1} along ℓ (see Figure 2(b)). Let $c_0, c_1, \dots, c_{l-1}$ be the arcs crossed by τ in the order they are encountered going from $s(v_i)$ to $t(v_i)$ along τ; denote by y_j and z_j the endpoints of c_j, with y_j to the left of z_j ($0 \leq j \leq l-1$). Notice that y_j is to the left of y_{j+1} and z_j is to the

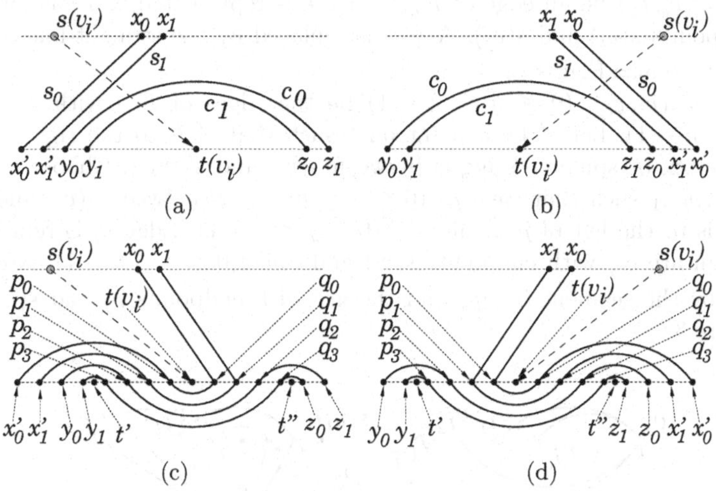

Fig. 2. Transformation of the shape of the edges intersected by the trajectory of v_i

right of z_{j+1}. Refer to Figures 2(c) and 2(d). Let t' and t'' be two points of ℓ' such that t', $t(v_i)$ and t'' appear in this left-to-right order along ℓ' and no vertex or spine crossing is between t' and $t(v_i)$ and between $t(v_i)$ and t'' on ℓ'. Choose $h+l$ points $p_0, p_1, \dots, p_{h-1}, p_h, \dots, p_{h+l-1}$ such that each p_j $(0 \le j \le h+l-1)$ is between t' and $t(v_i)$ on ℓ' and p_j is to the right of p_{j+1} on ℓ'. Choose $h+l$ points $q_0, q_1, \dots, q_{h-1}, q_h, \dots, q_{h+l-1}$ such that each q_j $(0 \le j \le h+l-1)$ is between $t(v_i)$ and t'' on ℓ' and q_j is to the left of q_{j+1} on ℓ'. If **Case a** holds (see Figures 2(c)), replace each segment $s_j = \overline{x_j x'_j}$ $(0 \le j \le h-1)$ with: (i) an arc with endpoints x'_j and p_j; (ii) an arc with endpoints p_j and q_j; (iii) a straight-line segment $\overline{q_j x_j}$. If **Case b** holds (see Figures 2(d)), replace each segment $s_j = \overline{x_j x'_j}$ $(0 \le j \le h-1)$ with: (i) an arc with endpoints x'_j and q_j; (ii) an arc with endpoints q_j and p_j; (iii) a straight-line segment $\overline{p_j x_j}$. Replace each arc c_j $(0 \le j \le l-1)$ whose endpoints are y_j and z_j with: (i) an arc with endpoints y_j and p_{h+j}; (ii) an arc with endpoints p_{h+j} and q_{h+j}; (ii) an arc with endpoints q_{h+j} and z_j.

Transformation of the shape of the edges incident on v_i: We partition the edges incident on v_i into four sets. The set $E_{t,l}$ (respectively, $E_{b,l}$) contains the edges $e = (v_j, v_i)$ such that $j < i$ and the arc of e incident on v_i is in the top (respectively, bottom) page of Γ. Analogously we can define sets $E_{t,r}$ and $E_{b,r}$ for the edges (v_i, v_j) with $i < j$.

Let $e = (v_j, v_i)$ be an edge of $E_{t,l}$ or $E_{b,l}$. When we move v_i, v_j has already been moved to ℓ' (because $j < i$) and therefore when going from v_j to v_i along e in Γ_i we find the (possibly empty) lower sequence σ_l of e, the inter-spine segment s_e of e, and the (possibly empty) upper sequence σ_u of e. Let x' be the endpoint of s_e on ℓ'. Replace s_e and σ_u with an arc whose endpoints are x' and $t(v_i)$.

Let $e = (v_i, v_j)$ be an edge of $E_{b,r}$. Edge e is represented in Γ_i as an arc c_e with endpoints $s(v_i)$ and $s(v_j)$. Arc c_e is replaced by the straight-line segment $\overline{t(v_i)s(v_j)}$; see also Figure 3.

Let $e_j = (v_i, v_{i_j})$ $(0 \leq j \leq h-1)$ be the edges of $E_{t,r}$ with $i_j < i_{j+1}$ $(0 \leq j < h-1)$. Let s' be a point on ℓ such that s' is to the right of $s(v_i)$ and no vertex or spine crossing is between $s(v_i)$ and s' on ℓ. Choose h points $p_0, p_1, \ldots, p_{h-1}$ such that each p_j $(0 \leq j \leq h-1)$ is between $s(v_i)$ and s' on ℓ and p_j is to the left of p_{j+1} along ℓ $(0 \leq j < h-1)$. Edge e_j is represented in Γ_i as an arc c_{e_j} with endpoints $s(v_i)$ and $s(v_{i_j})(0 \leq j \leq h-1)$. Arc c_{e_j} is replaced by the segment $\overline{t(v_i)p_j}$ and the arc with endpoints p_j and $s(v_{i_j})$; see also Figure 3.

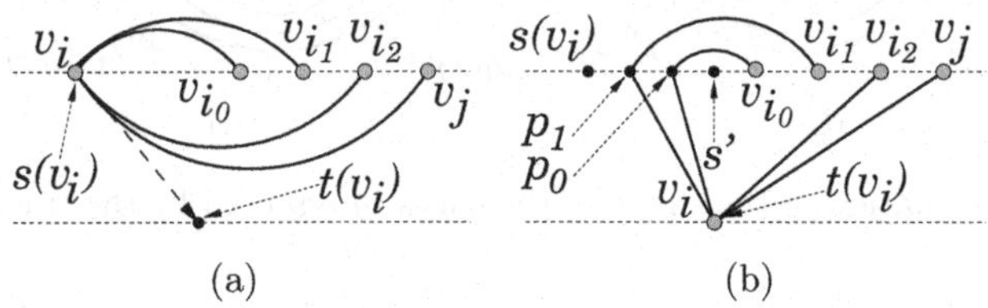

Fig. 3. Transformation of the shape of the edges incident on v_i

After n steps have been executed, and hence all vertices have been moved to their target positions, we obtain a drawing Γ_n where all vertices are aligned and have a left-to-right order coincident with λ. It can be proved that the drawing Γ_{i+1} obtained after the execution of Step i is a 2-spine drawing of G. It follows that Γ_n is a 2-spine drawing (and hence a topological book embedding) of G. □

By means of Lemma 3 and Theorem 2 the following results can be proved.

Lemma 4. *Let G be an n-colored planar graph with n vertices and let σ be an n-colored sequence compatible with G. G admits an augmenting n-colored Hamiltonian path consistent with σ and inducing at most $3n-3$ flat division vertices and at most 2 pointy division vertices per edge.*

Theorem 4. *Let G be a k-colored planar graph with n vertices such that $1 \leq k \leq n$ and let S be a k-colored set of points compatible with G. There exists an $O(n^2 \log n)$-time algorithm that computes a k-colored point-set embedding of G on S having curve complexity at most $3n+2$.*

Since by Theorem 1 k-colored point-set embeddings can have a linear number of edges each requiring a linear number of bends, the upper bound on the curve complexity expressed by Theorem 4 is asymptotically tight. However, as the next theorem shows, there can be special colorings of the input graph which guarantee a curve complexity that depends on k and does not depend on n.

Theorem 5. *Let G be a k-colored planar graph with n vertices such that: (i) $1 \leq k < n$; (ii) $|V_i| = 1$ for every $0 \leq i \leq k-2$; (iii) $|V_{k-1}| = n-k+1$. Let S*

be a k-colored set of points compatible with G. There exists an $O(n^2 \log n)$-time algorithm that computes a k-colored point-set embedding of G on S having curve complexity at most $9k - 1$.

References

1. Abellanas, M., Garcia, J., Hernández-Peñver, G., Noy, M., Ramos, P.: Bipartite embeddings of trees in the plane. Discr. Appl. Math. 93(2-3), 141–148 (1999)
2. Akiyama, J., Urrutia, J.: Simple alternating path problem. Discrete Mathematics 84, 101–103 (1990)
3. Badent, M., Di Giacomo, E., Liotta, G.: Drawing colored graphs on colored points. Technical Report RT-005-06, DIEI, Univ. Perugia (2006) `http://www.diei.unipg.it/rt/RT-005-06-Badent-DiGiacomo-Liotta.pdf`
4. Bose, P.: On embedding an outer-planar graph on a point set. Computational Geometry: Theory and Applications 23, 303–312 (2002)
5. Bose, P., McAllister, M., Snoeyink, J.: Optimal algorithms to embed trees in a point set. Journal of Graph Algorithms and Applications 2(1), 1–15 (1997)
6. Brandes, U., Erlebach, T. (eds.): Network Analysis: Methodological Foundations. LNCS, vol. 3418. Springer, Heidelberg (2005)
7. Di Battista, G., Eades, P., Tamassia, R., Tollis, I.G.: Graph Drawing. Prentice-Hall, Upper Saddle River, NJ (1999)
8. Di Giacomo, E., Didimo, W., Liotta, G., Meijer, H., Trotta, F., Wismath, S.K.: k-colored point-set embeddability of outerplanar graphs. In: Kaufmann, M., Wagner, D. (eds.) GD 2006. LNCS, vol. 4372, pp. 318–329. Springer, Heidelberg (2007)
9. Di Giacomo, E., Didimo, W., Liotta, G., Wismath, S.K.: Curve-constrained drawings of planar graphs. Computational Geometry 30, 1–23 (2005)
10. Di Giacomo, E., Liotta, G., Trotta, F.: On embedding a graph on two sets of points. Int. J. of Foundations of Comp. Science 17(5), 1071–1094 (2006)
11. Gritzmann, P., Mohar, B., Pach, J., Pollack, R.: Embedding a planar triangulation with vertices at specified points. Am. Math. Monthly 98(2), 165–166 (1991)
12. Halton, J.: On the thickness of graphs of given degree. Inf. Sc. 54, 219–238 (1991)
13. Kaneko, A., Kano, M.: Straight line embeddings of rooted star forests in the plane. Discrete Applied Mathematics 101, 167–175 (2000)
14. Kaneko, A., Kano, M.: Discrete geometry on red and blue points in the plane - a survey. In: Discrete & Computational Geometry, pp. 551–570. Springer, Heidelberg (2003)
15. Kaneko, A., Kano, M., Suzuki, K.: Path coverings of two sets of points in the plane. In: Pach, J. (ed.) Towards a Theory of Geometric Graphs. Contemporary Mathematics, vol. 342, American Mathematical Society, Providence, RI (2004)
16. Kaufmann, M., Wiese, R.: Embedding vertices at points: Few bends suffice for planar graphs. Journal of Graph Algorithms and Applications 6(1), 115–129 (2002)
17. Pach, J., Wenger, R.: Embedding planar graphs at fixed vertex locations. Graph and Combinatorics 17, 717–728 (2001)
18. Tamassia, R., Di Battista, G., Batini, C.: Automatic graph drawing and readability of diagrams. IEEE Trans. on Syst., Man, and Cyber. 18(1), 61–79 (1988)

Discrepancy-Sensitive Dynamic Fractional Cascading, Dominated Maxima Searching, and 2-d Nearest Neighbors in Any Minkowski Metric

Mikhail J. Atallah[1], Marina Blanton[1],
Michael T. Goodrich[2], and Stanislas Polu[3]

[1] Dept. of Computer Sciences, Purdue Univ.,
{mja,mbykova}@cs.purdue.edu
[2] Dept. of Computer Science, Univ. of California, Irvine
last-name@acm.org
[3] École Polytechnique
first-name.last-name@polytechnique.fr

Abstract. This paper studies a discrepancy-sensitive approach to dynamic fractional cascading. We provide an efficient data structure for dominated maxima searching in a dynamic set of points in the plane, which in turn leads to an efficient dynamic data structure that can answer queries for nearest neighbors using any Minkowski metric.

1 Introduction

Discrepancy theory deals with the degrees to which point sets differ from their expected uniformity (e.g., see Chazelle [8,9]). This theory is usually applied globally, for entire sets, but we are interested in local notions of discrepancy, dealing with how sets differ from their expected uniformity in small intervals. This interest is motivated from *dynamic fractional cascading* [10,11,17].

In fractional cascading [10,11], we are given a bounded-degree[1] *catalog graph* G, such that each vertex v of G stores a catalog $C(v) \subset U$, for a total order U. Given a value x belonging to the total order for a path P in G, a query for x in P searches for x in the catalog $C(v)$ for each vertex v in P. If insertions and deletions are allowed in the $C(v)$'s, then we have the "dynamic fractional cascading" [17] problem. Static fractional cascading solutions due to Chazelle and Guibas [10,11] allow for queries to be performed in a path of length k in time $O(\log n + k)$, where n is the total size of all the catalogs, and dynamic fractional cascading solutions due to Mehlhorn and Näher [17] show that such queries can be done in a dynamic setting in $O(\log n + k \log\log n)$ time, with updates taking $O(\log n \log\log |U|)$ amortized time. The reduced efficiency of dynamic fractional cascading seems to come from its need to dynamically handle discrepancy. Our interest in this paper, therefore, is to address discrepancy head on—to design a scheme for dynamic fractional cascading that is *discrepancy sensitive.*

[1] We note that a catalog graph of degree $d > 3$ can be transformed into a degree-3 catalog graph by replacing high-degree nodes with complete binary trees.

F. Dehne, J.-R. Sack, and N. Zeh (Eds.): WADS 2007, LNCS 4619, pp. 114–126, 2007.

Previous Related Work. For prior results in discrepancy theory, for example, please see the excellent book by Chazelle [9]. Subsequent to the introduction of fractional cascading by Chazelle and Guibas [10,11] and its dynamic implementation by Mehlhorn and Näher [17], there have been many specific uses for this technique, as well as a generalization, due to Sen [21], based on randomized skip lists, and an extension for I/O efficiency due to Yap and Zhu [24].

The prior work on nearest neighbor structures is vast; for more detailed reviews, see the surveys by Alt [1] or Clarkson [12]. For static data, there are several ways to achieve $O(\log n)$ time for nearest-neighbor queries in the plane, including constructing a planar point location data structure "on top" of a Voronoi diagram (e.g., see [20]). For uniformly distributed data, Bentley, Weide, and Yao [5] give optimal algorithms for static data, and Bentley [4] gives an optimal algorithm for the semidynamic (deletion only) case. For approximate nearest-neighbor queries, Arya *et al.* [3] give an optimal static structure, and Eppstein *et al.* [14] give an optimal dynamic structure. Finally, for general exact nearest-neighbor queries, Chan [7] gives a dynamic method that achieves polylogarithmic expected times for updates and queries. In addition, there has been some work on nearest-neighbors in non-Euclidean settings for "reasonably separated" uniform point sets (e.g., see [6,16,15]), but this does work does not apply efficiently to Euclidean metrics on point sets taken from continuous uniform distributions.

Our Results. In this paper, we introduce a study of a discrepancy-sensitive approach to dynamic fractional cascading. Unlike the Mehlhorn-Näher approach, which assumes a worst-case distribution for the discrepancies between adjacent catalogs, our approach is sensitive to these differences. That is, it runs faster through low-discrepancy neighbors and slower through high-discrepancy neighbors. We show, for example, that a search for a value x in a collection of catalogs, of size at most n, stored in vertices of a path P can be done in time $O(\log n + \sum_{(v,w)\in P} \log \delta_{v,w}(x))$, where $\delta_{v,w}(x)$ is the relative local discrepancy at x of the catalogs stored at the nodes v and w in G. Such a discrepancy-sensitive result is useful in a number of real-world scenarios, as we show that there are several practical distributions such that the sum of the relative local discrepancies in the catalogs belonging to a path of length k is $O(k)$ with high probability. For example, we use this approach to provide an efficient data structure for dominated maxima searching in a dynamic set of uniformly distributed points in the plane. This, together with the known fact that the expected number of maxima points in an uniformly distributed set S of n points in $\mathbb{R}^2$ is $O(\log n)$, shows that we can construct a dynamic data structure that can answer queries for nearest neighbors in S using any Minkowski metric, where insertions and deletions run in $O(\log^2 n)$ expected time and queries run in $O(\log n)$ expected time, as well. These expectations assume a uniform distribution, but even with real-life (not uniformly distributed) data we experimentally observe it to hold.

2 Discrepancy-Sensitive Dynamic Fractional Cascading

Weisstein [23] defines a notion for *local discrepancy*, which, for an interval I, gives a measure of how much the number of points intersecting I differs from the normalized length of I. We are, however, interested in the application to dynamic fractional cascading, which involves comparing adjacent catalogs to each other, not arbitrary intervals to catalogs. Suppose, therefore, that (v, w) is an edge in G and that $C(v)$ and $C(w)$ are the catalogs stored respectively at the vertices v and w in G. Let us assume, without loss of generality, that $C(v)$ and $C(w)$ both store sentinel values, "$-\infty$" and "$+\infty$," which are respectively the smallest and the largest elements in the common total order to which all catalog elements belong. For any value x, and vertex v in G, let $\text{pred}_v(x)$ denote the predecessor of x in $C(v)$, that is, the largest element in $C(v)$ less than or equal to x. Likewise, let $\text{succ}_v(x)$ denote the successor of x in $C(v)$, that is, the smallest element in $C(v)$ greater than or equal to x. For any edge (v, w) in G, we define the *relative local discrepancy* from $C(v)$ to $C(w)$ at x as follows:

$$\delta_{v,w}(x) = |[a, b] \cap C(v)| + |[a, b] \cap C(w)|,$$

where $a = \min\{\text{pred}_v(x), \text{pred}_w(x)\}$ and $b = \max\{\text{succ}_v(x), \text{succ}_w(x)\}$, i.e.,, the relative local discrepancy from $C(v)$ to $C(w)$ at x is the number of items of $C(v)$ and $C(w)$ falling in the closed interval $[a, b] = [\text{pred}_v(x), \text{succ}_v(x)] \cup [\text{pred}_w(x), \text{succ}_w(x)]$. It is a measure of how different $C(v)$ and $C(w)$ are in the vicinity of x. Note that $\delta_{v,w}(x) \geq 2$, even if $C(v) = C(w)$.

Augmenting a Catalog Graph to Support Searches and Updates. Let us first give some intuition about our augmentation. Imagine that we have a deterministic skip list [19] built "on top" of the elements in $C(v)$ and that the nodes in this structure are all colored black. Likewise, imagine that we have a deterministic skip list built "on top" of the elements in $C(w)$ and that the nodes in this structure are all colored white. These structures allow for both top-down and bottom-up searches and updates to be performed in $O(\log n)$ time [19]. Now imagine further that we merge these two structures into a common structure by having each black node "cut" any white edge (i.e., interval of white nodes) that it is contained in and having each white node "cut" any black edge that it is contained in. Let us then link the roots of all the remaining bottom-level skip lists. The remaining structure is the "fractionally-cascaded" merge of $C(v)$ and $C(w)$ and this is the structure that we will maintain dynamically.

More formally, our structure is defined so that we maintain the following substructures for each edge (v, w) in G (see Fig. 1):

- We maintain in a "black" deterministic skip list each maximal contiguous interval of $C(v)$ that contains no elements of $C(w)$.
- We maintain in a "white" deterministic skip list each maximal contiguous interval of $C(w)$ that contains no elements of $C(v)$.
- We maintain black-white links between the roots of these skip lists.
- Each bottom-level skip-list interval that is cut by a skip list of the other color has a link to and from the root of that skip list.

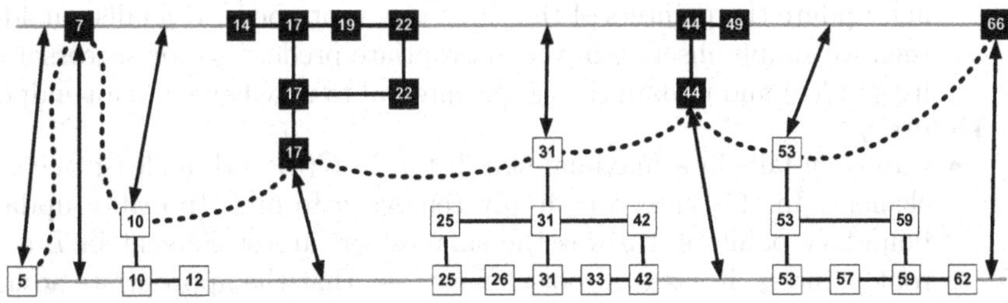

Fig. 1. An example of the fractionally-cascaded structures that join a "black" $C(v)$ to a "white" $C(w)$. Skip-list edges are shown in bold, with those cut by a sublist of the opposite colored gray. The links between skip-list roots are shown dashed and the arrowed lines show the links between bottom-level skip-list edges and the roots of the opposite-color skip lists that cut that edge.

Searches. A search in a catalog graph G consists of an element x for which we would like to find $\text{pred}_v(x)$ in $C(v)$ for each node v in a given path $P = (v_1, v_2, \ldots, v_k)$. We assume that we have a complete deterministic skip list for the first node, v_1, of P. This allows us to locate $\text{pred}_{v_1}(x)$ in $O(\log n)$ time, where n is the maximum size of any catalog. For locating x in $C(v_{i+1})$, for $i = 1, \ldots, k-1$, we start from a pointer to $\text{pred}_{v_i}(x)$, which we will have found inductively. There are two cases at this point:

- Case 1: x falls inside a maximal skip list in $C(v_i)$. In this case, we traverse up the skip list for this interval in $C(v_i)$ to its root and then follow the pointer from the root to the interval in $C(v_{i+1})$ containing x.
- Case 2: x falls outside a maximal skip list in $C(v_i)$. In this case, we follow the pointer from the "cut" interval in $C(v_i)$ containing x to the root of the skip list in $C(v_{i+1})$ falling in this interval. We then search down this skip list to locate the predecessor of x in $C(v_{i+1})$.

Note that, in either case, each step i of the search, after the first, runs in $O(\log \delta_{v_i,v_{i+1}}(x))$ time, since the size of the skip list we search in for either case is $O(\delta_{v_i,v_{i+1}}(x))$.

Updates. Let us consider how to perform an update in our structure, that is, an insertion or deletion in a $C(v)$ list, assuming we have already located the place in $C(v)$ where the update is to occur (let us account separately for the time needed to find this location). We perform the necessary updates for each edge (v, w), of which there are only a constant number, according to the following cases:

- **Insert** y:
  - Case 1: y falls inside a maximal skip list L in $C(v)$. In this case, we simply insert y in L.
  - Case 2: y falls outside a maximal skip list in $C(v)$. In this case, we follow the interval pointer from the (gray) interval in $C(v)$ containing y to the skip list L in $C(w)$ and search down for y in this list. If y falls in the interior of L then we split L at y, set up y as its own skip list in $C(v)$

and update the pointers of the three new root nodes. If y falls outside L, then we simply insert y in the appropriate predecessor or successor skip list in $C(v)$ and update the (gray) interval to now have y as an endpoint.

- **Delete** y:
  - Case 1: y falls in a maximal skip list L in $C(v)$ with at least one other element. In this case, we simply remove y from L (possibly updating boundary pointers if y was the smallest or largest element in L or the root pointers, if y was a root element—so that the appropriate adjacent pointers now point to the new root of L).
  - Case 2: y is the only element of its skip list in $C(v)$. In this case, we follow the pointers from y's (root) node to the two skip lists in $C(w)$ that y separates, and we perform a splice of these two structures, updating the root pointers as needed.

Note that in either an insertion or a deletion, the time needed to perform all the necessary local searching, insertions, deletions, splits, and/or splices is $O(\log \delta_{v,w}(y))$.

Theorem 1. *A catalog graph G, with maximum catalog size n, can be augmented with additional structures so as to support searches for an element x in the catalogs in a path P in G in time $O(\log n + \sum_{(v,w)\in P} \log \delta_{v,w}(x))$. Likewise, a sequence of updates for an element y in catalogs in a path P in G can be done in these structures in time $O(\log n + \sum_{(v,w)\in P} \log \delta_{v,w}(y))$.*

Uniform data. Suppose that each catalog in G contains n points chosen independently and uniformly at random from the interval $[0, 1]$. In this case, the set of points in a catalog $C(v)$ define a set of order statistics, and the distribution of the length of consecutive spacings therefore follows the Beta distribution with parameters 1 and n (e.g., see [2,13]). Thus, the expected interval length is $1/(n+1)$. Having fixed such an interval in $C(v)$, the number of points in $C(w)$ that falls in this interval follows a Binomial distribution, with probability equal to the length of the interval. Thus, the distribution of each $\delta_{v,w}(v)$ follows the Beta-Binomial distribution, with parameters 1 and n, which has expected value $\mu = n/(n+1)$ [22].

The performance of searching and updating our augmented structures at an element x along a path $P = (v_1, \ldots, v_k)$ in a catalog graph G depends on the random variable,

$$T_P = \sum_{(v_i,v_{i+1})\in P} \log \delta_{v_i,v_{i+1}}(x).$$

Unfortunately, the relative local discrepancies for consecutive edges in P are not necessarily independent. Even so, we can write

$$T_P = \sum_{(v_i,v_{i+1})\in P,\ \text{odd } i} \log \delta_{v_i,v_{i+1}}(x) + \sum_{(v_i,v_{i+1})\in P,\ \text{even } i} \log \delta_{v_i,v_{i+1}}(x), \quad (1)$$

and we note that each term in the separate sums are independent. Thus, we can bound the degree to which T_P differs from its expectation by adding bounds on the two sums. Combining this with the expected value of the associated Beta-Binomial distribution given above, we can use a Chernoff bound twice (e.g., see [18]) to prove the following (we give the proof in the final version):

Theorem 2. *Given a catalog graph G such that each catalog is a set of $O(n)$ independent, uniform random points in the interval $[0,1]$, then for any path P of length k in G, $\sum_{(v,w)\in P} \log \delta_{v,w}(x)$ is $O(k)$ with probability $1 - 1/2^k$.*

Using this result, we can take the dynamic range searching structure of Mehlhorn and Näher [17], which is based on range trees (e.g., see [20], and replace their dynamic fractional cascading solution with ours, which gives us the following:

Theorem 3. *We can maintain a dynamic range searching data structure for a set of points taken uniformly at random in the unit cube so as to support point insertions and deletions in $O(\log n)$ time w.h.p. and the reporting of all the points in a rectangular query range $[x_1, x_2] \times [y_1, y_2]$ in $O(\log n + k)$ time w.h.p., where k is the number of points returned by the query.*

3 Dynamic Dominated Maxima

This section describes a scheme for dynamically maintaining a set S of points drawn from a uniform distribution in a rectangle, so that a *dominated maxima* query can be done in $O(\log n)$ expected time: Given a query point q, the query returns the set of maximal elements among the points of S that are dominated by q; note that the expected size of the output is itself $O(\log n)$ (because of the uniform distribution). The expected time for an update will be shown to be $O(\log^2 n)$.

We shall find it necessary to maintain 4 such data structures, one for each of the 4 possible sets of coordinate axes obtained by reversing the direction of {neither,one,both} of the x and y axes – having all 4 such structures makes it possible to achieve the bounds we claim but imposes only a constant factor of 4 on the complexity bounds.

In order to more explicitly define the 4 above-mentioned problems, and also to facilitate the understanding of our algorithm, we will consider the smallest origin-centered square containing the whole set S for a given state of S. We position four coordinate systems, one at each of the four corners of the square, with the origin being at the corresponding corner and the directions of the axes pointing from the origin along the edges of the square. We call these four coordinate systems *South-West* (abbreviated as *SW*), *South-East* (*SE*), *North-West* (*NW*), *North-East* (*NE*). For a point $q \in S$, we use $x_{SW}(q)$ (resp., $y_{SW}(q)$) to denote the x (resp., y) coordinate of q in the SW coordinate system. A similar notation is used for the other three coordinate systems.

The 4 problems mentioned above are then the following: (i) A South-West problem that pertains to the subset of S that is dominated by the query point q_0 in the SW coordinate system, i.e., the subset "below and to the left of q_0"; (ii) a South-East problem that pertains to the subset of S that is dominated by the query point q_0 in the SE coordinate system (the subset "below and to the right of q_0"); (iii) a North-East problem that pertains to the subset of S that is dominated by the query point q_0 in the NE coordinate system (the subset "above and to the right of q_0"); and (iv) a North-West problem that pertains to the subset of S that is dominated by the query point q_0 in the NW coordinate system (the subset "above and to the left of q_0").

Recall that a point q is *maximal* in the set S relative to the SW coordinate system iff for every other point $q' \in S$ at least one of the following inequalities holds:

$$x_{SW}(q') \leq x_{SW}(q) \qquad y_{SW}(q') \leq y_{SW}(q),$$

which, in words, can be stated as: "no other point of S dominates q in the SW coordinate system." For a point q and a set S we also define the notion of a maximal set in the SW coordinate system with respect to q. This set, denoted by $M_{SW}(S, q)$, is computed by first considering only those points in S that are dominated by q in the SW coordinate system (i.e., the subset of S below and to the left of q) and then computing the maximal points of that subset. All points in $M_{SW}(S, q)$ are assumed to be sorted by increasing x coordinates. A similar notation is used for the other three coordinate systems.

In the rest of our discussion we focus on the South-West problem. All of our solutions for this South-West problem can be translated into similar ones for the South-East, North-East, and North-West problems.

The Data Structure. Let T_x be an n-node search tree structure whose nodes are the n points of S ordered by their x coordinates. T_x verifies the following properties, v being a node of T_x :

- T_x is a weight balanced binary search tree
- All nodes in the right subtree of v have greater x value than v
- All nodes in the left subtree of v have lesser value than v

For each node v in T_x, we use Sl_v to denote the subset of S that lies in the subtree of v and have x coordinate lesser or equal to v's one. Each such Sl_v is itself organized as a dynamic search structure according to the y coordinates of the points in it. The T_x tree and its associated Sl_v's are organized as the dynamic fractional cascading structure described above. With this structure in place, for every path $\mathcal{P}$ in T_x, searching for y_0 in Sl_v for every $v \in \mathcal{P}$ can be done in $O(\log n + |\mathcal{P}|)$ expected time.

An update to this structure due to insertion or deletion of a point consists of adding or removing a node of T_x, updating all the Sl_v sets from that node to the root and finally then rebalancing T_x. Note that the insertion of a point (x_0, y_0) does not cause the creation of a new node in T_x if there exists already a point with x_0 coordinates, but only an update in the underlying dynamic fractional cascading structure. We have the equivalent property for deletion. Rebalancing the tree implies $O(1)$ rotations. A rotation associated with three node v, v', v'' implies the reconstruction of the underlying sets Sl_v, Sl'_v, Sl''_v, that is, $O(|Sl_v|)$ insertions and deletions in the dynamic fractional cascading structure. Since T_x

is a weight balanced search tree, the amortized value of $|Sl_v|$ is $\log n$. Thus an update to this structure takes $O(\log n)$ amortized time.

In addition to the above, each copy of a point q in Sl_v stores the following:

- $l_{SW}(v,q)$ = the leftmost (hence, highest) point in $M_{SW}(S_v,q)$.
- $r_{SW}(v,q)$ = the rightmost (hence, lowest) point in $M_{SW}(S_v,q)$.
- $l_{SE}(v,q)$ = the leftmost (hence, lowest) point in $M_{SE}(S_v,q)$.
- $r_{SE}(v,q)$ = the rightmost (hence, highest) point in $M_{SE}(S_v,q)$.
- $l_{NW}(v,q)$ = the leftmost (hence, lowest) point in $M_{NW}(S_v,q)$.
- $r_{NW}(v,q)$ = the rightmost (hence, highest) point in $M_{NW}(S_v,q)$.
- $l_{NE}(v,q)$ = the leftmost (hence, highest) point in $M_{NE}(S_v,q)$.
- $r_{NE}(v,q)$ = the rightmost (hence, lowest) point in $M_{NE}(S_v,q)$.

The above quantities will be shown to facilitate a query, but they also impose the burden of dynamically updating them. We need to describe how a query is processed, and how to dynamically update all of the above quantities.

Processing a Query. The query processing consists of, given a query point q_0, returning the maximal elements of the subset of S dominated by q_0 in the SW coordinate system. (The query point is arbitrary and need not be in S.)

More formally, to process a query for a point q_0 with the coordinates (x_0, y_0), we do the following:

1. First we locate the node which has greatest x value lesser or equal to x_0 in T_x, thereby defining a root-to-leaf path $\mathcal{P}$ in T_x. Let $v_1, \ldots, v_t$ be (in left to right order) the nodes whose right sibling is on $\mathcal{P}$. We henceforth refer to these nodes as the *fringe of x_0 in T_x*. Note that $t \leq \log n$, and that every point in $\bigcup_{i=1}^{t} Sl_{v_i}$ has an x coordinate that is $\leq x_0$ and that there is no other such points.
2. Within every Sl_{v_i}, $1 \leq i \leq t$, let y_i' be the largest y coordinate that is $\leq y_0$. Computing all the y_i's involves locating y_0 in every Sl_{v_i}. Using the dynamic fractional cascading search structure, the computation of all the y_i's can be done in $O(\log n + t)$ expected time, which is $O(\log n)$.
3. Let $Y_1, \ldots, Y_t$ be defined inductively as follows:
   (a) $Y_t = -\infty$
   (b) $Y_{k-1} = \max\{Y_k, y_k'\}$ for $k = t-1, t-2, \ldots, 1$.
   In words, Y_k ($k < t$) is the largest y coordinate among the points in $\bigcup_{i=k+1}^{t} Sl_{v_k}$.
4. Enumerate the points in $M_{SW}(S,q)$. Before explaining how this enumeration done, we point out that the point of S that constitutes the South-West solution must belong to $M_{SW}(S,q)$, which is easy to prove by contradiction. We also point out that the expected number of points in $M_{SW}(S,q)$ is $O(\log |S|)$, hence $O(\log n)$. Thus, the $O(\log n)$ average query performance would be achieved if we could somehow enumerate the points of $U = M_{SW}(S,q)$ in time $O(|U|)$. We do this by first observing that the subset of S from which the maximal points are computed consists of the subset of $\bigcup_{i=1}^{t} Sl_{v_k}$ having y coordinates $< y_0$. Our strategy will be to enumerate, in the order $k = 1, \ldots, t$ the maximal points of Sl_{v_k} that belong to U, call their set U_k,

stopping as soon as the about-to-be-enumerated y coordinate drops below Y_k. (If we did not stop at that point, we would be enumerating points that do not belong to U.) This enumeration of U_k is done as follows:
   (a) Let q_k be the point with the y coordinate y'_k (that is, q_k is the highest point of Sl_{v_k} whose y coordinate is $\leq y_0$).
   (b) While the y coordinate of q_k is $\geq Y_k$, we (i) include q_k as a member of U_k, and then (ii) set $q_k = r_{SE}(v, q_k)$, which is the rightmost (hence, highest) point in $M_{SE}(S_v, q_k)$.

   Of course, in the above, U is the concatenation of $U_1, \dots, U_t$.
5. Since we have not checked the points with y-coordinate equal to y_0 in $M_{SW}(S, q)$, we need to add them to U. This can be done by searching for y_0 in the fringe of x_0 which takes $O(\log n)$ expected time using the fractional cascading structure.

As argued above, the average complexity of the above query processing is $O(\log n)$. We now turn our attention to the dynamic updates. We begin with the case of insertions.

Processing an Insertion. Let $q_0 = (x_0, y_0)$ be the point being inserted. We already argued that the fractional cascading structure can be updated in $O(\log n)$ expected time as a result of this insertion. The main task we face now is how to update the quantities $l_{SW}(v,q)$, $r_{SW}(v,q)$, $l_{SE}(v,q)$, $r_{SE}(v,q)$, $l_{NW}(v,q)$, $r_{NW}(v,q)$, $l_{NE}(v,q)$, and $r_{NE}(v,q)$, for each $q = (x,y) \in S$ and each v that is ancestor of x in T_x. We explain how to update only $r_{SE}(v,q)$ for all v's that are ancestors of x in T_x; similar updating can be repeated for each of the seven other quantities (relative to their own frame of reference).

We begin with the updating of the $r_{SE}(v,q)$'s for all points other than q_0 (i.e., the points in $S - \{q_0\}$). And we will explain how to compute the $r_{SE}(v, q_0)$ separately.

The first step is to compute, as a query that is processed just as in the previous section (except that the coordinate system is different), the set $U = M_{NE}(S, q_0)$, where, as before, the expected size of U is $O(\log n)$. The only points q of S whose $r_{SE}(v,q)$ may change are in U. For each point q of U, we update its (at most $\log n$) $r_{SE}(v,q)$ values. This is done in constant time for each value, by checking whether q_0 can cause an improvement when v is ancestor of q_0. The total update time for doing this is therefore $O(|U| \log n)$, which is $O(\log^2 n)$ on average.

To compute the $r_{SE}(v, q_0)$, we first compute $U' = M_{SE}(S, q_0)$ as a query, hence in $O(\log n)$ expected time. We then walk along the path from x_0 to the root in T_x, and at each node v along this path we set $r_{SE}(v, q_0)$ equal to the highest point of U' that is in Sl_v. Note that this whole walk can be done in time $O(\log n)$ because of monotonicity: The Sl_v's of the nodes on that walk to the root monotonically "swallow" U' in left-to-right order (hence, by increasing y coordinates). Thus we end up going through U' only once (not $\log n$ times).

Processing a Deletion. Let $q_0 = (x_0, y_0)$ be the point being deleted. We already argued that the fractional cascading structure can be updated in $O(\log n)$ expected time as a result of this deletion. Now we need to show how to update the quantities $l_{SW}(v,q)$, $r_{SW}(v,q)$, $l_{SE}(v,q)$, $r_{SE}(v,q)$, $l_{NW}(v,q)$, $r_{NW}(v,q)$, $l_{NE}(v,q)$, and $r_{NE}(v,q)$, for each $q = (x,y) \in S$ and each v that is ancestor of x_0 in T_x. We explain how to do it for $r_{SE}(v,q)$ for all v that are ancestors of x_0 in T_x, all other values are updated similarly (relative to their own frame of reference).

First, we compute each of the sets $U = M_{NW}(S, q_0)$ and $U' = M_{SW}(S, q_0)$ as queries (and, hence, in $O(\log n)$ expected time). The only points q of S whose $r_{SE}(v,q)$ may change as a result of the deletion are in U. Moreover, for each such point q whose $r_{SE}(v,q)$ changes, its new $r_{SE}(v,q)$ is either in U' or it is the old $r_{SE}(v,q_0)$. The best candidate from U' for each $q \in U$ need not be done in isolation; rather, it can be done for all the points of U together. This can be performed in a manner reminiscent of the way two sorted lists are merged, by walking simultaneously along U and U'. This has to be done only once (not repeated for the $r_{SE}(v,q)$ of every ancestor v of x_0). On the other hand, the comparison of the old $r_{SE}(v,q)$ with the two new candidates, which are the old $r_{SE}(v,q_0)$, and the point of U' determined during the above-mentioned merge-like procedure, needs to be done for every v and q. Hence, the overall time for a deletion is $O(\log^2 n)$ on average.

4 Dynamic Nearest Neighbors in Minkowski Metrics

Given a nearest-neighbor query for a point q_0, in a set S of uniformly-distributed points in an axis-aligned rectangle, we partition the problem into four subproblems: (i) a South-West problem that consists of computing the nearest neighbor from among the subset of S that is dominated by the query point q_0 in the *SW* coordinate system, i.e., the subset "below and to the left of q_0"; (ii) a South-East problem that consists of computing the nearest neighbor from among the subset of S that is dominated by the query point q_0 in the SE coordinate system (the subset "below and to the right of q_0"); (iii) a North-East problem that consists of computing the nearest neighbor from among the subset of S that is dominated by the query point q_0 in the NE coordinate system (the subset "above and to the right of q_0"); and (iv) a North-West problem that consists of computing the nearest neighbor from among the subset of S that is dominated by the query point q_0 in the NW coordinate system (the subset "above and to the left of q_0"). We solve all of (i)–(iv) and choose, as the solution to the nearest-neighbor query, the best from the four answers they return. Our performance bounds for this problem therefore immediately follow from those we established in the previous section for the dynamic dominated maxima problem: $O(\log n)$ expected query time, and $O(\log^2 n)$ expected time for an update (insertion or deletion).

5 Experimental Results

Local Discrepancy on a Range Tree. In this section we explore the distributions of the local discrepancy in the catalogs of the nodes of a range tree, augmented using our dynamic fractional cascading structure.

To evaluate the distributions of the local discrepancy along a path in the range tree we use, we have inserted the points of the real data set S in such a range tree and chose random query points (x, y). For each point, we calculated the local discrepancy relative to y for each edge on the path from the leaf associated with x to the root of the tree. We also did the same work with the same number of evenly distributed points (see Fig. 2).

As we see in Fig. 2, the distributions of local discrepancy for the real data set is very close to the distributions of local discrepancy in the case of evenly distributed points. Their plot in logarithmic scale indicates that they are very close to exponential distributions, which shows that the demonstration for theorem 3 still holds in the case of the real data set.

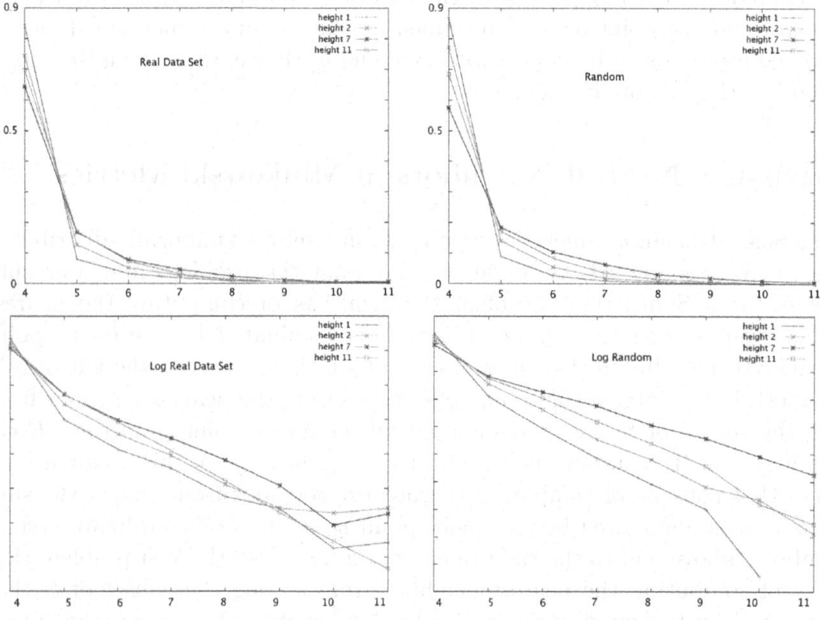

Fig. 2. Distributions of the local discrepancy along top-down path in a range tree using real data set in the upper-left corner and evenly distributed points on the upper-right corner. The distribution $height_k$ represents the distribution of local discrepancy for edges between nodes at height $k-1$ and k containing respectively 2^{k-1} and 2^k points in their catalogs. The two plots below show the same distributions on a log scale.

Acknowledgments

Portions of this work were supported by Grants IIS-0325345, CCR-0312760, and CNS-0627488 from the National Science Foundation, and by sponsors of the Center for Education and Research in Information Assurance and Security.

References

1. Alt, H.: The nearest neighbor. In: Computational Discrete Mathematics. LNCS, vol. 2122, pp. 13–24. Springer, Heidelberg (2001)
2. Arnold, B.C., Balakrishnan, N., Nagaraja, H.N.: A First Course in Order Statitics. Wiley-Interscience, Chichester (1992)
3. Arya, S., Mount, D.M., Netanyahu, N.S., Silverman, R., Wu, A.: An optimal algorithm for approximate nearest neighbor searching in fixed dimensions. J. ACM 45, 891–923 (1998)
4. Bentley, J.L.: K-d trees for semidynamic point sets. In: SCG '90. Proceedings of the sixth annual symposium on Computational geometry, pp. 187–197. ACM Press, New York (1990)
5. Bentley, J.L., Weide, B.W., Yao, A.C.: Optimal expected-time algorithms for closest point problems. ACM Trans. Math. Softw. 6(4), 563–580 (1980)
6. Beygelzimer, A., Kakade, S., Langford, J.: Cover trees for nearest neighbor. In: ICML '06. Proceedings of the 23rd international conference on Machine learning, pp. 97–104 (2006)
7. Chan, T.M.: A dynamic data structure for 3-d convex hull and 2-d nearest neighbor queries. In: Proceedings of the seventeenth ACM-SIAM symposium on Discrete algorithm, pp. 1196–1202. ACM Press, New York (2006)
8. Chazelle, B.: Geometric complexity and the discrepancy method. In: Abstracts 15th European Workshop Comput. Geom., pp. 21–23. INRIA Sophia-Antipolis (1999)
9. Chazelle, B.: The Discrepancy Method. Cambridge Univ. Press, Cambridge (2002)
10. Chazelle, B., Guibas, L.J.: Fractional cascading: I. A data structuring technique. Algorithmica 1(3), 133–162 (1986)
11. Chazelle, B., Guibas, L.J.: Fractional cascading: II. Applications. Algorithmica 1, 163–191 (1986)
12. Clarkson, K.L.: Nearest-neighbor searching and metric space dimensions. In: Shakhnarovich, G., Darrell, T., Indyk, P. (eds.) Nearest-Neighbor Methods for Learning and Vision: Theory and Practice, pp. 15–59. MIT Press, Cambridge (2006)
13. David, H.A., Nagaraja, H.N.: Order Statitics, 3rd edn. Wiley-Interscience, Chichester (2003)
14. Eppstein, D., Goodrich, M.T., Sun, J.Z.: The skip quadtree: A simple dynamic data structure for multidimensional data. In: SCG. 21st ACM Symp. on Computational Geometry, pp. 296–305. ACM Press, New York (2005)
15. Karger, D.R., Ruhl, M.: Finding nearest neighbors in growth-restricted metrics. In: STOC '02. Proceedings of the thiry-fourth annual ACM symposium on Theory of computing, pp. 741–750. ACM Press, New York (2002)
16. Krauthgamer, R., Lee, J.R.: Navigating nets: simple algorithms for proximity search. In: SODA. Proceedings of the 15th ACM-SIAM Symposium on Discrete Algorithms, pp. 798–807. ACM Press, New York (2004)
17. Mehlhorn, K., Näher, S.: Dynamic fractional cascading. Algorithmica 5, 215–241 (1990)

18. Motwani, R., Raghavan, P.: Randomized Algorithms. Cambridge University Press, New York (1995)
19. Munro, J.I., Papadakis, T., Sedgewick, R.: Deterministic skip lists. In: SODA. Proc. Annual ACM-SIAM Symposium on Discrete Algorithms, pp. 367–375 (1992)
20. Preparata, F.P., Shamos, M.I.: Computational Geometry: An Introduction. Springer-Verlag, New York (1985)
21. Sen, S.: Fractional cascading revisited. J. Algorithms 19(2), 161–172 (1995)
22. Weisstein, E.W.: Beta binomial distribution. In: MathWorld—A Wolfram Web Resource. Wolfram (2007) http://mathworld.wolfram.com/BetaBinomialDistribution.html
23. Weisstein, E.W.: Local discrepancy. In: MathWorld—A Wolfram Web Resource. Wolfram (2007) http://mathworld.wolfram.com/LocalDiscrepancy.html
24. Yap, C., Zhu, Y.: Yet another look at fractional cascading: B-graphs with application to point location. In: CCCG'01. Proceedings of the 13th Canadian Conference on Computational Geometry, pp. 173–176 (2001)

Priority Queues Resilient to Memory Faults

Allan Grønlund Jørgensen[1,⋆], Gabriel Moruz[1], and Thomas Mølhave[1,⋆⋆]

BRICS[⋆⋆⋆], MADALGO[†], Department of Computer Science,
University of Aarhus, Denmark
{jallan,gabi,thomasm}@daimi.au.dk

Abstract. In the faulty-memory RAM model, the content of memory cells can get corrupted at any time during the execution of an algorithm, and a constant number of uncorruptible registers are available. A resilient data structure in this model works correctly on the set of uncorrupted values. In this paper we introduce a resilient priority queue. The deletemin operation of a resilient priority queue returns either the minimum uncorrupted element or some corrupted element. Our resilient priority queue uses $O(n)$ space to store n elements. Both insert and deletemin operations are performed in $O(\log n + \delta)$ time amortized, where δ is the maximum amount of corruptions tolerated. Our priority queue matches the performance of classical optimal priority queues in the RAM model when the number of corruptions tolerated is $O(\log n)$. We prove matching worst case lower bounds for resilient priority queues storing only structural information in the uncorruptible registers between operations.

1 Introduction

Memory devices continually become smaller, work at higher frequencies and lower voltages, and in general have increased circuit complexity [1]. Unfortunately, these improvements come at the cost of reliability [2,3]. A number of factors, such as alpha particles, infrared radiation, and cosmic rays, can cause *soft memory errors* where a bit flips and as a consequence the value stored in the corresponding memory cell is corrupted. An unreliable memory can cause problems in most software ranging from the harmless to the very serious, such as breaking cryptographic protocols [4,5], taking control of a Java Virtual Machine [6] or breaking smart-cards and other security processors [7,8,9]. Furthermore, many modern computing centers consist of relatively cheap of-the-shelf components, and the large number of individual memories involved in these clusters substantially increase the frequency of memory corruptions in the system. Hence it is crucial that the software running on these machines is robust. Since the amount

⋆ Supported in part by an Ole Roemer Scholarship.

⋆⋆ Supported in part by an Ole Roemer Scholarship from the Danish National Science Research Council and by a Scholarship from the Oticon Foundation.

⋆⋆⋆ Basic Research in Computer Science, research school.

† Center for Massive Data Algorithmics, a Center of the Danish National Research Foundation.

F. Dehne, J.-R. Sack, and N. Zeh (Eds.): WADS 2007, LNCS 4619, pp. 127–138, 2007.

of cosmic rays increases dramatically with altitude, soft memory errors are of special concern in fields like avionics or space research. Furthermore, soft memory error rates are expected to rise for both DRAM and SRAM memories [2].

At the hardware level, the soft memory errors can be handled by means of error detection mechanisms such as parity checking, redundancy or Hamming codes. Unfortunately, implementing these mechanisms incur penalties with respect to performance, size and money. Therefore, memories using these technologies are rarely found in large scale computing clusters or ordinary workstations. On the software level, a series of low-level techniques have been proposed for dealing with the soft memory errors, many of them coping with corrupted instructions. Examples include algorithm based fault tolerance [10], assertions [11], control flow checking [12], or procedure duplication [13].

Traditionally, the work within the algorithmic community has focused on models where the integrity of the memory system is not an issue. In these models, the corruption of even a single memory cell can have a dramatic effect on the output. For instance, a single corrupted value can induce as much as $\Theta(n^2)$ inversions in the output of a standard implementation of mergesort [14]. Replication can help in dealing with corruptions, but is not always feasible, since the time and space overheads are not negligible.

A multitude of algorithms that deal with unreliable information in various ways were developed during the last decades. Aumann and Bender [15] introduced *fault tolerant pointer-based data structures*. In their model, error detection is done upon access, *i.e.* accessing a faulty pointer yields an error message. Obviously, this is not always the case in practice, since a pointer might get corrupted to a valid value and thus an error is not reported. Furthermore, their algorithms allow a certain amount of the data structure to be lost upon corruptions, and this is not accepted in many practical applications. The *liar model* considers algorithms in a comparison model where the result of a comparison is unreliable. Work in this model include fundamental problems such as sorting and searching [16,17,18]. A standard technique used in the design of algorithms in the liar model is query replication, which is not of much help when memory cells, and not comparisons, are unreliable. Kutten and Peleg [19,20] introduced the concept of *fault local mending* in the context of distributed networks. A problem is fault locally mendable if there exists a correction algorithm whose running time depends only on the (unknown) number of faults. Some other works studying network fault tolerance include [21,22,23,24,25,26,27].

Finocchi and Italiano [14] introduced the *faulty-memory random access machine*, which is a random access machine where the content of memory cells can get corrupted at *any time* and at *any location*. Corrupted cells cannot be distinguished from uncorrupted cells. The model is parametrized by an upper bound δ on the number of corruptions occurring during the lifetime of an algorithm. It is assumed that $O(1)$ reliable memory cells are provided, a reasonable assumption since CPU registers are considered reliable. Also, copying an element is considered an atomic operation, *i.e.* the elements are not corrupted while being copied. An algorithm is *resilient* if it is able to achieve a correct output at least for the uncorrupted values. This is the best one can hope for, since the output can get corrupted just after the algorithm finishes its execution. For instance a resilient sorting algorithm guarantees that there are no inversions between the uncorrupted elements in the output sequence.

Several important results has been achieved in the faulty-memory RAM. In the original paper, Finocchi and Italiano [14] proved lower bounds and gave (non-optimal) resilient algorithms for sorting and searching. Algorithms matching the lower bounds for sorting and searching(expected time) were presented in [28]. An optimal resilient sorting algorithm takes $\Theta(n \log n + \delta^2)$ time, whereas optimal searching is performed in $\Theta(\log n + \delta)$ time. Furthermore, in [29] a resilient search tree that performs searches and updates in $O(\log n + \delta^2)$ time amortized was developed. Finally, in [30] it was shown that resilient sorting algorithms are of practical interest.

Results. In this paper we design and analyze a priority queue in the faulty-memory RAM model. It uses $O(n)$ space for storing n elements and performs both INSERT and DELETEMIN in $O(\log n+\delta)$ time amortized. Our priority queue matches the bounds for an optimal comparison based priority queue in the RAM model while tolerating $O(\log n)$ corruptions. It is a significant improvement over using the resilient search tree in [29] as a priority queue, since it uses $O(\log n+\delta^2)$ time amortized per operation and thus only tolerates $O(\sqrt{\log n})$ corruptions to preserve the $O(\log n)$ bound per operation. Our priority queue is the first resilient data structure allowing $O(\log n)$ corruptions, while still matching optimal bounds in the RAM model. Our priority queue does not store elements in reliable memory between operations, only structural information like pointers and indices. We prove that any comparison based resilient priority queue behaving this way requires worst case $\Omega(\log n+\delta)$ time for either INSERT or DELETEMIN.

The resilient priority queue is based on the cache-oblivious priority queue by Arge *et al.* [31]. The main idea is to gather elements in large sorted groups of increasing size, such that expensive updates do not occur too often. The smaller groups contain the smaller elements, so they can be retrieved faster by DELETEMIN operations. We extensively use the resilient merging algorithm in [28] to move elements among the groups. Due to the large sizes of the groups, the extra work required to deal with corruptions in the merging algorithm becomes insignificant compared to the actual work done.

Outline. The remainder of the paper is structured as follows. In Section 2 we define the resilient priority queue and introduce some notation. We give a detailed description of the resilient priority queue in Section 3, while in Section 4 we prove its correctness and complexity bounds. Finally, in Section 5 we prove matching lower bounds for resilient priority queues.

2 Preliminaries

In this section we define the resilient priority queue and introduce some notation used throughout the paper.

Given two sequences X and Y, we let XY denote the *concatenation* of X and Y. A sequence X is *faithfully ordered* if its uncorrupted keys appear in non-decreasing order. Finally, a *reliable value* is a value stored in unreliable memory which can be retrieved reliably in spite of possible corruptions. This is achieved by replicating the given value $2\delta+1$ times. Retrieving a reliable value takes $O(\delta)$ time using the majority algorithm in [32], which scans the $2\delta+1$ values keeping a single majority candidate and a counter in reliable memory.

Definition 1. *A resilient priority queue maintains a set of elements under the operations* INSERT *and* DELETEMIN. *An* INSERT *adds an element and a* DELETEMIN *deletes and returns the minimum uncorrupted element or a corrupted one.*

We note that our definition of a resilient priority queue is consistent with the resilient sorting algorithms introduced in [14]. Given a sequence of n elements, inserting all of them into a resilient priority queue followed by n DELETEMIN operations yields a faithfully ordered sequence.

3 Fault Tolerant Priority Queue

In this section we introduce the resilient priority queue. It resembles the cache-oblivious priority queue by Arge *et al.* [31]. The elements are stored in faithfully ordered lists and are moved using two fundamental primitives, PUSH and PULL, based on faithful merging. We describe the structure of the priority queue in Section 3.1 and then introduce the PUSH and PULL primitives in Section 3.2. Finally, in Section 3.3, we describe the INSERT and DELETEMIN operations.

3.1 Structure

The resilient priority queue consists of an insertion buffer I together with a number of layers $L_0, \ldots, L_k$, with $k = O(\log n)$. Each layer L_i contains an up-buffer U_i and a down-buffer D_i, represented as arrays. Intuitively, the up-buffers contain large elements that are on their way to the upper layers in the priority queue, whereas the down-buffers contain small elements, on their way to lower layers. The buffers in the priority queue are stored as a doubly linked list $U_0, D_0, \ldots, U_k, D_k$, see Figure 1. For each up and down buffer we reliably store the pointers to their adjacent buffers in the linked list and their size. In the reliable memory we store pointers to I, U_0 and D_0, together with $|I|$. Since the position of the first element in U_0 and D_0 is not always the first memory cell of the corresponding buffer, we also store the index of the first element in these buffers in reliable memory. The insertion buffer I contains up to $b = \delta + \log n + 1$ elements. For layer L_i we define the threshold s_i by $s_0 = 2 \cdot (\delta^2 + \log^2 n)$ and $s_i = 2s_{i-1} = 2^{i+1} \cdot (\delta^2 + \log^2 n)$, where n is the number of elements in the priority queue. We use these thresholds to decide whether an up buffer contains too many elements or whether a down buffer has too few. For the sake of simplicity, the up and down buffers are grown and shrunk as needed during the execution such that they don't use any extra space.

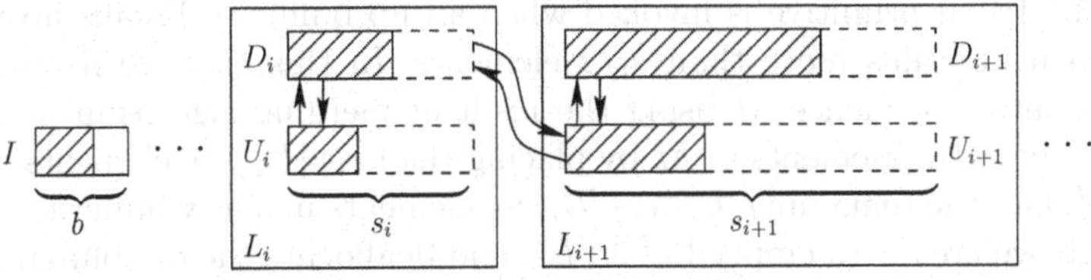

Fig. 1. The structure of the priority queue. The buffers are stored in a doubly linked list using reliably stored pointers. Additionally, the size of each buffer is stored reliably.

To structure the priority queue, we maintain the following invariants for the up and down buffers.

- *Order invariants:*
  1. All buffers are faithfully ordered.
  2. $D_i D_{i+1}$ and $D_i U_{i+1}$ are faithfully ordered, for $0 \leq i < k$.
- *Size invariants:*
  3. $s_i/2 \leq |D_i| \leq s_i$, for $0 \leq i < k$.
  4. $|U_i| \leq s_i/2$, for $0 \leq i < k$.

By maintaining all the up and down buffers faithfully ordered, it is possible to move elements between neighboring layers efficiently, using faithful merging. By invariant 2, all uncorrupted elements in D_i are smaller than all uncorrupted elements in both D_{i+1} and U_{i+1}. This ensures that small elements belong to the lower layers of the priority queue. We note that there is no assumed relationship between the elements in the up and down buffers in the same layer. Finally, the size invariants allow the sizes of the buffers to vary within a large range. This way, $\Omega(s_i)$ INSERT or DELETEMIN operations occur between two operations on the same buffer in L_i, yielding the desired amortized bounds.

Since the s_i values depend on n, whenever the size of the priority queue increases or decreases by $\Theta(n)$, we perform a global rebuilding. This rebuilding is done by collecting all elements, sorting them with an optimal resilient sorting algorithm [28], and redistributing the output into the down buffers of all the layers starting with L_0. After the global rebuilding, the up buffers are empty and the down buffers full, except possibly the last down buffer.

3.2 Push and Pull Primitives

We now introduce the two fundamental primitives used by the priority queue. The PUSH primitive is invoked when an up buffer contains too many elements, breaking invariant 4. It "pushes" elements upwards, repairing the size invariants locally. The PULL operation is invoked when a down buffer contains too few elements, breaking invariant 3. It fills this down buffer by "pulling" elements from the layer above, again locally repairing the size invariants. Both operations faithfully merge consecutive buffers in the priority queue and redistribute the resulting sequence among the participating buffers. After merging, we deallocate the old buffers and allocate new arrays for the new buffers.

Push. The PUSH primitive is invoked when an up buffer U_i breaks invariant 4, *i.e.* when it contains more than $s_i/2$ elements. In this case we merge U_i, D_i and U_{i+1} into a sequence M using the resilient merging algorithm in [28]. We then distribute the elements in M by placing the first $|D_i| - \delta$ elements in a new buffer D_i', and the remaining $|U_{i+1}|+|U_i|+\delta$ elements in a new buffer U_{i+1}'. After the merge, we create an empty buffer, U_i', and deallocate the old buffers. If U_{i+1}' contains too many elements, breaking invariant 4, the PUSH primitive is invoked on U_{i+1}'. When L_i is the last layer, we fill D_i' with the first elements of M and create a new layer L_{i+1} placing the remaining elements of M into D_{i+1}' instead of U_{i+1}'. Since $|D_i'|$ is smaller than $|D_i|$, it could violate invariant 3. This situation is handled by using the PULL operation and is described after introducing PULL.

Unlike the priority queue in [31], the PUSH operation decreases the size of a down buffer. This is required to preserve invariant 2, in spite of corruptions. After a PUSH call, D_i' can contain elements from $U_i \cup U_{i+1}$. Since there is no assumed relationship between elements in $U_i \cup U_{i+1}$ and those in $D_{i+1} \cup U_{i+2}$, we need to ensure that each element in D_i' originating from $U_i \cup U_{i+1}$ is faithfully smaller than the elements in $D_{i+1} \cup U_{i+2}$. Assume the size of D_i is preserved, *i.e.* $|D_i'| = |D_i|$. Consider a corruption that alters an element in D_i to some large value before the PUSH. This corrupted value could be placed in U_{i+1}' and, since $|D_i'| = |D_i|$, an element from $U_i \cup U_{i+1}$ must be placed in D_i'. This new element in D_i' potentially violates invariant 2.

Pull. The PULL operation is called on a down buffer D_i when it contains less than $s_i/2$ elements, breaking invariant 3. In this case, the buffers D_i, U_{i+1}, and D_{i+1} are merged into a sequence M using the resilient merging algorithm in [28]. The first s_i elements from M are written to a new buffer D_i', and the next $|D_{i+1}|-(s_i-|D_i|)-\delta$ elements are written to D_{i+1}'. The remaining elements of M are written to U_{i+1}'. A PULL is invoked on D_{i+1}', if it is too small.

Similar to the PUSH operation, the extra δ elements lost by D_{i+1} ensure that the order invariants hold in spite of possible corruptions. That is, a corruption of an element in $D_i \cup D_{i+1}$ to a very large value may cause an element from U_{i+1} to take the place of the corrupted element in D_{i+1}' and this element is possibly larger than some uncorrupted element in $D_{i+2} \cup U_{i+2}$.

After the merge, U_{i+1}' contains δ more elements than U_{i+1} had before the merge, and thus it is possible that it has too many elements, breaking invariant 4. We handle this situation as follows. Consider a maximal series of subsequent PULL invocations on down buffers $D_i, D_{i+1}, \ldots, D_j$, $0 \leq i < j < k$. After the first PULL call on D_i and before the call on D_{i+1} we store a pointer to D_i in the reliable memory. After all the PULL calls we investigate all the affected up buffers, by simply following the pointers between the buffers starting from D_i, and invoke the PUSH primitive wherever necessary. The case when PUSH operations cause down buffers to underflow is handled similarly.

3.3 Insert and Deletemin

An element is inserted in the priority queue by simply appending it to the insertion buffer I. If I gets full, its elements are added to U_0 by first faithfully sorting I and then faithfully merging I and U_0. If U_0 breaks invariant 4, we invoke the PUSH primitive. If L_0 is the only layer of the priority queue and D_0 violates the size constraint, we faithfully merge the elements in I with D_0 instead.

To delete the minimum element in the priority queue, we first find the minimum of the first $\delta+1$ values in D_0, the minimum of the first $\delta+1$ values in U_0, and the minimum element in I. We then take the minimum of these three elements, delete it from the appropriate buffer and return it. After deleting the minimum, we right-shift all the elements in the affected buffer from the beginning up to the position of the minimum. This way we ensure that elements in any buffer are stored consecutively. If D_0 underflows, we invoke the PULL primitive on D_0, unless L_0 is the only layer in the priority queue. If U_0 or D_0 contains $\Theta(\log n+\delta)$ empty cells, we create a new buffer and copy the elements from the old buffer to the new one.

4 Analysis

In this section we analyze the resilient priority queue. We prove the correctness in Section 4.1 and analyze the time and space complexity in Section 4.2.

4.1 Correctness

To prove correctness of the resilient priority queue, we show that the DELETEMIN operation returns the minimum uncorrupted value or a corrupted value. We first prove that the order invariants are maintained by the PULL and PUSH operations.

Lemma 1. *The* PULL *and* PUSH *primitives preserve the order invariants.*

Proof. Recall that in a PULL invocation on buffer D_i, the buffers D_i, U_{i+1}, and D_{i+1} are faithfully merged into a sequence M. The elements in M are then distributed into three new buffers D'_i, U'_{i+1}, and D'_{i+1}, see Figure 2. To argue that the order invariants are satisfied we need to show that the elements of the down buffer on layer L_j, for $0 \leq j < k$, are faithfully smaller than the elements of the buffers on layer L_{j+1}, where k is the index of the last layer. The invariants hold trivially for unaffected buffers. The faithful merge guarantees that $D'_iD'_{i+1}$ as well as $D'_iU'_{i+1}$ are faithfully ordered, and thus the individual buffers are also faithfully ordered. Since invariant 2 holds for the original buffers all uncorrupted elements in D_{i+1} and U_{i+1} are larger than the uncorrupted elements in D_i, guaranteeing that $D_{i-1}D'_i$ is faithfully ordered. Finally, we now show that $D_{i+1}D_{i+2}$ and $D_{i+1}U_{i+2}$ are faithfully ordered.

Let m be the minimum uncorrupted element in $D_{i+2} \cup U_{i+2}$. We need to show that all uncorrupted elements in D'_{i+1} are smaller than m. If no uncorrupted element from U_{i+1} is placed in D'_{i+1}, the invariant holds by the order invariants before the operation. Otherwise, assume that an uncorrupted element $y \in U_{i+1}$ is moved to D'_{i+1}. Since $|U'_{i+1}| = |U_{i+1}|+\delta$ and y is moved to D'_{i+1}, at least $\delta+1$ elements originating from $D_i \cup D_{i+1}$ are contained in U'_{i+1}. Since there can be at most δ corruptions, there exists at least one uncorrupted element, x, among

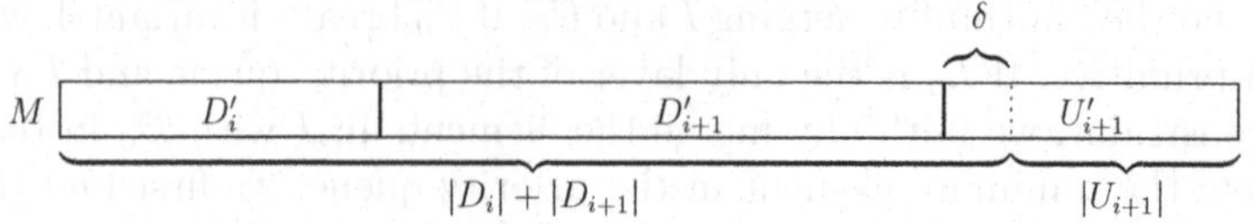

Fig. 2. The distribution of M into buffers

these. By faithful merging, all uncorrupted elements in D'_{i+1} are smaller than x, which means that $y \leq x$. Since x originates from $D_i \cup D_{i+1}$, it is smaller than m. We obtain $y \leq m$.

A similar argument proves correctness of the PUSH operation. We conclude that both order invariants are preserved by PULL and PUSH operations. □

Having proved that the order invariants are maintained at all times, we now prove the correctness of the resilient priority queue.

Lemma 2. *The* DELETEMIN *operation returns the minimum uncorrupted value in the priority queue or a corrupted value.*

Proof. We recall that the DELETEMIN operation computes the minimum of the first $\delta+1$ elements of U_0 and D_0. It compares these values with the minimum of I, found in a scan, and returns the smallest of these elements. Since U_0 and D_0 are faithfully ordered, the minimum of their first $\delta + 1$ elements is either the minimum uncorrupted value in these buffers, or a corrupted value even smaller. Furthermore, according to the order invariants, all the values in layers $L_1, \ldots, L_k$ are faithfully larger than the minimum in D_0. Therefore, the element reported by DELETEMIN is the minimum uncorrupted value or a corrupted value. □

4.2 Complexity

In this section we show that our resilient priority queue uses $O(n)$ space and that INSERT and DELETEMIN take $O(\log n + \delta)$ amortized time. We first prove that the PULL and PUSH primitives restore the size invariants.

Lemma 3. *If a size invariant is broken for a buffer in L_0, invoking* PULL *or* PUSH *on that buffer restores the invariants. Furthermore, during this operation* PULL *and* PUSH *are invoked on the same buffer at most once. No other invariants are broken before or after this operation.*

Proof. Assume that PUSH is invoked on U_0, and that it is called iteratively up to some layer L_l. By construction of PUSH, the size invariants for all the up buffers now hold. Since a PUSH steals δ elements from the down buffers, the layers $L_0, \ldots, L_l$ are traversed again and PULL is invoked on these as needed. The last of these PULL operations might proceed past layer L_l. Similarly, a PULL may cause an up buffer to overflow. However, since the cascading PUSH operations left $|U_i| = 0$ for $i \leq l$, any new PUSH are invoked on up buffers only on layer L_{l+1} or higher, thus PUSH is invoked on each buffer at most once. A similar argument works for the PULL operation. □

Lemma 4. *The resilient priority queue uses $O(n+\delta)$ space to store n elements.*

Proof. The insertion buffer always uses $O(\log n + \delta)$ space. We prove that the remaining layers use $O(n)$ space. For each layer we use $O(\delta)$ space for storing structural information reliably. In all layers, except the last one, the down buffer contains $\Omega(\delta^2)$ elements by invariant 3. This means that for each of these layers the elements stored in the down buffer dominate the space complexity. The structural information of the last layer requires additional $O(\delta)$ space. □

The space complexity of the priority queue can be reduced to $O(n)$ without affecting the time complexity, by storing the structural information of L_0 in safe memory, and by doubling or halving the insertion buffer during the lifetime of the algorithm such that it always uses $O(|I|)$ space.

Lemma 5. *Each* Insert *and* DeleteMin *takes $O(\log n + \delta)$ amortized time.*

Proof. We define the potential function:

$$\Phi = \sum_{i=1}^{k} (c_1 \cdot (\log n - i) \cdot |U_i| + c_2 \cdot i \cdot |D_i|)$$

We use Φ to analyze the amortized cost of a Push operation. In a Push operation on U_i, buffers U_i, D_i, and U_{i+1} are merged. The elements are then distributed into new buffers U'_i, D'_i, and U'_{i+1}, such that $|U'_i| = 0$, $|D'_i| = |D_i| - \delta$, and $|U'_{i+1}| = |U_{i+1}| + |U_i| + \delta$. This gives the following change in potential $\Delta\Phi$:

$$\begin{aligned}\Delta\Phi &= -|U_i| \cdot c_1 \cdot (\log n - i) - \delta \cdot c_2 \cdot i + (|U_i| + \delta) \cdot c_1(\log n - (i+1)) \\ &= -c_1 \cdot |U_i| + \delta(-c_2 \cdot i + c_1 \cdot \log n - c_1 \cdot i - c_1) .\end{aligned}$$

Since the Push is invoked on U_i, invariant 4 is not valid for U_i and therefore $|U_i| \geq \frac{s_i}{2} = 2^i (\log^2 n + \delta^2)$. Thus:

$$\Delta\Phi \leq -c_1 \cdot |U_i| + c_1 \cdot \delta \cdot \log n \leq -c_1 \cdot 2^i \cdot (\log^2 n + \delta^2) + c_1 \cdot \delta \cdot \log n \leq -c_1 \cdot c' \cdot |U_i| , \quad (1)$$

for some constant $c' > 0$.

Since faithfully merging two sequences of size n takes $O(n+\delta^2)$ time [28], the time used for a Push on U_i is upper bounded by $c_m \cdot (|U_i| + |D_i| + |U_{i+1}| + \delta^2)$, where c_m depends on the resilient merge. This includes the time required for retrieving reliably stored variables. Adding the time and the change in potential we are able to get the amortized cost less than zero by tweaking c_1 based on equation (1). This is because $|U_i|$ is $\Omega(\delta^2)$ and at most a constant fraction smaller than the participants in the merge.

A similar analysis works for the Pull primitive. We now calculate the amortized cost of Insert and DeleteMin. We ignore any Push or Pull operations since their amortized costs are negative. The amortized time for inserting an element in I, sorting I, and merging it with U_0 is $O(\log n + \delta)$ per operation. The change in potential when adding elements to L_0 is $O(\log n)$ per element.

The time needed to find the smallest element in a DELETEMIN is $O(\log n + \delta)$, and the change in potential when an element is deleted from L_0 is negative.

The cost of global rebuilding is dominated by the cost of sorting, which is $O(n \log n + \delta^2)$. There are $\Theta(n)$ operations between each rebuild, which leads to $O(\log n + \delta)$ time per operation, since $\delta \leq n$, and this concludes the proof. □

Theorem 1. *The resilient priority queue takes $O(n)$ space and uses amortized $O(\log n + \delta)$ time per operation.*

5 Lower Bound

In this section we prove that any resilient priority queue takes $\Omega(\log n + \delta)$ time for either INSERT or DELETEMIN in the comparison model, under the assumption that no elements are stored in reliable memory between operations. This implies optimality of our resilient priority queue under these assumptions. We note that the reliable memory may contain any structural information, e.g. pointers, sizes, indices.

Theorem 2. *A resilient priority queue containing n elements, with $n > \delta$, uses $\Omega(\log n + \delta)$ comparisons to perform* INSERT *followed by* DELETEMIN.

Proof. Consider a priority queue Q with n elements, with $n > \delta$, that uses less than δ comparisons for an INSERT followed by a DELETEMIN. Also, Q does not store elements in reliable memory between operations. Assume that no corruptions have occurred so far. Without loss of generality we assume that all the elements in Q are distinct. We prove there exists a series of corruptions C, $|C| \leq \delta$, such that the result of an INSERT of an element e followed by a DELETEMIN returns the same element regardless of the choice of e.

Let $k < \delta$ be the number of comparisons performed by Q during the two operations. We force the result of each comparison to be the same regardless of e by suitable corruptions. In all the comparisons involving e, we ensure that e is the smallest. We do so by corrupting the value which e is compared against if necessary, by adding some positive constant $c \geq e$ to the other value. If two elements different than e are compared, we make sure the outcome is the same as if no corruptions had happened. If one of them was corrupted, adding c to the other one reestablishes their previous ordering. If both of them were corrupted by adding c, their ordering is unchanged and no corruptions are needed. Forcing any comparison to give the desired outcome requires at most one corruption, and therefore $|C| \leq k < \delta$.

We now consider the value e' returned by DELETEMIN on Q. If $e = e'$ then we choose e to be larger than some element $x \in Q$ not affected by a corruption in C. Such a value exists because the size of the priority queue is larger than δ. Since $e = e' > x$, Q returned an uncorrupted element that was not the minimum uncorrupted element in Q. If $e \neq e'$ we choose e to be smaller than any element in Q. With such a choice of e, no corruptions are required and the value returned by Q was not corrupted, but still larger than e. This proves Q is not resilient.

Adding the classical $\Omega(\log n)$ bound for priority queues in the comparison model the result follows. □

Acknowledgments

We would like to thank Gerth S. Brodal and Lars Arge for their very helpful comments. We would also like to thank the anonymous reviewers for their valuable comments, especially suggestions for simplifying the proof of Lemma 1.

References

1. Constantinescu, C.: Trends and challenges in VLSI circuit reliability. IEEE micro 23(4), 14–19 (2003)
2. Tezzaron Semiconductor: Soft errors in electronic memory - a white paper (2004) `http://www.tezzaron.com/about/papers/papers.html`
3. van de Goor, A.J.: Testing Semiconductor Memories: Theory and Practice. ComTex Publishing, Gouda, The Netherlands (1998)
4. Boneh, D., DeMillo, R.A., Lipton, R.J.: On the importance of checking cryptographic protocols for faults. In: Fumy, W. (ed.) EUROCRYPT 1997. LNCS, vol. 1233, pp. 37–51. Springer, Heidelberg (1997)
5. Xu, J., Chen, S., Kalbarczyk, Z., Iyer, R.K.: An experimental study of security vulnerabilities caused by errors. In: Proc. International Conference on Dependable Systems and Networks, pp. 421–430 (2001)
6. Govindavajhala, S., Appel, A.W.: Using memory errors to attack a virtual machine. In: IEEE Symposium on Security and Privacy, pp. 154–165. IEEE Computer Society Press, Los Alamitos (2003)
7. Anderson, R., Kuhn, M.: Tamper resistance - a cautionary note. In: Proc. 2nd Usenix Workshop on Electronic Commerce, pp. 1–11 (1996)
8. Anderson, R., Kuhn, M.: Low cost attacks on tamper resistant devices. In: Christianson, B., Lomas, M. (eds.) Security Protocols. LNCS, vol. 1361, pp. 125–136. Springer, Heidelberg (1997)
9. Skorobogatov, S.P., Anderson, R.J.: Optical fault induction attacks. In: Kaliski Jr., B.S., Koç, Ç.K., Paar, C. (eds.) CHES 2002. LNCS, vol. 2523, pp. 2–12. Springer, Heidelberg (2003)
10. Huang, K.H., Abraham, J.A.: Algorithm-based fault tolerance for matrix operations. IEEE Transactions on Computers 33, 518–528 (1984)
11. Rela, M.Z., Madeira, H., Silva, J.G.: Experimental evaluation of the fail-silent behaviour in programs with consistency checks. In: Proc. 26th Annual International Symposium on Fault-Tolerant Computing, pp. 394–403 (1996)
12. Yau, S.S., Chen, F.-C.: An approach to concurrent control flow checking. IEEE Transactions on Software Engineering SE-6(2), 126–137 (1980)
13. Pradhan, D.K.: Fault-tolerant computer system design. Prentice-Hall, Inc, Englewood Cliffs (1996)
14. Finocchi, I., Italiano, G.F.: Sorting and searching in the presence of memory faults (without redundancy). In: Proc. 36th Annual ACM Symposium on Theory of Computing, pp. 101–110. ACM Press, New York (2004)
15. Aumann, Y., Bender, M.A.: Fault tolerant data structures. In: Proc. 37th Annual Symposium on Foundations of Computer Science, Washington, DC, p. 580. IEEE Computer Society Press, Los Alamitos (1996)
16. Borgstrom, R.S., Kosaraju, S.R.: Comparison-based search in the presence of errors. In: Proc. 25th Annual ACM symposium on Theory of Computing, pp. 130–136. ACM Press, New York (1993)

17. Lakshmanan, K.B., Ravikumar, B., Ganesan, K.: Coping with erroneous information while sorting. IEEE Transactions on Computers 40(9), 1081–1084 (1991)
18. Ravikumar, B.: A fault-tolerant merge sorting algorithm. In: Ibarra, O.H., Zhang, L. (eds.) COCOON 2002. LNCS, vol. 2387, pp. 440–447. Springer, Heidelberg (2002)
19. Kutten, S., Peleg, D.: Fault-local distributed mending. Journal of Algorithms 30(1), 144–165 (1999)
20. Kutten, S., Peleg, D.: Tight fault locality. SIAM Journal on Computing 30(1), 247–268 (2000)
21. Diks, K., Pelc, A.: Optimal adaptive broadcasting with a bounded fraction of faulty nodes (extended abstract). In: Burkard, R.E., Woeginger, G.J. (eds.) ESA 1997. LNCS, vol. 1284, pp. 118–129. Springer, Heidelberg (1997)
22. Gasieniec, L., Pelc, A.: Broadcasting with a bounded fraction of faulty nodes. Journal of Parallel and Distributed Computing 42(1), 11–20 (1997)
23. Hastad, J., Leighton, T.: Fast computation using faulty hypercubes. In: Proc. 21st Annual ACM Symposium on Theory of Computing, pp. 251–263. ACM Press, New York (1989)
24. Hastad, J., Leighton, T., Newman, M.: Reconfiguring a hypercube in the presence of faults. In: Proc. 19th Annual ACM Symposium on Theory of Computing, pp. 274–284. ACM Press, New York (1987)
25. Kaklamanis, C., Karlin, A.R., Leighton, F.T., Milenkovic, V., Raghavan, P., Rao, S., Thomborson, C.D., Tsantilas, A.: Asymptotically tight bounds for computing with faulty arrays of processors (extended abstract). In: Proc. 31st Annual Symposium on Foundations of Computer Science, pp. 285–296 (1990)
26. Leighton, F.T., Maggs, B.M.: Expanders might be practical: Fast algorithms for routing around faults on multibutterflies. In: Proc. 30th Annual Symposium on Foundations of Computer Science, pp. 384–389 (1989)
27. Park, S., Bose, B.: All-to-all broadcasting in faulty hypercubes. IEEE Transactions on Computers 46(7), 749–755 (1997)
28. Finocchi, I., Grandoni, F., Italiano, G.F.: Optimal resilient sorting and searching in the presence of memory faults. In: Bugliesi, M., Preneel, B., Sassone, V., Wegener, I. (eds.) ICALP 2006. LNCS, vol. 4051, pp. 286–298. Springer, Heidelberg (2006)
29. Finocchi, I., Grandoni, F., Italiano, G.F.: Resilient search trees. In: Proc. 18th ACM-SIAM Symposium on Discrete Algorithms (to appear)
30. Petrillo, U.F., Finocchi, I., Italiano, G.F.: The price of resiliency: a case study on sorting with memory faults. In: Azar, Y., Erlebach, T. (eds.) ESA 2006. LNCS, vol. 4168, pp. 768–779. Springer, Heidelberg (2006)
31. Arge, L., Bender, M.A., Demaine, E.D., Holland-Minkley, B., Munro, J.I.: Cache-oblivious priority queue and graph algorithm applications. In: Proc. 34th Annual ACM Symposium on Theory of Computing, pp. 268–276. ACM Press, New York (2002)
32. Boyer, R.S., Moore, J.S.: MJRTY: A fast majority vote algorithm. In: Automated Reasoning: Essays in Honor of Woody Bledsoe, pp. 105–118 (1991)

Simple and Space-Efficient Minimal Perfect Hash Functions*

Fabiano C. Botelho[1], Rasmus Pagh[2], and Nivio Ziviani[1]

[1] Dept. of Computer Science, Federal Univ. of Minas Gerais, Belo Horizonte, Brazil
{fbotelho,nivio}@dcc.ufmg.br

[2] Computational Logic and Algorithms Group, IT Univ. of Copenhagen, Denmark
pagh@itu.dk

Abstract. A *perfect hash function* (PHF) $h : U \to [0, m-1]$ for a key set S is a function that maps the keys of S to unique values. The minimum amount of space to represent a PHF for a given set S is known to be approximately $1.44n^2/m$ bits, where $n = |S|$. In this paper we present new algorithms for construction and evaluation of PHFs of a given set (for $m = n$ and $m = 1.23n$), with the following properties:

1. Evaluation of a PHF requires constant time.
2. The algorithms are simple to describe and implement, and run in linear time.
3. The amount of space needed to represent the PHFs is around a factor 2 from the information theoretical minimum.

No previously known algorithm has these properties. To our knowledge, any algorithm in the literature with the third property either:

- Requires exponential time for construction and evaluation, or
- Uses near-optimal space only asymptotically, for extremely large n.

Thus, our main contribution is a scheme that gives low space usage for realistic values of n. The main technical ingredient is a new way of basing PHFs on random hypergraphs. Previously, this approach has been used to design simple PHFs with superlinear space usage.

1 Introduction

Perfect hashing is a space-efficient way of associating unique identifiers with the elements of a static set S. We will refer to the elements of S as *keys*. A *perfect hash function* (PHF) maps $S \subseteq U$ to unique values in the range $[0, m-1]$. We let $n = |S|$ and $u = |U|$ — note that we must have $m \geq n$. A *minimal perfect hash function* (MPHF) is a PHF with $m = n$. For simplicity of exposition, we consider in this paper the case $\log u \ll n$. This allows us to ignore terms in the space usage that depend on u.

* This work was supported in part by GERINDO Project–grant MCT/CNPq/CT-INFO 552.087/02-5, and CNPq Grants 30.5237/02-0 (Nivio Ziviani) and 142786/2006-3 (Fabiano C. Botelho).

F. Dehne, J.-R. Sack, and N. Zeh (Eds.): WADS 2007, LNCS 4619, pp. 139–150, 2007.

In this paper we present a simple, efficient, near space-optimal, and practical family $\mathcal{F}$ of algorithms for generating PHFs and MPHFs. The algorithms in $\mathcal{F}$ use r-uniform random hypergraphs given by function values of r hash functions on the keys of S. An r-uniform hypergraph is the generalization of a standard undirected graph where each edge connects $r \geq 2$ vertices. The idea of basing perfect hashing on random hypergraphs is not new, see e.g. [14], but we will proceed differently to achieve a space usage of $O(n)$ bits rather than $O(n \log n)$ bits. (As in previous constructions based on hypergraphs we assume that the hash functions used are uniformly random and have independent function values. However, we argue that our scheme can also be realized using explicitly defined hash functions using small space.) Evaluation time for all schemes considered is constant. For $r = 2$ we obtain a space usage of $(3 + \epsilon)n$ bits for a MPHF, for any constant $\epsilon > 0$. For $r = 3$ we obtain a space usage of less than $2.7n$ bits for a MPHF. This is within a factor of 2 from the information theoretical lower bound of approximately $1.4427n$ bits. More compact, and even simpler, representations can be achieved for larger m. For example, for $m = 1.23n$ we can get a space usage of $1.95n$ bits. This is slightly more than two times the information theoretical lower bound of around $0.89n$ bits. The bounds for $r = 3$ assume a conjecture about the emergence of a 2-core in a random 3-partite hypergraph, whereas the bounds for $r = 2$ are fully proved. Choosing $r > 3$ does not give any improvement of these results.

We will argue that our method is far more practical than previous methods with proven space complexity, both because of its simplicity, and because the constant factor of the space complexity is more than 6 times lower than its closest competitor, for plausible problem sizes. We verify the practicality experimentally, using heuristic hash functions, and slightly more space than in the mentioned theoretical bounds.

2 Related Work

In this section we review some of the most important theoretical and practical results on perfect hashing. Czech, Havas and Majewski [4] provide a more comprehensive survey.

2.1 Theoretical Results

Fredman and Komlós [9] proved that at least $n \log e + \log\log u - O(\log n)$ bits are required to represent a MPHF (in the worst case over all sets of size n), provided that $u \geq n^\alpha$ for some $\alpha > 2$. Logarithms are in base 2. Note that the two last terms are negligible under the assumption $\log u \ll n$. In general, for $m > n$ the space required to represent a PHF is around $(1 + (m/n - 1)\ln(1 - n/m))\, n \log e$ bits. A simpler proof of this was later given by Radhakrishnan [18].

Mehlhorn [15] showed that the Fredman-Komlós bound is almost tight by providing an algorithm that constructs a MPHF that can be represented with at most $n \log e + \log\log u + O(\log n)$ bits. However, his algorithm is far from practice because its construction and evaluation time is exponential in n.

Schmidt and Siegel [19] proposed the first algorithm for constructing a MPHF with constant evaluation time and description size $O(n + \log \log u)$ bits. Their algorithm, as well as all other algorithms we will consider, is for the *Word RAM* model of computation [10]. In this model an element of the universe U fits into one machine word, and arithmetic operations and memory accesses have unit cost. From a practical point of view, the algorithm of Schmidt and Siegel is not attractive. The scheme is complicated to implement and the constant of the space bound is large: For a set of n keys, at least $29n$ bits are used, which means a space usage similar in practice to the best schemes using $O(n \log n)$ bits. Though it seems that [19] aims to describe its algorithmic ideas in the clearest possible way, not trying to optimize the constant, it appears hard to improve the space usage significantly.

More recently, Hagerup and Tholey [11] have come up with the best theoretical result we know of. The MPHF obtained can be evaluated in $O(1)$ time and stored in $n \log e + \log \log u + O(n(\log \log n)^2 / \log n + \log \log \log u)$ bits. The construction time is $O(n + \log \log u)$ using $O(n)$ words of space. Again, the terms involving u are negligible. In spite of its theoretical importance, the Hagerup and Tholey [11] algorithm is also not practical, as it emphasizes asymptotic space complexity only. (It is also very complicated to implement, but we will not go into that.) For $n < 2^{150}$ the scheme is not well-defined, as it relies on splitting the key set into buckets of size $\hat{n} \leq \log n/(21 \log \log n)$. If we fix this by letting the bucket size be at least 1, then buckets of size one will be used for $n < 2^{300}$, which means that the space usage will be at least $(3 \log \log n + \log 7)\, n$ bits. For a set of a billion keys, this is more than 17 bits per element. Thus, the Hagerup-Tholey MPHF is not space efficient in practical situations. While we believe that their algorithm has been optimized for simplicity of exposition, rather than constant factors, it seems difficult to significantly reduce the space usage based on their approach.

2.2 Practical Results

We now describe some of the main "practical" results that our work is based on. They are characterized by simplicity and (provably) low constant factors.

The first two results assume uniform random hash functions to be available for free. Czech et al [14] proposed a family of algorithms to construct MPHFs based on r-uniform hypergraphs (i.e., with edges of size r). The resulting functions can be evaluated in $O(1)$ time and stored in $O(n \log n)$ bits. Botelho, Kohayakawa and Ziviani [3] improved the space requirement of one instance of the family considering $r = 2$, but the space requirement is still $O(n \log n)$ bits. In both cases, the MPHF can be generated in expected $O(n)$ time. It was found experimentally in [3] that their construction procedure works well in practice.

Pagh [16] proposed an algorithm for constructing MPHFs of the form $h(x) = (f(x) + d[g(x)]) \bmod n$, where f and g are randomly chosen from a family of universal hash functions, and d is a vector of "displacement values" that are used to resolve collisions that are caused by the function f. The scheme is simple and evaluation of the functions very fast, but the space usage is $(2 + \epsilon)n \log n$ bits, which is suboptimal. Dietzfelbinger and Hagerup [5] improved [16], reducing from the space usage to $(1+\epsilon)n\log n$ bits, still using simple hash functions. Woelfel [20] has shown how to decrease the space usage further, to $O(n\log\log n)$ bits asymptotically, still with a quite simple algorithm. However, there is no empirical evidence on the practicality of this scheme.

2.3 Heuristics

Fox et al. [7,8] presented several algorithms for constructing MPHFs that in experiments require between 2 and 8 bits per key to be stored. However, it is shown in [4, Section 6.7] that their algorithms have exponential running times in expectation. Also, there is no warranty that the number of bits per key to store the function will be fixed as n increases. The work by Lefebvre and Hoppe [13] has the same problem. They have designed a PHF method to specifically represent sparse spatial data and the resulting functions requires more than 3 bits per key to be stored.

3 A Family of Minimal Perfect Hashing Methods

In this section we present our family $\mathcal{F}$ of algorithms for constructing near space-optimal MPHFs. The basic idea is as follows. The first step, referred to as the *Mapping Step*, maps the key set S to a set of $n = |S|$ edges forming an acyclic r-partite hypergraph $G_r = (V, E)$, where $|E(G_r)| = n$, $|V(G_r)| = m$ and $r \geq 2$. Note that each key in S is associated with an edge in $E(G_r)$. Also in the Mapping Step, we order the edges of G_r into a list L such that each edge e_i contains a vertex that is not incident to any edge that comes after e_i in L. The next step, referred to as the *Assigning Step*, associates uniquely each edge with one of its r vertices. Here, "uniquely" means that no two edges may be assigned to the same vertex. Thus, the Assigning Step finds a PHF for S with range $V(G_r)$. If we desire a PHF with a smaller range ($n < |V(G_r)|$), we subsequently map the assigned vertices of $V(G_r)$ to $[0, n-1]$. This mapping is produced by the *Ranking Step*, which creates a data structure that allows us to compute the *rank* of any assigned vertex of $V(G_r)$ in constant time.

For the analysis, we assume that we have at our disposal r hash functions $h_i : U \to [i\frac{m}{r}, (i+1)\frac{m}{r} - 1]$, $0 \leq i < r$, which are independent and uniformly distributed function values. (This is the "uniform hashing" assumption, see Section 6 for justification.) The r functions and the set S define, in a natural way, a random r−partite hypergraph. We define $G_r = G_r(h_0, h_1 \ldots, h_{r-1})$ as the hypergraph with vertex set $V(G_r) = [0, m-1]$ and edge set $E(G_r) = \{\{h_0(x), h_1(x), \ldots, h_{r-1}(x)\} \mid x \in S\}$. For the Mapping Step to work, we need G_r to be simple and acyclic, i.e., G_r should not have multiple edges and cycles. This is handled by choosing r new hash functions in the event that the Mapping Step fails. The PHF $p : S \to V(G_r)$ produced by the Assigning Step has the form

$$p(x) = h_i(x), \text{ where } i = (g(h_0(x)) + g(h_1(x)) + \cdots + g(h_{r-1}(x))) \bmod r. \quad (1)$$

The function $g : V(G_r) \to \{0, 1, \ldots, r\}$ is a labeling of the vertices of $V(G_r)$. We will show how to choose the labeling such that p is 1-1 on S, given that G_r is acyclic. In addition, $g(y) \neq r$ if and only if y is an assigned vertex, i.e., exactly when $y \in p(S)$. This means that we get a MPHF for S as follows:

$$h(x) = \text{rank}(p(x)) \tag{2}$$

where rank : $V(G_r) \to [0, n-1]$ is a function defined as:

$$\text{rank}(u) = |\{v \in V(G_r) \mid v < u \wedge g(v) \neq r\}|. \tag{3}$$

The Ranking Step produces a data structure that allows us to compute the rank function in constant time.

Figure 1 presents a pseudo code for our family of minimal perfect hashing algorithms. If we omit the third step we will build PHFs with $m = |V(G_r)|$ instead. We now describe each step in detail.

```
procedure Generate (S, r, g, rankTable)
   Mapping (S, G_r, L);
   Assigning (G_r, L, g);
   Ranking (g, rankTable);
```

Fig. 1. Main steps of the family of algorithms

3.1 Mapping Step

The Mapping Step takes the key set S as input, where $|S| = n$, and creates an acyclic random hypergraph G_r and a list of edges L. We say that a hypergraph is *acyclic* if it is the empty graph, or if we can remove an edge with a node of degree 1 such that (inductively) the resulting graph is acyclic. This means that we can order the edges of G_r into a list $L = e_1, \ldots, e_n$ such that any edge e_i contains a vertex that is not incident to any edge e_j, for $j > i$. The list L is obtained during a test which determines whether G_r is acyclic, which runs in time $O(n)$ (see e.g. [14]).

Let Pr_a denote the probability that G_r is acyclic. We want to ensure that this is the case with constant probability, i.e., $Pr_a = \Omega(1)$. Define c by $m = cn$. For $r = 2$, we can use the techniques presented in [12] to show that $Pr_a = \sqrt{1-(2/c)^2}$. For example, when $c = 2.09$ we have $Pr_a = 0.29$. This is very close to 0.294 that is the value we got experimentally by generating 1,000 random bipartite 2-graphs with $n = 10^7$ keys (edges). For $r > 2$, it seems to be technically difficult to obtain a rigorous bound on Pr_a. However, the heuristic argument presented in [4, Theorem 6.5] also holds for our $r-$partite random hypergraphs. Their argument suggests that if $c = c(r)$ is given by

$$c(r) = \begin{cases} 2+\varepsilon, \varepsilon > 0 & \text{for } r = 2 \\ r\left(\min_{x>0}\left\{\frac{x}{(1-e^{-x})^{r-1}}\right\}\right)^{-1} & \text{for } r > 2, \end{cases} \tag{4}$$

then the acyclic random r-graphs dominate the space of random r-graphs. The value $c(3) \approx 1.23$ is a minimum value for Eq. (4). This implies that the acyclic r-partite hypergraphs with the smallest number of vertices happen when $r = 3$. In this case, we have got experimentally $Pr_a \approx 1$ by generating $1,000$ 3-partite random hypergraphs with $n = 10^7$ keys (hyperedges).

It is interesting to remark that the problems of generating acyclic r-partite hypergraphs for $r = 2$ and for $r > 2$ have different natures. For $r = 2$, the probability Pr_a varies continuously with the constant c. But for $r > 2$, there is a phase transition. That is, there is a value $c(r)$ such that if $c \leq c(r)$ then Pr_a tends to 0 when n tends to ∞ and if $c > c(r)$ then Pr_a tends to 1. This phenomenon has also been reported by Majewski et al [14] for general hypergraphs.

3.2 Assigning Step

The Assigning Step constructs the labeling $g : V(G_r) \to \{0, 1, \ldots, r\}$ of the vertices of G_r. To assign values to the vertices of G_r we traverse the edges in the reverse order $e_n, \ldots, e_1$ to ensure that each edge has at least one vertex that is traversed for the first time. The assignment is created as follows. Let *Visited* be a boolean vector of size m that indicates whether a vertex has been visited. We first initialize $g[i] = r$ (i.e., each vertex is unassigned) and $\mathit{Visited}[i] = \mathit{false}$, $0 \leq i \leq m - 1$. Then, for each edge $e \in L$ from tail to head, we look for the first vertex u belonging to e not yet visited. Let j, $0 \leq j \leq r - 1$ be the index of u in e. Then, we set $g[u] = (j - \sum_{v \in e \wedge \mathit{Visited}[v] = \mathit{true}} g[v]) \bmod r$. Whenever we pass through a vertex u from e, if it has not yet been visited, we set $\mathit{Visited}[u] = \mathit{true}$. As each edge is handled once, the Assigning Step also runs in linear time.

3.3 Ranking Step

The Ranking Step obtains MPHFs from the PHFs presented in Section 3.2. It receives the labeling g as input and produces the rankTable as output. It is possible to build a data structure that allows the computation in constant time of function rank presented in Eq. (3) by using $o(m)$ additional bits of space. This is a well-studied primitive in succinct data structures (see e.g. [17]).

Implementation. We now describe a practical variant that uses $\epsilon\, m$ additional bits of space, where ϵ can be chosen as any positive number, to compute the data structure rankTable in linear time. Conceptually, the scheme is very simple: store explicitly the rank of every kth index in a rankTable, where $k = \lfloor \log(m)/\epsilon \rfloor$. In the implementation we let the parameter k to be set by the users so that they can trade off space for evaluation time and vice-versa. In the experiments we set k to 256 in order to spend less space to store the resulting MPHFs. This means that we store in the rankTable the number of assigned vertices before every 256th entry in the labeling g.

Evaluation. To compute rank(u), where u is given by Eq. (1), we look up in the rankTable the rank of the largest precomputed index $v \leq u$, and count the number of assigned vertices from position v to $u-1$. To do this in time $O(1/\epsilon)$ we use a lookup table that allows us to count the number of assigned vertices in $\Omega(\log m)$ bits in constant time. Such a lookup table takes $m^{\Omega(1)}$ bits of space.

In the experiments, we have used a lookup table that allows us to count the number of assigned vertices in 8 bits in constant time. Therefore, to compute the number of assigned vertices in 256 bits we need 32 lookups. Such a lookup table fits entirely in the cache because it takes 2^8 bytes of space.

We use the implementation just described because the smallest hypergraphs are obtained when $r = 3$ (see Section 3.1). Therefore, the most compact and efficient functions are generated when $r = 2$ and $r = 3$. That is why we have chosen these two instances of the family to be discussed in the following sections.

4 The 2-Uniform Hypergraph Instance

The use of 2-graphs allows us to generate the PHFs of Eq.(1) that give values in the range $[0, m-1]$, where $m = (2+\varepsilon)n$ for $\varepsilon > 0$ (see Section 3.1). The significant values in the labeling g for a PHF are $\{0, 1\}$, because we do not need to represent information to calculate the ranking (i.e., $r = 2$). Then, we can use just one bit to represent the value assigned to each vertex. Therefore, the resulting PHF requires m bits to be stored. For $\varepsilon = 0.09$, the resulting PHFs are stored in approximately $2.09n$ bits.

To generate the MPHFs of Eq. (2) we need to include the ranking information. Thus, we must use the value $r = 2$ to represent unassigned vertices and now two bits are required to encode each value assigned to the vertices. Then, the resulting MPHFs require $(2+\epsilon)m$ bits to be stored (remember that the ranking information requires ϵm bits), which corresponds to $(2+\epsilon)(2+\varepsilon)n$ bits for any $\epsilon > 0$ and $\varepsilon > 0$. For $\epsilon = 0.125$ and $\varepsilon = 0.09$ the resulting functions are stored in approximately $4.44n$ bits.

4.1 Improving the Space

The range of significant values assigned to the vertices is clearly [0,2]. Hence we need log(3) bits to encode the value assigned to each vertex. Theoretically we use arithmetic coding as block of values. Therefore, we can compress the resulting MPHF to use $(\log(3) + \epsilon)(2+\varepsilon)n$ bits of storage space by using a simple packing technique. In practice, we can pack the values assigned to every group of 5 vertices into one byte because each assigned value comes from a range of size 3 and $3^5 = 243 < 256$. Thus, if $\epsilon = 0.125$ and $\varepsilon = 0.09$, then the resulting functions are stored in approximately $3.6n$ bits.

We now sketch another way of improving the space to just over 3 bits per key, adding a little complication to the scheme. Use $m = (2+\epsilon/2)n$ for $\epsilon > 0$. Now store separately the set of assigned vertices in a bit array T of size m, such that rank operations are efficiently supported using $(\epsilon/2)n$ bits of extra space. Then, store for each vertex $v \in V(G_r)$ the bit $g(v)$ (must be 0 or 1), which costs m bits. Now we can create a compressed representation g' that uses only n bits and enables us to compute any bit of g in constant time by using rank on the set of assigned vertices represented by T. That is, $g'[\text{rank}(v)] = g[v]$. This is possible since $\text{rank}(v)$ is 1-1 on elements in $V(G_r)$. In conclusion, we can replace g by g' and reduce the space usage to $(3+\epsilon)n$ bits.

5 The 3-Uniform Hypergraph Instance

The use of 3−graphs allows us to generate more compact PHFs and MPHFs at the expense of one more hash function h_2. An acyclic random 3−graph is generated with probability $\Omega(1)$ for $m \geq c(3)n$, where $c(3) \approx 1.23$ is the minimum value for $c(r)$ (see Section 3.1). Therefore, we will be able to generate the PHFs of Eq. (1) so that they will produce values in the range $[0, (1.23+\varepsilon)n - 1]$ for any $\varepsilon \geq 0$. The values assigned to the vertices are drawn from $\{0, 1, 2, 3\}$ and, consequently, each value requires 2 bits to be represented. Thus, the resulting PHFs require $2(1.23+\varepsilon)n$ bits to be stored, which corresponds to $2.46n$ bits for $\varepsilon = 0$.

We can generate the MPHFs of Eq. (2) from the PHFs that take into account the special value $r = 3$. The resulting MPHFs require $(2+\epsilon)(1.23+\varepsilon)n$ bits to be stored for any $\epsilon > 0$ and $\varepsilon \geq 0$, once the ranking information must be included. If $\epsilon = 0.125$ and $\varepsilon = 0$, then the resulting functions are stored in approximately $2.62n$ bits.

5.1 Improving the Space

For PHFs that map to the range $[0, (1.23+\varepsilon)n-1]$ we can get still more compact functions. This comes from the fact that the only significant values assigned to the vertices that are used to compute Eq. (1) are $\{0, 1, 2\}$. Then, we can apply the arithmetic coding technique presented in Section 4.1 to get PHFs that require $\log(3)(1.23+\varepsilon)n$ bits to be stored, which is approximately $1.95n$ bits for $\varepsilon = 0$. For this we must replace the special value $r = 3$ to 0.

6 The Full Randomness Assumption

The full randomness assumption is not feasible because each hash function $h_i : U \to [i\frac{m}{r}, (i+1)\frac{m}{r} - 1]$ for $0 \leq i < r$ would require at least $n \log \frac{m}{r}$ bits to be stored, exceeding the space for the PHFs. From a theoretical perspective, the full randomness assumption is not too harmful, as we can use the "split and share" approach of Dietzfelbinger and Weidling [6]. The additional space usage is then a lower order term of $O(n^{1-\Omega(1)})$. Specifically, the algorithm would split S into $O(n^{1-\delta})$ buckets of size n^δ, where $\delta < 1/3$, say, and create a perfect hash function for each bucket using a pool of $O(r)$ simple hash functions of size $O(n^{2\delta})$, where each acts like truly random functions on each bucket, with high probability. From this pool, we can find r suitable functions for each bucket, with high probability. Putting everything together to form a perfect hash function for S can be done using an offset table of size $O(n^{1-\delta})$.

Implementation. In practice, limited randomness is often as good as total randomness [19]. For our experiments we choose h_i from a family $\mathcal{H}$ of universal hash functions proposed by Alon, Dietzfelbinger, Miltersen and Petrank [1], and we verify experimentally that the schemes behave well (see Section 7). We use a function h' from $\mathcal{H}$ so that the functions h_i are computed in parallel. For that, we impose some upper bound L on the lengths of the keys in S. The function h' has the following form: $h'(x) = Ax$, where $x \in S \subseteq \{0,1\}^L$ and A is a $\gamma \times L$ matrix in which the elements are randomly chosen from $\{0,1\}$. The output is a bit string of an a priori defined size γ. Each hash function h_i is computed by $h_i(x) = h'(x)[a,b] \bmod (\frac{m}{r}) + i(\frac{m}{r})$, where $a = \beta i$, $b = a + \beta - 1$ and β is the number of bits used from h' for computing each h_i. In [2] it is shown a tabulation idea that can be used to efficiently implement h' and, consequently, the functions h_i. The storage space required for the hash functions h_i corresponds to the one required for h', which is $\gamma \times L$ bits.

7 Experimental Results

In this section we evaluate the performance of our algorithms. We compare them with the main practical minimal perfect hashing algorithms we found in the literature. They are: Botelho, Kohayakawa and Ziviani [3] (referred to as BKZ), Fox, Chen and Heath [7] (referred to as FCH), Majewski, Wormald, Havas and Czech [14] (referred to as MWHC), and Pagh [16] (referred to as PAGH). For the MWHC algorithm we used the version based on 3-graphs. We did not consider the one that uses 2-graphs because it is shown in [3] that the BKZ algorithm outperforms it. We used the linear hash functions presented in Section 6 for all the algorithms.

The algorithms were implemented in the C language and are available at `http://cmph.sf.net` under the GNU Lesser General Public License (LGPL). The experiments were carried out on a computer running the Linux operating system, version 2.6, with a 3.2 gigahertz Intel Xeon Processor with a 2 megabytes L2 cache and 1 gigabyte of main memory. Each experiment was run for 100 trials. For the experiments we used two collections: (i) a set of randomly generated 4 bytes long IP addresses, and (ii) a set of 64 bytes long (on average) URLs collected from the Web.

To compare the algorithms we used the following metrics: (i) The amount of time to generate MPHFs, referred to as Generation Time. (ii) The space requirement for the description of the resulting MPHFs to be used at retrieval time, referred to as Storage Space. (iii) The amount of time required by a MPHF for each retrieval, referred to as Evaluation Time. For all the experiments we used $n = 3,541,615$ keys for the two collections. The reason to choose a small value for n is because the FCH algorithm has exponential time on n for the generation phase, and the times explode even for number of keys a little over.

We now compare our algorithms for constructing MPHFs with the other algorithms considering generation time and storage space. Table 1 shows that our algorithm for $r = 3$ and the MWHC algorithm are faster than the others to

Table 1. Comparison of the algorithms for constructing MPHFs considering generation time and storage space, and using $n = 3,541,615$ for the two collections

Algorithms		Generation Time (sec)		Storage Space	
		URLs	IPs	Bits/Key	Size (MB)
Our	$r = 2$	19.49 ± 3.750	18.37 ± 4.416	3.60	1.52
	$r = 3$	9.80 ± 0.007	8.74 ± 0.005	2.62	1.11
BKZ		16.85 ± 1.85	15.50 ± 1.19	21.76	9.19
FCH		5901.9 ± 1489.6	4981.7 ± 2825.4	3.66	1.55
MWHC		10.63 ± 0.09	9.36 ± 0.02	26.76	11.30
PAGH		52.55 ± 2.66	47.58 ± 2.14	44.16	18.65

generate MPHFs. The storage space requirements for our algorithms with $r = 2$, $r = 3$ and the FCH algorithm are 3.6, 2.62 and 3.66 bits per key, respectively. For the BKZ, MWHC and PAGH algorithms they are $\log n$, $1.23 \log n$ and $2.03 \log n$ bits per key, respectively.

Now we compare the algorithms considering evaluation time. Table 2 shows the evaluation time for a random permutation of the n keys. Although the number of memory probes at retrieval time of the MPHF generated by the PAGH algorithm is optimal [16] (it performs only 1 memory probe), it is important to note in this experiment that the evaluation time is smaller for the FCH and our algorithms because the generated functions fit entirely in the L2 cache of the machine (see the storage space size for our algorithms and the FCH algorithm in Table 1). Therefore, the more compact a MPHF is, the more efficient it is if its description fits in the cache. For example, for sets of size up to 6.5 million keys of any type the resulting functions generated by our algorithms will entirely fit in a 2 megabyte L2 cache. In a conversely situation where the functions do not fit in the cache, the MPHFs generated by the PAGH algorithm are the most efficient (because of lack of space we will not show this experiment).

Table 2. Comparison of the algorithms considering evaluation time and using the collections IPs and URLs with $n = 3,541,615$

Algorithms		Our		BKZ	FCH	MWHC	PAGH
		$r = 2$	$r = 3$				
Evaluation Time (sec)	IPs	1.35	1.36	1.45	1.01	1.46	1.43
	URLs	2.63	2.73	2.81	2.14	2.85	2.78

Now, we compare the PHFs and MPHFs generated by our family of algorithms considering generation time, storage space and evaluation time. Table 3 shows that the generation times for PHFs and MPHFs are almost the same, being the algorithms for $r = 3$ more than twice faster because the probability to obtain an acyclic 3-graph for $c(3) = 1.23$ tends to one while the probability for a 2-graph where $c(2) = 2.09$ tends to 0.29 (see Section 3.1). For PHFs with $m = 1.23n$

Table 3. Comparison of the PHFs and MPHFs generated by our algorithms, considering generation time, evaluation time and storage space metrics using $n = 3,541,615$ for the two collections. For packed schemes see Sections 4.1 and 5.1.

r	Packed	m	Generation Time (sec)		Eval. Time (sec)		Storage Space	
			IPs	URLs	IPs	URLs	Bits/Key	Size (MB)
2	no	$2.09n$	18.32 ± 3.352	19.41 ± 3.736	0.68	1.83	2.09	0.88
2	yes	n	18.37 ± 4.416	19.49 ± 3.750	1.35	2.63	3.60	1.52
3	no	$1.23n$	8.72 ± 0.009	9.73 ± 0.009	0.96	2.16	2.46	1.04
3	yes	$1.23n$	8.75 ± 0.007	9.95 ± 0.009	0.94	2.14	1.95	0.82
3	no	n	8.74 ± 0.005	9.80 ± 0.007	1.36	2.73	2.62	1.11

instead of MPHFs with $m = n$, then the space storage requirement drops from 2.62 to 1.95 bits per key. The PHFs with $m = 2.09n$ and $m = 1.23n$ are the fastest ones at evaluation time because no ranking or packing information needs to be computed.

8 Conclusions

We have presented an efficient family of algorithms to generate near space-optimal PHPs and MPHFs. The algorithms are simpler and has much lower constant factors than existing theoretical results for $n < 2^{300}$. In addition, it outperforms the main practical general purpose algorithms found in the literature considering generation time and storage space as metrics.

Acknowledgment. We thank Djamal Belazzougui for suggesting arithmetic coding to generate more compact functions and Yoshiharu Kohayakawa for helping us with acyclic r-partite random r-graphs.

References

1. Alon, N., Dietzfelbinger, M., Miltersen, P.B., Petrank, E., Tardos, G.: Linear hash functions. Journal of the ACM 46(5), 667–683 (1999)
2. Alon, N., Naor, M.: Derandomization, witnesses for Boolean matrix multiplication and construction of perfect hash functions. Algorithmica 16(4-5), 434–449 (1996)
3. Botelho, F.C., Kohayakawa, Y., Ziviani, N.: A practical minimal perfect hashing method. In: Nikoletseas, S.E. (ed.) WEA 2005. LNCS, vol. 3503, pp. 488–500. Springer, Heidelberg (2005)
4. Czech, Z.J., Havas, G., Majewski, B.S.: Fundamental study perfect hashing. Theoretical Computer Science 182, 1–143 (1997)
5. Dietzfelbinger, M., Hagerup, T.: Simple minimal perfect hashing in less space. In: Meyer auf der Heide, F. (ed.) ESA 2001. LNCS, vol. 2161, pp. 109–120. Springer, Heidelberg (2001)

6. Dietzfelbinger, M., Weidling, C.: Balanced allocation and dictionaries with tightly packed constant size bins. In: Caires, L., Italiano, G.F., Monteiro, L., Palamidessi, C., Yung, M. (eds.) ICALP 2005. LNCS, vol. 3580, pp. 166–178. Springer, Heidelberg (2005)
7. Fox, E.A., Chen, Q.F., Heath, L.S.: A faster algorithm for constructing minimal perfect hash functions. In: Proc. of the 15th Annual International ACM SIGIR Conference on Research and Development in Information Retrieval, pp. 266–273 (1992)
8. Fox, E.A., Heath, L.S., Chen, Q., Daoud, A.M.: Practical minimal perfect hash functions for large databases. Communications of the ACM 35(1), 105–121 (1992)
9. Fredman, M.L., Komlós, J.: On the size of separating systems and families of perfect hashing functions. SIAM Journal on Algebraic and Discrete Methods 5, 61–68 (1984)
10. Fredman, M.L., Komlós, J., Szemerédi, E.: Storing a sparse table with O(1) worst case access time. Journal of the ACM 31(3), 538–544 (1984)
11. Hagerup, T., Tholey, T.: Efficient minimal perfect hashing in nearly minimal space. In: Ferreira, A., Reichel, H. (eds.) STACS 2001. LNCS, vol. 2010, pp. 317–326. Springer, Heidelberg (2001)
12. Janson, S.: Poisson convergence and poisson processes with applications to random graphs. Stochastic Processes and their Applications 26, 1–30 (1987)
13. Lefebvre, S., Hoppe, H.: Perfect spatial hashing. ACM Transactions on Graphics 25(3), 579–588 (2006)
14. Majewski, B.S., Wormald, N.C., Havas, G., Czech, Z.J.: A family of perfect hashing methods. The Computer Journal 39(6), 547–554 (1996)
15. Mehlhorn, K.: Data Structures and Algorithms 1: Sorting and Searching. Springer, Heidelberg (1984)
16. Pagh, R.: Hash and displace: Efficient evaluation of minimal perfect hash functions. In: Dehne, F., Gupta, A., Sack, J.-R., Tamassia, R. (eds.) WADS 1999. LNCS, vol. 1663, pp. 49–54. Springer, Heidelberg (1999)
17. Pagh, R.: Low redundancy in static dictionaries with constant query time. SIAM Journal on Computing 31(2), 353–363 (2001)
18. Radhakrishnan, J.: Improved bounds for covering complete uniform hypergraphs. Information Processing Letters 41, 203–207 (1992)
19. Schmidt, J.P., Siegel, A.: The spatial complexity of oblivious k-probe hash functions. SIAM Journal on Computing 19(5), 775–786 (1990)
20. Woelfel, P.: Maintaining external memory efficient hash tables. In: Díaz, J., Jansen, K., Rolim, J.D.P., Zwick, U. (eds.) APPROX 2006 and RANDOM 2006. LNCS, vol. 4110, pp. 508–519. Springer, Heidelberg (2006)

A Near Linear Time Approximation Scheme for Steiner Tree Among Obstacles in the Plane

Matthias Müller-Hannemann and Siamak Tazari*

Darmstadt University of Technology, Dept. of Computer Science
{muellerh,tazari}@algo.informatik.tu-darmstadt.de

Abstract. We present a polynomial time approximation scheme (PTAS) for the Steiner tree problem with polygonal obstacles in the plane with running time $O(n \log^2 n)$, where n denotes the number of terminals plus obstacle vertices. To this end, we show how a planar spanner of size $O(n \log n)$ can be constructed that contains a $(1+\epsilon)$-approximation of the optimal tree. Then one can find an approximately optimal Steiner tree in the spanner using the algorithm of Borradaile et al. (2007) for the Steiner tree problem in planar graphs. We prove this result for the Euclidean metric and also for all uniform orientation metrics, i.e. particularly the rectilinear and octilinear metrics.

Keywords: Steiner Tree, Obstacles, PTAS, Euclidean Metric, Uniform Orientation Metric, Spanner, Banyan, Planar Graph.

1 Introduction

We consider the following network design problem: given a set of points in the plane and a set of disjoint polygonal obstacles, find the shortest network interconnecting the points and avoiding the interior of the obstacles. We refer to the given points as *terminals* and to the obstacle vertices as *corners*. We let n be the total number of terminals *and* corners. The shortest interconnecting network of the terminals will be a tree, a so-called *Steiner tree*, and it might use corners and additional vertices called *Steiner points* (note that we use this term only to refer to points that do not coincide with terminals or corners). This problem is called the *obstacle-avoiding Steiner minimum tree* problem (SMTO) or ESMTO when we are using the Euclidean metric.

Uniform orientation metrics are derived from λ-*geometries*. In a λ-geometry, one is allowed to move only along $\lambda \geq 2$ orientations building consecutive angles of π/λ. The *rectilinear* or *Manhattan* metric corresponds to the 2-geometry and the *octilinear* metric to the 4-geometry. We call the corresponding SMT problems λ-SMT or, when obstacles are to be avoided, λ-SMTO. In this case, the obstacle edges must obey the restrictions of the given orientations, too.

It has been a long-standing open problem whether these SMT problems *among obstacles* admit a polynomial time approximation scheme (PTAS). With the

* This author was supported by the Deutsche Forschungsgemeinschaft (DFG), grant MU1482/3-1.

F. Dehne, J.-R. Sack, and N. Zeh (Eds.): WADS 2007, LNCS 4619, pp. 151–162, 2007.

recent result of Borradaile et al. [1,2] about Steiner trees in planar graphs, this question can now be answered affirmatively by combining a number of results in the literature (see Section 1.1). However, to obtain a near linear running time, new ideas and more sophisticated techniques are required; this is the main contribution of our work. Our approach is based on constructing a planar graph of size $O(n \log n)$ that contains a $(1+\epsilon)$-approximation of the solution and then find an approximate solution in that graph. The total running time will be $O(n \log^2 n)$. Along the way, we prove a number of spanner results and other properties of SMTOs both for the Euclidean and uniformly oriented case.

The SMT problem and its many variations are of high theoretical (see below) and practical relevance. The applications reach from all kinds of network design to phylogenetic trees. Especially the geometric case with obstacles is very important in VLSI design, since there are usually regions in the plane that may not be crossed by wire. Also, it is often only allowed to route the tree along a rectilinear or octilinear grid and so, SMTs in uniformly oriented metrics are required.

1.1 Related Work

The ESMTO problem is clearly $\mathcal{NP}$-hard since it contains the Steiner minimum tree problem without obstacles as a special case [3]. For the SMT problem without obstacles, Arora [4] and Mitchell [5] were the first to present a PTAS. Rao and Smith [6] improved the running time of Arora's algorithm from $O(n(\frac{1}{\epsilon}\log n)^{O(1/\epsilon)})$ to $O(2^{\text{poly}(1/\epsilon)}n + n\log n)$ using a certain spanner graph they call a "banyan" and this is the best running time known so far. However, none of these algorithms are applicable to the case with obstacles since a so-called "patching lemma" that is necessary for these approaches, fails to hold. Provan [7] has shown how to approximate ESMTO by an SMT problem in graphs and derived an FPTAS for the special case when the terminals lie on a constant number of "boundary polygons" and interior points.

The PTASs discussed above also apply to λ-SMTs for all $\lambda \geq 2$. The rectilinear and octilinear case have been shown to be $\mathcal{NP}$-complete in [8,9]. For general fixed λ no proof has been published so far, though it is widely believed that these problems are hard, too. Properties of uniformly oriented SMTs have been studied by Brazil et al. [10]. Approximation algorithms for rectilinear SMTO have been proposed by Ganley and Cohoon [11] and for the octilinear case by Müller-Hannemann and Schulze [9,12]. For rectilinear SMTO with a constant number of obstacles, Liu et al. [13] presented a PTAS based on Mitchell's [5] approach. The SMT problem with length restrictions on obstacles has been studied by Müller-Hannemann and Peyer [14] in the rectilinear case, and by Müller-Hannemann and Schulze [12] in the octilinear case, and constant-factor approximation algorithms have been proposed.

The SMT problem in graphs has also been studied widely in the literature. It has been shown to be $\mathcal{APX}$-complete [15] and thus, no PTAS exists unless $\mathcal{P} = \mathcal{NP}$. The best approximation factor known so far is $1.55 + \epsilon$ [16]. The case of planar graphs has very recently been shown to admit a PTAS by Borradaile et al. [1,2]. This results immediately in a PTAS for rectilinear, octilinear, and

Euclidean SMTO using the following results from the literature: the so-called Hanan-grid [11,17] for the rectilinear case, the result of Müller-Hannemann and Schulze [12] for the octilinear case, and Provan's construction [7] together with the planar spanner result of Arikati et al. [18] for the Euclidean case. However, in all these cases, the PTAS of Borradaile et al. has to be run on a graph of size $O(n^2)$ and thus, the total running time will be $O(n^2 \log n)$. In this work, we show alternative constructions with running time $O(n \log^2 n)$.

1.2 On Spanners and Banyans

A *t-spanner* of a set of points P is a graph that contains a path between any two points of P that is at most a factor of t longer than the shortest path between them. Spanners have been vastly studied in the literature [19] and have been often used in the design of PTASs [6]. Of particular interest to us are spanners of the *visibility graph* among obstacles in the plane. The visibility graph contains all straight line connections between terminals and corners that do not cross the interior of any obstacle. Clarkson [20] showed how to construct a $(1+\epsilon)$-spanner of linear size of the visibility graph in $O(n \log n)$ time. A linear-sized *planar* spanner for both the rectilinear and Euclidean metric has been shown to exist and also to be computable in $O(n \log n)$ time by Arikati et al. [18,21]. We will show how to extend these ideas to derive sparse planar spanners in the same time bound for all uniform orientation metrics.

Rao and Smith [6] introduced the notion of *banyans*. A banyan is a graph that contains a $(1+\epsilon)$-approximation of the SMT of a given set of points[1] and whose weight is at most a constant factor larger than the SMT. Rao and Smith showed how to construct a banyan of size $O(n)$ in time $O(n \log n)$ in the obstacle-free case.

1.3 Our Contribution and Outline

One of the main results of our work is to show how to construct a *planar banyan for SMTO* of size $O(n \log n)$ in $O(n \log^2 n)$ time by building on the framework of Rao and Smith using new ideas and combining other results from the literature, especially the banyan-result for planar graphs contained in [1]. An approximate Steiner tree can then be obtained on this planar graph using [1]. Since the algorithm in [1,2] is exponential in $1/\epsilon$, so is our resulting algorithm.

The main difficulties that arise when obstacles are present are to deal with visibility and the fact that only a subset of corners is included in the SMT, i.e. we do not know which vertices of a spanner will be part of the SMT. In particular, a spanner might include arbitrary short edges between corners that are not part of the SMT and this causes an important proof idea of Rao and Smith to fail. Roughly speaking, they show that in the obstacle-free case, there is always a "long enough" spanner edge near non-negligible SMT edges and so,

[1] Their construction was in fact more powerful as it included an approximate SMT for *any* subset of the terminals.

they introduce a grid of candidate Steiner points in a neighborhood around every spanner edge to capture these SMT vertices. Our main new algorithmic idea is to use $O(\log n)$ layers of candidate Steiner points around each spanner edge, so that we are guaranteed to find such appropriate points even when our spanner edges are short. Another important difference is that we use planar spanners, so that afterwards, we can use the algorithm of [1] instead of building on Arora's approach [4] to obtain our PTAS. We present our algorithm in Section 2 and then present two proofs for the correctness of our algorithm for the Euclidean case in Section 3: one using an analog of the hexagon property [3] and another one using a generalization of the empty ball lemma [6]. Even though our proofs follow the lines of the proofs of Rao and Smith, they differ conceptually at some key points and other techniques have to be used, see Section 3.1.

Afterwards, in Section 4, we turn our attention to uniform orientation metrics and argue how the presented proofs can be modified to work for these cases, too. Along the way, we have to argue that a lemma by Provan [7] about δ-grids among obstacles in the plane still holds true for all uniform orientation metrics. At last, we prove a variation of Arikati et al.'s planar spanner result [18] to apply to uniform orientation metrics.

Due to space limitations, several details and proofs had to be shortened or omitted. We refer the interested reader to the full version of this paper.

2 The Algorithm

The main result of our work is the following theorem:

Theorem 2.1. *The Steiner minimum tree problem among disjoint polygonal obstacles in the plane admits a PTAS in the Euclidean metric and in all uniform orientation metrics. The running time is $O(n \log^2 n)$, where n is the total number of terminals and obstacle corners.*

Our algorithm is summarized in Alg. 1. We are given a set of terminals Z and a set of disjoint polygonal obstacles O as described in the introduction. In the first step, we find a $(1+\epsilon_1)$-spanner G_1 of the visibility graph of $Z \cup O$. In the Euclidean case, this can be done using the algorithm of Clarkson [20] or Arikati et al. [18,21] in $O(n \log n)$ time. Arikati et al. also provide an algorithm for the rectilinear case and we will show in Section 4.2 how to extend this result to other uniform orientation metrics. Note that these algorithms construct a spanner without having to build the full visibility graph. The graph G_1 will have $O(n)$ edges. Around each such edge, we place $\lceil \log_2 n \rceil$ circles with doubling radii and place a grid of constant size inside each of them. Here, we make use of a constant κ that depends on the metric being used. For $\epsilon_1 \leq 1$, in the Euclidean metric, κ can be chosen to be ≤ 226, and in the rectilinear metric ≤ 50. This introduces a set P_0 of $O(n \log n)$ "candidate Steiner points" from which we remove the ones that lie inside obstacles. This can be done using a sweep-line algorithm in $O(k \log k) = O(n \log^2 n)$-time with $k = n + |P_0| = O(n \log n)$. Let G_2 be the visibility graph of $Z \cup P_0 \cup O$ and let G_3 be a *planar* $(1+\epsilon_2)$-spanner of G_2. G_3 can

Algorithm 1. A PTAS for SMTO

Input : a set of terminals Z and a set of disjoint polygonal obstacles O in the plane and the desired accuracy $0 < \epsilon \leq 1$.
Output : a $(1 + \epsilon)$-approximation of the obstacle-avoiding Steiner minimum tree of the terminals.
Note : κ is a constant and can be ≤ 226 in the Euclidean case and ≤ 50 in the rectilinear case, ϵ_1 and ϵ_2 have to be chosen appropriately, e.g. $\epsilon_1 = \epsilon_2 = \frac{\epsilon}{22}$.

begin
  find a $(1 + \epsilon_1)$-spanner G_1 of the visibility graph of $Z \cup O$;
  let $P_0 = \emptyset$;
  for *each edge e in* G_1 **do** `// let` ℓ `be the length of` e
    for $i = 0$ *to* $\lceil \log_2 n \rceil$ **do**
      let $r = \kappa 2^i \ell / \epsilon_1$;
      let C be a circle of radius r around the midpoint of e;
      place a grid with spacing $\delta(r) = r\epsilon_1^3/\kappa^2$ inside C;
      `// the grid has` $\leq 4\kappa^4/\epsilon_1^6 = O(1)$ `points`
      add these points to P_0;
  remove all the points from P_0 that lie inside obstacles;
  `// let` G_2 `be the visibility graph of` $Z \cup P_0 \cup O$
  find a planar $(1 + \epsilon_2)$-spanner G_3 of G_2;
  find a $(1 + \epsilon/3)$-approximate SMT T of Z in G_3;
  `// using the PTAS of Borradaile et al. [1]`
  return T;
end

be found using Arikati et al.'s algorithm or our extension of it for other uniform orientation metrics. Let $k = O(n \log n)$ be the number of vertices of G_2. The spanner algorithms need $O(k \log k)$ time and introduce $O(k)$ additional Steiner points to achieve planarity. Thus, G_3 can be constructed in $O(n \log^2 n)$ time and has $O(n \log n)$ vertices (note that G_2 is not constructed explicitly). Now we find a $(1 + \epsilon/3)$-approximate Steiner minimum tree of the terminals Z in G_3 using the PTAS of Borradaile et al. [1,2] for the Steiner tree problem in planar graphs. The time needed for this step is $O(k \log k)$ and hence, the total runtime of our algorithm is $O(n \log^2 n)$.

Note that the first step of the PTAS of Borradaile et al. is to determine a subgraph G_4 of G_3 that contains a $(1 + \epsilon/3)$-approximation of the SMT of G_3 and has weight at most a constant times the weight of the SMT of G_3. Hence, G_4 is a planar banyan of the terminal set Z and so, our algorithm also delivers a planar banyan of a set of terminals among obstacles in the plane.

A note on the running time. Of course, the constants hidden in the O-notations above all depend on $1/\epsilon$. Our algorithm builds the planar graph G_3 in time $O(\frac{\kappa^4}{\epsilon^{11}} n \log^2 n)$ and its size is more precisely $O(\frac{\kappa^4}{\epsilon^{11}} n \log n)$. The PTAS of Borradaile et al. takes time singly exponential in $1/\epsilon$ [2].

3 Correctness

We present two proofs for the correctness of Alg. 1. The first one results in better constants but does not work in the rectilinear case. The other one is more general and can even be partly extended to give us some structural information about SMTOs in higher dimensions but uses much larger constants. The proof technique and the generalization of the empty ball lemma used in the second proof might also be interesting in their own right. In the next section, we discuss uniform orientation metrics where we include a simpler proof for the rectilinear case that results in small constants.

3.1 Key Differences

The main new idea in our algorithm compared to that of Rao and Smith's is the use of $O(\log n)$ layers of grids around each edge of the spanner G_1. We had to do this because in our case, we do not have the so-called spanner path property (Lemma 34 in [6]), that essentially says that two vertices that are connected in the SMT by an edge of length L, can not be connected in a spanner by a chain of "tiny" edges of length $< L$. In our case, two terminals and/or corners *can* be connected by a path consisting entirely of "tiny" edges, finding their way among obstacles. But we know that any two vertices in the spanner are connected by a short path with at most n edges and one of them can be made "long enough" by multiplying it with a power of 2 if necessary.

Also, we have the additional problem that many vertices of our spanner need not be part of the optimal Steiner tree: we do not know which corners will be included in the SMT. And there is the issue of visibility. To deal with these problems, we formulate and prove Lemma 3.1, which is a technical lemma that is very important in both of our proofs. The reason it works well is that since we do not make use of the spanner path property, it is not necessary for us that the vertices that we find close to a Steiner point A be in fact close to A *in the SMT* or even be part of the SMT at all. This enabled us to prove our generalizations of the hexagon property and the empty ball lemma and use them to prove our main theorem.

3.2 First Proof

We use the notation $d(A, B)$ for the length of the shortest obstacle-avoiding path between two points A and B and the notation $d_G(A, B)$ for the shortest path between A and B in a graph G. We start by mentioning some well known facts about Euclidean Steiner minimum trees [22,7,23] (recall that we use the term Steiner point for vertices of the tree that do not coincide with terminals or corners):

Fact 0: The SMT is a tree that includes all terminals as vertices. It might include corners or Steiner points as additional internal vertices.
Fact 1: Every Steiner point has 3 incident edges making angles of 120°.

Fact 2: A Steiner point may not occur on the boundary of some obstacle.
Fact 3: Every terminal and corner has degree at most 3 in the SMT.
Fact 4: If there are k terminals, there are at most $k-2$ Steiner points.
Fact 5: Two edges of the SMT meet only at a common endpoint, i.e. the SMT is not self-intersecting.
Fact 6: ($120°$ wedge property) If s is a Steiner point of the SMT, then in any closed $120°$ wedge with apex s, there exists a terminal or corner v and an SMT path sv that lies entirely inside the wedge.

The following lemma is of central importance for our work:

Lemma 3.1. *Let S be a closed convex region of the plane and let $A \in S$ be a point that is not contained in the interior of any obstacle. Then, we have*

(i) a terminal or corner in S that is visible to A; or
(ii) the maximal visible area to A in S is a closed convex region $S' \subseteq S$ that contains no terminal or corner and that shares its border with S except for finitely many straight line segments, where its border may consist of obstacle edges. Furthermore, any obstacle-avoiding path contained in S and connected to A is contained in S'.

Consider an SMT edge AB and some fixed distance D. Let H_A be the regular hexagon of side length D that has A as a corner, does not contain AB, and builds two $120°$ angles with AB, i.e. if we extend AB, it would cut H_A in half. Furthermore, we define an SMT edge AB to be *locally D-bounded* if when walking from A or B for at most 3 SMT edges or until we encounter a terminal or corner (whichever comes first), all edges we pass have length at most D. We have the following property:

Lemma 3.2 (Generalization of the hexagon property). *Let AB be a locally D-bounded SMT edge. Then the regular hexagon H_A of side length D defined above contains a terminal or a corner that is visible to A (this terminal or corner could be A itself).*

Let AB be an SMT edge of length L. For given constants $c \geq 1$ and $\epsilon_1 > 0$, we define AB to be *locally long* if it is locally cL/ϵ_1-bounded. Otherwise we call it *locally short.* The next lemma builds the heart of our first proof of Theorem 2.1. It assures that near every locally long SMT edge AB, we find a spanner edge of G_1 that is long enough, so that a layer of grids around it will enclose the points A and B; and short enough, so that the grid spacing does not introduce too large an error.

Lemma 3.3. *Let AB be a locally long SMT edge of length L as defined above with some constants $c \geq 1$ and $0 < \epsilon_1 \leq 1$ to be specified. Consider Alg. 1 with a constant $\kappa \geq 8c+2$. Then there exists an edge e of length ℓ in the $(1+\epsilon_1)$-spanner G_1 and an integer $0 \leq i \leq \lceil \log_2 n \rceil$, so that $L \leq 2^i \ell \leq \kappa L/\epsilon_1$ and so that A and B are included in a circle of radius $\kappa 2^i \ell/\epsilon_1$ around the midpoint of e.*

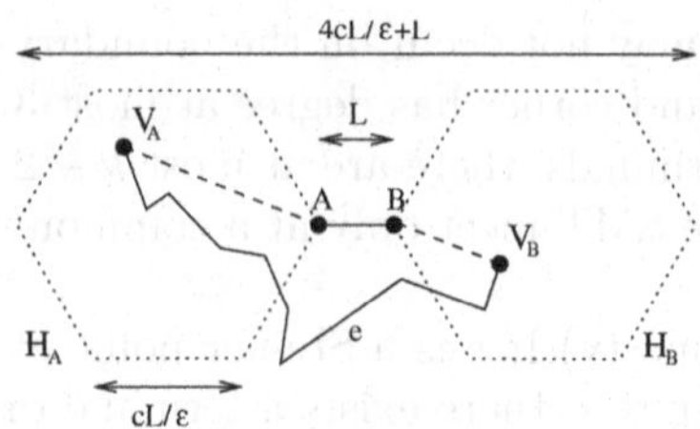

Fig. 1. Proof of Lemma 3.3; $L \leq d(V_A, V_B) \leq 4cL/\epsilon_1 + L$

Proof. By the hexagon property above with $D = cL/\epsilon_1$, we know that there exists a terminal or corner V_A inside H_A and a terminal or corner V_B inside H_B, so that V_A is visible to A and V_B is visible to B (note that V_A resp. V_B could be equal to A resp. B). Then we know that $L \leq d(V_A, V_B) \leq L + 4cL/\epsilon_1 =: M$ (see Fig. 1). Now consider the shortest path between V_A and V_B in the spanner G_1. It consists of at most n edges and its length is at least L and at most $(1+\epsilon_1)M = ((4c+1)\epsilon_1 + 4c + \epsilon_1^2)L/\epsilon_1 \leq \kappa L/\epsilon_1$ if we choose $\kappa \geq 8c+2$. Also, this path lies entirely inside a circle Q of radius $R := (1+\epsilon_1/2)M$ around the midpoint of the edge AB, since otherwise it would be too long for G_1 to be a $(1+\epsilon_1)$-spanner. Hence, there exists an edge e of length ℓ on this path inside Q, so that $L/n \leq \ell \leq \kappa L/\epsilon_1$. If $\ell < L$, one can choose an integer $0 \leq i \leq \lceil \log_2 n \rceil$ so that $L \leq 2^i \ell \leq 2L \leq \kappa L/\epsilon_1$ otherwise choose $i = 0$. Also, since e is inside Q, the distance between the midpoint of e to A and B is at most $R + L/2 = ((2c+1.5)\epsilon_1 + 4c + \epsilon_1^2/2)L/\epsilon_1 \leq \kappa 2^i \ell/\epsilon_1$.

Proof (First proof of Theorem 2.1 for the Euclidean metric). Let us denote the length of a tree T by $\ell(T)$. Let $T^\star$ be an optimal obstacle-avoiding Steiner tree of the terminal set and let T be the tree returned by Alg. 2.1. We show that the graph G_3 contains a Steiner tree T' with $\ell(T') \leq (1+\epsilon/2)\ell(T^\star)$. Then we know that $\ell(T) \leq (1+\epsilon/3)\ell(T') \leq (1+\epsilon)\ell(T^\star)$ and we are done.

We partition the edges of $T^\star$ into locally long and locally short edges as defined above and construct the tree T' as follows: for every locally long edge in $T^\star$, we find appropriate endpoints and a short path in G_3 to add to T'; then we get a number of connected components and interconnect them with a minimum forest in G_3. We now analyze the length of T'.

Let AB be a locally long edge of $T^\star$ of length L. By Lemma 3.3, we find an edge e of length ℓ in G_1 and a circle C of radius $r = \kappa 2^i \ell/\epsilon_1$ around the midpoint of e for some integer $0 \leq i \leq \lceil \log_2 n \rceil$, so that A and B are included in C. The grid inside C has spacing[2] $\delta = r\epsilon_1^3/\kappa^2 = 2^i \ell \epsilon_1^2/\kappa \leq L\epsilon_1$ since $2^i \ell \leq \kappa L/\epsilon_1$ by Lemma 3.3. A technical lemma of Provan (Lemma 3.2 in [7]) says that for a given Steiner point A in a δ-grid among obstacles, we can always find a terminal, corner or grid point A' that is visible to A, so that $d(A, A') \leq 2\delta$. Let A' and B' be vertices of G_3 that are visible to and closest to A and B, respectively. Add the shortest path between A' and B' in G_3 to T'. We have

[2] The grid spacing in [6] is $r\epsilon_1^2/\kappa^2$ but we believe that the exponent of ϵ_1 should be 3.

$$
\begin{aligned}
d(A', B') &\leq d(A', A) + d(A, B) + d(B, B') \\
&\leq L + 4\delta \leq L + 4L\epsilon_1 = (1 + 4\epsilon_1)L
\end{aligned} \tag{1}
$$

and thus, $d_{G_3}(A', B') \leq (1 + \epsilon_2)(1 + 4\epsilon_1)L$.

We leave the detailed analysis of locally short edges for the full paper; one can show that the overhead caused by ignoring locally short edges is at most equal to the total length of all locally short edges and can be upper bounded[3] by $(1 + \epsilon_2)\epsilon_1 \ell(T^\star)$. So, we get that

$$
\ell(T') \leq (1 + \epsilon_2)(1 + 4\epsilon_1)\ell(T^\star) + (1 + \epsilon_2)\epsilon_1 \ell(T^\star) \leq (1 + \epsilon/2)\ell(T^\star) \tag{2}
$$

if ϵ_1 and ϵ_2 are chosen appropriately, e.g. $\epsilon_1 = \epsilon_2 = \frac{\epsilon}{22}$.

Second Proof. Due to space limitations, we leave our second proof for the full paper. Here, we just state our generalization of the empty ball lemma:

Lemma 3.4 (Generalization of the empty ball lemma). *Let S_1 and S_2 be closed convex regions in the plane whose interiors are free of terminals and obstacle edges but whose borders may partly consist of obstacle-edges. Denote the parts of their borders that are not obstacle-edges as the* free border. *Assume that S_2 encloses S_1 and that the distance between every point on the free border of S_1 to any point on the free border of S_2 is at least $\gamma > 0$. Then, for any obstacle-avoiding SMT, the number of Steiner points inside S_1 is bounded by a constant $s_0 \leq (96e)^8$ (where e is the base of the natural logarithm).*

4 Uniform Orientation Metrics

We briefly discuss how our proofs adapt to uniform orientation metrics and provide a somewhat different proof for the rectilinear case that results in much better constants. Afterwards, we prove our generalization of Arikati et al.'s [18] planar spanner result to the cases with $\lambda \geq 3$.

4.1 Adapting the Proofs

The Cases $\lambda \geq 3$. Brazil et al. [10] showed that for $\lambda \geq 3$, there always exists an SMT, such that the minimum angle at a Steiner point is $90° \leq \alpha_{\min} \leq 120°$ and the maximum angle is $120° \leq \alpha_{\max} \leq 150°$. For these cases, we get an $\alpha_{\max}$-wedge property like Fact 6 of the Euclidean case and we can use it to prove an analog of the Hexagon property (Lemma 3.2). Also, using this $\alpha_{\max}$-wedge property one can generalize Provan's lemma [7] to ensure that in a δ-grid around a Steiner point A, one can always find a grid point, terminal, or corner A' that is visible to A, so that $d(A, A') \leq \delta / cos\frac{\alpha_{\max}}{2}$. Using these two results, one can generalize both of our proofs from Section 3 straightforwardly to all λ-geometries with $\lambda \geq 3$, where again, the first proof results in much better constants (but possibly different ones from the Euclidean case).

[3] To achieve this, we have to set $c = 28$ and thus, we can choose $\kappa = 8c + 2 = 226$.

The Rectilinear Case. In the rectilinear case, we do not have an α-wedge property for an $\alpha < 180°$; in fact, $180°$-angles can occur at any Steiner point. But instead, the structure of rectilinear Steiner trees is well-studied. Particularly, one can derive the following lemma using results from [24,25]:

Lemma 4.1. *For a given set of terminals and disjoint rectilinear obstacles in the plane, there exists an obstacle-avoiding RSMT that has the following properties: (i) for any two Steiner points A and B that are connected by a horizontal line-segment, B is not connected to a third Steiner point by a vertical line segment; (ii) if there is a grid with spacing δ around a Steiner point A, then there exists a grid point, terminal, or corner A' that is visible to A, so that $d(A, A') \leq 2\delta$.*

Using this lemma, both of our proofs from the last section adapt straightforwardly to the rectilinear case. Furthermore, one can choose $c = 12$ and also $\kappa = 4c + 2 = 50$.

4.2 Planar Spanners

Consider a λ-geometry and let $\omega = \pi/\lambda$ be the smallest allowed angle. Before we start with the construction of our spanner, we need the following technical lemma:

Lemma 4.2. *Consider a λ-geometry with smallest allowed angle ω and let a set of disjoint polygonal obstacles be given whose edges are parallel to the allowed directions. Consider two points A and B in the plane. Then there exists a shortest path (with respect to the metric of the λ-geometry) between A and B that passes through $A = v_0, v_1, v_2, \ldots, v_k = B$, so that each v_i with $0 < i < k$ is a corner and so that the path between each v_i and v_{i+1} is either a straight line in an allowed direction or is comprised of two straight lines in allowed directions that build an angle of $\pi - \omega$ with each other.*

Arikati et al. [18,21] showed how to find a planar rectilinear $(1+\epsilon)$-spanner of the visibility graph among disjoint polygonal obstacles in the plane that uses at most $O(n)$ Steiner points in time $O(n \log n)$. This spanner might include obstacle-edges that are not rectilinear but their length is measured in the rectilinear metric. We first show that one can rotate the axes of the coordinate system to build an arbitrary angle and still obtain such a spanner:

Lemma 4.3. *Given a set of terminals and a set of disjoint polygonal obstacles in the plane, one can find a planar $(1 + \epsilon)$-spanner of the visibility graph of size $O(n)$ in $O(n \log n)$ time that uses only edges in two directions d_1 and d_2 building an angle ω with each other (except for parts of the spanner that coincide with obstacle edges).*[4]

Now we can use a similar trick to the one used by Arikati et al. to obtain their Euclidean spanner: let $d_1, d_2, \ldots, d_\lambda$ be the allowed directions, so that two

[4] The spanning property is with respect to the metric induced by d_1 and d_2.

consecutive ones build an angle of $\omega = \pi/\lambda$ with each other. Find $(1+\epsilon)$-spanners $G_1, \ldots, G_\lambda$, so that G_i uses only edges parallel to d_i and d_{i+1} (or d_λ and d_1) using Lemma 4.3. Let G be the graph obtained by superimposing all these graphs on each other, i.e. putting them on each other and adding all intersection points as new vertices to the graph. A straightforward adaption of the proof of Arikati et al. for the Euclidean case (published in the thesis of Zeh [21]) shows that G will still have $O(n)$ vertices[5] and can be obtained in $O(n \log n)$ time. Also, by Lemma 4.2, G is indeed a $(1+\epsilon)$-spanner of the visibility graph (an approximate shortest path between each v_i and v_{i+1} of the lemma lies entirely in a spanner G_j), i.e. we have

Theorem 4.4. *Consider a λ-geometry and let a set of terminals and a set of disjoint polygonal obstacles whose edges are in the allowed directions be given, so that the total number of terminals and corners is n. Then one can find a planar $(1+\epsilon)$-spanner (with respect to the metric of the λ-geometry) of the visibility graph of size $O(n)$ in $O(n \log n)$ time that uses only edges in the allowed directions.*

References

1. Borradaile, G., Kenyon-Mathieu, C., Klein, P.N.: A polynomial-time approximation scheme for Steiner tree in planar graphs. In: SODA '07: Proceedings of the 18th Annual ACM-SIAM Symposium on Discrete Algorithms, pp. 1285–1294 (2007)
2. Borradaile, G., Kenyon-Mathieu, C., Klein, P.N.: Steiner tree in planar graphs: An $O(n \log n)$ approximation scheme with singly exponential dependence on epsilon. In: WADS '07: Proceedings of the 10th Workshop on Algorithms and Data Structures. LNCS, Springer, Heidelberg (2007)
3. Garey, M., Graham, R., Johnson, D.: The complexity of computing Steiner minimal trees. SIAM Journal on Applied Mathematics 32, 835–859 (1977)
4. Arora, S.: Polynomial time approximation schemes for the Euclidean traveling salesman and other geometric problems. Journal of the ACM 45, 753–782 (1998)
5. Mitchell, J.S.B.: Guillotine subdivisions approximate polygonal subdivisions: A simple polynomial-time approximation scheme for geometric TSP, k-MST, and related problems. SIAM Journal on Computing 28, 1298–1309 (1999)
6. Rao, S.B., Smith, W.D.: Approximating geometrical graphs via "spanners" and "banyans". In: STOC '98: Proceedings of the 30th annual ACM Symposium on Theory of Computing, pp. 540–550. ACM Press, New York (1998)
7. Provan, J.S.: An approximation scheme for finding steiner trees with obstacles. SIAM Journal on Computing 17, 920–934 (1988)
8. Garey, M., Johnson, D.: The rectilinear Steiner tree problem is $\mathcal{NP}$-complete. SIAM Journal on Applied Mathematics 32, 826–834 (1977)
9. Müller-Hannemann, M., Schulze, A.: Hardness and approximation of octilinear Steiner trees. In: Deng, X., Du, D.-Z. (eds.) ISAAC 2005. LNCS, vol. 3827, pp. 256–265. Springer, Heidelberg (2005)

[5] This is done by utilizing the special structure of the spanners; each spanner is based on a division of the plane into $O(n)$ regions, each containing a grid of constant size; it is shown that each region of each spanner can overlap with at most a constant number of regions of every other spanner.

10. Brazil, M., Thomas, D., Winter, P.: Minimum networks in uniform orientation metrics. SIAM Journal on Computing 30, 1579–1593 (2000)
11. Ganley, J.L., Cohoon, J.P.: Routing a multi-terminal critical net: Steiner tree construction in the presence of obstacles. In: Proceedings of the International Symposium on Circuits and Systems, pp. 113–116 (1994)
12. Müller-Hannemann, M., Schulze, A.: Approximation of octilinear Steiner trees constrained by hard and soft obstacles. In: Arge, L., Freivalds, R. (eds.) SWAT 2006. LNCS, vol. 4059, pp. 242–254. Springer, Heidelberg (2006)
13. Liu, J., Zhao, Y., Shragowitz, E., Karypis, G.: A polynomial time approximation scheme for rectilinear Steiner minimum tree construction in the presence of obstacles. In: 9th IEEE International Conference on Electronics, Circuits and Systems. vol. 2, pp. 781–784 (2002)
14. Müller-Hannemann, M., Peyer, S.: Approximation of rectilinear Steiner trees with length restrictions on obstacles. In: Dehne, F., Sack, J.-R., Smid, M. (eds.) WADS 2003. LNCS, vol. 2748, pp. 207–218. Springer, Heidelberg (2003)
15. Bern, M., Plassmann, P.: The Steiner problem with edge lengths 1 and 2. Information Processing Letters 32, 171–176 (1989)
16. Robins, G., Zelikovsky, A.: Improved Steiner tree approximation in graphs. In: SODA '00: Proceedings of the 11th Annual ACM-SIAM Symposium on Discrete Algorithms, pp. 770–779 (2000)
17. Hanan, M.: On Steiner's problem with rectilinear distance. SIAM Journal on Applied Mathematics 14, 255–265 (1966)
18. Arikati, S.R., Chen, D.Z., Chew, L.P., Das, G., Smid, M.H.M., Zaroliagis, C.D.: Planar spanners and approximate shortest path queries among obstacles in the plane. In: Díaz, J. (ed.) ESA 1996. LNCS, vol. 1136, pp. 514–528. Springer, Heidelberg (1996)
19. Eppstein, D.: Spanning trees and spanners. In: Sack, J.R., Urrutia, J. (eds.) Handbook of Computational Geometry, pp. 425–461. Elsevier, Amsterdam (2000)
20. Clarkson, K.: Approximation algorithms for shortest path motion planning. In: STOC '87: Proceedings of the 19th Annual ACM Symposium on Theory of Computing, pp. 56–65. ACM Press, New York (1987)
21. Zeh, N.: I/O-Efficient Algorithms for Shortest Path Related Problems. PhD thesis, Carleton University, Ottawa, Canada (2002)
22. Gilbert, E.N., Pollak, H.O.: Steiner minimal trees. SIAM Journal on Applied Mathematics 16, 1–29 (1968)
23. Zachariasen, M., Winter, P.: Obstacle-avoiding Euclidean Steiner trees in the plane: An exact approach. In: Goodrich, M.T., McGeoch, C.C. (eds.) ALENEX 1999. LNCS, vol. 1619, pp. 282–295. Springer, Heidelberg (1999)
24. Hwang, F.: On Steiner minimal trees with rectilinear distance. SIAM Journal on Applied Mathematics 30, 104–114 (1976)
25. Prömel, H., Steger, A.: The Steiner Tree Problem: A Tour through Graphs, Algorithms, and Complexity. Advanced Lectures in Mathematics, Vieweg (2002)

A Pseudopolynomial Time $O(\log n)$-Approximation Algorithm for Art Gallery Problems

Ajay Deshpande[1], Taejung Kim[2], Erik D. Demaine[1], and Sanjay E. Sarma[1]

[1] Massachusetts Institute of Technology, Cambridge, MA 02139 USA
{ajayd,edemaine,sesarma}@MIT.EDU
[2] Dankook University, Hanam-Dong, Seoul, 140-714 Korea
taejungkim@dankook.ac.kr

Abstract. In this paper, we give a $O(\log c_{opt})$-approximation algorithm for the point guard problem where c_{opt} is the optimal number of guards. Our algorithm runs in time polynomial in n, the number of walls of the art gallery and the spread Δ, which is defined as the ratio between the longest and shortest pairwise distances. Our algorithm is pseudopolynomial in the sense that it is polynomial in the spread Δ as opposed to polylogarithmic in the spread Δ, which could be exponential in the number of bits required to represent the vertex positions. The special subdivision procedure in our algorithm finds a finite set of potential guard-locations such that the optimal solution to the art gallery problem where guards are restricted to this set is at most $3c_{opt}$. We use a set cover cum VC-dimension based algorithm to solve this restricted problem approximately.

1 Introduction

The art gallery problem addresses the following question [7]: How many guards are required to guard an art gallery with n walls? This problem was first posed by Victor Klee in 1973 [8]. Chvátal showed that $\lfloor \frac{n}{3} \rfloor$ guards are always sufficient and occasionally necessary [10]. Since then, numerous variations of this problem have been studied including mobile guards, guards with limited visibility, guarding rectilinear polygons, etc., see, e.g., [8,9,7]. In this paper, we study one version of the art gallery problem, also known as the *point-guard problem.* The point-guard problem involves finding the minimum number of points and their positions so that guards located at these points *cover* (i.e. *see*) every point in the interior of the art gallery.

Lee and Lin show that the point-guard problem is NP-hard [11]. Eidenbenz, Stamm and Widmayer prove that even finding a $(1+\epsilon)$-approximation for this problem for any $\epsilon > 0$ is *NP-hard* [4]. They also show that the problem of art gallery with holes can not be approximated by a polynomial time algorithm with ratio $(\frac{1-\epsilon}{12}) \ln n$ for any $\epsilon > 0$, unless $NP \subseteq TIME(n^{O(\log \log n)})$. Brodén, Hammar and Nisson prove that the point-guard problem even for a special class of art galleries, which are *2-link polygons*, is *APX-hard* [12].

F. Dehne, J.-R. Sack, and N. Zeh (Eds.): WADS 2007, LNCS 4619, pp. 163–174, 2007.

In [5], Ghosh proposes an $O(\log n)$-approximation algorithm for the minimum *vertex-guard problem* where guards can be be located only at the vertices of the art gallery. González-Banos and Latombe [3] consider another version of the art gallery problem in which guards have range and incidence constraints and are required to cover only the walls of the art gallery. They choose a set of uniformly randomly selected points from the art gallery as potential guard-locations and solve this new problem. They argue that their algorithm computes with high probability a solution whose size is at most a factor $O(\log n \cdot \log (c \log n))$ times the size of the optimal solution c. In [15], Efrat and Har-Peled consider another variant of the art gallery problem where guards are restricted to be placed on the points of a dense grid and propose a randomized algorithm which with high probability yields the approximation ratio within $O(\log c')$, where c' is the optimal solution size for the modified problem. In the same paper [15], Efrat and Har-Peled propose an exact algorithm for the point-guard problem with running time at most $O((nc)^{3(2c+1)})$, where is c is the size of the optimal solution. This is the first known exact solution to the problem, although the running time is exponential in the size of the optimal solution.

Our result. We give a pseudopolynomial time $O(\log c_{opt})$-approximation algorithm for the point-guard problem, where c_{opt} is the size of the optimal solution which can be as large as $\Theta(n)$ in some cases. Our algorithm is pseudopolynomial in the sense that it is polynomial in the number of walls n of the art gallery and the *spread* Δ of the vertices of the art gallery. The spread of a set of points is defined as the ratio of the longest and shortest pairwise distances [13,14]. In the worst case, the spread Δ could possibly be exponential in the number of bits required to represent positions of the vertices of the art gallery. To the best of our knowledge, this is the first pseudopolynomial time algorithm that yields a solution with a guaranteed approximation ratio.

Our basic approach involves using a special subdivision procedure to obtain a finite set of potential guard-locations. We then consider a new problem of choosing the minimum number of guards from this finite set. We devise our algorithm such that the new problem has an optimal solution at most three times the optimal solution to the original point-guard problem. We solve the new problem using a set cover cum VC-dimension-based algorithm. Our overall algorithm can be summarized in the following 3 steps:

- **Step 1:** Generate an initial triangulation of the art gallery based on the *visibility cell decomposition.*
- **Step 2:** Subdivide the initial triangulation such that each triangle in the final triangulation satisfies a special property – the region that is visible to any point in a triangle is always a subset of the region simultaneously visible to the three vertices of the triangle.
- **Step 3:** Formulate the set cover problem and solve it approximately using the VC-dimension-based algorithm of González-Banos and Latombe [3].

2 Basic Terminology

Most of the definitions and notation we present in this section have been borrowed from [1,2]; however, we reformulate some of these and define new ones for our convenience. Most of the notions we describe below are illustrated in Figure 1.

For the sake of simplicity, we consider the case of an art gallery without holes. At the end of the paper, we comment about the case of an art gallery with holes. An art galley without holes can be represented as a simple polygon. Here, we consider the boundary also as a part of the polygon. Let P be a simple polygon with n vertices. Some of these are *reflex* vertices that subtend an angle greater than 180^0 inside P. We say two points in P *see* each other if the line segment between them does not intersect with the exterior of P.

The *visibility polygon* $V(x)$ for any point $x \in P$, is the polygon consisting of all the points in P that are visible from x. Note that some of the edges of $V(x)$ coincide with those of the original polygon P and some are newly introduced as shown in Figure 1(a). A new edge is introduced at a reflex vertex of P that blocks the view of x. We call this reflex vertex a *blocking reflex vertex*. The other end-point of the new edge which lies on the boundary of P is referred to as an *image of x through the blocking reflex vertex*. To remove any ambiguities, we assume that for P and $V(x)$, no two consecutive edges are collinear.

For any point $x \in P$, we say that x *sees* an edge of P, if it sees a point on the edge. If x cannot see either of the end-points of a visible edge of P, we say that x sees the edge *partially*. We call the corresponding edge of P a *partial edge with respect to x*. We say that x *sees* a visible edge of P *non-partially*, if it sees at least one of its end-points. We call the corresponding edge of P a *non-partial edge with respect to x*. If we join every vertex of $V(x)$ to x, we get a triangulation of $V(x)$. We call each triangle as a *visibility sector* of x. The edge of a visibility sector that is a part of an edge of P is referred to as a *base* of the visibility sector. Depending upon the type of the edge of P corresponding to the base of a visibility sector, we classify the visibility sector into *non-partial-edge sector* or *partial-edge sector*.

3 Initial Triangulation Using Visibility Cell Decomposition

In this section, we define a particular subdivision of the polygon – the *visibility cell decomposition*. Then we show how to triangulate this subdivision to generate the initial triangulation in Step 1 of our algorithm.

The *visibility cell decomposition* of P is a subdivision induced by visibility polygons of all the vertices of P. We call each component of the subdivision a *visibility cell*. We state without proofs the following properties of the visibility cell decomposition that are useful both in the construction and analysis of our algorithm. We refer interested readers to the papers by Bose *et al.* [1] and Guibas *et al.* [2] for further details.

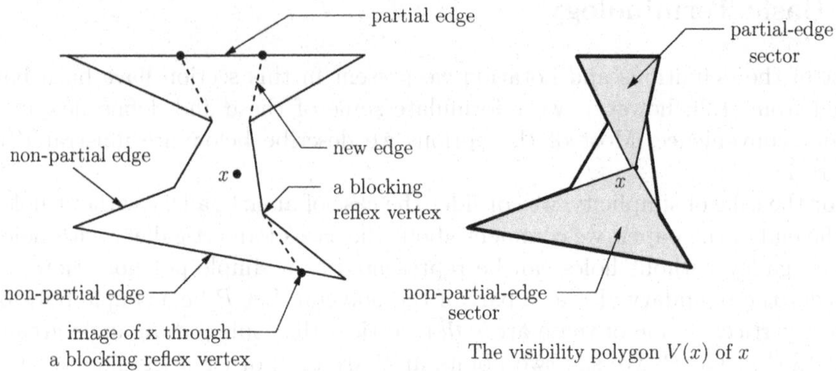

Fig. 1. Visibility polygon and visibility sectors

- Each visibility cell is a convex polygon.
- The total number of visibility cells in the visibility cell decomposition is $O(n^3)$.
- By definition, any two points in a visibility cell see the same set of vertices of P. Furthermore any two points in the same visibility cell see the same set of non-partial edges and the same set of partial edges of P.

Step 1 of our algorithm can be summarized as follows:

Construct the visibility cell decomposition of the polygon. Triangulate each visibility cell simply by joining its one particular vertex to every other vertex.

4 Further Subdivision of the Initial Triangulation

In this section, we describe Step 2 of our algorithm. We give a procedure to subdivide the initial triangulation in such a way that each triangle in the final triangulation satisfies a special visibility property – the region that is visible to any point in a triangle is covered by the visibility polygons of the three vertices of the triangle.

4.1 Vertex-Visibility Property and Vertex-Pair-Visibility Property

We first define the desirable property which each triangle in the final triangulation is required to satisfy.

Definition 1. *Let $\triangle abc$ be a triangle in the polygon P. We say $\triangle abc$ satisfies the* vertex visibility property, *if for any point $x \in \triangle abc$, $V(x) \subseteq V(a) \cup V(b) \cup V(c)$.*

Covering the visibility polygon of a point is equivalent to covering every visibility sector of the point. This motivates the following definition.

Definition 2. *A triangle in a visibility cell satisfies the* vertex-visibility property with respect to a particular edge *of the polygon, if the corresponding visibility sector of any point in the triangle is a subset of the union of the visibility polygons of the vertices of the triangle.*

The vertex-visibility property is not directly useful in the construction of our algorithm. We define a more convenient property.

Definition 3. *A triangle in a visibility cell satisfies the* vertex-pair-visibility property with respect to a particular edge *of the polygon, if the visibility sectors of any two vertices of the triangle overlap on the edge.*

Consider the images of two points in a visibility cell through a blocking reflex vertex on an edge of the polygon. We call the portion of the edge between the two images as a *span* of the two points corresponding to the blocking reflex vertex. Note that the image of a point on the segment joining these two points lies in the span by one-to-one mapping. Now consider the images of the three vertices of a triangle in a visibility cell through a blocking reflex vertex on an edge of the polygon. One of the three images lies between the other two. We call the portion of the edge between the two extreme images as a *span of the triangle* through the blocking reflex vertex.

Lemma 1. *For any point in a triangle in a visibility cell, its image through a blocking reflex vertex always lies in the span of the triangle through the blocking reflex vertex.*

Proof. The image of any point on a segment lies in the span of the two endpoints of the segment corresponding to a blocking vertex. Thus, the image of any point on the perimeter of a triangle lies in the span of the triangle. Now consider any point in the interior of the triangle. The image of this point is same as the image of the point on the perimeter of the triangle where the line segment joining this point, the blocking reflex vertex and its image intersects the perimeter. Hence the image of any point in the triangle is in its span. □

Theorem 1. *A triangle in a visibility cell satisfies the vertex-pair-visibility as well as the vertex-visibility property with respect to a non-partial edge.*

Proof. Let $\triangle abc$ be a triangle in a visibility cell C. Let e be a non-partial edge. As we have already seen, at least one of the end-points of a non-partial edge is visible from any point in a visibility cell. Depending on whether one or both the end-points of a non-partial edge are visible, we make two cases and deal with each case separately.

Case 1: Both the end-points of e are visible from any point in C. In this case, by definition, $\triangle abc$ satisfies the vertex-pair-visibility property. Let u and v be the end-points of e. Consider the convex hull of a, b, c, u and v. Since $\triangle abc$ is on one side of e, line segment uv must be one of the edges of the convex hull. Therefore, the convex hull can also be formed by considering the union of $\triangle abc$ and the visibility sectors of a, b and c. Note that the convex hull is a subset of $V(a) \cup V(b) \cup V(c)$ and the visibility sector of any point $x \in \triangle abc$ is a subset of this convex hull. Therefore, $\triangle abc$ also satisfies the vertex-visibility property with respect to e.

Case 2: In this case, only one end-point of e is visible from any point in C. Let u be the visible end-point. Let r be a blocking reflex vertex. Again by definition $\triangle abc$ satisfies the vertex-pair-visibility property because u is a common visible point. Now, consider any point x in $\triangle abc$. The visibility sector of x with respect to e consists of two triangles, $\triangle xur$ and $\triangle urx'$, where x' is the image of x through r. By similar arguments as in the first case, we can prove that $\triangle xur$ is a subset of $V(a) \cup V(b) \cup V(c)$. By Lemma 1, x' lies in the span of the image of $\triangle abc$ through r. Thus, at least one of a, b or c cover $\triangle rux'$. Therefore, $\triangle abc$ satisfies the vertex-visibility property with respect to e. □

Theorem 2. *If a triangle in a visibility cell satisfies the vertex-pair-visibility property with respect to a partial edge e, then it also satisfies the vertex-visibility property with respect to e.*

Proof. Let $\triangle abc$ be a triangle in a visibility cell C such that it satisfies the vertex-pair-visibility property with respect to the partial edge e. Let r_1 and r_2 be the two blocking reflex vertices. Consider vertices a and b. The visibility sectors of a and b overlap on e. Let a_1 and a_2 be images of a through r_1 and r_2 respectively. Let b_1 and b_2 be images of b through r_1 and r_2 respectively. Since a_1a_2 and b_1b_2 overlap on e, at least one of b_1 and b_2 lies in between a_1 and a_2. Since e is a partial edge, a_1b_1 and a_2b_2 do not overlap on e. In other words, the spans of a and b with respect to r_1 and r_2 do not overlap on e. By extending this argument to the three vertices, a, b and c, the spans of any two vertices with respect to r_1 and r_2 do not overlap. This implies that the spans of $\triangle abc$ also do not overlap on e because if they do, the previous condition of pairwise vertices having non-overlapping spans is violated for at least one pair. The portion of e that is simultaneously visible to a, b and c consists of the spans of $\triangle abc$ through r_1 and r_2 and the patch between the two spans. By Lemma 1, for any point x in $\triangle abc$, the two images of x through r_1 and r_2 lie in the spans of $\triangle abc$ through r_1 and r_2 respectively. Thus, the portion of e that is visible to x is contained in the portion that is visible to a, b and c. Therefore, the visibility sector of x is a subset of $V(a) \cup V(b) \cup V(c)$. □

The theorem we prove below is useful in the analysis of the algorithm. Let *subtriangle* be a triangle that is contained within a triangle.

Theorem 3. *If a triangle in a visibility cell satisfies the vertex-pair-visibility property with respect to a partial edge e, then any subtriangle also satisfies the vertex-pair-visibility property with respect to e.*

Proof. Let $\triangle abc$ be a triangle in a visibility cell C such that it satisfies the vertex-pair-visibility property with respect to the partial edge e. Let r_1 and r_2 be the two blocking reflex vertices. We already proved in the proof of Theorem 2 that the spans of $\triangle abc$ through r_1 and r_2 do not overlap on e because it satisfies the vertex-pair-visibility property. For any two points x and y in $\triangle abc$, the spans of x and y through r_1 and r_2 do not overlap on e because they are contained in the spans of $\triangle abc$ through r_1 and r_2. Therefore, the visibility sectors of x and y overlap on e. Therefore, any $\triangle xyz$ in $\triangle abc$ satisfies the vertex-pair-visibility property. □

The above theorem allows us to further subdivide the visibility cell without affecting already existent vertex-pair visibility property with respect to a partial-edge visibility sector.

4.2 Further Subdivision

In this subsection, we give a procedure to further subdivide the initial triangulation obtained in Step 1 of our algorithm. The subdivision procedure described below generates the final triangulation where every triangle satisfies the vertex-visibility property. This property is required so that we can reduce the art gallery problem to a problem with guaranteed approximation ratio. Using the results of Theorem 1 and Theorem 2, we achieve this by developing a subdivision procedure which is based on a stronger condition, the vertex-pair-visibility property.

First we define a notion that is useful in the description of our algorithm. Let a and b be two points in a visibility cell such that the visibility sectors of a and b do not overlap on a partial edge. Let r_1 and r_2 be the corresponding blocking reflex vertices. Consider the convex hull of a, b, r_1 and r_2. We call a triangle obtained by taking set difference between the convex hull and the union of the visibility sectors of a and b as a *dark triangle* of segment ab. An example of a dark triangle is shown in Figure 2(a).

Step 2 of our algorithm can be summarized as follows.

For every $\triangle abc$ in the initial triangulation obtained in Step 1, repeat the following procedure:

1. Construct a set S of partial edges for which $\triangle abc$ does not satisfy the vertex-pair-visibility property. Repeat the following procedure for every edge $e \in S$:
   (a) Construct a dark triangle of every edge of $\triangle abc$.
   (b) For each dark triangle whose interior is not disjoint with $\triangle abc$, invoke *SUBDIVIDE-DARK-TRIANGLE.*
   (c) Intersect with $\triangle abc$, the subdivisions of all such dark triangles on which the function *SUBDIVIDE-DARK-TRIANGLE* is invoked in the above step to generate a new subdivision of $\triangle abc$.
2. Intersect all the subdivisions of $\triangle abc$ corresponding to every edge $e \in S$ to generate the final subdivision. Triangulate the final subdivision in the similar way as in Step 1 of our algorithm and return the final triangulation of $\triangle abc$.

Function *SUBDIVIDE-DARK-TRIANGLE*:

Input: A dark triangle $\triangle aob$ corresponding to the two blocking reflex vertices r_1 and r_2

Procedure: Let a_1b_1 and a_2b_2 be the two spans of ab through r_1 and r_2 respectively on the partial edge. Construct a line joining the reflex vertex r_2 and the image a_1 of a through r_1 and another line joining the reflex vertex r_1 and the image b_2 of b through r_2. Depending upon whether the two lines intersect inside or outside $\triangle aob$, choose one of the following two steps.

(**Case 1**) The two lines meet outside $\triangle aob$: Return the new subdivision of $\triangle aob$ induced by the two lines (Figure 2(a)). Terminate the function.

(**Case 2**) The two lines meet in $\triangle aob$: Return the new subdivision of $\triangle aob$ without $\triangle a'o'b'$, where o' is the point of intersection of the two lines, and a' and b' are the points of intersection of the two lines with the segment ab. Check if $\triangle a'o'b'$ satisfies the vertex-pair-visibility property. If it does not, invoke *SUBDIVIDE-DARK-TRIANGLE* on $\triangle a'o'b'$. (Figure 2(b))

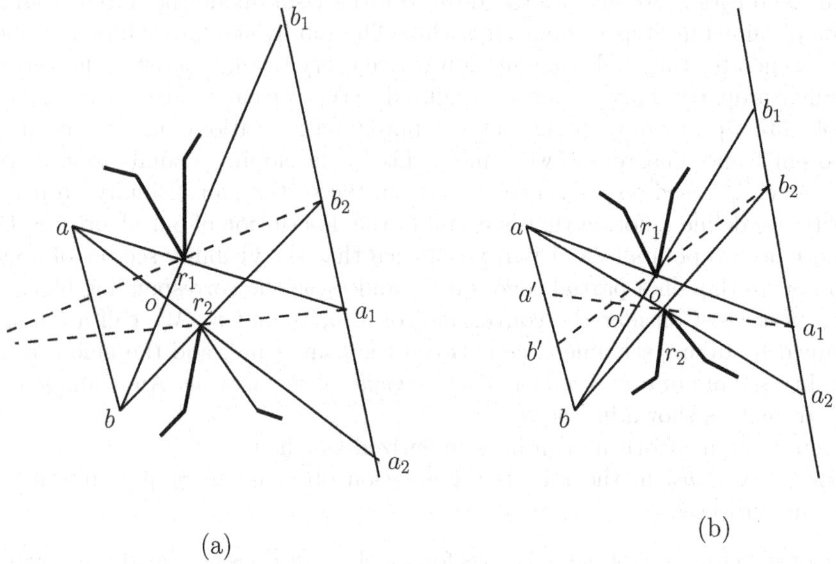

Fig. 2. $\triangle aob$ is a dark triangle. Two cases in *SUBDIVIDE-DARK-TRIANGLE*: (a)lines a_1r_2 and b_2r_1 meet outside $\triangle aob$ (b) lines a_1r_2 and b_2r_1 meet in $\triangle aob$.

As a result of Theorem 1 and Theorem 2, in our subdivision procedure, we need to subdivide a triangle only if it does not satisfy the vertex-pair-visibility property with respect to a partial edge. The result of our subdivision procedure is the final triangulation where every triangle satisfies the vertex-pair visibility property and in turn, the vertex-visibility property. Now we prove this result.

As we have already mentioned, using the results of Theorem 1 and Theorem 2, we check whether a triangle in the initial triangulation satisfies the vertex-pair-visibility with respect to partial edges only. As a result of Theorem 3, the subdivision procedure of a triangle with respect to one edge is 'independent' of the subdivision procedure with respect to another edge. This allows us to subdivide a triangle in the edge-by-edge fashion.

Lemma 2. *Consider the partial-edge visibility sector of a point in a visibility cell. Any triangle that lies in the visibility sector as well as the visibility cell always satisfies the vertex-pair-visibility property.*

Proof. Let x be a point in a visibility cell C. Let r_1 and r_2 be the blocking reflex vertices corresponding to the partial edge. Let x_1 and x_2 be the images of point x through r_1 and r_2 respectively. Any point a that lies in the visibility sectors of x as well as in the same visibility cell C, sees the line segment x_1x_2. Therefore, by definition, any triangle that lies in the visibility sector of x as well as in C satisfies the vertex-pair-visibility property. □

Let $\triangle abc$ be a triangle in the initial triangulation. Suppose that it does not satisfy the vertex-pair-visibility property with respect to a partial edge. Consider the convex hull of a, b, c, r_1 and r_2. The convex hull can also be obtained by taking union of the visibility sectors of a, b and c and the dark triangles of all the edges of $\triangle abc$. By Lemma 2, the portions of $\triangle abc$ that lie in the visibility sector of any of the vertices satisfies the vertex-pair-visibility property. The remaining part of $\triangle abc$ is a subset of the union of the dark triangles. Therefore, in our subdivision procedure in Step 2, we just subdivide the dark triangles.

Now we prove correctness of the function *SUBDIVIDE-DARK-TRIANGLE* with reference to Figure 2

Theorem 4. *In the first case, the subdivision of $\triangle aob$ satisfies the vertex-pair-visibility property.*

Proof. Consider line a_1r_2. It subdivides $\triangle aob$ into two part. a_1 is always visible from any point in one part. Therefore that always satisfies the vertex-pair visibility property. Similarly line b_2r_1 subdivides $\triangle aob$ in two parts out of which one part always satisfies the vertex-pair-visibility property because b_2 is the common visible point from that part. In the first case lines a_1r_2 and b_2r_1 meet outside $\triangle abc$. Both the parts of mentioned above that satisfy the vertex-pair-visibility property cover $\triangle aob$ in the first case. Therefore, the subdivision of $\triangle aob$ satisfies the vertex-pair-visibility property. □

Theorem 5. *In the second case, the subdivision of $\triangle aob$ except $\triangle a'o'b'$ satisfies the vertex-pair-visibility property.*

The proof of the above theorem is similar to the proof of Theorem 4. $\triangle a'o'b'$ may not satisfy the vertex-pair-visibility property. In that case, we subdivide $\triangle a'o'b'$ by again invoking the function *SUBDIVIDE-DARK-TRIANGLE.* The first case is the termination case for the recursion in *SUBDIVIDE-DARK-TRIANGLE.* In the next section, we show that *SUBDIVIDE-DARK-TRIANGLE* indeed terminates. Thus, subdivision generated by *SUBDIVIDE-DARK-TRIANGLE* always satisfies the vertex-pair-visibility property.

The function *SUBDIVIDE-DARK-TRIANGLE* in the subdivision procedure described above is recursive. Here, we address the question after how many steps this recursion ends. We define *spread* Δ of the vertices of the art gallery as the ratio of the longest and shortest pairwise distances [13,14]. Now we prove the following theorem.

Theorem 6. *The recursive function SUBDIVIDE-DARK-TRIANGLE ends in $O(\Delta)$ steps.*

Proof. Let L be the longest and let ϵ be the shortest pairwise distances among the vertices of the art gallery. Thus, $\Delta = L/\epsilon$. The length of each subdivision of the partial edge at the end of the recursive procedure is at most ϵ. Since the length of any partial edge can be at most L, the total number of subdivisions does not exceed Δ. □

5 Set Cover Formulation and Approximate Solution

In this section, we describe Step 3 of our algorithm. We choose all the vertices of the final triangulation obtained in Step 2 as the potential guard-locations and formulate the set cover problem. The set cover problem is then solved approximately using a VC-dimension-based algorithm.

Step 3 of our algorithm can be summarized in the following way:

1. Construct a set G consisting of all the vertices of the final triangulation obtained in Step 2 of our algorithm. Let $|G| = m$.
2. Construct the visibility polygon for every $g_i \in G$ and generate the new subdivision of the polygon. Enumerate all the cells in the new subdivision and group them in the set $X = \{1, 2, .., l\}$. For each $g_i \in G$, construct a set R_i of cells visible from g_i, that is, $R_i = \{x \in X | x \in V(g_i)\}$. Build the set family, $R = \{R_1, R_2 ..., R_m\}$. Group X and R together to form the set system (X, R).
3. Invoke the function *SET-COVER* on the set system (X, R) to obtain a near-optimal covering of X from the set family R.

The function *SET-COVER* used in the above procedure is based on the algorithm proposed by Brönnimann and Goodrich [6] for finding set covers for set systems with finite VC-dimension. Here, we do not give details of the function *SET-COVER*. Instead, we refer interested readers to [3] for further details.

6 Analysis of the Algorithm

In this section, we analyze the bound on the approximation ratio and running time of our algorithm.

6.1 Bound on the Approximation Ratio of Our Algorithm

Consider the set system (X, R) that we construct in Step 3 of our algorithm. Let T_x, where $x \in X$, be a set consisting of all the sets in R that contain x. We define the *dual set system* (Y, S) of (X, R) by setting $Y = R$ and $S = \{T_x | x \in X\}$ [6,3]. Y corresponds to the set of candidate locations for guards. An element in S corresponds to a cell and is a set of candidate guard-locations that are visible from every point in the cell. We can also write this dual set system as $(G, \{G \cap V(x) \mid x \in P\})$, where G consists of the set of all candidate guard-locations. Valtr showed that the VC-dimension of the more general set system $(P, \{P \cap V(x) \mid x \in P\})$ is bounded by 23 [16][1]. Using the definition of the VC-dimension it is easy to prove that the VC-dimension of the dual set system (Y, S) is also bounded by 23.

The result from [6] implies that it is possible to compute an approximate solution to the set cover problem with the approximation ratio $O(d \log(dc))$, where d is the VC-dimension and c is the size of the optimal solution. The constant bound on the VC-dimension in this case implies that we obtain $O(\log c_{opt})$-approximate solution, where c_{opt} is the size of the optimal solution. c_{opt} can be as large as $\Theta(n)$ in some cases.

6.2 Analysis of the Running Time of the Algorithm

Theorem 7. *The running time of our algorithm is polynomial in the number of walls, n and the spread Δ of the vertices of the art gallery.*

Proof. Here we provide only the sketch of the proof. In Step 1, the initial triangulation can be generated in $O(n^4)$ time and consists of $O(n^4)$ triangles [1,2]. In Step 2, for each triangle in the initial triangulation, we can check in $O(n)$ time whether it satisfies the vertex-pair visibility property. In the worst case, the recursive subdivision procedure for each triangle with respect to a partial edge may run in $O(\Delta)$ time as shown in Theorem 6 and may generate $O(\Delta)$ line segments to form the subdivision. This ensures that the number of triangles in the final subdivision is polynomial in n and Δ. In Step 3, *SET-COVER* runs in $O(|X|)$ time [6,3], where $|X|$ is the total number of cells. □

Δ can be at most exponential in the input size. Thus, our algorithm runs in pseudopolynomial time.

6.3 Art Gallery with Holes

When the art gallery has holes, our algorithm can still be used. Guibas et al. [2] extend the visibility cell decomposition to a polygon with holes; except that in this case, the vertices of the holes also act as the blocking vertices. The subdivision procedure of our algorithm is still valid in this case. Valtr prove that for this case of an art gallery with holes the VC-dimension is bounded by $O(\log h)$, where h is the number of holes [16]. Thus, in this case our algorithm yields a solution with the approximation ratio $O(\log h \cdot \log(c_{opt} \log h))$.

---

[1] In the earlier draft of this paper, we had used $O(\log n)$ bound on the VC-dimension. Csaba Toth pointed us to the constant VC-dimension bound in [16].

7 Conclusions

In this paper, we have presented a pseudopolynomial time algorithm for the point guard problem with guaranteed $O(\log n)$ approximation ratio. The imminent question is whether we can improve the running time of our algorithm. An interesting topic for future research is to investigate whether our subdivision procedure can be applied to other variants of the art gallery problems, particularly for the case when guards have limited range.

Acknowledgments

We thank Csaba Toth for pointing to us a better VC-dimension bound (see Footnote 1). We thank the anonymous referees for their useful comments and bringing to our attention the reference [15]. The first author would like to thank Prahladh Harsha for many useful discussions and feedback on the earlier drafts of the paper.

References

1. Bose, P., Lubiw, A., Munro, J.I.: Efficient visibility queries in simple polygons. In: Proc. 4th Canad. Conf. Comput. Geom. pp. 23–28 (1992)
2. Guibas, L.J., Motwani, R., Raghavan, P.: The robot localization problem. SIAM J. Comput. 26(4), 1120–1138 (1997)
3. González-Banos, H., Latombe, J.: A randomized art-gallery algorithm for sensor placement. In: Proc. 17th Symp. Comput. Goem, pp. 232–240 (2001)
4. Eidenbenz, S., Stamm, C., Widmayer, P.: Inapproximability Results for Guarding Polygons and Terrains. Algorithmica 31, 79–113 (2001)
5. Ghosh, S.: Approximation algorithm for art gallery problems. In: Canad. Information Processing Soc. Congress (1987)
6. Brönnimann, H., Goodrich, M.T.: Almost optimal set covers in finite VC-dimension. In: Proc. 10th Symp. Comp. Geom. pp. 293–302 (1994)
7. Urrutia, J.: Art Gallery and Illumination Problems. In: Sack, J. R., Urrutia, J. (eds.) Handbook of Computational Geometry (2000)
8. O'Rourke, J.: Art Gallery Theorems and Algorithms (1987)
9. Shermer, T.: Recent results in art galleries. Proc. IEEE. 80, 1384–1399 (1992)
10. Chvátal, V.: A combinatorial theorem in plane geometry. J. Combinat. Theory B 18, 39–41 (1975)
11. Lee, D.T., Lin, A.K.: Computational complexity of art gallery problems. IEEE Trans. Info. Theory. IT-32, 276–282 (1986)
12. Brodén, B., Hammar, M., Nilsson, B.J.: Guarding lines and 2-link polygons is APX-hard. In: Proc. 13th Canad. Conf. Comp. Geom. pp. 45–48 (2001)
13. Erickson, J.: Nice point sets can have nasty Delaunay triangulations. In: Proc. 17th Symp. Comp. Geom. pp. 96–105 (2001)
14. Erickson, J.: Dense point sets have sparse Delaunay triangulations: or "...but not too nasty". In: Proc. 13th Symp. Disc. Algo. pp. 125–134 (2002)
15. Efrat, A., Har-Peled, S.: Guarding galleries and terrains. Info. Proc. Lett. 100(6), 238–245 (2006)
16. Valtr, P.: Guarding galleries where no point sees a small area. Israel J. Math. 104, 1–16 (1998)

Optimization for First Order Delaunay Triangulations*

Marc van Kreveld, Maarten Löffler, and Rodrigo I. Silveira

Department of Information and Computing Sciences,
Utrecht University, 3508TB Utrecht, The Netherlands
{marc,loffler,rodrigo}@cs.uu.nl

Abstract. This paper discusses optimization of quality measures over first order Delaunay triangulations. Unlike most previous work, our measures relate to edge-adjacent or vertex-adjacent triangles instead of only to single triangles. We give efficient algorithms to optimize certain measures, whereas other measures are shown to be NP-hard. For two of the NP-hard maximization problems we provide for any constant $\varepsilon > 0$, factor $(1-\varepsilon)$ approximation algorithms that run in $2^{O(1/\varepsilon)} \cdot n$ and $2^{O(1/\varepsilon^2)} \cdot n$ time (when the Delaunay triangulation is given). For a third NP-hard problem the NP-hardness proof provides an inapproximability result. Our results are presented for the class of first-order Delaunay triangulations, but also apply to triangulations where every triangle has at most one flippable edge. One of the approximation results is also extended to k-th order Delaunay triangulations.

1 Introduction

Triangulation is a well-studied topic in computational geometry. The input is a point set or planar straight line graph in the plane, and the objective is to generate a subdivision where all faces are triangles, except for the outer face. In some cases extra points are allowed, in which case we speak of a Steiner triangulation. Since a point set (or planar straight line graph) allows many different triangulations, one can try to compute one that optimizes a criterion. For example, one could maximize the minimum angle used in any triangle, or minimize the total edge length (minimum weight triangulation). The former optimization is solved with the Delaunay triangulation in $O(n \log n)$ time for n points. The latter optimization is NP-hard [18].

Several other optimization measures exist. In finite element methods, triangular meshes with various quality constraints are used, and Steiner points may be used to achieve this; Bern and Plassmann [4] give a survey. Other optimization measures arise if the triangulation represents a terrain (called a polyhedral terrain in computational geometry): all vertices have a specified height, and the

* This research has been partially funded by the Netherlands Organisation for Scientific Research (NWO) under FOCUS/BRICKS grant number 642.065.503 (GADGET) and under the project GOGO.

F. Dehne, J.-R. Sack, and N. Zeh (Eds.): WADS 2007, LNCS 4619, pp. 175–187, 2007.

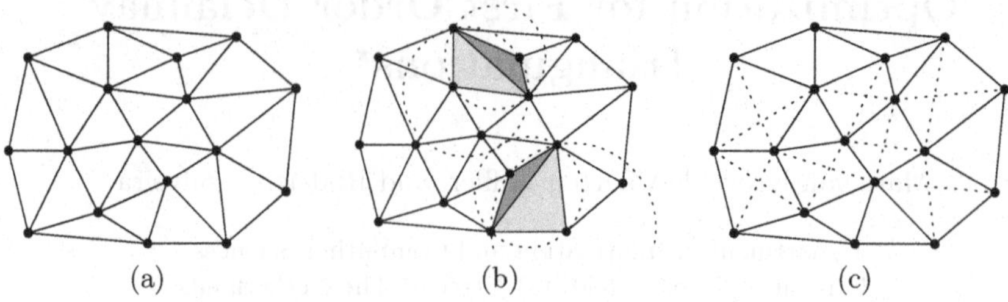

Fig. 1. (a) Delaunay triangulation (zero-th order). (b) Second order Delaunay triangulation (light grey triangles are first order, darker triangles are second order). (c) For first order Delaunay triangulations, one of every pair of dotted edges must be chosen.

height of points on edges and on triangles is obtained by linear interpolation. Such a terrain representation is common in GIS and is called a TIN [6,22].

Bern et al. [3] show that measures like maxmin triangle height, minmax slope, and minmax eccentricity of any triangle can be optimized with a technique called *edge insertion* in $O(n^3)$ or $O(n^2 \log n)$ time. Other measures such as minmax angle [9] and minmax edge length can also be optimized in polynomial time [8]. Interestingly, the Delaunay triangulation optimizes several measures simultaneously: maxmin angle, minmax circumscribed circle, minmax enclosing circle, and minimum integral of the gradient squared (e.g. [3]). For terrain modeling in GIS, Steiner points cannot be used because their elevation would not be known. Terrain modeling leads to a number of optimization criteria, both to yield good rendering of the terrain for visualization, and to make it suitable for modeling processes like water runoff and erosion [13,17]. Slope characteristics are especially important. Furthermore, local minima and artificial dams, which may be artifacts due to the creation of the triangulation, should be avoided [7,14,21].

The *Delaunay triangulation* of a set P of points is defined as the triangulation where all vertices are points of P and the circumcircle of the three vertices of any triangle does not contain any other point of P. If no four points of P are cocircular, then the Delaunay triangulation is uniquely defined. Gudmundsson et al. [10] generalize this to *higher order Delaunay triangulations.* A triangulation is *k-th order Delaunay* if the circumcircle of the three vertices of any triangle contains at most k other points (see Figure 1). For higher order Delaunay triangulations, fewer results are known. Minimizing local minima in a terrain becomes NP-hard for orders higher than n^ε, for constant ε. Experiments showed that low order Delaunay triangulations can reduce the number of local minima significantly [7].

First order Delaunay triangulations have a special structure. All edges that are in any first order Delaunay triangulation form a subdivision that only has triangles and convex quadrilaterals (see Figure 1(c)). In the quadrilaterals, both diagonals are possible to obtain a first order Delaunay triangulation. We call these quadrilaterals and diagonals *flippable.* Due to this structure, measures like the number of local minima or extrema can be minimized in $O(n \log n)$ time.

Table 1. Optimization problems and complexity results for first order Delaunay triangulations. d is the maximum vertex degree in the Delaunay triangulation.

Triangles incident to	Opt. worst local measure (minmax)	Result	Opt. # occurrences	Result
edge	area ratio	$O(n \log n)$	max #convex edges	NP-hard
	angle of outward normals	$O(n \log n)$		
vertex	area ratio	$O(nd \log n)$	max #convex vertices	$O(n \log n)$
	angle of outward normals	$O(nd \log n)$	min #local minima	$O(n \log n)$ [10]
	vertex degree	NP-hard	min #mixed vertices	NP-hard

The same holds for minimizing the maximum area triangle, minimizing the total edge length, and various other measures [10]. On the other hand, minimizing the maximum vertex degree was only approximated by a factor of roughly 3/2.

Many of the measures mentioned above are measures for single triangles. Exceptions are total edge length, number of local minima or extrema, and maximum vertex degree. In this paper, we consider measures that depend on pairs of triangles that are edge-adjacent, and measures that depend on groups of triangles that are vertex-adjacent. Note that a single flip in a first order Delaunay triangulation influences five pairs of edge-adjacent triangles and four vertex-adjacent groups. We consider objectives of the maxmin or minmax type, and objectives where the number of undesirable situations must be minimized. Examples of minmax objectives for edge-adjacent triangles include minimizing the maximum ratio of edge-adjacent triangle areas, which is relevant for numerical methods on meshes, or minimizing the maximum spatial angle of the normals of edge-adjacent triangles in polyhedral terrains, which is important for flow modeling. Geomorphologists classify parts of mountains or hills as footslopes, hillslopes, valley heads, etc. [13]. If we know that a part of a terrain is a valley head, we should maximize the number of convex edges or convex vertices in that part. A vertex of a polyhedral terrain is *convex* if there is a plane through that vertex such that all of its neighbors are on or below that plane, and at least one strictly below. A vertex is *mixed* if every plane containing it has neighbors strictly above and below the plane. We study maximization of convex edges, maximization of convex vertices, and minimization of mixed vertices.

Given a planar point set P with or without elevation, we study the complexity of optimizing measures over all first order Delaunay triangulations. Measures we consider and results are shown classified in Table 1. The optimization of other worst local measures for edge-adjacent triangles can also be solved in $O(n \log n)$ time with the same technique, like minimizing the largest minimum enclosing circle of any two edge-adjacent triangles. Our proof of NP-hardness of minimizing the maximum vertex degree justifies the factor 3/2 approximation algorithm given before in [10]. The proof yields inapproximability beyond a constant greater than 1 in polynomial time unless P=NP. It was already known that triangulating a biconnected planar graph while minimizing the maximum degree is NP-hard [15]. The NP-hard problems of maximizing convex edges and maximizing non-mixed vertices can be approximated within a factor $1 - \varepsilon$ in

$2^{O(1/\varepsilon)} \cdot n$ and $2^{O(1/\varepsilon^2)} \cdot n$ time, if the Delaunay triangulation is given. The NP-hardness results show that, despite the simple structure of first order Delaunay triangulations, optimization of various measures is hard.

2 Exact Algorithms

We start this section with a problem that turns out to be surprisingly easy to solve, namely, maximizing the number of convex vertices over all possible first order Delaunay triangulations. Let P be a set of n points in the plane, where each point has a height value. As observed before, if we take the Delaunay triangulation T of P, it has a number of edges that are in any first order Delaunay triangulation, and a number of flippable edges, and no two flippable edges bound the same Delaunay triangle [10]. The Delaunay triangulation and its flippable edges can be determined in $O(n \log n)$ time.

For any flippable quadrilateral, one diagonal is reflex and the other diagonal is convex in 3-dimensional space, unless the four vertices of the quadrilateral are co-planar. Consider a convex vertex v in T. If it is incident to a flippable quadrilateral where the convex diagonal is present, then v will remain convex if we use the reflex diagonal instead (regardless of which diagonal is incident to v). In other words: using only reflex edges in flippable quadrilaterals does not cause any vertex to become non-convex. At the same time, it may turn non-convex vertices into convex ones. It follows that the maximization problem on the given point set P can be solved in $O(n \log n)$ time.

2.1 Measures on Edge-Adjacent Triangles

In this section we show how to optimize a measure function M defined for a triangulation T, over all first order Delaunay triangulations of P. The function M should be of the shape $M(T) = \max_{q \in T} \mu(q)$ for q a (not necessarily flippable) quadrilateral, and we wish to minimize $M(T)$ over all first order Delaunay triangulations T. We also use $\mu(e)$ for any edge e in a triangulation to denote $\mu(q)$, where e is the diagonal of q. A first order Delaunay triangulation has four types of edges: between two fixed triangles, between a fixed triangle and a flippable quadrilateral, between two flippable quadrilaterals, and flippable edges. As a consequence, there are only $O(n)$ possible values for $M(T)$, and we can determine and sort them in $O(n \log n)$ time.

We solve the min $M(T)$ problem by transforming it into a series of 2-SAT instances. We will use 2-SAT to answer the following question: Is there a first order Delaunay triangulation T such that $M(T) \leq \mu_0$? Since there are $O(n)$ interesting values for μ_0, we can apply binary search to find the smallest one.

Let S be the subdivision that is the Delaunay triangulation of P with all flippable edges removed, and let μ_0 be given. For every edge e of S between a triangle and a quadrilateral, decide which of the two diagonals of the quadrilateral has $\mu(e) > \mu_0$. If neither does, then we can answer the question immediately with "no". If only one diagonal has $\mu(e) > \mu_0$, then we fix the other diagonal in

S. Otherwise, we continue with the next edge between a triangle and a quadrilateral. This step may have made flippable quadrilaterals into two fixed triangles in S. Next we test the possible diagonals of each quadrilateral of S. If both diagonals give $\mu(.) > \mu_0$, then we answer with "no" again. If only one diagonal gives $\mu(.) > \mu_0$, then we fix the other diagonal to make two new triangles in S. Next we test all edges of S between adjacent triangles. If any such edge does not satisfy $\mu(e) > \mu_0$, then we answer the question with "no" again.

It remains to solve the problem for edges between quadrilaterals of S. For every quadrilateral q we introduce a Boolean variable x_q, and let one diagonal choice represent TRUE and the other FALSE. Let e be an edge of S between two quadrilaterals q and r. For each choice of diagonals in q and r that gives $\mu(e) > \mu_0$, for example the one with TRUE in q and FALSE in r, we make a clause $(\neg x_q \vee x_r)$. We get at most four clauses for any edge between two quadrilaterals, so $O(n)$ clauses overall. The conjunction of all clauses is a 2-SAT instance, which we can solve in linear time with the algorithm of Aspvall et al. [1]. The binary search must try $O(\log n)$ values for μ_0 until we find the one minimizing $M(T)$. Hence, the whole algorithm takes $O(n \log n)$ time.

2.2 Measures on Vertex-Adjacent Triangles

The algorithm described in the previous section can easily be extended to minimize measure functions of the form $M(T) = \max_{t,t' \in T} \mu(t, t')$ for t and t' triangles in T with a common vertex. The set of possible values of $M(T)$ induced by pairs of triangles incident to a vertex v is $\binom{d(v)}{2}$, where $d(v)$ denotes the degree of v. Since the sum of the degrees of all vertices is $O(n)$, the total number of possible values of $M(T)$ is at most $\sum_{v \in T} d(v)^2 = d \cdot \sum_{v \in T} d(v) = O(dn)$, where d is the maximum degree of any vertex in the triangulation.

Theorem 1. *A first order Delaunay triangulation that minimizes the maximum area ratio of edge adjacent triangles can be computed in $O(n \log n)$ time. If the triangulation represents a polyhedral terrain, the same result holds for minimizing the maximum angle of outward normals. If we consider these ratio measures over pairs of vertex adjacent triangles, the algorithms take $O(nd \log n)$ time, where d is the maximum vertex degree in the Delaunay triangulation.*

3 NP-Hardness Results

We show NP-hardness for three different optimization problems on first order Delaunay triangulations. The proof for the first problem, minimization of the number of mixed vertices in a terrain, is treated in detail. The other two NP-hardness results are only stated; the proofs can be found in the full paper [23].

3.1 Mixed Vertices

In a terrain, we call a vertex *mixed* if every plane through it has neighboring vertices above and below the plane. In some types of terrains, such vertices are

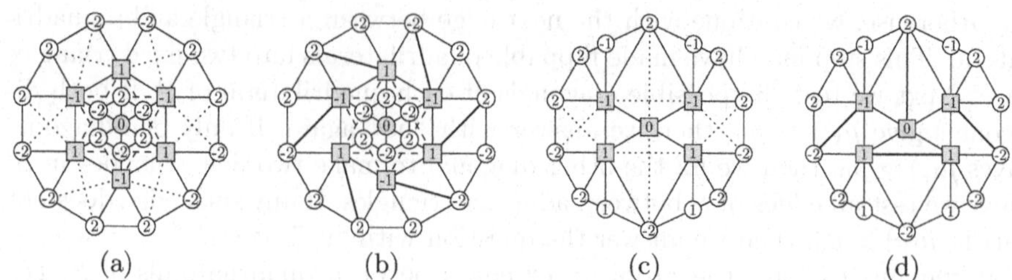

Fig. 2. (a) A fan. (b) One of the solutions: a left-turning fan. Similarly a right-turning fan is possible. (c) An inverter gadget. (d) One of the two solutions.

uncommon, so we may want to minimize their number. Given a set of points with height information, we study the problem of constructing a first order Delaunay triangulation of this point set such that the number of mixed vertices is minimal. This problem is NP-hard. We prove this by reduction from planar 3-SAT [16].

We represent the variables of a 3-SAT instance by *fan gadgets*, see Figure 2(a). A fan gadget consists of 25 points with elevations. In the figure, all possible first order Delaunay edges are shown. Solid edges are in every first order Delaunay triangulation; dashed and dotted edges are flippable. The square nodes and the dotted edges are the most important part. We observe that a square vertex is mixed if and only if both incident dotted edges are in the triangulation.

We construct the gadget in such a way that the state of the round vertices does not depend on any of the dotted edges. The white round vertices are always non-mixed, even if all incident edges would be in the triangulation; the grey round vertices are always mixed, already if only the fixed edges are in the triangulation. Hence the number of mixed vertices is only affected by square vertices, and can

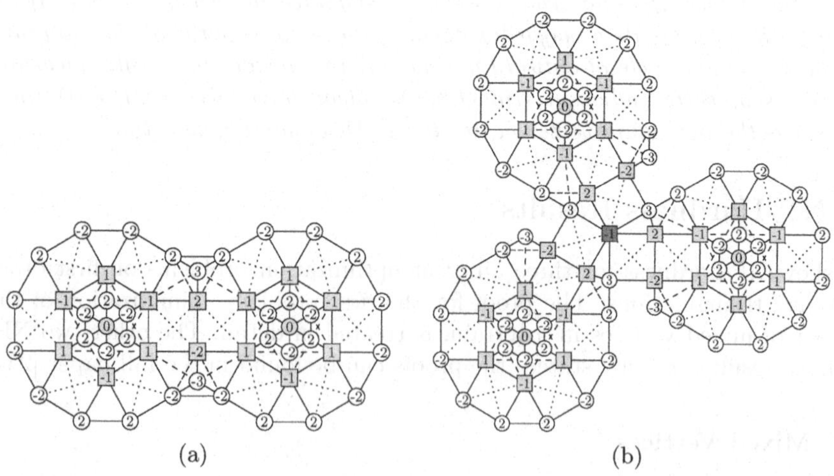

Fig. 3. (a) Connecting variables. (b) Three variables come together in a clause.

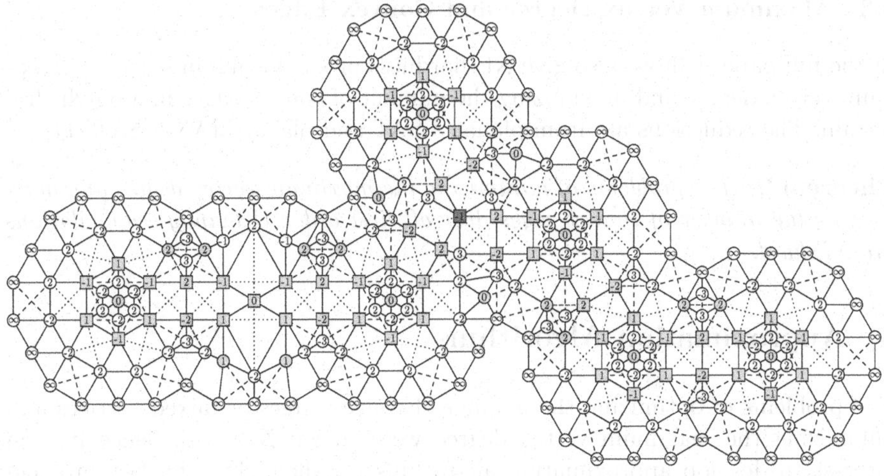

Fig. 4. A coating to shield the construction from the outside

only be minimal if there are never two dotted edges at the same square vertex. A fan gadget therefore has two possible states, see Figure 2(b).

We can connect fans together to form larger chains that are all in the same state, see Figure 3(a). We turn two more vertices into squares, and if the left fan is left-turning, the right fan must also be left-turning and the other way around. We can connect up to three fans to an existing fan, so chains can also split. We also make negations in chains using the inverter gadget in Figure 2(c). Here, if the leftmost square has its positive sloping diagonal in the triangulation, the rightmost square must have its negative sloping diagonal and the other way around, see Figure 2(d). We use an inverter gadget in a chain to negate a variable.

We represent the clauses occurring in the 3-SAT instance by a special clause vertex, see Figure 3(b). Here three fan chains come together at one square vertex in a darker shade of grey. This vertex has a slightly different property than the other square vertices. A clause vertex is mixed if and only if all three incident dotted edges are in the triangulation.

So, the clause can be satisfied if at least one of the three fans is *not* right-turning, and by including inverters at the appropriate places this can represent any Boolean clause of a 3-SAT formula. With these gadgets we can build the whole planar 3-SAT instance. Finally, we need to triangulate the remaining gaps, so we need to ensure that the vertices on the boundary really have a fixed value. We add an extra layer of sufficiently high vertices (labeled ∞ in Figure 4). These vertices are all non-mixed, and the properties of the vertices that are not on the boundary can be checked locally.

Theorem 2. *Minimizing the number of mixed vertices over all first order Delaunay triangulations is NP-hard.*

3.2 Maximum Vertex Degree and Convex Edges

In the full paper [23] we also give NP-hardness proofs for minimizing the maximum vertex degree and maximizing the number of convex edges in a polyhedral terrain. The reductions are from planar 3-SAT and planar MAX-2-SAT [11].

Theorem 3. *The problems of minimizing the maximum vertex degree and maximizing the number of convex edges over all first order Delaunay triangulations are NP-hard.*

4 Approximation Algorithms

The problems of optimizing the number of convex edges or mixed vertices and minimizing the maximum vertex degree were shown NP-hard; hence it is of interest to develop approximation algorithms for them. For the last problem there is already a 1.5-approximation [10], and our NP-hardness proof shows that no polynomial time approximation scheme exists unless P=NP. For the other two problems we present polynomial time approximation schemes. We also sketch an extension to k-th order Delaunay triangulations for maximizing convex edges.

The general idea is as follows. First we transform the problem into a graph problem on some planar graph that can be obtained from the Delaunay triangulation without flippable edges. The resulting graph is partitioned into layers of outerplanarity at most λ. For each choice of i, where $0 \leq i < \lambda$, we delete every $(j\lambda + i)$-th layer of vertices, where $j = 0, 1, 2, \ldots$. The resulting "thick" layers are independent. For each thick layer, we compute a tree decomposition of width at most $3\lambda - 1$ and solve the problem optimally on this decomposition in $2^{O(\lambda)}n$ time, using dynamic programming. Finally, the union of the solutions of all the thick layers for a given i yields a solution to the original problem. We simply choose i such that the size of the solution is the maximal, and return the corresponding triangulation as output. Such an approach gives a $(1 - \varepsilon)$-approximation if λ is chosen suitably, depending on the problem and ε [2,12].

4.1 Maximizing the Number of Convex Edges

We build a graph G that has a vertex (called q-vertex) for each flippable quadrilateral, and an edge between two q-vertices if and only if their corresponding quadrilaterals share an edge. The rest of the input (all the fixed triangles) are not explicitly represented, see Figure 5(b). Each q-vertex has two possible states, *convex* or *reflex*, depending on the choice of the diagonal. It also has a value that depends on its state and represents the number of convex edges among the flippable edge and any edges that the quadrilateral shares with fixed triangles when the q-vertex is in that state (from 0 to 5). Furthermore, every edge in G has a value that depends on the states of both incident q-vertices. The goal of the algorithm is to find a state for each q-vertex such that the sum of the values (total number of convex edges) is maximized.

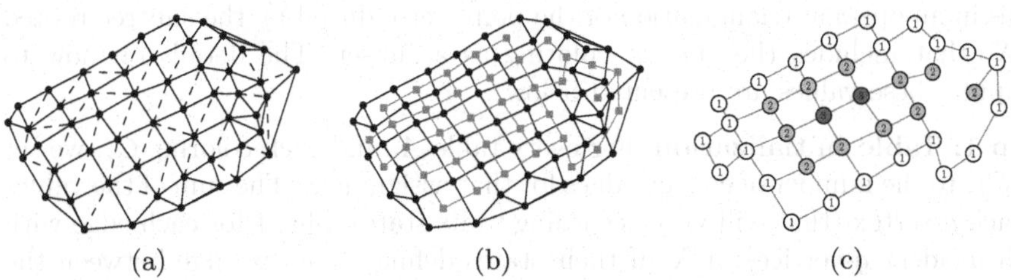

Fig. 5. (a) Initial triangulation (solid edges are fixed). (b) Graph (in gray) where each vertex represents a flippable quadrilateral. (c) The same graph showing the outerplanarity layers.

To create the independent thick layers from the graph we will remove the edges that connect two consecutive layers $j\lambda + i$ and $j\lambda + i + 1$ in G, where $j = 0, 1, 2, \ldots$, for all choices of $0 \leq i < \lambda$. The layers created after removing one set of layers of edges are independent, so if we optimize them separately and then join them by adding the removed edges, the number of convex edges after the join cannot decrease. Some edges are not considered for every i, but only in $\lambda - 1$ out of λ solutions. We get a $(1 - \varepsilon)$-approximation algorithm by taking $\lambda = \lceil \frac{1}{\varepsilon} \rceil$, due to the pigeonhole principle [2,12].

Once we have the thick layers, each layer is solved optimally by using a tree decomposition approach. Since each layer is a λ-outerplanar graph, a tree decomposition with treewidth at most $3\lambda - 1$ can be computed in time linear in the number of nodes of the graph [5]. Once we have this decomposition we can apply one of the standard techniques to deal with problems on graphs of small treewidth. The technique consists of building tables of partial solutions in the nodes of the tree decomposition [5,19].

Definition 1. *(from [19], originally in [20]) Let $G = (V, E)$ be a graph. A* tree decomposition *of G is a pair $\langle \{X_i | i \in I\}, T \rangle$ where each X_i is a subset of V, called a* bag, *and T is a tree with the elements of I as nodes. The following three properties must hold:*

1. *$\bigcup_{i \in I} X_i = V$;*
2. *for every edge $\{u, v\} \in E$, there is an $i \in I$ such that $\{u, v\} \subseteq X_i$; and*
3. *for all $i, j, k \in I$, if j is on the path between i and k in T then $X_i \cap X_k \subseteq X_j$.*

The width *of $\langle \{X_i | i \in I\}, T \rangle$ equals* $\max\{|X_i| \mid i \in I\} - 1$. *The* treewidth *of G is the minimum ω such that G has a tree decomposition of width ω.*

We will make T rooted by choosing any node to be the root. For each bag X_i, we will store a table A_i ($i \in I$). Tables will be created in a bottom up fashion as follows. For each bag X_i, the table A_i has 2^{n_i} rows and $n_i + 1$ columns, where $n_i = |X_i|$. Each row represents a *coloring*, which is an assignment of a state (*reflex*/*convex*) to each q-vertex (flippable quadrilateral) in X_i. All the different possible colorings for the bag are represented in the table. Furthermore, for each coloring C_j an extra value $m_i(C_j)$ is stored, containing the number of convex edges in an optimal triangulation of the point set induced by the subtree rooted at X_i that includes the current coloring as a subset. The details on how to compute these values are presented below.

Step 1: Table initialization. For every table A_i and each coloring C_j, we set $m_i(C_j)$ to the number of convex edges for that assignment: The sum of the values of each q-vertex (that will vary according to its state), plus 1 for each edge with both incident q-vertices in X_i if their states define a convex edge between the corresponding quadrilaterals (with diagonals chosen).

Step 2: Table update. Next the tree is traversed, starting from the leaves, finishing at the root. For each node, the column m_i of A_i is updated based on its children. Let i be the parent of node j. Bags X_i and X_j have some q-vertices in common. We sort both tables first by the columns of the shared q-vertices, and second by m_i. Then we scan A_i row by row, and for each coloring C_l we update $m_i(C_l)$ based on the highest value that $m_j()$ has for that combination of the shared q-vertices. For later reconstruction of the triangulation we also store a pointer to the corresponding row in A_j. When a node X_i has several children, we update A_i against each child, one at a time, in the same way. Once the root node is updated, the number of convex edges in an optimal triangulation will be in the last column of one of the rows of its table. The final triangulation can be computed by following the pointers in the tables.

The correctness of the method follows from the definition and properties of tree decompositions, and the arguments are identical to the ones that hold for other well-known problems where the same technique has been used, such as vertex cover or dominating set (see [19]).

The running time is dominated by the computation and merging of the tables. The sorting of each table can be done in time $O(2^\omega \omega)$ (because all but one column have only two states). The time for updating a table based on another one is linear in the size of the largest one, so $O(2^\omega)$. The number of tables is linear in the number of nodes $|I|$ of tree T, hence the total running time is $O(2^\omega \omega \cdot |I|)$. Since the graph is λ-outerplanar we can compute a tree decomposition of width $\omega \leq 3\lambda - 1$ and $|I| = O(n)$ nodes [5,19]. We apply this algorithm to the λ different values of i to get an approximation scheme, so the worst-case running time is $O(\lambda 2^\omega \omega \cdot |I|) = O(\lambda^2 8^\lambda \cdot n) = O(\frac{1}{\varepsilon^2} 8^{\frac{1}{\varepsilon}} \cdot n) = 2^{O(1/\varepsilon)} \cdot n$.

Theorem 4. *For any $\varepsilon > 0$, a $(1-\varepsilon)$-approximation algorithm for maximizing the number of convex edges over all first order Delaunay triangulations exists that takes $2^{O(1/\varepsilon)} \cdot n$ time (if the Delaunay triangulation is given).*

4.2 Maximizing the Number of Non-mixed Vertices

Using a similar approach as above, we can also maximize the number of non-mixed vertices of a terrain. Because the mixed/non-mixed state of a vertex is determined by a large (possibly non-constant) number of neighboring quadrilaterals, several adaptations are needed. We now construct a graph with vertices for both the vertices and the quadrilaterals of the terrain. We remove the graph vertices that represent terrain vertices that have a fixed state, a high degree, or that can always be satisfied without disturbing the others. Of the remaining graph, we create λ-thick layers again, and we compute a tree decomposition of every layer, which we blow up such that every vertex contains all its neighbors in some bag. We solve the problem in each layer optimally by dynamic programming. More details are in the full paper [23]. We achieve:

Theorem 5. *For any $\varepsilon > 0$, a $(1-\varepsilon)$-approximation algorithm for maximizing the number of non-mixed vertices over all first order Delaunay triangulations exists that takes $2^{O(1/\varepsilon^2)} \cdot n$ time (if the Delaunay triangulation is given).*

4.3 Maximizing the Number of Convex Edges, k-th Order

The approximation algorithm for maximizing convex edges extends to k-th order Delaunay triangulations. To assure that every k-order Delaunay edge with its incident triangles is considered as a potentially convex edge in enough subproblems, we need layers with thickness proportional to k/ε. To use tree decompositions with bounded treewidth for maximizing convex edges, we also need to assure that the four vertices involved in two adjacent k-order Delaunay triangles appear in some bag of the tree decomposition. We show in the full paper [23]:

Lemma 1. *If $\langle\{X_i | i \in I\}, T\rangle$ is a tree decomposition of the Delaunay triangulation of a set of points with width ω, then a tree decomposition of width at most $2^{O(k)}\omega^2$ exists where every pair of adjacent k-th order Delaunay triangles appears in some bag.*

The number of states of a bag is exponential in the treewidth, and combining two bags takes time nearly linear in their number of states. This leads to:

Theorem 6. *For any $\varepsilon > 0$, a $(1-\varepsilon)$-approximation algorithm for maximizing the number of convex edges over all k-th order Delaunay triangulations exists that takes $2^{2^{O(k)}/\varepsilon^2} \cdot n$ time (if the Delaunay triangulation is given).*

5 Discussion

We analyzed the algorithmic complexity of optimizing various measures that apply to triangulations, and terrains represented by triangulations. The class of triangulations over which optimization is done is the first order Delaunay triangulations. We gave efficient algorithms for four measures, NP-hardness proofs for three other measures, and polynomial time approximation schemes for two measures that were shown NP-hard. One approximation algorithm could be extended to k-th order Delaunay triangulations.

Other measures related to terrain modeling in GIS may be of interest to optimize. Also, certain measures that have efficient, optimal algorithms for first order Delaunay triangulations may become harder for second and higher order Delaunay triangulations. These are interesting topics for further research. It is also unknown how to generalize the approximation algorithm for maximizing non-mixed vertices to higher order Delaunay triangulations. Finally, improving on the doubly-exponential dependency on the order k in the approximation algorithm for maximizing convex edges is worthwhile.

Acknowledgements. The authors thank Hans Bodlaender and René van Oostrum for helpful discussions.

References

1. Aspvall, B., Plass, M., Tarjan, R.: A linear-time algorithm for testing the truth of certain quantified boolean formulas. Inf. Proc. Lett. 8, 121–123 (1979)
2. Baker, B.S.: Approximation algorithms for NP-complete problems on planar graphs. J. ACM 41, 153–180 (1994)
3. Bern, M., Edelsbrunner, H., Eppstein, D., Mitchell, S., Tan, T.S.: Edge insertion for optimal triangulations. Discrete Comput. Geom. 10(1), 47–65 (1993)
4. Bern, M., Plassmann, P.: Mesh generation. In: Sack, J., Urrutia, J. (eds.) Handbook of Computational Geometry, pp. 291–332. Elsevier, Amsterdam (1997)
5. Bodlaender, H.L.: A partial k-arboretum of graphs with bounded treewidth. Theoretical Computer Science 209, 1–45 (1998)
6. de Floriani, L., Magillo, P., Puppo, E.: Applications of computational geometry in Geographic Information Systems. In: Sack, J., Urrutia, J. (eds.) Handbook of Computational Geometry, pp. 333–388. Elsevier, Amsterdam (1997)
7. de Kok, T., van Kreveld, M., Löffler, M.: Generating realistic terrains with higher-order Delaunay triangulations. Comput. Geom. Th. Appl. 36, 52–65 (2007)
8. Edelsbrunner, H., Tan, T.S.: A quadratic time algorithm for the minmax length triangulation. SIAM J. Comput. 22, 527–551 (1993)
9. Edelsbrunner, H., Tan, T.S., Waupotitsch, R.: $O(N^2 \log N)$ time algorithm for the minmax angle triangulation. SIAM J. Sci. Stat. Comput. 13, 994–1008 (1992)
10. Gudmundsson, J., Hammar, M., van Kreveld, M.: Higher order Delaunay triangulations. Comput. Geom. Theory Appl. 23, 85–98 (2002)
11. Guibas, L.J., Hershberger, J.E., Mitchell, J.S.B., Snoeyink, J.S.: Approximating polygons and subdivisions with minimum link paths. Internat. J. Comput. Geom. Appl. 3(4), 383–415 (1993)
12. Hochbaum, D.S., Maass, W.: Approximation schemes for covering and packing problems in image processing and VLSI. J. ACM 32, 130–136 (1985)
13. Huggett, R.J.: Fundamentals of Geomorphology. Routledge (2003)
14. Jenson, S.K., Trautwein, C.M.: Methods and applications in surface depression analysis. In: Proc. Auto-Carto, vol. 8, pp. 137–144 (1987)
15. Kant, G., Bodlaender, H.L.: Triangulating planar graphs while minimizing the maximum degree. Inform. Comput. 135, 1–14 (1997)
16. Lichtenstein, D.: Planar formulae and their uses. SIAM J. Cmp. 11, 329–343 (1982)
17. Maidment, D.R.: GIS and hydrologic modeling. In: Goodchild, M., Parks, B., Steyaert, L. (eds.) Environmental modeling with GIS, pp. 147–167. Oxford University Press, New York (1993)
18. Mulzer, W., Rote, G.: Minimum weight triangulation is NP-hard. In: Proc. 22nd Annu. ACM Sympos. Comput. Geom., pp. 1–10 (2006)
19. Niedermeier, R.: Invitation to Fixed-Parameter Algorithms. Oxford University Press, New York (2006)
20. Robertson, N., Seymour, P.D.: Graph minors II. Algorithmic aspects of tree width. J. Algorithms 7, 309–322 (1986)

21. Theobald, D.M., Goodchild, M.F.: Artifacts of TIN-based surface flow modelling. In: Proc. GIS/LIS, pp. 955–964 (1990)
22. van Kreveld, M.: Geographic Information Systems. In: Goodmann, J.E., O'Rourke, J. (eds.) Handbook of Discrete and Computational Geometry, ch. 58, pp. 1293–1314. Chapman & Hall/CRC, Boca Raton (2004)
23. van Kreveld, M., Löffler, M., Silveira, R.I.: Optimization for first order Delaunay triangulations. Technical Report UU-CS-2007-011, Utrecht University, Institute of Information and Computing Sciences (2007)

Constant Factor Approximations for the Hotlink Assignment Problem

Tobias Jacobs

Department of Computer Science, University of Freiburg,
Georges-Köhler-Allee 79, 79110 Freiburg, Germany
jacobs@informatik.uni-freiburg.de

Abstract. An approach for reducing the navigation effort for the users of a web site is to enhance its hyperlink structure with additional *hotlinks*. We address the task of adding at most one such additional outgoing edge to each page of a tree-like site, minimizing the *path length*, i.e. the expected number of "clicks" necessary for the user to reach his destination page. Another common formulation of that problem is to maximize the *gain*, i.e. the path length reduction achieved by the assignment.

In this work we analyze the natural greedy strategy, proving that it reaches the optimal gain up to the constant factor of 2. Considering the gain, we also prove the existence of a PTAS. Finally, we give a polynomial time 2-approximation which constitutes the first constant factor approximation in terms of the path length. The algorithms' performance analyses are made possible by a set of three new basic operations for the transformation of hotlink assignments.

Keywords: Hotlink Assignment, Approximation Algorithms, Graph Theory, Greedy Algorithms, Dynamic Programming.

1 Introduction

Due to the extensive growth of the Internet as a huge information source, the task of making an increasing amount of information accessible in a user-friendly way is becoming more and more important. The value of any information is closely related to its accessibility. Therefore, the effort spent by users searching for a specific piece of information, or trying to get an overview of some subset of the available information, should be minimized.

In this work we address the concept of improving the design of large web directories or similar structures by assigning additional *hotlinks* to its pages. By taking access frequencies into account, hotlinks can especially reduce the access times of popular pages, while the site's original structure is preserved. A considerable amount of research has been spent on this approach, see e.g. [1,2,3,4,5,6,7,8,9,10,11,12,13]. It can be applied in a number of additional scenarios, e.g. knowledge bases, file systems, menus of computer applications, and, as observed by Bose et. al. in [15], even in asymmetric communication protocols.

Problem definition: A hierarchical web site can be modeled as a *weighted tree* $T = (V, E, \omega)$ where (V, E) is a tree. Let $L \subseteq V$ be the set of leaves of T. The

F. Dehne, J.-R. Sack, and N. Zeh (Eds.): WADS 2007, LNCS 4619, pp. 188–200, 2007.

weight function $\omega : L \to \mathbb{R}_0^+$ assigns a non-negative weight to each leaf. The weights can be interpreted as access frequencies or, if normalized to sum up to 1, as access probabilities. Let $\text{ch}(v)$ denote the set of children and let $\text{desc}(v)$ ($\text{anc}(v)$) denote the set of proper descendants[1] (ancestors) of a node v.

We assume that information is stored in the leaves only. In order to access a leaf l, the user has to traverse the directed path from the root to l. A *hotlink assignment (HLA)* is a set $A \subset V \times V$ of additional edges, providing shortcuts for the user. The elements of A are called *hotlinks.* For $(u, v) \in A$ we refer to v as u's *hotchild.* We further refer to u as v's *hotparent* and say that the hotlink *starts* in u and *ends* v.

In this work we assume that the user only knows about the outgoing hotlinks of the nodes he has already visited and always takes any hotlink that leads him closer to his destination leaf. This is referred to as the *greedy user model.* In contrast, in the *clairvoyant user model* users take the shortest path in $(V, E \cup A)$.

The greedy user assumption leads us to demand that any hotlink assignment A satisfies three *feasibility properties* (referring to [8]).

(i) For every edge $(u, v) \in A$, $v \in \text{desc}(u)$ in T.
(ii) Let $(u, v), (u', v') \in A$ and let $u' \in \text{desc}(u) \cap \text{anc}(v)$. Then $v' \in \text{anc}(v)$.
(iii) For every node $u \in V$ there is at most one edge $(u, v) \in A$.

Properties (i) and (ii) are consequences of the greedy user assumption, since hotlinks violating these properties would never be taken by a greedy user. Property (iii) demands that each node may have at most one hotchild. This reflects the fact that the number of hotlinks on a concise web page must be somehow limited. Relaxations of property (iii) are considered in Section 6.

For a given weighted tree T, the *path length* of a hotlink assignment A is defined as

$$p(A) = \sum_{l \in L} \omega(l) \text{dist}^A(r, l) ,$$

where L is the set of leaves, r is the root of T and $dist^A(u, v)$ determines the number of edges and hotlinks a greedy user traverses when traveling from u to v. The *Hotlink Assignment Problem* denotes the problem of finding a hotlink assignment for a given tree, minimizing the path length. An alternative formulation of that problem is to maximize the *gain*

$$g(A) = p(\emptyset) - p(A) .$$

The two problem formulations are equivalent in the sense that a HLA maximizes the gain if and only if it achieves a minimum path length. They are however not equivalent when we are interested in approximation ratios. We will see in the later sections that there are in fact algorithms that approximate the gain (path length) up to a constant ratio but do not guarantee any constant ratio in terms of the path length (gain).

[1] All descendants (ancestors) a node v, not including v itself, are proper descendants (ancestors) of v.

Related work: The concept of assigning hotlinks to web sites was suggested by Perkowitz and Etzoni in [14]. Bose et. al. have shown in [1] that the problem is NP-hard when considering general DAGs instead of trees. They have proposed algorithms for full binary trees with special probability distributions on the leaves. The first strategies for assigning at most a fixed number of k hotlinks to each node of full d-ary trees have been given in [2].

In [1] the authors have also proven a lower bound for the path length using coding theory: No assignment of at most k hotlinks to each node can result in a path length smaller than $H(p)/\log(d+k)$, where $H(p) = \sum_{l \in L} \omega(l) \log(\frac{1}{\omega(l)})$ is the entropy of the probability distribution over the leaves. Efficient algorithms achieving a path length of $O(H(p))$ have been published in [4,12] and most recently in [13]. Dynamic maintenance of such HLAs has also been studied in [12]. These algorithms hold constant approximation ratios for trees of fixed degree.

For the clairvoyant user model, Matichin and Peleg have proven in [9] that the natural greedy strategy holds an approximation ratio of 2 even in DAGs. In [11] the same authors have given a 2-approximation for arbitrary trees in the more realistic greedy user model. Their algorithm assigns only hotlinks from nodes to one of their grandchildren.

Gerstel et. al. ([7]) and Pessoa et. al. ([8]) have independently discovered an optimal algorithm whose running time is exponential in the depth of the tree and thus polynomial for trees of logarithmic depth. It is yet unknown if the optimal solution for arbitrary trees can be computed in polynomial time.

A number of experimental papers on hotlink assignment have been published ([3,6,10]) and a software tool for assigning hotlinks to web sites has been developed ([5]). A surprising application of hotlink assignment in asymmetric communication protocols has been suggested by Bose et. al. in [15].

Our contribution: We present two constant factor approximations and one PTAS for the hotlink assignment problem on arbitrary trees and thus substantially improve the best approximation ratios previously known for both the path length and the gain.

- The natural greedy algorithm GR, which always adds a hotlink achieving the greatest gain, has exhibited the best performance among the approximation algorithms studied experimentally in [3,6,10][2]. So far, no bound for its ratio in the greedy user model was known. We show that GR is a 2-approximation in terms of the gain.
- We prove the existence of a PTAS in terms of the gain. More specific, we give an upper bound for the loss of gain caused by restricting the length of hotlinks to a fixed value. Then we modify the PATH-algorithm of Pessoa, Laber and Souza ([8]) such that it achieves the gain of any length-restricted assignment in polynomial time.
- We present the first algorithm that yields a constant approximation ratio in terms of the path length for trees of unbounded degree. Our approach decomposes the tree into subtrees called *heavy centipedes*. Optimal hotlink

[2] GR is called *recursive* in [3,6] and *greedyBFS* in [10].

assignments for such subtrees can be computed efficiently by dynamic programming. The resulting approximation ratio is 2.

The foundation of all the algorithms' performance analyses is formed by three basic operations for HLA modification. These are introduced in Section 2 together with some further notation. The algorithms are then presented and analyzed in the following three sections, which can be read independently from each others. In Section 6 we briefly discuss if and how our algorithms can be generalized for relaxations of feasibility property (iii). Section 7 concludes. Omitted proofs appear in the long version of the paper, as well as lower bounds for all algorithms' approximation ratios and detailed descriptions of the generalized algorithms.

2 Further Notation and Basic Operations

In this section we introduce some further notation for our problem and subsequently define and analyze three new operations for the modification of hotlink assignments.

For nodes v in T we denote by $T(v)$ the maximum subtree of T rooted at v. For any set V' of nodes, let $T(-V')$ be the tree obtained by omitting from T all maximum subtrees rooted at a node in $V \cap V'$. Let further $T(v)(-A) = T(v)(-\{v' \in \text{desc}(v) \mid \exists(u,v') \in A : u \in \text{anc}(v)\})$ be the subtree rooted at v where the maximum subtrees rooted at the hotchildren of v's ancestors are omitted. Finally, for any subtree T' of T, we define $A|T' = \{(u,v) \in A \mid u, v \text{ are nodes in } T'\}$.

We extend the weight function ω to be defined also for inner nodes. The weight of such a node u is calculated by summing up the weights of all leaves l where u is on a greedy user's path from r to l. Given an assignment A, these are exactly the leaves in $T(u)(-A)$, i.e.

$$\omega^A(u) = \begin{cases} \omega(u) & \text{if } u \text{ is a leaf} \\ \sum_{v \in \text{ch}^A(u)} \omega^A(v) & \text{otherwise} \end{cases},$$

where $\text{ch}^A(u)$ denotes the set of u's children and hotchildren in $T(u)(-A)$.

The gain of a hotlink assignment A can be formulated as the sum over the path shortening contributions of its hotlinks, i.e.

$$g(A) = \sum_{(u,v) \in A} \omega^A(v)(\text{dist}^{\emptyset}(u,v) - 1) .$$

The gain of a single hotlink of A is defined as $g^A(u,v) = \text{dist}((u,v) - 1) \cdot \omega^A(v)$.

Concerning the empty assignment we use the following abbreviations: $\omega^{\emptyset} = \omega$, $\text{dist}^{\emptyset} = \text{dist}$ and $g^{\{(u,v)\}}(u,v) = g(u,v)$. For $T = (V, E, \omega)$ we sometimes write $v \in T$ instead of $v \in V$.

We proceed introducing three basic operations that modify hotlink assignments. Note that none of these operations is actually applied by any algorithm given in this paper. We need them however for the performance analyses in Sections 3, 4 and 5.

Definition 1. *Let A be a hotlink assignment for a weighted tree T and let u be a node in T. The operation* PUSHDOWN(u) *is defined as follows:*

If u has no hotchild under A, do nothing. If u's hotchild is one of its children or grandchildren, delete the corresponding hotlink. Otherwise, let v be u's hotchild and let u_1 be the child of u that is located on the path from u to v. Apply PUSHDOWN(u_1) *and subsequently replace (u, v) by (u_1, v) in the assignment.*

After applying PUSHDOWN(u) to a hotlink assignment, the node u is guaranteed to have no hotchild.

Lemma 1. *Let A be a HLA for a weighted tree T and let u be a node in T. The operation* PUSHDOWN(u) *causes a decrease in gain of at most $\omega^{A\setminus\{(u,v)\}}(u_1)$.* □

Note that in most of the analyses in the later sections it suffices to use a weaker version of Lemma 1, assuming a maximum decrease in gain of $\omega^A(u)$.

Definition 2. *Let A be a hotlink assignment for a weighted tree T, let u be an inner node in T that has no hotchild under A and let v be a node in $T(u)(-A)$. The operation* FREE-INSERT(u, v) *is defined as follows:*

Let $uu_1 \dots u_n v$ be the path from u to v, let $u_{k(1)}, \dots, u_{k(m)}$ be the list of nodes on this path that have a hotchild in $T(v)$, ordered by decreasing depth. Let further $v_{k(i)}$ be the hotchild of $u_{k(i)}$, $1 \leq i \leq m$.

For $i = 1, \dots, m$, apply PUSHDOWN(v) *and subsequently replace $(u_{k(i)}, v_{k(i)})$ by $(v, v_{k(i)})$. After these m iterations, add (u, v) to the assignment.*

Lemma 2. *No decrease in gain occurs when applying* FREE-INSERT. □

Definition 3. *Let A be a HLA for a weighted tree T, let $(u, v) \in A$ and let $v' \in desc(u) \cap anc(v)$. The operation* SHORTEN-HL(u, v') *is defined as follows: Temporarily allow u to have a second hotchild and perform* FREE-INSERT(u, v'). *Then apply* PUSHDOWN(v'). *Finally, replace (u, v) by (v', v).*

Due to feasibility condition (ii), v' is also a node in $T(u)(-A)$, which ensures that the operation is always applicable.

Lemma 3. *Let A' be the HLA obtained from A by applying* SHORTEN-HL(u, v'). *Then $g(A) - g(A') \leq \omega^{A'}(v')$.*

Proof. The claim follows straightforward from Lemma 1 and 2. □

3 The Natural Greedy Strategy

The natural greedy strategy, denoted as GR in this paper, initializes A with $A = \emptyset$ and always adds a hotlink (u, v) to A that maximizes $g^A(u, v)$. The procedure terminates when any further hotlink would violate a feasibility condition.

GR can be formulated in a number of alternative ways. One is to traverse the nodes in a topological order[3] and for each node u add a hotlink (u, v) that

[3] In a topological order each node appears before its children.

causes a maximum gain. Another equivalent formulation is to add a maximum gain hotlink (r, v) that starts in the root and recursively apply the algorithm to the subtrees $T(u_1)(-A), \ldots, T(u_d)(-A)$ and $T(v)$, where $u_1, \ldots, u_d$ are the children of r.

In the latter formulation GR has been introduced by Czyzowics et al. in [3], where it is called *recursive.* In that work the authors have also considered another greedy algorithm *greedyBFS*[4]. The difference to GR is that greedyBFS does not take the hotlinks it has already assigned into account for the calculation of the inner node's weights, i.e. it works with $\omega^\emptyset$ instead of ω^A. The weight function $\omega^\emptyset$ is also used by *recursive* in that work, but in that formulation of the algorithm it makes no difference since hotlinks are only added to empty assignments. The experiments in [3,6] examine in fact a slightly better performance of GR.

Theorem 1. *GR holds an approximation ratio of 2 in terms of the gain.*

Proof. We show how to transform an optimal HLA $\tilde{A}$ into a GR-assignment A^{GR} loosing not more than half of the gain. Let A be our working assignment, initialized by $A = \tilde{A}$. We traverse the nodes of T in some topological order. For each node u we delete the hotlink $(u, v) \in A$ if it exists. Let $(u, v') \in A^{\text{GR}}$. Due to the topological traversal, v' is a node in $T(u)(-A)$. Thus, we are able to apply FREE-INSERT(u, v') to A, preserving the gain of the assignment due to Lemma 2. Furthermore, the topological traversal ensures that $\omega^A(v')$ will not decrease later in the transformation process. From the definition of GR follows that the decrease caused by the deletion of (u, v) is not greater than the gain guaranteed by (u, v').

So, after the complete transformation, we have lost a total amount of gain that is not greater than the gain of the resulting assignment. □

4 An Approximation Scheme for the Gain

In this section we present an algorithm that computes a $(1 + \epsilon)$-approximation in time $O(n3^{\frac{1}{\epsilon}})$. In [11], Matichin and Peleg have shown that restricting hotlinks to length 2 at most halves the possible gain. They have given an algorithm that computes a best length 2 hotlink assignment and is thus a 2-approximation.

On a high level, our approach can be interpreted as a generalization to that of Matichin et al. However, both our proof technique and our algorithmic idea are completely different. Let the length of the hotlink (u, v) be defined as $\text{dist}(u, v)$. We prove that one looses at most $\frac{1}{h}$ of the gain when restricting hotlinks to a maximum length of h. Then we show how to compute an assignment achieving at least the gain of the best length h HLA.

Lemma 4. *For any tree T and any integer $h > 1$ there is a HLA A^h with $dist(u, v) \leq h$ for each $(u, v) \in A^h$ and $\frac{h}{h-1} g(A^h) \geq g(\tilde{A})$, where $\tilde{A}$ is an optimal HLA for T.*

[4] In contrast, the algorithm called *greedyBFS* in [10] is actually GR.

Proof sketch: $\tilde{A}$ can be transformed into a length h assignment A^h by cutting each long hotlink at length h using SHORTEN-HL. Not more than $1/h$ of $\tilde{A}$'s gain is lost due to these operations. □

Now we show how to efficiently compute optimal length-restricted HLAs. The PATH algorithm presented in [8] computes a best HLA A that satisfies $\text{dist}^A(r,u) < h$ for any node u. We give a modified version LPATH which computes an optimal HLA A under the restriction that $dist^{A\setminus\{(u,v)\}}(u,v) \leq h$ for any $(u,v) \in A$. Since the latter restriction is weaker than demanding $dist^{\emptyset}(u,v) \leq h$, the HLA computed by LPATH is at least as good as any assignment of hotlinks of maximum length h. Thus, Lemma 4 implies an approximation ratio of $\frac{h}{h-1}$ for LPATH.

Algorithmic idea: The main idea is to determine the concrete hotchildren of the nodes as late as possible. Assume we are given a subtree T' together with a number of hotlinks that have already been decided to end in T'. One of these hotlinks could end in the root r of T'. For each of the others, we have to decide in which subtree of T' it ends.

Detailed description of the algorithm: The structure of the subproblems is exactly the same as for PATH. Let T be the input tree. Subproblems are defined by a triple (q, a, T'), where $q = q_1 \dots q_n$ is a directed path, $a = a_1 \dots a_n a_{n+1} b \in \{0,1\}^{n+2}$ is a binary vector and T' is a subtree of T. It represents the tree qT' obtained by appending the root r of T' to q_n. The vector a represents a number of restrictions concerning hotlink assignments for qT'. Path nodes q_i are only allowed to have a hotchild if $a_i = 1$. They are not allowed to be hotchildren themselves. The node r is only allowed to be a hotfather if $a_{n+1} = 1$, and may only be a hotchild if $b = 1$. Observe that the original problem can also be described as such a subproblem.

We proceed describing how LPATH computes a best HLA for (q, a, T'), i.e. maximizes $\tilde{g}(q, a, T')$. Let r be an inner node. Then we distinguish between two cases.

Case I: There is a hotlink that ends in r. The hotfather of r is some node q_i with $a_i = 1$. Due to feasibility condition (ii), there can be no hotlink starting in any q_j with $j > i$. Moreover, hotlinks (q_k, v) with $k < i$ and $v \in \text{desc}(r)$ will decrease the gain of (q_i, r) by $(\text{dist}(q_i, r) - 1) \cdot \omega(v)$. This is equivalent to saying that the gain of (q_i, r) is fixed and the distance between q_k and v is reduced by $(\text{dist}(q_i, r) - 1)$. So the gain of an optimal assignment for (q, a, T') is calculated by

$$\tilde{g}_{\text{I}}(q, a, T') = \max_{a_i=1}\{(n-i)\cdot\omega(r) + \tilde{g}(q_1 \dots q_i, a_1 \dots a_{i-1} 0 a_{n+1} 0, T')\} \ .$$

Case II: No hotlink ends in r. The hotchild of any q_i with $a_i = 1$ must be either in some subtree rooted at the first child u_1 of r (assuming any order), or in a subtree rooted at another child of r. The same holds for r if $a_{n+1} = 1$. We use a vector $c \in \{0,1\}^{n+1}$ to describe the distribution of these hotchildren among $T(u_1)$ and $T(-\{u_1\})$. The set C contains all feasible vectors for the subproblem, i.e. $C = \{c \in \{0,1\}^{n+1} \mid c_i + a_i \leq 1 \forall i = 1, \dots, n+1\}$. The gain of an optimal HLA is calculated by

$$\tilde{g}_{\text{II}}(q, a, T') = \max_{c \in C}\{\tilde{g}(q_1 \dots q_n r, c_1 \dots c_{n+1} 11, T'(u_1)) + \tilde{g}(p, \bar{c}0, T'(-\{u_1\}))\} ,$$

where $\bar{c}0$ represents the vector $(1 - c_1, \dots, 1 - c_{n+1}, 0)$.

As we want to bound the relative length of hotlinks to h, we cut off the first $n-h$ components of q and a whenever $n > h$. This constitutes the only difference between PATH and LPATH, as PATH would set the gain to $-\infty$ in case of $n > h$. PATH and LPATH behave identically in the basic case when T' is a leaf.

So the formula used by LPATH for calculating the gain of an optimal HLA for the subproblem (q, a, T') is

$$\tilde{g}(q, a, T') = \begin{cases} \max(\{0\} \cup \{n - i \mid a_i = 1\}) \cdot \omega(r) & \text{if r is a leaf} \\ 0 & \text{if } \text{ch}(r) = \emptyset \\ \tilde{g}(q_{n-h} \dots q_n, a_{n-h} \dots a_{n+1} b, T') & \text{if } n > h \\ \tilde{g}_{\text{II}}(q, a, T') & \text{if } b = 0 \\ \max\{\tilde{g}_{\text{I}}(q, a, T'), \tilde{g}_{\text{II}}(q, a, T')\} & \text{otherwise .} \end{cases}$$

Analysis of LPATH: LPATH uses dynamic programming, maintaining a table with an entry for each configuration of a and T'. The performance analysis is the same as for PATH, we therefore refer the reader to [8]. The memory requirements are in $O(n2^h)$, while the runtime is in $O(n3^h)$.

Theorem 2. *For any $\epsilon > 0$, LPATH computes a $(1+\epsilon)$-approximation in time $O(n3^{\frac{1}{\epsilon}})$ and space $O(n2^{\frac{1}{\epsilon}})$.*

Proof. For a given $\epsilon > 0$, choose h such that $\frac{h}{h-1} \leq 1 + \epsilon < \frac{h-1}{h-2}$. According to Lemma 4, the left inequality implies that LPATH with parameter h guarantees an approximation ratio of $(1+\epsilon)$.

For the runtime and memory analysis we assume $(1+\epsilon) = \frac{h-1}{h-2}$, as a smaller ϵ only implies weaker runtime and memory demands. Thus, $h = 2 + \frac{1}{\epsilon}$ and the runtime of LPATH is in $O(n3^{2+\frac{1}{\epsilon}}) = O(n3^{\frac{1}{\epsilon}})$, while the algorithm uses $O(n2^{\frac{1}{\epsilon}})$ space. □

5 A 2-Approximation in Terms of the Path Length

In this section we develop the first constant approximation in terms of the path length. We give a lower bound $p_{\min}$ for the path length of any hotlink assignment. Then we show that this bound can be reached up to the constant factor of 2 by a HLA that satisfies the *centipede property* (see Definition 5). Finally we give a polynomial time algorithm for computing the best centipede assignment.

Definition 4. *Let T be a weighted tree rooted at r. The function $p_{min} : T \to \mathbb{R}$ is recursively defined as follows:*

$$p_{min}(T) = \begin{cases} 0 & \textit{if } r \textit{ is a leaf} \\ (\omega(r) - \max_{u \in ch(r)} \omega(u)) + \sum_{u \in ch(r)} p_{min}(T(u)) & \textit{otherwise .} \end{cases}$$

Intuitively, $p_{\min}$ is the sum over the weights of all nodes having a heavier sibling.

Lemma 5. *$p_{min}(T)$ is a lower bound for the path length that any hotlink assignment can achieve.*

Proof. We prove the lemma by induction over the depth of the tree. The basic case is obvious. For the induction step we consider a HLA A for a tree T with depth n. Let r be the root of that tree and let $u_1, \ldots, u_d$ be the children of r where w.l.o.g. u_1 is an ancestor of r's hotchild. By applying PUSHDOWN(r) to A, due to Lemma 1, we increase the path length by at most $\omega(u_1)$. Let A' be the resulting assignment. Then

$$\begin{aligned} p(A) &\geq p(A') - \omega(u_1) \\ &\geq \textstyle\sum_{u \in \mathrm{ch}(r)} (\omega(u) + p(A'|T(u))) - \max_{u \in \mathrm{ch}(r)} \omega(u) \\ &\geq \ \omega(r) + \textstyle\sum_{u \in \mathrm{ch}(r)} p_{\min}(T(u)) - \max_{u \in \mathrm{ch}(r)} \omega(u) = \ l_{\min} \ . \end{aligned}$$

Here $p(A'|T(u))$ is the path length of A' in $T(u)$. The last inequality is implied by the induction hypothesis and the fact that the weights of r's children sum up to $\omega(r)$. □

Definition 5. *A HLA A for a tree T satisfies the centipede property, iff*

$$(u, v) \in A, u' \in desc(u) \cap anc(v) \Rightarrow u' \succ u'' \ \forall \ siblings \ u'' \ of \ u' \ ,$$

where "$\succ$" is some total order of the children of a node with $\omega(u') > \omega(u'') \Rightarrow u' \succ u''$.

Lemma 6. *For each weighted tree T there exists a hotlink assignment A^c, where $p(A^c) \leq 2 \cdot p(\tilde{A})$ for any HLA $\tilde{A}$, and A^c satisfies the centipede property.*

Proof. We show how to transform any HLA $\tilde{A}$ into a centipede assignment A^c. For each node u' where there exists a $u'' \succ u'$ and which has ancestors that are hotparents of descendants of u' under $\tilde{A}$, perform the following operation: Let u be among these ancestors the one having the shortest distance to the root. We apply SHORTEN-HL(u, u'), increasing the path length by at most $\omega(u')$.

No hotlink starting in an ancestor of u' is added or modified during the operation (although some of these hotlinks might be deleted). Thus, for any ancestor of u' satisfying the centipede property before the the application of SHORTEN-HL, that property will also hold afterwards. So if we consider the nodes u' in a topological order, the centipede condition will hold for the resulting HLA. The total increase in path length caused by the SHORTEN-HL-operations is bounded by $p_{\min}(T)$, which implies the lemma. □

It remains to show that the best HLA satisfying the centipede property can be computed in polynomial time. Assume that we want to compute a centipede assignment for a tree T. For any node u having a sibling $u' \succ u$, there can be no hotlink from an ancestor of u to a descendant of u. This fact has two implications. The first is that, when computing the hotchildren of u's ancestors, we can consider the tree T' obtained from T by deleting u's descendants and transforming u into a leaf of weight $\omega(u)$. The second implication is that the partial assignment $A|T(u)$ can be computed independently from $T(-\{u\})$. So the subtrees T' and $T(u)$ can be considered separately.

By applying this observation to every node u with $u' \succ u$ for some sibling u' of u, we split the tree into the set of *heavy centipedes* $C_1, \ldots, C_k$, where C_i is a *centipede tree* for $1 \leq i \leq k$.

Definition 6. *A centipede tree is a tree whose inner nodes have at most one non-leaf child. Let h be the depth and r be the root of a centipede tree. Then*

$$\text{lev}(v) = \begin{cases} h - dist(v, r) & \textit{if } v \textit{ is an inner node} \\ h - dist(v, r) + 1 & \textit{if } v \textit{ is a leaf, but not the root} \\ 1 & \textit{otherwise} . \end{cases}$$

The definition of lev implies that in a non-trivial centipede tree each level consists of exactly one inner node and the leaf children of that node.

In the remainder of the section we show that for the heavy centipedes $C_1, \ldots, C_k$ it is possible to efficiently compute best HLAs $A_1, \ldots, A_k$. Then the union $\bigcup_{1 \leq i \leq k} A_i$ of these assignments is a best centipede-HLA for T and, due to Lemma 6, holds an approximation ratio of 2. We give two structural properties of optimal hotlink assignments for centipede trees and subsequently formulate a dynamic programming algorithm that takes advantage of these properties.

Lemma 7. *Let $\tilde{A}$ be an optimal HLA for a centipede tree C rooted at r. If $(r, l) \in \tilde{A}$ and l is a leaf, then there is no $(u', l') \in A$ with $\omega(l') > \omega(l)$.* □

Lemma 8. *For any centipede tree C rooted at r there is an optimal hotlink assignment $\tilde{A}$ satisfying the following property:*

$(r, l) \in \tilde{A}, l$ *is a leaf* $\Rightarrow \text{lev}(l) < \text{lev}(l')$ *for all l' with $\omega(l') = \omega(l)$.* □

Lemma 9. *An optimal hotlink assignment for centipede trees can be computed in polynomial time.*

Proof. We extend "$\succ$" (see Definition 5) to be some total order of all leaves of a centipede tree with

$$\Big(\omega(l) > \omega(l')\Big) \vee \Big(\omega(l) = \omega(l') \wedge \text{lev}(l) < \text{lev}(l')\Big) \Rightarrow l \succ l' .$$

From Lemma 7 and 8 follows that an optimal assignment containing (r, l) contains no hotlink (u, l') with $l' \succ l$.

Let C be a centipede tree of depth h. For any leaf l in C and $h \geq x \geq \text{lev}(l) \geq y \geq 1$, the tree $C[x, y, l]$ is defined as the maximum subtree of C satisfying the following properties:

(a) It contains only nodes v with $x \geq \text{lev}(v) \geq y$.
(b) It contains no leaf $l' \succ l$.

The gain of an optimal hotlink assignment for $C[x,y,l]$ can recursively calculated by the following formulae:

$$\tilde{g}(C[x,y,l]) = \begin{cases} \max(G_L(x,y,l) \cup G_N(x,y,l)) & \text{if } x-y \geq 1 \\ 0 & \text{otherwise} \end{cases}$$

$$G_L(x,y,l) = \max\{g(v_x,l') + \tilde{g}(C[x-1,y,l'']) \mid l',l'' \in C[x-1,y,l], l'' \prec l'\}$$

$$G_R(x,y,l) = \max\{(x-k-1)\omega^{[x,y,l]}(v_k) + \tilde{g}(C[x-1,k+1,l_1]) + \tilde{g}(C[k,y,l_2]) \mid l_1 \preceq l, l_2 \preceq l, x > \text{lev}(l_1) > k \geq \text{lev}(l_2) \geq y\}\ .$$

where v_i is the inner node at level i and $\omega^{[x,y,l]}(u)$ is $\omega(u)$ in $C[x,y,l]$. G_L represents the HLAs where the hotchild of the subtree's root is a leaf, while G_N represents those where that hotchild is an inner node.

Let n be the number of nodes in C. There are $O(n^3)$ different configurations of (x,y,l). Thus by dynamic programming we can calculate an optimal hotlink assignment for C with space requirements $O(n^4)$.

For a fixed l' we can restrict the choice of l'' to the greatest (with respect to "$\succ$") possible leaf. The same holds for l_1 and l_2, when k is fixed. So for each configuration of (x,y,l) we have to compare $O(n)$ different possibilities and thus the runtime of the algorithm is in $O(n^4)$. □

Theorem 3. *There is a polynomial time 2-approximation in terms of the remaining path length for the hotlink assignment problem.*

Proof. The claim follows directly from Lemma 6 and 9. □

6 Generalization to Multiple Hotlink Assignment

In this section we briefly discuss how to generalize our algorithms in order to, for any fixed k, assign at most k hotlinks to each node of a weighted tree. See the long version of the paper for details.

The generalized version of GR preserves the approximation ratio of 2 by traversing the tree in a topological order, assigning the currently best set of k hotchildren to each node. The best known implementation of that strategy leads to a running time of $O(n^4k^2)$.

A natural generalization of LPATH computes a $(1+\epsilon)$-approximation in terms of the gain in time $O(n(\frac{1}{2}(k+1)(k+2))^{\frac{1}{\epsilon}})$ and space $O(nk^{\frac{1}{\epsilon}})$.

There seems to be no natural generalization of the centipede algorithm which has both a polynomial running time and a constant approximation ratio.

7 Final Remarks

All approximation ratios given in this paper are tight, and each of the presented algorithms holds a constant ratio only for one optimization term (path length or gain). Instances proving this proposition are given in the long version of the paper. Nevertheless, one could obtain a constant factor approximation for both optimization terms by choosing the best HLA from those generated e.g. by GR and the centipede algorithm.

Open problems: While we know that the hotlink assignment problem is NP-hard for general DAGs (see [1]), it is still an open question if that complexity holds when the input graph is a tree. Besides the computation of the optimal solution, the question arises if there are algorithms that exhibit better approximation ratios and/or lower resource requirements than the known strategies. Finally, it would be interesting to know about the practical performance of the newer algorithms given in [11,12,13] and this work.

References

1. Bose, P., Czyzowicz, J., Gasienicz, L., Kranakis, E., Krizanc, D., Pelc, A., Martin, M.V.: Strategies for hotlink assignments. In: Lee, D.T., Teng, S.-H. (eds.) ISAAC 2000. LNCS, vol. 1969, Springer, Heidelberg (2000)
2. Fuhrmann, S., Krumke, S.O., Wirth, H.-C.: Multiple hotlink assignment. In: Brandstädt, A., Le, V.B. (eds.) WG 2001. LNCS, vol. 2204, Springer, Heidelberg (2001)
3. Czyzowicz, J., Kranakis, E., Krizanc, D., Pelc, A., Martin, M.V.: Evaluation of hotlink assignment heuristics for improving web access. In: Proceedings of the 2nd International Conference on Internet Computing (ICOMP) (2001)
4. Kranakis, E., Krizanc, D., Shende, S.: Approximate hotlink assignment. In: Eades, P., Takaoka, T. (eds.) ISAAC 2001. LNCS, vol. 2223, Springer, Heidelberg (2001)
5. Kranakis, E., Krizanc, D., Martin, M.V.: The hotlink optimizer. In: Proceedings of the 3rd International Conference on Internet Computing (ICOMP) (2002)
6. Czyzowicz, J., Kranakis, E., Krizanc, D., Pelc, A., Martin, M.V.: Enhancing hyperlink structure for improving web performance. Journal of Web Engineering 1(2), 93–127 (2003)
7. Gerstel, O., Kutten, S., Matichin, R., Peleg, D.: Hotlink enhancement algorithms for web directories. In: Ibaraki, T., Katoh, N., Ono, H. (eds.) ISAAC 2003. LNCS, vol. 2906, Springer, Heidelberg (2003)
8. Pessoa, A., Laber, E., de Souza, C.: Efficient algorithms for the hotlink assignment problem: The worst case search. In: Fleischer, R., Trippen, G. (eds.) ISAAC 2004. LNCS, vol. 3341, Springer, Heidelberg (2004)
9. Matichin, R., Peleg, D.: Approximation algorithm for hotlink assignments in web directories. In: Dehne, F., Sack, J.-R., Smid, M. (eds.) WADS 2003. LNCS, vol. 2748, Springer, Heidelberg (2003)
10. Pessoa, A., Laber, E., de Souza, C.: Efficient implementation of a hotlink assignment algorithm for Web sites. In: Proceedings of the 6th Workshop on Algorithm Engineering and Experiments (ALENEX) and the First Workshop on Analytic Algorithmics and Combinatorics (ANALCO) (2004)
11. Matichin, R., Peleg, D.: Approximation algorithm for hotlink assignment in the greedy model. In: Kralovic, R., Sýkora, O. (eds.) SIROCCO 2004. LNCS, vol. 3104, Springer, Heidelberg (2004)
12. Douïeb, K., Langerman, S.: Dynamic Hotlinks. In: Dehne, F., López-Ortiz, A., Sack, J.-R. (eds.) WADS 2005. LNCS, vol. 3608, Springer, Heidelberg (2005)

13. Douïeb, K., Langerman, S.: Near-entropy hotlink assignments. In: Azar, Y., Erlebach, T. (eds.) ESA 2006. LNCS, vol. 4168, Springer, Heidelberg (2006)
14. Perkowitz, M., Etzioni, O.: Towards adaptive web sites: Conceptual framework and case study. Computer Networks 31(11-16), 1245–1258 (1999)
15. Bose, P., Krizanc, D., Langerman, S., Morin, P.: Asymmetric communication protocols via hotlink assignments. In: Proceeding of the 9th Colloquium on Structural Information and Communication Complexity (SIROCCO) (2002)

Approximation Algorithms for the Sex-Equal Stable Marriage Problem

Kazuo Iwama[1,*], Shuichi Miyazaki[2,**], and Hiroki Yanagisawa[1,3]

[1]Graduate School of Informatics, Kyoto University,
Yoshida-honmachi, Sakyo-ku, Kyoto 606–8501, Japan
{iwama,yanagis}@kuis.kyoto-u.ac.jp

[2]Academic Center for Computing and Media Studies, Kyoto University,
Yoshida-honmachi, Sakyo-ku, Kyoto 606–8501, Japan
shuichi@media.kyoto-u.ac.jp

[3]Tokyo Research Laboratory, IBM,
1623-14, Shimo-tsuruma, Yamato-shi, Kanagawa 242–8502, Japan
yanagis@jp.ibm.com

Abstract. The stable marriage problem is a classical matching problem introduced by Gale and Shapley. It is known that for any instance, there exists a solution, and there is a polynomial time algorithm to find one. However, the matching obtained by this algorithm is man-optimal, that is, the matching is preferable for men but unpreferable for women, (or, if we exchange the role of men and women, the resulting matching is woman-optimal). The sex-equal stable marriage problem posed by Gusfield and Irving asks to find a stable matching "fair" for both genders, namely it asks to find a stable matching with the property that the sum of the men's score is as close as possible to that of the women's. This problem is known to be strongly NP-hard.

In this paper, we give a polynomial time algorithm for finding *a near optimal* solution in the sex-equal stable marriage problem. Furthermore, we consider the problem of optimizing additional criterion: among stable matchings that are near optimal in terms of the sex-equality, find a minimum egalitarian stable matching. We show that this problem is NP-hard, and give a polynomial time algorithm whose approximation ratio is less than two.

Keywords: the stable marriage problem, the sex-equal stable marriage problem, approximation algorithms.

1 Introduction

An instance I of the stable marriage problem consists of n men, n women, and each person's preference list. A preference list is a totally ordered list including all members of the opposite sex depending on his/her preference. For a matching

* Supported in part by KAKENHI (16092101, 16092215, 16300002).
** Supported by Grant-in-Aid for Scientific Research, MEXT 16092215 and 17700015.

F. Dehne, J.-R. Sack, and N. Zeh (Eds.): WADS 2007, LNCS 4619, pp. 201–213, 2007.

M between men and women, a pair of a man m and a woman w is called a *blocking pair* if both prefer each other to their current partners. A matching with no blocking pair is called *stable*. Gale and Shapley showed that every instance admits at least one stable matching, and proposed a linear time algorithm to find one, which is known as the Gale-Shapley algorithm [4]. However, in general, there are many different stable matchings for a single instance, and the Gale-Shapley algorithm finds only one of them (*man-optimal* or *woman-optimal*) with an extreme property: In the man-optimal stable matching, each man is matched with his best possible partner, while each woman gets her worst possible partner, among all stable matchings. Hence, it is natural to try to obtain a matching which is not only stable but also "good" in some criterion.

There are three major optimization criteria for the quality of stable matchings. Let $p_m(w)$ ($p_w(m)$, respectively) denote the position of woman w in man m's preference list (the position of man m in woman w's preference list, respectively). For a stable matching M, define a *regret cost* $r(M)$ to be

$$r(M) = \max_{(m,w)\in M} \max\{p_m(w), p_w(m)\}.$$

Also, define an *egalitarian cost* $c(M)$ to be

$$c(M) = \sum_{(m,w)\in M} p_m(w) + \sum_{(m,w)\in M} p_w(m),$$

and a *sex-equalness cost* $d(M)$ to be

$$d(M) = \sum_{(m,w)\in M} p_m(w) - \sum_{(m,w)\in M} p_w(m).$$

The minimum regret stable marriage problem (*the minimum egalitarian stable marriage problem* and *the sex-equal stable marriage problem*, respectively) is to find a stable matching M with minimum $r(M)$ ($c(M)$ and $|d(M)|$, respectively) [6]. Note that the number of stable matchings for one instance grows exponentially in general (see [8], e.g.). Nevertheless, for the first two problems, Gusfield [5], and Irving, Leather and Gusfield [9], respectively, proposed polynomial time algorithms by exploiting a lattice structure which is of polynomial size but contains information of all stable matchings.

In contrast, it is hard to obtain a sex-equal stable matching. The question of its complexity was posed by Gusfield and Irving [6], and was later proved to be strongly NP-hard by Kato [11]. Thus, the next step should be its approximability for which we have no knowledge so far.

Our Contribution. In this paper, we consider finding *near optimal* solutions for the sex-equal stable marriage problem. Let M_0 and M_z be the man-optimal and the woman-optimal stable matchings, respectively. Note that $d(M_0) \leq d(M) \leq d(M_z)$ for any stable matching M (see Fig. 1). Our goal is to obtain a stable matching M such that $-\epsilon\Delta \leq d(M) \leq \epsilon\Delta$ for a given constant ϵ, where $\Delta = \min\{|d(M_0)|, |d(M_z)|\}$. Namely, we define the following problem called *Near SexEqual* (*NSE* for short). Given a stable marriage instance I and a positive constant ϵ, it asks to find a stable matching M such that $|d(M)| \le \epsilon\Delta$ if such M exists, or answer "No" otherwise. We give a polynomial time algorithm for NSE, which runs in time $O(n^{3+\frac{1}{\epsilon}})$.

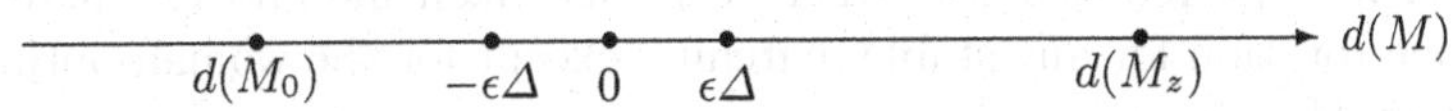

Fig. 1. The sex-equalness costs of stable matchings

NSE asks to find an *arbitrary* stable matching whose sex-equalness cost lies within some range. However, we may want to find a good one if there are several solutions in the range. In fact, there is an instance I that has two stable matchings M and M' such that $d(M) = d(M') = 0$ but $c(M) \ll c(M')$. This motivates us to consider the following corresponding optimization problem MinESE (*Minimum Egalitarian Sex-Equal stable marriage problem*): Given a stable marriage instance I and a positive constant ϵ, find a stable matching M which minimizes $c(M)$ under the condition that $|d(M)| \le \epsilon\Delta$, (or answer "No" if none exists). We show that MinESE is NP-hard, and give a polynomial time $(2-(\epsilon-\delta)/(2+3\epsilon))$-approximation algorithm for an arbitrary δ such that $0 < \delta < \epsilon$, whose running time is $O(n^{4+\frac{1+\epsilon}{\delta}})$. Here, an Algorithm A is said to be a c-approximation algorithm if $A(I)/OPT(I) \le c$ holds for any input I, where $A(I)$ and $OPT(I)$ are the costs of A's solution and optimal solution, respectively.

Our results also hold for the *weighted* versions of the above problems, in which $p_m(w)$ ($p_w(m)$, respectively) represents not simply a rank of w in m's preference list, but an arbitrary score of m for w (of w for m), where $p_m(w) > 0$ ($p_w(m) > 0$) and $p_m(w) < p_m(w')$ if and only if m prefers w to w' ($p_w(m) < p_w(m')$ if and only if w prefers m to m') for all m and w.

Related Results. As mentioned above, the minimum regret stable marriage problem and the minimum egalitarian stable marriage problem can be solved in polynomial time [5,9,6], but the sex-equal stable marriage problem is strongly NP-hard [11]. If we allow ties in preference lists, all these problems become hard, even to approximate, if we seek for optimal *weakly* stable matching: For each problem, there exists a positive constant δ such that there is no polynomial-time approximation algorithm with approximation ratio δn unless P=NP [7].

2 Rotation Poset

In this section, we explain a *rotation poset (partially-ordered set)*, originally defined in [8], which is an underlying structure of stable matchings. Here, we give only a brief sketch necessary for understanding the algorithms given later. Readers can refer to [6] for further details.

We fix an instance I. Let M be a stable matching for I. For each such M, we can associate a *reduced list*, which is obtained from the original preference lists by removing entries by some rule. One property of the reduced list associated with M is that, in M, each man is matched with the first woman in the reduced list, and each woman is matched with the last man. A *rotation exposed in* M is an ordered list of pairs $\rho = (m_0, w_0), (m_1, w_1), \ldots, (m_{r-1}, w_{r-1})$ such that, for every i $(0 \le i \le r-1)$, m_i and w_i are matched in M, and w_{i+1} is at the second position in m_i's reduced list, where $i+1$ is taken modulo r. There exists at least one rotation for any stable matching except for the woman-optimal stable matching M_z.

For a stable matching M and a rotation $\rho = (m_0, w_0), (m_1, w_1), \ldots, (m_{r-1}, w_{r-1})$ exposed in M, *eliminating* ρ *from* M means to replace m_i's partner from w_i to w_{i+1} for each i $(0 \le i \le r-1)$, (and to update a reduced list accordingly). Note that by eliminating a rotation, men become worse off while women become better off. The resulting matching is denoted by M/ρ. It is well known that M/ρ is also stable for I. If a rotation is exposed in M/ρ, then we can similarly obtain another stable matching by eliminating it.

Now, let $\mathcal{M}$ be the set of all stable matchings for I, and Π be the set of rotations ρ such that ρ is exposed in some stable matching in $\mathcal{M}$. Then, it is known that $|\Pi| \le n^2$. The *rotation poset* $(\Pi, \prec)$, which is uniquely determined for instance I, is the set Π with a partial order $\prec$ defined for elements in Π. For two rotations ρ_1 and ρ_2 in Π, $\rho_1 \prec \rho_2$ intuitively means that ρ_1 must be eliminated before ρ_2, or ρ_2 is never exposed until ρ_1 is eliminated. It is known that the rotation poset can be constructed in $O(n^2)$ time.

A *closed subset* R of the rotation poset $(\Pi, \prec)$ is a subset of Π such that if $\rho \in R$ and $\rho' \prec \rho$ then $\rho' \in R$. There is a one-to-one correspondence between $\mathcal{M}$ and the set of closed subsets of $(\Pi, \prec)$: Let R be a closed subset. Starting from the man-optimal stable matching M_0, if we eliminate all rotations in R successively in a proper order defined by $\prec$, then we can obtain a stable matching. Conversely, any stable matching can be obtained by this procedure for some closed subset. We denote the stable matching corresponding to a closed subset R by M_R. For simplicity, we sometimes write $c(R)$ and $d(R)$ instead of $c(M_R)$ and $d(M_R)$, respectively. Especially, the empty subset corresponds to the man-optimal stable matching M_0, and the set Π itself corresponds to the woman-optimal stable matching M_z. From M_0, if we eliminate all rotations according to the order $\prec$, then we eventually reach M_z.

For a rotation $\rho = (m_0, w_0), (m_1, w_1), \ldots, (m_{r-1}, w_{r-1})$, we define $w_c(\rho)$ and $w_d(\rho)$, which represent the cost change of egalitarian and sex-equalness, respectively, by eliminating ρ:

$$w_c(\rho) = \sum_{i=0}^{r-1}(p_{m_i}(w_{i+1}) - p_{m_i}(w_i)) + \sum_{i=0}^{r-1}(p_{w_i}(m_{i-1}) - p_{w_i}(m_i)),$$

$$w_d(\rho) = \sum_{i=0}^{r-1}(p_{m_i}(w_{i+1}) - p_{m_i}(w_i)) - \sum_{i=0}^{r-1}(p_{w_i}(m_{i-1}) - p_{w_i}(m_i)).$$

Here, note that $w_d(\rho) > 0$ for all ρ since by eliminating a rotation, some men become worse off, and a some women become better off, and other people remain matched with the same partners. Now, let ρ be a rotation exposed in a stable matching M. Then, it is obvious from the definition that $c(M/\rho) = c(M) + w_c(\rho)$ and $d(M/\rho) = d(M) + w_d(\rho)$. Also, it is easy to see that for any closed subset R,

$$c(M_R) = c(M_0) + \sum_{\rho \in R} w_c(\rho) \quad \text{and} \quad d(M_R) = d(M_0) + \sum_{\rho \in R} w_d(\rho).$$

Hence, the minimum egalitarian stable marriage problem (the sex-equal stable marriage problem, respectively) is equivalent to the problem of finding a closed subset R such that $c(M_0) + \sum_{\rho \in R} w_c(\rho)$ ($|d(M_0) + \sum_{\rho \in R} w_d(\rho)|$, respectively) is minimum. For example, the algorithm for finding a minimum egalitarian stable matching in [9] efficiently finds such R by exploiting network flow.

3 The Sex-Equal Stable Marriage Problem

Recall that M_0 is the man-optimal stable matching and M_z is the woman-optimal stable matching. Note that any stable matching M satisfies $d(M_0) \leq d(M) \leq d(M_z)$. Thus, this problem is trivial if $d(M_0) \geq 0$ or $d(M_z) \leq 0$, namely, if $d(M_0) \geq 0$, M_0 is optimal, while if $d(M_z) \leq 0$, M_z is optimal. Therefore, we consider the case where $d(M_0) < 0 < d(M_z)$. Recall that $\Delta = \min\{|d(M_0)|, |d(M_z)|\}$. In the following, we assume without loss of generality that $|d(M_0)| \leq |d(M_z)|$ since otherwise, we can exchange the role of men and women. Hence, $\Delta = \min\{|d(M_0)|, |d(M_z)|\} = |d(M_0)|$.

We first briefly give the underlying idea of our algorithm presented in this section. Recall that, for a given instance I and ϵ, we are to find a stable matching M such that $-\epsilon\Delta \leq d(M) \leq \epsilon\Delta$ if any. As an easy case, assume that all rotations ρ of I satisfy $w_d(\rho) \leq 2\epsilon\Delta$. Now, we construct a rotation poset $(\Pi, \prec)$ of I, and starting from M_0, we eliminate rotations in an order of any linear extension of $\prec$. Recall that by eliminating a rotation, the sex-equalness cost increases, but by at most $2\epsilon\Delta$ by assumption. Note that $d(M_0) < 0 < d(M_z)$, and recall that if we eliminate all rotations from M_0, we eventually reach M_z. Then, in this sequence, we certainly meet a desirable stable matching at some point.

However, this procedure fails if there is a rotation with large sex-equalness cost: If we eliminate such a rotation, then we may "jump" from M to M' such that $d(M) < -\epsilon\Delta$ and $d(M') > \epsilon\Delta$ even if there is a feasible solution. To resolve this problem, we will try all combinations of selecting such "large" rotations, and treat "small" rotations in the above manner. To evaluate the time complexity, we show that the number of large rotations is limited.

Before giving a description of our algorithm, we give a couple of notations. Let R be any (not necessarily closed) subset of a poset $(\Pi, \prec)$. Then $R_{\min} = R \cup \{\rho \mid \text{there exists a } \rho' \text{ such that } \rho' \in R \text{ and } \rho \prec \rho'\}$. That is, $R_{\min}$ is the minimal closed subset of Π satisfying $R_{\min} \supseteq R$. Similarly, $R_{\max} = R \cup \{\rho \mid$ there exists a ρ' such that $\rho' \in R$ and $\rho' \prec \rho\}$.

Algorithm 1

1. Construct the rotation poset $(\Pi, \prec)$.

2. Let R^L be the set of rotations ρ such that $w_d(\rho) > 2\epsilon\Delta$, and R^S be $\Pi \setminus R^L$.

3. For each set R in 2^{R^L} such that $|R| \le \frac{1+\epsilon}{2\epsilon}$, do,

(a) If $R_{\min} \cap (R^L \setminus R)_{\max} \ne \emptyset$, then go to **3** and choose the next R.

(b) If $-\epsilon\Delta \le d(R_{\min}) \le \epsilon\Delta$, then output $M_{R_{\min}}$.

(c) Fix an arbitrary order $\rho_1, \rho_2, \cdots, \rho_k \in R^S \setminus (R_{\min} \cup (R^L \setminus R)_{\max})$ which is consistent with $\prec$.

(d) For $i = 1$ to k, if $-\epsilon\Delta \le d(R_{\min} \cup \{\rho_1, \rho_2, \cdots, \rho_i\}) \le \epsilon\Delta$, then output $M_{R_{\min} \cup \{\rho_1, \rho_2, \cdots, \rho_i\}}$ and halt.

4. Output "No," and halt.

Theorem 1. *There is an algorithm for NSE whose running time is* $O(n^{3+\frac{1}{\epsilon}})$.

Proof. **Correctness Proof.** Clearly, if there is no M such that $-\epsilon\Delta \le d(M) \le \epsilon\Delta$, then the algorithm answers "No." On the other hand, suppose that there is M_X such that $-\epsilon\Delta \le d(M_X) \le \epsilon\Delta$, where X is the set of rotations corresponding to M_X. Let $X^L = X \cap R^L$ and $X^S = X \cap R^S$. Then, $d(X^L) \le d(M_X) \le \epsilon\Delta$. Because $w_d(\rho) > 2\epsilon\Delta$ for any rotation $\rho \in X^L$, $|X^L| < \frac{d(X^L) - d(M_0)}{2\epsilon\Delta} \le \frac{|d(M_0)| + \epsilon\Delta}{2\epsilon\Delta} = \frac{1+\epsilon}{2\epsilon}$. So, Algorithm 1 selects X^L at Step 3 as R, and we consider this particular execution of Step 3.

First, note that $d((X^L)_{\min}) \le \epsilon\Delta$ since otherwise, $d(M_X) \ge d((X^L)_{\min}) > \epsilon\Delta$, a contradiction. If $-\epsilon\Delta \le d((X^L)_{\min}) \le \epsilon\Delta$, then Algorithm 1 outputs $M_{(X^L)_{\min}}$ at Step 3(b). Finally, suppose that $d((X^L)_{\min}) < -\epsilon\Delta$. Note that $d((X^L)_{\min} \cup \{\rho_1, \rho_2, \cdots, \rho_k\}) \ge d(M_X) \ge -\epsilon\Delta$ and that any rotation ρ_i $(1 \le i \le k)$ satisfies $w_d(\rho_i) \le 2\epsilon\Delta$. Hence there must be j $(1 \le j \le k)$ such that $-\epsilon\Delta \le d((X^L)_{\min} \cup \{\rho_1, \rho_2, \cdots, \rho_j\}) \le \epsilon\Delta$.

Time Complexity. Steps 1 and 2 can be performed in $O(n^2)$. Inside the loop of Step 3 can be performed in $O(n^2)$ since the number of rotations is at most $O(n^2)$. Clearly, Step 4 can be performed in constant time.

We consider the number of repetitions of Step 3, i.e., the number of R satisfying the condition at Step 3. Let this number be t. Recall that the number of rotations is at most n^2 as mentioned in Sec. 2. So, $|R^L| \le n^2$. Since $|R| \le \frac{1+\epsilon}{2\epsilon}$,

$$t = \sum_{k=1}^{\lfloor \frac{1+\epsilon}{2\epsilon} \rfloor} \binom{n^2}{k} \le \sum_{k=1}^{\lfloor \frac{1+\epsilon}{2\epsilon} \rfloor} \frac{(n^2)^{\lfloor \frac{1+\epsilon}{2\epsilon} \rfloor}}{k!} = O(n^{\frac{1+\epsilon}{\epsilon}}).$$

Hence the time complexity of Algorithm 1 is $O(n^2) \cdot O(n^{\frac{1+\epsilon}{\epsilon}}) = O(n^{3+\frac{1}{\epsilon}})$. □

4 The Minimum Egalitarian Sex-Equal Stable Marriage Problem

In NSE, we are asked to find a stable matching whose sex-equalness cost is in some range close to 0. However, if there are several stable matchings satisfying the condition, there might be good ones and bad ones. In fact, there is an instance I that has two stable matchings M and M' whose sex-equalness costs are the same (0), but egalitarian costs are significantly different. (Because of the space restriction, we omit the construction of this instance.) This motivates us to consider the following problem, *MinESE* (*the Minimum Egalitarian Sex-Equal stable marriage problem*): Given an instance I and a constant ϵ such that $0 < \epsilon < 1$, find a stable matching M with minimum $c(M)$, under the condition that $|d(M)| \leq \epsilon\Delta$, (or answer "No" if no such solution exists). First, in Sec. 4.1, we show that MinESE is NP-hard. Then, in Sec. 4.2, we give an approximation algorithm for MinESE.

4.1 NP-Hardness of MinESE

It turned out that there is a polynomial-time algorithm for obtaining a stable matching M such that (a) $-\epsilon\Delta \leq d(M) \leq \epsilon\Delta$ *or* (b) $c(M)$ is minimum. Interestingly, it is hard to obtain M satisfying (a) *and* (b).

Theorem 2. *MinESE is NP-hard.*

Proof. Because of the space restriction, we give only a rough idea of the proof, and omit the details.

We show that MinESE is NP-hard by a reduction from the k-clique problem. In this problem, we are given a graph $G(V, E)$ and an integer k, and asked if there exists a clique of size k. This problem is NP-complete.

Given a graph $G = (V, E)$ and an integer k, we first construct a poset $(\Pi, \prec)$ in a similar manner as the construction in [10]. Let Π be $V \cup E$, and define the precedence relation $\prec$ as follows: $v \prec e$ if and only if $v \in V$ is incident to $e \in E$ in $G(V, E)$. We then give weights to each element in $(\Pi, \prec)$: We give some negative weight to each element corresponding to an edge, and positive weight to an element corresponding to a vertex. Note that if we want to select a closed subset with smaller weight, we want to select many elements corresponding to edges, but to make the subset closed, we need to select elements corresponding to adjacent vertices, which may increase the weight. We give weights to vertices and edges appropriately, so that only closed subsets corresponding to k-cliques can have a desirable (negative) weight.

Next, we construct an instance I of MinESE from the poset $(\Pi, \prec)$ using a similar construction as [11], so that the rotation poset of I is exactly $(\Pi, \prec)$. In the construction, we ensure that the egalitarian cost of each rotation is exactly the same to the weight of the corresponding element defined above. We also adjust sex-equalness cost of rotations so that if R is a closed subset corresponding to a k-clique of G, then $d(M_R)$ lies between $-\epsilon\Delta$ and $\epsilon\Delta$. In summary, our reduction satisfies the following: G has a k-clique if and only if there is a stable matching M in I such that $-\epsilon\Delta \leq d(M) \leq \epsilon\Delta$ and $c(M) < c(M_0)$.

We can show that MinESE is NP-hard as follows: Given an instance of the clique problem, we construct a MinESE instance I by the above reduction. Then, we find an optimal solution M and the man-optimal stable matching M_0. Finally, we compare $c(M_0)$ and $c(M)$: If $c(M) < c(M_0)$, the answer to the clique problem is "yes," otherwise, "no." □

Remark. Although details are omitted, the reduction in the NP-hardness proof produces an instance (I, ϵ) of MinESE such that $|d(M_0)| = |d(M_z)|$ in I, and ϵ is any constant such that $0 < \epsilon < 1$. Observe that if $|d(M_0)| = |d(M_z)|$ and $\epsilon = 1$, then MinESE is equivalent to the minimum egalitarian stable marriage problem, which can be solved in polynomial time.

Remark. We can modify the reduction so that it preserves the gap of the Dense Subgraph Problem (DSP) (see [3,1], e.g.). Although some PTASs are known for DSP in some settings of parameters, existence of PTAS is not known for general case. Feige [2] and Khot [12] provided evidence that DSP may be hard to approximate within some constant factor. Therefore, we conjecture that MinESE either does not have a PTAS.

4.2 Approximation Algorithms for MinESE

Here, we give a $(2 - (\epsilon - \delta)/(2 + 3\epsilon))$-approximation algorithm for MinESE for an arbitrary δ such that $0 < \delta < \epsilon$. Similarly as Sec. 3, we assume that $|d(M_0)| \leq |d(M_z)|$. In this section, we prove two simple but important lemmas that link the egalitarian cost and the sex-equalness cost, whose proofs are given later. (i) For any stable matching M, $|d(M)| < c(M)$ (Lemma 1). (ii) For any stable matching M and a rotation ρ exposed in M, by eliminating ρ from M, the cost change in the egalitarian cost is at most the cost change in the sex-equalness cost (Lemma 2).

To illustrate an idea of the algorithm, we first consider a restricted case and show that our algorithm achieves 2-approximation. For a fixed $\delta > 0$, suppose that all rotations satisfy $w_d(\rho) \leq \delta\Delta$. Given I and ϵ, we first find a minimum egalitarian stable matching M_{eg}, which can be done in polynomial time. If $-\epsilon\Delta \leq d(M_{eg}) \leq \epsilon\Delta$, then we are done since M_{eg} is an optimal solution for MinESE. If $d(M_{eg}) < -\epsilon\Delta$, then we eliminate rotations one by one as Algorithm 1 until the sex-equalness cost first becomes $-\epsilon\Delta$ or larger. If $d(M_{eg}) > \epsilon\Delta$, then we "add" rotations one by one until the sex-equalness cost first becomes $\epsilon\Delta$ or smaller. Here, "adding a rotation" means the reverse operation of eliminating a rotation. If we do not reach a feasible solution by this procedure, then we can conclude that there is no feasible solution, by a similar argument as in Sec. 3. If we find a stable matching M such that $-\epsilon\Delta \leq d(M) \leq \epsilon\Delta$, then we can show that this is a 2-approximation, namely, $c(M) \leq 2c(M_{eg})$ using (i) and (ii) above (note that the optimal cost is at least $c(M_{eg})$): Suppose, for example, that $d(M_{eg}) < -\epsilon\Delta$ (see Fig. 2). Then, by (ii), $c(M) - c(M_{eg}) \le d(M) - d(M_{eg})$, and by (i), $|d(M_{eg})| < c(M_{eg})$. Also, since the costs of rotations are at most $\delta\Delta$, and since M is the first feasible solution found by this procedure, $d(M) \le -(\epsilon - \delta)\Delta < 0$. Putting these together, we have that $c(M)/c(M_{eg}) < 2$.

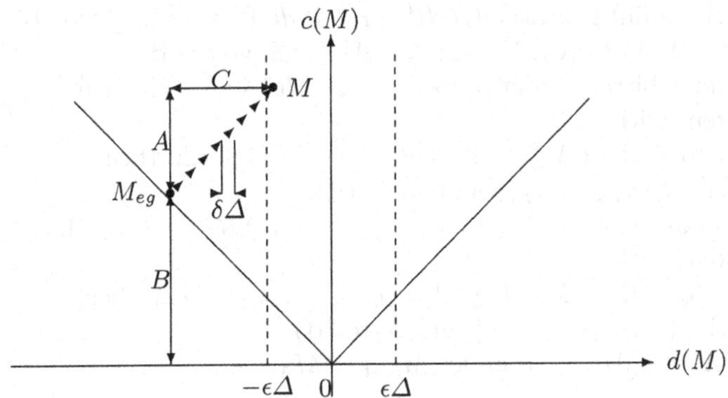

Fig. 2. $C \le B$ by (i) and $A \le C$ by (ii). Hence $A + B \le C + B \le 2B$.

However, we may have rotations of large costs. Then we take a similar approach as in Sec. 3: Let R^L be the set of such large rotations. Then, for any partition R_1 and R_2 of R^L ($R_1 \cup R_2 = R^L$ and $R_1 \cap R_2 = \emptyset$), we want to obtain a minimum egalitarian stable matching whose corresponding closed subset A contains all rotations in R_1 but none in R_2. For this purpose, we need to solve the following problem: Given an instance I and disjoint subsets of rotations R_1 and R_2 of R^L, find a minimum egalitarian stable matching M_A under the condition that the corresponding closed subset A satisfies $A \supseteq R_1$ and $A \cap R_2 = \emptyset$. For this problem, we can use the same algorithm for the minimum egalitarian stable marriage problem in [6]. We denote this procedure by minEgalitarian(R_1, R_2). First, we review the following proposition described in [6]:

Proposition 1. *[6] Given a poset $(\Pi, \prec)$, there is an $O(n^4)$-time algorithm which finds a minimum-weight closed subset of $(\Pi, \prec)$ with respect to the egalitarian cost.*

Our procedure minEgalitarian(R_1, R_2) is as follows: Without loss of generality, assume that there are no elements such that $r_2 \prec r_1$ ($r_1 \in R_1$ and $r_2 \in R_2$) since there exists no solution in such a case. Construct the poset $(\Pi', \prec)$ by removing all the rotations in $(R_1)_{\min}$ and $(R_2)_{\max}$ from $(\Pi, \prec)$ (recall the definitions of $R_{\min}$ and $R_{\max}$ given before Algorithm 1), and let R' be the subset obtained by using Proposition 1 to $(\Pi', \prec)$. Then, it is easy to see that $(R_1)_{\min} \cup R'$ is an optimal solution for minEgalitarian(R_1, R_2). Now, we are ready to give the algorithm for MinESE.

Algorithm 2
1. Construct the rotation poset $(\Pi, \prec)$.
2. Let $M_{best} = NULL$.
3. Let R^L be the set of rotations ρ such that $w_d(\rho) > \delta\Delta$, and R^S be $\Pi \setminus R^L$.
4. For each set R in 2^{R^L} such that $|R| \leq \frac{1+\epsilon}{\delta}$, do,

(a) Let $A = \text{minEgalitarian}(R, R^L \setminus R)$. If $d(A) < -\epsilon\Delta$, go to **(b)**. If $d(A) > \epsilon\Delta$, go to **(c)**. If $-\epsilon\Delta \leq d(A) \leq \epsilon\Delta$, go to **(d)**.
(b) Fix an arbitrary order $\rho_1, \rho_2, \cdots, \rho_k \in R^S \setminus (A \cup (R^L \setminus R)_{\max})$ which is consistent with $\prec$.
For $i = 1$ to k, if $-\epsilon\Delta \leq d(A \cup \{\rho_1, \rho_2, \cdots, \rho_i\}) \leq \epsilon\Delta$, then let $A = A \cup \{\rho_1, \rho_2, \cdots, \rho_i\}$ and go to **(d)**.
(c) Fix an arbitrary order $\rho_1, \rho_2, \cdots, \rho_k \in (A \cap R^S) \setminus R_{\min}$ which is consistent with $\prec$.
For $i = k$ to 1, if $-\epsilon\Delta \leq d(A \setminus \{\rho_i, \rho_{i+1}, \cdots, \rho_k\}) \leq \epsilon\Delta$, then let $A = A \setminus \{\rho_i, \rho_{i+1}, \cdots, \rho_k\}$ and go to **(d)**.
(d) If $c(A) < c(M_{best})$, then let $M_{best} = M_A$.

5. If $M_{best} \neq NULL$, then output M_{best}, otherwise output "No," and halt.

Theorem 3. *There is a* $(2 - (\epsilon - \delta)/(2 + 3\epsilon))$*-approximation algorithm for MinESE whose running time is* $O(n^{4+\frac{1+\epsilon}{\delta}})$ *for an arbitrary* δ *such that* $0 < \delta < \epsilon$.

Proof. **Correctness Proof.** Clearly, if there is no M such that $-\epsilon\Delta \leq d(M) \leq \epsilon\Delta$, then the algorithm answers "No." On the other hand, suppose that there is a feasible solution, and let M_{opt} be an optimal solution. We first show that Algorithm 2 finds a feasible solution. Let OPT be the rotation set corresponding to M_{opt}, and $OPT^L = OPT \cap R^L$. Then, $d(OPT^L) \leq d(M_{opt}) \leq \epsilon\Delta$. Because $w_d(\rho) > \delta\Delta$ for any rotation $\rho \in OPT^L$, $|OPT^L| < \frac{d(OPT^L) - d(M_0)}{\delta\Delta} \leq \frac{|d(M_0)| + \epsilon\Delta}{\delta\Delta} = \frac{1+\epsilon}{\delta}$. So, Algorithm 2 selects OPT^L at Step 4 as R, and we consider this particular execution of Step 4. We show that in this execution, Algorithm 2 finds a feasible solution. Let $A_{opt} = \text{minEgalitarian}(OPT^L, R^L \setminus OPT^L)$. There are three cases:

(i) $-\epsilon\Delta \leq d(A_{opt}) \leq \epsilon\Delta$**.** $M_{A_{opt}}$ is selected as M_{best} at Step 4(d).
(ii) $d(A_{opt}) < -\epsilon\Delta$**.** Note that $d(A_{opt} \cup \{\rho_1, \rho_2, \cdots, \rho_k\}) \geq d(M_{opt}) \geq -\epsilon\Delta$ and that any rotation ρ_i $(1 \leq i \leq k)$ satisfies $w_d(\rho_i) \leq \delta\Delta$. Hence there must be j $(1 \leq j \leq k)$ such that $-\epsilon\Delta \leq d(A_{opt} \cup \{\rho_1, \rho_2, \cdots, \rho_j\}) \leq -(\epsilon - \delta)\Delta$. (See Fig. 3.)
(iii) $d(A_{opt}) > \epsilon\Delta$**.** Note that $d(A_{opt} \setminus \{\rho_1, \rho_2, \cdots, \rho_k\}) \leq d(M_{opt}) \leq \epsilon\Delta$ and that any rotation ρ_i $(1 \leq i \leq k)$ satisfies $w_d(\rho_i) \leq \delta\Delta$. Hence there must be j $(1 \leq j \leq k)$ such that $(\epsilon - \delta)\Delta \leq d(A_{opt} \setminus \{\rho_j, \rho_{j+1}, \cdots, \rho_k\}) \leq \epsilon\Delta$.
Next, we analyze the approximation ratio. Let M^* be the matching found in this particular execution of Step 4. We show that $c(M^*) \leq (2 - (\epsilon - \delta)/$

$(2+3\epsilon))c(M_{opt})$, which gives a proof for the approximation ratio. We first prove the following two lemmas:

Lemma 1. *For any stable matching M, $|d(M)| < c(M)$.*

Proof. If $d(M) \geq 0$, then $c(M) - |d(M)| = 2\sum_{(m,w)\in M} p_w(m) > 0$. Otherwise, $c(M) - |d(M)| = 2\sum_{(m,w)\in M} p_m(w) > 0$. □

Lemma 2. *Let $R = \{\rho_1, \ldots, \rho_{r-1}\}$ be a set of rotations and let $M_1, \cdots, M_r$ be stable matchings such that $M_{i+1} = M_i/\rho_i$ for $1 \leq i < r$. Then, $|c(M_r)-c(M_1)| \leq d(M_r) - d(M_1)$.*

Proof. Suppose that $m = M_i(w) = M_{i+1}(w')$ and $w = M_i(m) = M_{i+1}(m')$ for a fixed i. By the properties of the rotation [6], m prefers w to w' and w prefers m' to m. Let $d(m) = p_m(w') - p_m(w)$ and $d(w) = p_w(m) - p_w(m')$. Then $d(m) \geq 0$ and $d(w) \geq 0$, and it follows that

$$|c(M_{i+1})-c(M_i)| = \left|\sum_m d(m) - \sum_w d(w)\right| \leq \left|\sum_m d(m) + \sum_w d(w)\right| = d(M_{i+1})-d(M_i).$$

By summing up the above inequality for all i, we have

$$|c(M_r)-c(M_1)| \leq \sum_{i=1}^{r-1} |c(M_{i+1})-c(M_i)| \leq \sum_{i=1}^{r-1}(d(M_{i+1})-d(M_i)) = d(M_r)-d(M_1).$$

□

Note that A_{opt} = minEgalitarian($OPT^L, R^L \setminus OPT^L$). So, $c(A_{opt}) \leq c(M_{opt})$ since OPT, the rotation set corresponding to M_{opt}, is one of the candidates for A_{opt}. We will consider the following four cases (note that $d(A_{opt}) \geq -\Delta$ for any stable matching M):

Case (i): $-\epsilon\Delta \leq d(A_{opt}) \leq \epsilon\Delta$. In this case, $M^* = M_{A_{opt}}$, which is an optimal solution since $c(A_{opt}) = c(M_{opt})$.

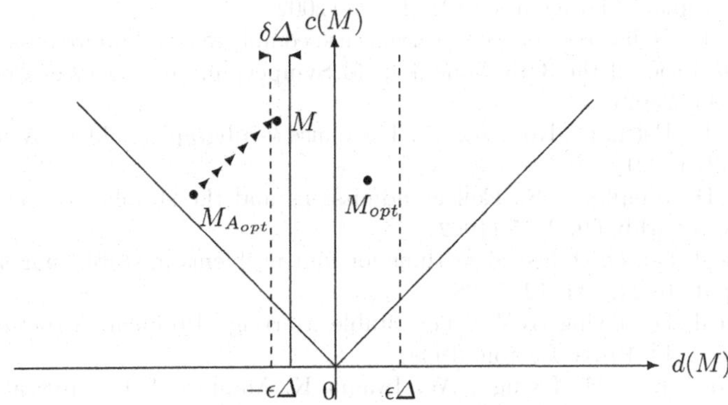

Fig. 3. Finding a feasible solution

Case (ii): $\epsilon\Delta < d(A_{opt}) \leq (2+3\epsilon)\Delta$. In this case, Step 4(b) of Algorithm 2 is executed. We have $|c(A_{opt}) - c(M^*)| \leq d(A_{opt}) - d(M^*)$ by Lemma 2. Since $d(M^*) \geq (\epsilon - \delta)\Delta$ and (ii) hold, $|c(A_{opt}) - c(M^*)| \leq (1 - (\epsilon - \delta)/(2+3\epsilon))d(A_{opt})$. Since $|d(A_{opt})| < c(A_{opt})$ by Lemma 1 and $c(A_{opt}) \leq c(M_{opt})$, $c(M^*) < (2 - (\epsilon - \delta)/(2+3\epsilon))c(M_{opt})$.

Case (iii): $(2+3\epsilon)\Delta < d(A_{opt})$. Since both M_{opt} and M^* can be obtained by repeatedly eliminating rotations from M_0, $|c(M_{opt}) - c(M_0)| \leq d(M_{opt}) - d(M_0)$ and $|c(M^*) - c(M_0)| \leq d(M^*) - d(M_0)$ by Lemma 2. Since both $d(M_{opt})$ and $d(M^*)$ are at most $\epsilon\Delta$, $c(M^*) - c(M_{opt}) \leq 2(1+\epsilon)\Delta$ (note that $|d(M_0)| = \Delta$). It follows that $c(M^*) - c(M_{opt}) \leq 2(1+\epsilon)d(A_{opt})/(2+3\epsilon) = (2 - \epsilon/(2+3\epsilon))d(A_{opt})$. Since we have $|d(A_{opt})| < c(A_{opt})$ by Lemma 1 and $c(A_{opt}) \leq c(M_{opt})$, $c(M^*) \leq (2 - \epsilon/(2+3\epsilon))c(M_{opt})$.

Case (iv): $-\Delta \leq d(A_{opt}) < -\epsilon\Delta$. The same as Case (ii).

Time Complexity. Steps 1, 2, 3, and 5 can be executed in $O(n^2)$ time. Step 4(a) is performed in the same time complexity as finding a minimum egalitarian stable matching, namely, $O(n^4)$. We can see that Steps 4(b) through 4(d) can be performed in time $O(n^2)$ by a similar analysis of Algorithm 1. The number of repetitions of Step 4 can be analyzed in the same way as the proof of Theorem 1, which is $O(n^{\frac{1+\epsilon}{\delta}})$. Hence the time complexity of Algorithm 2 is $O(n^{4+\frac{1+\epsilon}{\delta}})$. □

5 Concluding Remarks

In this paper, we gave a polynomial time algorithm for finding near optimal sex-equal stable matching. Furthermore, we proved NP-hardness and developed a polynomial time approximation algorithm whose approximation ratio is less than 2 for MinESE. Our future work is to improve the approximation ratio of MinESE.

References

1. Asahiro, Y., Hassin, R., Iwama, K.: Complexity of finding dense subgraphs. Discrete Applied Mathematics 121, 15–26 (2002)
2. Feige, U.: Relations between average case complexity and approximation complexity. In: Proc. of the 34th Annual ACM Symposium on Theory of Computing, pp. 534–543 (2002)
3. Feige, U., Peleg, D., Kortsarz, G.: The dense k-subgraph problem. Algorithmica 29, 410–421 (2001)
4. Gale, D., Shapley, L.S.: College admissions and the stability of marriage. Amer. Math. Monthly 69, 9–15 (1962)
5. Gusfield, D.: Three fast algorithms for four problems in stable marriage. SIAM J. Comput. 16(1), 111–128 (1987)
6. Gusfield, D., Irving, R.W.: The Stable Marriage Problem: Structure and Algorithms. MIT Press, Boston (1989)
7. Halldórsson, M.M., Irving, R.W., Iwama, K., Manlove, D.F., Miyazaki, S., Morita, Y., Scott, S.: Approximability results for stable marriage problems with ties. Theoretical Computer Science 306, 431–447 (2003)

8. Irving, R.W., Leather, P.: The complexity of counting stable marriages. SIAM J. Comput. 15(3), 655–667 (1986)
9. Irving, R.W., Leather, P., Gusfield, D.: An efficient algorithm for the "optimal" stable marriage. J. ACM 34(3), 532–543 (1987)
10. Johnson, D.S., Niemi, K.A.: On knapsacks, partitions, and a new dynamic programming technique for trees. Mathematics of Operations Research 8(1), 1–14 (1983)
11. Kato, A.: Complexity of the sex-equal stable marriage problem. Japan Journal of Industrial and Applied Mathematics (JJIAM) 10, 1–19 (1993)
12. Khot, S.: Ruling out PTAS for graph min-bisection, densest subgraph and bipartite clique. In: Proc. of the 45th Annual IEEE Symposium on Foundations of Computer Science, pp. 136–145 (2004)

A Stab at Approximating Minimum Subadditive Join

Staal A. Vinterbo[1,2,3]

[1] Decision Systems Group, Brigham and Women's Hospital, Boston
[2] Harvard Medical School, Boston
staal@dsg.harvard.edu
[3] Harvard-MIT, Division of Health Sciences and Technology, Boston

Abstract. Let $(L, *)$ be a semilattice, and let $c: L \to [0, \infty)$ be monotone and increasing on L. We state the Minimum Join problem as: given size n sub-collection X of L and integer k with $1 \leq k \leq n$, find a size k sub-collection $(x'_1, x'_2, \ldots, x'_k)$ of X that minimizes $c(x'_1 * x'_2 * \cdots * x'_k)$. If $c(a * b) \leq c(a) + c(b)$ holds, we call this the Minimum Subadditive Join (MSJ) problem and present a greedy $(k-p+1)$-approximation algorithm requiring $O((k-p)n + n^p)$ joins for constant integer $0 < p \leq k$. We show that the MSJ Minimum Coverage problem of selecting k out of n finite sets such that their union is minimal is essentially as hard to approximate as the Maximum Balanced Complete Bipartite Subgraph (MBCBS) problem. The motivating by-product of the above is that the privacy in databases related k-ambiguity problem over L with subadditive information loss can be approximated within $k - p$, and that the k-ambiguity problem is essentially at least as hard to approximate as MBCBS.

1 Introduction

In this paper we will often talk about finite collections of elements from some set. What we mean by a collection is an ordered multiset. Let $I_n = \{1, 2, \ldots, n\}$, and let U be a set. We can then represent a collection of size n of elements from U as a function $\mathcal{X} : I_n \to U$. We will denote the set of all collections of elements from U of size n as $\mathcal{C}_n(U)$. We can now formally define the Minimum Subadditive Join[1] (MSJ) problem as follows.

Problem 1 (Minimum Subadditive Join). Let $(L, *)$ be a semilattice where $*$ is polynomial time computable, and let $c: S \to [0, \infty)$ be a monotone, increasing and polynomial time computable function on L such that

$$c(a * b) \leq c(a) + c(b) \tag{1}$$

holds for $a, b \in L$. Given $\mathcal{X}_n \in \mathcal{C}_n(L)$ an integer k with $1 \leq k \leq n$, find a subset $S = \{s_1, s_2, \ldots, s_k\} \subset I_n$ that minimizes $c(\mathcal{X}_n(s_1) * \mathcal{X}_n(s_2) * \cdots * \mathcal{X}_n(s_k))$. We denote an instance of this problem as $((L, *), \mathcal{X}_n, c, k)$.

[1] We could equally well have chosen to call the operation $*$ "meet", but motivated by the behavior of cardinality over $(2^X, \cup)$, we chose "join".

F. Dehne, J.-R. Sack, and N. Zeh (Eds.): WADS 2007, LNCS 4619, pp. 214–225, 2007.

If (1) is not required to hold, we call the problem the Minimum Join (MJ) problem.

Our study of the Minimum Join problem is principally motivated by its relation to privacy in databases. Biomedical research is dependent on sharing of data [1,2]. Two immediate reasons for this are the principle of reproducibility of research, and the need of comparative retrospective data analysis. However, we are ethically [3], and legally [4] bound to protect the privacy of individuals, and the consequences of loss of trust in the preservation of privacy can be dire [5].

With the advent of large collections of data relatively easily available in electronic form on the web, privacy is endangered through the potential ability of linking a combination of individually seemingly "safe" data items across tables.

Consider the following data matrix in Table 1. Let each row represent what we know about a particular individual. We want to release the data contained

Table 1. Example data set

	a	b	c	d
1	1	0	1	1
2	0	1	1	1
3	0	0	0	1
4	0	1	1	0

in Table 1 without endangering the privacy rights of the individuals in question. As we don't know what a potential adversary is capable of, we postulate the existence of a "linking machine" ϕ that when given a row in Table 1, yields the identity of the individual. Introducing ambiguity in the data is a countermeasure [6,7,8,9] that can be applied to reduce the applicability of ϕ. A way to introduce this ambiguity is to define the value $\top$ to be indistinguishable from both 0 and 1, in effect representing both of these values. A row is then made indistinguishable from others by substituting values with $\top$. Consider row 1. It differs from row 2 in attributes a and b. We collect the sets of attributes in which row 1 differs from all the others it in a collection $C_1 = (\{a,b\},\{a,c\},\{a,b,d\})$. If we need to make row 1 indistinguishable from k rows, we can do this by selecting $k-1$ elements from C_1 and substituting $\top$ for the values in row 1 identified by the union of the selected elements. Say $k=3$, we can then choose the sets $\{a,b\}$ and $\{a,b,d\}$ corresponding to rows 2 and 4, respectively. The result for row 1 is $(\top\top1\top)$.

The above is an example of k-ambiguity [9] by cell suppression. A particular flavor of k-ambiguity is k-anonymity [6], in that k-anonymity requires that every record in the transformed data is equivalent to k records in the *transformed* data. The problem of introducing k-anonymity by a minimum number of cell suppressions was shown to be approximable within $O(k \log k)$ [10], and subsequently a more general version of the k-anonymity problem was shown to be approximable within $\max\{2k-1, 3k-5\}$ [11]. We will in the following

- present a polynomial time $(k-p+1)$-approximation algorithm for the MSJ problem and show that this algorithm fails to provide any bound for non-subadditive instances of MJ, and
- prove that a specialization Minimum Coverage (MinC) of the MSJ problem is essentially as hard to approximate as the known NP-hard Maximum Balanced Complete Bipartite Subgraph (MBCBS) problem,
- use the connection between MJ and k-ambiguity to present a $(k-p)$-approximation algorithm for subadditive information loss instances of the k-ambiguity problem. We also show that negative approximation results obtained for MSJ are applicable to the k-ambiguity problem.

2 A Polynomial Time Approximation Algorithm

The algorithm GREEDY-MSJ in Algorithm 1 is a simple greedy algorithm that essentially works by iteratively adding the unused element that minimizes the added cost to the solution. We also include that by pre-computing an optimal solution for $p \leq k$, we can improve the upper approximation bound. Let $V^* = \mathcal{X}_n(v_1) * \mathcal{X}_n(v_2) * \cdots * \mathcal{X}_n(v_m)$ for any $V = \{v_1, v_2, \ldots, v_m\} \subseteq I_n$.

Algorithm 1. The greedy $(k-p+1)$-approximation algorithm

```
GREEDY-MSJ(X_n, c, k, p)
C ← I_n
S ← ∅
(i_1, i_2, ..., i_p) ← arg min_{1≤j_1<j_2<···<j_p≤n} c({j_1, j_2, ..., j_p}*)
S ← {i_1, i_2, ..., i_p}
z ← c(S*)
C ← C − S
while k > p
    i ← arg min_{j∈C} c(z * X_n(j))
    S ← S ∪ {i}
    z ← z * X_n(i)
    C ← C − {i}
    k ← k − 1
return S
```

Theorem 1. *For constant integer $0 < p \leq k$ we have that* GREEDY-MSJ *is a polynomial time $(k-p+1)$-approximation algorithm for the Minimum Subadditive Join problem that can be implemented to run in $O(((k-p)n + n^p)t(*))$ time, where $t(*)$ is the time it takes to compute $*$.*

Proof. Let

- $S(i) = \{s_1, s_2, \ldots, s_i\}$ be the solution returned by GREEDY-MSJ$(\mathcal{X}_n, c, i, p)$, with s_j being added to the solution before s_{j+1} for all $j > p$, and

– $O(i) = \{{}_io_1, {}_io_2, \ldots, {}_io_i\}$, be an optimal solution for instance $(\mathcal{X}_n, c, i)$, and let $O(i)$ be ordered such that any elements in $O(i)$ occurring in $S(i)$ come first and in the same order as in $S(i)$. This ensures that

$$ {}_io_i \notin S(i-1) \tag{2} $$

holds.

First of all, note that by design of the algorithm, we have that

$$ c(O(p)^*) = c(S(p)^*) \; . \tag{3} $$

Next we note that because of monotonicity of c and (1) we have that for any $i \geq p$

$$ c(O(i)^*) = 0 \Rightarrow c(S(i)^*) = 0 \; , \tag{4} $$

and GREEDY-MSJ produces an optimal solution in this case. Therefore now assume that $c(O(i)^*) > 0$. By the greediness of GREEDY-MSJ, (2), and (1) we also have that that for $i \geq p$

$$ \begin{aligned} c(S(i+1)^*) &= c(S(i)^* * \mathcal{X}_n(s_{i+1})) \\ &\leq c(S(i)^* * \mathcal{X}_n({}_{i+1}o_{i+1})) \\ &\leq c(S(i)^*) + c(\mathcal{X}_n({}_{i+1}o_{i+1})) \end{aligned} \tag{5} $$

holds. By (3) we get

$$ c(S(p)^*) + c(\mathcal{X}_n({}_{p+1}o_{p+1})) = c(O(p)^*) + c(\mathcal{X}_n({}_{p+1}o_{p+1})) \; . \tag{6} $$

Further, we have that $c(O(i)^*) \geq c(\{{}_io_1, {}_io_2, \ldots, {}_io_{i-1}\}^*) \geq c(O(i-1)^*)$ by monotonicity of c and optimality of $O(i-1)$. This, combined with (5) and (6), means that

$$ \begin{aligned} \frac{c(S(p+1)^*)}{c(O(p+1)^*)} &\leq \frac{c(O(p)^*) + c(\mathcal{X}_n({}_{p+1}o_{p+1}))}{c(O(p+1)^*)} \\ &= \frac{c(O(p)^*)}{c(O(p+1)^*)} + \frac{c(\mathcal{X}_n({}_{p+1}o_{p+1}))}{c(O(p+1)^*)} \leq 1 + 1 \leq 2 \; . \end{aligned} \tag{7} $$

Furthermore, using (5) and $c(O(i)^*) \geq c(O(i-1)^*)$, we get that for $i > p$

$$ \begin{aligned} \frac{c(S(i)^*)}{c(O(i)^*)} &\leq \frac{c(S(i-1)^*) + c(\mathcal{X}_n({}_io_i))}{c(O(i)^*)} \\ &= \frac{c(S(i-1)^*)}{c(O(i)^*)} + \frac{c(\mathcal{X}_n({}_io_i))}{c(O(i)^*)} \\ &\leq \frac{c(S(i-1)^*)}{c(O(i-1)^*)} + \frac{c(\mathcal{X}_n({}_io_i))}{c(O(i)^*)} \end{aligned} $$

holds. By monotonicity of c we have that $c(\mathcal{X}_n({}_io_i)) \leq c(O(i)^*)$, by which we get

$$ \frac{c(S(i)^*)}{c(O(i)^*)} \leq \frac{c(S(i-1)^*)}{c(O(i-1)^*)} + 1 \; . \tag{8} $$

Using (7) and induction on (8), we get

$$\frac{c(S(k)^*)}{c(O(k)^*)} \leq k - p + 1 . \tag{9}$$

The computation of an optimal solution for $i = p$ in line 3 can be implemented as p nested loops over C, which yields the $O(n^p t(*))$ term. The subsequent steps are $O((k-p)nt(*))$, making the algorithm implementable in $O(((k-p)n + n^p)t(*))$ time. □

We now investigate the non-subadditive case. We will in the following show that for any $\beta > 1$, we can construct a MJ instance such that GREEDY-MSJ produces a solution worse than β times the optimal.

Proposition 1. GREEDY-MSJ *fails to approximate MJ within any given bound* $\beta > 1$.

Proof. Let $X = \{s_1, s_2, \ldots, s_k, o_1, o_2, \ldots, o_k\} = L$, and let $*$ be defined such that the Hasse diagram of $\leq$ is Figure 1. Further, let

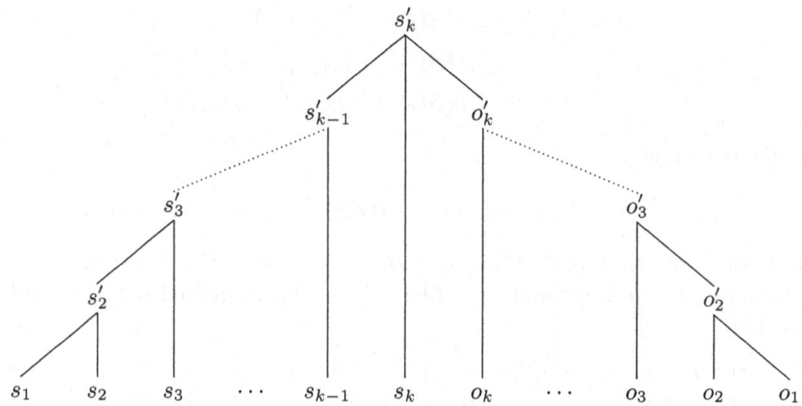

Fig. 1. Example of GREEDY-MSJ failure

- $c(s_1) = \frac{1}{3k}$ and $c(s_i) = \frac{1}{2k}$ for $2 \leq i \leq k$,
- $c(o_i) = \frac{1}{k}$ for $1 \leq i \leq k$,
- $c(s'_i) = \sum_{j=1}^{i} c(s_j)$ for $1 \leq i < k$,
- $c(o'_i) = \sum_{j=1}^{i} c(o_j)$ for $1 \leq i \leq k$, and
- $c(s'_k) = (\beta + \epsilon)$,

for some $\epsilon > 0$ and $\beta > 1$. Let $\mathcal{X}_n$ represent X, then we have that

$$\frac{c(\text{GREEDY-MSJ}(\mathcal{X}_n, c, i, 1)^*)}{c(O_i)} = 1, \text{ for } 1 \leq i < k, \text{ and}$$

$$\frac{c(\text{GREEDY-MSJ}(\mathcal{X}_n, c, k, j)^*)}{c(O_k)} > \beta$$

for $j < k$ if we let O_i be an optimal solution for $k = i$. □

3 Minimum Coverage

Consider the problem of selecting k out of n finite subsets of some set such that their union is minimal. It is clear that this problem can be formulated as follows.

Problem 2 (Minimum Coverage). Let U be a finite set, and let $\mathcal{X}_n \in \mathcal{C}_n(2^U)$. Find $S = \{s_1, s_2, \ldots, s_k\} \subset I_n$ that minimizes $|\mathcal{X}_n(s_1) \cup \mathcal{X}_n(s_2) \cup \cdots \cup \mathcal{X}_n(s_k))|$. We denote an instance of this problem as $(\mathcal{X}_n, k)$.

We call the above Problem 2 Minimum Coverage (MinC) since the problem of *maximizing* $|\mathcal{X}_n(s_1) \cup \mathcal{X}_n(s_2) \cup \cdots \cup \mathcal{X}_n(s_k))|$ in Problem 2 is known as the Maximum Coverage problem [12].

It is also clear that since $(2^U, \cup)$ is a semi-lattice, that $|X \cup Y| \geq |X|$ and $|X \cup Y| \leq |X| + |Y|$ for all $X, Y \in 2^U$, and that both $\cup$ and $||$ are computable in polynomial time, we have that a Minimum Coverage instance $(\mathcal{X}_n, k)$ is the MSJ instance $((2^U, \cup), \mathcal{X}_n, c, k)$ where $c(X) = |X|$. We formulate the above as the following proposition.

Proposition 2. *The Minimum Coverage instance $(\mathcal{X}_n, k)$ is the Minimum Subadditive Join instance $((2^U, \cup), \mathcal{X}_n, c, k)$ where $c(X) = |X|$.*

Proposition 3. *Let $(L, \vee, \wedge, ^-, 0, 1)$ be a finite Boolean algebra such that $\vee$, $\wedge$, and $^-$ all are computable in polynomial time. Let polynomial time computable function $g : L \to [0, \infty)$ be such that $g(x) = g(1) - g(\overline{x})$ for all $x \in L$, and let $h(x) = g(1) - g(x)$. Also, let $\overline{\mathcal{X}_n}$ be such that $\overline{\mathcal{X}_n}(i) = \overline{\mathcal{X}_n(i)}$. Then MSJ instances $((L, \vee), \mathcal{X}_n, g, k)$ and $((L, \wedge), \overline{\mathcal{X}_n}, h, k)$ are polynomial time equivalent.*

Proof. We start by noting that by applying DeMorgan's laws we have that

$$\mathcal{X}_n(i) \vee \mathcal{X}_n(j) = \overline{\overline{\mathcal{X}_n(i)} \wedge \overline{\mathcal{X}_n(j)}} = \overline{\overline{\mathcal{X}_n}(i) \wedge \overline{\mathcal{X}_n}(j)} \ .$$

Using this we have that

$$g(\mathcal{X}_n(i) \vee \mathcal{X}_n(j)) = g(\overline{\overline{\mathcal{X}_n}(i) \wedge \overline{\mathcal{X}_n}(j)}) = g(1) - g(\overline{\mathcal{X}_n}(i) \wedge \overline{\mathcal{X}_n}(j)) = h(\overline{\mathcal{X}_n}(i) \wedge \overline{\mathcal{X}_n}(j)) \ .$$

We also have that

$$g(x) = g(1) - g(\overline{x}) = h(\overline{x}) \ .$$

The proof then follows from the polynomial time computability of the Boolean algebra operators. □

Now consider the problem of finding k out of n finite subsets of some set such that their intersection is maximal. This problem we call the Maximum Intersection (MaxI) problem and define it formally as follows.

Problem 3 (Maximum Intersection). Let U be a finite set, let and let $\mathcal{X}_n \in \mathcal{C}_n(2^U)$. Find $S = \{s_1, s_2, \ldots, s_k\} \subset I_n$ that maximizes $|\mathcal{X}_n(s_1) \cap \mathcal{X}_n(s_2) \cap \cdots \cap \mathcal{X}_n(s_k))|$. We denote an instance of this problem as $(\mathcal{X}_n, k)$.

Proposition 4. *Maximum Intersection and Minimum Coverage are polynomial time equivalent.*

Proof. We begin by noting that for a finite set U we have that $(2^U, \cup, \cap, ^-, \emptyset, U)$ is a Boolean algebra with polynomial time computable operations. Let $g(X) = |X|$, we then have that $g(X) = g(U) - g(\overline{X})$, and that $g(X \cup Y) \geq g(X)$ and $g(X \cup Y) \leq g(X) + g(Y)$ for all $X, Y \in 2^U$. Let $h(X) = g(U) - g(X)$. Then by Proposition 3 we have that MSJ instances $((2^U, \cup), \mathcal{X}_n, g, k)$ and $((2^U, \cap), \overline{\mathcal{X}_n}, h, k)$ are polynomially equivalent. We finish the proof by recognizing that $((2^U, \cup), \mathcal{X}_n, g, k)$ is the Minimum Coverage instance $(\mathcal{X}_n, k)$ and since maximizing $g(x) = |X|$ is equivalent to minimizing $|\overline{X}| = h(X)$, we recognize $((2^U, \cap), \overline{\mathcal{X}_n}, h, k)$ as the Minimum Intersection instance $(\overline{\mathcal{X}_n}, k)$. □

4 Hardness

Consider the Maximum Balanced Complete Bipartite Subgraph (MBCBS) problem defined as follows.

Problem 4 (Maximum Balanced Complete Bipartite Subgraph). Given a bipartite graph $G = ((V, W), E)$ find a maximum biclique H in G such that $|H \cap V| = |H \cap W|$.

This problem is known to be NP-hard as the associated decision problem is NP-complete [13].

We start by noting that any vertex in a bipartite graph that has no incident edges is of no interest in the context of the MBCBS problem and can be removed in polynomial time. Hence, we can assume that any instance $G = ((V, W), E)$ of the MBCBS problem can be represented by $\mathcal{X}_n \in \mathcal{C}_n(2^W)$. Further, let $\mathcal{A}$ be an algorithm for the MaxI problem such that $\mathcal{A}(\mathcal{X}_n, k)$ is the solution given by $\mathcal{A}$ for the MaxI instance $(\mathcal{X}_n, k)$. Now consider the MBCBS algorithm given as Algorithm 2.

Algorithm 2. The MBCBS algorithm using the MaxI algorithm $\mathcal{A}$

```
MBCBS(𝒳_n)
i ← n
while i > 0
    S' ← 𝒜(𝒳_n, i)
    s ← |∩_{x∈S'} 𝒳_n(x)|
    if s ≥ i
        return S'
    i ← i − 1
return ∅
```

Lemma 1. *Let H_o be an optimal solution to MBCBS instance $\mathcal{X}_n$, and let H be the solution found by applying* MBCBS($\mathcal{X}_n$). *Then*

$$\frac{|H_o|}{|H|} \leq r \tag{10}$$

where r is the performance ratio of the best known algorithm $\mathcal{A}$ for MaxI.

Proof. Let $\mathcal{A}_o$ be an optimal algorithm for the MaxI problem, let $a_o(i) = |\bigcap_{x \in \mathcal{A}_o(\mathcal{X}_n,i)} \mathcal{X}_n(x)|$, and let $a(i) = |\bigcap_{x \in \mathcal{A}(\mathcal{X}_n,i)} \mathcal{X}_n(x)|$. Note that

$$a_o(i) \geq a_o(i+1) \ .$$

Note that we can, without loss of generality, assume that

$$a(i) \geq a(i+1) \ .$$

This because we can construct a MaxI algorithm $\mathcal{A}'$ that for a given i in polynomial time finds $n \geq j \geq i$ such that $a(j) - a(i)$ is maximal, and return any i size subset S'' of the corresponding solution. We then have that $|S''^{\cap}| \geq a(l)$, for any $l > i$, and we can use $\mathcal{A}'$ instead of $\mathcal{A}$ in MBCBS.

Now let i_o be the largest i in MBCBS such that $a_o(i) \geq i$, and let $i_{\mathcal{A}}$ be the largest i in MBCBS such that $a(i) \geq i$. Inspecting the MBCBS algorithm we have that

$$\begin{aligned} a_o(i_o+1) &< i_o+1 \\ a_o(i_o) &\geq i_o \\ a(i_{\mathcal{A}}+1) &< i_{\mathcal{A}}+1 \\ a(i_{\mathcal{A}}) &\geq i_{\mathcal{A}} \ . \end{aligned}$$

Now note that if $i_o = i_{\mathcal{A}}$ we have that $\frac{|H_o|}{|H|} = 1$. Then the theorem holds as 1 is a lower bound for performance ratios under our (implicitly assumed standard) computational model. Therefore in the following assume that $i_o \neq i_{\mathcal{A}}$, which means that $i_o > i_{\mathcal{A}}$. Combining the above we then get

$$a(i_{\mathcal{A}}) \geq i_{\mathcal{A}} \geq a(i_{\mathcal{A}}+1) \geq a(i_o) \ ,$$

and

$$a_o(i_{\mathcal{A}}) \geq a_o(i_{\mathcal{A}}+1) \geq a_o(i_o) \geq i_o \ .$$

Using this we get that

$$\frac{|H_o|}{|H|} = \frac{i_o}{i_{\mathcal{A}}} \leq \frac{a_o(i_o)}{a(i_o)} \leq r \ . \tag{11}$$

□

Theorem 2. *MinC is NP-hard.*

Proof. We first note that if $\mathcal{A}$ runs in polynomial time, then MBCBS runs in polynomial time. Then the NP-hardness of MaxI follows from Lemma 1, as we can solve the NP-hard MBCBS problem optimally in polynomial time if we have an optimal polynomial time algorithm for MaxI. The theorem then follows from Proposition 4. □

Corollary 1. *MJ and MSJ are NP-hard.*

We now state the main theorem of this section, that MinC is essentially as hard to approximate as the MBCBS problem.

Theorem 3. *Let n be the number of vertices in a MBCBS problem instance graph, and let g be a monotonically increasing function. Let the MBCBS problem be hard to approximate within a factor of $g(n)$ unless proposition P holds. Then MinC is hard to approximate within $g(n)$ unless proposition P holds.*

Proof. This follows from Lemma 1 and Proposition 4. □

Corollary 2. *MJ and MSJ are essentially at least as hard to approximate as MBCBS.*

Corollary 3. *Let $\epsilon > 0$ be an arbitrarily small constant. Then there is no polynomial time algorithm for MinC that achieves an approximation ratio $n^{\epsilon'}$ where $\epsilon' = \frac{1}{2^{O(1/\epsilon \log(1/\epsilon))}}$ unless there exists probabilistic algorithm for SAT that runs in time 2^{n^ϵ} on an instance of size n. Also, assuming that $NP \not\subseteq \text{BPTIME}(2^{n^\epsilon})$ MinC allows no PTAS.*

Proof. This follows from Theorem 3 and hardness results established by Khot [14]. □

5 Approximation of Minimum Loss k-Ambiguity

Informally, k-ambiguity is a property of a transformation Δ of a data set X such that every point in the transformed data set $\Delta(X)$ is a generalization of at least k elements in X. Following Vinterbo [9] we express this formally in Problem 5.

Problem 5 (Minimum Loss k-Ambiguity). Let V be a set and let $\leq$ be a partial order on V with a single maximal element. For $v \in V$ and $\mathcal{X}_n \in \mathcal{C}_n(V)$ define $\sigma(v) = \{i \in I_n | \mathcal{X}_n(i) \leq v\}$ to be the "meaning" of element v with respect to $\mathcal{X}_n$. Associate with V a measure $\lambda : V \to [0, \infty)$ of information loss computable in polynomial time and let the measure λ be monotone and increasing in our partial order $\leq$. Given finite $\mathcal{X}_n \in \mathcal{C}_n(V)$ and a positive integer $k \leq n$ find $\Delta : V \to V$ such that for each $x \in I_n$ both

$$x \in \sigma(\Delta(\mathcal{X}_n(x))) \tag{12}$$

$$|\sigma(\Delta(\mathcal{X}_n(x)))| \geq k \tag{13}$$

hold and $\lambda(\Delta(\mathcal{X}_n(x)))$ is minimized.

Vinterbo [9] calls the requirements (12) the preservation of meaning, and (13) k-ambiguity.

Theorem 4. *Any instance $\mathcal{X} = (V, \leq, \mathcal{X}_n, \lambda, k)$ of Problem 5 such that $(V, \leq)$ is a join-semilattice $(V, \vee)$ and for which*

$$\lambda(x \vee y) \leq \lambda(x) + \lambda(y)$$

holds for all $x, y \in V$, is approximable within $k - p$ in $O((2n(k - p) + p + (n - 1)^p)t(\vee))$ time where $p \leq k$ is a positive integer.

Proof. Since we have that the join $\vee(x_1, x_2, \ldots, x_k)$ for any k elements $x_1, x_2, \ldots, x_k \in V$ can be written as a join $\vee(x_1 \vee x_2, x_1 \vee x_3, \ldots, x_1 \vee x_k)$ of $k-1$ elements, we have that all joins of size k elements from $\mathcal{X}_n$ containing one fixed element $x = \mathcal{X}_n(j)$ can be written as all joins of $k-1$ elements from $\mathcal{X}_{n-1}^x = \{(i, x \vee \mathcal{X}_n(y)) | y \in I_n - \{j\} \wedge i = y - I(y > j)\}$, where I is the Boolean indicator function. Computing $\mathcal{X}_{n-1}^x$ can be done in polynomial time as $\vee$ is polynomial time computable.

Let $\mathcal{Y}_x = ((V, \vee), \mathcal{X}_{n-1}^x, \lambda, k-1)$ be an instance of MSJ, and let for for all $x \in X$ $\Delta(x) = $ GREEDY-MSJ$(\mathcal{Y}_x, \lambda, k-1, p)^{\vee}$. From the argument above and Theorem 1, we can conclude the proof. □

Problem 6 (k-Ambiguity by Cell Suppression). Let X be a set, let $\top \notin X$, and $V = (X \cup \{\top\})^m$. Define $*$ on V such that $(x_1 x_2 \cdots x_m) * (y_1 y_2 \cdots y_m) = (z_1 z_2 \cdots z_m)$ where $z_i = x_i$ if $x_i = y_i$ and $z_i = \top$ otherwise, and let $\mathcal{X}_n \in \mathcal{C}_n(V)$. The class of Problem 5 instances $(V, *, \mathcal{X}_n, \lambda, k)$, where $\lambda(x_1 x_2 \cdots x_m) = \sum_{i=1}^{m} I(x_i = \top)$ where I is the Boolean indicator function, is called the k-ambiguity by cell suppression problem.

We see that the example of k-ambiguity given in Section 1 is an instance of k-ambiguity by cell suppression over $V = \{0, 1, \top\}^4$. We can use the connection between k-ambiguity by cell suppression and MinC to show that the former is as hard to approximate as the latter.

Theorem 5. *k-ambiguity by cell suppression is essentially as hard to approximate as the MBCBS problem.*

Proof. Let $(C = (S_1, S_2, \ldots, S_n), k)$ be an instance of the Minimum Coverage problem. We can in polynomial time construct a binary data table with $n+1$ rows in which row 1 differs from row j in attributes corresponding exactly to S_{j-1}. Let S be the set of suppressed entries in row 1 returned by a k-ambiguity by cell suppression algorithm. Then we know that S is a superset of at least k of the sets in C. We can in polynomial time find k of these. The non-approximation results then follow from Theorem 3. □

Corollary 4. *Problem 5 is essentially at least as hard to approximate as the MBCBS problem.*

6 Discussion

The above results suggest that finding a PTAS for the Minimum Join problem is highly unlikely. However, it seems likely that the analysis of particular NP-Hard specializations of MSJ might yield better approximation algorithms for these than the generic GREEDY-MSJ algorithm.

The proof of GREEDY-MSJ failure on non-subadditive MJ instances begs the question whether such examples of failure can be constructed for larger classes of algorithms.

Even though our analysis of MSJ was primarily motivated by the connection to privacy in databases, there are other motivating factors as well. Minimum Coverage, in addition to being a complement to Maximum Coverage analyzed by Hochbaum et al. [12], and its equivalent problem, Maximum Intersection, can be used to gain insight into problems from other areas as well.

One such (arguably esoteric) example is the problem of finding k out of n integers that share the most factors. As each square-free integer in a finite set can be represented by a finite set containing its prime factors, the problem restriction to square-free integers corresponds to the Maximum Intersection problem. This means that it is for instance highly unlikely that there exists a PTAS for the square-free instance and hence the problem in general.

Another example might be a location selection type problem. Assume we have k factories each of which we wish to place in one of n locations. Each placement has a set of requirements that needs to be met. The requirements are such that if it is met for one location, it is also met for all other locations. Assuming that the the fulfillment of each requirement has a positive cost associated with it, it is natural to seek out the k locations that minimize this cost.

Acknowledgments. Thanks go to Stephan Dreiseitl for fruitful input. This work was funded by NIH grant R01 LM007273-04A1.

References

1. Melton, L.: The threat to medical-records research. N. Engl. J. Med. 337(20), 1466–1470 (1997)
2. Dick, R.S., Steen, E.B., Detmer, D.E.: The Computer Based Patient Record: An Essential Technology for Health Care, Revised Edition. Institute of Medicine (1997)
3. Hippocrates: The oath and law of hippocrates. The Harvard Classics, vol. 38. P.F. Collier & Son, New York (1909-1914)
4. United States Department of Health and Human Services: 45 CFR Parts 160 and 164 RIN 0991-AB14, Standards for Privacy of Individually Identifiable Health Information. Federal Register 67(157) (August 2002)
5. Walton, J., Doll, R., Asscher, W., Hurley, R., Langman, M., Gillon, R., Strachan, D., Wald, N., Fletcher, P.: Consequences for research if use of anonymised patient data breaches confidentiality. BMJ 319(7221) 1366 (November 1999)
6. Sweeney, L.: k-anonymity: a model for protecting privacy. International Journal on Uncertainty, Fuzziness and Knowledge-based Systems 10(5), 557–570 (2002)
7. Hundepool, A.J., Willenborg, L.C.R.J.: Mu- and tau-argus: Software for statistical disclosure control. In: 3rd International Seminar on Statistical Confidentiality at Bled. (1996)
8. Øhrn, A., Ohno-Machado, L.: Using boolean reasoning to anonymize databases. Artif. Intell. Med. 15(3), 235–254 (1999)
9. Vinterbo, S.A.: Privacy: A machine learning view. IEEE Transactions on Knowledge and Data Engineering 16(8), 939–948 (2004)
10. Meyerson, A., Williams, R.: On the complexity of optimal k-anonymity. In: PODS '04: Proceedings of the twenty-third ACM SIGMOD-SIGACT-SIGART symposium on Principles of database systems, pp. 223–228. ACM Press, New York (2004)

11. Aggarwal, G., Feder, T., Kenthapadi, K., Motwani, R., Panigrahy, R., Thomas, D., Zhu, A.: Approximating algorithms for k-anonymity. Journal of Privacy Technology 1, 1–18 (2005)
12. Hochbaum, D., Pathria, A.: Analysis of the greedy approach in covering problems. Naval Research Quarterly 45, 615–627 (1998)
13. Garey, M.R., Johnson, D.S.: Computers and Intractability, A Guide to the Theory of NP-Completeness. W.H. Freeman and Company, New York (1979)
14. Khot, S.: Ruling out ptas for graph min-bisection, densest subgraph and bipartite clique. In: FOCS, pp. 136–145. IEEE Computer Society Press, Los Alamitos (2004)

Algorithmic Challenges for Systems-Level Correlational Analysis: A Tale of Two Datasets*

Michael A. Langston

Department of Computer Science, University of Tennessee, Knoxville, TN 37996–3450, USA

I will discuss novel algorithmic, combinatorial and correlational tools for the analysis of complex natural systems. A pair of illustrative but widely divergent applications will be described. Despite huge differences in data acquisition methodologies, the algorithmic missions for both problems are similar, and help to highlight the rich interplay between data quality and effective computation.

The first application centers on determining the effects of environment on man. As a case study, we search for biological pathways relevant to the human allergic response. We exploit well-designed studies and quantitative data generated with state-of-the-art technologies. We extract putative relationships from the simultaneous expression of vast numbers of genes, under the premise that genes encoding proteins functioning in a common pathway often exhibit correlated levels of expression. Thus the identities and ontologies of these genes can be used to pinpoint existing and assimilate new functional pathway elements. Armed with advanced technologies and high-quality data, we seek to elucidate genetic components relevant to allergic rhinitis, asthma and eczema.

The second application focuses on the rather complementary problem of determining the effect of man on environment. As a case study, we analyze quantifiable variables of significance to oceanic ecosystems. These variables encompass a huge variety of biotic and abiotic factors, and tend to possess differing periodicities and other diverse properties. Only heuristic experimental designs and incomplete and sometimes dubious historical data is available. We labor to uncover temporal, spatial and other meaningful patterns on an immense scale, and to shed light on inflection points, putative regime changes and other complex relationships. Data quality and missing or corrupted values are significant, as is the mining of information at multiple levels of granularity. Armed with powerful technologies but highly challenging data, we seek to establish dependencies upon which we can draw conclusions about the impact of man and other agents upon the sea.

* This research has been supported in part by the U.S. National Institutes of Health under grants 1-P01-DA-015027-01, 5-U01-AA-013512 and 1-R01-MH-074460-01, by the U.S. Department of Energy under the EPSCoR Laboratory Partnership Program, by the Australian Research Council, and by the European Commission under the Sixth Framework Programme.

F. Dehne, J.-R. Sack, and N. Zeh (Eds.): WADS 2007, LNCS 4619, p. 226, 2007.

Flooding Countries and Destroying Dams*

Rodrigo I. Silveira and René van Oostrum

Department of Information and Computing Sciences
Utrecht University, 3508 TB Utrecht, The Netherlands
{rodrigo, rene}@cs.uu.nl

Abstract. In many applications of terrain analysis, pits or local minima are considered artifacts that must be removed before the terrain can be used. Most of the existing methods for local minima removal work only for raster terrains. In this paper we consider algorithms to remove local minima from polyhedral terrains, by modifying the heights of the vertices. To limit the changes introduced to the terrain, we try to minimize the total displacement of the vertices. Two approaches to remove local minima are analyzed: lifting vertices and lowering vertices. For the former we show that all local minima in a terrain with n vertices can be removed in the optimal way in $\mathcal{O}(n \log n)$ time. For the latter we prove that the problem is NP-hard, and present an approximation algorithm with factor $2 \ln k$, where k is the number of local minima in the terrain.

1 Introduction

Digital terrain analysis is an important area of GIS. In many cases, when the terrains are used for purposes concerning land erosion, landscape evolution or hydrology, it is generally accepted that the majority of the depressions present in the terrains are likely to be spurious features. The sources of such artifacts can be many, including low-quality input data, interpolation errors during the generation of the terrain model and truncation of interpolation values [12]. As a result, it is standard in many applications of terrain analysis, particularly in hydrologic applications such as automatic drainage analysis, to do some kind of preprocessing of the terrain to remove these spurious sinks [21,18]. This is because this kind of artifact can severely hinder flow routing. Several related terms have been used before to refer to these features, such as depressions, sinks, pits and local minima. In this paper, following the computational geometry literature, we use the term *local minimum.*

The most widely used type of *digital terrain model*, or simply *terrain*, is the square grid digital elevation model (raster DEM), mainly due to its simplicity. Another common type of terrain is the triangulated irregular network (TIN), which is a triangulation of a set of points with elevation. It involves a more complex data structure because it is necessary to store its irregular topology, but also has several advantages, such as variable density and continuity.

* This research has been partially funded by the Netherlands Organisation for Scientific Research (NWO) under the project GOGO.

F. Dehne, J.-R. Sack, and N. Zeh (Eds.): WADS 2007, LNCS 4619, pp. 227–238, 2007.

Regarding the removal of local minima, most of the literature in GIS has focused on algorithms for (raster) DEMs. Most of the proposed methods are some type of "pit filling" technique [2,12,21]. They consist in raising the local minimum to the elevation of its lowest neighbor. This type of method implicitly assumes that most of the spurious local minima result only from underestimation errors, neglecting the ones caused by overestimation. Not many papers address the problem in the opposite way, removing local minima by lowering a neighboring vertex to a lower height. An example of such a technique was proposed by Rieger [14] and is also part of the "outlet breaching" algorithm of Martz and Garbrecht [13]. Even though pit filling is the most widely implemented method for local minima removal, recent studies have shown that the lowering methods perform significantly better than the depression filling techniques, in terms of the impact on the terrain attributes [10].

When the terrain is modeled as a TIN, a few algorithms have been presented to deal with the problem of local minima. Theobald and Goodchild [19] show experimental results on the number of local minima produced by different methods to extract TINs. Liu and Snoeyink [11] present an algorithm to simulate the flooding of a TIN, a problem that, although different, is related to removing local minima by pit filling. A different approach against local minima is the one followed by de Kok et al. [3] and Gudmundsson et al. [7]. Instead of modifying the elevation of the points, they choose the edges of the triangulation in such a way that the number of local minima is minimized. They optimize over a particular class of well-shaped triangulations, the *higher-order Delaunay triangulations* [7].

In this paper we present algorithms to remove local minima from TINs by modifying the heights of the vertices. We study both lifting points (pit filling) and lowering points (breaching). In both cases we want to remove local minima while modifying the terrain as little as possible. To formalize this second goal, we introduce a cost function that is applied to each point or vertex whose height is modified. The objective is to minimize the total cost of the removal. There is no obvious choice for this measure of the cost, and many of them are reasonable. The one adopted throughout most of this paper is the total displacement of the vertices. A few other measures are discussed in Section 3.3. To our knowledge, no previous paper deals with optimization for local minima removal.

The different possibilities for the cost function give rise to different problems. Furthermore, another source of variants of the problem is choosing *what* local minima to remove. Possible options are: removing *a given subset* of the local minima, removing *all* of them, or removing the *cheapest* k, for k a parameter. When the removal method is lifting, the three options can be solved on a one-by-one basis, that is, by removing each of the local minima separately. This is possible because the removal of one minimum does not affect the removal of the others. For lowering, the situation is different, because the way a local minimum is removed may affect the cost of removing other minima.

We study both approaches, lifting and lowering, independently. Some comments on their combination are made in Section 4. For the lifting approach, we show how the use of *contour trees* allows to remove all the local minima in

$\mathcal{O}(n \log n)$ time, by facilitating the location of the vertices that must be used to remove each minimum. The lowering approach turns out to be much harder than the lifting version. We start by showing that removing all local minima is NP-hard, and then propose an approximation algorithm to solve the problem, based on an existing algorithm for the Node-Weighted Steiner Tree Problem.

We begin by studying the simplest of the two, the lifting approach, and then we focus on the lowering technique.

2 Removing Local Minima by Lifting

In this section we present an algorithm to remove local minima by increasing the elevation of some of the vertices. This can be seen as a flooding or pit filling technique for TINs. We begin with a few basic definitions that will be also used in the next sections. A *polyhedral terrain* T, or just *terrain*, is a triangulated point set in the plane where each point or vertex v has a height, denoted $h(v)$. Any terrain has an associated graph, G_T. Sometimes we will refer to both the terrain and the associated graph as the *terrain*.

A (local) *minimum* is a maximally connected set of vertices $M \subset T$ such that all the vertices in M have the same height and no vertex in M has a neighbor with lower height. Even though a minimum can be made of more than one vertex, for the purpose of this paper it is more convenient to treat each minimum as consisting of only one vertex. For example, if a minimum at vertex u is lifted to height h, we will assume that also all the other vertices of the minimum u belongs to are lifted in the same way. We do the same with the definition of *saddle*: below we define it as being one vertex, but in practice it can be a connected set of them. This does not affect our algorithms or their running times. These considerations apply to the whole paper. From now on, we treat each minimum or saddle as consisting of one vertex only.

A vertex is a *saddle* if and only if it has some neighboring vertices around it that are higher, lower, higher, lower, in cyclic order around it. To simplify the presentation of the algorithms, we will assume the terrain has only one global minimum, and we will adopt the convention that when we refer to *local* minima, we do not include the *global* minimum.

The cheapest way to remove a local minimum at a vertex v, with height $h(v)$, is by lifting it to $h(w)$, where w is the lowest neighbor of v. However, this may turn w into a local minimum. To fix this, the lifting procedure must be propagated until no new local minimum exists.

Conceptually, the idea is as follows. We explain how to compute a list $S = \{s_1, \ldots, s_k\}$ of vertices that must be lifted to remove a local minimum at v. Initially, $S = \{s_1 = v\}$, and it is expanded every time the lifting must be propagated. Let $U = \{u_1, \ldots, u_m\}$ be the union of the neighbors of all vertices in S that are not in S themselves, and denote the vertex in U with lowest height with $u_{\min}$. We raise all vertices in S such that $h(s_i) = h(u_{\min})$ for $i \in \{1, \ldots, k\}$. If $u_{\min}$ is a saddle vertex of the terrain, then it is connected to another lower vertex and we are done. Otherwise, we remove $u_{\min}$ from U and add it to S as

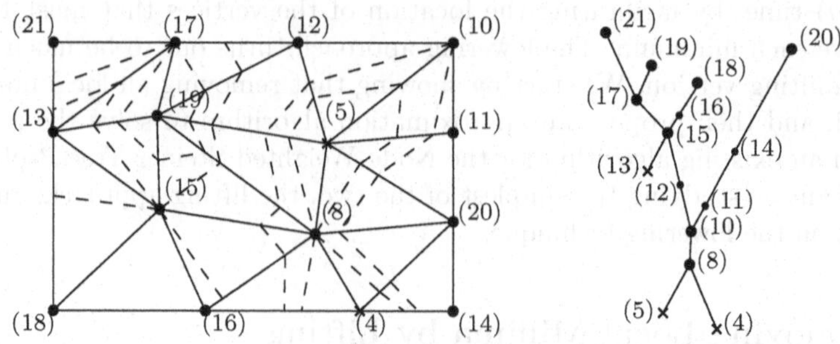

Fig. 1. Left: example of a terrain showing the elevation of the vertices (between parenthesis) and some of the contour lines. Right: the augmented contour tree of the terrain.

s_{k+1}, and we set k to $k+1$. Next, we add the neighbors of the new s_k that are not already in S to U. After the changes made to S and U, $u_{\min}$ refers to the new lowest vertex in U. This iterative approach lifts the whole basin of the local minimum, in a bottom-up fashion, until it reaches its lowest saddle vertex.

The propagation of the lifting is facilitated by the variation of the *contour tree* used by van Kreveld et al. [20]. Contour trees have been previously used in image processing and GIS research [4,6,16,17] and are also related to the *Reeb graph* used in Morse Theory [15]. Minima and maxima in the terrain are represented by leaves in the contour tree, and saddle vertices in the terrain correspond to nodes of degree three or higher. Ordinary, non-critical vertices appear as vertices of degree two. See Figure 1 for an example. The contour tree for a terrain with n vertices can be computed in $\mathcal{O}(n \log n)$ time [1,20].

To remove a given local minimum at v in the terrain by propagated lifting, we look at the corresponding leaf v' in the contour tree. Let w' be the node of degree three or more in the contour tree that is closest to v'. It corresponds to a saddle w in the terrain, the first one encountered when "flooding" v. To remove the local minimum at v, we must lift all the vertices in the terrain that correspond to nodes on the path from v' to w' in the contour tree, including v', but excluding w', to the height of the saddle vertex.

After this change we must update the contour tree to reflect the changes. The branch that ended at v' disappears, and the nodes on it become ordinary (non-critical) ones, at the same height as the saddle node. If necessary, we can store any relevant information about the lifted vertices, like total displacement, together with the saddle node. If before the lifting step the saddle had only one downwards branch, then after the lifting it became a new local minimum, and the lifting needs to be propagated until the lowest of the saddle ancestors is reached (there could be more than one). If before the lifting step the saddle had two or more downwards branches, then the saddle might continue as a saddle or become an ordinary node. In both cases this lifting step is over. In any case, during the removal of the local minima every node involved is processed only once, because after lifting it, it becomes "part" of the saddle node.

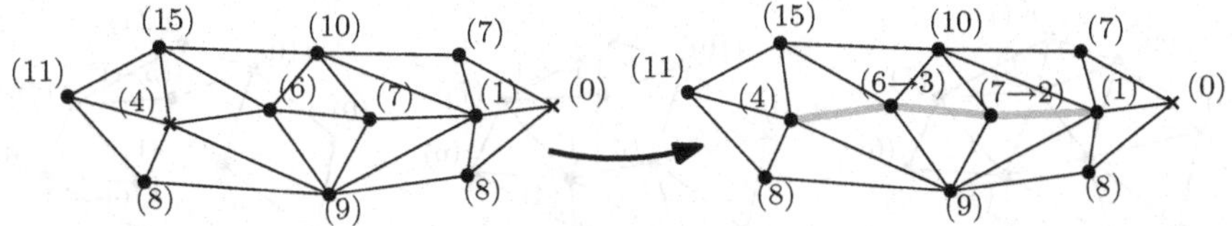

Fig. 2. Removing the local minimum of height 4 by lowering. Left: initial heights (between parenthesis). The change in height of the lowered vertices is shown with an arrow. Right: first a vertex is lowered from 6 to 3, turning it into a new local minimum. To remove it, a neighboring vertex is lowered from 7 to 2. A breaching path from the vertex of height 4 to the one of height 1 is highlighted.

Removing all the local minima, given the contour tree, can be done in linear time. Hence the total running time equals the time needed to build the tree.

Theorem 1. *Given a terrain T with n vertices, k of them local minima, a lifting of the vertices that removes all local minima while minimizing the total displacement of the vertices can be computed in $\mathcal{O}(n \log n)$ time.*

3 Removing Local Minima by Lowering

In this section we study how to remove local minima by lowering the heights of some vertices. We begin with some definitions and basic observations.

Any vertex in a terrain can be *lowered*, meaning that its height can be decreased. The *cost* of lowering a vertex is defined as the difference between its original height and the new one. Some other definitions are discussed in Section 3.3. Recall that each local minimum is treated as consisting of exactly one vertex.

A given local minimum at u is removed by lowering some neighboring vertex to a height less than or equal to $h(u)$. Since the lowered vertex can become a local minimum itself, and we do not want to create new local minima (or make an existing one worse), a propagation takes place, until the original and the newly created local minima are removed. Figure 2 shows an example. Observe that any given local minimum can be removed in this way (recall that we do not consider the global minimum to be a local minimum). The same can be done for any set of local minima that does not include the lowest minimum in the terrain.

The following definitions formalize the basic ideas related to lowering.

Definition 1. *Let T be a terrain, and let u and v be two vertices of T, with $h(u) > h(v)$. A* breaching path *from u to v is a tuple $\rho = (P, D)$, where P is a list of vertices that induce a path between u and v, that is, $P = \{u, w_1, w_2, \cdots, w_\eta, v\}$, and $D = \{0, d_1, d_2, \cdots, d_\eta, 0\}$ is a list of height displacements for the intermediate vertices of the path, such that $h(u) \geq h(w_1) + d_1$, $h(w_i) + d_i \geq h(w_{i+1}) + d_{i+1}$ for every $1 \leq i < \eta$, and $h(w_\eta) + d_\eta \geq h(v)$. The* cost *of a breaching path is $Cost(\rho) = \sum_i |d_i|$.*

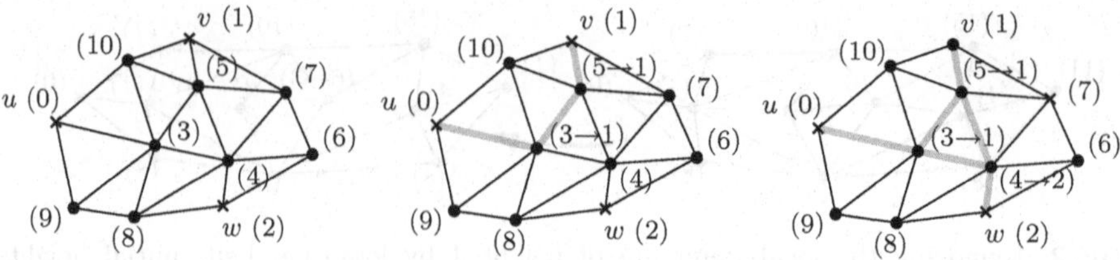

Fig. 3. From left to right: a terrain with three local minima: at u,v and w; a breaching path from v to u; a breaching graph of minimum cost that connects u,v and w

Intuitively, a breaching path is a path used to remove a local minimum at u, by connecting it to a lower vertex v, by modifying the heights of all the vertices in between (see Figure 2). A natural extension of the concept of breaching path is the *breaching graph.*

Definition 2. *Let T be a terrain. A* breaching graph *is a tuple $\psi = (P, D)$, where P is a set of vertices that induce a subgraph of T, such that all the vertices of degree one are minima, and each vertex with degree higher than one has a height displacement in D such that for each connected component, ψ includes a breaching path connecting each of its local minima to the lowest minimum of the component. The cost of the breaching graph is defined as $Cost(\psi) = \sum_i |d_i|$.*

Note that since we aim at removing *all* local minima, the breaching graph must be connected, because all local minima must be connected to the global minimum. A minimum cost breaching graph may contain cycles, but it can be turned into a tree by discarding some edges, without changing its cost. Hence we will sometimes refer to a minimum cost breaching *tree.* Throughout this paper we refer to *connecting* two minima u and v, meaning that the minimum at u or v (the highest one) is removed by creating a breaching path connecting u and v.

Since we are minimizing the total displacement, each vertex will be lowered as little as possible. Hence if a vertex w_i needs to be lowered to connect a local minimum at u to a lower vertex v, the new height of w_i will be $h(u)$.

In a breaching graph it can occur that an intermediate vertex w_i is part of more than one breaching path. See for example the vertex of initial height 3 in Figure 3. In that case the new height of the vertex must be set to the height of the lowest of those minima. We say that u *pays* for the lowering of a vertex w_i if u is the lowest local minimum that is removed by a breaching path through w_i. It is interesting to look at the structure of an optimal solution, in relation to how the payment of the lowering is distributed. Figure 4 shows an example.

3.1 Removing All Local Minima

It is easy to verify that one given local minimum u can be removed optimally by computing the shortest paths between u and each of the lower minima in a directed graph based on the terrain. The vertices and edges of this graph are the same as in the terrain, but edges are made directed and the weights are related

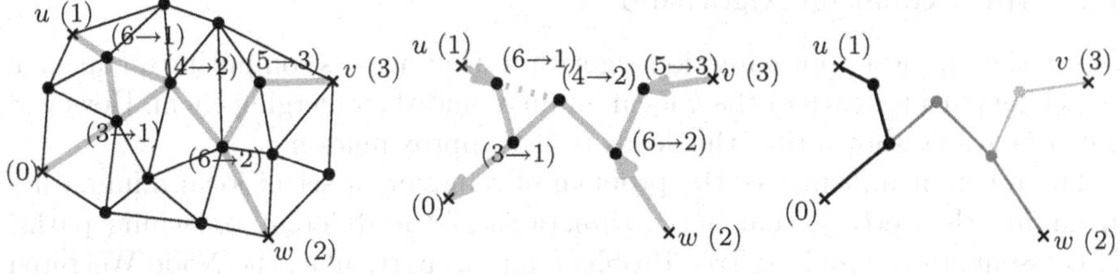

Fig. 4. Left: terrain with four minima, with the breaching graph of an optimal solution (only relevant heights shown). Center: breaching graph seen as a tree. Right: the black vertices are paid by u, the medium gray vertices by w, and the light gray vertex by v.

to the vertical distance between neighboring vertices. Therefore a breaching path of minimum cost removing u can be found in $\mathcal{O}(n \log n)$ time.

Extending this idea to several local minima involves dealing with the possible ways to combine the different breaching paths. For a non-constant number of local minima, this leads to an exponential time algorithm. Moreover, in this section we show that the problem of removing all local minima from a terrain by lowering, minimizing the total displacement, is NP-hard. We use a reduction from Planar Connected Vertex Cover (PCVC), which is known to be NP-hard [5]. The optimization version of PCVC consists in given a planar graph $G = (V, E)$, finding a set of vertices of minimum size such that every edge has at least one end among the selected vertices, and the subgraph induced by the set is connected.

We show how to solve any instance of PCVC, given by a planar graph $G = (V, E)$, with an algorithm for our problem. We create a new graph G' as follows. Take G as the initial graph G'. Assign height 1 to each existing vertex. Then for each edge $e \in E$, create a vertex "in the middle" of the edge at height 0 (these will be all minima). Finally, change the height of an arbitrary local minimum to -1, to create a global minimum. The resulting graph has exactly $|E|$ minima. Each of them corresponds to an edge in G that must be covered by the vertex set. To turn the graph into a terrain, we compute a first arbitrary triangulation, and for every edge added during the triangulation, we add a vertex on its midpoint at height $+\infty$. The resulting non-triangular faces are triangulated in some arbitrary way. This guarantees that all the new edges have one endpoint with a vertex at height $+\infty$, so they will never be part of an optimal solution. It is straightforward to see that the whole construction can be done in polynomial time.

Removing all local minima in G' induces a set of vertices that must be lowered. The fact that all local minima have been removed implies that all edges have one end in the chosen set, so it is indeed a solution to Vertex Cover, and it is a tree, hence also connected. It is easy to verify that if the local minima are removed optimally, the vertex cover is also optimal.

Theorem 2. *Given a terrain T with n vertices, it is NP-hard to compute a lowering of the vertices that removes all local minima, that minimizes the total displacement of the vertices.*

3.2 Approximation Algorithm

One of the simplest approximation algorithms that arise is computing an optimal breaching path for each of the k local minima, and then merging them. However, it can be easily shown that this leads to a k-approximation.

The inherent difficulty of the problem of removing a set of local minima lies in finding the vertices that act as junctions of the different breaching paths. This resembles the Steiner Tree Problem, and in particular, the Node-Weighted Steiner Tree (NWST) problem in networks, which is a more general version of the problem where the costs are assigned to the vertices. Even though constant factor approximation algorithms are known for the standard Steiner Tree problem, no approximation algorithms with factor less than logarithmic exist for the NWST problem, unless $NP \subseteq DTIME[n^{\mathcal{O}(polylog\, n)}]$ [9]. Our problem is still different from the NWST problem because the cost paid for using a vertex is not fixed; it depends on the heights of the local minima that are being removed through it.

In order to make an algorithm for NWST work for our problem, non-trivial adaptations are needed. In this section we present an approximation algorithm, with factor $2 \ln k$, which is an adaptation of the NWST approximation algorithm of Klein and Ravi [9]. The general idea at each step is to connect some minima in some simple way, using *spiders*, as to minimize the ratio of the cost of the spider to the number of minima it connects. We begin with some definitions.

Definition 3. *(Adapted from [9]) A* spider *is a tree with at most one vertex of degree greater than 2. A* center *of a spider is a vertex from which there are edge-disjoint paths to the leaves of the spider. A spider has a number of* feet*, comprised by its leaves and, if the spider contains at least 3 leaves, its center. A* nontrivial spider *is one with at least two feet.*

Definition 4. *(from [9]) Let G be a graph, and let M be a subset of its vertices. A* spider decomposition *of M in G is a set of vertex-disjoint nontrivial spiders in G such that the union of the feet of the spiders contains M.*

We relate each spider to a breaching graph, and define its *cost* as follows:

Definition 5. *Let T be a terrain and let S be a spider in T, with center v and feet F, where at least two of its feet are minima of T. The* cost *of S, $Cost(S)$, is the minimum total displacement required to remove all (but the lowest) minima in F by lowering vertices of S.*

The algorithm. The goal is to find up to k spiders that cover the minima and connect them with each other. This is done through an iterative process, that at each iteration finds one spider and uses it to remove a number of local minima. A spider is specified by a center vertex c and a set of leaves F. The elements of F will be minima, which can always be removed (except for the global minimum) if they are all connected to c, and c is connected to the lowest minimum of F.

The algorithm begins by computing the shortest (breaching) paths from each minimum to all the other vertices. It then computes for each vertex, a list with the k minima sorted by increasing distance to the vertex (in the breaching path sense). All this preprocessing takes $\mathcal{O}(kn \log n)$ time.

Then up to k iterations take place. At each iteration, a spider is found that connects at least two minima, hence removing at least one. The spider is chosen as one that minimizes the ratio $C(F_i)/|F_i|$, where F_i is the set of minima connected by the spider, $|F_i|$ its size and $C(F_i)$ the cost of the spider. To find such an optimal spider, the algorithm needs to find a center vertex c_i and the set of minima that will be connected to the center, F_i. This is done as follows.

Every possible vertex is tried as the center c_i. For each center candidate, all the possible second lowest minima in F_i are tried. There are $\mathcal{O}(nk)$ of these pairs. For a pair of center c_i and second lowest minimum $v_i^{(2)}$, we still need to find the other elements of F_i. The lowest element of F_i, $v_i^{(1)}$, is set to the *nearest* local minimum lower than $v_i^{(2)}$, where the distance equals the displacement needed to remove $v_i^{(2)}$ by connecting it to $v_i^{(1)}$ (going through c_i). After this we have $F_i = \{v_i^{(1)}, v_i^{(2)}\}$. By construction this is the optimal choice for this pair and $|F_i| = 2$. Next we can start augmenting F_i by adding, one by one, the local minimum *nearest* to c_i, among the ones higher than $v_i^{(2)}$. By *nearest* we mean the one with the minimum cost breaching path to c_i. This results in optimal choices for each value of $|F_i|$, because the fact that all the local minima that are added are higher than $v_i^{(2)}$ guarantees that the total cost of the removal, $C(F_i)$, is increased only by the cost of connecting the added vertices to c_i (the connection from c_i to $v_i^{(1)}$ is paid by $v_i^{(2)}$). We choose the minimum ratio combination over all the ones considered in the current iteration.

Since each iteration removes at least one local minimum, the total number of iterations is $\mathcal{O}(k)$, yielding a $\mathcal{O}(k^3 n + kn \log n)$ running time.

Approximation factor. Our proof of the approximation factor follows the proof in [9]. We first define some notation. T_i is the terrain just after iteration i. The number of local minima (not yet removed) in T_i is denoted by ϕ_i. The number of minima connected at iteration i (which is one more than the number of local minima removed at that iteration) is denoted h_i. The cost of the lowering done at iteration i is C_i. Finally, OPT is the cost of an optimal solution. Since any solution induces a breaching graph of T, which can be seen as a tree, we will sometimes refer to the *optimal tree*, meaning one of the trees associated with an optimal solution. The following lemma relates the ratio of the combination chosen at step i to the ratio of an optimal solution.

Lemma 1. *At any iteration i of the algorithm,*

$$\frac{C_i \phi_{i-1}}{OPT} \le h_i \tag{1}$$

Proof. Let T^* be a tree of an optimal solution. Some of the vertices in T^* may correspond to local minima that have been already removed in the current terrain (T_{i-1}). Let T_i^* be a tree based on T^* where all the leaves that correspond to local minima that have been removed in T_{i-1} have been deleted, together with all the paths that connected them to the rest of the tree (we keep everything just as to keep the remaining minima connected). If there is a removed local minimum in T^* that is not a leaf, we treat that vertex as a normal one — non-local minimum).

Let $\text{Cost}(T_i^*)$ be the cost of T_i^*, in the same way as it is defined for breaching graphs: the cost of removing all local minima in the tree by lowering only vertices of the tree. Observe that $\text{Cost}(T_i^*)$ can be different from $\text{Cost}(T^*) = OPT$, because the removal of some leaves of the tree may change the minimum that pays for the lowering of some intermediate vertex. However, since the minimum paying is always the lowest one, the new minimum paying for the intermediate vertex will be higher than the previous one, and the displacement will decrease, hence we have $\text{Cost}(T_i^*) \leq OPT$.

Given a tree and a subset of its vertices M, $|M| \geq 2$, there is always a spider decomposition of M contained in the tree [9]. Thus we can compute a spider decomposition of T_i^*, where M are the minima. Let $c_1, ..., c_r$ be the centers of the spiders in the decomposition. For a spider with only two leaves (a path), we pick any vertex in the path as its center. Let the cost of the spider S_j, centered at c_j, be $\text{Cost}(S_j)$, and let n_j be the number of minima that it connects.

During main step i of the algorithm, vertex c_j (for any j), will be considered as a possible center vertex to remove a set of local minima. The quotient of this vertex was defined as to minimize the ratio between the cost of connecting the minima and the number of minima that it connects. The c_j with the minimum ratio will be selected. That ratio can never be more than the ratio of the spider with center c_j. Notice that it could be lower, if for example some of the vertices that must be lowered have been already partially lowered in a previous iteration. Then for each spider S_j in the decomposition we have $\frac{C_i}{h_i} \leq \frac{\text{Cost}(S_j)}{n_j}$.

Rewriting and summing over all the spiders in the decomposition we get

$$\frac{C_i}{h_i} \sum_j n_j \leq \sum_j \text{Cost}(S_j) \tag{2}$$

Now we argue that $\sum_j \text{Cost}(S_j) \leq \text{Cost}(T_i^*)$. The cost of some spider S_i, $\text{Cost}(S_i)$, can be different from the cost of the associated subgraph in T_i^*. Some vertices may have been lowered in a different way. This is because when T_i^* is divided into subtrees (each corresponding to one spider), it might occur that the lowest vertex in one of the subtrees changes (because the original one is now in some other subtree). This causes a series of changes in the way the local minima of the subtree are removed, because another vertex must act as the global minimum of the component. However, the new global minimum must be higher than the previous one, hence using the same arguments used to claim that $\text{Cost}(T_i^*) \leq OPT$, it follows that $\text{Cost}(S_i)$ cannot be higher than the cost of the associated subgraph of T_i^*.

Going back to Equation (2), and using that $\text{Cost}(T_i^*) \leq OPT$ and that the sum $\sum_j n_j$ equals the number of minima at the beginning of the current iteration, ϕ_{i-1}, we get $(C_i/h_i)\phi_{i-1} \leq OPT$, which is (after rewriting) the result claimed.

To get the approximation factor, exactly the same arguments used in [9] can be used to conclude that if p is the total number of iterations of the algorithm, then $\sum_{j=1}^{p} C_j \leq 2 \ln k \cdot OPT$. An example showing that the approximation factor is nearly tight can be constructed.

Theorem 3. *Given a terrain T with n vertices, k of them minima, a lowering of the vertices that removes all the local minima, that minimizes the total displacement of the vertices, can be computed in $\mathcal{O}(k^3 n + kn \log n)$ time, where the total displacement is at most $2 \ln k$ times the minimum one.*

3.3 Other Measures

Even though most of this paper considered the total displacement as the measure being optimized, there are several other measures that are also interesting. Two of them, similar to the one studied here, are the number of vertices lowered and the total volume reduction. Both can be shown to be NP-hard for removing all local minima, and our approximation algorithm can be adapted to work for them. On the other hand, if the goal is to minimize the maximum displacement, that is, the maximum height that any vertex is moved, then we can do it in polynomial time, because for this measure simply merging the optimal breaching paths to remove each minimum leads to an optimal solution.

4 Conclusions and Future Work

This paper studied optimization problems related to the removal of local minima from triangulated terrains, by modifying the heights of the vertices. Two techniques were analyzed, lifting and lowering, with the objective of removing all local minima while minimizing the total displacement of the vertices. For the lifting technique, we showed how to use contour trees to facilitate finding which vertices need to be lifted to remove each local minimum. For the lowering technique, we showed that one local minimum can be removed efficiently, but in the general case the problem is NP-hard. With that in mind, we proposed an approximation algorithm with factor $2 \ln k$.

There are many directions for further research, specially for the lowering approach. Approximation algorithms with better factors are one of them. There are some better approximation algorithms for the NWST problem, like the ones of Guha and Khuller [8], that improve the $2 \ln k$ factor of [9] to $1.5 \ln k$ or even $(1.35 + \epsilon) \ln k$, but it is unclear how to adapt them to our problem. Another aspect worth studying is the combination of lifting and lowering, that sometimes can result in a smaller total displacement. Finally, many other variants, like removing *the cheapest k local minima* may pose interesting challenges.

Acknowledgments. We thank Marc van Kreveld for proposing the problem and for many helpful suggestions.

References

1. Carr, H., Snoeyink, J., Axen, U.: Computing contour trees in all dimensions. Comput. Geom. Theory Appl. 24, 75–94 (2003)
2. Charleux-Demargne, J., Puech, C.: Quality assessment for drainage networks and watershed boundaries extraction from a digital elevation model. In: Proc. 8th ACM Symp. on Advances in GIS, pp. 89–94. ACM Press, New York (2000)
3. de Kok, T., van Kreveld, M., Löffler, M.: Generating realistic terrains with higher-order Delaunay triangulations. Comput. Geom. Th. Appl. 36, 52–65 (2007)
4. Freeman, H., Morse, S.P.: On searching a contour map for a given terrain profile. J. of the Franklin Institute 248, 1–25 (1967)
5. Garey, M.R., Johnson, D.S.: Computers and Intractability: A Guide to the Theory of NP-Completeness. W. H. Freeman, New York (1979)
6. Gold, C., Cormack, S.: Spatially ordered networks and topographic reconstructions. In: Proc. 2nd Internat. Sympos. Spatial Data Handling, pp. 74–85 (1986)
7. Gudmundsson, J., Hammar, M., van Kreveld, M.: Higher order Delaunay triangulations. Comput. Geom. Theory Appl. 23, 85–98 (2002)
8. Guha, S., Khuller, S.: Improved methods for approximating node weighted steiner trees and connected dominating sets. Inform. Comput. 150, 57–74 (1999)
9. Klein, P., Ravi, R.: A nearly best-possible approximation algorithm for node-weighted steiner trees. J. Algorithms 19, 104–115 (1995)
10. Lindsay, J.B., Creed, I.F.: Removal of artifact depressions from digital elevation models: towards a minimum impact approach. Hydr. Proc. 19, 3113–3126 (2005)
11. Liu, Y., Snoeyink, J.: Flooding triangulated terrain. In: Proc. 11th Int. Symp. on Spatial Data Handling, pp. 137–148 (2004)
12. Mark, D.: Network models in geomorphology. In: Anderson, M.G. (ed.) Modelling Geomorphological Systems, ch. 4, pp. 73–97. John Wiley & Sons, West Sussex (1988)
13. Martz, L.W., Garbrecht, J.: An outlet breaching algorithm for the treatment of closed depressions in a raster dem. Comp. & Geosciences 25, 835–844 (1999)
14. Rieger, W.: A phenomenon-based approach to upslope contributing area and depressions in DEMs. Hydrological Processes 12, 857–872 (1998)
15. Shinagawa, Y., Kunii, T.L.: Constructing a Reeb graph automatically from cross sections. IEEE Comput. Graph. Appl. 11, 44–51 (1991)
16. Sircar, J.K., Cerbrian, J.A.: Application of image processing techniques to the automated labelling of raster digitized contours. In: Proc. 2nd Internat. Sympos. Spatial Data Handling, pp. 171–184 (1986)
17. Takahashi, S., Ikeda, T., Shinagawa, Y., Kunii, T.L., Ueda, M.: Algorithms for extracting correct critical points and constructing topological graphs from discrete geographical elevation data. In: Eurographics'95, vol. 14, pp. C–181–C–192 (1995)
18. Temme, A., Schoorl, J., Veldkamp, A.: Algorithm for dealing with depressions in dynamic landscape evolution models. Comp. & Geosciences 32, 452–461 (2006)
19. Theobald, D., Goodchild, M.: Artifacts of tin-based surface flow modeling. In: Proc. GIS/LIS' 90, pp. 955–964 (1990)
20. van Kreveld, M., van Oostrum, R., Bajaj, C., Pascucci, V., Schikore, D.: Contour trees and small seed sets for isosurface generation. In: Rana, S. (ed.) Topological Data Structures for Surfaces, ch. 5, pp. 71–85. Wiley, New York (2004)
21. Zhu, Q., Tian, Y., Zhao, J.: An efficient depression processing algorithm for hydrologic analysis. Comp. & Geosciences 32, 615–623 (2006)

I/O-Efficient Flow Modeling on Fat Terrains

Mark de Berg[1], Otfried Cheong[2], Herman Haverkort[1], Jung Gun Lim[2], and Laura Toma[3]

[1] TU Eindhoven, The Netherlands, mdberg@win.tue.nl, cs.herman@haverkort.net
[2] KAIST, Korea, otfried@kaist.ac.kr, araste@gmail.com
[3] Bowdoin College, USA, ltoma@bowdoin.edu

Abstract. We study the flow of water on *fat terrains*, that is, triangulated terrains where the minimum angle of any triangle is bounded from below by a positive constant. We give improved bounds for the worst-case complexity of river networks on fat terrains, and show how to compute the river network and other flow-related structures I/O-efficiently.

1 Introduction

In this paper we study one of the most important problems on terrains, analyzing the flow of water. The basic questions in flow analysis are to identify the river network and, for any given point q, its watershed (the part of the terrain from which water flows to q). Acquiring real flow data for a terrain is tedious, time-consuming and often impossible. Fortunately high-resolution elevation data is now widely available. As a result, flow modeling and analysis based on elevation data is a popular topic for researchers in GIS (geographic information systems).

One common representation of terrain data in a GIS is the TIN (*triangulated irregular network*). A TIN—in computational-geometry terms: a triangulated polyhedral terrain—is obtained by triangulating a collection of irregularly spaced sample points and then giving each triangulation vertex the elevation of the corresponding sample point. When working with very large terrains, the data is too large to fit into the computer's main memory. Most of the data must therefore reside on disk during the computation, making I/O (moving data between main memory and disk) the bottleneck of the computation. This leads us to the topic of our paper: the study of river networks and watersheds on TINs, and the design of I/O-efficient algorithms for computing these structures.

We analyze our algorithms with the model introduced by Aggarwal and Vitter [2], which has become the standard model for I/O-efficient algorithms. In this model, a computer has an internal memory of size M and an arbitrarily large external memory (disk) where data is stored in blocks of size B. The I/O-*complexity* of an algorithm in this model is measured in terms of the number of I/O's—reading or writing a block from or to external memory—it performs. In this model, scanning (reading a set of n consecutive items from disk) takes $Scan(n) = \Theta(n/B)$ I/O's, and sorting $Sort(n) = \Theta((n/B)\log_{M/B}(n/B))$ I/O's.

The previous work on modeling flow on TINs falls into two classes. Most GIS papers [9,15,16,17,18] adopt a *discrete* approach and route flow from a triangle

F. Dehne, J.-R. Sack, and N. Zeh (Eds.): WADS 2007, LNCS 4619, pp. 239–250, 2007.

to one of its three neighbour triangles using the direction of steepest descent. This approach is appealing because of its simplicity; it is problematic, however, because it discretizes flow and tends to lead to inconsistencies when the triangles in the TIN differ a lot in size [10,19]. The approach taken in the computational-geometry literature considers the TIN as a *continuous* surface on which water always flows in the direction of steepest descent. De Berg et al. [6], McAllister [11,12], McAllister and Snoeyink [13] and Yu and Snoeyink [19] study the structure and the complexity of the river network and other structures on TINs under this model. In particular, De Berg et al. prove that the complexity of the river network—see Section 2 for a formal definition—on a TIN of n vertices is $\Theta(n^3)$ in the worst case: there can be $\Theta(n)$ separate rivers, each with complexity $\Theta(n^2)$. None of the papers mentioned above provide I/O-efficient algorithms.

I/O-efficient flow modeling was first studied—on grids, the other type of data representation in GIS—by Arge et al. [4]. Their system, `Terraflow`, has become the state of the art in flow modeling on massive grids. `Terraflow` uses a discrete approach which can be easily extended to TINs. However, discrete flow is only an approximation of real flow. Thus, the challenge is to develop I/O-efficient algorithms to model continuous flow on TINs, which is ultimately more accurate.

The main step in the computation of flow, and at the same time the bottleneck in the I/O-model, is tracing paths of steepest descent across the triangles that they intersect—in particular, any river is such a path of steepest descent. While in internal memory a path of size k can be traced in $O(k)$ time, the best known I/O-bound is $O(k/\log B)$ I/O's on planar graphs [1]. This results in a straightforward bound of $O((r+n)/\log B)$ I/O's for the computation of a river network of size r, but this would be prohibitively expensive.

Moreover, De Berg et al. [6] showed that r is $\Theta(n^3)$ in the worst case. However, the worst case is a construction that is unlikely to occur in real life. In computational geometry such discrepancies between worst case and practice have led to the study of input models that resemble realistic inputs better. Moet et al. [14] studied visibility and distance problems on *realistic terrains*. In this paper we consider flow modeling on *fat terrains*, that is, terrains where the minimum angle of any triangle is bounded from below by a positive constant[1]. Our notion of a fat terrain is less restrictive than the notion of realistic terrains from Moet et al.

Our results. In this paper we give improved bounds for the complexity of the river network on a fat terrain and show how to compute a number of flow-related structures I/O-efficiently. The main ingredient in our solution is to represent the terrain by a directed graph, which we call the *descent graph*, $\mathcal{G}_{\text{desc}}$. The nodes of $\mathcal{G}_{\text{desc}}$ represent the edges of the triangulation, and we define the arcs of $\mathcal{G}_{\text{desc}}$ such that following a path of steepest descent on the terrain corresponds to following a path in $\mathcal{G}_{\text{desc}}$. Unfortunately, in its basic form $\mathcal{G}_{\text{desc}}$ can have cycles, and a path of steepest descent can visit the same edge more than once. In fact this is exactly the reason why the complexity of a river in an arbitrary terrain can be $\Theta(n^2)$: it can visit a linear number of edges each a linear number of times [6].

[1] Here the angle of a triangle is measured in space. Our results also hold if the angles are measured in the projection on the xy-plane, or if all triangles are non-obtuse.

In the I/O-model, a descent graph with cycles does not only signify a potentially problematic output size, but it also constitutes an algorithmic problem, because it is not known how to store such a graph on disk such that any path of length k can be traced using $O(k/B)$ I/O's [1].

The cornerstone of this paper is an idea that solves both problems at the same time: we subdivide each edge of the triangulation into a number of segments, in such a way that the descent graph, defined on these segments instead of the original edges, is *acyclic*. Moreover we show that for fat terrains a constant number of segments per edge suffices. This implies that any path of steepest descent can visit each segment at most once, and hence its worst-case complexity is $\Theta(n)$. This in turn implies an $O(n^2)$ bound on the complexity s of the *strip map*: the subdivision $\mathcal{S}$ of the terrain induced by the paths of steepest ascent and descent from all vertices. It also follows that the complexity r of the river network on a fat terrain is $O(n^2)$, which is a linear factor smaller than the $O(n^3)$ bound for river networks on general terrains. In the full version of this paper we shows that our bounds are tight in the worst case.

The acyclicity of the descent graph allows us to apply time-forward processing [3,7] and traverse paths in a batched I/O-efficient manner. By applying and refining ideas from McAllister [11] and Yu et al. [19], we obtain the following algorithms and data structures, all computable in $O(Sort(s))$ I/O's: (1) an algorithm to compute the river network, with a piecewise quadratic function of $O(r+n)$ pieces whose value is the area of the watershed for each point of the network; (2) a data structure that reports the boundary of the watershed of any query point q in $O(l + w/B)$ I/O's, where l is the number of I/O's needed to locate q in $\mathcal{S}$ and w is the complexity of the reported watershed; (3) a structure that reports the flow path from any point q (the course of water flowing from q) in $O(l + c/B)$ I/O's, where c is the complexity of the path.

One of the open questions posed by De Berg et al. [6] was if one could prove an $O(n^2)$ bound on the complexity of river networks in Delaunay triangulations. In the full version of this paper we answer this question negatively and show that we can construct Delaunay triangulations with river networks of size $\Theta(n^3)$.

2 Preliminaries

Let $\mathcal{T}$ be a TIN defined on n vertices. To model flow we assume that water always runs downhill in the direction of steepest descent. Furthermore, we assume that the direction of steepest descent is unique for any point in the terrain (so there are no horizontal triangles); no water flows off the terrain, and no edge is parallel to the direction of steepest descent on an adjacent triangle. We discuss how to do without the last three assumptions in the long version of this paper.

Following Yu et al. [19], we distinguish three types of edges in $\mathcal{T}$: *transfluent edges* are edges that receive water from one adjacent triangle which continues its way down the other triangle; *channels* are edges that receive water from both adjacent triangles; *ridges* are edges that do not receive water from any triangle.

The direction of steepest descent from a vertex may be along an incident edge, or on an incident triangle orthogonal to its contour lines. To capture this, we define the *slope profile* of a point p on the terrain as the function $s_p : S^1 \to \mathbb{R}$ such that $s_p(\theta)$ is the slope of the path that leaves p in the direction θ. The interesting directions for a vertex v of $\mathcal{T}$ correspond to the local maxima and minima in the slope profile of v. Note that these directions include the directions of channels and ridges incident to v, and also the directions of steepest descent and ascent. A vertex v is a *pit* if its slope profile is entirely positive.

We define up-paths and down-paths as paths of locally steepest ascent or descent as follows. An *up-path* from p is a path that starts at p, goes into a direction θ that is a positive local maximum in the slope profile of p leading onto the interior of an incident triangle, and then follows the steepest ascent until a vertex or a ridge of $\mathcal{T}$ is reached. (The requirement that the path leads onto the interior of the triangle incident to p prevents an up-path from leaving p along a ridge.) Similarly, a *down-path* from p is a path that starts at p, goes into a direction θ that is a negative local minimum in the slope profile of p leading onto the interior of an incident triangle, and then follows the steepest descent until a vertex or a channel of $\mathcal{T}$ is reached. An up-path from p to q is a down-path from q to p. Therefore, we may sometimes refer to a down-path as an up-path or the other way around, depending on our point of view.

The *strip map* $\mathcal{S}$ of $\mathcal{T}$ is the subdivision of $\mathcal{T}$ induced by the channels, ridges, up-paths and down-paths from all vertices of $\mathcal{T}$. We call the $O(n)$ faces of this map *strips*. Each strip is bounded by a portion of a ridge, a portion of a channel (the foot), and two possibly empty chains of up-paths. Note that from every point at the foot of a strip, the up-path through the strip has the same combinatorial structure: it crosses the same triangles and leads to the same ridge.

The *watershed* $W(q)$ of a point q on $\mathcal{T}$ is the set of all points on $\mathcal{T}$ from which the water flows to q. The *river network* of $\mathcal{T}$ is the set of all points on $\mathcal{T}$ with watersheds of non-zero area, or in other words, the set of points whose watersheds are two-dimensional regions. De Berg et al. [6] argue that the river network of $\mathcal{T}$ is the union of the channels of $\mathcal{T}$ and the paths of steepest descent that start from the lower endpoints of channels.

3 Modeling Flow with a Descent Graph

In this section we describe how to model flow on a triangulated terrain $\mathcal{T}$ using a *descent graph* $\mathcal{G}_{\text{desc}}$. Our goals are to define $\mathcal{G}_{\text{desc}}$ such that it is acyclic and such that any path of steepest descent in $\mathcal{T}$ corresponds to a path in $\mathcal{G}_{\text{desc}}$. For α-fat terrains, that is, terrains $\mathcal{T}$ where the minimum angle of each triangle (measured in the plane supporting the triangle) is at least a constant $\alpha > 0$, we show how to construct $\mathcal{G}_{\text{desc}}$ with only $O(n/\alpha^2)$ nodes. This leads to the following bounds on the size of down-paths, up-paths, river networks and strip maps:

Theorem 1. *Any down-path or up-path in an α-fat terrain has $O(n/\alpha^2)$ vertices, and the total complexity of its river network or the strip map is $O(n^2/\alpha^2)$.*

To define $\mathcal{G}_{\text{desc}}$, assume that the edges of $\mathcal{T}$ have been subdivided into *segments*, as will be described below. Then the descent graph $\mathcal{G}_{\text{desc}}$ of $\mathcal{T}$ contains a node for each vertex and each segment in $\mathcal{T}$; let $seg(v)$ denote the vertex/segment corresponding to node v. To ensure that the sets $seg(v)$ are disjoint, we consider a segment to include its upper endpoint (unless this is a vertex of $\mathcal{T}$), and to exclude its lower endpoint. $\mathcal{G}_{\text{desc}}$ contains a directed arc from node a to node b if and only if one of the following applies: (1) $seg(a)$ is the upper endpoint of $seg(b)$, and $seg(b)$ is a ridge or channel; (2) $seg(b)$ is the lower endpoint of $seg(a)$, and $seg(a)$ is a ridge or channel; (3) $seg(a)$ and $seg(b)$ are on the boundary of the same triangle Δ, and there is a path of steepest descent across the interior of Δ from $seg(a)$ to $seg(b)$. It can be seen that any path of steepest descent in $\mathcal{T}$ corresponds to a path in $\mathcal{G}_{\text{desc}}$.

To ensure that $\mathcal{G}_{\text{desc}}$ is acyclic we impose the following requirements on the subdivision of the edges of $\mathcal{T}$. Channels/ridges of $\mathcal{T}$ are not subdivided: each channel/ridge is one segment. For a segment s on a transfluent edge e and a triangle Δ incident to e, we have: (1) If s is not incident to a vertex, then the (open) smallest enclosing sphere of s (that is, the sphere with s as a diameter) does not intersect any other edge of Δ; (2) If s is incident to a vertex p of Δ, then the (open) sphere centered at p with s as a radius does not intersect any segment s' on an edge of Δ, where s' is not incident to p and s' is not separated from s by a line of steepest descent through p on Δ. We call a subdivision of the edges of $\mathcal{T}$ that meets these requirements *compliant*.

We denote the upper endpoint, midpoint and lower endpoint of $seg(v)$ by $up(v)$, $md(v)$ and $lw(v)$. If $seg(v)$ is a vertex we have $up(v) = md(v) = lw(v) = seg(v)$. We define the anchor $an(v)$ and relative position $rp(v)$ of v as follows:

- if $seg(v)$ is a channel, then $an(v) = lw(v)$ and $rp(v) = 1$ (above the anchor);
- if $seg(v)$ is a vertex or a segment of a transfluent edge, then $an(v) = md(v)$ and $rp(v) = 0$ (on the anchor);
- if $seg(v)$ is a ridge, then $an(v) = up(v)$ and $rp(v) = -1$ (below the anchor).

We now prove that any descent graph $\mathcal{G}_{\text{desc}}$ based on a compliant subdivision is acyclic. To this end, we define a partial order $\succ$ as follows: For two nodes a and b in $\mathcal{G}_{\text{desc}}$, we define $a \succ b$ if and only if $z(an(a)) > z(an(b))$, or $z(an(a)) = z(an(b))$ and $rp(a) > rp(b)$, where $z(p)$ is the elevation of p.

Lemma 1. *Given $\mathcal{G}_{desc}$ of a compliant subdivision, if $seg(a)$ and $seg(b)$ are on the boundary of the same triangle Δ, $seg(b)$ is not a channel, and there is a path of steepest descent across the interior of Δ from $seg(a)$ to $seg(b)$, then $z(up(a)) \geq z(up(b))$, and equality can only hold if $seg(a)$ is incident to $up(b)$.*

Proof. Let $p \in seg(a)$ and $q \in seg(b)$ such that the segment pq lies in Δ and follows the direction of steepest descent. If $seg(b)$ is a vertex of $\mathcal{T}$, then $z(up(a)) \geq z(p) > z(q) = z(up(b))$. Since $seg(b)$ is not a ridge or a channel, the only remaining case is that $seg(b)$ is a segment on a transfluent edge.

Consider now the plane Γ supporting Δ. Let S be the open half-plane on Γ containing p, bounded by the line through $seg(b)$. Let ℓ be the intersection of Γ

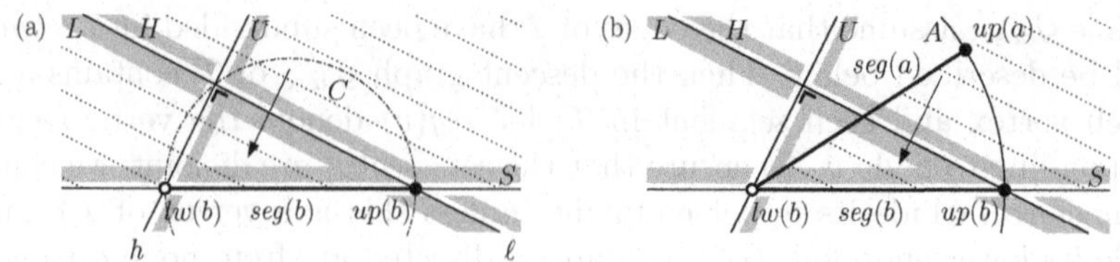

Fig. 1. Proof of Lemma 1

with the horizontal plane through $up(b)$, and let h be the line on Γ orthogonal to ℓ through $lw(b)$. Note that ℓ is a contour line of Γ, the line h follows the direction of steepest descent on Γ, and pq is parallel to h. Let L and H be the closed lower and the open upper half-planes of Γ bounded by ℓ, and let U be the open half-plane of Γ bounded by h containing $seg(b)$. Let C be the minimum circumscribed circle C of $seg(b)$ on Γ. See Fig. 1 (a) for an illustration. Since $q \in seg(b)$, we must have $p \in U \cap S$. We distinguish three cases:

First, if $seg(b)$ is not incident to a vertex of $\mathcal{T}$, then by compliance condition (1) C does not intersect $seg(a)$. By Thales' theorem, the rectangular triangle $U \cap S \cap L$ lies in C, and so $p \notin L$. This implies $z(up(a)) \geq z(p) > z(up(b))$.

Second, if $lw(b)$ is a vertex of $\mathcal{T}$, then let A be the open disk centered at $lw(b)$ through $up(b)$. If $seg(a)$ is not incident to $lw(b)$, then by compliance condition (2) it does not intersect A (since $p \in U \cap seg(a)$, $seg(a)$ is not separated from $seg(b)$ by h). Since $C \subset A$, we again have $z(up(a)) \geq z(p) > z(up(b))$. If $seg(a)$ is incident to $lw(b)$ (Fig. 1 (b)), then by compliance condition (2) $up(a)$ must lie in $U \setminus A$, and hence in H, implying $z(up(a)) > z(up(b))$.

Finally, if $up(b)$ is a vertex of $\mathcal{T}$, then let A' be the open disk centered at $up(b)$ through $lw(b)$. If $seg(a)$ is incident to $up(b)$, then clearly $z(up(a)) \geq z(up(b)))$, otherwise by condition (2) $seg(a)$ lies outside A', and p lies again in H. □

Corollary 1. *If $seg(a)$ and $seg(b)$ are on the boundary of the same triangle Δ, $seg(a)$ is not a ridge, and there is a path of steepest descent across the interior of Δ from $seg(a)$ to $seg(b)$, then $z(lw(a)) \geq z(lw(b))$, and equality can only hold if $seg(b)$ is incident to $lw(a)$.*

Lemma 2. *Given $\mathcal{G}_{desc}$ of a compliant subdivision, if $\mathcal{G}_{desc}$ contains an arc from a to b, then $a \succ b$.*

Proof. We can verify that if $seg(a)$ is the upper endpoint of $seg(b)$, or if $seg(b)$ is the lower endpoint of $seg(a)$, then $a \succ b$. It remains to discuss the case where there is a path of steepest descent from some point $p \in seg(a)$ to some point $q \in seg(b)$ across the interior of their common triangle Δ. We note that $seg(a)$ cannot be a channel, and $seg(b)$ cannot be a ridge. We have $z(up(a)) \geq z(p) > z(q) \geq z(lw(b))$. We now distinguish four cases, proving $z(an(a)) > z(an(b))$ and thus $a \succ b$ in each: First, if $seg(a)$ is a ridge and $seg(b)$ is a channel, then $z(an(a)) = z(up(a)) > z(lw(b)) = z(an(b))$. Second, if $seg(a)$ is a ridge and $seg(b)$ is a vertex or a segment on a transfluent edge, then $z(an(a)) = z(up(a)) \geq z(up(b))$ (by Lemma 1). Since $z(up(a)) > z(lw(b))$, then $z(an(a)) > z(md(b)) = z(an(b))$.

Third, if $seg(a)$ is a vertex or a segment and $seg(b)$ is a channel, it is handled symmetrically. Finally, if both $seg(a)$ and $seg(b)$ are segments of transfluent edges, then we have both $z(up(a)) \geq z(up(b))$ and $z(lw(a)) \geq z(lw(b))$. We cannot have equality in *both* inequalities, since if $seg(a)$ and $seg(b)$ would share both $up(b)$ and $lw(a)$, that would imply $up(b) = lw(a)$ and thus $z(up(a)) = z(lw(b))$, contradicting $z(p) > z(q)$. Hence equality holds in at most one of the above comparisons, so $z(an(a)) = z(md(a)) > z(md(b)) = z(an(b))$. □

It follows that $\mathcal{G}_{\text{desc}}$ of a compliant subdivision is a directed acyclic graph. We now sketch how to construct a compliant subdivision of size $O(n)$ for an α-fat terrain. Consider a vertex p of $\mathcal{T}$, and let r be the distance from p to the nearest edge not incident on p. We create a segment of length $r/2$ on each transfluent edge incident to p. Since the ratio of the longest and shortest edge in a fat triangle is bounded by a constant, and assuming the degree of p is constant, each of these segments covers a constant fraction of its edge. We can subdivide the remainder of the transfluent edges into a constant number of segments satisfying condition (1). In the long version of this paper we prove that the assumption on the vertex degree is not necessary and we analyse the dependency on α.

Theorem 2. *Any α-fat terrain has a compliant subdivision of size $O(n/\alpha^2)$.*

4 Computing the Rivers and the Watershed Area Map

This section shows how the acyclic descent graph can be used to construct the river network and the strip map of the terrain I/O-efficiently. We also show how to compute, for every point on the river network, the area of its watershed.

First we need the following. In Section 3 we showed that an α-fat terrain can be modeled by a descent graph $\mathcal{G}_{\text{desc}}$ of $O(n/\alpha^2)$ nodes and arcs. We have:

Lemma 3. *$\mathcal{G}_{desc}$ of an α-fat terrain can be computed in $O(Sort(n/\alpha^2))$ I/O's.*

Recall that the river network consists of the channels in the terrain and the paths of steepest descent that start at the lower endpoints of channels. The channels can be extracted from the terrain in a straightforward way. The challenge is tracing paths of steepest descent I/O-efficiently: we need to trace each such path from its upper endpoint across the triangles that it intersects, one segment at a time. We would like that a path that crosses k triangles can be computed in $O(Sort(k))$ I/O's. A general solution for this problem is not known in the I/O-model. This is where $\mathcal{G}_{\text{desc}}$ comes to rescue. The key idea is that every segment of a path of steepest descent is captured by an arc in the descent graph. Thus, instead of tracing such paths on the original terrain, we trace them in $\mathcal{G}_{\text{desc}}$. Doing this path by path would not be any faster, since even for planar directed acyclic graphs it is not known how to preprocess them for fast path traversals. Instead, we trace *all* paths of steepest descent in parallel while traversing $\mathcal{G}_{\text{desc}}$ in topological order (highest nodes first). In this way we compute the segments of the paths of steepest descent I/O-efficiently in a batched way.

More precisely, our approach is as follows. We put the arcs of $\mathcal{G}_{desc}$ on a stack A sorted by $\succ$-order of their nodes of origin, with the nodes that appear first in $\succ$-order on top. Furthermore, we initialize an I/O-efficient priority queue Q that stores pairs of the type (v, p), where v is a node in $\mathcal{G}_{desc}$, and p is a point of $seg(v)$. The queue is also organized by $\succ$-order. Initially we fill Q with all pairs (v, p) where p is a vertex of a channel, and v is the corresponding node in $\mathcal{G}_{desc}$.

We now repeat the following until Q is empty. From Q we extract the pair (v, p_1) with highest priority, and all further pairs $(v, p_2), (v, p_3), \ldots$ with the same priority. From A we pop arcs $\overrightarrow{uw_1}$ until $u = v$ or until $v \succ u$. In the latter case, no arcs that lead out of v are found, and all paths of steepest descent that reach v end there—we proceed to extracting the next pair (v, p_1) from Q. Otherwise at least one arc leads out of v. We also pop any remaining arcs $\overrightarrow{vw_2}, \overrightarrow{vw_3}, \ldots$ that originate from v from A. For each pair (v, p_i), we now select the arc $\overrightarrow{vw_j}$ that captures the path of steepest descent from p_i to a point q_i on $seg(w_j)$. If $seg(w_j)$ is not incident to p_i, we output the line segment from p_i to q_i as a river segment. If $seg(w_j)$ is not a channel or its bottom endpoint, we insert (w_j, q_i) in Q so that the path of steepest descent is traced further. After handling all pairs $(v, p_1), (v, p_2), \ldots$, we proceed by extracting the next pair from Q. We get:

Theorem 3. *The river network of an α-fat terrain can be computed in $O(Sort(r + n/\alpha^2))$ I/O's, where r is the size of the river network.*

The above approach can be extended to compute the subdivision of $\mathcal{T}$ induced by *all* channels, ridges, up-paths and down-paths starting from vertices of $\mathcal{T}$, that is, the strip map, in $O(Sort(s+n/\alpha^2))$ I/O's, where s is the size of the map.

Let q be a point on a channel segment. The watershed $W(q)$ of q contains parts of the at most two strips that have q at their feet [19]; the area of each part is given by a quadratic function of the position of q that is determined by scanning the list of triangles that intersect the strip. All other strips lie either inside or outside $W(q)$; more precisely, a strip lies in $W(q)$ if and only if its lowest point lies at q or on the river network upstream of q.

To compute the piecewise quadratic function whose value is the watershed area for every point on the river network, we process the river network from the leaves to the root and collect, for every edge of the river network, the area functions for the strips to the left and to the right and the total area of the strips that drain into the subtree below (that is, upstream of) that edge. This can be done with standard techniques (e.g. a post-order traversal of the river network).

Theorem 4. *Given an α-fat terrain $\mathcal{T}$ we can compute in $O(Sort(s + n/\alpha^2))$ I/O's a piecewise quadratic function of $O(r+n)$ pieces whose value is the watershed area of every point in $\mathcal{T}$, where r is the size of the river network and s is the size of the strip map of $\mathcal{T}$.*

5 A Data Structure for Flow Path and Watershed Queries

In this section we describe an I/O-efficient data structure for fast flow path and watershed boundary queries: given a point q on the terrain, we want to report

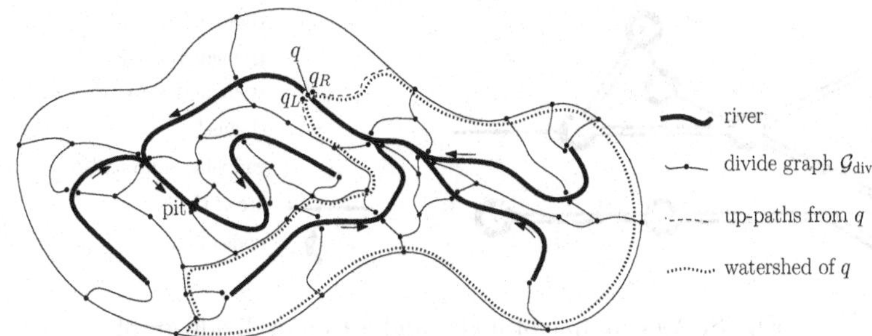

Fig. 2. Finding the watershed of q

the flow path starting at q and the boundary of the watershed $W(q)$ of q. As explained in the previous section, the watershed of a point q on the river network can be found by traversing the subtree of the river network upstream of q, and collecting the strips that drain into that subtree and parts of the strips that have q at their feet. However, this would yield all strip boundaries that lie in the interior of the watershed, whose total size may be much bigger than the boundary of the complete watershed given as a simple polygon. To address this problem McAllister [11] suggested to build a divide graph, $\mathcal{G}_{\text{div}}$, that consists of sections of ridges and certain up-paths in $\mathcal{T}$. The watershed boundary of a query point q can then be found by adding the up-paths from q to $\mathcal{G}_{\text{div}}$. This will divide the face of $\mathcal{G}_{\text{div}}$ that contains q into two subfaces: the subface upstream of q is the watershed of q (see Fig. 2). Below we describe a refined version of $\mathcal{G}_{\text{div}}$ and sketch how to store it face by face in a way that makes it possible to report watershed boundaries I/O-efficiently.

For every vertex v of $\mathcal{T}$, and for every endpoint v of an up-path or a down-path from a vertex of $\mathcal{T}$, consider an infinitesimally small circle centered at the projection of v on the horizontal plane. Cut this circle where it is crossed by the projections of up-paths that start from v, channels that lead down to v, or the path of steepest descent from v. Note that we do not cut the circle at every down-path from v, but only at the down-path that descends steepest from v. Every piece of the circle that results constitutes a node of $\mathcal{G}_{\text{div}}$, which is not (directly) connected to the other pieces of the circle. The arcs of $\mathcal{G}_{\text{div}}$ correspond to (i) the ridges of $\mathcal{T}$ and (ii) two copies of each up- or down-path starting from a vertex v of $\mathcal{T}$. The two copies are assumed to lie at an infinitesimally small distance from the real course of the path, such that one copy runs to the left of the path, and the other to the right, leaving a narrow corridor between them.

The arcs of $\mathcal{G}_{\text{div}}$ are connected to the nodes of $\mathcal{G}_{\text{div}}$ as follows. Every arc that corresponds to a path π that has v as an endpoint, is connected to the node whose piece of the circle around v is crossed by the projection of π on the plane. When an up- or down-path passes between two pieces of the circle on its way to or from v, the copy of the path that is offset to its left is connected to the piece of the circle to its left, and the copy that is offset to its right is connected to the piece of the circle to its right (Fig. 3). Note that arcs are incident to the same

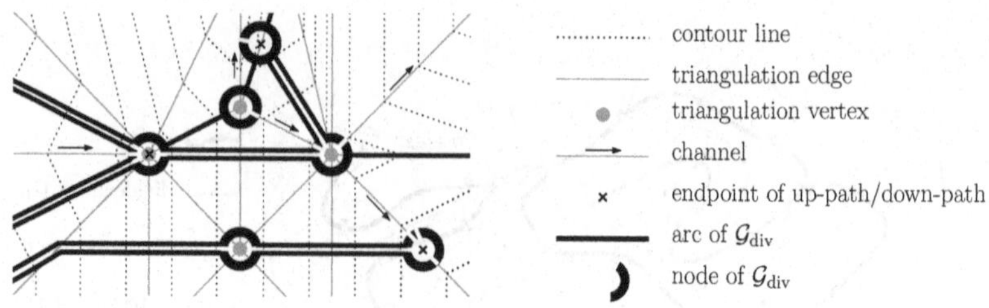

Fig. 3. An example of nodes and arcs in a divide graph

node—that is, the same piece of a circle—if and only if the geometric entities to which these arcs correspond are not separated by any path of steepest descent. Thus, no watershed is divided by face boundaries of $\mathcal{G}_{\mathrm{div}}$. The projection of a vertex v on the horizontal plane always lies in the *interior* of a face of $\mathcal{G}_{\mathrm{div}}$, and never on a node or arc of $\mathcal{G}_{\mathrm{div}}$. Thus every vertex of $\mathcal{T}$ lies inside a well-defined watershed of a pit of the terrain.

We now find the watershed boundary of a point q as follows. If q is a pit, we report the boundary of the face $F(q)$ in $\mathcal{G}_{\mathrm{div}}$ that contains q. Otherwise, let q_L and q_R be the nodes in $\mathcal{G}_{\mathrm{div}}$ corresponding to the circular pieces around q immediately counterclockwise and clockwise, respectively, of the direction of steepest descent from q. The area $W(q)$ that drains through q is the area enclosed by the path in $\mathcal{G}_{\mathrm{div}}$ that follows the boundary of $F(q)$ clockwise from q_R to q_L (see Fig. 2).

We now sketch the data structure that makes it possible to trace the up-paths and the flow path from q efficiently and to trace the boundary of $W(q)$ without going down and back up arcs of $\mathcal{G}_{\mathrm{div}}$ in the interior of $W(q)$. Our solution consists of four ingredients: (1) the divide graph $\mathcal{G}_{\mathrm{div}}$, stored face by face as explained below; (2) the river network, preprocessed for fast downstream traversals (with Hutchinson et al. [8]); (3) the strip map, stored strip by strip to facilitate fast traversals of steepest-ascent and steepest-descent paths and to provide pointers into the divide graph and the river network; (4) a point location structure on the strip map so that we can locate, for each query point q, the strip that contains it (with Arge and Vahrenhold [5]). Below we sketch our solution for storing a face of the divide graph. Details can be found in the long version of this paper.

Let p be a pit. The boundary of the face $F(p)$ of $\mathcal{G}_{\mathrm{div}}$ that represents $W(p)$ can be seen as a clockwise cycle γ with trees protruding to its right, into the watershed of p. Pick any node r that corresponds to a circular piece around p; and let π be the path from r to a node $s(p)$ of the boundary γ, such that $s(p)$ is the first and only node of γ on π. Our idea is to "cut" the graph at r, and convert the boundary of $F(p)$ to a tree (see Fig. 4). To do this, we start by splitting all edges and vertices on π in a left copy and a right copy. Every right vertex is connected to the right edges incident to it, and to any trees that protrude into the watershed of p to the right of π. Furthermore, the right copy $s_R(p)$ of $s(p)$ is connected to the edge that follows $s(p)$ in γ. Every left vertex is connected to the left edges incident to it, and to any trees that protrude into the watershed of p to the left of π. Furthermore, the left copy $s_L(p)$ of $s(p)$ is connected to the

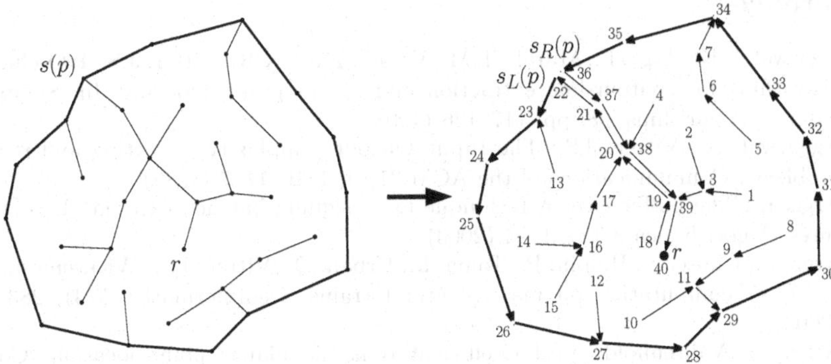

Fig. 4. Transforming the boundary of a face into a rooted tree. In the right figure, the edges are directed from child to parent, and numbers represent postorder ranks.

edge that precedes $s(p)$ in γ. We root the resulting tree at the right copy of r. We break up arcs that represent up-paths or down-paths in segments: one segment for each triangle crossed by the path, with vertices at the points where the path crosses a transfluent edge of the triangulation. Let the resulting tree be $\mathcal{B}(p)$; we store with each node its rank in a postorder traversal of the tree, and preprocess the tree for fast leaf-to-root traversals with the method of Hutchinson et al. [8].

To find the watershed boundary of a query point q, we briefly discuss the (interesting) case when q is in the interior of a channel. Assume we have the point location structure in [5] on the strip map so that we can locate the strips that have q at their feet. We follow the up-paths from q until they meet $\mathcal{G}_{\text{div}}$ in points u_L and u_R on the ridges at the top of these strips. We then report the path between them that encloses the area upstream from q; from our construction, this is the path that connects u_L and u_R in $\mathcal{B}(pit(q))$, where $pit(q)$ is the pit in the face of $\mathcal{G}_{\text{div}}$ that contains q. The triangulation edges crossed by the up-paths from q to u_L and u_R are stored in the strips on each side of the channel, and they can be retrieved in a linear number of I/O's. The arcs of $\mathcal{B}(pit(q))$ that contain u_L and u_R are stored with the strip. The path between u_L and u_R is obtained from $\mathcal{B}(pit(q))$ in a linear number of I/O's by tracing it from u_L and u_R up to their lowest common ancestor, which is easily recognized using the post-order numbering of the nodes.

Let us denote $q(s) = O((s/B)\log_B s)$ and $l(s) = O(\log_B^2 s)$ the number of I/O's to build and query, respectively, a point location structure as described by Arge and Vahrenhold [5]. We have:

Theorem 5. *A data structure of size $O(s)$ for answering flow path and watershed boundary queries on an α-fat terrain $\mathcal{T}$ can be computed in $O(q(s) + Sort(n/\alpha^2))$ I/O's, where s is the size of the strip map of $\mathcal{T}$. The structure reports the watershed boundary or the flow path of any query point q in $O(l(s) + k/B)$ I/O's, where k is the complexity of the answer.*

References

1. Agarwal, P.K., Arge, L., Murali, T.M., Varadarajan, K.R., Vitter, J.S.: I/O-efficient algorithms for contour-line extraction and planar graph blocking. In: Symp. on Discrete Algorithms '98, pp. 117–126 (1998)
2. Aggarwal, A., Vitter, J.S.: The Input/Output complexity of sorting and related problems. Communications of the ACM 31(9), 1116–1127 (1988)
3. Arge, L.: The buffer tree: A technique for designing batched external data structures. Algorithmica 37(1), 1–24 (2003)
4. Arge, L., Chase, J., Halpin, P., Toma, L., Urban, D., Vitter, J.S., Wickremesinghe, R.: Flow computation on massive grid terrains. GeoInformatica 7(4), 283–313 (2003)
5. Arge, L., Vahrenhold, J.: I/O-efficient dynamic planar point location. Comp. Geom. 29, 147–162 (2004)
6. de Berg, M., Bose, P., Dobrint, K., van Kreveld, M., Overmars, M., de Groot, M., Roos, T., Snoeyink, J., Yu, S.: The complexity of rivers in triangulated terrains. In: Canad. Conf. Comp. Geom. '96, pp. 325–330 (1996)
7. Chiang, Y.-J., Goodrich, M.T., Grove, E.F., Tamassia, R., Vengroff, D.E., Vitter, J.S.: External-memory graph algorithms. In: Symp. on Discr. Alg. '95, pp. 139–149 (1995)
8. Hutchinson, D., Maheshwari, A., Zeh, N.: An external memory data structure for shortest path queries. In: ACM-SIAM Computing and Combin. Conf. '99, pp. 51–60 (1999)
9. Jones, N.L., Wright, S.G., Maidment, D.R.: Watershed delineation with triangle-based terrain models. Journal of Hydraulic Engineering 116(10), 1232–1251 (1990)
10. van Kreveld, M.: Digital elevation models and TIN algorithms. In: van Kreveld, M., Roos, T., Nievergelt, J., Widmayer, P. (eds.) Algorithmic Foundations of Geographic Information Systems. LNCS, vol. 1340, pp. 37–78. Springer, Heidelberg (1997)
11. McAllister, M.: The computational geometry of hydrology data in geographic information systems. PhD th., Univ. of British Columbia (1999)
12. McAllister, M.: A watershed algorithm for triangulated terrains. Canad. Conf. Comp. Geom. '99 (1999)
13. McAllister, M., Snoeyink, J.: Extracting consistent watersheds from digital river and elevation data. Ann. Conf. Amer. Soc. for Photogrammetry and Remote Sensing'99 (1999)
14. Moet, E.: M. v. Kreveld, and A. F. v/d Stappen. On realistic terrains. In: Symp. on Computational Geom. '06, pp. 177–186 (2006)
15. Palacios-Velez, O., Gandoy-Bernasconi, W., Cuevas-Renaud, B.: Geometric analysis of surface runoff and computation of unit elements in distributed hydrological models. J. Hydrology 211, 266–274 (1998)
16. Silfer, A.T., Kinn, G.J., Hassett, J.M.: A geographic information system utilizing the triangulated irregular network as a basis for hydrologic modeling. Auto-Carto 8, 129–136 (1987)
17. Theobald, D.M., Goodchild, M.F.: Artifacts of TIN-based surface flow modeling. In: Proc. GIS/LIS'90, pp. 955–964 (1990)
18. Tucker, G., Lancaster, S., Gasparini, N., Rybarczyk, S.: An object-oriented framework for hydrology and geomorphic modeling using TINs. Computers and Geosc. 27(8), 959–973 (2001)
19. Yu, S., van Kreveld, M., Snoeyink, J.: Drainage Queries in TINS: from local to global and back again. In: Symp. on Spatial Data Handling '96, pp. 1–14 (1996)

Computing the Visibility Map of Fat Objects*

Mark de Berg and Chris Gray

Department of Computing Science, TU Eindhoven
{mdberg,cgray}@win.tue.nl

Abstract. We give an output-sensitive algorithm for computing the visibility map of a set of n constant-complexity convex fat polyhedra or curved objects in 3-space. Our algorithm runs in $O((n + k) \text{polylog}\, n)$ time, where k is the combinatorial complexity of the visibility map. This is the first algorithm for computing the visibility map of fat objects that does not require a depth order on the objects and is faster than the best known algorithm for general objects. It is also the first output-sensitive algorithm for curved objects that does not require a depth order.

1 Introduction

Hidden-surface removal is an important and well-studied computational-geometry problem with obvious applications in computer graphics. The problem is to find those portions of objects in a scene that are visible from a given viewpoint. There are two main approaches to the hidden-surface removal problem: the *image-space approach* and the *object-space approach*. In the former, one calculates the visible object for each pixel of the image; the well known Z-buffer algorithm is the standard example of this. In the latter, one computes the so-called *visibility map* of the scene, which gives an exact description of the visible part of each object; this is the approach taken in computational geometry.

Formally, the visibility map of a set $\mathcal{P}$ of objects in $\mathbb{R}^3$ with respect to a viewpoint p is defined as the subdivision of the viewing plane into maximal regions such that in each region a single object in $\mathcal{P}$ is visible from p, or no object is visible. We will assume in this paper, as is usual, that the objects are disjoint. The visibility map of a set of n constant-complexity objects can be computed in $O(n^2)$ time [17]. Since the (combinatorial) complexity of the visibility map can be $\Omega(n^2)$—a set of n long and thin triangles that form a grid-like pattern when projected on the viewing plane is an example—this is optimal in the worst case. In most cases, however, the complexity of the visibility map is much smaller than quadratic. Therefore the main challenge in the design of algorithms for computing visibility maps has been to obtain *output-sensitive* algorithms: algorithms whose running time depends not only on the complexity of the input, n, but also on the complexity of the output (that is, the visibility map), k. Ideally the running time should be near-linear in n and k.

* This research was supported by the Netherlands' Organisation for Scientific Research (NWO) under project no. 639.023.301.

F. Dehne, J.-R. Sack, and N. Zeh (Eds.): WADS 2007, LNCS 4619, pp. 251–262, 2007.

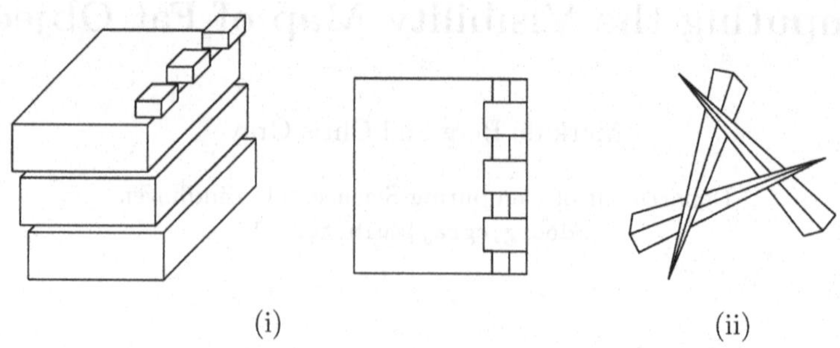

Fig. 1. (i) The visibility map of fat boxes can have quadratic complexity. Left: the scene. Right: the visibility map for $p = (0, 0, \infty)$. (ii) The visibility map of a scene with cyclic overlap.

The first output-sensitive algorithms for computing visibility maps only worked for polygons parallel to the viewing plane or for the slightly more general case that a depth order on the objects exists and is given [10,13,14,19,20,21]. Unfortunately a depth order need not exist since there can be cyclic overlap among the objects—see Figure 1 (ii). De Berg and Overmars [7] (see also [3]) developed a method to obtain an output-sensitive algorithm that does not need a depth order. When applied to axis-parallel boxes (or, more generally, c-oriented polyhedra) it runs in $O((n+k)\log n)$ time [7] and when applied to arbitrary triangles it runs in $O(n^{1+\varepsilon} + n^{2/3+\varepsilon}k^{2/3})$ time [1]. Unfortunately, the running time for the algorithm when applied to arbitrary triangles is not near-linear in n and k; for example, when $k = n$ the running time is $O(n^{4/3+\varepsilon})$. For general curved objects no output-sensitive algorithm is known,[1] not even when a depth order exists and is given.

In this paper we study the hidden-surface removal problem for so-called *fat objects*—see the next section for a definition of fatness. As illustrated in Figure 1, the complexity of the visibility map of fat objects can still be $\Theta(n^2)$, so also here the main challenge is to obtain an output-sensitive algorithm. Fat objects have received ample attention over the past decade or so, both from a combinatorial and from an algorithmic point of view, and many problems can be solved much more efficiently for fat objects than for general objects. Since hidden-surface removal has been widely studied in computational geometry, it is not surprising that it has also been studied for fat objects: Katz *et al.* [15] gave an algorithm with running time $O((U(n) + k)\log^2 n)$, where $U(m)$ denotes the maximum complexity of the union of the projection onto the viewing plane of any subset of m objects. Since $U(m) = O(m \log\log m)$ for fat polyhedra [18] and $U(m) = O(\lambda_{s+2}(m)\log^2 m)$ for fat curved objects [5], their algorithm is near-linear in n and k. (Here $\lambda_{s+2}(n)$ is the maximum length of an $(n, s+2)$ Davenport-Schinzel sequence; $\lambda_{s+2}(n)$ is almost linear in n.) However, the algorithm only works if a depth order exists and is given. This leads to the main question we wish

[1] Some of the algorithms can be generalized to curved objects using standard techniques. The resulting algorithms are not very efficient, however, and typically have running time close to quadratic even when the visibility map has linear complexity.

to answer: is it possible to obtain an output-sensitive hidden-surface removal algorithm for fat objects that is near-linear in n and k and does not need a depth order on the objects? We answer this question affirmatively by giving an algorithm with running time $O((n+k)\,\mathrm{polylog}\,n)$ for fat convex objects of constant-complexity. More precisely, the running time is $O((n \log n (\log\log n)^2 + k)\log^3 n)$ when the objects are polyhedra, and it is $O((n \log^{5+\varepsilon} n + k)\log^3 n)$ when the objects are curved.

The main difficulty we have to overcome is that the only known method for output-sensitive hidden-surface removal that can handle objects without depth order [3,7] needs an auxiliary data structure for ray shooting in so-called *curtains*—these are semi-infinite surfaces, extending downward from the edges of the input objects—and it appears to be difficult to profit from the fact that the objects are fat when implementing this data structure. This also explains why there is no (efficient) output-sensitive algorithm for hidden-surface removal in curved objects: there are no efficient data structures known for ray shooting (with curved rays, in this case) in curved curtains. Our method therefore works differently: instead of ray shooting in curtains, we trace the rays on several two-dimensional planes by performing many simultaneous and coordinated sweeps on these planes. To obtain a suitable set of planes, we use a suitably augmented variant of the BSP for ray shooting that was recently introduced by De Berg [4].

2 Preliminaries

Let $\mathcal{P}$ be a set of disjoint convex objects in $\mathbb{R}^3$. We assume the objects are β-fat according to the following definition of fatness [9]: an object o in $\mathbb{R}^d$ is *β-fat* if for any ball b whose center lies in o and that does not fully contain o, we have $vol(b \cap o) \geq \beta \cdot vol(b)$, where $vol(o)$ denotes the volume of o. (For convex objects this definition is equivalent, up to constant factors, to other definitions of fatness that have been proposed.)

We define size(o), the *size* of an object o, to be the radius of the smallest enclosing ball of o. The *density* of a set S of objects is defined as the smallest number λ such that any ball b is intersected by at most λ objects $o \in S$ such that $size(o) \geq size(b)$. The following well-known lemma [9] relates the density of a group of objects to the fatness constant.

Lemma 1. [9] *A set of disjoint β-fat objects has density λ where $\lambda = O(1/\beta)$.*

For a curve e in $\mathbb{R}^3$ define the *curtain* of e, denoted curt(e), as the ruled surface constructed by taking a vertical ray pointing downward and moving its starting point from one end of e to the other. Thus, if e is a segment then curt(e) is an infinite polygon defined by e and two unbounded edges, each parallel to the z-axis. For a set E of curves we let $\mathrm{curt}(E) := \{\mathrm{curt}(e) | e \in E\}$.

Next we define some notation and terminology relating to visibility maps. We assume from now on that we are looking at the scene from above with the viewpoint at $z = \infty$—hence, we are dealing with a parallel view. As already mentioned, the visibility map $\mathcal{M}(\mathcal{P})$ of $\mathcal{P}$ is the subdivision of the viewing plane into maximal regions such that in each region a single object in $\mathcal{P}$ is visible from the viewpoint p, or no object is visible. We assume without loss of generality that the viewing plane is the xy-plane.

Consider an object $o \in \mathcal{P}$. We denote the projection of o onto the viewing plane by $\mathrm{proj}(o)$. Since o is convex, the boundary of $\mathrm{proj}(o)$ consists of the projection of all points of vertical tangency of o. Let $\sigma(o)$ denote the curve[2] on the boundary of o that projects onto the boundary of $\mathrm{proj}(o)$. Note that if o is polyhedral, $\sigma(o)$ consists of certain edges of o. We cut $\sigma(o)$ into two pieces at the points of minimum and maximum x-coordinate; we can assume without loss of generality that these points are unique. We call these pieces *silhouette curves*—note that for polyhedral objects a silhouette curve consists of multiple edges of the object—and their endpoints *vertices*.

$\mathcal{M}(\mathcal{P})$ is a plane graph whose *nodes* are intersection points of projected silhouette curves and whose *arcs* are portions of projected silhouette curves. Arcs of the visibility map will be denoted by a, and silhouette curves by e. The curve whose projection contains the arc a is denoted $e(a)$. It will be convenient to also consider the projections of visible endpoints of silhouette curves (that is, visible vertices) as nodes. Since we cut $\sigma(o)$ into two pieces when it changes direction with respect to the x-axis, the arcs of $\mathcal{M}(\mathcal{P})$ are x-monotone.

The existing output-sensitive hidden-surface removal algorithm from [3] works as follows. It sweeps over the viewing plane from left to right, detecting the arcs of the visibility map along the way. Note that the left endpoint of an arc is one of two types. It is either the projection of a visible vertex of a silhouette curve or it is the right endpoint of some other arc.

To detect left endpoints of the first type we need a data structure to determine for a vertex v whether it is visible or not. If it is, two new arcs start at $\mathrm{proj}(v)$, which are contained in the projections of the two silhouette curves incident to v. Detecting if v is visible can be done by vertical ray shooting: shoot a ray from v vertically upwards; if no object is hit then v is visible.

We also need to be able to detect the right endpoint of an arc (and thereby the left endpoints of the second type). An arc a can end for two reasons. One is that the silhouette curve $e(a)$ projecting onto a ends. The other is that $\mathrm{proj}(e(a))$ intersects some other projected silhouette curve $\mathrm{proj}(e')$ such that either $e(a)$ becomes invisible or e' becomes visible—see Figure 2 (i). These two latter events are detected using a ray shooting operation in a set of curtains, as explained next. When $e(a)$ becomes invisible because it disappears below some object o, then the ray along $e(a)$ must hit the curtain hanging from one of o's silhouette curves. When some other silhouette curve e' becomes visible, something similar holds. To this end, we define a ray[3] $\rho(a)$ for an arc a of the visibility map as follows. Let q be the point on $e(a)$ projecting onto the left endpoint of a. Project the portion of $e(a)$ to the right of q onto the object $o(q)$ immediately below q. (If there is no such object, we project onto a plane below all objects.) This gives us

[2] For simplicity of presentation we assume o does not have any vertical facets, so that $\sigma(o)$ is uniquely defined. It is easy to adapt the definitions to the general case.

[3] Note that in case of curved objects, the ray will be curved.

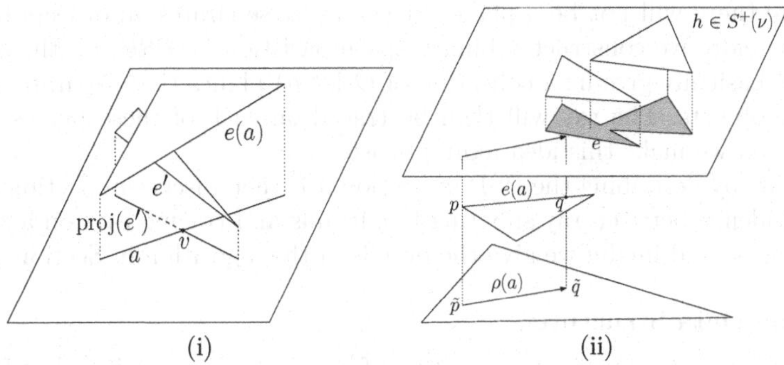

Fig. 2. (i) The node v in the visibility map is made by the intersection of proj(e) and proj(e'). (ii) $\rho(a)$ hits a curtain in curt(E) at point q when its projection intersects a silhouette curve of a union stored at $S^+(\nu)$. Note that the objects pictured here are not fat under our definition, but could be the top surfaces of fat polyhedra. We draw the objects in this way to ease visualization.

a ray on the surface of $o(q)$ whose projection contains a. It can be argued [3] that the point where $\rho(a)$ hits curt(e') corresponds to the point where the silhouette curve e' becomes visible. (Note that $\rho(a)$ is about to leave $o(q)$ when it reaches a silhouette curve of $o(q)$; in this case $\rho(a)$ hits the curtain hanging from that silhouette curve, which is then the curve that becomes visible.) Since any curtain hit by the ray along $e(a)$ is also hit by $\rho(a)$—after all, $\rho(a)$ is below $e(a)$—we can detect events where $e(a)$ becomes invisible by shooting along $\rho(a)$ as well.

The next lemma summarizes the discussion above.

Lemma 2 ([3]). *Let E be the set of silhouette curves of the objects in $\mathcal{P}$. The right endpoint of an arc a of $\mathcal{M}(\mathcal{P})$ is the leftmost of the following event points:*

- *The projection of the right endpoint of $e(a)$.*
- *The projection of the first intersection of $\rho(a)$ with a curtain in* curt(E).

3 The Algorithm

As mentioned in the introduction, it seems hard to implement a structure for ray shooting in curtains that profits from the fact that the objects are fat. Therefore we use the following idea.

Suppose that all objects are above a plane h and the query ray $\rho(a)$ is below h. Then we can project all objects and the ray onto h, and shoot with the projected ray in the union of the projected objects; the point where the ray hits a curtain then corresponds to the point where the projected ray hits the union. This is true because in our application the ray will always be visible, so the projected ray cannot start inside the union. Unfortunately two-dimensional ray shooting is still too costly. If, however, we have to answer many queries, then we can project all of them onto h, and perform a sweep to detect when they intersect the union.

Of course there will not be a plane h that nicely separates all objects from all rays. Therefore we construct a binary space partition (a *BSP*) on the objects. This will basically give us a collection of $O(\log n)$ planes that separate any ray from the objects. The ray will then be traced on each of these planes. In the next section we make this idea more precise.

We start by describing the BSP in Section 3.1, then discuss in Section 3.2 the correspondence between ray shooting in curtains and tracing rays on a suitable set of planes, and finally we give the details of the algorithm in Section 3.3.

3.1 The Data Structure

A *balanced aspect ratio tree* (or *BAR-tree* for short) is a special type of BSP for storing points. It was introduced by Duncan [11,12]. The variant known as the *object BAR-tree* [8] stores objects rather than points and has proved especially useful in designing data structures for fat objects. It has been used as a basis for vertical ray shooting [4,6] as well as approximate range searching and nearest neighbor searching [8].

We denote the region associated with a node ν in the object BAR-tree for $\mathcal{P}$ by region(ν), and we let $\mathcal{P}_\nu$ denote the set of all objects $o \in \mathcal{P}$ intersecting region(ν), clipped to region(ν). The following lemma states the properties of the object BAR-tree we will need.

Lemma 3. [8] *Let $\mathcal{P}$ be a set of n β-fat disjoint convex objects in $\mathbb{R}^d$. An object BAR-tree on $\mathcal{P}$ is a BSP tree $\mathcal{T}$ for $\mathcal{P}$ with the following properties:*

(i) the tree has $O(n)$ leaves; each leaf region intersects $O(1/\beta)$ objects from $\mathcal{P}$;
(ii) the depth of the tree is $O(\log n)$;
(iii) for each node ν, region(ν) *has constant complexity and fatness.*

De Berg [4] has shown how to augment an object BAR-tree $\mathcal{T}$ with secondary structures, so that vertical ray shooting can be performed efficiently. The augmentation is as follows.

- For each leaf node μ of $\mathcal{T}$, we store the set $\mathcal{P}_\mu$ in a list $\mathcal{L}_\mu$.
- For an internal node ν, let h_ν denote the splitting plane stored at ν.
  - If h_ν is vertical, then we store the set $\{h_\nu \cap o : o \in \mathcal{P}_\nu\}$—that is, the cross-sections of the polyhedra in $\mathcal{P}_\nu$ with h_v—in a structure $\mathcal{T}_\nu$, which is an optimal point-location structure [16] on the trapezoidal map defined by $h_\nu \cap \mathcal{P}_\nu$.
  - If h_ν is not vertical, then ν has two associated data structures, $\mathcal{T}_\nu^+$ and $\mathcal{T}_\nu^-$, defined as follows.
    Let $\mathcal{P}_\nu^+$ denote the set of object parts from $\mathcal{P}_\nu$ lying above h_ν. Thus $\mathcal{P}_\nu^+ = \mathcal{P}_\mu$, where μ is the child of ν corresponding to the region above h_ν. Let proj($\mathcal{P}_\nu^+$) denote the set of vertical projections of the objects in $\mathcal{P}_\nu^+$ onto h_ν. Then $\mathcal{T}_\nu^+$ is an optimal point-location structure for U(proj($\mathcal{P}_\nu^+$)), the union of proj($\mathcal{P}_\nu^+$). In our application, we not only store the point-location structure for U(proj($\mathcal{P}_\nu^+$)), but also an explicit list of all union edges.
    The associated structure $\mathcal{T}_\nu^-$ is defined similarly, but this time for the object parts below h_ν.

Recall that we want to use the structure to answer ray shooting queries in curtains, where the query rays are projections of (parts of) silhouette curves onto the object immediately below. A problem with this approach is that an object may be cut into many pieces,[4] and we would then have to spend time whenever the ray goes from one piece to the next.

To avoid this problem we need some extra information. In particular, for each object $o \in \mathcal{P}$ we need to store the union of the projection of a certain subset $\mathcal{P}(o) \subset \mathcal{P}$ onto $\partial^+(o)$, the top surface of o. (The top surface of o is the part of the boundary of o visible from above.) The subset $\mathcal{P}(o)$ is defined as follows.

Call an object o *large* at a node ν of $\mathcal{T}$ if o intersects region(ν) and the following two conditions are met: (i) size(o) $<$ size(region(parent(ν))) and (ii) either size(o) $\geq$ size(region(ν)) or ν is a leaf. Now we define

$$\mathcal{P}(o) := \{\, o' \in \mathcal{P} : \text{there is a node } \nu \text{ such that } o \text{ is large at } \nu, \; o' \text{ intersects region}(\nu) \text{ and } o' \text{ is above } o \,\}$$

Besides these extra unions on the top surface of each object o, we also need the union of the projections of all the objects in $\mathcal{P}$ onto the xy-plane. (The xy-plane can be seen as a dummy object added below the whole scene, which is large at the root of $\mathcal{T}$.)

Next we analyze the cost of the additional information.

Lemma 4. *Any object $o \in \mathcal{P}$ is large at $O(\log n)$ nodes, and at any node ν there are $O(1/\beta)$ large objects.*

Proof. By Lemma 3 we know that every cell of $\mathcal{T}$ is $O(1)$-fat. This means that any collection of disjoint cells has density $O(1)$. Therefore, since the cells at any level of the BAR-tree are disjoint, the number of nodes ν in any level of the BAR-tree intersecting some $o \in \mathcal{P}$ with size(region(ν)) $\geq$ size(o) is $O(1)$. An object o can only be large at the node ν if size(region(parent(ν))) $\geq$ size(o). Thus, the number of cells per level at which o can be large is $O(1)$. Finally we know that $\mathcal{T}$ has $O(\log n)$ levels by Lemma 3, proving the first part of the lemma.

A set of disjoint β-fat objects has density $O(1/\beta)$—see Lemma 1—which, together with Lemma 3(i), implies the second part. □

From Lemma 4 we derive:

$$\begin{aligned}\textstyle\sum_o |\mathcal{P}(o)| &\leq \textstyle\sum_\nu \{(\#\text{ large objects at } \nu) \cdot (\#\text{ objects intersecting region}(\nu))\} \\ &\leq O(1/\beta) \cdot \textstyle\sum_\nu |\mathcal{P}_\nu| \;\leq\; O((1/\beta) \cdot n \log n),\end{aligned}$$

where the last inequality follows from [4]. Together with the known bounds on the union of fat objects [5,18] this is easily seen to imply that the total amount of

[4] The fact that an object is cut into many pieces also prevents us from applying the following simple strategy: compute the object BAR-tree, use it to find a depth order on the resulting set of pieces, and apply the algorithm of Katz *et al.* [15]. The problem is that the visibility map of the pieces may be much more complex than the visibility map of the original objects.

storage and preprocessing time for the unions of the projections of $\mathcal{P}(o)$ onto the top surfaces $\partial^+(o)$ does not increase the total amount of storage or preprocessing asymptotically, and the bounds we get are the same as in [4]. (The constants in the O-notation depend on the fatness factor β.)

Lemma 5. *Let β be a fixed constant. The data structure above for convex β-fat polyhedral objects requires $O(n \log^3 n (\log\log n)^2)$ storage and $O(n \log^4 n (\log\log n)^2)$ preprocessing time, and $O(n \log^{7+\varepsilon} n)$ storage and $O(n \log^{8+\varepsilon} n)$ preprocessing time for convex β-fat curved objects. With this structure, we can answer vertical ray shooting queries in $O(\log^2 n)$ time.*

3.2 Tracing an Arc

Recall that the right endpoint of an arc a can be found by shooting with $\rho(a)$ in $curt(E)$. Next we explain how to find the right endpoint of a using the unions stored in $\mathcal{T}$ and additional unions described above. The key is to find a collection of $O(\log n)$ unions such that the first point where $\rho(a)$ hits a curtain corresponds to the first point where one of the unions is hit.

To this end we first define for a node ν a collection $S^+(\nu)$ of $O(\log n)$ splitting planes, which consists of those splitting planes $h_{\nu'}$ such that ν' is an ancestor of ν and region(ν) is below $h_{\nu'}$. Now let $e(a)$ be the silhouette curve defining an arc a, and let $p \in e(a)$ be the point projecting onto the left endpoint of a. Recall that $\rho(a)$ is a ray on the top surface of the object o directly below p. We denote the projection of p onto o by $\tilde{p}$.

The first curtain hit by $\rho(a)$ can now be found using the following lemma.

Lemma 6. *Let $\rho(a)$ be a ray on the top surface of an object $o \in \mathcal{P}$, let $\tilde{p}$ be the starting point of $\rho(a)$, and let ν be the node in $\mathcal{T}$ such that $\tilde{p} \in$ region(ν) and o is large at ν. Then $\rho(a)$ hitting a curtain from $curt(E)$ inside region(ν) corresponds to (a suitable projection of) $\rho(a)$ hitting either the union of (the projection of) $\mathcal{P}(o)$ on o or a union on one of the splitting planes in $S^+(\nu)$.*

Proof. Note that the node ν referred to in the lemma is unique and must exist, since we consider the xy-plane to be a dummy object below the whole scene.

Let $\tilde{q}$ be the first point where $\rho(a)$ intersects a curtain in $curt(E)$, let e be the silhouette curve defining the curtain, and let $q \in e$ be the point directly above $\tilde{q}$. If $q \in$ region(ν) then the object containing the silhouette curve e is a member of $\mathcal{P}(o)$ and we are done. Otherwise there is a splitting plane $h_{\nu'}$ stored at some ancestor ν' of ν with q above $h_{\nu'}$ and $\tilde{q}$ below $h_{\nu'}$. Then the relevant portion of e must be part of the union stored at the first such node ν' (as seen from the root of $\mathcal{T}$). See Figure 2 (ii).

Conversely, since all the unions considered are generated by (parts of) objects above o, we know that $\rho(a)$ cannot hit such a union before it hits a curtain. □

3.3 Details of the Algorithm

We now describe a space-sweep algorithm for computing the visibility map of a set $\mathcal{P} = \{o_1, \ldots, o_n\}$ of convex, constant-complexity, β-fat objects. We move a sweep plane h parallel to the yz-plane from left to right through space. The space sweep induces a plane sweep for each of the unions stored in $\mathcal{T}$. Thus, instead of thinking about the algorithm as a 3D sweep, one may also think about it as a number of coordinated 2D sweeps. That is, while we sweep $\mathbb{R}^3$ with h, we also sweep each (non-vertical) splitting plane h_ν with the line $h \cap h_\nu$. This 2D sweep is performed to detect intersections of the union on h_ν with certain rays (projected onto h_ν). The same holds for the unions stored for each object: while we sweep $\mathbb{R}^3$ with h, we sweep the top surface $\partial_{\mathrm{top}}(o)$ of each object o with the curve $h \cap \partial_{\mathrm{top}}(o)$. Finally, the sweep of h induces a sweep on the viewing plane. As in the algorithm from [3], the visibility map will be computed as we go, so that at the end of the sweep the entire visibility map has been computed.

The space-sweep algorithm is supported by the following data structures:

- There is a global event queue Q, where the priority of an event is its x-coordinate. Initially, all vertices of the objects (that is, all endpoints of silhouette curves) are placed into Q. In addition, all vertices of any of the unions stored in $\mathcal{T}$ are placed into Q. During the sweep, new event points will be inserted into Q, for example endpoints of arcs of the visibility map. It is also possible that events will be removed before they are handled.
- For every splitting plane h_ν (and the top surface of every object o) we maintain a balanced binary tree, which we will call the *intersection-detection data structure*. This tree will store the edges of the union on the splitting plane (resp. $\partial_{\mathrm{top}}(o)$) that intersect the sweep line $h \cap h_\nu$ (resp. $h \cap \partial_{\mathrm{top}}(o)$) as well as the rays traced on it that intersect the sweep line; the edges and rays are stored in order of their intersection with the sweep line. Thus we are essentially running the standard line-segment intersection algorithm of Bentley and Ottmann [2] on the union edges and rays.

Next we discuss the events that can take place, and how they are handled.

(i) *The sweep reaches the left endpoint of an arc a.*
Let $e(a)$ be the silhouette curve defining a, and let $p \in e(a)$ be the point whose projection is the left endpoint of a. Let o be the first object that a vertical ray downward from p hits, and ν be the node where o is large in region(ν) and $\tilde{p} \in$ region(ν). Determine $S^+(\nu)$, and insert the portion of $e(a)$ starting at p into each of the intersection-detection data structures associated with the splitting planes in $S^+(\nu)$. (More precisely, the projection of the silhouette curve on the plane is added.) Also add the projection of the silhouette curve onto $\partial_{\mathrm{top}}(o)$ to the intersection-detection structure for o. Determine any new events using these data structures in the standard way (that is, by checking new pairs of adjacent elements); add any new events to Q. Finally, add the following three events to Q: the right endpoint of $e(a)$, the (first) intersection of $\rho(a)$ with the boundary of region(ν), and the (first) intersection of $\rho(a)$ with the silhouette of o.

(ii) *The sweep reaches the right endpoint of an arc a.*
Determine ν and o as above. Remove a from all intersection-detection data structures in $S^+(\nu)$ and the intersection-detection data structure

associated with o. Remove all events associated with a from Q. Check for new events in each of the intersection-detection data structures; add any new events to Q. Output a as an arc of $\mathcal{M}$. (Note that the right endpoint of an arc may be the left endpoint of one or two other arcs; in this case the left endpoints will be separate events, which are handled according to case (i).)

(iii) *The sweep reaches the left vertex v of a silhouette curve.*
(In other words, we reach the leftmost point of an object $o \in \mathcal{P}$.) Determine if v is visible by shooting a ray vertically up from it. If v is visible, two arcs start at the projection of v onto the viewing plane. Run the actions from case (i) for each of these arcs.

(iv) *The sweep reaches the right vertex v of a silhouette curve it is currently tracing.*
Run the actions from case (ii) for the arc ending at the projection of v.

(v) *The sweep reaches the intersection point of the union boundary on some splitting plane (or top surface of an object) and an arc a traced on the plane (or top surface).*
This case corresponds to a hitting a curtain in curt(E). Now the arc a ends. Run the actions from case (ii) for a. One or two new arcs may start at this point, at most one along the silhouette curve $e(a)$, and one along the silhouette curve corresponding to the curtain that is hit. Run the action from case (i) for the new arc(s).

(vi) *The sweep reaches a point p where the projection of a currently visible silhouette curve onto the object o below hits the boundary of a cell ν where o is large.*
Remove a from all the intersection-detection data structures in $S^+(\nu)$ and all events associated with a from Q. Run the action for case (i) for the continuation of the arc a defined by the silhouette curve. (The only thing that happens here is that the set $S^+(\cdot)$ changes, because the ray that we are tracing moves out of a cell where the object o on which the ray is traced is large.)

(vii) *The sweep reaches the point where the object o immediately below a currently visible silhouette curve changes.*
Now p is the right endpoint of an arc a. Run the actions from case (ii) for a. Two new arcs start at p, one that is the continuation of a, and one that is along a silhouette curve of o (which became visible). Run the actions for case (i) on both curves.

(viii) *The sweep reaches a point on a splitting plane (or top surface of an object), where a union edge starts or ends.*
In this case we only have to update the relevant intersection-detection data structure, check for new events in the intersection-detection data structures, and add any new events to Q.

Lemma 7. *The number of events of type (i)–(vii) is $O(n + k \log n)$, where k is the complexity of $\mathcal{M}$, and the total number of events of type (viii) is $O(n \log^3 n (\log \log n)^2)$ for fat polyhedra and $O(n \log^{7+\varepsilon} n)$ for fat curved objects.*

Proof. Clearly, the number of events of types (i), (ii), (iv), (v), and (vii) is $O(k)$, since they can be charged to a vertex of $\mathcal{M}$. The number of events of type (iii) is $O(n)$. It remains to bound the number of events of type (vi). Consider the portion of a silhouette curve $e(a)$ defining some arc a. This portion has a unique object o immediately below it. Since o is large at $O(\log n)$ cells by Lemma 4 and the projection of $e(a)$ onto o can leave any cell only a constant number of times, we can conclude that there are only $O(\log n)$ type (vi) events for any arc a, this giving $O(k \log n)$ such events in total.

The bound on the number of events of type (viii) follows immediately from Lemma 5. □

Lemma 8. *The time taken for each event of type (i)–(vii) is $O(\log^2 n)$, and the time taken for each event of type (viii) is $O(\log n)$.*

Proof. In all event types, we may need to perform several actions: vertical ray shooting, updating intersection-detection data structures, determining a set $S^+(\nu)$, and updating Q.

By Lemma 5, the time taken for the vertical ray shooting is $O(\log^2 n)$. Each event needs to do only a constant number of ray shooting queries, so this is $O(\log^2 n)$ in total. The intersection-detection data structures are balanced binary trees, so updates take $O(\log n)$ time. At each event we have to update $O(\log n)$ intersection-detection data structures, so the total time taken for updating is $O(\log^2 n)$. Determining new events in the intersection-detection data structures takes $O(1)$ per data structure, so the total amount of time taken for events of type (iii) is $O(\log^2 n)$. Determining a set $S^+(\nu)$ can be done in $O(\log n)$ time by searching in $\mathcal{T}$. At each event we may have to remove $O(\log n)$ event points from Q, each removal taking $O(\log n)$ time. Hence, all events of type (i)–(vii) can be handled in $O(\log^2 n)$ time, as claimed.

The events of type (viii) require $O(\log n)$ time, since they involve a constant number of operations on a single intersection-detection data structure. □

The correctness of the algorithm follows from Lemmas 2 and 6 as well as the correctness of the algorithm in [3]. We conclude with the following theorem.

Theorem 1. *The visibility map of a set of n disjoint constant-complexity convex β-fat polyhedra in $\mathbb{R}^3$ can be computed in time $O((n \log n (\log \log n)^2 + k) \log^3 n)$, where k is the complexity of the visibility map. When the objects are curved (and disjoint, constant-complexity, convex, and β-fat) the visibility map can be computed in time $O((n \log^{5+\varepsilon} n + k) \log^3 n)$.*

References

1. Agarwal, P.K., Matoušek, J.: Ray shooting and parametric search. SIAM Journal on Computing 22(4), 794–806 (1993)
2. Bentley, J., Ottmann, T.: Algorithms for reporting and counting geometric intersections. IEEE Transactions on Computers 28, 643–647 (1979)
3. de Berg, M.: Ray Shooting, Depth Orders and Hidden Surface Removal. LNCS, vol. 703. Springer, Heidelberg (1993)

4. de Berg, M.: Vertical ray shooting for fat objects. In: Proc. 21st Annual Symposium on Computational Geometry, pp. 288–295 (2005)
5. de Berg, M.: Improved bounds for the union complexity of fat objects. In: Ramanujam, R., Sen, S. (eds.) FSTTCS 2005. LNCS, vol. 3821, pp. 116–127. Springer, Heidelberg (2005)
6. de Berg, M., Gray, C.: Vertical ray shooting and computing depth orders for fat objects. In: Proc. 17th Annual Symposium on Discrete Algorithms, pp. 494–503 (2006)
7. de Berg, M., Overmars, M.H.: Hidden-surface removal for c-oriented polyhedra. Comput. Geom. Theory Appl. 1, 247–268 (1992)
8. de Berg, M., Streppel, M.: Approximate range searching using binary space partitions. In: Proc. 24th Conference on Foundations of Software Technology and Theoretical Computer Science, pp. 110–121 (2004)
9. de Berg, M., van der Stappen, A.F., Vleugels, J., Katz, M.J.: Realistic input models for geometric algorithms. Algorithmica 34(1), 81–97 (2002)
10. Bern, M.: Hidden surface removal for rectangles. J. Comp. Syst. Sciences 40, 49–69 (1990)
11. Duncan, C.: Balanced Aspect Ratio Trees. PhD thesis, Johns Hopkins University (1999)
12. Duncan, C., Goodrich, M., Kobourov, S.: Balanced aspect ratio trees: Combining the advantages of k-d trees and octtrees. In: Proc. 10th Annual ACM-SIAM Sympos. on Discrete Algorithms, pp. 300–309 (1999)
13. Goodrich, M.T., Atallah, M.J., Overmars, M.H.: An input-size/output-size tradeoff in the time-complexity of rectilinear hidden surface removal. In: Paterson, M.S. (ed.) Automata, Languages and Programming. LNCS, vol. 443, pp. 689–702. Springer, Heidelberg (1990)
14. Güting, R.H., Ottmann, T.: New algorithms for special cases of the hidden line elimination problem. Comp. Vision, Graphics and Image Processing 40, 188–204 (1987)
15. Katz, M.J., Overmars, M., Sharir, M.: Efficient hidden surface removal for objects with small union size. Computational Geometry: Theory and Applications 2, 223–234 (1992)
16. Kirkpatrick, D.: Optimal search in planar subdivisions. SIAM J. Comput. 12, 28–35 (1983)
17. McKenna, M.: Worst-Case Optimal Hidden Surface Removal. ACM Trans. Graphics 6, 19–28 (1987)
18. Pach, J., Tardos, G.: On the boundary complexity of the union of fat triangles. SIAM J. Comput. 31, 1745–1760 (2002)
19. Preparata, F.P., Vitter, J.S., Yvinec, M.: Computation of the axial view of a set of isothetic parallellipipes. ACM Trans. Graphics 9, 278–300 (1990)
20. Reif, J., Sen, S.: An efficient out-sensitive hidden surface removal algorithm and its parallelization. In: Proc. 4th Annual Symposium on Computational Geometry, pp. 193–200 (1988)
21. Sharir, M., Overmars, M.H.: A simple method for output-sensitive hidden surface removal. ACM Trans. Graphics 11, 1–11 (1992)

Independent Sets in Bounded-Degree Hypergraphs

Magnús M. Halldórsson[1,*] and Elena Losievskaja[2,*]

[1] Dept. of Computer Science, Reykjavík University, Iceland
magnusmh@gmail.com

[2] Dept. of Computer Science, University of Iceland, Iceland
elenal@hi.is

Abstract. In this paper we analyze several approaches to the Maximum Independent Set problem in hypergraphs with degree bounded by Δ. We propose a general technique that reduces the worst case analysis of certain algorithms to their performance in the case of ordinary graphs. This technique allows us to show that the greedy algorithm that corresponds to the classical greedy set cover algorithm has a performance ratio of $(\Delta + 1)/2$. It also allows us to apply results on local search algorithms of graphs to obtain a $(\Delta + 1)/2$ approximation for a weighted case and $(\Delta + 3)/5 - \epsilon$ approximation for an unweighted case. We improve the bound in the weighted case to $\lceil(\Delta + 1)/3\rceil$ using a simple partitioning algorithm. Finally, we show that another natural greedy algorihthm, that adds vertices of minimum degree, achieves only a ratio of $\Delta - 1$, significantly worse than on ordinary graphs.

1 Introduction

In this paper we consider the independent set problem in hypergraphs. A *hypergraph* H is a pair (V, E), where $V = \{v_1, \ldots, v_n\}$ is a set of vertices and $E = \{e_1, \ldots, e_m\}$ is a collection of subsets of V, or (hyper)edges. An *independent set* in H is a subset of V that doesn't properly contain any edge of H. Let MIS denote the problem of finding a maximum independent set in hypergraphs.

MIS is of fundamental interest, both in practical and theoretical aspects. It arises in various applications in data mining, image processing, database design, parallel computing and many others. MIS is intimately related with classical covering problems. The vertices not contained in a weak independent set form a vertex cover, or a *hitting set*. Moreover, a set cover in the dual of a hypergraph (replacing each set by a vertex and including a set for the incidences of each original node) is equivalent to a hitting set in the original hypergraph. Thus, in terms of optimization, MIS is equivalent to the Hitting Set and the Set Cover problems.

Numerous results are known about independent sets in hypergraphs, including approximation algorithms for MIS in [15] and [18]. The focus of the current work

* Research funded by grants of the Icelandic Research Fund and the Research Fund of the University of Iceland, work done at University of Iceland.

F. Dehne, J.-R. Sack, and N. Zeh (Eds.): WADS 2007, LNCS 4619, pp. 263–274, 2007.

is on bounded-degree hypergraphs, where each vertex is of degree at most Δ. Given that MIS generalizes the independent set problem in graphs, the problem is NP-hard to approximate within a factor $\Delta/2^{O(\sqrt{\log\Delta})}$ unless $P = NP$ [23].

Various approximation algorithms have been given for MIS in graphs. Halldórsson and Radhakrishnan [13] showed that the minimum-degree greedy algorithm approximates the size of an unweighted MIS within a factor of $\frac{\Delta+2}{3}$. A simple partitioning algorithm due to Halldórsson and Lau [12] gives a $(\Delta+2)/3$ approximations of weighted MIS. A better approximation ratio for an unweighted MIS is $(\Delta+3)/5$ obtained by Berman and Fujito [3] using a local search algorithm. Another local search algorithm by Berman [2] approximates weighted MIS in $(d+1)$-claw free graphs within a factor of $(d+1)/2$, which implies also a $\Delta/2$-approximation. For large values of Δ, the best approximation is obtained by using semi-definite programming, with a ratio of $O(\Delta \log\log\Delta/\log\Delta)$ due to Vishwanathan [24] (and also in the weighted case, due to Halldórsson [11] and Halperin [14]).

One of the best studied heuristics of all times is the greedy set cover algorithm, which repeatedly adds to the cover the set with the largest number of uncovered elements. In spite of its simplicity, it is in various ways also one of the most effective ones. Johnson [17] and Lovász [20] showed that it approximates the Set Cover problem with $H_n \leq \ln n + 1$ factor, which was shown by Feige [9] to be the best possible up to a lower order term. Generalizations to weights [7] and submodular functions [25] also yield equivalent ratios. And under numerous variations on the objective function does it still achieve the best known/possible performance ratio, e.g. Sum Set Cover [10], Entropy Set Cover [5], and Test Set.

Bazgan, Monnot, Paschos and Serrière [1] studied the *differential approximation ratio* of the greedy set cover algorithm, this ratio measures how many sets are *not* included in the cover. When viewed on the dual hypergraph, this is equivalent to studying the performance ratio the greedy set cover algorithm for MIS. They proved that when modified with a post-processing phase, it has a performance ratio of at most $\Delta/1.365$ and at least $(\Delta+1)/4$. Caro and Tuza [6] showed that the greedy set cover algorithm applied to MIS in r-uniform hypergraphs always finds a weak independent set of size at least $\Theta\left(n/\Delta^{\frac{1}{r-1}}\right)$. Thiele [22] extended their result to non-uniform hypergraphs and gave a lower bound on the size of an independent set found by the greedy set cover algorithm as a complex function of the number of edges of different sizes incident on each vertex in a hypergraph.

Another popular algorithm design technique is local search. This technique is based on the concept of a *neighborhood* - a set of solutions close to a given solution S. The idea is to start with some (arbitrary) solution S and iteratively replace S by a better solution found in the neighborhood of S. Local search gives the best approximations of weighted and unweighted MIS in bounded-degree graphs for small values of Δ, due to Berman [2], Fujito [3] and Halldórsson [11]. Bazgan, Monnot, Paschos and Serrière [1] considered a simple 2-OPT local search algorithm to approximate MIS in hypergraphs and proved a tight bound of $(\Delta+1)/2$.

A yet simpler approach in approximation algorithm design is partitioning. The strategy is to break the problem into a set of easier subproblems, solve each subproblem and output the largest of the found solutions. Halldórsson [11] applied this approach to obtain $\lceil(\Delta+1)/3\rceil$ approximations to the weighted MIS in graphs.

In this paper we describe a general technique that reduces the worst case analysis of certain algorithms to their performance in the case of ordinary graphs. Given an approximation algorithm A, this technique, called *shrinkage reduction*, truncates a hypergraph H to a graph G such that an optimal solution on H is also an optimal solution in G, and A produces the same worst approximate solution on H and G, proving that the performance ratio of A in hypergraphs is no worse than in graphs.

We also present three different approaches to approximate weighted and unweighted MIS in bounded-degree hypergraphs. First, we apply shrinkage reduction to extend a factor $(\Delta+3)/5$ algorithm of Berman and Fujito [3] for unweighted MIS and a factor $\Delta/2$ algorithm of Berman [2] for weighted MIS. We improve the bound in the weighted case to $\lceil(\Delta+1)/3\rceil$ using a simple partitioning algorithm. Finally, using our technique we give a tight analysis of the classical greedy set cover algorithm applied to MIS. It starts with the set of all vertices, and greedily removes vertices of maximum degree until feasibility is achieved. The performance ratio of this algorithm is exactly $\frac{\Delta+1}{2}$, improving the bounds obtained by Bazgan et al. [1]. In addition, while their analysis required a post-processing maximalization phase, our bound applies to the greedy algorithm alone. A second natural greedy algorithm acts in an opposite way of greedily adding vertices of minimum degree to an initial empty solution. This algorithm has a performance ratio at most $\Delta-1$, which is nearly tight.

The paper is organized as follows. After giving essential definitions of various hypergraph properties, we describe shrinkage reduction technique in Section 3, local search and partitioning approaches in Section 4 and 5, and conclude with the analysis of two greedy algorithms in Section 6.

2 Definitions

Given a hypergraph $H=(V,E)$, let n and m be the number of vertices and edges in H. The *degree* of a vertex v is the number of edges incident on v. We denote by $\Delta(\overline{d})$ the maximum (average) degree in the hypergraph. In a *bounded-degree* hypergraph Δ is a constant. A hypergraph is *Δ-regular* if all vertices have the same degree Δ.

The *rank* r of a hypergraph H is the maximum edge size in H. A hypergraph is *r-uniform* if all edges have the same cardinality r.

A vertex u is a *neighbor* of a vertex v, if there exist an edge $e \in E$ that includes both u and v. A *hyperclique* is a hypergraph in which each vertex is a neighbor of all other vertices. Note that a hyperclique need not be a uniform hypergraph. By analogy with the graph being a 2-uniform hypergraph, a clique is a 2-uniform hyperclique.

An *n-star* is a tree on $n+1$ nodes with one node having the degree n (the *root* of the star) and the other having degree 1 (the *endpoints* of the star).

In the remainder we let $\mathcal{H}$ ($\mathcal{G}$) be the collection of all hypergraphs (graphs) and H (G) be a hypergraph (graph) in $\mathcal{H}$ ($\mathcal{G}$). By *cover* we mean a hitting set (a vertex cover) in H (G).

3 Shrinkage Reduction

Shrinkage reduction is a general technique that allows to apply the worst case analysis of algorithms on graphs to the hypergraphs case. It is based on a *shrinkage* hypergraph, or *shrinkage* for short.

Definition 1. *A hypergraph H' is a* shrinkage *of H if $V(H') = V(H)$ and for any edge $e \in E(H)$ there exist an edge $e' \in E(H')$ such that $e' \subseteq e$, in other words, the edges of H might be truncated in H' into sets of smaller size (and at least 2).*

Shrinkage reduction works for *hereditary* optimization problems. Given an instance I, an optimization problem consists of a set of feasible solutions $\mathcal{S}_I$ and a function $w : \mathcal{S}_I \rightarrow R^+$ assigning a non-negative cost to each solution $S \in \mathcal{S}_I$.

Definition 2. *An optimization problem on hypergraphs is* hereditary, *if for a shrinkage H' of a hypergraph H it satisfies $\mathcal{S}_{H'} \subseteq \mathcal{S}_H$.*

Many problems on hypergraphs are hereditary, including the Minimum Hitting Set, the Maximum Independent Set, the Minimum Coloring, the Shortest HyperPath, etc. An example of non hereditary problem is the Longest HyperPath. Given a hereditary problem, the essence of shrinkage reduction is the following.

Proposition 1. *Let A be an approximation algorithm for a hereditary problem. Suppose we can construct a shrinkage graph G of a hypergraph H such that an optimal solution in H is also an optimal solution in G and A produces the same worst approximate solution on H and G, then the performance ratio of A in hypergraphs is no worse than in graphs.*

Note that Proposition 1 applies also to non-deterministic (or randomized) approximation algorithms.

It is not easy to give a general rule on how to construct a shrinkage for an arbitrary approximation algorithm. In the following sections we describe reductions for greedy set cover and local search algorithms for weighted and unweighted MIS in bounded-degree hypergraphs.

4 Local Search

The idea of the local search approach is to start with an initial solution and continually replace it by a better solution found in its neighborhood while possible. We need formal definitions to determine what a "better solution" and a "neighborhood" mean.

A *neighborhood function* Γ maps a solution $S \in \mathcal{S}_I$ into a set of solutions $\Gamma_I(S) \subseteq \mathcal{S}_I$, called the *neighborhood* of S. A feasible solution $\tilde{S}$ is *locally optimal w.r.t.* Γ, or Γ*-optimal* for short, if for a minimization (maximization) problem such solution satisfies $w(\tilde{S}) \leq w(S)$ ($w(\tilde{S}) \geq w(S)$) for all $S \in \Gamma_I(S)$. A feasible solution S^* is *globally optimal*, or *optimal* for short, if for a minimization (maximization) problem such solution satisfies $w(S^*) \leq w(S)$ ($w(S^*) \geq w(S)$) for all $S \in \mathcal{S}$. To specify more precisely the neighborhood functions used in our local search algorithms, we need the following definition.

Definition 3. *A neighborhood function Γ is said to be* edge-monotone *for a hereditary problem on hypergraphs if for any shrinkage H' of a given hypergraph H and any solution $S \in \mathcal{S}_{H'}$ the neighborhood of S satisfies $\Gamma_{H'}(S) \subseteq \Gamma_H(S)$.*

In other words, the edge-monotonicity means that edge reduction can only decrease the neighborhood size.

A Γ*-optimal algorithm* is a local search algorithm that given an instance I, starts with a (arbitrary) solution S and repeatedly replaces it by a better solution found in $\Gamma_I(S)$ until S is Γ-optimal. The *approximation ratio* $\varrho_{\Gamma,I}$ of a Γ-optimal algorithm on a instance I is the maximum ratio between the sizes of Γ-optimal and optimal solutions over all Γ-optimal solutions on I, i.e. $\varrho_{\Gamma,I} = \max_{\forall \tilde{S} \in \mathcal{S}_I} \frac{|\tilde{S}|}{|S^*|} \left(\varrho_{\Gamma,I} = \max_{\forall \tilde{S} \in \mathcal{S}_I} \frac{|S^*|}{|\tilde{S}|} \right)$ for a minimization (maximization) problem. The *performance ratio* $\rho_{\Gamma,\mathcal{I}}$ of a Γ-optimal algorithm is the worst approximation ratio over all instances I in the class of instances $\mathcal{I}$.

Theorem 1. *Given an edge-monotone neighborhood function Γ and a hypergraph H with an optimal cover S^* and a Γ-optimal cover $\tilde{S}$, there exists a shrinkage graph G of H on which S^* and $\tilde{S}$ are also optimal and Γ-optimal covers, respectively.*

Proof. Given H, S^* and $\tilde{S}$, we construct a shrinkage G as follows. From each edge e in $E(H)$, we arbitrarily pick vertices u and v such that $\{u, v\} \cap \tilde{S} \neq \emptyset$ and $\{u, v\} \cap S^* \neq \emptyset$, and add (u, v) to $E(G)$.

Any edge in $E(G)$ contains at least one vertex from $\tilde{S}$ and at least one vertex from S^*, and so $\tilde{S}$ and S^* are covers in G, i.e. $\tilde{S}, S^* \in \mathcal{S}_G$. Given that Γ is edge-monotone, $\mathcal{S}_G \subseteq \mathcal{S}_H$ by definition. Since $w(S^*) \leq w(S)$ for all $S \in \mathcal{S}_H$, we have $w(S^*) \leq w(S)$ for all $S \in \mathcal{S}_G$, and so S^* is an optimal cover in G. The same argument also applies to the local optimality of $\tilde{S}$ in G.

Corollary 1. *If a neighborhood function Γ is edge-monotone for MIS, then $\rho_{\Gamma,\mathcal{H}} \leq \rho_{\Gamma,\mathcal{G}}$.*

Proof. Given a hypergraph $H(V, E)$, the vertices not contained in a weak independent set I form a vertex cover S in H, i.e. $I = V \backslash S$. Given an edge-monotone neighborhood function Γ for MIS, we define a new neighborhood function $\Gamma'(S) = \{S' : V \backslash S' \in \Gamma(V \backslash S)\}$. Note that $\Gamma'(S)$ is edge-monotone for the Hitting Set problem. Moreover, if I^* and $\tilde{I}$ are optimal and Γ-optimal weak independent sets in H, then $S^* = V \backslash \tilde{I^*}$ and $\tilde{S} = V \backslash \tilde{I}$ are optimal and Γ-optimal covers in H. The claim then follows from Theorem 1.

The simplest local search algorithm for MIS is t-Opt, which repeatedly tries to extend the current solution by deleting t elements while adding $t+1$ elements. It is easy to verify that the corresponding neighborhood function $\Gamma(S) = \{S' \in \mathcal{S}_H : |S \oplus S'| \leq t\}$ defined on $\mathcal{S}_H$ in is edge-monotone. Then, the following theorems are straightforward from Corollary 1 and the results of Hurkens and Schrijver on graphs [16].

Theorem 2. *t-Opt approximates an unweighted MIS within $\Delta/2 + \epsilon$, where $\lim_{t\to\infty} \epsilon(t) = 0$.*

Theorem 3. *2-Opt approximates an unweighted MIS within $(\Delta + 1)/2$.*

Theorem 4. *For every $\epsilon > 0$ an unweighted MIS can be approximated within $(\Delta + 3)/5 + \epsilon$ for even Δ and within $(\Delta + 3.25)/5 + \epsilon$ for odd Δ.*

Proof. We extend the algorithm $SIC_{\Delta,k}$ of Berman and Fürer [4] for MIS in bounded degree graphs to the hypergraph case. Given a hypergraph $H(V, E)$ and a weak independent set A in H, let B_A equal $V - A$ if the maximum degree of H is three, and otherwise equal the set of vertices that have at least two incident edges with vertices in A. Let $Comp(A)$ be the subhypergraph induced by B_A.

```
ALGORITHM HSIC(H, Δ, k)
  If Δ ≤ 2 then compute MIS exactly and stop
  Let A be any maximal weak independent set
  Repeat
    Do all possible local improvements of size O(k log n)
    If Δ = 3 then l = 1 else l = 2
    Recursively apply HSIC(Comp(A), Δ − l, k)
    and select the resulting weak independent set if it is bigger
  Until no A has no improvements
```

There are two neighborhood functions in $HSIC$. The first function, which maps a solution A to a set of all possible local improvements of size $O(k \log n)$, is t-optimal with $t = O(k \log n)$ and so edge-monotone. The second function, which maps a solution A to a set of weak independent sets in $Comp_H(A)$, is also edge-monotone. This is because shrinking H to H' reduces the degree of some vertices, implying $B_A(H') \subseteq B_A(H)$. Consequently, a weak independent set in $Comp_{H'}(A)$ is also a weak independent set in $Comp_H(A)$. Thus, both neighborhood functions are edge-monotone and the performance ratio of $HSIC$ is no worse than the performance ratio of $SIC_{\Delta,k}$ by Corollary 1.

Theorem 5. *Weighted MIS can be approximated within $(\Delta + 1)/2$ in hypergraphs of a constant rank r.*

Proof. We extend the algorithm $SquareIMP$ of Berman [2] for a weighted MIS in bounded degree graphs to the hypergraph case. Given a hypergraph $H(V, E)$ and a vertex $v \in V$, we denote by $N(v)$ a set of neighbors of v. Similarly, given a set $U \subseteq V$, we denote by $N(U)$ a set of neighbors of vertices in U. Let S be a weak independent set in H. We define (A, B) to be an improvement of S, if there is a vertex $v \in S$ such that $A \subseteq N(v) \cap (V \backslash S)$, $B \subseteq N(A) \cap S$, $(S \backslash B) \cup A$ is a weak independent set and $w^2((S \backslash B) \cup A) > w^2(S)$.

ALGORITHM $HSquareIMP(H)$
  $S \leftarrow \emptyset$
  While there exist an improvement (A, B) of S
    $S \leftarrow (S \backslash B) \cup A$
  Output S

The neighborhood function in $HSquareIMP$ is edge-monotone. Shrinking H to H' reduces the degree of some vertices and so every improvement A, B of S in H' is also an improvement of S in H. Hence, the performance ratio of $HSquareIMP$ is no worse than the performance ratio of $SquareIMP$ by Corollary 1.

Note that finding an improvement (A, B) takes $O(n2^{\Delta^2(r-2)(r-1)})$ steps. Because in the worst case we check every vertex $v \in S$ for a possible improvement, and given v we consider every possible subset $A \subseteq N(v) \cap (V \backslash S)$ and every possible subset $B \subseteq N(A) \cap S$ to check whether $(S \backslash B) \cup A$ is a weak independent set and $w^2((S \backslash B) \cup A) > w^2(S)$. Since $|N(v) \cap (V \backslash S)| \leq \Delta(r-2)$, there are at most $2^{\Delta(r-2)}$ possible A sets. Similarly, since $|N(A) \cap S| \leq \Delta(r-2)(\Delta(r-1)-1)$, there are at most $2^{\Delta(r-2)(\Delta(r-1)-1)}$ possible B sets. In total we consider at most $2^{\Delta^2(r-2)(r-1)}$ possible pairs (A, B) for every vertex in $v \in S$ until an improvement is found.

5 Partitioning

The idea of the partitioning approach is to split a given hypergraph into k induced subhypergraphs so that MIS can be solved optimally on each subhypergraph in polynomial time. This is based on the strategy of [12] for ordinary graphs. Note that the largest of the solutions on the subhypergraphs is a k-approximation of MIS, since the size of any optimal solution is at most the sum of the sizes of the largest independent sets on each subhypergraph.

We extend a partitioning lemma of Lovász [19] to the hypergraph case.

Lemma 1. *The vertices of a given hypergraph can be partitioned into $\lceil (\Delta+1)/3 \rceil$ sets, where each set induces a subhypergraph of maximum degree at most two.*

Proof. Start with an arbitrary vertex partitioning into $\lceil (\Delta+1)/3 \rceil$ sets. While a set contains a vertex v with degree more than two, move v to another set that properly contains at most two edges incident on v. Such a set exists, because otherwise the total number of edges incident on v would be at least $3\lceil (\Delta+1)/3 \rceil \geq \Delta+1$. Any such move increases the number of edges between different sets, and so the process terminates with a partition where every vertex has at most two incident edges in its set.

The method can be implemented in time $O(\sum_{e\in E}|e|)$ by using an initial greedy assignment as shown in [12].

Lemma 2. *A weighted MIS in hypergraphs of maximum degree two can be solved optimally in polynomial time.*

Proof. Given a hypergraph $H(V,E)$ we consider the dual hypergraph $H'(E,V)$, whose vertices $e_1,\ldots,e_m$ correspond to the edges of H and the edges $v_1,\ldots,v_n$ correspond to the vertices of H, i.e. $v_i = \{e_j : v_i \in e_j \text{ in } H\}$. The maximum edge size in H' equals to the maximum degree of H, thus H' is a graph, possibly with loops. A vertex cover in H is an edge cover in H' (where an edge cover in H' is defined as a subset of edges that touches every vertex in H'), and a minimum weighted edge cover in graphs can be found in polynomial time via maximum weighted matching [8]. All edges not in a minimum cover in H' correspond to a maximum independent set of vertices in H.

The following result now is straightforward from Lemmas 1 and 2.

Theorem 6. *A weighted MIS can be approximated within* $\lceil(\Delta+1)/3\rceil$ *in polynomial time.*

6 Greedy

The idea of the greedy approach is to construct a solution by repeatedly selecting the best candidate on each iteration. There are two variations, called GreedyMAX and GreedyMIN, depending on whether we greedily add or greedily reject vertices.

6.1 The GreedyMAX Algorithm

The GreedyMAX algorithm constructs a cover S by adding a vertex of maximum degree, deleting it and its incident edges from the hypergraph, and iterating until the edge set is empty. It then outputs the remaining vertices as an independent set I.

Theorem 7. *The performance ratio of GreedyMAX is at most* $\frac{\Delta+1}{2}$.

Proof. Given a hypergraph $H(V,E)$, let S^* be a minimum cover. Then, the performance ratio of GreedyMAX is:

$$\rho = \max_{\forall H} \frac{n-|S^*|}{n-|S|} . \tag{1}$$

The proof has two parts. First we show that the worst case for GreedyMAX occurs on (ordinary) graphs. Namely, we show that we can reduce any hypergraph to a graph (actually, a multigraph) G for which GreedyMAX has a no better approximation ratio. We then show that the bound actually holds for (multi)graphs.

Lemma 3. *Given a hypergraph H with a minimum cover S^*, there exists a shrinkage multigraph G of H on which S^* is still a cover and where GreedyMAX constructs the same cover for G as for H.*

Proof. The proof is by induction on s, the number of iterations of GreedyMAX. For the base case, $s = 0$, the claim clearly holds for the unchanged empty graph.

Suppose now that the claim holds for all hypergraphs for which GreedyMAX selects $s - 1 \geq 0$ vertices. Let u_1 be the first vertex chosen by GreedyMAX, $E(u_1)$ be the set of incident edges, and H_1 be the remaining hypergraph after removing u_1 (with edge set $E(H) \setminus E(u_1)$). Based on $E(u_1)$, we form a set $E'(u_1)$ of ordinary edges as follows. If u_1 is contained in both S and S^*, then for each edge e in $E(u_1)$ pick an arbitrary vertex v from e and add (u_1, v) to $E'(u_1)$. If u_1 is only in S and not in S^*, then for each edge e in $E(u_1)$ we pick an arbitrary vertex u from e that is contained in S^* and add (u_1, u) to $E'(u_1)$; such a vertex u must exist, since e is covered by S^*. This completes the construction of $E'(u_1)$.

By the inductive hypothesis, there is a shrinkage multigraph G_1 of H_1 with a greedy cover of $S \setminus \{u_1\}$ that is still covered by S^*. We now form the multigraph G on the same vertex set as H with the edge set $E'(u_1) \cup E(G_1)$, and claim that it satisfies the statement of the lemma. Since G_1 is covered by S^* and all edges of $E'(u_1)$ are also covered by vertices of S^*, S^* covers all edges of G. The edge shrinkage only decreases degrees of vertices, but does not affect the degree of u_1. Therefore, u_1 remains the first vertex chosen by GreedyMAX and, by induction, the vertices chosen from G_1 are the same as those chosen from H_1. Hence, GreedyMAX outputs the same solution on G as on H, completing the lemma.

From Lemma 3 it follows immediately that GreedyMAX on hypergraphs has no worse performance ratio than on graphs.

Lemma 4. *The performance ratio of GreedyMAX on (multi)graphs is at most $\frac{\Delta+1}{2}$.*

Proof. Each vertex can cover at most Δ of the m edges of a graph. Therefore, any optimal cover S^* is of size at least

$$|S^*| \geq \frac{m}{\Delta} = \frac{\bar{d}n}{2\Delta} . \tag{2}$$

Sakai, Togasaki, and Yamazaki [21] obtained a lower bound on the size of weighted independent set produced by GreedyMAX on graphs. In unweighted case this bound reduces to a Caro-Wei improvement of the Turan bound on graphs:

$$|I| \geq \sum_{v \in V} \frac{1}{d(v)+1} . \tag{3}$$

It can be verified that the proof of [21] applies also to multigraphs, if we understand the neighborhood of a node to become a multiset.

We show that GreedyMAX attains its worst performance ratio in regular or almost regular graphs. For that we refine $\bar{d}$ as follows: let $k \in [0, 1]$ be the value so that kn vertices have degree Δ and the remaining $(1-k)n$ vertices have the average degree $d' \leq \Delta - 1$. Then

$$\bar{d} = k\Delta + (1-k)d' . \tag{4}$$

Using (4) we rewrite (2) and (3) as

$$|S^*| \geq \frac{n\left(k\Delta + (1-k)d'\right)}{2\Delta} \tag{5}$$

and

$$|I| \geq \frac{kn}{\Delta+1} + \frac{(1-k)n}{d'+1} . \tag{6}$$

Combining (1), (5) and (6) we obtain an upper bound on the performance ratio of GreedyMAX:

$$\rho = \max_{\forall H} \frac{n-|S^*|}{n-|S|} = \max_{\forall H} \frac{n-|S^*|}{|I|} \leq \frac{2\Delta - k\Delta - (1-k)d'}{2\Delta\left(\frac{k}{\Delta+1} + \frac{1-k}{d'+1}\right)} =$$

$$= \frac{(\Delta+1)(d'+1)}{2\Delta}\left(1 + \frac{\Delta - d' - 1}{\Delta + 1 - k(\Delta - d')}\right) \tag{7}$$

$$\leq \frac{(\Delta+1)(d'+1)}{2\Delta}\left(1 + \frac{\Delta - d' - 1}{\Delta + 1 - (\Delta - d')}\right) = \frac{\Delta+1}{2}, \tag{8}$$

where the first inequality in (8) is obtained by setting $k = 1$.

Note that multiple edges in a graph don't affect the performance ratio of GreedyMAX. The edge reduction in $E(H)$ might create multiples edges in $E(G)$, but the number of edges in G and H is not changed and the maximum degree in G and H also remains the same. Thus, from Lemmas 3 and 4 the performance ratio of GreedyMAX on hypergraphs is at most $\frac{\Delta+1}{2}$, completing the theorem.

Proposition 2. *There exist hypergraphs where the approximation ratio of GreedyMAX is* $\frac{\Delta+1}{2}$.

Proof. It remains to construct examples showing that the performance of GreedyMAX can actually reach this value, for each Δ. Consider the hypergraph $H_{\Delta+1,\Delta+1}$, formed by a complete bipartite graph missing a single perfect matching. GreedyMAX may remove vertices alternately from each side, until two vertices remains as a maximal independent set. The optimal solution consists of one of the bipartitions, of size $\Delta + 1$. By taking independent copies, this can be extended for arbitrarily large hypergraphs.

6.2 The GreedyMIN Algorithm

The GreedyMIN algorithm starts with an empty independent set I, then iteratively adds a vertex of minimum degree into I and deletes it from the hypergraph.

If the vertex deletion results in loops (edges containing only one vertex), then the algorithm also deletes the vertices with loops along with the edges incident on such vertices. The algorithm terminates when the vertex set is empty and outputs I.

Theorem 8. *The performance ratio of GreedyMIN is at most $\Delta - 1$.*

Proof. We split the sequence of the iterations of the algorithm into epochs, where a new epoch starts when the algorithm selects a vertex of degree Δ. Clearly, if the algorithm always selects a vertex of degree less than Δ, the whole sequence of iterations is just one epoch. We show that in any epoch the ratio between the sizes of the greedy and an optimal solutions is at most $\Delta - 1$.

In each iteration i, the algorithm selects a vertex v_i, whose set of neighbors in the edges of size 2 we denote by $N(v_i)$. The vertices of $N(v_i)$ are deleted in the iteration along with all incident edges. The maximum number of nodes removed in an iteration i that can belong to an optimal solution is at most the degree of v_i.

Suppose one of the deleted edges is incident on a vertex u outside of $N(v_i)$. Then, in iteration $i + 1$, the vertex u will have the degree at most $\Delta - 1$, and therefore, the degree of v_{i+1} is at most $\Delta - 1$. Thus, iteration $i + 1$ will be in the same epoch as i. The last iteration of an epoch occurs when a vertex is chosen whose neighborhood is contained in $N(v_i) \cup \{v_i\}$. This neighborhood then forms a hyperclique, because any vertex in $N(v_i)$ has at least the degree of v_i and all its neighbors are contained in $N(v_i) \cup \{v_i\}$. Notice that we may assume without loss of generality that the hypergraph is *simple*, namely that no edge is a proper subset of any other edge. Therefore, since the degree of v_i is at most Δ, any edge of the hyperclique contains at most $\Delta - 1$ vertices, and any optimal solution can contain at most $\Delta - 2$ vertices.

We see that the optimal solution can contain at most $\Delta - 2$ vertices from the vertices removed in the last iteration of an epoch, $\Delta - 1$ from the intermediate iterations and Δ from the first iteration of an epoch. On average, it can thus contain at most $\Delta - 1$ vertices per iteration, while the greedy algorithm contains exactly one.

Proposition 3. *The performance ratio of GreedyMIN is tight for $\Delta = 3$ and at least $\Delta - 2 + \frac{2}{\Delta+1}$ for any $\Delta \geq 4$.*

Proof. The details of the proof are omitted here due to the limited space.

References

1. Bazgan, C., Monnot, J., Paschos, V., Serrière, F.: On the differential approximation of MIN SET COVER. Theor. Comput. Sci. 332, 497–513 (2005)
2. Berman, P.: A $d/2$ approximation for Maximum Weight Independent Set in d-claw free graphs. In: Halldórsson, M.M. (ed.) SWAT 2000. LNCS, vol. 1851, pp. 214–219. Springer, Heidelberg (2000)
3. Berman, P., Fujito, T.: On Approximation Properties of the Independent Set Problem for Low Degree Graphs. Theory Comput. Syst. 32(2), 115–132 (1999)

4. Berman, P., Fürer, M.: Approximating maximum independent set in bounded degree graphs. SODA, pp. 365–371 (1994)
5. Cardinal, J., Fiorini, S., Joret, G.: Tight Results on Minimum Entropy Set Cover. In: Díaz, J., Jansen, K., Rolim, J.D.P., Zwick, U. (eds.) APPROX 2006 and RANDOM 2006. LNCS, vol. 4110, Springer, Heidelberg (2006)
6. Caro, Y., Tuza, Z.: Improved lower bounds on k-independence. J. of Graph Theory 15, 99–107 (1991)
7. Chvátal, V.: A greedy heuristic for the set-covering problem. Math. of Operat. Research. 3, 233–235 (1979)
8. Edmonds, J.: Paths, trees and flowers. Canadian J. of Math. 17, 449–467 (1965)
9. Feige, U.: A threshold of $\ln n$ for approximating set cover. J. of the ACM 45, 634–652 (1998)
10. Feige, U., Lovász, L., Tetali, P.: Approximating min-sum set cover. Algorithmica 40(4), 219–234 (2004)
11. Halldórsson, M.: Approximations of independent sets in graphs. In: Jansen, K., Rolim, J.D.P. (eds.) APPROX 1998. LNCS, vol. 1444, Springer, Heidelberg (1998)
12. Halldórsson, M.M., Lau, H.-C.: Low-degree graph partitioning via local search with applications to Constraint Satisfaction, Max Cut, and 3-Coloring. J. of Graph Alg. and Appl. 1(3), 1–13 (1997)
13. Halldórsson, M., Radhakrishnan, J.: Greed is good: Approximating independent sets in sparse and bounded-degree graphs. Algorithmica 18, 143–163 (1997)
14. Halperin, E.: Improved approximation algorithms for the vertex cover problem in graphs and hypergraphs. SODA (2000)
15. Hofmeister, Th., Lefmann, H.: Approximating maximum independent sets in uniform hypergraphs. In: Brim, L., Gruska, J., Zlatuška, J. (eds.) MFCS 1998. LNCS, vol. 1450, pp. 562–570. Springer, Heidelberg (1998)
16. Hurkens, C.A.J., Schrijver, A.: On the size of systems of sets every t of which have an SDR, with an application to the worst-case ratio of heuristics for packing problems. SIAM J. Disc.Math. 2(1), 68–72 (1989)
17. Johnson, D.S.: Approximation algorithms for combinatorial problems. J. Comput. Syst. Sci. 9, 256–278 (1974)
18. Krivelevich, M., Nathaniel, R., Sudakov, B.: Approximating coloring and maximum independent set in 3-uniform hypergraphs. J. of Algorithms 41, 99–113 (2001)
19. Lovász, L.: On decomposition of graphs. Stud. Sci. Math. Hung. 1, 237–238 (1966)
20. Lovász, L.: On the ratio of optimal integral and fractional covers. Disc. Math. 13, 383–390 (1975)
21. Sakai, S., Togasaki, M., Yamazaki, K.: A note on greedy algorithms for the maximum weighted independent set problem. Disc. Appl. Math. 126(2), 313–322 (2003)
22. Thiele, T.: A lower bound on the independence number of arbitrary hypergraphs. J. of Graph Theory 32, 241–249 (1999)
23. Trevisan, L.: Non-approximability results for optimization problems on bounded degree instances. In: Proc. of the 33rd ACM STOC (2001)
24. Vishwanathan, S.: Private communication (1998)
25. Wolsey, L.A.: An analysis of the greedy algorithm for the submodular set covering problem. Combinatorica 2, 385–393 (1982)

Steiner Tree in Planar Graphs: An $O(n \log n)$ Approximation Scheme with Singly-Exponential Dependence on Epsilon

Glencora Borradaile*, Philip N. Klein**, and Claire Mathieu

Brown University, Providence RI 02912, USA
{glencora,philip,claire}@cs.brown.edu

Abstract. We give an algorithm that, for any $\epsilon > 0$, any undirected planar graph G, and any set S of nodes of G, computes a $(1+\epsilon)$-optimal Steiner tree in G that spans the nodes of S. The algorithm takes time $O(2^{\text{poly}(1/\epsilon)} n \log n)$.

1 Introduction

The Steiner problem in graphs is a fundamental and well-studied optimization problem: given a graph with edge lengths and a set of terminals, find a minimum-length connected subgraph that includes all the terminals. The problem is NP-hard [11] (even in planar graphs [8]) and is max SNP-complete in general graphs [5]. Much work [22,14,24,23,15,26,2,25,17,12,10,21] has gone into obtaining constant-factor approximation algorithms. There has also been work [1,16,20] on approximation schemes for the case of Euclidean plane (or more generally low-dimensional Euclidean space). In [6], we gave an $O(n \log n)$ approximation scheme for Steiner tree in planar graphs; more precisely, we showed that for any $\epsilon > 0$, there is an $O(n \log n)$ algorithm that returns a solution whose length is at most $1 + \epsilon$ times optimal. However, the constant factor for this algorithm is triply exponential in $1/\epsilon$. In this paper, we give another $O(n \log n)$ approximation scheme, one for which the constant factor is singly exponential in a polynomial in $1/\epsilon$:

Theorem 1. *For any $\epsilon > 0$, there is an algorithm that, given a planar graph G with edge lengths and a set S of vertices of G, finds a Steiner tree that spans S and whose length is at most $1 + \epsilon$ times the length of the optimal Steiner tree spanning S. The running time is $O(2^{\text{poly}(1/\epsilon)} n + n \log n)$ where n is the number of vertices of G.*

The previous approximation scheme fit into a framework given in [13]. What the framework required to yield an approximation scheme (and what was provided in [6]) was a kind of "spanner" result: an algorithm that, for a planar graph G_{in}

* Supported by a Kanellakis Fellowship and a Brown University Dissertation Fellowship.

** Partially supported by NSF Grant CCF-0635089.

F. Dehne, J.-R. Sack, and N. Zeh (Eds.): WADS 2007, LNCS 4619, pp. 275–286, 2007.

and set of terminals Q, would return a "short" subgraph of G_{in} that approximately preserved the value of the optimum. The approximation scheme then follows rather directly from the framework. However, the spanner result had a doubly-exponential dependence on $1/\epsilon$, which led to the triply-exponential dependence of the final approximation scheme's running time.

We overcome this deficiency using two ideas. One of the exponentials came from a theorem in which we showed how to reduce the complexity of a subtree of the optimal Steiner tree without increasing its cost too much. In this paper, we prove the same theorem but with a polynomial instead of an exponential (see Theorem 4). This idea by itself, if plugged back into [6], would yield doubly exponential dependence on $1/\epsilon$.

The other idea is more global, and perhaps will turn out to be more generally applicable for obtaining approximation schemes in planar graphs. The first step of the spanner construction consisted of finding a short grid-like subgraph MG of the input graph G_{in} that contains every terminal. In this paper, we use the term *brick* to refer the subgraph consisting of a face of MG and the subgraph of G_{in} embedded inside it.

In the new approximation scheme, we also start by finding MG. We then decompose MG into "parcels" with short boundaries such that each parcel has low carving-width (a relative of tree-width). Of course, MG is not the original graph; it is missing the bricks. We add back the bricks but connect them to MG via a small number of "portal edges". We add some new terminals and, for each parcel-plus-bricks, find an optimal Steiner tree using dynamic programming (exploiting the low carving-width of the parcel and the small number of portal edges). We prove that the union of all these trees is not much longer than the optimal Steiner tree in the original graph. The base case for the dynamic program uses an exact algorithm by Erickson et al. [7] for the special case where all terminals are on the boundary of an embedded planar graph. We summarize their result in the following theorem:

Theorem 2. [7] *Let G be a planar embedded graph and Q be a set of k terminals that all lie on the boundary of a single face. Then there is an algorithm*[1] *to find an optimal Steiner tree of Q in G in time $O(nk^3 + (n \log n)k^2)$.*

We can replace the term $n \log n$ by n using the linear-time planarity-exploiting shortest-path algorithm of [9], obtaining a running time of $O(nk^3)$.

1.1 Notation

We assume without loss of generality that the input graph G_{in} is planar embedded and has degree at most three. The input graph has positive edge-lengths $\ell(\cdot)$. For a set A of edges, we use $\ell(A)$ to denote $\sum_{e \in A} \ell(e)$. We take Q as the set of terminal vertices.

[1] This algorithm has been generalized by Bern [3] and by Bern and Bienstock [4] to handle some additional special cases, e.g. where the terminals lie on a constant number of faces. Provan [19,18] used the same approach to give exact and approximate algorithms for some geometric special cases.

For notational convenience, we prove a slightly weaker version of main theorem (Theorem 1). We show that, for any $\epsilon > 0$, there is an $O(n \log n)$ algorithm to find a Steiner tree whose length is at most $(1 + c\epsilon)\text{OPT}(G_{in}, Q)$, where c is a constant and $\text{OPT}(G, Q)$ denotes the length of the Steiner minimal tree that spans Q in graph G. To prove Theorem 1, for any given $\bar{\epsilon} > 0$, we set $\epsilon = \bar{\epsilon}/c$ and invoke the algorithm.

The *boundary of a face* of a planar embedded graph is the set of edges adjacent to the face; it does not always form a simple cycle (Figure 2(a)). The *boundary* ∂H of a planar embedded graph H is the set of edges bounding the infinite face. An edge is *strictly enclosed* by the boundary of H if the edge belongs to H but not to ∂H.

Graphs are identified with sets of edges, thus a subgraph H of a graph G is also considered a subset of the edges of G. The set of vertices that are endpoints of edges in H is denoted $V(H)$. For a tree T and vertices $x, y \in V(T)$, we denote the unique simple x-to-y path in T by $T[x, y]$. In particular, if T is a path then $T[x, y]$ is the x-to-y subpath. We denote the length of the shortest x-to-y path in G as $dist_G(x, y)$.

For a connected planar embedded graph G, there is another connected planar embedded graph denoted G^*. The faces of G are the vertices of G^*; the edges of the two graphs are identified. We refer to G as the *primal* graph and to G^* as the *dual*.

2 Algorithm

2.1 Mortar Graph

We first find a connected grid-like subgraph of the input graph G_{in}, based on the set Q of terminals and the given precision ϵ. The $O(n \log n)$-construction is given in [6]. We call the subgraph the *mortar graph* and denote it MG (see Figure 1(b)). The mortar graph spans every terminal in Q and has length at most $5\epsilon^{-1} \cdot \text{OPT}(G_{in}, Q)$. We defer the remaining properties of MG to Section 3, where they are needed for the proof of correctness.

Step 1: Construct the mortar graph, MG.

2.2 Bricks

Each face f of the mortar graph that strictly encloses at least one edge of G_{in} defines a graph called a *brick*. The brick consists of the edges of G_{in} that are enclosed by the boundary ∂f of f. This boundary is a cycle of edges, possibly with repetition if some edges occur twice in the boundary (an example of such a situation is shown on Figure 2). We duplicate the repeated edges as follows:

Cut the original graph G_{in} along ∂f, duplicating the edges you cut along (and replicating the vertices), and define the brick to be the subgraph of G_{in} embedded inside that cycle, including the boundary edges according to their

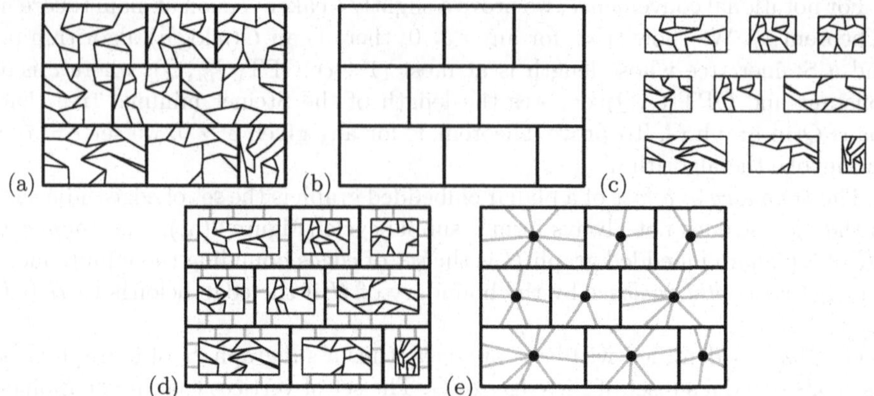

Fig. 1. (a) A fragment of an input graph G_{in}. The bold edges belong to the mortar graph *MG*, the corresponding fragment of which is shown in (b). The corresponding bricks are shown in (c), and the corresponding fragment of the portal-connected graph, $\mathcal{B}^+(MG)$, appears in (d). The portal edges are grey. (e) $\mathcal{B}^+(MG)$ with the bricks contracted to *brick vertices*.

multiplicity in ∂f. That is, if an edge occurs twice in the boundary of the face, then there are two copies of that edge in the corresponding brick.

Step 2: Compute the set of bricks, $\mathcal{B}$.

It is easy to see that Step 2 takes $O(n)$.

The boundary ∂B of a brick B is the simple cycle of boundary edges. The corresponding face of *MG* is called the *mortar boundary* of B. Each edge of the mortar graph occurs at most twice in the disjoint union of the boundaries of the bricks. Since we defined bricks corresponding only to non-empty faces, every brick contains at least one edge not belonging to *MG*. Figure 1(c) is an example of the set of bricks corresponding to the mortar graph of Figure 1(b). The construction of a brick is illustrated in Figures 2(a) and (b).

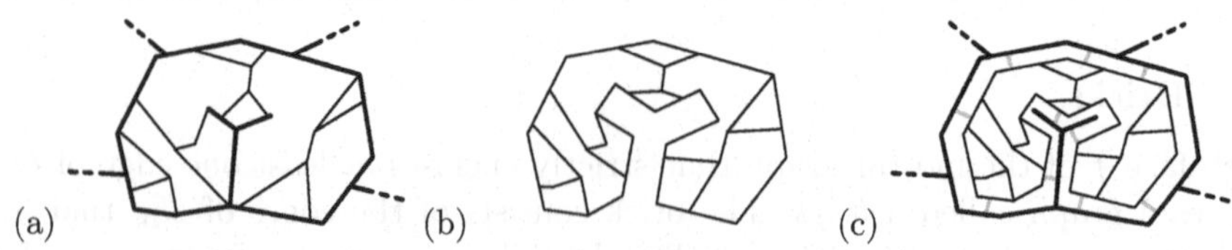

Fig. 2. Construction of a brick: (a) The boundary of a face f of *MG* is a cycle of edges (thick edges), possibly with repetition (i.e. an edge can occur twice in the boundary). The light edges are those in the interior of f in G_{in}. (b) We obtain the corresponding *brick* via Step 2. The resulting brick B has boundary ∂B. (c) A brick, copied.

2.3 Portals

For portal selection, we use a parameter $\theta(\epsilon) = 2\alpha(\epsilon)5\epsilon^{-2}$ that depends on a value $\alpha(\epsilon)$ that in turn comes out of Theorem 3. Portals are selected greedily as in [6] to satisfy:

Property 1. For any vertex x on ∂B, there is a portal y such that the x-to-y subpath of ∂B has length at most $\ell(\partial B)/\theta$.

Step 3: For each brick B, designate θ vertices of ∂B as *portals.*

We additionally require that one portal be the endpoint of an edge that is strictly enclosed in the brick (this, in addition to the assumption that our input graph is degree three, allows us to build a binary recursion tree for the dynamic program).

2.4 Portal-Connected Graph and the Operation $\mathcal{B}^+$

In preparation for stating our Structure Theorem, we define an operation called *brick insertion.* For any subgraph MG of G, we derive a planar embedded graph $\mathcal{B}^+(G)$ as follows. For each face f of G corresponding to a brick B, embed a copy of B inside the face f, and, for each portal vertex v of B, connect v in the brick to the corresponding vertex in f, using a zero-length artificial edge (Figure 2(c)). We refer to the artificial edges as *portal edges.* This step is illustrated in Figure 1(d). We refer to $\mathcal{B}^+(MG)$ as the *portal-connected graph,* and we denote it by $\mathcal{B}^+(MG)$. Intuitively, this graph is almost the same as the input graph G_{in}, except that artificial cost-zero separations have been added so that paths that connect vertices strictly enclosed by faces of the mortar graph to outside vertices are forced to go through the portals.

If a vertex of MG is a terminal of Q, we do not consider its copy on the brick to be a terminal vertex. Thus a brick has no terminals; this helps in the design of the dynamic program (Section 2.7).

The following lemma follows directly from the fact that each portal edge in $\mathcal{B}^+(MG)$ connects a vertex of a brick to the corresponding vertex of MG.

Lemma 1. *If A is a connected subgraph of $\mathcal{B}^+(MG)$ that spans Q, then $A - \{portal\ edges\}$ is a connected subgraph of G_{in} that spans Q.*

The following theorem, proved in Section 3, is central to the proof of correctness of the spanner construction and the approximation scheme. Indeed, taken together, Lemma 1 and Theorem 3 provide a reduction from the Steiner tree problem on G_{in} to the Steiner tree problem on $\mathcal{B}^+(MG)$.

Theorem 3 (Structure Theorem). *There exists a constant $\theta(\epsilon)$ depending polynomially on $1/\epsilon$ such that, for any choice of portals satisfying the Coverage Property, the corresponding portal-connected graph $\mathcal{B}^+(MG)$ satisfies*

$$OPT(\mathcal{B}^+(MG), Q) \leq (1 + c\epsilon) OPT(G_{in}, Q)$$

where c is an absolute constant.

2.5 Parcels

First we further decompose MG into subgraphs called *parcels*, using an integer parameter $\eta(\epsilon) = \epsilon^{-2}$.

Step 4(a): Do breadth-first search in the planar dual MG^* starting from r.

Define the *level* of a vertex of MG^* (face of MG) as its distance from r. Let E_i denote the set of edges whose two endpoints are at levels i and $i+1$.

Step 4(b): For $k = 0, 1, \ldots, \eta - 1$, let $\mathcal{E}_k = E_k \cup E_{k+\eta} \cup E_{k+2\eta} \cup \ldots$.
Let k^* be the index that minimizes $\ell(\mathcal{E}_k)$.

Let $\mathcal{Y}$ denote the set of connected components of $MG^* - \mathcal{E}_{k^*}$. For each $Y \in \mathcal{Y}$, let H_Y denote the subgraph of MG consisting of the boundaries of faces in $V(Y)$ The set of *parcels* of MG is $\mathcal{H} = \{H_Y : Y \in \mathcal{Y}\}$.

Step 4(c): Find the set $\mathcal{H}$ of parcels of MG.

Lemma 2. *The parcel decomposition has the following two properties:*

Radius Property: *The planar dual of each parcel has a spanning tree of depth at most $\eta + 1$.*
Boundary-Length Property: *The sum of the lengths of the boundaries of the parcels is at most $2\ell(MG)/\eta$.*

2.6 New Terminals

The next step is to select the new terminals. These new terminals will ensure that the Steiner trees we find later will combine to form a connected subgraph. The parcel-boundary length property ensures that connecting to these new terminals does not increase the lengths of the optimal parcel solutions by much.

Step 5: For each parcel H and for each connected component C of the boundary of H, if $\mathcal{B}^+(MG) - V(C)$ disconnects some terminals, then designate a vertex of C as a new terminal.

Note that the new terminals are vertices of the mortar graph, not of the bricks. We omit the $O(n)$ implementation of Step 5.

Lemma 3. *The new terminals have the following two properties:*

Spannable Property: *Let T be a tree in $\mathcal{B}^+(MG)$ that spans the original terminals and let H be a parcel. Then $T \cup \{$parcel boundary edges$\}$ contains a tree in $\mathcal{B}^+(H)$ that spans the original and new terminals in H.*
Connecting Property: *For each parcel H that contains a terminal, let T_H be a tree in $\mathcal{B}^+(H)$ spanning the original and new terminals belonging to H. Then $\bigcup_H T_H$ is a connected subgraph of $\mathcal{B}^+(MG)$.*

2.7 Optimal Solution Within the Parcels

> *Step 6:* For each parcel H, if H contains an original or new terminal then find an optimal Steiner tree in $\mathcal{B}^+(H)$ spanning the original and new terminals in H.

This step is solved by a $O(c^{\theta\eta}m)$-time dynamic programming algorithm where m is the number of edges in H and c is a constant. We briefly sketch the idea. Lemma 2 states that the planar dual of H has a spanning tree T^* of depth at most $\eta+1$. When we apply $\mathcal{B}^+$ to H (inserting the bricks), we connect each brick to the corresponding face boundary using at most θ portal edges. Suppose we then contract the bricks in $\mathcal{B}^+(H)$, turning them into *brick vertices* as shown in Figure 1(e). Each brick vertex is connected to MG by at most θ portal edges. In the dual, these portal edges form a cycle encircling the brick vertex. Add to T^* all these edges except the one that in the primal graph is incident to an internal brick edge. Let $\widehat{T^*}$ be the resulting spanning tree. Its depth is at most $\theta(\eta+1)$.

Let $\widehat{T}$ be the set of edges in the contracted graph that do not belong to $\widehat{T^*}$. A classical result in planarity states that the complement of a spanning tree of the dual is a spanning tree of the primal, so $\widehat{T}$ is a spanning tree of the primal. The (primal) input graph had degree three. For each brick, $\widehat{T}$ has one edge connecting the brick vertex to a vertex v of the mortar graph. In the input graph there was an edge incident to v that is no longer present in the contracted graphs, so $\widehat{T}$ has degree at most three.

Root $\widehat{T}$ at a non-brick-vertex of degree at most two, and use the rooted tree to guide a dynamic-programming algorithm for Steiner tree in $\mathcal{B}^+(H)$. For each vertex v of $\widehat{T}$, the subtree rooted at v corresponds to a subgraph of $\mathcal{B}^+(H)$ (replace each brick vertex by the corresponding brick). We show that this subgraph connects with the rest of $\mathcal{B}^+(H)$ via few edges. Suppose v is not the root, and let e be the edge connecting v to its parent. In the dual, e is not an edge of $\widehat{T^*}$, so it forms a cycle with the simple path in $\widehat{T^*}$ between its endpoints. Since $\widehat{T^*}$ has depth at most $\theta(\eta+1)$. the cycle has at most $2\theta(\eta+1)+1$ edges. This shows that, in the primal, the subgraph connects to the rest of $\mathcal{B}^+(H)$ via at most $2\theta(\eta+1)+1$ edges, which enables us to do dynamic programming. Each vertex corresponds to a subproblem. The size of the table for this subproblem is $d^{2\theta(\eta+1)+1}$ where d is a constant. Because each vertex of $\widehat{T}$ has at most two children, only two subproblems need to be combined at a time. If v is a brick vertex, then v is a leaf in $\widehat{T}$, and the subproblem corresponding to v can be solved used the algorithm of Erickson et al. (Theorem 2).

> *Step 7:* Take the union of the edge-sets of all the Steiner trees found in Step 6 (not including portal edges), together with the edges of S, and return the connected component containing the terminals.

This completes the description of the approximation scheme. Lemma 3 shows that the output is a feasible solution. Lemmas 2 and 3, together with the definition of η, show that the length of the output solution is at most

$(1+d\epsilon)\mathrm{OPT}(\mathcal{B}^+(MG), Q)$, and is therefore (by Theorem 3) at most $(1+d\epsilon)(1+c\epsilon)\mathrm{OPT}(G_{in}, Q)$.

3 Proof of the Structure Theorem (Theorem 3)

The construction of what we here call a brick decomposition was given in [6]. Step (i) of the construction involved *cutting open* the input graph along a 2-approximate Steiner tree, obtaining a graph G_1. Step (ii) used shortest paths to decompose G_1 into "strips." Step (iii) found some shortest paths within each strip, and Step (iv) designated some of the shortest paths found in Step (iii) as *supercolumns*. For this paper, we define the *mortar graph* MG of G_{in} to be the planar embedded subgraph consisting of (A) the edges of the 2-approximate Steiner tree of Step (i), (B) the edges of the shortest paths used in Step (ii), and (C) the edges of the supercolumns. The choice of supercolumns in Step (iv) involves a parameter κ; choosing $\kappa(\epsilon) = 8\epsilon^{-2}(1+\epsilon^{-1})$ yields Lemma 5 (below). In [6], we showed the following properties:

Lemma 4. *The boundary of a brick B, in counterclockwise order, is the concatenation of four paths $W_B \cup S_B \cup E_B \cup N_B$ such that:*

1. *Every terminal of Q that is in B is on N_B or on S_B.*
2. *N_B and S_B are ϵ-short.*
3. *$V(S_B)$ has a κ-element subset $\{s_0, \ldots, s_\kappa\}$ (in left-to-right order) where for any i and any vertex $x \in S_B[s_i, s_{i+1})$, $dist_{S_B}(x, s_i) < \epsilon dist_B(x, N_B)$.*

Lemma 5. *Summing over all bricks B, $\sum_B \ell(E_B) + \ell(W_B) \le \epsilon\,\mathrm{OPT}(G_{in}, Q)$.*

3.1 Structural Property of Bricks

This decomposition into bricks is useful because there exists a near-optimal Steiner tree that crosses the boundary of each face of MG a small number of times. For H a subgraph of a graph G and P a path in H, a *joining vertex of H with P* is a vertex of P that is the endpoint of an edge of $H - P$.

Theorem 4 (Structural Property of Bricks). *Let B be a plane graph with boundary $N \cup E \cup S \cup W$ and satisfying the brick properties of Lemma 4. Let F be a set of edges of B. There is a forest $\tilde{F}$ of B with the following properties:*

1. *If two vertices of $N \cup S$ are connected in F then they are connected in $\tilde{F}$.*
2. *The number of joining vertices of F with both N and S is at most $\alpha(\epsilon)$.*
3. *$\ell(\tilde{F}) \le (1+c\epsilon)\ell(F)$.*

In the above, $\alpha(\epsilon) = o(\epsilon^{-5.5})$ and c is a fixed constant.

"N", "E", "S", and "W" stand for north, east, south, and west.

In [6], the analogous theorem appeared as Theorem 3.1 with $\alpha(\epsilon) = 2^{\mathrm{poly}(1/\epsilon)}$ instead. In order to prove Theorem 4, we use the following lemmas.

Lemma 6. *Let G be a planar embedded graph, let P be an ϵ-short path that is a subpath of the boundary ∂G of G, and let T be a tree in G rooted at a vertex r whose leaves are exactly $V(T) \cap V(P)$. There is another tree $\widehat{T}$ rooted at r spanning $V(T) \cap V(P)$ whose total length is at most $(1 + c_1 \cdot \epsilon)\ell(T)$ such that $\widehat{T}$ has at most $c_2 \cdot \epsilon^{-1.45}$ joining vertices with P.*

Proof. Following [6], as long as there is a vertex u with at least 3 children, do this: choosing u to be closest to the root, replace the subtree rooted at u with (1) the minimal subpath P' of P containing all leaves of that subtree, and (2) the shortest u-to-P' path. The resulting tree T' has $\ell(T') \le (1+\epsilon)\ell(T)$.

Define the *level* of a vertex v to be the number of degree-3 vertices on the r-to-v path of T'. Let $\mathcal{U}$ be the set of degree-3 vertices having level k (we will choose k later). A *super-edge* is a maximal descending path in T' whose internal vertices have degree 2, and its *level* is the level of its first vertex.

For each $u \in \mathcal{U}$, replace the subtree T'_u of T' rooted at u with another tree T''_u rooted at u that is the union of the shortest subpath P_u of P spanning the vertices of $T'_u \cap P$ and the shortest u-to-P_u path (Figure 3.1(a)). Let T'' be the result. Analysis is as follows.

For a degree-3 vertex u, let Q_u be the path in T' between u's leftmost and rightmost descendent leaves. For each i, let $E_i = \bigcup\{Q_u : u \text{ has level } i\}$, and let $L_i = \bigcup\{\text{level-}i \text{ super-edges}\} - E_{i-1}$. See Figure 3.1 (b). Let $S_i = \cup_{j=i}^{\infty} L_j$.

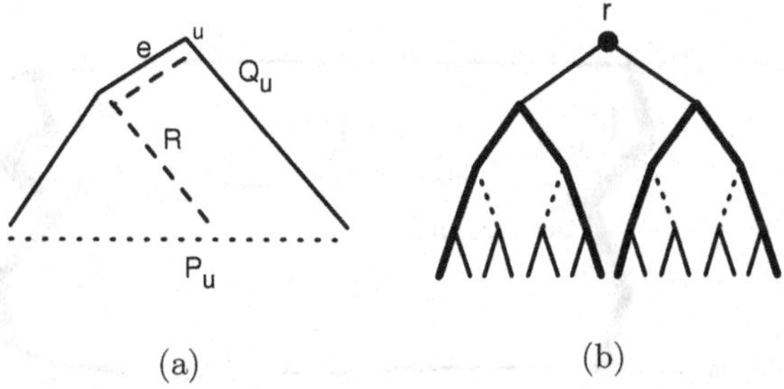

Fig. 3. (a) Solid line is Q_u, dashed line is R, dotted line is P_u. Root of tree shown is u, and left child edge is e.(b) Bold edges: E_1, dotted edges: L_2.

Let k' be the first level i for which $\ell(L_i) \le \ell(S_{i+2})$ (if there is no such level, let $k' = \infty$). Let $k = \min\left(k', \lceil \log_\Phi(\sqrt{5}(1/\epsilon - 1)) \rceil\right)$, where Φ is the golden ratio. Since $k \le \lceil \log_\Phi(\sqrt{5}(1/\epsilon + 1)) \rceil$, the number of level-$k$ vertices is $\le 2^k \le 11 \cdot \epsilon^{-1.45}$ (for $\epsilon < 1$), which leads to the bound on joining vertices. It remains to show that $\ell(T'') \le (1+\epsilon)\ell(T')$.

Let u be a vertex in level k. Let e be the unique super-edge in L_k whose parent is u (as illustrated in Figure 3.1 (a). Let R be the path from u to P that traverses e and subsequently uses only edges of $E_{k+1} - E_k$.

$$\ell(T''_u) = \ell(P_u) + dist_G(u, P_u) \le (1+\epsilon)\ell(Q_u) + \ell(R) \tag{1}$$

Case 1: $k = k'$**.** S_{k+2} is disjoint from Q_u and R, so the RHS of (1) is at most $(1+\epsilon)[\ell(T'_u)+\ell(e)-\ell(S_{k+2}\cap T'_u)] < (1+\epsilon)[\ell(T'_u)+\ell(L_k\cap T'_u)-\ell(S_{k+2}\cap T'_u)]$. Summing over all level-i vertices u, $\ell(T'') < (1+\epsilon)[\ell(T')+\ell(L_k)-\ell(S_{k+2})] < (1+\epsilon)\ell(T')$ since $\ell(L_k) \le \ell(S_{k+2})$.

Case 2: $k \ne k'$**.** The RHS of (1) is $\le (1+\epsilon)\ell(Q_u \cup R)+\ell(e) \le (1+\epsilon)\ell(T'_u)+\ell(e)$. Summing over all $u \in \mathcal{U}$, $\ell(T'') \le (1+\epsilon)\ell(T')+\ell(L_k) \le (1+\epsilon)\ell(T')+\ell(S_k)$. Note that S_i is the disjoint union of L_i and S_{i+1}, so $\ell(S_i) = \ell(L_i)+\ell(S_{i+1})$. Since $\ell(L_i) > \ell(S_{i+2})$ for every $i \le k$, we have $\ell(S_i) \ge \ell(S_{i+2})+\ell(S_{i+1})$. It follows that $\ell(S_1) \ge k^{th}$ Fibonacci number $\cdot\, \ell(S_k)$. Then by choice of k, $\ell(S_k) \le \epsilon\ell(S_1) \le \epsilon\ell(T')$. □

Lemma 7. *Let G be a planar embedded graph and let T be a tree in G with leaves on an ϵ-short path P that is a subpath of the boundary ∂G of G. Let p and q be two vertices of T. There is another tree $\widehat{T}$ spanning p, q, and the vertices of $T \cap P$ whose total length is at most $(1+c_1\cdot\epsilon)\ell(T)$ such that $\widehat{T}$ has at most $c_2\cdot\epsilon^{-2.45}$ joining vertices with P.*

The proof for Lemma 7 can be derived from the section titled *Achieving the Third Property* in [6]. It builds on Lemma 6.

Proof idea for Theorem 4. We use the term *bases* to refer to the vertices $s_0,\ldots,s_k$ of Part 3 of Lemma 4. We select S-to-N paths $P_0, P_1, \ldots$, modifying F as we go, as follows. (Let F' denote the modified F.)

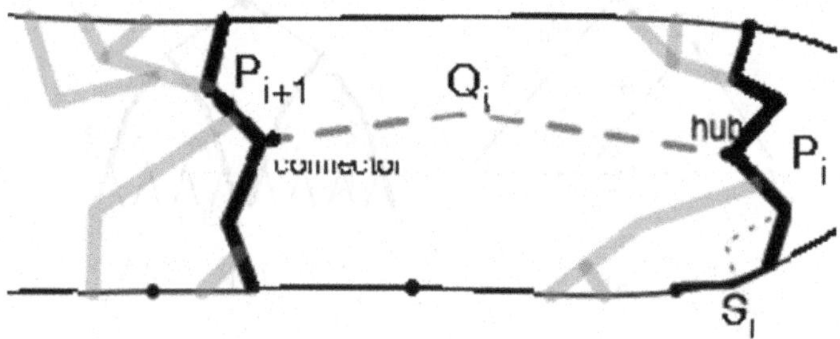

Assume inductively that P_i has been selected, and let x_i be its first vertex. Let S_i be the subpath of S going west from x_i to the first base encountered. Note that $\ell(S_i) \le \epsilon\ell(P_i)$. We add S_i to F, possibly creating cycles. To fix this, remove an edge not in $P_i \cup S_i$ from a cycle until no cycles remain. Next, let P_{i+1} be the eastmost S-to-N path in F that starts from a vertex west of x_i and that is vertex-disjoint from $P_i \cup S$. By acyclicity, there is at most one path Q_i from a vertex of P_{i+1} to a vertex of $P_i \cup S_i$. If there is such a path, designate its first vertex as a *connector* of P_{i+1} and its last vertex as a *hub* of P_i. Note that the hub has the following property: in the component of $F' - Q_i - Q_{i-1}$ containing P_i, every S-to-N path goes through the hub. If there is no such path Q_i, we define $Q_i = \emptyset$ and arbitrarily select as the hub of P_i any vertex satisfying that property.

Next, we transform F', obtaining another forest F''. For each i, consider the component of $F' - Q_i - Q_{i-1}$ that contains P_i. Decompose this component into two trees: the southern tree consists of the paths from the hub to vertices in S, and the northern tree consists of the paths from the hub to vertices in N. We apply Lemma 6 to whichever of these trees does not contain a connector (taking r=the hub), and we apply Lemma 7 to whichever does contain a connector (taking p=the hub and q=the connector). □

3.2 Completion of the Proof of Theorem 3

The Structure Theorem (Theorem 3) states that $\text{OPT}(\mathcal{B}^+(MG), Q) \leq (1 + c\epsilon)\text{OPT}(G_{in}, Q)$. We now give the proof using Theorem 4.

Proof. We start from an optimal solution T to the Steiner tree problem in G_{in} and gradually transform it into a solution $\widehat{T}$ to the Steiner tree problem in $\mathcal{B}^+(MG)$, while approximately preserving its length. First, let T_1 be the union of T with the east and west boundaries (E_B and W_B) for every brick B in G. Using Lemma 5, we have $\ell(T_1) \leq \text{OPT} + \epsilon\text{OPT}(G_{in}, Q)$. Remove edges (other than east/west boundary edges) to break cycles.

Next, apply Theorem 4 to the subgraph of T_1 that is contained in each brick, obtaining T_2 such that $\ell(T_2) \leq (1 + c'\epsilon)\ell(T_1)$.

Next, we must obtain a solution in $\mathcal{B}^+(MG)$. Let T_2^a be the set of brick-boundary edges of T_1, and let T_2^b the other edges of T_2. Let T_3 be the set of edges of $\mathcal{B}^+(MG)$ consisting of (a) the edges of MG corresponding to edges of T_3^a, and (b) the edges of T_3^b. Note that T_3 is not a connected subgraph of $\mathcal{B}^+(MG)$. A path in T_2 might pass from the interior of a brick to the boundary; the corresponding sequence of edges in T_3 would have a gap: the "path" would stop at a vertex of the brick boundary, and resume at the corresponding vertex of the mortar boundary. To close the gap, we must add paths connecting each to the nearest portal vertex associated with that brick and then add the corresponding portal edge. The resulting graph T_4 is connected.

It remains to bound the length of all these detours. For brick B, the distance to the nearest portal is at most $\ell(\partial B)/\theta$, so the length of the detour is at most $2\ell(\partial B)/\theta$. By Theorem 4, the number of detours for this brick is at most α, so the length of all these detours is at most $2\alpha\ell(\partial B)/\theta$. Summing over all bricks and using the bound from Section 2.1 on the length of the mortar graph, we obtain a bound of $10\alpha\epsilon^{-1} \cdot \text{OPT}(G_{in}, Q)/\theta$. The choice of θ ensures that this is at most $\epsilon \cdot \mathit{opt}(G_{in}, Q)/\theta$.

□

References

1. Arora, S.: Polynomial-time approximation schemes for euclidean TSP and other geometric problems. JACM 45(5), 753–782 (1998)
2. Berman, P., Ramaiyer, V.: Improved approximations for the Steiner tree problem. Journal of Algorithms 17, 381–408 (1994)

3. Bern, M.: Faster exact algorithms for Steiner trees in planar networks. Networks 20, 109–120 (1990)
4. Bern, M., Bienstock, D.: Polynomially solvable special cases of the Steiner problem in planar networks. Annals of Operations Research 33, 405–418 (1991)
5. Bern, M., Plassmann, P.: The Steiner problem with edge lengths 1 and 2. IPL 32, 171–176 (1989)
6. Borradaile, G., Kenyon-Mathieu, C., Klein, P.: A polynomial-time approximation scheme for Steiner tree in planar graphs. In: 18th SODA, pp. 1285–1294 (2007)
7. Erickson, R., Monma, C., Veinott, A.: Send-and-split method for minimum-concave-cost network flows. Mathematics of Operations Research 12, 634–664 (1987)
8. Garey, M., Johnson, D.: The rectilinear Steiner tree problem is NP-complete. SIAM J. Appl. Math. 32(4), 826–834 (1977)
9. Henzinger, M., Klein, P., Rao, S., Subramanian, S.: Faster shortest-path algorithms for planar graphs. J. Comput. System Sci. 55(1), 3–23 (1997)
10. Hougardy, S., Prömel, H.J.: A 1.598 approximation algorithm for the Steiner problem in graphs. In: 10th SODA, pp. 448–453 (1999)
11. Karp, R.: On the computational complexity of combinatorial problems. Networks 5, 45–68 (1975)
12. Karpinski, M., Zelikovsky, A.: New approximation algorithms for the Steiner tree problem. Journal of Combinatorial Optimization 1, 47–65 (1997)
13. Klein, P.: A linear-time approximation scheme for planar weighted TSP. In: 46th FOCS, p. 647 (2005)
14. Kou, L., Markowsky, G., Berman, L.: A fast algorithm for Steiner trees. Acta Informatica 15, 141–145 (1981)
15. Mehlhorn, K.: Approximation algorithm for the Steiner problem in graphs. IPL 27(3), 125–128 (1988)
16. Mitchell, J.: Guillotine subdivisions approximate polygonal subdivisions: A simple polynomial-time approximation scheme for geometric tsp, k-mst, and related problems. SIAM J. Comput. 28(4), 1298–1309 (1999)
17. Prömel, H.J., Steger, A.: RNC approximation algorithms for the Steiner problem. In: 39th STOC. LNCS, vol. 1200, pp. 559–570. Springer, Heidelberg (1997)
18. Provan, J.: An approximation scheme for finding Steiner trees with obstacles. SIAM J. Comput. 17, 920–934 (1988)
19. Provan, J.: Convexity and the Steiner tree problem. Networks 18, 55–72 (1988)
20. Rao, S., Smith, W.: Approximating geometrical graphs via spanners and banyans. In: 30th STOC, pp. 540–550 (1998)
21. Robins, G., Zelikovsky, A.: Tighter bounds for graph Steiner tree approximation. SIAM J. Discret. Math. 19(1), 122–134 (2005)
22. Takahashi, H., Matsuyama, A.: An approximate solution for the Steiner problem in graphs. Mathematica Japonicae 24, 571–577 (1980)
23. Widmayer, P.: A fast approximation algorithm for Steiner's problem in graphs. In: Tinhofer, G., Schmidt, G. (eds.) WG 1986. LNCS, vol. 246, pp. 17–28. Springer, Heidelberg (1987)
24. Wu, Y., Widmayer, P., Wong, C.: A faster approximation algorithm for the Steiner problem in graphs. Acta informatica 23(2), 223–229 (1986)
25. Zelikovsky, A.: Better approximation bounds for the network and Euclidean Steiner tree problems. Technical Report CS-96-06, University of Virginia (1994)
26. Zelikovsky, A.: An 11/6-approximation algorithm for the network Steiner problem. Algorithmica 9, 463–470 (1999)

Computing a Minimum-Depth Planar Graph Embedding in $O(n^4)$ Time*

Patrizio Angelini, Giuseppe Di Battista, and Maurizio Patrignani

Dipartimento di Informatica e Automazione – Università di Roma Tre
{angelini,gdb,patrigna}@dia.uniroma3.it

Abstract. Consider an n-vertex planar graph G. We present an $O(n^4)$-time algorithm for computing an embedding of G with minimum distance from the external face. This bound improves on the best previous bound by an $O(n \log n)$ factor. As a side effect, our algorithm improves the bounds of several algorithms that require the computation of a minimum depth embedding.

1 Introduction

As pointed out in [4,9,8], the quality of a planar embedding of a planar graph can be measured in terms of maximum distance of its vertices from the external face f_e. Such a distance can be given in terms of different incidence relationships between vertices and faces. For example, if we say that two faces are adjacent if they share a vertex, the maximum distance to f_e is called *radius* [10]. If two vertices are adjacent if they are endpoints of an edge, the maximum distance to f_e is called *width* [7]. If two vertices are adjacent if they are on the same face and f_e is adjacent to all its vertices, the maximum distance to f_e is called *outerplanarity* [2]. If two faces are adjacent if they share an edge, the maximum distance to f_e is called *depth* [3].

The algorithms that compute a planar embedding such that the vertices have minimum maximum distance to the external face have several applications. The algorithm by Dolev, Leighton, and Trickey for drawing planar graphs with asymptotically optimal area [7] requires the computation of the embedding with minimum width. In [2] Baker gives approximation algorithms on planar graphs for many NP-complete problems, including maximum independent set and minimum vertex cover. The time complexity and the optimality bounds of such algorithms depend on the outerplanarity of the graph. Finally, the algorithm by Di Giacomo et al. [6] for constructing crossing-free minimum radial drawings of planar graphs is based on the computation of their outerplanarity.

In [4] Bienstock and Monma present an algorithm to compute the planar embedding of an n-vertex planar graph G with minimum maximum distance to the

* Work partially supported by EC - Fet Project DELIS - Contract no 001907 and by MUR under Project "MAINSTREAM: Algoritmi per strutture informative di grandi dimensioni e data streams".

F. Dehne, J.-R. Sack, and N. Zeh (Eds.): WADS 2007, LNCS 4619, pp. 287–299, 2007.

external face in $O(n^5 \log n)$ time. The distance they consider is the depth. However, the other distances listed above can be computed with simple variations of the algorithm. The algorithm is based on the decomposition of G into its bi- and tri-connected components. The general approach is the one of selecting a positive integer k and to check if an embedding exists with depth k. A binary search is performed to determine the optimal value of k. For each selected k the decomposition of G is visited associating to each component μ a left and a right weight, corresponding to the distances of μ to the external face of G, which is independent on the embedding of μ. The components are then visited to check if their weights can be composed to construct en embedding with depth k. The space complexity of the algorithm is not analyzed in the paper. In [9] Pizzonia and Tamassia present an algorithm for solving in $O(n)$ time an analogous problem where the depth of the embedding is expressed in terms of biconnected components traversed to reach the external face and the biconnected components are "rigid", in the sense that their embedding cannot be changed.

In this paper we present an algorithm that improves the time bound of [4] to $O(n^4)$ time. As a side effect, we improve also the time bound of the algorithms listed above that need to compute a planar embedding with minimum maximum distance to the external face. Our approach is inspired by the methods in [4], and develops on top of such methods several new techniques. As in [4], we decompose the graph into bi- and tri-connected components, using BC-trees and SPQR-trees [5]. However, we are able to solve the problem on each biconnected component with a given edge on the external face in $O(n^3)$ time. Then, we use techniques analogous to those in [4] for assembling the results on each biconnected component into a general solution. Among the techniques presented in this paper, a key issue, that might have other applications, is the ability of representing implicitly and with reasonable size all the possible values of depth of each triconnected component. The space complexity of the algorithm is $O(n^3)$.

The paper is organized as follows. Section 2 gives basic definitions. Section 3 deals with the combinatorial structure of the depth of planar embeddings and develops a theory of the set of integer pairs that is exploited by the algorithm. Section 4 presents the algorithm for biconnected graphs and Section 5 extends such an algorithm to general connected graphs. In Section 6 we give concluding remarks and further compare our approach with the one in [4]. Because of space limitation some proofs are omitted and can be found in [1].

2 Background

A graph $G(V, E)$ is *connected* if every pair of vertices of G is connected by a path. A *separating k-set* of a graph G is a set of k vertices whose removal increases the number of connected components of G. Separating 1-sets and 2-sets are called *cutvertices* and *separation pairs*, respectively. A connected graph is *biconnected* if it has no cutvertices. The maximal biconnected subgraphs of a graph are its *blocks*. Observe that each edge of G falls into a single block of G, while cutvertices are shared by different blocks. The *block cutvertex tree*, or

BC-tree, of a connected graph G has a B-node for each block of G and a C-node for each cutvertex of G. Edges in the BC-tree connect each B-node μ to the C-nodes associated with the cutvertices in the block of μ. The BC-tree of G may be thought as rooted at a specific block ν.

The *SPQR-tree* $\mathcal{T}$ of a biconnected graph G describes the arrangement of its triconnected components. Here we provide an intuitive sketch of the definitions and properties related to SPQR-trees. For more details, see [5,1]. The nodes of $\mathcal{T}$ are of four types: S-, P-, Q-, and R-nodes, and represent a recursive decomposition of the graph G into structural components, each one attached to the rest of the graph by means of a separation pair or a pair of adjacent vertices, called *poles*. A *Q-node* is a component corresponding to an edge of G and is the base case of the decomposition. An *S-node* (*P-node*, *R-node*, respectively) is a component that corresponds to a more complex portion of the graph which can be in its turn decomposed as a series of components (a parallel of components, or an arrangement of components which is not a series nor a parallel, respectively). Given a node μ of $\mathcal{T}$, the *skeleton* of μ, denoted by $sk(\mu)$, is a graph showing how its $\delta(\mu)$ lesser components, represented by *virtual edges*, are arranged into the current component. The poles are always placed on the external face of the skeleton. Starting from a virtual edge (u, v) of $sk(\mu)$, or from the skeleton of the corresponding child node ν, by recursively replacing each virtual edge with the skeleton of the corresponding component, it can be obtained a subgraph of G called the *pertinent graph* of (u, v) and denoted by $pertinent(\nu)$.

The SPQR-tree $\mathcal{T}$ of a graph G with n vertices and m edges has m Q-nodes and $O(n)$ S-, P-, and R-nodes. Also, the total number of vertices of the skeletons stored at the nodes of $\mathcal{T}$ is $O(n)$.

3 The Combinatorial Structure of Planar Embeddings and Their Depths

A biconnected graph G is planar if and only if the skeletons of all the nodes of the SPQR-tree $\mathcal{T}$ of G are planar. An SPQR-tree $\mathcal{T}$ rooted at a given Q-node can be used to represent all the planar embeddings of G having the reference edge (associated with the Q-node at the root) on the external face. Namely, any embedding can be obtained by selecting one of the two possible flips of each skeleton around its poles and selecting a permutation of the children of each P-node with respect to their common poles.

Let G be a biconnected planar graph and let μ_i be a component of the SPQR-tree decomposition $\mathcal{T}$ of G rooted at edge e. Observe that any embedding Γ_G of G with e on the external face corresponds to an embedding Γ_{G_i} of the pertinent graph G_i of μ_i with poles u_i and v_i on the external face. Also, the external face of Γ_{G_i} corresponds to two faces of Γ_G, which can be arbitrarily called *left and right external faces of* G_i and denoted by $f_l^{\mu_i}$ and $f_r^{\mu_i}$. Following the approach of [4], we give a definition of $f_l^{\mu_i}$ and $f_r^{\mu_i}$ which is independent on Γ_G and only depends on Γ_{G_i}. Let (u_i, v_i) be the virtual edge of μ_i that represents in μ_i the portion of G containing e and denote by G_i^+ the graph obtained by adding edge

(u_i, v_i) to the pertinent graph G_i of μ_i. The (u_i, v_i)-*dual* of G_i is obtained by computing the dual of G_i^+ and removing the edge corresponding to (u_i, v_i). The faces incident to the removed edge are $f_l^{\mu_i}$ and $f_r^{\mu_i}$.

An SPQR-tree component μ_i *satisfies* the pair of non-negative integers $\langle x, y\rangle$ if its pertinent graph G_i admits an embedding Γ_{G_i}, with its poles on the external face, where it is possible to find a partition of the set of its internal faces into two sets, denoted by F_l and F_r, such that all faces in F_l have distance from $f_l^{\mu_i}$ less or equal than x and all faces in F_r have distance from $f_r^{\mu_i}$ less or equal than y. Obviously, if μ_i satisfies $\langle x, y\rangle$, then it satisfies any pair $\langle w, z\rangle$ with $w \geq x$ and $z \geq y$. The infinite set of integer pairs satisfied by component μ_i is the *admissible set of* μ_i and is denoted by $A(\mu_i)$.

In order to efficiently represent the admissible set of μ_i, we need to investigate its combinatorial properties. Hence, we provide a definition of a "precedence" relationship between integer pairs and explore the combinatorial properties of sets of integer pairs. Such properties can be also expressed in terms of poset theory or inclusion relationships between geometric curves.

We say that pair $\langle x_1, y_1\rangle$ *precedes wrt* x (*precedes wrt* y) a pair $\langle x_2, y_2\rangle$ when $x_1 \leq x_2$ ($y_1 \leq y_2$). We denote this relationship by $\preceq_x$ ($\preceq_y$). For example $\langle 3, 1\rangle \preceq_y \langle 3, 5\rangle$. We say that pair $\langle x_1, y_1\rangle$ *precedes* a pair $\langle x_2, y_2\rangle$ when $\langle x_1, y_1\rangle$ precedes $\langle x_2, y_2\rangle$ both wrt x and wrt y. We denote this relationship by $\preceq$. For example $\langle 3, 4\rangle \preceq \langle 3, 5\rangle$. Two pairs $\langle x_1, y_1\rangle$ and $\langle x_2, y_2\rangle$ are *incomparable* if none of them precedes the other, i.e., if $\langle x_1, y_1\rangle \npreceq \langle x_2, y_2\rangle$ and $\langle x_2, y_2\rangle \npreceq \langle x_1, y_1\rangle$. The incomparability relationship is denoted by $\nsim$, as, for example, in $\langle 3, 4\rangle \nsim \langle 2, 5\rangle$. Based on the above definition, if μ_i satisfies $\langle x, y\rangle$, then it satisfies any pair $\langle w, z\rangle$ such that $\langle x, y\rangle \preceq \langle w, z\rangle$.

A set S of pairs of non-negative integers $\langle x, y\rangle$ is *succinct* if the pairs of S are pairwise incomparable. Given two sets S and S' of pairs of integers, S' *precedes* S if for any pair $p \in S$ there exists at least one pair $p' \in S'$ such that $p' \preceq p$. For example $\{\langle 0, 4\rangle\langle 3, 3\rangle\langle 5, 4\rangle\} \preceq \{\langle 0, 5\rangle\langle 4, 5\rangle\}$. Also, S' *reduces* S if $S' \preceq S$ and $S' \subseteq S$. Further, if S' is succinct and reduces S, S' is a *gist* of S. For example $\{\langle 0, 4\rangle\langle 3, 3\rangle\}$ is a gist of $\{\langle 0, 4\rangle\langle 3, 3\rangle\langle 5, 4\rangle\}$. The gist S' of a set S is unique and is the smallest set preceding S.

Consider a set S and two integer pairs $p_1 = \langle x_1, y_1\rangle, p_2 = \langle x_2, y_2\rangle \in S$. Denote by $x^{\max}(S)$ ($y^{\max}(S)$) the maximum value of x_i (y_i) in any pair $\langle x_i, y_i\rangle \in S$.

Property 1. If S is succinct, then $p_1 \preceq_x p_2 \Leftrightarrow p_2 \preceq_y p_1$. Hence, the relationship $\preceq_x$ induces a total order on S, which is an inverse total order with respect to relationship $\preceq_y$. Also, $|S| \leq x^{\max}(S)$ and $|S| \leq y^{\max}(S)$.

Let S_j, $j = 1, \ldots, k$, be k sets of integer pairs and let S'_j be their gists. Suppose that each S'_j is known and sorted with respect to the $\preceq_x$ relationship.

Lemma 1. *The gist of* $S_1 \cap S_2 \cap \cdots \cap S_k$ *and the gist of* $S_1 \cup S_2 \cup \cdots \cup S_k$*, sorted with respect to the* $\preceq_x$ *relationship, can be computed in* $O(\sum_{j=1}^{k}(x^{\max}(S'_j)))$ *or, equivalently, in* $O(\sum_{j=1}^{k}(|S'_j|))$ *time.*

Given the previously described precedence relationship, it is possible to represent the infinite set $A(\mu_i)$ of integer pairs satisfied by component μ_i with its gist, which is denoted by $\hat{A}(\mu_i)$ and assumed ordered wrt the $\preceq_x$ relationship.

Let μ_i be a component, G_i its pertinent graph and n_i the number of nodes of G_i. If $\langle x, y\rangle \in \hat{A}(\mu_i)$, then $\langle y, x\rangle \in \hat{A}(\mu_i)$. Also, $\hat{A}(\mu_i)$ contains exactly one pair $\langle x, y\rangle$ with $x = 0$ and one pair $\langle x', y'\rangle$ with $y' = 0$. In such pairs $x = y' = O(n_i)$. Hence, $\hat{A}(\mu_i)$ is finite and $|\hat{A}(\mu_i)|$ is $O(n_i)$.

4 Computing a Minimum-Depth Embedding of a Biconnected Planar Graph

The minimum-depth embedding of a biconnected planar graph G can be found by applying for each edge e of G the algorithm presented in this section, which computes the minimum-depth embedding of G with e on the external face. Such a computation is performed by means of two traversals of its SPQR-tree $\mathcal{T}$ rooted at e. The first traversal is a bottom-up traversal. Its purpose is to label each virtual edge e_i, corresponding to node μ_i, with suitable values that describe the properties with respect to the depth of all possible embeddings of the pertinent graph G_i. Such values are the gist of the admissible set of μ_i and the distance between $f_l^{\mu_i}$ and $f_r^{\mu_i}$ in the (u_i, v_i)-dual of G_i, which is called the *thickness* of μ_i and is denoted by $t(\mu_i)$. In [4], where the concept of thickness was also used, it is shown that $t(\mu_i)$ is independent on the embedding of the pertinent graph G_i of μ_i. At the end of the first bottom-up traversal, the child component of the root of the SPQR-tree is labeled with the gist of the admissible set of G, from which an optimal pair can be selected and used in a top-down traversal of $\mathcal{T}$ to provide a suitable embedding for each skeleton of its nodes.

4.1 Labeling an SPQR-Tree with Minimum-Depth Embedding Descriptors

For each component μ_i, we compute $t(\mu_i)$ and $\hat{A}(\mu_i)$ based on the thickness and the gist computed for its children during the bottom-up traversal of $\mathcal{T}$.

The embeddings Γ_μ^j of $sk(\mu)$, provided that $sk(\mu)$ admits more than one, induce a partition on the embeddings of G_μ. Hence, in order to compute $A(\mu)$ through all the possible embeddings of G_μ, we first compute the admissible sets $A^j(\mu)$, restricted to those embeddings of G_μ corresponding to a single embedding Γ_μ^j of $sk(\mu)$, and then perform the union of the $A^j(\mu)$.

Given an embedding Γ_{G_μ} of the pertinent graph G_μ of μ we distinguish two types of faces. We call *children faces* the faces of Γ_{G_μ} that are also faces of some Γ_{G_ν}, with ν child of μ. We call *skeleton faces* all the other faces. Essentially, "shrinking" each pertinent graph of the children of μ into a single (virtual) edge, the skeleton faces of Γ_{G_μ} transform into the faces of an embedding Γ_μ^j of $sk(\mu)$.

Observe that, once the embedding Γ_μ^j of $sk(\mu)$ has been fixed, the distances of each skeleton face of any embedding of Γ_{G_μ} from f_l^μ and f_r^μ depend on the values $t(\nu_1), \ldots, t(\nu_k)$ only, which, in turn, are independent from the embedding of each child component of μ. Hence, each face f of Γ_μ^j can be labeled with its depths $d_l(f)$ and $d_r(f)$, which correspond to the above distances.

We provide for $sk(\mu)$ definitions analogous to those given for pertinent(μ), and say that Γ_μ^j *satisfies* the pair of non-negative integers $\langle x, y\rangle$ if it is possible to find a partition of its faces into two sets, denoted by F_l and F_r, such that each face $f \in F_l$ has $d_l(f) \leq x$ and each face $f \in F_r$ has $d_r(f) \leq y$. The infinite set of integer pairs satisfied by Γ_μ^j is the *admissible set of* Γ_μ^j, and is denoted by $A(\Gamma_\mu^j)$. Once the embedding Γ_μ^j of $sk(\mu)$ has been fixed, the admissible set $A^j(\mu)$ can be computed starting from $A(\Gamma_\mu^j)$, the admissible set of each child ν_i, with $i = 1, \dots, \delta(\mu)$, and the values of the depths of its left and right external face $f_l^{\nu_i}$ and $f_r^{\nu_i}$. Namely, μ satisfies the integer pair $\langle x, y\rangle$ if $\langle x, y\rangle \in A(\Gamma_\mu^j)$ and each child ν_i satisfies a pair $\langle x^i, y^i\rangle$ such that: *i)* $x^i + d_l(f_l^{\nu_i}) \leq x$ or $x^i + d_r(f_l^{\nu_i}) \leq y$, and *ii)* $y^i + d_r(f_r^{\nu_i}) \leq y$ or $y^i + d_l(f_r^{\nu_i}) \leq x$.

Hence, in order to obtain $A(\mu)$, we compute for each child ν_i of μ the set of integer pairs that are satisfied by ν_i when embedded into μ, that is, the set of integer pairs that verify the conditions above. Namely, let μ be a node of the SPQR-tree, let ν be a child of μ, and let $\langle x, y\rangle$ be a pair of non-negative integers. Node ν *satisfies* $\langle x, y\rangle$, *nested into* μ, if the pertinent graph G_μ admits an embedding Γ_{G_μ} where it is possible to find a partition of the set of the children faces corresponding to the internal faces of ν into two sets, denoted by F_l and F_r, such that all faces in F_l have distance from f_l^μ less or equal than x and all faces in F_r have distance from f_r^μ less or equal than y.

The infinite set of integer pairs satisfied by component ν nested into μ is the *admissible set of* ν *into* μ, and is denoted by $A(\nu|\mu)$. The gist of $A(\nu|\mu)$ is denoted by $\hat{A}(\nu|\mu)$ and assumed ordered with respect to the $\preceq_x$ relationship.

Lemma 2. *Given an embedding Γ_μ^j of $sk(\mu)$, the admissible set $A^j(\mu)$ of G_μ (restricted to those embeddings of G_μ corresponding to Γ_μ^j) can be obtained by intersecting the $\delta(\mu)$ sets $A(\nu_i|\mu)$ and $A(\Gamma_\mu^j)$.*

As said above, the admissible set $A(\mu)$ can be easily obtained as the union of the admissible sets $A^j(\mu)$ computed for any embedding Γ_μ^j of $sk(\mu)$.

The Series Case. In the series case, $sk(\mu)$ has exactly one embedding, and such an embedding has no internal face. Hence, in order to compute $\hat{A}(\mu)$, it is not necessary to compute $A(\Gamma_\mu^j)$ and it is sufficient, by Lemma 2, to intersect the gists $\hat{A}(\nu_i|\mu)$ of the admissible sets of the components ν_i nested into μ.

Given an S-node μ and one of its children ν, we suitably build a set S in $O(n(\nu))$ time, starting from $\hat{A}(\nu)$ and $t(\mu)$, that is the gist $\hat{A}(\nu|\mu)$ of the admissible set of ν nested into μ. First, initialize S with $\hat{A}(\nu)$. Observe that $\hat{A}(\nu)$ contains the two pairs $p_{first} = \langle 0, y_{max}\rangle$ and $p_{last} = \langle x_{max}, 0\rangle$, with $y_{max} = x_{max}$. For each pair $p^k = \langle x_k, y_k\rangle$ of $\hat{A}(\nu)$ define $p_{first}^k = \langle 0, max(y_k, x_k + t(\mu))\rangle$. Denote by $\overline{p}_{first}$ the p_{first}^k with minimum y and by $\overline{p}_{last}$ the pair obtained from $\overline{p}_{first}$ swapping the two elements x and y. If $\overline{p}_{first} \preceq p_{first}$ insert $\overline{p}_{first}$ into S and remove from S any pair p^* such that $\overline{p}_{first} \preceq p^*$. If $\overline{p}_{last} \preceq p_{last}$ append $\overline{p}_{last}$ to S and remove from S any pair p^* such that $\overline{p}_{last} \preceq p^*$.

Let μ be an S-node with children ν_i, for $i = 1, \ldots, \delta(\mu)$ and let $n(\nu_i)$ be the number of vertices of ν_i. Since $t(\mu) = \min_i(t(\nu_i))$, for $i = 1, \ldots, \delta(\mu)$, we have:

Lemma 3. *The thickness $t(\mu)$ can be computed in $O(\delta(\mu))$ time.*

Lemma 4. *Starting from $\hat{A}(\nu_i)$ and $t(\nu_i)$, for $i = 1, \ldots, \delta(\mu)$, the gist $\hat{A}(\mu)$ can be computed in time $O(\sum_{i=1}^{\delta(\mu)} n(\nu_i))$.*

The Rigid Case. In the rigid case, since $sk(\mu)$ is a 3-connected component, it admits exactly two embeddings, Γ_μ^1 and Γ_μ^2, which only differ for a flipping around its poles. Hence, it is possible to consider one of the two embeddings only, say Γ_μ^1, compute the admissible set $A^1(\mu)$ of μ restricted to Γ_μ^1, and obtain the admissible set $A^2(\mu)$ of μ restricted to Γ_μ^2 by swapping, for every pair of $A^1(\mu)$, elements x and y. The gist $\hat{A}(\mu)$ is given by the union of the two sets. By Lemma 2, $A^1(\mu)$ can be obtained by intersecting the gist $\hat{A}(\Gamma_\mu^1)$ of the admissible set of $sk(\mu)$ and the gists $\hat{A}(\nu_i|\mu)$ of the admissible sets of the components ν_i nested into μ, both computed on Γ_μ^1.

In order to compute $\hat{A}(\nu_i|\mu)$ and $\hat{A}(\Gamma_\mu^1)$ it is useful to label each face f of Γ_μ^1 with its depths $d_l(f)$ and $d_r(f)$. This can be done in linear time performing a single-source shortest path from the two external faces f_l^μ and f_r^μ.

Given an R-node μ and one of its children ν, we suitably build a set S in $O(n(\nu))$ time, starting from $\hat{A}(\nu)$, $t(\mu)$, and the depths $d_l(f)$ and $d_r(f)$ of each face f of Γ_μ^1, that is the gist $\hat{A}(\nu|\mu)$ of the admissible set of ν nested into μ. First, build a set S' containing a pair $\langle x_k + d_l(f_l^\nu), y_k + d_r(f_r^\nu)\rangle$ for each pair $\langle x_k, y_k\rangle \in \hat{A}(\nu)$, and a set S'' containing a pair $\langle y_k + d_l(f_r^\nu), x_k + d_r(f_l^\nu)\rangle$ for each pair $\langle x_k, y_k\rangle \in \hat{A}(\nu)$. Initialize $S = S' \cup S''$. Observe that S contains the two pairs $p_{first} = \langle 0, y_{max}\rangle$ and $p_{last} = \langle x_{max}, 0\rangle$. For each pair $p^k = \langle x_k, y_k\rangle$ of $\hat{A}(\nu)$ define $p^k_{first} = \langle 0, max(y_k, x_k + t(\mu))\rangle$. Denote by $\overline{p}_{first}$ the p^k_{first} with minimum y and by $\overline{p}_{last}$ the pair obtained from $\overline{p}_{first}$ swapping the two elements x and y. If $\overline{p}_{first} \preceq p_{first}$ insert $\overline{p}_{first}$ into S and remove from S any pair p^* such that $\overline{p}_{first} \preceq p^*$. If $\overline{p}_{last} \preceq p_{last}$ append $\overline{p}_{last}$ to S and remove from S any pair p^* such that $\overline{p}_{last} \preceq p^*$.

Given an R-node μ and the values of the depths $d_l(f)$ and $d_r(f)$ of each face $f \in \Gamma_\mu^j$, we build the gist $\hat{A}(\Gamma_\mu^j)$ of the admissible set of $sk(\mu)$ in $O(\delta(\mu))$ time. First, initialize $\hat{A}(\Gamma_\mu^j) = \emptyset$. Then, for increasing values of x, consider the partition of the internal faces of Γ_μ^j into F_l and F_r such that all faces at distance less or equal than x from f_l^μ are into F_l and the other ones are into F_r. Then, a pair $\langle x, y\rangle$ such that all faces in F_r are at distance less or equal than y from f_r^μ is created and inserted into $\hat{A}(\Gamma_\mu^j)$ if it is incomparable with the last pair inserted.

Let μ be an R-node with children ν_i, for $i = 1, \ldots, \delta(\mu)$ and let $n(\nu_i)$ be the number of vertices of ν_i. By performing a shortest-path on the (u, v)-dual of G_μ between the external faces f_l^μ and f_r^μ, we have:

Lemma 5. *The thickness $t(\mu)$ can be computed in $O(\delta(\mu))$ time.*

Lemma 6. *Starting from $\hat{A}(\nu_i)$ and $t(\nu_i)$, for $i = 1, \ldots, \delta(\mu)$, the gist $\hat{A}(\mu)$ can be computed in time $O(\sum_{j=1}^{\delta(\mu)} \sum_{i=1}^{j} n(\nu_i))$.*

The Parallel Case. In the parallel case $sk(\mu)$ is composed of two vertices, u and v, with $\delta(\mu)$ parallel edges between them and admits a factorial number of embeddings, corresponding to all the possible permutations of its $\delta(\mu)$ edges. Hence, according to Lemma 2, the gist $\hat{A}(\mu)$ can be obtained by performing the union between $\delta(\mu)!$ sets, where each set $A^j(\mu)$ corresponds to a different embedding Γ_μ^j of $sk(\mu)$. Also, $A^j(\mu)$ can be computed by intersecting $\hat{A}(\Gamma_\mu^j)$ and the gists $\hat{A}(\nu_i|\mu)$ of the admissible sets of the components ν_i nested into μ. Hence, a naïve computation of $\hat{A}(\mu)$ employs a factorial number of steps. We reduce the number of permutations to be analyzed by exploiting the following properties and considerations.

Let μ be a P-node with children ν_i, for $i = 1, \ldots, \delta(\mu)$. First, consider a pair $\langle x, y\rangle \in \hat{A}(\mu)$ and an embedding Γ_{G_μ} of G_μ satisfying $\langle x, y\rangle$, i.e., whose internal faces can be partitioned into two sets F_l and F_r such that all faces in F_l (F_r) have distance from f_l^μ (f_r^μ) less or equal than x (y).

Lemma 7. *Let Γ_{G_μ} be an embedding of G_μ satisfying pair $\langle x, y\rangle \in \hat{A}(\mu)$. There exists a partition F_l' and F_r' such that: (i) all faces in F_l' have distance from f_l^μ less or equal than x, (ii) all faces in F_r' have distance from f_r^μ less or equal than y, (iii) there exists at most one component ν_c whose internal faces belong to both F_l' and F_r', and (iv) each child component at the left(right) of ν_c has its internal faces in $F_l'(F_r')$.*

The unique component ν_c, if any, whose faces belong to both F_l' and F_r' is called the *center* of the permutation. Intuitively, Lemma 7 states that we could restrict to consider those partitions F_l and F_r of the internal faces of Γ_{G_μ} such that each child component different from ν_c has its internal faces into the same set F_l or F_r. In other words, for each child component ν_i different from ν_c, $\hat{A}(\nu_i)$ can be assumed to contain the two pairs $\langle x, 0\rangle$ and $\langle 0, y\rangle$ only.

The gist $\hat{A}(\mu)$ can be computed by choosing, one by one, each child component ν_c as the center of the permutation and inserting the other components either to the left or the right of ν_c until a complete permutation is obtained. Each subsequence σ of components is associated with the gist $\hat{A}(\sigma)$ of its admissible set $A(\sigma)$, which is properly updated when a component is inserted. This approach would obtain the same permutation $\delta(\mu)$ times, exploring $O(\delta(\mu) \cdot \delta(\mu)!)$ sequences. Hence, at first glance, the computational complexity is augmented. However, we show in the following that focusing on ν_c can greatly help to reduce the number of permutations to be considered.

Lemma 8. *Let σ be a sequence of child components of μ and let $\nu_i \notin \sigma$ be a child component with $\langle 0, y_i\rangle, \langle x_i, 0\rangle \in \hat{A}(\nu_i)$. Adding ν_i to the left(right) of σ we obtain a sequence $\sigma'(\sigma'')$. The set S' containing a pair $\langle \max(x + t(\nu_i), x_i), y\rangle$ for each pair $\langle x, y\rangle \in \hat{A}(\sigma)$ is such that $S' \preceq A(\sigma')$ and $S' \subset A(\sigma')$. Analogously, the set S'' containing a pair $\langle x, \max(y + t(\nu_i), y_i)\rangle$ for each pair $\langle x, y\rangle \in \hat{A}(\sigma)$ is such that $S'' \preceq A(\sigma'')$ and $S'' \subset A(\sigma'')$.*

Let ν_i be a component. We introduce function $l(\nu_i) = t(\nu_i) - x_i$, where x_i is such that pair $\langle x_i, 0\rangle \in \hat{A}(\nu_i)$.

Lemma 9. *Let ν_c, ν', and ν'' be three child components of μ, and let Γ_μ^1 and Γ_μ^2 be two embeddings of $sk(\mu)$ corresponding to two permutations of its components which only differ for the swapping of two components ν' and ν'' lying on the same side of ν_c. If $l(\nu') < l(\nu'')$, then $A^2(\mu) \preceq A^1(\mu)$.*

Intuitively, any permutation with a component ν' further from ν_c than a second component ν'' with $l(\nu'') < l(\nu')$ can be ignored since its admissible set is preceded by that computed with another permutation. Therefore, the number of analyzed permutations can be reduced by ordering the components by decreasing values of $l(\nu)$ and, once the center ν_c of the permutation has been chosen, by adding the other components, ordered wrt l, either to its left or to its right. To do so, we build a rooted tree $T(\nu_c)$ of height $\delta(\mu)+1$. Each node p at distance d from the root is a pair $\langle x_p, y_p\rangle$ of non-negative integers and is associated with a sequence σ_p of d child components of μ such that $\langle x_p, y_p\rangle \in \hat{A}(\sigma_p)$. The nodes at distance d from the root are the incomparable pairs of integers satisfied by some sequence of length d. Hence, the set of nodes at distance $\delta(\mu)$ from the root is $\hat{A}(\mu)$ restricted to the permutations having ν_c as the center.

Tree $T(\nu_c)$ is built as follows. The root is pair $\langle 0, 0\rangle$ and corresponds to the empty sequence. The root is added as many children as many pairs in the gist $\hat{A}(\nu_c)$ of ν_c, each one associated with the sequence composed by ν_c only. The following levels are obtained by considering one by one the other components in decreasing order of l. When the k-th component ν_k is processed, each node p at depth $k-1$ is added two children p_l and p_r, corresponding to the sequences $\nu_k \cdot \sigma_p$ and $\sigma_p \cdot \nu_k$, respectively. Pairs p_l and p_r are computed, starting from p, with the function presented in Lemma 8. From the set of pairs introduced at level k all those preceded by a pair of the same level can be removed, pruning the tree. The gist $\hat{A}(\mu)$ can be obtained as the union of the gists $\hat{A}_{\nu_c}(\mu)$, for each child component ν_i chosen as the center of the permutation ν_c.

To efficiently build $T(\nu_c)$, we keep the nodes at each level ordered wrt the $\preceq_x$ relationship. When adding the k-th level, we produce the set $P_l(P_r)$ of nodes obtained by concatenating ν_k to the left(right) of σ_p, for each p at distance $k-1$ from the root. Since the nodes of the $(k-1)$-th level are ordered wrt the $\preceq_x$ relationship, and due to the formula used to compute nodes in $P_l(P_r)$, such a set can be kept ordered wrt the $\preceq_x$ relationship and succinct by comparing the pair to be inserted with the last pair only. The set of nodes of the k-th level, ordered wrt the $\preceq_x$ relationship and succinct, is the union of P_l and P_r.

Since $t(\mu) = \sum_{i=1}^{\delta(\mu)} t(\nu_i)$, we have:

Lemma 10. *The thickness $t(\mu)$ can be computed in $O(\delta(\mu))$ time.*

Lemma 11. *Starting from $\hat{A}(\nu_i)$ and $t(\nu_i)$, the gist $\hat{A}_{\nu_c}(\mu)$ of the admissible set restricted to all the permutations having ν_c as center can be computed in time $O(\sum_{k=1}^{\delta(\mu)} \sum_{i=1}^{k} n(\nu_i))$.*

Lemma 12. *Starting from $\hat{A}(\nu_i)$ and $t(\nu_i)$, for $i = 1, \dots, \delta(\mu)$, the gist $\hat{A}(\mu)$ can be computed in time $O(\delta(\mu) \cdot \sum_{k=1}^{\delta(\mu)} \sum_{i=1}^{k} n(\nu_i))$.*

4.2 Computing the Minimum Depth and the Minimum-Depth Embedding

At the end of the bottom-up traversal of $\mathcal{T}$, the value of the minimum depth can be computed starting from the gist of the admissible set of the component μ that is the child of the root e of $\mathcal{T}$. Namely, for each pair $\langle x_h, y_h \rangle \in \hat{A}(\mu)$, denote by m_h the maximum between x_h and y_h. The minimum depth is the minimum of the m_h.

Lemma 13. *Let μ be the child of the root of the SPQR-tree of an n-vertex graph G. Starting from $\hat{A}(\mu)$, the minimum depth of G can be computed in $O(n)$ time.*

Theorem 1. *Let G be an n-vertex biconnected planar graph and let $\mathcal{T}$ be the SPQR-tree of G rooted at e. The minimum depth of an embedding of G with e on the external face can be computed in $O(n^3)$ time and $O(n^2)$ space.*

Proof: Consider the three sets S, R and P containing the series, rigid and parallel nodes of $\mathcal{T}$, respectively. For each series component $\mu_s \in S$, by Lemmas 3 and 4, $t(\mu_s)$ and $\hat{A}(\mu_s)$ can be computed in $O(\sum_{i=1}^{\delta(\mu_s)} n(\nu_i))$ time. Hence, the overall complexity for all the series nodes is $O(\sum_{\mu_s \in S} \sum_{i=1}^{\delta(\mu_s)} n(\nu_i))$. Since the number of the series nodes is $O(n)$, the above sum is $O(n^2)$.

For each rigid component $\mu_r \in R$, by Lemmas 5 and 6, $t(\mu_r)$ and $\hat{A}(\mu_r)$ can be computed in $O(\sum_{j=1}^{\delta(\mu_r)} \sum_{i=1}^{j} n(\nu_i))$. Hence, the overall complexity for all the rigid nodes is $O(\overbrace{\sum_{\mu_r \in R} \sum_{j=1}^{\delta(\mu_r)}}^{O(n)} \overbrace{\sum_{i=1}^{j} n(\nu_i)}^{O(n)})$, which is $O(n^2)$, since the total number of children of all the rigid nodes is $O(n)$.

For each parallel component $\mu_p \in P$, by Lemmas 10 and 12, $t(\mu_p)$ and $\hat{A}(\mu_p)$ can be computed in $O(\delta(\mu_p) \sum_{k=1}^{\delta(\mu_p)} \sum_{i=1}^{k} n(\nu_i))$ time. Hence, the overall complexity for all the parallel nodes is $O(\overbrace{\sum_{\mu_p \in P} \delta(\mu_p) \sum_{k=1}^{\delta(\mu_p)}}^{O(n^2)} \overbrace{\sum_{i=1}^{k} n(\nu_i)}^{O(n)})$, which is $O(n^3)$, since the total number of children of all the parallel nodes is $O(n)$. The time complexity of the bottom-up traversal is $O(n^2) + O(n^2) + O(n^3) = O(n^3)$. Starting from the gist of the admissible set of the root, the minimum depth is computed, by Lemma 13, in $O(n)$ time.

The space bound can be obtained by considering that there are $O(n)$ components in $\mathcal{T}$ and that the size of the gists of their admissible sets is $O(n)$. □

To produce a minimum-depth embedding of G with an edge e on the external face we need some additional information to be added to each component during the bottom-up traversal of $\mathcal{T}$, meant to describe how the components must be attached together in order to obtain an embedding satisfying each pair of the gist of the admissible set.

Namely, for each node μ and for each pair $p \in \hat{A}(\mu)$ we attach an "embedding descriptor" composed of (i) a Boolean variable b_μ specifying whether μ must be attached to its parent component ν with $f_l(\mu)$ corresponding to $f_l(\nu)$ or not, and (ii) an integer pair p_i for each child component ν_i of μ specifying how ν_i must be, in its turn, embedded in order to obtain an embedding of μ satisfying p. In addition, if μ is a parallel component, we record the needed ordering of its child components ν_i.

The minimum-depth embedding is computed with a top-down traversal of the SPQR-tree $\mathcal{T}$ rooted at e, using the above described additional structures, by replacing each virtual edge with the skeleton of the corresponding component.

Theorem 2. *Let G be an n-vertex biconnected planar graph. The minimum-depth embedding of G can be computed in $O(n^4)$ time and $O(n^3)$ space.*

Proof: For each edge e of G, compute the SPQR-tree rooted at e in $O(n)$ time and the minimum-depth embedding with e on the external face in $O(n^3)$ time. The cubic space bound is due to the fact that, for each component and for each integer pair of the gist of its admissible set, it must be recorded an integer pair for each child and, eventually, an ordering of the children. □

5 Extension to General Planar Graphs

The minimum-depth embedding of a simply-connected planar graph G, described by its BC-tree, can be found with an approach similar to that used in [4]. The key point of such an approach is that the algorithm to compute a minimum-depth embedding of a biconnected graph with a specified edge on the external face can be suitably modified in order to be applied to each block μ_i, taking into account the depth of the blocks that are attached to the cutvertices of μ_i, and maintaining the $O(n_i^3)$ time complexity, where n_i is the number of vertices of μ_i (see [1]). Each child block μ_j, sharing the cutvertex v_j with μ_i, will be embedded with v_j on its external face. Hence, we apply the modified algorithm using as reference edge each one of the edges incident to v_j and choose the embedding with minimum depth.

We start by choosing a root block for the BC-tree, and a reference edge inside such a block. We traverse bottom-up the BC-tree applying the modified algorithm. This computation has to be performed for each edge of each block chosen as the root block. The overall $O(n^4)$ complexity can be obtained by considering that the modified algorithm has to be launched at most three times for each edge of G. Namely, we launch the algorithm on each edge e of G when such an edge is chosen to be on the external face, taking into account the depths of all the attached blocks, and we launch the algorithm on each edge e incident to a cutvertex v (hence, at most two times for each e) taking into account the depths of all the attached blocks with the exception of those attached to v. Therefore, the following theorem follows.

Theorem 3. *Let G be an n-vertex connected planar graph. The minimum-depth embedding of G can be computed in $O(n^4)$ time and $O(n^3)$ space.*

6 Conclusions

We presented an $O(n^3)$-time algorithm for computing a minimum-depth embedding of a biconnected planar graph with a given edge on the external face. Then, we exploited such a result to solve the problem on a general planar graph in $O(n^4)$ time.

Since our approach is inspired by that in [4], it is useful to stress the similarities ad the differences between the two contributions. We take from [4] the fundamental idea of decomposing the graph into components and to separately consider each component. Also, the concept of *thickness* is the same as in [4]. In both papers there is the idea of equipping each component with pairs of integers, representing their distance from the external face. However, in [4] a pair represents the result of a "probe" that says that a certain component is feasible with that depth. In our case a set of pairs represents implicitly all the admissible values of depth of the component. The combinatorial structure of such pairs and their nice computational properties are a key ingredient of our paper. The techniques for combining the components used in the two papers are similar. However, in the critical problem of dealing with parallel compositions we develop an approach that has many new features.

The natural problem that remains open is to fill the gap from our $O(n^4)$ time to the linear time obtained in [9] for a simplified version of the problem.

References

1. Angelini, P., Di Battista, G., Patrignani, M.: Computing a minimum-depth planar graph embedding in $O(n^4)$ time. Tech. Rep. RT-DIA-116-2007, Dept. of Comp. Sci., Univ. Roma Tre (2007) `http://web.dia.uniroma3.it/ricerca/rapporti/rapporti.php`
2. Baker, B.S.: Approximation algorithms for NP-complete problems on planar graphs. J. of the Ass. for Comp. Mach. 41, 153–180 (1994)
3. Bienstock, D., Monma, C.L.: On the complexity of covering vertices by faces in a planar graph. SIAM-J. on Comp. 17, 53–76 (1988)
4. Bienstock, D., Monma, C.L.: On the complexity of embedding planar graphs to minimize certain distance measures. Algorithmica 5(1), 93–109 (1990)
5. Di Battista, G., Tamassia, R.: On-line planarity testing. SIAM J.C. 25(5), 956–997 (1996)
6. Di Giacomo, E., Didimo, W., Liotta, G., Meijer, H.: Computing radial drawings on the minimum number of circles. In: Pach, J. (ed.) GD 2004. LNCS, vol. 3383, pp. 250–261. Springer, Heidelberg (2005)

7. Dolev, D., Leighton, F.T., Trickey, H.: Planar embedding of planar graphs. Adv. in Comp. Res. 2 (1984)
8. Pizzonia, M.: Minimum depth graph embeddings and quality of the drawings: An experimental analysis. In: Healy, P., Nikolov, N.S. (eds.) GD 2005. LNCS, vol. 3843, pp. 397–408. Springer, Heidelberg (2006)
9. Pizzonia, M., Tamassia, R.: Minimum depth graph embedding. In: Paterson, M.S. (ed.) ESA 2000. LNCS, vol. 1879, pp. 356–357. Springer, Heidelberg (2000)
10. Robertson, N., Seymour, P.D.: Graph minors. III. Planar tree-width. J. Comb. Theory, Ser. B 36(1), 49–64 (1984)

On a Family of Strong Geometric Spanners That Admit Local Routing Strategies*

Prosenjit Bose, Paz Carmi, Mathieu Couture, Michiel Smid, and Daming Xu

School of Computer Science, Carleton University, Herzberg Building
1125 Colonel By Drive, Ottawa, Ontario, K1S 5B6 Canada
{jit,paz,mathieu,michiel,daming}@cg.scs.carleton.ca
http://www.scs.carleton.ca

Abstract. We introduce a family of directed geometric graphs, denoted G_λ^θ, that depend on two parameters λ and θ. For $0 \leq \theta < \frac{\pi}{2}$ and $\frac{1}{2} < \lambda < 1$, the G_λ^θ graph is a strong t-spanner, with $t = \frac{1}{(1-\lambda)\cos\theta}$. The out-degree of a node in the G_λ^θ graph is at most $\lfloor 2\pi/\min(\theta, \arccos\frac{1}{2\lambda})\rfloor$. Moreover, we show that routing can be achieved locally on G_λ^θ. Next, we show that all strong t-spanners are also t-spanners of the unit disk graph. Simulations for various values of the parameters λ and θ indicate that for random point sets, the spanning ratio of G_λ^θ is better than the proven theoretical bounds.

1 Introduction

A graph G whose vertices are points in the plane and edges are segments weighted by their length is a *geometric graph*. A geometric graph G is a t-spanner (for $t \geq 1$) when the weight of the shortest path in G between any pair of points a, b does not exceed $t \cdot |ab|$ where $|ab|$ is the Euclidean distance between a and b. Any path from a to b in G whose length does not exceed $t \cdot |ab|$ is a t-spanning path. The smallest constant t having this property is the *spanning ratio* of the graph. A t-spanning path from a to b is *strong* if the length of every edge in the path is at most $|ab|$. The graph G is a *strong* t-spanner if there is a strong t-spanning path between every pair of vertices.

The spanning properties of various geometric graphs have been studied extensively in the literature (see the book by Narasimhan and Smid [7] for a comprehensive survey on the topic). We are particularly interested in spanners that are defined by some proximity measure or emptiness criterion (see for example Bose *et al.* [2]). Our work was initiated by Chavez *et al.* [5] who introduced a new geometric graph called Half-Space Proximal (HSP). Given a set of points in the plane, HSP is defined as follows. There is an edge oriented from a point p to a point q provided there is no point r in the set that satisfies the following conditions: (a) $|pr| < |pq|$, (b) there is an edge from p to r and (c) q is closer to r than to p.

* Research supported in part by NSERC, MITACS, MRI, and HPCVL.

F. Dehne, J.-R. Sack, and N. Zeh (Eds.): WADS 2007, LNCS 4619, pp. 300–311, 2007.

The authors show that this graph has maximum out-degree[1] at most 6. The authors also claim that HSP has a spanning ratio[2] of $2\pi+1$ and that this bound is tight[3]. Unfortunately, we found statements made in their proofs for the latter two claims to be erroneous or incomplete as we outline in Section 3. However, in reviewing their experimental results as well as running some of our own, although their proofs are incomplete, we felt that the claimed results might be correct. Our attempts at finding a correct proof to their claims was the starting point of this work. Although we have been unable to find a correct proof of their claims, we introduce a family of directed geometric graphs that approach HSP asymptotically and possess several other interesting characteristics outlined below.

In this paper, we define a family G_λ^θ of graphs. These are directed geometric graphs that depend on two parameters λ and θ. We show that each graph in this family has bounded out-degree and is a strong t-spanner, where both the out-degree and t depend on λ and θ. Furthermore, graphs in this family admit local routing algorithms that find strong t-spanning paths. Finally, we show that all strong t-spanners are also spanners of the unit-disk graph, which are often used to model adhoc wireless networks.

The remainder of this paper is organized as follows. In Section 2, we introduce the G_λ^θ graph and prove its main theoretical properties. In Section 3, we outline the inconsistencies within statements of the proof of the upper and lower bounds on the stretch factor of HSP given in Chavez *et al.* [5]. In Section 4, we show that by intersecting the G_λ^θ graph with the unit disk graph, we obtain a spanner of the unit disk graph. In Section 5, we present some simulation results about the G_λ^θ graph.

2 The Family G_λ^θ of Graphs

In this section, we define the G_λ^θ graph and prove that it is a strong t-spanner of bounded out-degree. We first introduce some notation. Let P be a set of points in the plane, $0 \le \theta < \frac{\pi}{2}$ and $\frac{1}{2} < \lambda < 1$.

Definition 1. *The* θ-cone(p,r) *is the cone of angle* 2θ *with apex* p *and having the line through* p *and* r *as bisector.*

Definition 2. *The* λ-half-plane(p,r) *is the half-plane containing* r *and having as boundary the line perpendicular to* $\overline{pr}$ *and intersecting* $\overline{pr}$ *at distance* $\frac{1}{2\lambda}|pr|$ *from* p.

Definition 3. *The* destruction region *of* r *with respect to* p*, denoted* $K(p,r)$*, is the intersection of the* θ*-cone*(p,r) *and the* λ*-half-plane*(p,r) *(see Figure 1(a)).*

The directed graph $G_\lambda^\theta(P)$ is obtained by the following algorithm. For every point $p \in P$, do the following:

[1] Theorem 1 in [5].
[2] Theorem 2 in [5].
[3] Construction in Figure 2 in [5].

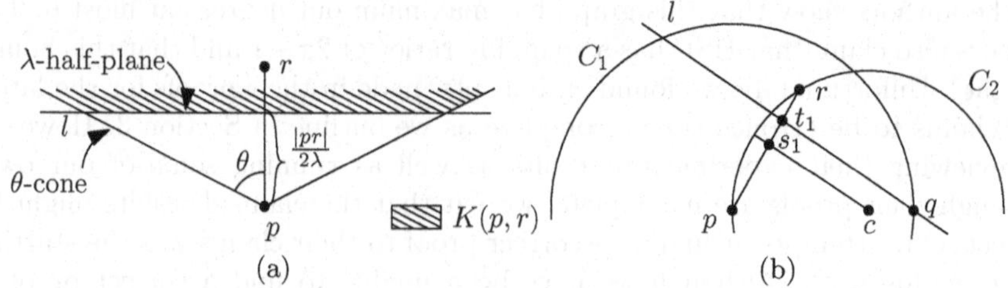

Fig. 1. (a) The destruction region of r with respect to p, and (b) the location of a point r destroying (p, q)

1. Let $N(p)$ be the set $P \setminus \{p\}$.
2. Let r be the point in $N(p)$ which is closest to p.
3. Add the directed edge (p, r) to $G_\lambda^\theta(P)$.
4. Remove all $q \in K(p, r)$ from $N(p)$ (i.e., $N(p) \leftarrow N(p) \setminus K(p, r)$).
5. If $N(p)$ is not empty go to 2.

Note that the graph computed by this algorithm has the following property:

Lemma 1. *There is an edge $(p, q) \in G_\lambda^\theta(P)$ iff there is no point $r \in P$, such that $|pr| \leq |pq|$, (p, r) is an edge of $G_\lambda^\theta(P)$ and $q \in K(p, r)$, (ties on the distances are broken arbitrarily). Such a point r is said to be a* destroyer *of the edge (p, q).*

2.1 Location of Destroyers

What prevents the directed pair (p, q) from being an edge in G_λ^θ? It is the existence of one point acting as a destroyer. Given two points p, q, where can a point lie such that it acts as the destroyer of the edge (p, q)? In this subsection, we describe the region containing the points r such that $q \in K(p, r)$. This region is denoted $\overline{K}(p, q)$. In other words, $\overline{K}(p, q)$ is the description of all the locations of possible destroyers of an edge (p, q).

Proposition 1. *Let $R(p, q, \lambda)$ be the intersection of the disks C_1 centered at p with radius $|pq|$ and C_2 centered at $c = p + \lambda(q - p)$ with radius $\lambda|pq|$. If $q \in K(p, r)$ and $|pr| \leq |pq|$, then $r \in R(p, q, \lambda)$.*

Proof: If r destroyed (p, q), then $|pr| \leq |pq|$. Therefore, r is in C_1. To complete the proof, we need to show that r is in C_2. We begin by considering the case when q lies on the line l which is the boundary of λ-half-plane(p, r) (see Figure 1(b)). Let s_1 be the midpoint of $\overline{pr}$, t_1 the intersection of l with $\overline{pr}$ and c' the intersection of $\overline{pq}$ with the bisector of $\overline{pr}$. Since the triangles Δpt_1q and $\Delta ps_1c'$ are similar, this implies that $|pc'| = |pq|\frac{|ps_1|}{|pt_1|} = |pq|\frac{|pr|}{2|pt_1|} = |pq|\frac{2\lambda|pr|}{2|pr|} = \lambda|pq| = |pc|$. Therefore, $c' = c$, which implies that $|cr| = |cp|$ thereby proving that r is on the boundary of C_2. In the case when q is not on l, then we have $|pc'| < |pc|$ and r lies on a circle centered at c' going through p. Therefore, r is contained in C_2, which completes the proof. □

From the definition of $\overline{K}(p,q)$ and Proposition 1, we have:

Proposition 2. *Let $\overline{K}(p,q)$ be the intersection of $R(p,q,\lambda)$ with the θ-cone(p,q). If $q \in K(p,r)$ and $|pr| \le |pq|$, then $r \in \overline{K}(p,q)$.*

2.2 The Spanning Ratio of G_λ^θ

Theorem 1. *For $0 \le \theta < \frac{\pi}{2}$ and $\frac{1}{2} < \lambda < 1$, the G_λ^θ graph is a strong t-spanner, with $t = \frac{1}{(1-\lambda)\cos\theta}$.*

Proof: Let P be a set of points in the plane, $p, q \in P$ and $d_G(p,q)$ be the length of the shortest path from p to q in $G_\lambda^\theta(P)$. By induction on the rank of $|pq|$, we show $d_G(p,q) \le t|pq|$.

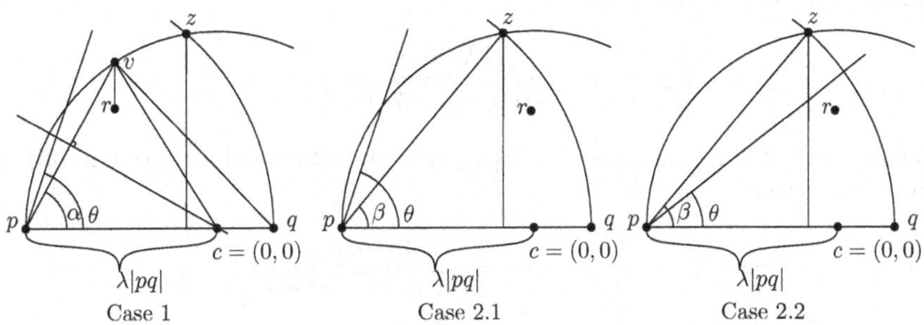

Fig. 2. Cases for the proof of Theorem 1

Base case: If p and q form a closest pair, then the edge (p,q) is in $G_\lambda^\theta(P)$ by definition. Therefore, $d_G(p,q) = |pq| \le t|pq|$.

Inductive case: If the edge (p,q) is in $G_\lambda^\theta(P)$, then $d_G(p,q) = |pq| \le t|pq|$ as required. We now address the case when (p,q) is *not* in $G_\lambda^\theta(P)$. By Proposition 2, there must be a point $r \in \overline{K}(p,q)$ with $|pr| < |pq|$ that is destroying (p,q) and such that the edge (p,r) is in $G_\lambda^\theta(P)$. Since $r \in \overline{K}(p,q)$ and $|pr| < |pq|$, we have that $|rq| < |pq|$. By the inductive hypothesis, we have $d_G(r,q) \le t|rq|$.

Let z be the intersection of the boundaries of the disks C_1 and C_2 defined in Proposition 1. We assume, w.l.o.g., that c is the origin and that points p, q are on the x-axis with p to the left of q as depicted in Figure 2. The remainder of the proof addresses two cases, depending on whether or not $r_x \le z_x$ (the notation p_x denotes the x-coordinate of a point p).

Case 1: $r_x \le z_x$. Let $v \in \overline{K}(p,q)$ be the point with the same x-coordinate as r and having the greatest y-coordinate. In other words, v is the highest point in $\overline{K}(p,q)$ that is strictly above r. We have:

$$d_G(p,q) \le |pr| + d_G(r,q) \le |pr| + t|rq| \le |pv| + t|vq|.$$

Now, let $\alpha = \angle vpq \leq \theta$ We express $|pv|$ and $|vq|$ as a function of $\cos\alpha$. Consider the triangle $\Delta(pvc)$ and note that $|vc| = |pc|$ by construction. Since $\cos\alpha = |pv|/2\lambda|pq|$, we have:

$$|pv| = 2\lambda|pq|\cos\alpha.$$

From the law of cosines, we see:

$$|vq|^2 = |pv|^2 + |pq|^2 - 2|pv||pq|\cos\alpha = |pq|^2(4\lambda^2\cos^2\alpha - 4\lambda\cos^2\alpha + 1)$$

which implies that:

$$\begin{aligned} d_G(p,q) &\leq 2\lambda|pq|\cos\alpha + t|pq|\sqrt{4\lambda^2\cos^2\alpha - 4\lambda\cos^2\alpha + 1} \\ &= |pq|(2\lambda\cos\alpha + t\sqrt{4\lambda^2\cos^2\alpha - 4\lambda\cos^2\alpha + 1}). \end{aligned}$$

Therefore, we have to show that

$$t \geq \frac{2\lambda\cos\alpha}{1 - \sqrt{4\lambda^2\cos^2\alpha - 4\lambda\cos^2\alpha + 1}}.$$

Since $\alpha \leq \theta < \pi/2$ implies $\cos\theta \leq \cos\alpha$, by straightforward algebraic manipulation we have that:

$$\frac{1}{(1-\lambda)\cos\alpha} \geq \frac{2\lambda\cos\alpha}{1 - \sqrt{4\lambda^2\cos^2\alpha - 4\lambda\cos^2\alpha + 1}}.$$

Case 2: $r_x > z_x$**.** Let $\beta = \angle zpq$. We first compute the value of $\cos\beta$. From the definition of C_1 and C_2, we have $z_x^2 + z_y^2 = \lambda^2|pq|^2$ and $(z_x - p_x)^2 + z_y^2 = |pq|^2$. Therefore, since $p_x = -\lambda|pq|$, we have $z_x = \frac{|pq|(1-2\lambda^2)}{2\lambda}$ which implies

$$\cos\beta = \frac{\lambda|pq| + z_x}{|pq|} = \lambda + \frac{1-2\lambda^2}{2\lambda} = \frac{1}{2\lambda}.$$

We need to consider two subcases, depending on whether or not $\beta \leq \theta$.

Case 2.1: $\beta \leq \theta$**.** In this case, we have:

$$d_G(p,q) \leq |pr| + d_G(r,q) \leq |pr| + t|rq| \leq |pz| + t|zq| = |pq| + t|zq|.$$

By the law of cosines, we have

$$|zq|^2 = |pq|^2(2 - \frac{1}{\lambda})$$

which implies

$$d_G(p,q) \leq |pq| + t|pq|\sqrt{2 - \frac{1}{\lambda}}.$$

Therefore, we have to show that

$$t \geq \frac{1}{1 - \sqrt{2 - \frac{1}{\lambda}}}.$$

Since $\beta \leq \theta$, we have $\cos\beta \geq \cos\theta$, and then

$$t = \frac{1}{(1-\lambda)\cos\theta} \geq \frac{1}{(1-\lambda)\cos\beta} = \frac{1}{(1-\lambda)(1/2\lambda)} = \frac{2\lambda}{1-\lambda} \geq \frac{1}{1 - \sqrt{2 - \frac{1}{\lambda}}}$$

where the last inequality holds because it is equivalent to $(1-\lambda)^2 \geq 0$.

Case 2.2: $\beta > \theta$**.** By the law of cosines we have

$$d_G(p,q) \leq |pq| + t|pq|\sqrt{2 - 2\cos\theta}$$

which means that we have to show that

$$t \geq \frac{1}{1 - \sqrt{2 - 2\cos\theta}}.$$

But $\frac{1}{1-\sqrt{2-2\cos\theta}} \geq \frac{1}{(1-\lambda)\cos\theta}$ since $\beta > \theta$ implies $\cos\theta > \cos\beta = \frac{1}{2\lambda}$. This completes the last case of the induction step. Note that the resulting t-spanning paths found in this inductive proof are strong since both $|pr|$ and $|rq|$ are shorter than $|pq|$. □

The above proof provides a simple local routing algorithm. A routing algorithm is considered local provided that the only information used to make a decision is the 1-neighborhood of the current node as well as the location of the destination (see [4] for a detailed description of the model). The routing algorithm proceeds as follows. To find a path from p to q, if the edge (p,q) is in $G_\lambda^\theta(P)$, then take the edge. If the edge (p,q) is not in $G_\lambda^\theta(P)$, then take an edge (p,r) where r is a destroyer of the edge (p,q). Recall that r is a destroyer of the edge (p,q) if $r \in \overline{K}(p,q)$. This can be computed solely with the positions of p, q and r. Therefore, determining which of the neighbors of p in $G_\lambda^\theta(P)$ destroyed the edge (p,q) is a local computation.

Proposition 3. *The out-degree of a node in* G_λ^θ *is at most* $\lfloor \frac{2\pi}{\min(\theta, \arccos\frac{1}{2\lambda})} \rfloor$.

Proof: Let (p,r) and (p,s) be two edges of the G_λ^θ graph. W.l.o.g., $|pr| \leq |ps|$. Let l be the line perpendicular to $\overline{pr}$ through $p + \frac{1}{2\lambda}(r-p)$. Then either $\angle spr \geq \theta$ or s lies on the same side of l as p. In the latter case, the angle $\angle spr$ is at least $\arccos\frac{1}{2\lambda}$ (see Figure 3). The angle $\angle spr$ is then at least $\min(\theta, \arccos\frac{1}{2\lambda})$, which means that p has at most $\lfloor 2\pi / \min(\theta, \arccos\frac{1}{2\lambda}) \rfloor$ outgoing edges. □

Corollary 1. *If* $\theta \geq \pi/3$ *and* $\lambda > \frac{1}{2\cos(2\pi/7)}$, *then the out-degree of a node in the* G_λ^θ *graph is at most six.*

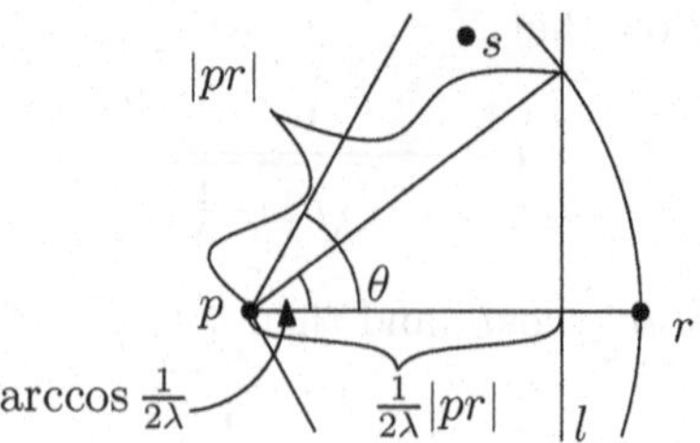

Fig. 3. The G_λ^θ graph has bounded out-degree

3 Half-Space Proximal

In this section, we outline the inconsistencies within statements of the proof of the upper and lower bounds on the stretch factor of HSP given in Chavez *et al.* [5].

In the proof of the upper bound (Theorem 2 of Chavez *et al.* [5]), claim 4 states that *all vertices* $u_0, u_1, u_2, \ldots, u_k$ *are either in clockwise or anticlockwise order around* v. The claim is that this situation exists when (u, v) is not an edge of HSP and no neighbor of u is adjacent to v. However, as stated, this is not true. A counter-example to this claim is shown in Figure 4. There is a unique path from u to v, namely uu_1u_2v, but this path is neither clockwise nor counter-clockwise around v. We believe that this situation may exist *in the worst case.* However, a characterization of the worst case situation must be given and it must be proven that the worst case situation has the claimed property.

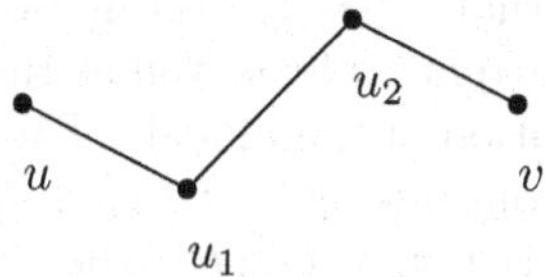

Fig. 4. Counter-example to the proof of Theorem 2 of [5]

For the lower bound, the authors also claim that the spanning ratio of HSP can be arbitrarily close to $2\pi+1$. However, the proof they provide to support that claim is a construction depicted in Figure 5 (reproduced from [5]). The claim is that the path from u to v can have length arbitrarily close to $(2\pi + 1)|uv|$. Although this may be true for the path that they highlight, this path is *not* the only path from u to v in HSP. The authors neglected the presence of the edge (u_1u_k) in their construction, which provides a shortcut that makes the distance between u and v much less than $2|uv|$.

One of the main reasons we believe the claims made in Chavez *et al.* [5] may be true is that in the simulations, all the graphs have small spanning ratio. In fact, the spanning ratio seems to be even smaller than $2\pi+1$. However, at this

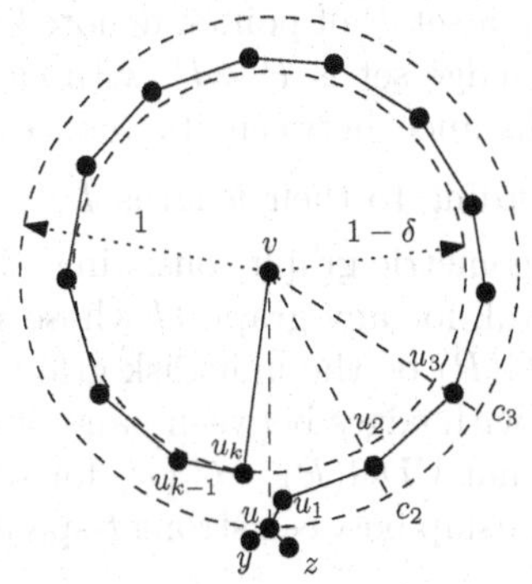

Fig. 5. The illustration of the lower bound on the spanning ratio of [5]

point, no proof that HSP is a constant spanner is known. We provide a lower bound of $3-\epsilon$ on the spanning ratio of HSP:

Proposition 4. *The HSP graph has stretch factor at least* $3-\epsilon$.

Proof: Consider the set of 6 point as in Figure 6, put $\delta = \epsilon/6$. The length of the path between p and q via a and b is equal to the length of the path between p and q via c and d. The length of both of these paths is $3-6\delta$. Since the shortest path between p and q in the HSP graph is one of the above paths, the stretch factor is $3-6\delta = 3-\epsilon$. □

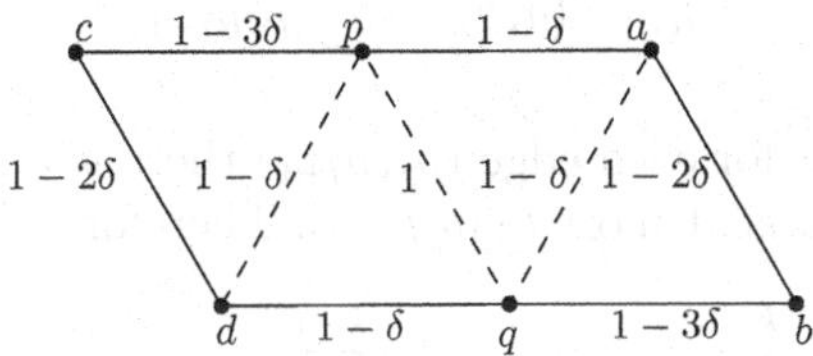

Fig. 6. Example of a 6 nodes HSP with a stretch factor of $3-\epsilon$. The solid edges are in HSP.

4 Unit Disk Graph Spanners

In Section 2, we showed that the G_λ^θ graph of a set of points in the plane is a strong t-spanner of the complete graph of these points, for a constant $t = \frac{1}{(1-\lambda)\cos\theta}$. We show in this section that strong t-spanners are also spanners of the unit disk graph. That is, the length of the shortest path between a pair of points in the graph resulting from the intersection of a strong t-spanner and a unit disk graph is not more than t times the length of the shortest path in the unit disk graph. Before proceeding, we need to introduce some notation.

For simplicity of exposition, we will assume that given a set P of points in the plane, no two pairs of points are at equal distance from each other. The *complete geometric graph* defined on a set P of points, denoted $C(P)$, is the graph whose vertex set is P and whose edge set is $P \times P$, with each edge having its weight equal to the Euclidean distance between its vertices. Let $e_1, \ldots, e_{\binom{n}{2}}$ be the edges of $C(P)$ sorted according to their lengths $L_1, \ldots, L_{\binom{n}{2}}$. For $i = 1 \ldots \binom{n}{2}$, we denote by $C_i(P)$ the geometric graph consisting of all edges whose length is no more than L_i. In general, for any graph G whose vertex set is V, we define G_i as $G \cap C_i(V)$. Let UDG(P) be the unit disk graph of P, which is the graph whose vertex set is P and with edges between pairs of vertices whose distance is not more than one. Note that UDG(P) $= C_i(P)$ for some i.

We now show the relationship between strong t-spanners and unit disk graphs.

Proposition 5. *If S is a strong t-spanner of $C(P)$, then for all $i = 1 \ldots \binom{n}{2}$ and all $j = 1 \ldots i$, S_i contains a t-spanning path linking the vertices of e_j.*

Proof: Let p and q be the vertices of e_j. Consider a strong t-spanner path in S between p and q. Each edge on this path has length at most $|pq| = L_j \leq L_i$. Therefore, each edge is in S_i. □

Proposition 6. *If S is a strong t-spanner of $C(P)$, then for all $i = 1 \ldots \binom{n}{2}$, S_i is a t-spanner of $C_i(P)$.*

Proof: Let a and b be any two points such that $d_{C_i(P)}(a,b)$ is finite. We need to show that in S_i there exists a path between a and b whose length is at most $t \cdot d_{C_i(P)}(a,b)$. Let $a = p_1, p_2, \ldots, p_k = b$ be a shortest path in $C_i(P)$ between a and b. Hence:

$$d_{C_i(P)}(a,b) = \sum_{j=1}^{k-1} |p_j p_{j+1}|.$$

Now, by proposition 5, for each edge (p_j, p_{j+1}) there is a path in S_i between p_j and p_{j+1} whose length is at most $t \cdot |p_j p_{j+1}|$. Therefore:

$$d_{S_i(P)}(a,b) \leq \sum_{j=1}^{k-1} t \cdot |p_j p_{j+1}| = t \sum_{j=1}^{k-1} |p_j p_{j+1}| = t \cdot d_{C_i(P)}(a,b)$$

which means that in S_i, there exists a path between a and b whose length is at most $t \cdot d_{C_i(P)}(a,b)$. □

Corollary 2. *If S is a strong t-spanner of $C(P)$, then $S \cap$ UDG(P) is a strong t-spanner of* UDG(P).

Proof: Just notice that UDG $= C_i$ for some i and the result follows from Proposition 6. □

Thus, we have shown sufficient conditions for a graph to be a spanner of the unit disk graph. We now show that these conditions are also necessary.

Proposition 7. *If S is a subgraph of $C(P)$ such that for all $i = 1 \ldots \binom{n}{2}$, S_i is a t-spanner of $C_i(P)$, then S is a strong t-spanner of $C(P)$.*

Proof: Let a, b be any pair of points chosen in P. We have to show that in S, there is a path between a and b such that

1. its length is at most $t \cdot |ab|$ and
2. every edge on the path has length at most $|ab|$.

Let $e_i = (a, b)$. We know that S_i is a t-spanner of $C_i(P)$. Since $C_i(P)$ contains e_i, $d_{C_i(P)}(a, b) = |ab|$. Hence, there is a path in S_i (and therefore in S) whose length is at most $t \cdot d_{C_i(P)}(a, b) = t|ab|$. Also, since it is in S_i, all of its edges have length at most $L_i = |ab|$. □

The two last results, together, allow us to determine whether or not given families of geometric graphs are also spanners of the unit disk graph. First, since the G_λ^θ graph is a strong t-spanner, we already know that it is also a spanner of the unit disk graph. Second, Bose *et al.* [3] showed that the Yao graph [8] and the Delaunay triangulation are strong t-spanners. Therefore, these graphs are also spanners of the unit disk graph. Third, the θ-graph [6] is not always a spanner of the unit disk graph. The reason for that is that in a cone, the edge you chose may not be the shortest edge. Hence, the path from a point p to a point q may contain edges whose length is greater than $|pq|$ (see Figure 7). Using Proposition 7, we thus know that the intersection of the θ-graph with the unit disk graph may not be a spanner of the unit disk graph. Indeed, the intersection of the θ-graph with the unit disk graph may not even be connected.

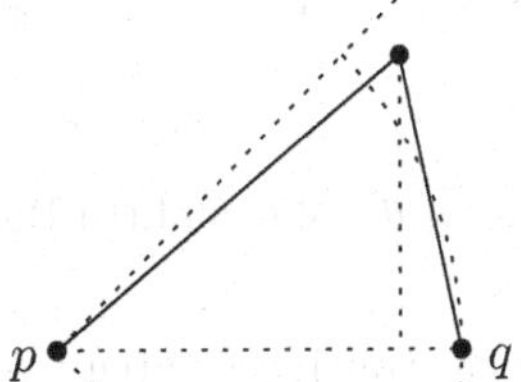

Fig. 7. The θ-graph is not a strong t-spanner

5 Simulation Results

In Section 2, we provided worst-case analysis of the spanning ratio of the G_λ^θ graph. Using simulation, we now provide estimates of the average spanning ratio of the G_λ^θ graph. Using a uniform distribution, we generated 200 sets of 200 points each and computed the spanning ratio for λ ranging from 0.5 to 1 and θ ranging from 5° to 90° (for $\theta = 0°$, the spanning ratio is exactly 1). For each graph, we then computed the spanning ratio and the local routing ratio. The *spanning ratio* is defined as the maximum, over all pair of points (p, q), of the length of the shortest from p to q path in the G_λ^θ graph divided by $|pq|$. The *local routing ratio* is defined as the maximum, over all pair of points (p, q), of the length of the path produced by using a local routing strategy in the G_λ^θ

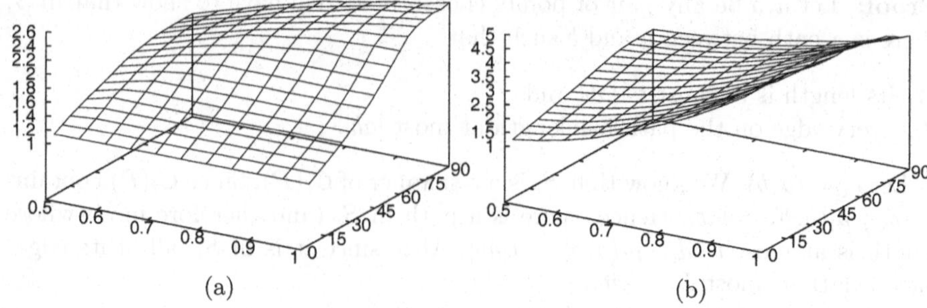

Fig. 8. (a) Spanning Ratio and (b) Local Routing Ratio for $\lambda = 0.5$ to 1 and $\theta = 5°$ to $90°$

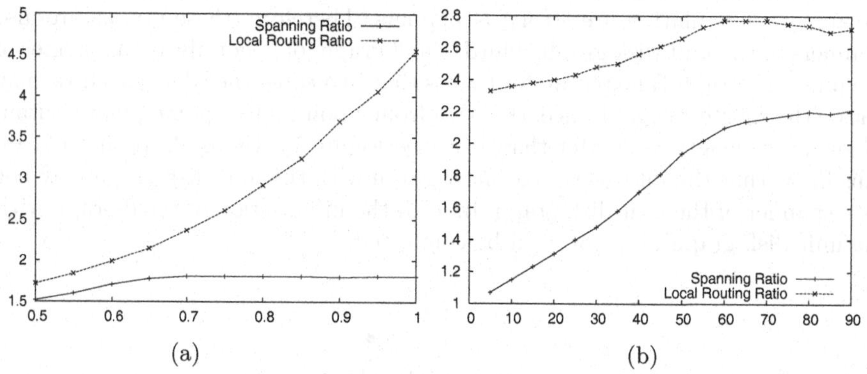

Fig. 9. (a) Ratios for $\theta = 45°$ and (b) Ratios for $\lambda = 0.75$

graph divided by $|pq|$. The local routing strategy we have used is the following: at each step, send the message to the neighbor which destroyed q. We also tried the strategy which consists in choosing the neighbor which is the nearest to q, and the results we obtained were the same.

Figure 8(a) shows the results we obtained for the spanning ratio. Figure 8(b) shows the results we obtained for the local routing ratio. Exact values of our simulations results can be found in the technical report version of this paper [1].

One interesting conclusion we can draw from these results is that for the spanning ratio, θ has a more decisive influence than λ. Figure 9(a) shows the simulation results for the cases where $\lambda = 0.75$. We see that even though both ratios generally increase when θ increase, the spanning ratio varies between 1.07 and 2.21 (107% variation), while the local routing ratio only varies between 2.33 and 2.77 (19% variation). For the local routing ratio, it is the other way around. It is λ which has a more decisive influence. Figure 9(b) shows the influence of λ when $\theta = 45°$. In that case, the local routing ratio varies between 1.72 and 4.55 (165% variation), while the spanning ratio only varies between 1.52 and 1.81 (19% variation).

6 Conclusion

We conclude with the problem that initiated this research. Determine whether or not HSP is a strong t-spanner for some constant t.

References

1. Bose, P., Carmi, P., Couture, M., Smid, M., Xu, D.: On a family of strong geometric spanners that admit local routing strategies. Technical Report TR-07-08, School of Computer Science, Carleton University (2007)
2. Bose, P., Devroye, L., Evans, W., Kirkpatrick, D.: On the spanning ratio of Gabriel graphs and β-skeletons. SIAM Journal of Disc. Math. 20(2), 412–427 (2006)
3. Bose, P., Maheshwari, A., Narasimhan, G., Smid, M., Zeh, N.: Approximating geometric bottleneck shortest paths. Comput. Geom. Theory Appl. 29(3), 233–249 (2004)
4. Bose, P., Morin, P.: Online routing in triangulations. SIAM J. Comput. 33(4), 937–951 (2004)
5. Chavez, E., Dobrev, S., Kranakis, E., Opatrny, J., Stacho, L., Tejeda, H., Urrutia, J.: Half-space proximal: A new local test for extracting a bounded dilation spanner. In: Anderson, J.H., Prencipe, G., Wattenhofer, R. (eds.) OPODIS 2005. LNCS, vol. 3974, Springer, Heidelberg (2006)
6. Keil, J.M., Gutwin, C.A.: Classes of graphs which approximate the complete euclidean graph. Discrete Comput. Geom. 7(1), 13–28 (1992)
7. Narasimhan, G., Smid, M.: Geometric Spanner Networks. Cambridge University Press, New York (2007)
8. Yao, A.C.-C.: On constructing minimum spanning trees in k-dimensional spaces and related problems. SIAM J. Comput. 11(4), 721–736 (1982)

Spanners for Geometric Intersection Graphs*

Martin Fürer** and Shiva Prasad Kasiviswanathan

Computer Science and Engineering, Pennsylvania State University
{furer,kasivisw}@cse.psu.edu

Abstract. A disk graph is an intersection graph of a set of disks with arbitrary radii in the plane. In this paper, we consider the problem of efficient construction of sparse spanners of disk (ball) graphs with support for fast distance queries. These problems are motivated by issues arising from topology control and routing in wireless networks.

We present the first algorithm for constructing spanners of ball graphs. For a ball graph in $\mathbb{R}^k$, we construct a $(1+\epsilon)$-spanner with $O(n\epsilon^{-k+1})$ edges in $O(n^{2\ell+\delta}\epsilon^{-k}\log^{\ell} S)$ expected time, using an efficient partitioning of the space into hypercubes and solving intersection problems. Here $\ell = 1-1/(\lfloor k/2 \rfloor+2)$, δ is any positive constant, and S is the ratio between the largest and smallest radius. For the special case where all the balls have the same radius, we show that the spanner construction has complexity almost equivalent to the construction of a Euclidean minimum spanning tree. Previously known constructions of spanners of unit ball graphs have time complexity much closer to n^2. Additionally, these spanners have a small vertex separator (hereditary), which is then exploited for fast answering of distance queries. The results on geometric graph separators might be of independent interest.

1 Introduction

Let $G = (V, E)$ be a weighted graph, and let $d_G(u, v)$ be the length of a shortest path between vertices u and v in G. For any fixed $\epsilon > 0$, a $(1 + \epsilon)$-stretch spanner of G is a subgraph G' such that for all pairs of vertices u and v, $d_{G'}(u, v)/d_G(u, v) \leq (1 + \epsilon)$. Spanner constructions have been widely investigated for general graphs and complete Euclidean graphs, also with additional properties like weight, diameter, degree [1].

We present a new method for producing spanners of geometric graphs based on a hierarchical decomposition of the plane into tiles of various sizes. Our constructions are also more general, as they are not restricted to complete Euclidean graphs, but extend to geometric disk graphs, as well as their higher dimensional versions, the ball graphs. In all cases, edge lengths are given by Euclidean distances, but not all edges have to be present in our graphs. The *difficulty* in constructing a spanner for the disk graph metric when compared to the metric

* A short abstract of this work appeared in the proceedings of FWCG 2006.

** This material is based upon work supported by the National Science Foundation under Grant No. CCR-0209099.

F. Dehne, J.-R. Sack, and N. Zeh (Eds.): WADS 2007, LNCS 4619, pp. 312–324, 2007.

induced by a complete Euclidean graph is that two points that are close in space are not necessarily close under the graph metric.

The intersection graphs of disks in the Euclidean plane has been studied for many years for its theoretical aspects as well as its applications. We consider weighted disk graphs where the weight of an edge is the Euclidean distance between centers. Such graphs have been used widely to model the communication between objects in the context of wireless networks [2,3]. For wireless networks they model the fact that two wireless nodes can directly communicate with each other only if they are within a certain distance.

Spanners are important for disk graphs because restricting the size of a network reduces the amount of routing information. Spanners are used in topology control for maintaining network connectivity, improving throughput, and optimizing network lifetime [3]. Geometric spanner constructions for disk-like graphs have been widely investigated in both theory and networking communities. Many constructions, both centralized and distributed, also with additional properties like planarity, power saving have been proposed [2,3]. However, all these constructions only work for some restricted cases of disk graphs (for example like for unit disk graphs).

We present the first algorithm for constructing spanners of general ball (disk) graphs. Additionally, for unit ball graphs, we show that constructing $(1+\epsilon)$-spanners has randomized complexity almost equivalent to the construction of a Euclidean minimum spanning tree. Therefore we cannot hope to find a faster algorithm for constructing spanners of unit ball graphs, unless we improve on some other well-studied problems [4]. The previously best known constructions of spanners of unit ball graphs were primarily based on Yao graph construction and therefore have time complexity $\Omega(n^{2-a(k)})$ for $a(k) = 2^{-k+1}$ in dimensions k greater than 3 [5,6].

Our spanners also have a small separator decomposition, which helps us to support fast answering of distance queries. Distance queries are important in ball graphs as they are widely used to determine coverage in wireless sensor networks, and for routing purposes [7]. Since complete Euclidean graphs are just a special case of (unit) ball graphs, our results also provides a new approach for constructing spanners with small separators in these graphs (a previous solution [8] turned out to be incorrect [9]).

In a companion paper [10], the authors used a much simpler hierarchical partitioning scheme to construct a data structure for answering distance queries in ball graphs. It is shown that a ball graph can be preprocessed in $O(mn^{1-1/k}\,\epsilon^{-k+1} + m\epsilon^{-k}\log S)$ time, producing a data structure of $O(n^{2-1/k}\,\epsilon^{-k+1} + n\epsilon^{-k}\log S)$ size, such that subsequent distance queries can be answered approximately in $O(n^{1-1/k}\,\epsilon^{-k+1} + \epsilon^{-k}\log S)$ time. Here S is the global scale factor (formally defined later). The resulting distance estimate is within an additive error which is less than ϵ times the longest edge on some shortest path. We build upon this foundation and introduce several new techniques to obtain our new results. Note that spanners are far more desirable (thus also tougher to construct) and the sparse graphs constructed in [10] are not spanners.

2 Preliminaries

Let $\mathcal{P} = \{p_1, \ldots, p_n\}$ be a set of points in $\mathbb{R}^k$ for any fixed dimension k. We assume w.l.o.g. that points are distinct. Let $\mathcal{D} = \{D_{p_1}, \ldots, D_{p_n}\}$ be a set of n balls such that (i) D_u is centered at $u \in \mathcal{P}$, and (ii) D_u has radius of r_u. Balls D_u and D_v intersect if $d(u,v) \leq (r_u + r_v)$, where $d(.,.)$ denotes the Euclidean metric. The ball graph $G = (\mathcal{P}, E)$ is a weighted graph where an edge between u and v with weight $d(u,v)$ exists iff D_u and D_v intersect. Let d_G denote the shortest path metric induced by the connected graph G on its vertices. The balls in $\mathcal{D}$ are rescaled such that the largest radius equals one. Now, the global scale factor (ratio between the largest and smallest radius) of $\mathcal{D}$ is defined as $\rho(\mathcal{D}) = 1/\min\{r_u \mid D_u \in \mathcal{D}\}$.

Our algorithms use a variant of 2^k-trees (quadtrees in the plane). For a node t, denote by $p(t)$ the parent of t in the tree. We use d_t to denote the depth of node t in the tree. A point (x,y) is *contained* in the node t representing a square with center (x_t, y_t) and length l_t in the quadtree iff $x_t - l_t/2 \leq x < x_t + l_t/2$ and $y_t - l_t/2 \leq y < y_t + l_t/2$. For a set of squares T in the quadtree a point is contained in T iff there exists $t \in T$, such that point is contained in t. The distance between two squares is the Euclidean distance between their centers.

Throughout the paper we refer to the vertices of a graph as *vertices* and vertices of a tree as *nodes*. We assume w.l.o.g. that ϵ^{-1} is a power of 2. Floors and ceilings are omitted throughout the paper, unless needed. Note that starting with a fixed ϵ we get a $(1 + c\epsilon)$-spanner for a fixed constant c. For simplicity we describe all the algorithms for $k = 2$ and then state the generalizations to higher k. We use the notation $\tilde{O}(f) \equiv O(f \text{ poly} \log f)$.

2.1 Our Contributions

For ball graphs in $\mathbb{R}^k$, we solve halfspace range searching problems to construct spanners with $O(n\epsilon^{-k+1})$ edges. For the interesting case when $\rho(\mathcal{D})$ is polynomially bounded by using the currently best algorithm of Agarwal and Matoušek [11] for halfspace range searching, we obtain a running time of $\tilde{O}(n^{2\ell+\delta}\epsilon^{-k})$, where $\ell = 1 - 1/(\lfloor k/2 \rfloor + 2)$ and δ is any positive constant.

In the case when all the balls have the same radius (unit ball graphs), we replace halfspace range searching problems by bichromatic closest pair problems. Therefore we get a running time of $\tilde{O}(n\epsilon^{-2})$ for $k = 2$. In higher dimensions using the currently best algorithm of Agarwal *et al.* [6] for solving the bichromatic closest pair problem, we get $\tilde{O}(n^{4/3}\epsilon^{-3})$ expected time for $k = 3$, and $O(n^{2-2/(\lceil k/2 \rceil+1)+\delta}\epsilon^{-k})$ expected time for $k \geq 4$.

The spanners constructed have an $O(n^{1-1/k}\epsilon^{-k+1/2} + \epsilon^{-2k+1} \log \rho(\mathcal{D}))$ vertex separator, which can be found in $O(n \log n)$ time. Using this separator we obtain fast algorithms for approximately answering distance queries in ball graphs. We show that the spanner can be preprocessed in $\tilde{O}(nf(n,\epsilon))$ time and space, such that subsequent distance queries under the d_G metric can be answered with $(1+\epsilon)$-stretch in $O(f(n,\epsilon))$ time and with $(2+\epsilon)$-stretch in $O(\log n)$ time. Here $f(n,\epsilon) = n^{1-1/k}\epsilon^{-k+1/2} + \epsilon^{-2k+1} \log \rho(\mathcal{D})$.

We now present an overview of some of the ideas that we use to obtain these results. All missing proofs and details, can be found in the full version [12].

2.2 Modified Yao Graph

A Yao graph [5] construction involves partitioning the space around each point into cones with a fixed opening angle and connecting the point to its nearest neighbor in each cone. Even though constructing the original Yao graph is costly (to the best of authors knowledge the currently best running time is from [6]), a variant of it called the θ-graph can be constructed in $\tilde{O}(n\epsilon^{-k})$ time [13]. In a θ-graph points inside a wedge are projected onto the angle bisector of a cone and the closest point in this distorted metric is used to add an edge. Even though θ-graphs are spanners of the complete Euclidean graphs, they fail to be spanners even for unit ball graphs due to the distortion.

It is well known that, for unit disk (ball) graphs the Yao graph with long edges removed is a spanner [14]. We now define a *modified* Yao graph, which forms a spanner of the general disk (ball) graph. Let $\mathcal{C}(p) = \{co_1(p), \ldots, co_{\epsilon^{-1}}(p)\}$ be a collection of ϵ^{-1} cones such that (i) each cone has its apex at $p \in \mathcal{P}$, (ii) each cone has an opening angle of $2\pi\epsilon$, and (iii) the union of these cones covers $\mathbb{R}^2$. We define a modified Yao graph Y in the following way. The vertices of Y are the points of $\mathcal{P}$. For each $p \in \mathcal{P}$ and $1 \le i \le \epsilon^{-1}$, add an edge from p to the point q contained in $co_i(p)$ if: (i) $r_q \ge r_p$, (ii) the edge (p, q) exists in G, and (iii) q is the closest point in $co_i(p)$ to p satisfying (i) and (ii).

Lemma 1. *Let (u, v) be an edge in the disk graph G and $\epsilon < 1/6$. Then there exists a path in the spanner Y such that $d_Y(u, v) \le (1 + \epsilon)d(u, v)$.*

From the above lemma by summing over all edges of a path in G we also get that graph Y is a $(1 + \epsilon)$-spanner of the disk graph G. For ball graphs in $\mathbb{R}^k$, the number of edges in Y can be bounded by $O(n\epsilon^{-k+1})$ using the results of Lukovszki [15]. We use the following ideas to efficiently construct a *variant* of the modified Yao graph.

2.3 Partitioning the Plane

A well-separated pair decomposition (introduced by Callahan and Kosaraju in [16]) for a given parameter s is a set of nonempty subsets of $\mathcal{P}$, $\{\{A_1, B_1\}, \ldots, \{A_m, B_m\}\}$, such that: (i) the sets A_i and B_i are disjoint, (ii) for any pair $p, q \in \mathcal{P}$, there is an unique pair $\{A_i, B_i\}$ such that $p \in A_i$ and $q \in B_i$, and (iii) for each pair $\{A_i, B_i\}$, there is a length r such that A_i and B_i can be enclosed by two r-balls, separated by a distance of at least sr. For a point set in the Euclidean space it is known that a well-separated pair decomposition with almost linear many pairs exists [16]. Given such a pair decomposition it can be easily converted into a $(1+\epsilon)$-spanner by picking a representative edge (an actual edge of the graph) from each pair into the spanner [17,18].

However, the disk graph metric being more general doesn't have the same nice properties as the complete Euclidean graph metric. For unit disk graphs, Gao and Zhang [7] have recently given a construction of a well-separated pair decomposition with $O(n \log n)$ pairs. They have also shown that for unit ball graphs in $\mathbb{R}^k$ at least $\Omega(n^{2-2/k})$ pairs are needed. Therefore a spanner constructed using *such a* pair decomposition has a super-linear number of edges. With disk graphs, the situation is even worse, as disk graphs do not have a sub-quadratic well-separated pair-decomposition [7].

Loosely speaking, we alter the notion of well-separated pair decomposition in two ways, (a) we relax the condition (ii) in the definition such that only pairs $p, q \in \mathcal{P}$ that needs to be covered are those that have an edge in the graph, and (b) we restrict the sets to only those that form a clique in G. Even though under this new notion we don't save on the number of pairs, but for constructing the spanner when combined with ideas of modified Yao graph, one can eliminate all but linear number of pairs. The final challenge is minimizing the time for finding the representative edges between set pairs.

3 Spanners of Disk Graphs

We first describe a high level idea of our algorithm and prove the claimed stretch factor, and then define an algorithmic version of the construction. Each disk is associated with a level, a disk D_u is of level l iff $2^{-l} \leq r_u < 2^{-l+1}$. Let l_m denote the largest level among disks in $\mathcal{D}$, i.e., $l_m = \lceil \log \rho(\mathcal{D}) \rceil$.

Quad-dissection: Our spanner construction involves recursively partitioning the plane using a simple variant of quadtrees. The input to quad-dissection is a set of disks in $\mathbb{R}^2$. Let $\mathcal{P}$ be the set of their centers. Define the *bounding box* to be the smallest axis-parallel rectangle enclosing $\mathcal{P}$. The left bottom corner of the bounding box is assumed to be the origin. An L-grid is defined by horizontal and vertical line segments drawn at $y \in L\mathbb{Z}$ and $x \in L\mathbb{Z}$ within the bounding box. A quad-dissection of the L-grid is a recursive partition into smaller squares. We view the resulting structure as a 4-ary forest with the root nodes as the non-empty squares in the L-grid. Each square is partitioned into four equal squares, which form its children. We continue partitioning all the non-empty squares until each disk center is contained in a separate square of size $\epsilon 2^{-l_m}$ or smaller. See Figure 1.

Constructing the forest: Let Γ denote the forest from the quad-dissection of the ϵ-grid. Let *Roots* initially be the non-empty squares of the ϵ-grid. Γ is a collection of disjoint trees, each of which is rooted at a node belonging to *Roots*. Note that the set of nodes at depth l in the forest corresponds to the set of non-empty squares defined by the $\epsilon 2^{-l}$-grid. We introduce a disk of level l only at depth l. For a node $t \in \Gamma$ of depth d_t, define $D(t)$ as the set of disks which are of level d_t or less and have their centers are contained in t. Let $C(t)$ be the set of disk centers of disks in $D(t)$. Note that if $u \in C(t)$ for an internal node t, then there exists a node t' with $t = p(t')$ and $u \in C(t')$.

We also add to *Roots* any node $t \in \Gamma$ with $C(t) \neq \emptyset$ whereas $C(p(t)) = \emptyset$. A node t of the forest is called *interesting* if $C(t) \neq C(p(t))$. By definition all

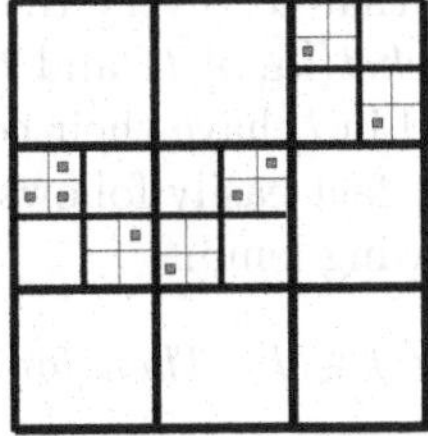
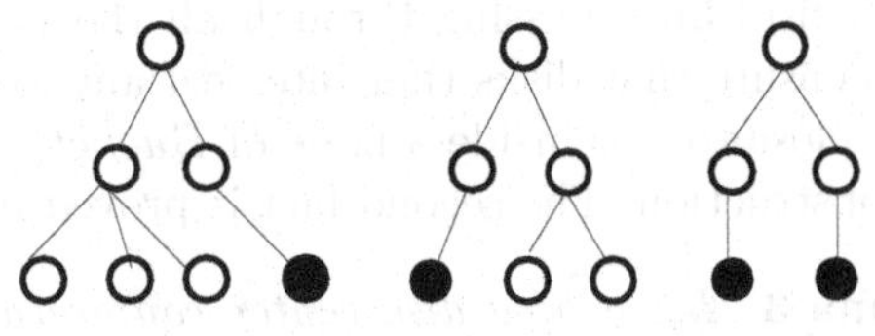

Fig. 1. The quad-dissection procedure for a set of disk centers and the corresponding forest Γ. Assume all the disks have the same radius. The donut shaped nodes are the interesting nodes in Γ.

nodes in *Roots* are interesting. It immediately follows that Γ has at most $2n$ interesting nodes.

Choosing the representatives: For every leaf node $t \in \Gamma$ we choose the disk center in $C(t)$ as its representative R_t. For an internal node t with non-empty $C(t)$, pick a child t_s satisfying $R_{t_s} \in C(t)$ and set $R_t = R_{t_s}$.

Neighborhood of nodes: For every interesting node t, define its *close neighborhood* ($N_c(t)$) as the set of all nodes at depth d_t which are within a distance of 2^{-d_t} from t.

We also define a *far neighborhood* ($N_f(t)$) for the nodes in *Roots*. To do so, we introduce some new definitions. For a node $t \in \Gamma$ and integers α, β define its $[\alpha, \beta]$-shift, ($S([\alpha, \beta], t)$) with $-2\epsilon^{-1} \leq \alpha, \beta \leq 4\epsilon^{-1}$ and $\max(|\alpha|, |\beta|) \geq (2\epsilon)^{-1}$, as the square t' obtained by shifting the x- and y-coordinates of all points in t by $\epsilon\alpha 2^{-d_t}$ and $\epsilon\beta 2^{-d_t}$ respectively. Informally, α tells the number of squares moved over in the horizontal direction and β does the same in the vertical direction. The bounds of α, β are chosen such that at least one of them is big. For every $t \in Roots$, construct $O(\epsilon^{-2})$ ordered buckets, where

$$bucket([\alpha, \beta], t) = \{S([\alpha, \beta], t), S([\alpha, \beta], p(t)), S([\alpha, \beta], p(p(t))), \ldots\},$$
$$Bucket(t) = \bigcup_{\alpha,\beta} bucket([\alpha, \beta], t).$$

Now for every $t \in Roots$ and every α, β-pair scan through $bucket([\alpha, \beta], t)$ to find the first node $t' \in bucket([\alpha, \beta], t)$ such that there exists $u \in C(t)$ and $v \in C(t')$ with edge (u, v) in G. Add t' to $N_f(t)$.

Edges in spanner G′: For every interesting node t, we add an edge between R_t and $R_{t'}$, where $t' \in N_c(t)$. Additionally, for every node $t \in Roots$, we pick an edge of G between a vertex in $C(t)$ and another in $C(t')$ into the spanner, where $t' \in N_f(t)$.

Lemma 2. *The graph G' has $O(n\epsilon^{-2})$ edges and is a subgraph of the graph G.*

The idea behind creating the *bucket* is: (i) to ensure that for every α, β-pair there exists lines passing through all the nodes of $bucket([\alpha, \beta], t)$ and t, and (ii) to ensure that disks that intersect any disk centered in t, have their centers either close to t or inside a node of $Bucket(t)$. The first fact easily follows from the construction. The second fact is proved in the following lemma.

Lemma 3. *Let u be a disk center contained in a node $t \in \Gamma$. Then for every edge (u, v) in G, v is contained in $Bucket(t) \cup N_c(t)$.*

Lemma 4. (Spanner Property). *Let (u, v) be an edge in the disk graph G. Then there exists a path in the spanner G' such that $d_{G'}(u, v) \leq (1 + c_1\epsilon)d(u, v)$ for some constant c_1.*

Sparsifying the spanner: Consider the set of cones $\mathcal{C}(p)$ for a point $p \in \mathcal{P}$ (as in Section 2.2). For each cone $co_i(p) \in \mathcal{C}(p)$, scan through all edges added to the spanner from p to a point contained in $co_i(p)$ and retain only the shortest such edge in the spanner. Since there are only ϵ^{-1} cones in $\mathcal{C}(p)$, we conclude that the number of edges in the spanner is $O(n\epsilon^{-1})$. Using the analysis of Lemma 1 along with Lemma 4, it follows that the graph is a $(1 + \epsilon)$-spanner. Henceforth, G' denotes this sparsified spanner.

3.1 Algorithmic Version

Constructing the forest: We construct a compressed forest in which we only introduce the interesting nodes of Γ and shortcut the degree one internal nodes. The construction of the compressed forest is omitted here, as it follows from simple modifications of known algorithms (like [19,16,20]) for generating compressed quadtrees. For details see [12].

Let Γ' denote this compressed forest. During the construction of Γ', we also maintain at every node a disk of the largest radius whose center is contained in the node. For any node $t \in \Gamma'$, if among the disks whose centers are contained in t the largest radius is greater than 2^{-d_t}, then the representative R_t is defined as the disk center of a largest radius disk. Additionally, if among the disks whose centers are contained in t the largest radius is less than 2^{-d_t}, then add nodes t' satisfying $t = p(t')$ and $R_{t'} \neq \emptyset$ to *Roots*.

Finding the close neighborhood: Finding the close neighborhood for nodes at depth 0 is straight-forward. Because we work with Γ' we construct a *close pseudo-neighborhood* ($N'_c(t)$) for a node t. A node $r \in \Gamma'$ is in $N'_c(t)$ if (i) r belongs to $N_c(t)$ or, (ii) in Γ the node r is the maximal depth ancestor of a node $r' \in N_c(t)$ with $r' \notin \Gamma'$.

Now assuming that we have constructed $N'_c(t)$, we describe the construction of $N'_c(t_s)$ for a child t_s of t in Γ'. We access all the nodes in $N'_c(t)$. We do a level-order traversal from these nodes with a modification that the subtree of any node b is accessed only if $d_b \leq d_{t_s}$ and R_b is at most $(1 + \sqrt{2}\epsilon)2^{-d_b}$ distance away from R_{t_s}. The nodes at which the traversal ends and whose representatives are at most $(1 + \sqrt{2}\epsilon)2^{-d_{t_s}}$ away from R_{t_s} define $N'_c(t_s)$.

Any node $g \in \Gamma'$ accessed during the traversal has a node $f \in \Gamma$ such that: (i) f is an ancestor of t_s in Γ, (ii) $d_g = d_f$, and (iii) the distance between f and g is $O(2^{-d_g})$. We charge the access of node g to the node f (f needn't be in Γ'). For this entire procedure we can charge all the accesses of node g to different nodes which are at the same depth as g and only $O(2^{-d_g})$ away (note there are only $O(\epsilon^{-2})$ such nodes). This implies that the time for finding close pseudo-neighborhoods for all nodes is $O(n\epsilon^{-2})$.

Far neighborhood for small global scale: If $\rho(\mathcal{D})$ is small, we use the observation from Gupta *et al.* [21] that maps the problem of reporting the intersection of a collection of disks with a query disk into halfspace range searching in two dimensions higher. For simplicity, we use a dynamic data structure for halfspace range searching. Agarwal and Matoušek [11] construct a data structure for $\mathbb{R}^k$ which for any parameter $n \leq m \leq n^{\lfloor k/2 \rfloor}$ and any positive constant δ, after $O(m^{1+\delta})$ space and time preprocessing answers halfspace queries in $\tilde{O}(n/m^{1/\lfloor k/2 \rfloor})$ time, and has $O(m^{1+\delta}/n)$ amortized update time. We construct this data structure bottom-up. The data structure for any node $t \in \Gamma'$ stores the points in $C(t)$. Among the children of t, let t_m be the node having the most number of points in its data structure. We update the data structure of t_m to construct a data structure for t. This is done by first deleting points from the data structure that are not in $C(t)$ and then by using siblings of t_m to insert the remaining points of $C(t)$. We then query the data structure with every disk in $C(t')$ with $t \in Bucket(t')$ to check for intersection.

If $\rho(\mathcal{D})$ is the global scale factor, then each $bucket([\alpha,\beta],t)$ is of size $O(\log \rho(\mathcal{D}))$. Therefore $|Bucket(t)| = O(\epsilon^{-2} \log \rho(\mathcal{D}))$ and the representation of $Bucket(t)$ in Γ' can be found as achieved for the close neighborhood. Each disk acts as a query disk $O(\epsilon^{-2} \log \rho(\mathcal{D}))$ many times, so total number of queries is at most $O(n\epsilon^{-2} \log \rho(\mathcal{D}))$.

To balance the total time for setting up the data structure at every node and the total query time, we assume the parameter m of [11] to be n^c for some $c \geq 1$. As we work in two dimensions higher, each query can be answered in $\tilde{O}(n^{1-c/2})$ time. The total time for answering queries is $\tilde{O}(n^{2-c/2}\epsilon^{-2} \log \rho(\mathcal{D}))$. The total time for setting up all the data structures of [11] by the procedure described above is $O(n^{c+\delta})$, where δ is any positive constant. On eliminating c, by balancing the query and construction times, we get a space and time bound of $O(n^{4/3+\delta}\epsilon^{-2} \log^{2/3} \rho(\mathcal{D}))$ for finding the far neighborhood of all nodes.

Far neighborhood for large global scale: In this case we use the adjacency list to place the edges into the right *bucket*. For a node $t \in Roots$, we only consider an edge (u,v) if it satisfies: (i) $u \in C(t)$, $v \notin C(t)$, and (ii) $r_u \leq r_v$. If this is the case, then we find points v' and v'' on the line segment connecting u to v such that $2d(u,v') = d(u,v'')$, v' is contained in $\bigcup_{\alpha,\beta} S([\alpha,\beta],t)$, and v'' is not contained in $\bigcup_{\alpha,\beta} S([\alpha,\beta],t)$. Let α', β' be such that v' is contained in $S([\alpha',\beta'],t)$. We put the edge (u,v) in $bucket([\alpha',\beta'],t)$. Once all edges have been put into their respective *bucket*, we pick the shortest edge in each *bucket* and add it to the spanner. The entire procedure can be implemented in $O(|E|)$ given the adjacency list.

3.2 Extension to Ball Graphs

The quadtrees become 2^k-trees in $\mathbb{R}^k$. The close neighborhood and far neighborhood of any node is of size $O(\epsilon^{-k})$, but again one can sparsify to get it to $O(\epsilon^{-k+1})$. Therefore the total number of edges is $O(n\epsilon^{-k+1})$. The sparsification can be done in $O(n\epsilon^{-k})$ time. In case of small global scale using results of range-searching from [11] we get a running time of $O(n^{2\ell+\delta}\epsilon^{-k}\log^{\ell}\rho(\mathcal{D}))$, where $\ell = 1 - 1/(\lfloor k/2 \rfloor + 2)$. In case of large global scale the running time is $O(|E| + n\epsilon^{-k})$ given the adjacency list of G.

We can summarize these results as follows:

Theorem 1. *Let G be a k-dimensional ball graph defined on $\mathcal{D}$. A $(1+\epsilon)$-spanner of G with $O(n\epsilon^{-k+1})$ edges can be constructed in $O(\min\{n^{2\ell+\delta}\epsilon^{-k}\log^{\ell}\rho(\mathcal{D}), Adj(\mathcal{D}) + n\epsilon^{-k}\})$ time, where δ is any positive constant, $\ell = 1 - 1/(\lfloor k/2 \rfloor + 2)$, and $Adj(\mathcal{D})$ is the time for constructing the adjacency list for G.*

3.3 Speeding up on Unit Ball Graphs

If G was defined on unit balls (disks) we can speed up the construction considerably. The algorithm remains the same until the part where we compute the far neighborhood. Here the set *Roots* contains only the non-empty squares of the ϵ-grid. For finding the far neighborhood of nodes in *Roots*, we solve a collection of bichromatic closest pair problems. If the disks corresponding to the closest pair intersect, we add the corresponding edge into the spanner. See Figure 2.

Let $f(n)$ be any function satisfying $\log f(n) = o(\log n)$, i.e., f grows slower than polynomial. We assume in the rest of discussion that for some $c \geq 1$, $n^c f(n)$ is an upper bound on the time for computing a bichromatic closest pair for a total of n points.

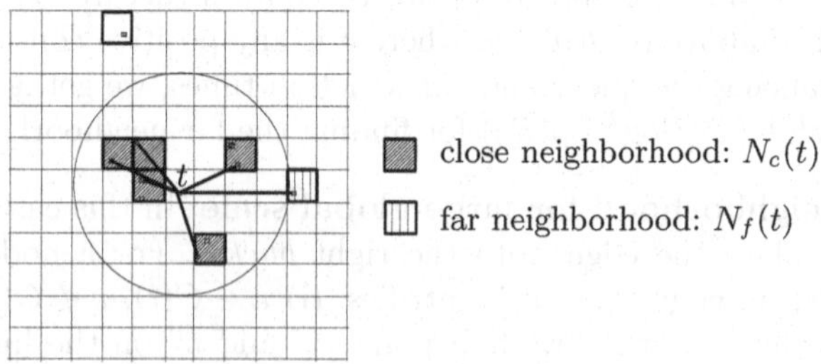

Fig. 2. The close and far neighborhood of a node t in *Roots* for a unit disk graph. All the squares within the circle are at most 2^{-d_t} distance away from t. Far neighborhood is determined by using bichromatic closest pair tests.

Lemma 5. *For unit ball graphs in $\mathbb{R}^k$, the worst case running time for finding the far neighborhood for all nodes in Roots is $O(n^c f(n)\epsilon^{-k})$.*

It is well known that the bichromatic closest pair problem in $\mathbb{R}^2$ can be solved in $O(n \log n)$ by using post-office queries [5]. The currently best algorithm of Agarwal *et al.* [6] for finding a bichromatic closest pair between points sets P and Q in $\mathbb{R}^k$ runs in $O((PQ \log P \log Q)^{2/3} + P \log^2 Q + Q \log^2 P)$ expected time for $k = 3$ and in $O((PQ)^{1-1/(\lceil k/2 \rceil+1)+\delta} + P \log Q + Q \log P)$ expected time for $k \geq 4$, where δ is any positive constant.

Theorem 2. *Let G be a unit ball graph in $\mathbb{R}^k$. A $(1+\epsilon)$-spanner of G with $O(n\epsilon^{-k+1})$ edges can be constructed in $O(n^c f(n)\epsilon^{-k})$ time.*

Corollary 1. *The spanner G' can be constructed in $\tilde{O}(n\epsilon^{-2})$ time for $k = 2$, in $\tilde{O}(n^{4/3}\epsilon^{-3})$ expected time for $k = 3$, and in $O(n^{2-2/(\lceil k/2 \rceil+1)+\delta}\epsilon^{-k})$ expected time for $k \geq 4$, where δ is any positive constant.*

The above result shows that finding a spanner of a unit ball graph in $\mathbb{R}^k$ is not much harder than computing a bichromatic closest pair for n points in $\mathbb{R}^k$. In the other direction from the results of Eppstein [1] and Chan [22], we know that the randomized expected time bounds for constructing a spanning forest of a unit ball graph in $\mathbb{R}^k$ and a Euclidean minimum spanning tree (or a bichromatic closest pair) in $\mathbb{R}^k$ are within constant factors. Once we have the spanner any graph traversal can be used to construct a spanning forest. To make this relation precise, define the *exponent* of a problem $\mathcal{A}$ with respect to input size n to be

$$\inf\{c \mid \text{there exists an algorithm for solving } \mathcal{A} \text{ with a running time of } O(n^c)\}.$$

Since the bichromatic closest pair and Euclidean minimum spanning tree problems have asymptotically same complexities [6,23], their exponents are also equal. The following theorem follows as a consequence of Theorem 2, and the above discussion.

Theorem 3. *The exponent of a $(1+\epsilon)$-spanner of a graph on n unit balls in $\mathbb{R}^k$ is the same as the exponent of a bichromatic closest pair for n points in $\mathbb{R}^k$.*

4 Separators in Spanner Graph

We show that G' has an $O(\sqrt{n}\epsilon^{-3/2} + \epsilon^{-3} \log \rho(\mathcal{D}))$-vertex separator, whose removal leaves two sets with each having at most 7/9 of the original vertices (called 7/9-split) with no edges going across sets. Furthermore, this separator can be found in $O(n \log n)$ time. The algorithm uses a similar approach to that used in [10], in that it uses line segments for partitioning the rectangles. But unlike the spanners, the graphs considered in [10] don't have short edges, and handling these short edges require significantly new approaches and insights. For lack of space we only provide the sketch of the algorithm. Missing details can be found in the full version [12].

We say a vertex crosses a line segment if any edge incident on it in G' crosses the line segment. During each step, the algorithm focuses on one rectangle, called the *active rectangle*. An active rectangle $\mathcal{R}$ has at least 2/3 of the total vertices inside it and there exists a set of $O(\sqrt{n}\epsilon^{-3/2}+\epsilon^{-3}\log\rho(\mathcal{D}))$ vertices which when removed ensures that no remaining vertex has an edge in G' that crosses the boundary of $\mathcal{R}$.

At every step, the algorithm uses two line segments (called *double line separators*) to divide the currently active rectangle. A *horizontal double line separator* of an active rectangle is a set of at most two horizontal line segments that partitions the active rectangle such that there exists a set of $O(\sqrt{n}\epsilon^{-3/2}+\epsilon^{-3}\log\rho(\mathcal{D}))$ vertices which when removed ensures that no remaining vertex crosses either of the vertical line segments (similarly define *vertical double line separator*). The algorithm recursively partitions an active rectangle alternatively with a horizontal or a vertical double line separator and stops when none of the new rectangles created contain enough vertices to become active. The initial active rectangle is the bounding box.

We now summarize the main result of this section in the following theorem.

Theorem 4. *Let G be a k-dimensional ball graph defined on $\mathcal{D}$ and G' be a $(1+\epsilon)$-spanner of G constructed as described above. An $O(n^{1-1/k}\epsilon^{-k+1/2}+\epsilon^{-2k+1}\log\rho(\mathcal{D}))$ vertex separator of G' with 7/9-split can be found in $O(n\log n)$ time.*

4.1 Approximate Proximity Problems Using Spanner

Gao and Zhang [7] discuss many approximate proximity problems for unit disk graphs. Using a well-separated pair decomposition, they show that a unit disk graph can be preprocessed in $O(n\sqrt{n\log n}\epsilon^{-3})$ time, such that subsequent distance queries can be answered with $(1+\epsilon)$-stretch in constant time.

We note that other than the standard advantages of using a sparse spanner for solving approximate proximity problems, the separator helps us to also support fast answering of distance queries in ball graphs. The idea is to use the algorithm of Arikati *et al.* [24] (originally used for answering distance queries in planar graphs) on the spanner G'. Using the same techniques we get the following

Corollary 2. *Let G be a k-dimensional ball graph defined on $\mathcal{D}$ and G' be a $(1+\epsilon)$-spanner of G constructed as described above. The graph G' can be preprocessed in $\tilde{O}(nf(n,\epsilon))$ time and space such that subsequent distance queries under the d_G metric can be answered with $(1+\epsilon)$-stretch in $O(f(n,\epsilon))$ time and with $(2+\epsilon)$-stretch in $O(\log n)$ time, where $f(n,\epsilon)=n^{1-1/k}\epsilon^{-k+1/2}+\epsilon^{-2k+1}\log\rho(\mathcal{D})$.*

Acknowledgments

We would like to Joachim Gudmundsson for telling us about [8] and Michiel Smid for pointing us to [14,4].

References

1. Eppstein, D.: Testing bipartiteness of geometric intersection graphs. In: SODA '04, SIAM, pp. 860–868 (2004)
2. Rajaraman, R.: Topology control and routing in ad hoc networks: A survey. SIGACT News 33, 60–73 (2002)
3. Li, X.Y.: Algorithmic, geometric and graphs issues in wireless networks. Wireless Communications and Mobile Computing 3(2), 119–140 (2003)
4. Erickson, J.: On the relative complexities of some geometric problems. In: CCCG '95, pp. 85–90 (1995)
5. Yao, A.C.C.: On constructing minimum spanning trees in k-dimensional spaces and related problems. SIAM Journal on Computing 11(4), 721–736 (1982)
6. Agarwal, P.K., Edelsbrunner, H., Schwarzkopf, O., Welzl, E.: Euclidean minimum spanning trees and bichromatic closest pairs. Discrete & Computational Geometry 6, 407–422 (1991)
7. Gao, J., Zhang, L.: Well-separated pair decomposition for the unit-disk graph metric and its applications. SIAM Journal on Computing 35(1), 151–169 (2005)
8. Gudmundsson, J.: Constructing sparse t-spanners with small separators. In: Lingas, A., Nilsson, B.J. (eds.) FCT 2003. LNCS, vol. 2751, pp. 86–97. Springer, Heidelberg (2003)
9. Gudmundsson, J.: Personal communication (2006)
10. Fürer, M., Kasiviswanathan, S.P.: Approximate distance queries in disk graphs. In: Erlebach, T., Kaklamanis, C. (eds.) WAOA 2006. LNCS, vol. 4368, pp. 174–187. Springer, Heidelberg (2007)
11. Agarwal, P.K., Matoušek, J.: Dynamic half-space range reporting and its applications. Algorithmica 13(4), 325–345 (1995)
12. Fürer, M., Kasiviswanathan, S.P.: Spanners for geometric intersection graphs (2007) Available at `http://www.cse.psu.edu/~kasivisw/spanner.pdf`
13. Ruppert, J., Seidel, R.: Approximating the d-dimensional complete euclidean graph. In: CCCG '91, pp. 207–210 (1991)
14. Bose, P., Maheshwari, A., Narasimhan, G., Smid, M.H.M., Zeh, N.: Approximating geometric bottleneck shortest paths. Comput. Geom 29(3), 233–249 (2004)
15. Lukovszki, T.: New results on geometric spanners and their applications. PhD. thesis, University of Paderborn (1999)
16. Callahan, P.B., Kosaraju, S.R.: A decomposition of multidimensional point sets with applications to K-nearest-neighbors and N-body potential fields. J. ACM 42(1), 67–90 (1995)
17. Callahan, P.B., Kosaraju, S.R.: Faster algorithms for some geometric graph problems in higher dimensions. In: SODA '93, SIAM, pp. 291–300 (1993)
18. Smid, M.: The well-separated pair decomposition and its applications. In: Gonzalez, T. (ed.) Handbook on Approximation Algorithms and Metaheuristics, Chapman & Hall/CRC (to appear)
19. Har-Peled, S.: Approximation algorithms in geometry. Available at: `http://valis.cs.uiuc.edu/~sariel/research/book/aprx/`
20. Bern, M.W., Eppstein, D., Teng, S.H.: Parallel construction of quadtrees and quality triangulations. Int. J. Computational Geometry & Applications 9(6), 517–532 (1999)
21. Gupta, P., Janardan, R., Smid, M.: Algorithms for some intersection searching problems involving circular objects. International Journal of Mathematical Algorithms 1, 35–52 (1999)

22. Chan, T.M.: Geometric applications of a randomized optimization technique. Discrete & Computational Geometry 22(4), 547–567 (1999)
23. Krznaric, D., Levcopoulos, C., Nilsson, B.J.: Minimum spanning trees in d dimensions. Nord. J. Comput. 6(4), 446–461 (1999)
24. Arikati, S.R., Chen, D.Z., Chew, L.P., Das, G., Smid, M.H.M., Zaroliagis, C.D.: Planar spanners and approximate shortest path queries among obstacles in the plane. In: Díaz, J. (ed.) ESA 1996. LNCS, vol. 1136, pp. 514–528. Springer, Heidelberg (1996)

On Generalized Diamond Spanners*

Prosenjit Bose, Aaron Lee, and Michiel Smid

School of Computer Science, Carleton University, Ottawa, Canada
{jit,michiel}@scs.carleton.ca, aaron@aaronlee.ca

Abstract. Given a set P of points in the plane and a set L of non-crossing line segments whose endpoints are in P, a *constrained* plane geometric graph is a plane graph whose vertex set is P and whose edge set contains L. An edge e has the α-visible diamond property if one of the two isosceles triangles with base e and base angle α does not contain any points of P visible to both endpoints of e. A constrained plane geometric graph has the d-good polygon property provided that for every pair x, y of visible vertices on a face f, the shorter of the two paths from x to y around the boundary has length at most $d \cdot |xy|$. If a constrained plane geometric graph has the α-visible diamond property for each of its edges and the d-good polygon property, we show it is a $\frac{8d(\pi-\alpha)^2}{\alpha^2 \sin^2(\alpha/4)}$-spanner of the visibility graph of P and L. This is a generalization of the result by Das and Joseph[3] to the constrained setting as well as a slight improvement on their spanning ratio of $\frac{8d\pi^2}{\alpha^2 \sin^2(\alpha/4)}$. We then show that several well-known constrained triangulations (namely the constrained Delaunay triangulation, constrained greedy triangulation and constrained minimum weight triangulation) have the α-visible diamond property for some constant α. In particular, we show that the greedy triangulation has the $\pi/6$-visible diamond property, which is an improvement over previous results.

1 Introduction

A graph G whose vertices are points in the plane and edges are segments weighted by their length is a t-spanner (for $t \geq 1$) provided that the shortest path in G between any two vertices x, y does not exceed $t|xy|$ where $|xy|$ is the Euclidean distance between x and y. The value t is the *spanning ratio* or *stretch factor* of the graph. The spanning properties of various geometric graphs has been studied extensively in the literature (see [7,14,10,13] for several surveys on the topic). Our work is a generalization of the result by [3] to the constrained setting. [3] showed that any graph possessing the *diamond* property and the *good polygon* property is a t-spanner where the constant t depends on parameters of each of the two properties.

Before we can state our results precisely, we outline what these properties are, what we mean by the constrained setting and how the spanning ratio of a geometric graph is measured in this setting. Throughout this paper, a graph will

* Research supported in part by NSERC.

F. Dehne, J.-R. Sack, and N. Zeh (Eds.): WADS 2007, LNCS 4619, pp. 325–336, 2007.

refer to a *geometric graph* whose vertex set is a set of points in the plane, and whose edge set is a set of line segments joining pairs of vertices. The edges are weighted by their length. Let P denote a set of points in the plane and L be a set of non-crossing line segments whose endpoints are in P. Two points p and q of P are *visible* with respect to L provided the segment pq does not properly intersect any segment of L. Two line segments intersect properly if they share a common interior point. The visibility graph of P constrained to L, denoted $Vis(P, L)$, is a geometric graph whose vertex set is P and whose edge set contains L as well as one edge for each visible pair of vertices (See Figure 1). A spanning subgraph of $Vis(P, L)$ whose edge set contains L is a geometric graph *constrained to* L. In such a graph, the set L is referred to as the *constrained edges* and all other edges are referred to as *unconstrained edges* or *visibility edges.*

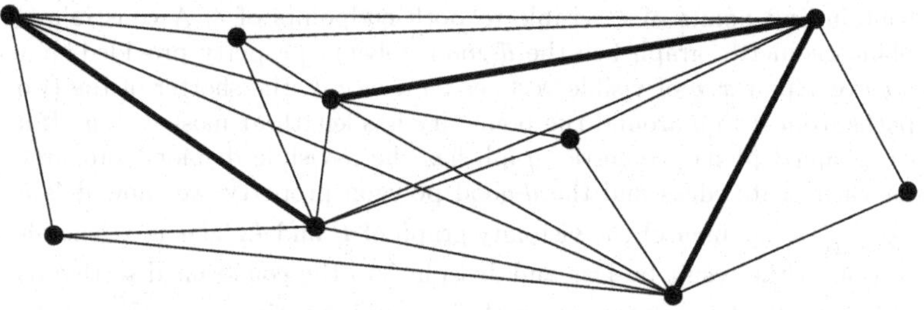

Fig. 1. The visibility graph $Vis(P, L)$ where segments of L are shown in bold

Definition 1. *Let* $t \geq 1$ *be a real number. A constrained geometric graph* $G(P, L)$ *is a* constrained t-spanner *provided that for every visibility edge* $[pq]$ *in* $Vis(P, L)$, *the length of the shortest path between* p *and* q *in* $G(P, L)$ *is at most* t *times the Euclidean distance between* p *and* q. *We refer to* t *as the* spanning ratio *or the* stretch factor *of* $G(P, L)$.

Note that if $G(P, L)$ is a constrained t-spanner, then for every pair of points p, q in P (not just visible edges), the shortest path from p to q in $G(P, L)$ is at most t times the shortest path from p to q in $Vis(P, L)$. We now define the two essential properties.

Definition 2. *Refer to Figure 2. Fix* $\alpha \in (0, \pi/2)$. *A constrained graph* $G(P, L)$ *is said to have the* visible α-diamond property *if, for every unconstrained edge* e *in the graph, at least one of the two isosceles triangles, with* e *as the base and base angle* α, *does not contain any points of* P *visible to the endpoints of* e. *We label this empty triangle as* $\triangle(e)$, *and the apex of* $\triangle(e)$ *as* $a(e)$.

Definition 3. *A constrained plane graph* $G(P, L)$ *has the* d-good polygon property *if for every visible pair of vertices* a *and* b *on a face* f, *the shortest distance from* a *to* b *around the boundary of* f *is at most* d *times the Euclidean distance between* a *and* b.

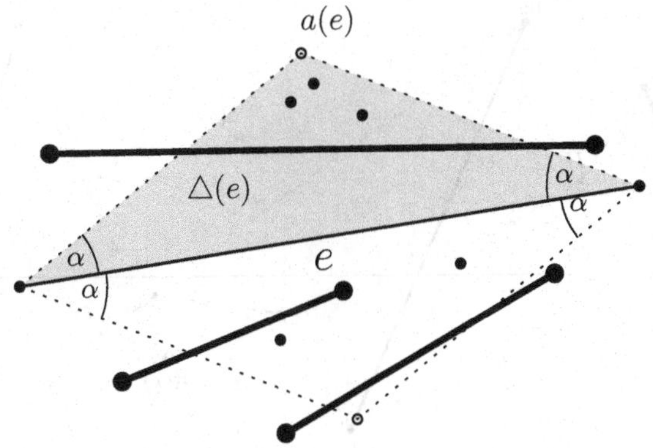

Fig. 2. The edge e has the visible α-diamond property

Our main results are the following:

Theorem 1. *Given fixed $\alpha \in (0, \pi/2)$ and $d \geq 1$, if a constrained plane graph $G(P, L)$ has both the visible α-diamond property and the d-good polygon property, then its stretch factor is at most $\frac{8(\pi-\alpha)^2 d}{\alpha^2 \sin^2(\alpha/4)}$.*

This is a generalization of the result in [3] to the constrained setting as well as a slight improvement on the spanning ratio from $\frac{8d\pi^2}{\alpha^2 \sin^2(\alpha/4)}$ to $\frac{8d(\pi-\alpha)^2}{\alpha^2 \sin^2(\alpha/4)}$.

Theorem 2. *The Constrained Greedy Triangulation has the visible $\frac{\pi}{6}$-diamond property.*

This is an improvement over the results of [3], [4] and [6] on this problem. In [3], they showed that the Greedy Triangulation has the $\pi/8$-diamond property. The results in [6], which is an extension of the results in [4], imply that the Greedy triangulation has the $\arctan(1/\sqrt{5})$-diamond property. Note that $\arctan\left(1/\sqrt{5}\right) \approx 24.1°$.

2 Constructing Spanner Paths

The proof of the main result is constructive. Consider a constrained plane graph $G(P, L)$ that has the visible α-diamond property and the d-good polygon property. Given a pair of points $a, b \in P$ that are visible with respect to L, we show how to construct a path from a to b in $G(P, L)$ whose length is at most $\frac{8d(\pi-\alpha)^2}{\alpha^2 \sin^2(\alpha/4)}$ times the Euclidean distance between a and b.

If $[ab]$ is an edge of $G(P, L)$ then such a path trivially exists. Therefore, assume that $[ab]$ is not an edge of the graph. In this case, either $[ab]$ intersects some edges of the graph or intersects no edges of the graph. In the latter case, this means that the segment $[ab]$ is a chord in a face of $G(P, L)$. The d-good polygon property ensures that the required path exists in this case. In the remainder of

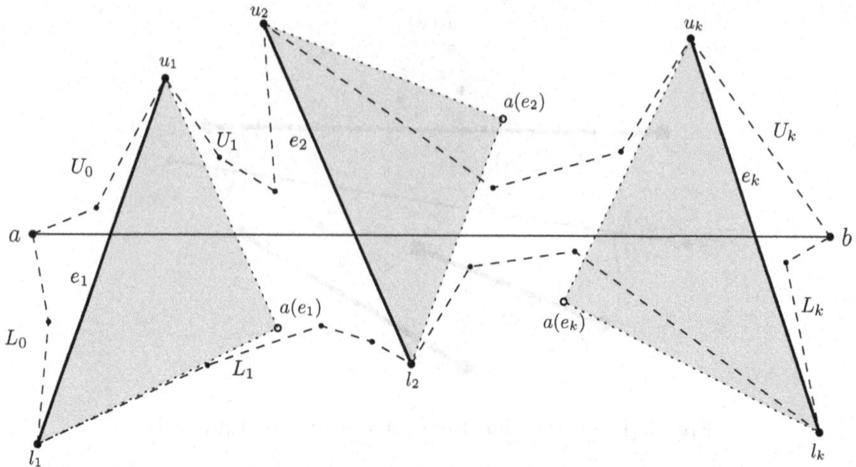

Fig. 3. Illustration of Structures

this section, we show that when $[ab]$ intersects some edges of $G(P, L)$, we can construct a spanner path from a to b.

Re-orient the coordinate system such that $[ab]$ lies on the x-axis. Let $e_1, e_2, \ldots, e_k$ be the edges of $G(P, L)$ that cross $[ab]$ in order from a to b. For simplicity of exposition, assume that none of these edges share a common endpoint. Sharing endpoints is a degenerate situation that only makes the proof simpler. Label the endpoint of e_i above $[ab]$ as u_i and the endpoint below $[ab]$ as l_i. The fact that a and b are visible with respect to L ensures that each of these edges is an unconstrained edge. Moreover, the visible α-diamond property implies that each of these edges is the base of a visibly empty triangle $\triangle(e_i)$ with apex $a(i)$, $1 \leq i \leq k$. Since $G(P, L)$ is a plane graph, e_i and e_{i+1} lie on a common face f. Let U_i be the shortest path from u_i to u_{i+1} in the face f. Since a and b are visible, this implies that U_i is a convex path. Similarly, let L_i be the shortest path from l_i to l_{i+1} in f. Note that the d-good polygon property ensures that there is path in $G(P, L)$ from u_i to u_{i+1} whose length does not exceed d times the length of U_i. Define U_0 (resp. U_k) to be the shortest path from a to u_1 (resp. u_k to b). L_0 and L_k are defined symmetrically. See Figure 3.

We have two different construction methods depending on where the apices of the empty triangles are with respect to $[ab]$. If all of the apices lie on the same side of the line through $[ab]$, we construct a path called a *one-sided* path, otherwise, we construct a *two-sided* path. We first show the construction of one-sided paths and bound their length.

2.1 One-Sided Paths

In this case, we assume that all the apices of the empty triangles lie below $[ab]$. The construction of the one-sided path starts with the union of the U_i. Now, each edge e in U_i can be approximated by a path in $G(P, L)$ whose length is at most $d \cdot e$. Therefore, the length of the one-sided path is at most $d(|U_0| + |U_1| + \ldots + |U_k|)$.

What remains to be shown is that this is a good approximation of $|ab|$ since a and b are visible. Let h be the line through $[ab]$ and h^- be the closed half-plane below h. To obtain a bound on $|U_0|+|U_1|+\ldots+|U_k|$, we consider the following structure $T = h^- \cup \bigcup_{i=1}^{k} \triangle(e_i)$ (See Figure 4). Denote by $T(a,b)$ the portion of the boundary of T between a and b. The fact that each of the triangles $\triangle(e_i)$ is empty of points visible to both endpoints of the edge e_i, all the apices of the empty triangles are below $[ab]$ and each of the U_i is formed by a shortest path imply that no edge of $T(a,b)$ can intersect any of the upper chains U_i. Since each of the upper chains U_i is a shortest path, we conclude the following:

Lemma 1. $\sum_{i=0}^{k} |U_i| \leq |T\,(a,b)\,|$

Before proceeding, we need a simple property about triangles that essentially follows from the sine law.

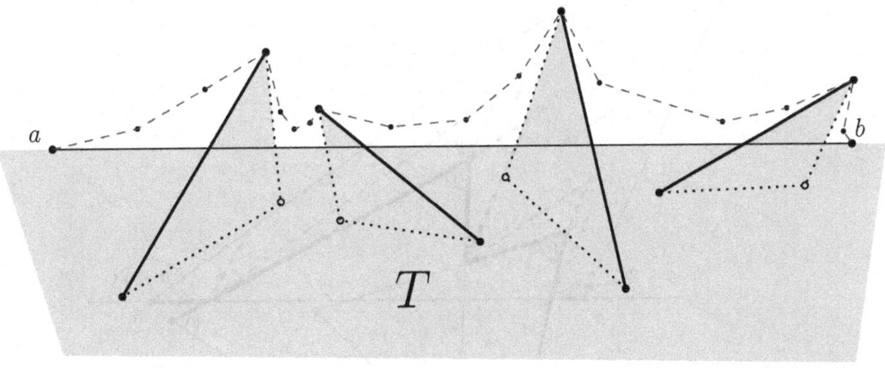

Fig. 4. Using the shaded region T to approximate the length of the upper chains

Lemma 2. *Given a triangle $\triangle(u,v,w)$ such that angle at vertex v is α, we have that $|uv| + |vw| \leq |uw| / \sin(\alpha/2)$.*

We now show how to bound the length of the one-sided path in terms of $|ab|$.

Lemma 3. *The length of a one-sided path from a to b in $G(P,L)$ is at most $|ab|\frac{2(\pi-\alpha)d}{\alpha \cdot \sin(\alpha/4)}$.*

Proof: By Lemma 1, and the d-good polygon property, it suffices to show that $|T\,(a,b)| \leq \frac{2(\pi-\alpha)|ab|}{\alpha \cdot \sin(\alpha/4)}$. Note that since the apex $a(e_i)$ of every empty triangle $\triangle(e_i)$ is below $[ab]$, only two sides of $\triangle(e_i)$ lie in the upper half-plane h^+. Hence, there is a well-defined left edge and right edge for the portion of $\triangle(e_i)$ that lies above $[ab]$.

To simplify the exposition, we assume that π is a multiple of α (a condition that can be easily removed). Partition the empty triangles $\triangle(e_1), \triangle(e_2), \ldots, \triangle(e_k)$ into $\frac{2(\pi-\alpha)}{\alpha}$ groups labelled $G_0, G_{\alpha/2}, \cdots, G_\theta, \cdots, G_{\pi-3\alpha/2}$, such that the left edges of the empty triangles in group G_θ make an angle in the range $\left[\theta, \theta + \frac{\alpha}{2}\right]$ with the x-axis. Since the base angle of the empty triangles is α, we see that the right edges will be in the range $\left[\theta + \alpha, \theta + \frac{3\alpha}{2}\right]$ with the x-axis. This is why the last group ends at $\pi - 3\alpha/2$.

Let T_θ be the union of all the triangles in G_θ with the half-plane h^- below the x-axis. Recall T from Lemma 1. Note that $T = T_0 \cup T_{\alpha/2} \cup \cdots \cup T_{\pi-3\alpha/2}$. Hence, it follows that the length of the boundary of T from a to b is bounded by $|T_0\,(a,b)| + \left|T_{\alpha/2}\,(a,b)\right| + \cdots + \left|T_{\pi-3\alpha/2}\,(a,b)\right|$

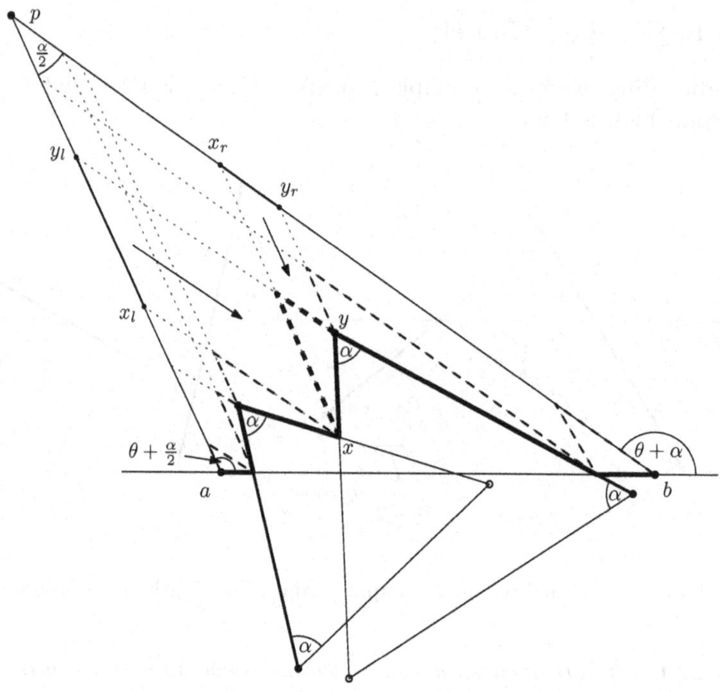

Fig. 5. The triangles of G_θ

Consider the group G_θ, as shown in Figure 5. The edges of the boundary $T_\theta\,(a,b)$ are shown in bold. Let p be the point such that $\measuredangle apb = \frac{\alpha}{2}$, $\measuredangle pab = \theta + \frac{\alpha}{2}$, and $p \in h^+$. By construction, the portion of the triangles in G_θ that lie above the x-axis (and thus $T_\theta\,(a,b)$) are completely contained inside $\triangle pab$.

$T_\theta\,(a,b)$ is a polygonal chain consisting of portions of left edges of empty triangles, portions of right edges of empty triangles, and portions of $[ab]$. Note that the angle restriction implies that the chain is monotone both in the direction pa and the direction pb. To bound the length of an edge xy of $T_\theta\,(a,b)$, project x and y onto pa by translating in a direction parallel to pb and denote the projected vertices on pa by x_l and y_l, respectively. Similarly, project xy onto pb

by translating in a direction parallel to pa resulting in projected vertices x_r and y_r, respectively. The triangle inequality guarantees that $|xy| \leq |x_l y_l| + |x_r y_r|$. Monotonicity guarantees that none of the projected edges of $T_\theta(a,b)$ overlap. Therefore, we have that $|T_\theta(a,b)| \leq |pa| + |pb|$.

Using Lemma 2, it follows that $|pa| + |pb| \leq \frac{|ab|}{\sin(\alpha/4)}$, regardless of the angle θ. Therefore, since there are $\frac{2(\pi-\alpha)}{\alpha}$ many groups, $|T(a,b)| \leq \frac{2(\pi-\alpha)|ab|}{\alpha \cdot \sin(\alpha/4)}$. □

2.2 Two-Sided Paths

We have seen that if every apex $a(e_1), a(e_2), \ldots, a(e_k)$ lies on the same side of the x-axis, then a one-sided path from a to b can be constructed, whose length is $\frac{2(\pi-\alpha)d}{\alpha \sin(\alpha/4)}$ times the length $|ab|$. We now outline the process of constructing a short path from a to b in the case where some apices $a(e_i)$ lie above the x-axis, while others lie below. The one-sided paths either followed the upper chain or the lower chain. In two-sided paths, we may need to cross over from upper chains to lower chains. Recall that U_i is the shortest path from u_i to u_{i+1} and L_i is the shortest path from l_i to l_{i+1}. For each pair of upper and lower chains, U_i and L_i, respectively, we add the unique edge on the shortest path from u_i to l_{i+1} that is not on either chain and the unique edge on the shortest path from l_i to u_{i+1} that is not on either chain. We refer to these two edges as *tangents* between the upper and lower chains.

Divide the set of edges $e_1, e_2, \ldots, e_k$ into two disjoint groups, U and L. U contains the edges that have their apex below $[ab]$, and L contains the edges whose apex is above $[ab]$. For the first group, define the region $T_U = h^- \cup \bigcup_{e \in U} \Delta(e)$. Correspondingly, define the region $T_L = h^+ \cup \bigcup_{e \in L} \Delta(e)$. Let $T_U(a,b)$ denote the upper boundary of T_U between a and b and similarly $T_L(a,b)$ for the lower boundary. Note that the length of $T_U(a,b)$ and $T_L(a,b)$ are each less than $|ab| \frac{2(\pi-\alpha)d}{\alpha \cdot \sin(\alpha/4)}$ as shown in Lemma 3. The two-sided path from a to b is constructed using disjoint portions of T_U, T_L and tangents. Since $|T_U(a,b)| + |T_L(a,b)| \leq 2|ab| \frac{2(\pi-\alpha)d}{\alpha \cdot \sin(\alpha/4)}$, we only need to bound the length of the tangents used.

The two-sided path from a to b is constructed as follows. Without loss of generality, assume that e_1 has its apex below $[ab]$. Let e_{i+1} be the first edge whose apex is above $[ab]$. If no subsequent edge has an apex below $[ab]$, then the path follows the upper chains from a to u_i, follows the path from u_i to l_{i+1} and follows the lower chains to b. This path has length at most $|T_U(a,b)| + |T_L(a,b)|$ since each of the two paths is one-sided and the length of the tangent is subsumed by the unused portions of the upper and lower chain.

The situation where a decision needs to be made on how to proceed is when there are two edges e_i and e_j (with $j > i+1$) having apices $a(e_i)$ and $a(e_j)$ below the x-axis and at least one edge e_k (with $i < k < j$) with apex $a(e_k)$ above the x-axis. We show how to construct a short path from u_i to u_j in this case. There are two possibilities in this case, either the path from u_i to u_j follows the upper chain, or it follows the path from u_i to l_{i+1}, continues on the lower chain until l_{j-1} and follows the path from l_{j-1} to u_j.

The decision whether or not to cross over from the upper chain to the lower chain depends on the tangents. Let $t_a = [u_a, l_a]$ be the tangent on the path from u_i to l_{i+1} and $t_b = [u_b, l_b]$ be the tangent on the path from l_{j-1} to u_j. Extend the tangents t_a and t_b until they intersect. Label their intersection point as C. Note that C may be below or above the x-axis. Label the smaller of the two angles between the two extended tangents as θ. If t_a and t_b are parallel, then C is a point at infinity, and $\theta = 0$. There are two cases to consider depending on the angle θ.

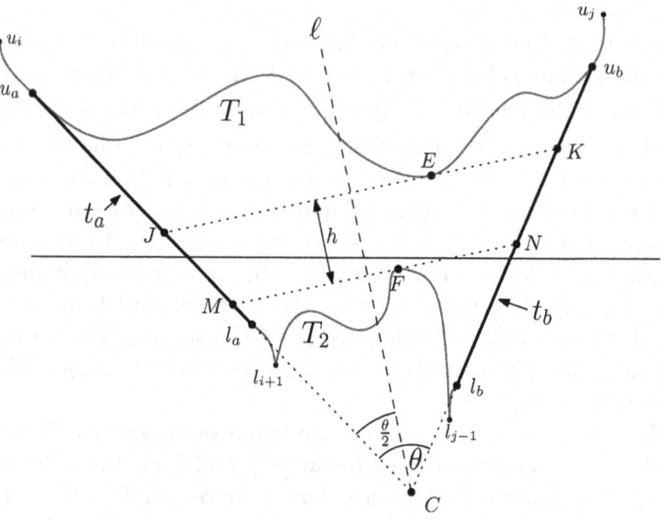

Fig. 6. Case 2: Construct an empty region

Case 1: $\theta \geq \alpha$. In this case, the angle between the tangents is relatively wide. Follow the path from u_i to l_{i+1} using the tangent t_a. Continue on the lower convex chains $L_{i+1} \cdots L_{j-1}$ from l_a until l_b, and cross back up along t_b to u_j.

Case 2: $\theta < \alpha$. Refer to Figure 6 for this case. Label the portion of $T_U(a, b)$ between the upper vertices u_a and u_b of the tangents as T_1. Similarly, label the portion of $T_L(a, b)$ between l_a and l_b as T_2. Consider the region bounded by the tangents t_a and t_b and the boundaries T_1 and T_2. By definition, this region is empty of vertices of G. Bisect the angle θ at C with a line labelled ℓ. (If t_a and t_b happen to be parallel, then ℓ is defined as the line parallel to the tangents, and halfway between them.) Let E be the point on T_1 whose orthogonal projection onto ℓ is lowest, and let F be the point on T_2 whose orthogonal projection onto ℓ is highest. Define h to be the distance between the orthogonal projections of E and F onto ℓ. Define points J and M on t_a, and K and N on t_b, with JEK and MFN perpendicular to ℓ. Let $w = \max(|JK|, |MN|)$.

a) If $h \leq \frac{w}{2\tan(\alpha/2)}$, then follow the same path from u_i to u_j as in Case 1.
b) If $h > \frac{w}{2\tan(\alpha/2)}$, then simply follow the upper chains $U_i, U_{i+1}, \ldots, U_{j-1}$ from u_i and u_j.

We now bound the lengths of the paths constructed in the context of the different cases stated above. Note that the main difficulty is in bounding the length of the tangents since we have a bound on the length of the upper and lower chains.

Lemma 4. *If $\theta \geq \alpha$, the length of the portion of the two-sided path from a to b that is between u_a and u_b is at most $\frac{2}{\sin(\alpha/2)}\big(|T_1| + |T_2|\big)$.*

Proof: Consider the triangle $\triangle(u_a, C, u_b)$. The path follows the two tangents and the lower chain. The length of the lower chain is $|T_2|$. We want to bound the length of the two tangents in terms of $|T_1|$ and $|T_2|$. If C is below $[ab]$, by Lemma 2, we have that $|u_aC| + |Cu_b| \leq |T_1|/\sin(\alpha/2)$. If C is above $[ab]$, by Lemma 2, we have that $|u_aC| + |Cu_b| \leq |T_2|/\sin(\alpha/2)$. Note that these above bounds on the tangents only hold when C does not lie on one of the tangents. Should C land on one of the tangents then an extra $|T_1|$ or $|T_2|$ term can bound the portion of the tangent that lies outside the triangle by the triangle inequality. Putting all the inequalities together completes the proof. □

Lemma 5 (Case 2*a*). *If $\theta < \alpha$, and $h \leq \frac{w}{2\tan(\alpha/2)}$, then the length of the portion of the two-sided path from a to b that is between u_a and u_b is at most $\frac{3}{\sin(\alpha/2)}\big(|T_1| + |T_2|\big)$.*

Proof: (Sketch): The path constructed in this case is identical to the path constructed in Case 1. The path follows the two tangents and the lower chains. The length of the lower chains is $|T_2|$. We need to bound the length of the tangents. Refer to Figure 6.

Note that the tangent t_a is decomposed into three segments: $[u_aJ]$,$[JM]$,$[Ml_a]$. Similarly t_b is decomposed into three segments: $[u_bK]$, $[KN]$, $[Nl_b]$. Since the angle at J in triangle $\triangle(u_aJE)$ is obtuse, we have that $|u_aJ|$ is shorter than the portion of T_1 from u_a to E. By this argument, we have that $|u_aJ| + |u_bK| < |T_1| + |T_2|$. To bound $|JM|$ and $|KN|$, we use the fact that $h \leq \frac{w}{2\tan(\alpha/2)}$. This allows us to show that $|JK|+|KN| \leq 2h/\cos(\alpha/2) \leq |T_1|/\sin(\alpha/2)$. Finally, by elementary trigonometry, we have that $|l_aM|+|l_bN| \leq (|l_aF|+|l_bF|)/\sin(\alpha/2) \leq |T_2|/\sin(\alpha/2)$. Combining the inequalities, we have that the length of the two-sided path is at most $2(|T_1| + |T_2|)/\sin(\alpha/2)$.

Note that this bound only holds when C does not lie on one of the tangents. However, C may lie on one of the two tangents. In this case, an extra $|T_1|/\sin(\alpha/2)$ or $|T_2|/\sin(\alpha/2)$ term needs to be added giving the stated bound[1]. □

A fairly lengthy argument along the same lines allows us to bound the path when $h > \frac{w}{2\tan(\alpha/2)}$.

[1] This is one of the cases that was omitted from the original proof by [3].

Lemma 6. *If $\theta < \alpha$, and $h > \frac{w}{2\tan(\alpha/2)}$, then the length of the portion of the two-sided path from a to b that is between u_a and u_b is at most $\frac{2(\pi-\alpha)}{\alpha\sin(\alpha/4)}(|T_1|+|T_2|)$.*

2.3 The Final Spanning Ratio

We now have all the pieces to prove Theorem 1. From the above lemmas, we have that the maximum length of the path between u_a and u_b is at most:

$$\max\left(\frac{2}{\sin\left(\frac{\alpha}{2}\right)}, \frac{3}{\sin\left(\frac{\alpha}{2}\right)}, \frac{2(\pi-\alpha)}{\alpha\sin\left(\frac{\alpha}{4}\right)}\right)\cdot\left(|T_1|+|T_2|\right) = \frac{2(\pi-\alpha)}{\alpha\sin\left(\frac{\alpha}{4}\right)}\cdot\left(|T_1|+|T_2|\right)$$

Since $|T_U(a,b)| + |T_L(a,b)| \leq 2|ab|\frac{2(\pi-\alpha)d}{\alpha\cdot\sin(\alpha/4)}$, we have that the path from a to b has length at most:

$$\frac{2(\pi-\alpha)}{\alpha\sin\left(\frac{\alpha}{4}\right)}\cdot\Big(|T_U(a,b)|+|T_L(a,b)|\Big) \leq \frac{2(\pi-\alpha)}{\alpha\sin\left(\frac{\alpha}{4}\right)}\cdot 2\left(\frac{2(\pi-\alpha)d|ab|}{\alpha\sin\left(\frac{\alpha}{4}\right)}\right) = \frac{8(\pi-\alpha)^2 d|ab|}{\alpha^2\sin^2\left(\frac{\alpha}{4}\right)}$$

proving Theorem 1.

3 Triangulations That Have the Diamond Property

In this section, we note that three constrained versions of classical triangulations have the visible α-diamond property, namely the constrained Delaunay triangulation, the constrained minimum weight triangulation and the constrained greedy triangulation. Since they are all triangulations, they satisfy the requirements for the d-good polygon property, for $d = 1$. Therefore, all three triangulations are constant spanners where the constant depends on α.

The first constrained triangulation considered is the constrained Delaunay triangulation (CDT), also called the Generalized Delaunay triangulation [11], and the Obstacle triangulation [2]. One of the important properties of the constrained Delaunay triangulation is that for every unconstrained edge e in the graph, there exists a circle C_e such that the endpoints of e lie on the boundary of this circle, and there are no vertices of S that are visible to both endpoints of the edge e [11] [2]. For such an edge e, we refer to C_e as its *visibly empty circle.* The existence of the visibly empty circle for each unconstrained edge implies that the edge has the visible $\frac{\pi}{4}$-diamond property.

Theorem 3. *The Constrained Delaunay Triangulation (CDT) has the visible $\frac{\pi}{4}$-diamond property.*

We note that techniques exploiting additional properties of the Delaunay triangulation have been used to reduce the spanning ratio from the one implied by the visible $\frac{\pi}{4}$-diamond property (see [5], [9] for the unconstrained setting, and [8], [1] for the constrained setting). Currently, the best known spanning ratio for the Constrained Delaunay triangulation is $\frac{4\pi\sqrt{3}}{9}$ as shown by [1]. It is conjectured that the spanning ratio for the Delaunay and Constrained Delaunay triangulation is $\pi/2$.

The second constrained triangulation that we consider is the constrained Greedy triangulation (CGT), which is a generalization of the standard Greedy triangulation [12]. The algorithm for computing such a triangulation is as follows: Sort the edges of $Vis(P, L)$ by length. First insert all the constrained edges to CGT. Next insert the unconstrained edges in sorted order into CGT as long as they do not introduce a crossing. In order to prove the result for the Constrained Greedy Triangulation (CGT), we make extensive use of the following.

Lemma 7. *Let x and y be points of P such that $xy \in Vis(P, L)$, but xy is not an edge of $CGT(P, L)$. Let e be the edge of $CGT(P, L)$ of shortest length that properly intersects the segment xy. Then $|e| \le |xy|$.*

Proof: Recall that the CGT is constructed by considering all possible edges of $Vis(P, L)$ in non-descending sorted order; an edge is inserted only if it does not intersect any previously inserted edge. If the lemma were false, then at the point in the algorithm when xy would be considered for insertion, none of the edges that intersect it would have been considered yet since they are all longer than xy. Hence, with no edges yet crossing xy, the segment joining x and y would be inserted CGT, which is in contradiction to the assumption that xy is not in the triangulation. □

Using this simple lemma, we can show that the Constrained Greedy Triangulation has the visible $\frac{\pi}{6}$-diamond property. The main approach to proving this theorem is by contradiction. If an edge xy of CGT does not have the visible $\frac{\pi}{6}$-diamond property, then both visible triangles adjacent to xy contain at least one point visible to both x and y. We use those points to show that the greedy process would have inserted a shorter edge intersecting xy, thereby contradicting that xy is part of the greedy triangulation. The analysis is similar in approach to the one presented in [3], however, by carefully reviewing each of the 6 cases in their analysis, we are able increase the angle from $\pi/8$ to $\pi/6$.

Theorem 4. *The Constrained Greedy Triangulation (CGT) has the visible $\frac{\pi}{6}$-diamond property.*

The size of the diamond in the above proof is an improvement over the original value of $\frac{\pi}{8}$ shown by [3]. It is also an improvement over the values shown by [4] and [6]. [4], prove that every edge e of the Greedy triangulation has a disc-shaped exclusion region centered at the midpoint of e, of radius $\frac{|e|}{\sqrt{5}} \approx 0.447|e|$. The size of this region is extended in [6] to include the tangents to the region. By basic trigonometry, it can be shown that the largest visible diamond inscribed in this region has $\alpha = \arctan\left(1/\sqrt{5}\right)$. Therefore, $\frac{\pi}{6}$ is currently the largest diamond for the Greedy and constrained Greedy triangulations. If a circle is inscribed inside the diamond of an edge e, its radius is $\frac{|e|}{2} = 0.5|e|$, an improvement on the size of a disc-shaped exclusion region for the Greedy triangulation.

An argument similar to the one for Theorem 4 shows that the constrained minimum weight triangulation has the visible $\pi/8$-diamond property.

Theorem 5. *The Constrained Minimum Weight Triangulation has the visible $\frac{\pi}{8}$-diamond property.*

References

1. Bose, P., Keil, M.: On the stretch factor of the constrained delaunay triangulation. In: Proc. of Voronoi Diagrams in Sci. and Eng., pp. 25–31 (2006)
2. Chew, L.P.: There is a planar graph almost as good as the complete graph. In: Proc. 2nd Annu. ACM Symp. Comp. Geom., pp. 169–177 (1986)
3. Das, G., Joseph, D.: Which triangulations approximate the complete graph. In: Djidjev, H.N. (ed.) Optimal Algorithms. LNCS, vol. 401, pp. 168–192. Springer, Heidelberg (1989)
4. Dickerson, M.T., Drysdale, R.L.S., McElfresh, S.A., Welzl, E.: Fast greedy triangulation algorithms. In: Proc. ACM Symp. Comp. Geom., pp. 211–220 (1994)
5. Dobkin, D.P., Friedman, S.J., Supowit, K.J.: Delaunay graphs are almost as good as complete graphs. In: Proc. IEEE Symp. Found. CS, pp. 20–26 (1987)
6. Drysdale, R.L., Rote, G., Aichholzer, O.: A simple linear time greedy triangulation algorithm for uniformly distributed points. Rep IIG-408, U. Graz (1995)
7. Eppstein, D.: Spanning trees and spanners. In: Handbook of Comp. Geom., pp. 425–461. Elsevier, Amsterdam (2000)
8. Karavelas, M.I.: Proximity Structures for Moving Objects in Constrained and Unconstrained Environments. Ph.D. thesis, Stanford University (2001)
9. Keil, J.M., Gutwin, C.A.: Classes of graphs which approximate the complete Euclidean graph. Disc. Comp. Geom. 7, 13–28 (1992)
10. Knauer, C., Gudmundsson, J.: Dilation and detour in geometric networks. In: Handbook of approximation algorithms and metaheuristics, CRC Press, Boca Raton (2007)
11. Lee, D.T.: Proximity and reachability in the plane. Report R-831, Dept. Elect. Engrg., Univ. Illinois, Urbana, IL (1978)
12. Manacher, G.K., Zobrist, A.L.: Neither the greedy nor the Delaunay triangulation of a planar point set approximates the optimal triangulation. Inform. Process. Lett. 9, 31–34 (1979)
13. Narasimhan, G., Smid, M.: Geometric Spanner Networks. Cambridge University Press, Cambridge (2007)
14. Smid, M.: Closest-point problems in computational geometry. In: Handbook of Comp. Geom., pp. 877–935. Elsevier, Amsterdam (2000)

The k-Resource Problem on Uniform and on Uniformly Decomposable Metric Spaces*

Marcin Bienkowski[1,**] and Jarosław Kutyłowski[2,***]

[1] Institute of Computer Science, University of Wroclaw, Poland
[2] International Graduate School of Dynamic Intelligent Systems, and Heinz Nixdorf Institute, University of Paderborn, Germany

Abstract. We define a natural generalization of the prominent k-server problem, the k-resource problem. It occurs in metric spaces with some demands and resources given at its points. The demands may vary with time, but the total demand may never exceed k. The goal of an online algorithm is to satisfy demands by moving resources, while minimizing the cost for transporting resources.

We give an asymptotically optimal $\mathcal{O}(\log(\min\{n, k\}))$-competitive randomized algorithm and an $\mathcal{O}(\min\{k, n\})$-competitive deterministic one for the k-resource problem on uniform metric spaces consisting of n points. This extends known results for paging to the more general setting of k-resource. Basing on the results for uniform metric spaces, we develop a randomized algorithm solving the k-resource and the k-server problem on metric spaces which can be decomposed into components far away from each other. The algorithm achieves a competitive ratio of $\mathcal{O}(\log(\min\{n, k\}))$, provided that it has some extra resources more than the optimal algorithm.

1 Introduction

In our work we define an extension of k-server, the k-resource problem. We introduce the definition of both problems for general metric spaces, whereas later we are addressing specific finite metric spaces, like the uniform or decomposable metric.

The k-server problem [7] is defined on a metric space $(\mathcal{X}, \xi)$, where $\xi(\cdot, \cdot)$ is a function measuring the distance between two points of space $\mathcal{X}$. An input consists of sequence of requests to points from $\mathcal{X}$. Upon seeing a request to point $x \in \mathcal{X}$, an algorithm has to move its servers, so that at least one server is placed at x. The costs of transports are defined by the distance function ξ. The goal

* Extended abstract. The full version of this paper is available online.

** Supported by MNiSW grant number N206 001 31/0436, 2006-2008.

*** Partially supported by the EU within the 6th Framework Programme under contract 001907 (DELIS) and by the DFG-Schwerpunktprogramm 1183: Organic Computing, project "Smart Teams: Local, Distributed Strategies for Self-Organizing Robotic Exploration Teams".

F. Dehne, J.-R. Sack, and N. Zeh (Eds.): WADS 2007, LNCS 4619, pp. 337–348, 2007.

of an algorithm solving the k-server problem is to minimize the total distance travelled by its servers in the runtime.

The setting of the k-server problem is powerful enough to model many applications in computer science, as well as in many other aspects of everyday life. On the other hand, the model has also some deficiencies. The k-resource problem, a very natural extension defined in this paper, has the advantage of dealing additionally with two of those: we allow requests to have different durations (whereas in k-server a request is regarded as processed after one unit time step) and we allow to request more than one server to a point of the metric. An application example for the k-resource could be a robotic scenario, where tasks appearing dynamically in a terrain impose certain demands (for some periods) on the number of robots that have to be sent to handle them.

We define the k-resource problem to be, similarly to k-server, based on a finite metric space $(\mathcal{X}, \xi)$. In each point of $\mathcal{X}$ we have some demand for resource units. The resources can be thought of as workers or computational resources. The demands are arbitrary integers, but their sum is guaranteed to be at most some fixed integer k. Time is slotted and divided into time steps. The input sequence contains changes of demand: in every time step, at exactly one place the demand decreases or increases by 1. Such assumption is typical also for other problems and models the scenarios in which demands may vary, but the pace of changes is restricted. At the end of any time step, the algorithm has to *cover* all the demands by its resource units, which means that the number of resources available at any node has to be at least the demand at this node. In order to do that, the algorithm may move the units between the points of $\mathcal{X}$, paying the sum of distances travelled by the units.

More formally, let the *resource vector* X_{t-1} denote the algorithm's distribution of resources to points in the metric space at the beginning of step t. In step t, the algorithm is given a *demand vector* d_t, which describes the demand for resources in all points of the metric. We allow only input sequences, for which $\sum_{i=1}^{n} |d_{t-1}[i] - d_t[i]| = 1$ and $|d_t| \leq k$ for all time steps t. Additionally, we assume that at the very beginning of the input sequence, X_0 is any but fixed and $|d_0| = 0$. At the end of a time step, an algorithm has to move resources between metric points, so that $X_t \geq d_t$, where X_t is the new resource vector. The algorithm pays $\xi(i, j)$ for moving a resource unit from i to j; its cost in step t is summed over all movements performed.

We consider the k-resource problem for two types of finite metrics. In the *uniform* metric space, the distances between two different points of the space are equal to 1. A metric space $(\mathcal{X}, \xi)$ is *uniformly decomposable* if it can be partitioned into a set Λ of disjoint components, each being a uniform metric space. We require that for any two distinct points x and y, $\xi(x, y) = 1$ if they belong to the same component, and $\xi(x, y) = \Theta(\Gamma)$ if they belong to different components. We will impose some restrictions on the *separation* between components given by Γ. Uniformly decomposable metrics are a natural extension of the uniform metric, capable of capturing some aspects of locality. In the following we will use the notions *metric space* and *graph* interchangeably. Whenever we speak of a graph, we mean a complete graph with nodes corresponding to points of $\mathcal{X}$, and with edge weights given by function ξ.

In this paper we are interested in online algorithms, i.e. the ones that have to make decisions in time step t solely on the basis of the input sequence up to time step t, and without any information about future requests. To analyze the efficiency of our algorithms we use competitive analysis [9], and — on any input sequence — compare the cost of our algorithm and the optimal offline schedule. For any algorithm A and any input sequence σ, we denote the cost of A by $\mathcal{C}_A(\sigma)$. We say that a deterministic algorithm A is α-competitive if there exists a constant β, such that for all sequences σ, it holds that $\mathcal{C}_A(\sigma) \leq \alpha \cdot \mathcal{C}_{\text{OPT}}(\sigma) + \beta$, where OPT denotes the optimal *offline* algorithm. If A is a randomized algorithm, then in the definition above we replace $\mathcal{C}_A(\sigma)$ by its expected value. The c-resource augmented version of the k-resource problem allows an online algorithm to use $(1 + c) \cdot k$ resources, whereas the offline schedule has to fulfill the same task using only k resource items.

Our contribution. We start our investigation of the k-resource problem by presenting its relation to the k-server problem. This yields lower bounds on the competitiveness of any algorithm solving k-resource, as shown in Section 2. The main results of the paper are divided into two parts. In Section 3, we consider the k-resource problem on uniform metric spaces consisting of n points. We present a deterministic and a randomized algorithm, both achieving asymptotically optimal competitive ratios: $\mathcal{O}(\min\{n, k\})$ and $\mathcal{O}(\log(\min\{n, k\}))$, respectively. In fact, our algorithms are able to solve some instances of a more general problem, the so-called *dynamic-resource* problem, where the number of available resources can change in each step.

The results above allow us to solve the k-resource problem on uniformly decomposable metric spaces with $\Gamma \geq k$. The key issue for this kind of metrics is to use resources available locally in the component and transport new ones from other components only rarely. In Section 4 we give a randomized algorithm with a competitive ratio $(1+1/c)\cdot\mathcal{O}(\log(\min\{n, k\}))$, where n is the size of the largest component of the metric space and c denotes the resource augmentation factor (see Theorem 3). By the relations between the k-resource and k-server problems, the same result holds for k-server. The competitiveness does not depend on the number of components in the decomposition.

Due to space limitations some proofs are only sketched and technical ones are omitted. All the proofs are available in the full version of the paper.

Related work. The k-server problem, introduced by Manasse et al in [7], has been prominent in the research on online problems from the advent of competitive analysis. Currently the best algorithm for general metrics, WFA [6], is $(2k - 1)$-competitive, which nearly matches the lower bound of k by [7].

Much effort was put in the search for randomized algorithms which could perform better. The results for general metric spaces are unsatisfactory: whereas there exists a lower bound on the competitive ratio of $\Omega(\log k/ \log\log k)$ [1,2], currently there is no randomized algorithm better than k-competitive. There are, however, several results for specific metric spaces that break the $o(k)$ barrier. A classical result [5] shows a randomized $\mathcal{O}(\log k)$-competitive algorithm for the uniform metric space. Recently, results on other metric spaces were obtained: the work by Csaba and Lodha [4] provides an $\mathcal{O}(n^{2/3} \log n)$-competitive algorithm for n equally spaced points placed on a line; Bartal and Mendel [3] generalized this work for growth-rate bounded graphs. Seiden [8] has related the k-server problem on decomposable metric spaces to unfair metrical task systems.

Our uniformly decomposable metric is a specific case of the metrics considered by Seiden [8]. The difference is that Seiden is dealing with non-uniformly recursively decomposable metrics with a restriction $\xi(x,y) \geq \Gamma$ for points x, y from distinct components. Additionally, his results do not require resource augmentation. On the other hand, for his general setting the presented competitive ratio is approximately $\mathcal{O}(z \log^2(k+z))$ where $z = |\Lambda|$.

Our result for k-server on uniformly decomposable metric spaces is one of the few, which relies on resource augmentation. Another successful applications are due to Young [10] for the weighted caching problem and Sleator and Tarjan [9] for paging.

2 Lower Bounds

It is quite straightforward that the k-resource problem is capable of modelling k-server instances.

Lemma 1. *Fix any metric space $(\mathcal{X}, \xi)$ and integers $k' \leq k$. If there exists an α-competitive (randomized) algorithm A for k-resource problem on $(\mathcal{X}, \xi)$, then there exists an α-competitive (randomized) algorithm A' for k'-server problem on $(\mathcal{X}, \xi)$.*

We may now apply known lower bounds for k-server mentioned in the introduction by relating the k-resource problem to the $(\min\{k, n-1\})$-server problem, where n is the number of points in $\mathcal{X}$.

Corollary 1. *Fix any space $(\mathcal{X}, \xi)$ consisting of n points, any integer k, and any α-competitive algorithm A for the k-resource problem. Let $f = \min\{k, n-1\}$. If A is deterministic, then $\alpha \geq f$; if A is randomized, then $\alpha = \Omega(\log f / \log\log f)$. Moreover, if $(\mathcal{X}, \xi)$ is a uniform metric, then for a randomized A, $\alpha = \Omega(\log f)$.*

3 Uniform Metric Spaces

In this section we investigate the k-resource problem on a uniform graph of $n \geq 2$ nodes. First, we present some definitions and notations. Then, we specify a very general class of *phase-based algorithms* and prove crucial properties about them. It appears that any (deterministic) algorithm from this class is $\mathcal{O}(\min\{n, k\})$-competitive. Finally, we show that by carefully using randomness, we are able to substantially reduce this factor to $\mathcal{O}(\log(\min\{n, k\}))$. By Corollary 1, our results are asymptotically optimal.

Preliminaries. Formally, by a *vector* we understand a vector of n integers. For any vector A, let $|A| = \sum_{i=1}^{n} A[i]$. We say that a vector A covers B (we write $A \geq B$) if $A[i] \geq B[i]$ for all $1 \leq i \leq n$. For any two vectors A and B we define the *distance* between them as $\delta(A, B) := \frac{1}{2} \cdot \sum_{i=1}^{n} |A[i] - B[i]|$. In these terms, in step t, we require that $X_t \geq d_t$ and the cost of the algorithm on the uniform metric is equal to $\delta(X_{t-1}, X_t)$. In the following we omit subscript t if it can be deduced from the context.

Dynamic-resource (DR) problem. Since we want to use our algorithms for the uniform spaces as building blocks in the construction for more complicated spaces, we have to slightly generalize the model. Precisely, at the beginning of each step t, the number of resources available may change. If it increases, the algorithm may choose where to place the resources; if it decreases, the algorithm may choose which units are removed. We also generalize the competitive analysis for this problem: the number of units available to the online algorithm in step t (denoted by k_t) is not necessarily equal to the number of units at optimal offline algorithm's disposal (k_t^{OPT}). Naturally, in each step $|d_t| \leq k_t$ and $|d_t| \leq k_t^{\mathrm{OPT}}$.

We call such extended problem the *dynamic-resource problem* (DR). In general, it seems to be much harder than k-resource, and we do not aim at constructing competitive algorithms for it. Instead, we construct algorithms which perform well on some instances of DR, in particular when $k_t \geq k_t^{\mathrm{OPT}}$, k_t^{OPT} is non-increasing, and k_t is non-decreasing. Note that the setting $k_t \equiv k_t^{\mathrm{OPT}} \equiv k$ corresponds to the original k-resource problem.

Division of input into phases. First, we show how to deterministically partition the whole input sequence σ into phases; each phase consist of several time steps. The procedure is described in Figure 1. In each phase we track the maximum demand occurring at any node, and we store this information in vector D_t. Within one phase, D is non-decreasing and since the change of d is restricted, only one entry of D can change, and only by 1. At the beginning of the phase P_q we store the current number of units, k_t, in the variable κ_q. Note that a phase P_q may end because either $k_t < \kappa_q$ or $|D_t| > \kappa_q$. If the former condition holds, we call such P_q *unfair*, otherwise P_q is *fair*. Note that any k-resource instance consists only of fair phases. As a byproduct, we get that within a phase $|d_t| \leq |D_t| \leq \kappa_q$. Note that the partitioning into phases depends only on the input sequence and can be computed online.

3.1 Phase-Based Algorithms for the DR Problem

In this section we construct a class of *phase-based* algorithms. Although the algorithms will always produce a feasible solution for the DR problem, we prove their their competitiveness in fair phases only. This immediately yields their competitiveness on k-resource instances.

The proof consists of two parts: in the first one we relate the cost of the optimal offline algorithm in each two consecutive phases P_{q-1} and P_q to the difference of the values of D at the end of these phases; in the second one we relate the cost of our algorithm to the same amount.

```
subroutine StartPhase (q, t)
    P_q ← {t};  κ_q ← k_t;  D_t ← d_t

q ← 1; StartPhase (q, 1)                              /* P_q = current phase */
for t ← 2 to |σ| do
    for each node i let D_t[i] ← max_{t'∈P_q} d_{t'}[i]
    if k_t < κ_q or |D_t| > κ_q
       then q ← q + 1;  StartPhase (q, t)            /* new phase starts now */
       else  P_q ← P_q ∪ {t}
return (P_1, P_2, ..., P_q)
```

Fig. 1. Division of input into phases

Algorithm construction. A phase-based algorithm PB bases its choices solely on the contents of vector D, namely it always tries to maintain an *invariant* $X \geq D$. This simplifies the analysis, as we may assume that the adversary manipulates vector D directly, instead of changing the demands. Moreover, in phase P_q, the algorithm uses only κ_q units, and we explicitly assume it in the analysis, i.e. $|X| = \kappa_q$. The surplus of units is left untouched by the algorithm and is stored in a fixed node v_S. If the real number of units decreases, the surplus decreases, and if surplus diminishes below zero, it implies the end of the phase because of the condition $k_t < \kappa_q$.

First, we describe the behavior of PB in the first step of a phase P_q. If the preceding phase was fair, then $\kappa_q \geq \kappa_{q-1}$. If this inequality is strict, then the algorithm has to increase the number of resources it is actually using. We simply assume that at the very beginning of P_q, the vector X increases by the surplus units, so that $|X| = \kappa_q$. This introduces no additional cost to the algorithm. If the preceding phase was unfair, then the situation is a little bit more complicated, because $\kappa_q < \kappa_{q-1}$, and the algorithm has to remove some units. In such case, in the first step a *reorganization* is performed, i.e. PB moves all the units, so that the demand is fulfilled exactly: $X[i] = d[i]$ for each node i but v_S, where all the remaining units are stored.

To precisely describe the behavior of PB within a phase, we divide the nodes into three groups. A node i is called *saturated* when $X[i] = D[i]$, *unsaturated* when $X[i] > D[i]$, and *infringing* when $X[i] < D[i]$. It usually makes a difference, when these conditions are checked: *before* or *after* the algorithm's decision in step t. In the former case, D_t is compared with X_{t-1}, and in the latter with X_t.

Phase-based algorithms are *lazy* (as introduced earlier for the k-server problem in [7]). It means that if the invariant holds at the beginning of time step t, i.e. if $X_{t-1} \geq D_t$, the algorithm does nothing ($X_t = X_{t-1}$). Otherwise, note that X_{t-1} covers D_{t-1} and since D may change only at one node, say j, there is at most one infringing node j. The algorithm may not move any of the units from saturated nodes without further violating the invariant, and therefore it considers the set of unsaturated nodes $\mathcal{U}_t = \{i \in V : X_{t-1}[i] > D_t[i]\}$. Since throughout the phase $|D_t| \leq \kappa_q = |X_t|$ and there exists one infringing node,

$\mathcal{U}_t$ is non-empty. PB chooses now one node $i \in \mathcal{U}_t$, according to some rule called $\mathcal{U}$*-rule*, and moves one unit from node i to j (we write that j *borrows* one unit from i). Note that PB is not fully defined, as we have some freedom in choosing the $\mathcal{U}$-rule. It appears that we may show some properties of PB and the bounds on its performance which hold for any $\mathcal{U}$-rule.

Basic observations and lower bound on OPT. We start with an easy observation on the structure of $\mathcal{U}_t$ sets and node saturation process.

Lemma 2. *For any two consecutive steps t and $t+1$ from one phase, $\mathcal{U}_{t+1} \subseteq \mathcal{U}_t$. Additionally, at the very end of a fair phase all nodes are saturated.*

Now fix any phase P_q. For the sake of analysis, if reorganization of X occurs at the end of the first step of P_q, we assume that it is performed at the very beginning of P_q, before the demands are presented to the algorithm in the first step. This assumption simplifies the notations below.

Let F_q be the resource vector X at the very beginning of P_q (but after the reorganization step, if any) and let F'_q be the resource vector at the very end of P_q. Obviously, it holds that $|F_q| = |F'_q| = \kappa_q$. If at some node we need a resource unit, we borrow it from a node which has too many units compared to its current demand. If we always borrowed from a "correct" node (one that will not need this unit anymore in this phase), we would pay exactly $\delta(F_q, F'_q)$. It appears that *on fair phases* no algorithm which uses few resources can perform asymptotically better.

Lemma 3. *For any two consecutive fair phases P_{q-1} and P_q, any algorithm for the DR problem, which has at most $\min\{\kappa_{q-1}, \kappa_q\}$ units in these phases, has to pay at least $\delta(F_q, F'_q)/2$.*

Tokens and cost in one phase. In an online setting it is usually not possible to borrow units in an optimal way. If — to fulfill the demand — node i borrows from some node j, which will need this unit in the future, we only get rid of the deficiency for a while and when the demand at node j increases appropriately, it will have to borrow from some other node.

To analyze the performance of our algorithm we introduce a concept of *tokens*. One may think of a token as a kind of currency used to pay for units sent from remote nodes. The phase begins with no tokens in the graph; at the end of a phase all tokens are removed. Whenever node i borrows a unit from node j it has to pay j, i.e. give j a token in exchange. If i has no tokens, then prior to the exchange it has to *produce* one. In effect, the cost paid by the algorithm in one phase is equal to the number of node changes of all the tokens produced in this phase. We denote the number of tokens at node i at the end of time step t by $T_t[i]$. On the basis of the description above, one can deduce that only at most $\delta(F_q, F'_q)$ tokens are produced during any phase P_q.

Bounds for the PB algorithms. The following lemma shows an essential upper bound on the cost of the PB algorithm in one phase.

Lemma 4. *Let* UNF_q *be a binary indicator variable denoting whether phase* P_q *was unfair. For any phase* P_q*, it holds that* $C_{\text{PB}}(P_q) \leq \min\{n-1, \kappa_q\} \cdot \delta(F_q, F'_q) + \kappa_q \cdot \text{UNF}_{q-1}$.

Proof. If the previous phase was unfair, then the cost of the reorganization is at most κ_q. By our earlier observation, the number of produced tokens is at most $\delta(F_q, F'_q)$ and therefore it is sufficient to show that each token is moved at most $\min\{n-1, \kappa_q\}$ times.

Fix any time step t. First, note that if node i borrows a unit from node j, then $i \notin \mathcal{U}_t$ and $j \in \mathcal{U}_t$. Therefore a token is sent from the node outside $\mathcal{U}_t$ to the node inside $\mathcal{U}_t$. Let y denote the size of set $\mathcal{U}_t$ when the first borrowing of the phase takes place. We have $y \leq n-1$ and $y \leq \kappa_q$. Since, by Lemma 2, $\mathcal{U}_{t+1} \subseteq \mathcal{U}_t$, a token can be sent only y times. This concludes the proof. □

When we combine Lemmas 3 and 4, we can relate the cost of the optimal algorithm in two consecutive fair phases P_{q-1} and P_q to the cost of the algorithm PB in P_q. We can extend this relation to whole contiguous sequences of fair phases. In particular, we get the following result.

Theorem 1. *Any (deterministic) phase-based algorithm* PB *is* $\mathcal{O}(\min\{n, k\})$*-competitive for the k-resource problem.*

3.2 Randomized Phase-Based Algorithm for DR Problem

In this section we show a randomized phase-based algorithm for the DR problem, which — on k-resource instances — is $\mathcal{O}(\log(\min\{k, n\}))$-competitive. We only have to specify the $\mathcal{U}$-rule: the algorithm chooses a node with the smallest number of tokens. If there is more than one such node, it is chosen uniformly at random. We denote the resulting algorithm by PBR.

Distribution of tokens. Our goal is to define the whole game between the adversary and the algorithm in terms of changes in token positions. For each time step t we introduce the notion of set $\mathcal{U}'_t = \{i \in V : X_t[i] > D_t[i]\}$, i.e. the set of all nodes which are unsaturated *after* the algorithm moves its unit. Similarly to the Lemma 2, we get that $\mathcal{U}'_{t+1} \subseteq \mathcal{U}_{t+1} \subseteq \mathcal{U}'_t$. Let ℓ_t denote the minimum number of tokens at a node from $\mathcal{U}'_t$. By a straightforward induction, we may show that each node from $\mathcal{U}'_t$ has either ℓ_t or $\ell_t + 1$ tokens.

The key to the analysis is the set $\mathcal{A}_t = \mathcal{U}'_t \cup \{i \in V : T_t[i] = \ell_t + 1\}$. Obviously, each node from $\mathcal{A}_t$ contains ℓ_t or ℓ_t+1 tokens. The following lemma redefines the whole game between the algorithm and the adversary in terms of token shifting.

Lemma 5. *In step* $t+1$*, there are two possible actions concerning tokens and set* $\mathcal{A}$ *that can be incurred by the adversary.*

A. A token is moved from $i \notin \mathcal{A}_t$ *to* $j \in \mathcal{A}_t$*. If* i *had no tokens, then one token was produced at* i*.*

B. A node i *is removed from set* $\mathcal{A}_t$*. If* i *had* $\ell_t + 1$ *tokens,* ℓ_t *tokens remain at* i *and one is moved to* $j \in \mathcal{A}_t$*.*

In both actions, node j is chosen uniformly at random amongst these nodes of $\mathcal{A}_t$ which have ℓ_t tokens. Additionally, as a result of Action A, set $\mathcal{A}$ may decrease.

The adversary can compute the set $\mathcal{A}_t$, but it does not know which nodes have ℓ_t tokens and which $\ell_t + 1$. Let r_t be the number of $\mathcal{A}_t$ nodes with $\ell_t + 1$ tokens; it is possible to show that such nodes are distributed uniformly in $\mathcal{A}_t$.

Number of token movements. In order to bound the expected number of token movements in one phase, we construct a potential function Φ, which relates this amount to the current distribution of tokens.

We fix any phase P_q and any time step $t \in P_q$. Let $L_t = \sum_{i \notin \mathcal{A}_t} T_t[i]$, the number of tokens outside set $\mathcal{A}_t$. Let

$$\Phi_t = |T_t| - \delta(F_q, F'_q) \cdot H_{|\mathcal{A}_t|} - L_t \ ,$$

where $H_i = \sum_{j=1}^{i} 1/j$. Below we show that the expected cost of PBR in one step is bounded by the change in the potential.

Lemma 6. *In time step $t+1$, $\mathbf{E}[\mathcal{C}_{\mathrm{PBR}}(t+1)] \leq \mathbf{E}[\Phi_{t+1} - \Phi_t]$.*

Proof. By Lemma 5, we have to consider two types of events.

Assume that Action A occurs. If a token is produced at node $i \notin \mathcal{A}_t$, both $|T|$ and L increase by 1, which does not change the potential. The cost of moving a token is paid by the decrease of L by 1. Finally, set $\mathcal{A}$ may decrease, but this can only increase the change in the potential.

Assume that Action B occurs. Then $\mathcal{A}_{t+1} = \mathcal{A}_t \setminus \{i\}$ and ℓ_t tokens are removed from set $\mathcal{A}$. $\mathbf{E}[\mathcal{C}_{\mathrm{PBR}}(t+1)] = r_t/|\mathcal{A}_t|$ because $r_t/|\mathcal{A}_t|$ is the probability of moving a token. Therefore, the expected change in the potential is equal to

$$\mathbf{E}[\Phi_{t+1} - \Phi_t] = \delta(F_q, F'_q) \cdot [H_{|\mathcal{A}_t|} - H_{|\mathcal{A}_{t+1}|}] - \ell_t = \frac{\delta(F_q, F'_q)}{|\mathcal{A}_t|} - \ell_t \leq \frac{r_t}{|\mathcal{A}_t|},$$

where the last inequality holds because the number of tokens in $\mathcal{A}_t$, $\ell_t \cdot |\mathcal{A}_t| + r_t$, is at most $\delta(F_q, F'_q)$. □

Since it always holds that $|\mathcal{A}_t| \leq \min\{n, \kappa_q\}$ and $|T_t| \leq \delta(F_q, F'_q)$, the absolute value of Φ is bounded by $\mathcal{O}(\log(\min\{n, \kappa_q\})) \cdot \delta(F_q, F'_q)$. By Lemma 6, the same bound holds for the expected number of token movements in phase P_q, and thus we get the following lemma.

Lemma 7. *For any phase P_q, $\mathbf{E}[\mathcal{C}_{\mathrm{PBR}}(P_q)] = \mathcal{O}(\log(\min\{n, \kappa_q\})) \cdot \delta(F_q, F'_q) + \kappa_q \cdot \mathrm{UNF}_{q-1}$.*

Theorem 2. PBR *is $\mathcal{O}(\log(\min\{n, k\}))$-competitive for the k-resource problem.*

4 Uniformly Decomposable Metric Spaces

Up to this point we have been considering the k-resource problem for uniform metric spaces. We now turn our attention to a metric space $(\mathcal{X}, d)$ which is uniformly decomposable into a set Λ of components, each being a uniform metric space. The distances between components are denoted by Γ and $z = |\Lambda|$.

Outline of the solution. To solve the k-resource problem on a uniformly decomposable metric space, we have to care about an appropriate resource assignment on two levels: the available resources must be split accordingly between the components of the metric space and they must be assigned to single points of the metric space within each of the components.

The resource assignment in each of the components is managed by separate instances of algorithms for the dynamic-resource problem (we use the algorithm PBR introduced in the previous section). Note that if we look only at the higher level and we treat each component as one point of the space, we have a usual k-resource game on a uniform metric. However, if we solve it naively, neglecting what really happens inside of particular components, we obtain an inefficient algorithm. For example, if the whole activity takes place in one component, but the total demand in this component is constant, the naive algorithm on higher level does nothing and may assign much fewer resources to this component than the optimal algorithm. Therefore, the algorithm PBR working inside this particular component has no chance to be competitive.

Thus, the PBR algorithms running inside the components have to give the higher level assignment algorithm *hints* in which components the resource demand changes in a significant manner. Nevertheless, even with these hints, the problem seems to be difficult and therefore we give to our algorithm a little bit more resources than to the optimal one (i.e. we use resource augmentation). This guarantees that there will be many phases in which PBR algorithms have at least as many resources as OPT. By the previous section, we know that the ratio between their costs in such phases is at most logarithmic in k and in the number of nodes in the specific component.

Upper level of DMR. We introduce an algorithm DMR. It works by using separate instances PBR_λ of the algorithm PBR in each component $\lambda \in \Lambda$. After the adversary changes the demands in component λ, the algorithm DMR checks whether more resources must be moved to component λ. If necessary, this movement is performed and afterwards algorithm PBR_λ is used to handle that request with the resources available locally in λ.

Let $F_q(\lambda)$, $D_t(\lambda)$, $I_q(\lambda)$, and $k_t(\lambda)$ denote, respectively, the variables F_q, D_t, I_q, and k_t for the algorithm PBR_λ (as introduced in the analysis of PBR). For any step t belonging to phase q we define a *padding* $g_t(\lambda)$ as

$$g_t(\lambda) = \sum_{i: D_t(\lambda)[i] > F_q(\lambda)[i]} (D_t(\lambda)[i] - F_q(\lambda)[i]) \ . \tag{1}$$

Let $w[\lambda]$ denote the total demand for resources in a component λ, i.e. $w_t[\lambda] = \sum_{p \in \lambda} d_t[p]$. The assignment of resources to components is performed by an algorithm which depends on changes of the padding g. In particular, we construct the following sequence of vectors:

$$W_t[\lambda] = \begin{cases} \max\{W_{t-1}[\lambda] + 1/\Gamma, w_t[\lambda]\} & \text{if } t > 2 \text{ and } g_{t-1}(\lambda) > g_{t-2}(\lambda), \\ \max\{W_{t-1}[\lambda], w_t[\lambda]\} & \text{otherwise.} \end{cases}$$

This recursive definition is still incomplete, since we have to give some start conditions. Note that by this definition only two elements of W can increase per time step. Using the values of W the algorithm DMR divides the input sequence σ into epochs $E_1, E_2, \ldots$ The first epoch starts with the first time step. At the beginning of each epoch we set $W = w$. An epoch ends at the end of a step t (after PBR executed all its actions in t) in which $|W_t| \geq (1+c) \cdot k - 2$. Since $|W|$ can increase by at most $1 + 1/\Gamma$ per step, at the very end of the epoch $|W| \leq (1+c) \cdot k$ when the epoch ends in t. For convenience we call each increase of W due to the condition $g_{t-1}(\lambda) > g_{t-2}(\lambda)$ a *hit* in component λ.

Let a vector B_t denote the number of resources available in the components. Obviously $|B_t| = (1+c) \cdot k$. In step t we *assign* exactly $\lfloor W_t[\lambda] \rfloor$ resources to component λ by setting $k_t(\lambda) = \lfloor W_t[\lambda] \rfloor$. If there are too few resources in λ (i.e. $\lfloor W_t[\lambda] \rfloor > B_{t-1}[\lambda]$), then a resource item is moved from another component. This is always possible since we have $(1+c) \cdot k$ resource units available and $\sum_{\lambda \in \Lambda} \lfloor W_t[\lambda] \rfloor \leq (1+c) \cdot k$ by the way epochs are constructed.

In effect, DMR constructs a part of the input sequence for the dynamic-resource problem solved by PBR_λ by changing the number of assigned resources. In any time step, the DMR algorithm moves resources between components prior to the algorithms PBR_λ. This assures that each PBR_λ has $\lfloor W_t[\lambda] \rfloor \geq w_t[\lambda]$ resources available to fulfill the demand. The PBR_λ algorithm may only use the assigned resources to fulfill the demand, and not all the resources really available in the component. Such construction implies that not all resources are always used; this appears counterproductive at first glance, but it keeps the division into epochs deterministic and dependent only on the input sequence.

Analysis of DMR. Using Lemma 7 and observing that with each hit $|W|$ increases by $1/\Gamma$, we may bound the cost of DMR in each epoch. A lower bound on the cost of OPT relies on the notion of *fair timespans*. In such a timespan, the algorithm PBR_λ has at least the same number of resources as OPT, and therefore the phases in this timespan are fair. Since DMR has $(1+c) \cdot k$ resources available, we can guarantee that there exist fair timespans of appropriate length during an epoch. In a fair timespan, the PBR_λ algorithm is competitive against OPT, and thus each hit implies that OPT has incurred some cost. This leads to an appropriate lower bound on the cost of OPT during the whole epoch, as shown in the following lemma.

Lemma 8. *Let $z \leq c \cdot \Gamma/4$ and $\Gamma \geq k$. Then for any epoch E_y,*

$$\mathbf{E}[\mathcal{C}_{\text{DMR}}(E_y)] \leq (1+c) \cdot k \cdot \Gamma \cdot \mathcal{O}(\log(\min\{n, k\})) ,$$

where the expectation is taken w.r.t. all random choices made by the algorithms PBR_λ. *On the other hand, $\mathcal{C}_{\text{OPT}}(E_y) = \Omega(c \cdot k \cdot \Gamma)$.*

The following theorem follows easily since each input sequence is deterministically partitioned into epochs and DMR is competitive for every finished epoch by the former lemma. The last potentially unfinished epoch contributes only an additive term.

Theorem 3. *Let $(\mathcal{X}, \xi)$ be a uniformly decomposable metric space with component set Λ. Then* DMR *is $\mathcal{O}((1 + 1/c) \cdot \log(\min\{n, k\}))$-competitive for the k-resource problem on $(\mathcal{X}, \xi)$ with c-resource augmentation, $|\Lambda| \leq c \cdot \Gamma/4$ and $\Gamma \geq k$. The same result holds for the k-server problem.*

References

1. Bartal, Y., Bollobás, B., Mendel, M.: A Ramsey-type theorem for metric spaces and its applications for metrical task systems and related problems. In: Proc. of the 42nd IEEE Symp. on Foundations of Computer Science (FOCS), pp. 396–405 (2001)
2. Bartal, Y., Linial, N., Mendel, M., Naor, A.: On metric Ramsey-type phenomena. In: Proc. of the 35th ACM Symp. on Theory of Computing (STOC), pp. 463–472 (2003)
3. Bartal, Y., Mendel, M.: Randomized k-server algorithms for growth-rate bounded graphs. Journal of Algorithms 55(2), 192–202 (2005)
4. Csaba, B., Lodha, S.: A randomized on-line algorithm for the k-server problem on a line. Random Structures and Algorithms 29(1), 82–104 (2006)
5. Fiat, A., Karp, R.M., Luby, M., McGeoch, L.A., Sleator, D.D., Young, N.E.: Competitive paging algorithms. Journal of Algorithms 12(4), 685–699 (1991)
6. Koutsoupias, E., Papadimitriou, C.H.: On the k-server conjecture. Journal of the ACM 42(5), 971–983 (1995) Also appeared. In: Proc. of the 26th STOC, pp. 507–511 (1994)
7. Manasse, M.S., McGeoch, L.A., Sleator, D.D.: Competitive algorithms for server problems. Journal of the ACM 11(2), 208–230 (1990) Also appeared as Competitive algorithms for on-line problems. In: Proc. of the 20th STOC, pp. 322–333 (1988)
8. Seiden, S.S.: A general decomposition theorem for the k-server problem. In: Meyer auf der Heide, F. (ed.) ESA 2001. LNCS, vol. 2161, pp. 86–97. Springer, Heidelberg (2001)
9. Sleator, D.D., Tarjan, R.E.: Amortized efficiency of list update and paging rules. Communications of the ACM 28(2), 202–208 (1985)
10. Young, N.E.: On-line file caching. Algorithmica 33(3), 371–383 (2002) Also appeared. In: Proc. of the 9th SODA, pp. 82–86 (1998)

On the Robustness of Graham's Algorithm for Online Scheduling*

Michael Gatto and Peter Widmayer

ETH Zurich, Switzerland
{gatto,widmayer}@inf.ethz.ch

Abstract. While standard parallel machine scheduling is concerned with good assignments of jobs to machines, we aim to understand how the quality of an assignment is affected if the jobs' processing times are perturbed and therefore turn out to be longer (or shorter) than declared. We focus on online scheduling with perturbations occurring at any time, such as in railway systems when trains are late. For a variety of conditions on the severity of perturbations, we present upper bounds on the worst case ratio of two makespans. For the first makespan, we let Graham's algorithm assign jobs to machines, based on the non-perturbed processing times. We compute the makespan by replacing each job's processing time with its perturbed version while still sticking to the computed assignment. The second is an optimal offline solution for the perturbed processing times. The deviation of this ratio from Graham's competitive ratio (of slightly less than 2) tells us about the "price of perturbations". For instance, we show a competitive ratio of 2 for perturbations decreasing the processing time of a job arbitrarily, and a competitive ratio of less than 2.5 for perturbations doubling the processing time of a job.

1 Introduction

The input to the parallel machine scheduling problem is a sequence of processing times of n jobs, each of which needs to be processed later on one of m identical machines. The objective is to minimize the completion time of the job terminating last (the *makespan*) [16]. A feasible solution is an assignment of jobs to machines; an optimal solution minimizes the makespan. The underlying assumption is as usual for all kinds of (offline or online) optimization problems: The input values accurately reflect the reality for which the planning takes place. In *online* parallel machine scheduling, the planning decision for each job, i.e., to which machine this job shall be assigned, must be made when the job is presented, before the next processing time (if any) is shown, and cannot be revoked later. Note that there is a clear distinction between the presentation of the processing times of the jobs for the purpose of planning, and processing the jobs on the planned machines. Therefore, the planned processing time of a job can

* This work was partially supported by the Future and Emerging Technologies Unit of EC (IST priority - 6th FP), under contract no. FP6-021235-2 (project ARRIVAL).

F. Dehne, J.-R. Sack, and N. Zeh (Eds.): WADS 2007, LNCS 4619, pp. 349–361, 2007.

sometimes be dramatically different from its real processing time, when processing takes place. For instance, this happens when a train ride turns out to take much longer than planned due to an engine breakdown.

In this paper, we aim at understanding to what extent a *limited number* of specific kinds of perturbations of the input affect the performance of online parallel machine scheduling. We assume that a probability distribution of perturbations is either not available, or does not tell a lot (as for engine breakdowns, which with very small probability may occur everywhere), and focus on a worst-case analysis. We make just a first step by studying the behavior under perturbations of the well-known algorithm by Graham [6] for online scheduling, the *List scheduling algorithm*. Graham's algorithm assigns the next presented job to the machine that will terminate earliest for the job sequence seen so far. While deterministic online algorithms have improved the competitive ratio for parallel machine scheduling to 1.9201 [5], and randomized algorithms such as [2] even more than that, Graham's algorithm remains, in its simplicity and with its competitive ratio of $2 - \frac{1}{m}$, a prime example of a good online algorithm.

The perturbations of the processing times are disclosed after the entire solution has been determined. The impact of perturbations is measured by the worst-case ratio of two makespans. The first is the makespan that we get by taking, for the original instance, Graham's assignment of jobs to machines, and by replacing, within this assignment, all original processing times with the perturbed ones. The second is the makespan of an optimal offline solution of the perturbed instance. This ratio tends to be larger than Graham's competitive ratio; the larger it is, the more Graham's algorithm suffers from perturbations.

A related but different question has been addressed in *sensitivity analysis*, asking for the amount of perturbation that can be tolerated before the structure of an optimal solution changes. Sensitivity analysis has been studied predominantly for linear programming [4], but also for many combinatorial problems such as network flows [1] and scheduling [8]. In this paper, we allow a small number of processing times to be perturbed a lot, but we will not change the structure of a solution and merely observe its change in quality, because the perturbation happens after the irrevocable assignment decision.

Related work. Robustness has been defined in a variety of ways in optimization theory. Exact input values have been replaced by probability distributions [18] in a probabilistic approach, or by intervals of possible values in a worst case scenario [13,9], or by uncertainty sets [3]. We are most interested in a notion of robustness that takes a worst-case view, but limits the number of perturbations that can actually happen. This view is motivated by everyday experience: A severe disturbance such as a locomotive breakdown can happen at any train ride, but we do not need to fear many such events simultaneously.

The effect of perturbations in offline scheduling algorithms has been studied under many different points of view, see [8,11] for an overview. These include sensitivity analysis for different parameters [8] and uncertainty in communication delays between inter-depending jobs scheduled on different machines [14,17].

The effect of perturbed processing times has also been addressed in different ways. Parallel machine scheduling where all jobs' processing times are accurate up to a factor $(1 \pm \varepsilon')$ of a declared value was analyzed in [15]. The quality of any algorithm deteriorates by a factor $(1 + \varepsilon)$ for the makespan, and by $\sqrt{1+\varepsilon}$ for the sum of completion times. We provide a better bound for the makespan, since our analysis exploits the structure of Graham's schedule. The performance of online scheduling algorithms where the processing times are drawn from a distribution, and the scheduling algorithm needs to schedule the jobs with the knowledge of the distributions only, has also been addressed in terms of average case analysis for the completion times [18] and of minimization of an objective in expectation, as in [12].

The tolerance to perturbations of Graham's List scheduling algorithm has been addressed both in terms of decrease in quality of the objective [6,7] and in the number of different (offline) schedules that arise from perturbing the processing time of one job and for different input sequences [10]. Graham [6,7] analyzed the effects of relaxations in a number of instance-defining parameters. For each parameter, he analyzed the worst-case ratio of the makespans of the relaxed instance with respect to the non-relaxed instance, for both instances scheduled with List. He showed a worst-case ratio of $2 - \frac{1}{n}$ for decreasing the processing times, and similar results were derived for the relaxation of other parameters.

Our contribution. We derive bounds on the competitive ratio of Graham's algorithm on m machines for the following scenarios. For integer r and arbitrary increase of the processing times of $r \leq n$ jobs, we show a competitive ratio of $2 + r - \frac{r+1}{m}$; for arbitrary decrease of the processing times of any number of jobs that turn out to be scheduled on $r \leq m$ machines in an optimal offline schedule, we show a competitive ratio of $2 + \frac{r-1}{m-r}$; for dividing the processing times of r jobs by a factor at most $x > 1, x \in \mathbb{Q}_0^+$, we show a competitive ratio of $2 + \frac{r(x-1)-1}{m}$; and for either dividing or multiplying (but not both) the processing times of an arbitrary number of jobs by a factor at most $x > 1$, we show a competitive ratio of $1 + x - \frac{1}{m}$. We also give (infinite families of) examples where these bounds are tight or come close.

Problem setting and notation. An instance is specified as a 3-tuple $(J, P, \tilde{P})$, where J is the sequence of jobs to be scheduled, P and $\tilde{P}$ specify the original processing time $p_j \in \mathbb{Q}_0^+$ and the perturbed processing time $\tilde{p}_j \in \mathbb{Q}_0^+$ of each job $j \in J$, respectively. Graham's algorithm schedules the *original instance*, that is, the job sequence J with processing times P, and produces a schedule $\text{List}(J, P)$.

Each instance is characterized by the perturbation against which we analyze robustness. The effect of the perturbation is reflected in the processing times $\tilde{P}$, which may increase or decrease; the perturbation may be of arbitrary size, or bounded for each job by a factor x of the job's original processing time. The latter setting is motivated by project scheduling, where the extent of the misjudgment of a task's processing time is often linked to the task's difficulty. We refer to the jobs J with processing times $\tilde{P}$ as the *perturbed instance*.

We denote by $\mathrm{OPT}(J, \tilde{P})$ the optimal offline schedule of the sequence of jobs J with processing times $\tilde{P}$. In the offline setting, the order in J is irrelevant. We denote by $\mathcal{L}(S, P)$ the makespan of the schedule S with processing times P. In this setting, we measure the robustness of Graham's online algorithm by computing the makespan of the schedule $\mathrm{List}(J, P)$ obtained with the original instance, but evaluated on the perturbed processing times $\tilde{P}$, that is, by evaluating $\mathcal{L}(\mathrm{List}(J, P), \tilde{P})$. We compare this makespan with the makespan $\mathcal{L}(\mathrm{OPT}(J, \tilde{P}), \tilde{P})$ of the optimal offline schedule $\mathrm{OPT}(J, \tilde{P})$. Note that comparing $\mathcal{L}(\mathrm{List}(J, P), \tilde{P})$ with $\mathcal{L}(\mathrm{List}(J, P), P)$ does not provide any useful information, since the total processing time is different for P and $\tilde{P}$: If the processing time of a job increases arbitrarily, *any* algorithm needs to process this job.

In this model, each machine processes its assigned jobs without pausing in between. The sum of the processing times of a machine is called *load.* When machine $i \in M$ is finished, it remains idle up to the makespan. We refer to the idle time as $s_i, i \in M$. We call the set of machines which process some perturbed jobs the *affected* machines, and denote them by $M^{\neq}$. Similarly, we call the set of machines which do not process any perturbed jobs *unaffected* machines, and denote them by $M^{=}$.

For perturbations increasing the jobs' processing times, we denote the perturbed processing times as $p_j^{\uparrow}, j \in J$, and the set of all increased processing times as $P^{\uparrow}$. Similarly, for decreases we use $p_j^{\downarrow}, j \in J$ and $P^{\downarrow}$. Changing the processing times of the jobs in a schedule also influences the idle times. Therefore, we refer to the idle time resulting after the perturbation as $s_i^{\uparrow}$ for increased processing times and as $s_i^{\downarrow}$ for decreases. In the analysis, we look at the perturbations of the jobs sequentially, in any order. In this way, we can specify the impact of the perturbation of each job on the idle time of each machine. When a job changes its processing time, the subsequent jobs shift accordingly in the schedule. This shift may shorten or lengthen the idle time of various machines. We denote the increase or decrease in idle time on machine $i \in M$ caused by perturbing job j by $\delta_j^i \in \mathbb{Q}$, which may be positive or negative.

For a makespan $\mathcal{L}$, the processing times of the jobs and the idle times of the m machines satisfy $m\mathcal{L} = \sum_{j \in J} p_j + \sum_{i=1}^{m} s_i$. For the sake of completeness, we revise some of the known lower bounds on the optimum for the classical parallel machine scheduling and for Graham's algorithm. Let $\mathcal{L}_{\mathrm{OPT}}$ be the optimal makespan. First, $p_j \leq \mathcal{L}_{\mathrm{OPT}}, \forall j \in J$, since each job must be scheduled non-preemptively. Furthermore, $\mathcal{L}_{\mathrm{OPT}} \geq \frac{\sum_{j \in J} p_j}{m}$, since no schedule can do better than distribute the total processing time evenly across all machines. Finally, consider an arbitrary job $\bar{\jmath}$ finishing at the makespan of Graham's schedule. Then, $s_i \leq p_{\bar{\jmath}}, \forall i \in M$, since otherwise Graham's algorithm would have scheduled job $\bar{\jmath}$ on the machine not satisfying the inequality. Finally, if we consider the instances where the perturbation increases the processing times, we have $\mathcal{L}_{\mathrm{OPT}} \leq \mathcal{L}(\mathrm{OPT}(J, P^{\uparrow}), P^{\uparrow})$, since the total processing time increases and the maximum processing time of the jobs can also only become larger.

2 Arbitrary Perturbations

In the following, we analyze robustness for arbitrary size perturbations of the processing times of jobs. First, we analyze the case of arbitrary decreases in processing times and then of arbitrary increases.

2.1 Arbitrary Decreases in Processing Time

First, we bound the best-possible quality of the solution of any online algorithm if the processing times may decrease arbitrarily.

Theorem 1. *No algorithm for online scheduling on m identical parallel machines on instances where the processing time of one job may decrease arbitrarily can have a competitive ratio smaller than* 2.

Proof. Assume such an algorithm exists, and consider the following sequence of jobs, all with processing time 1. First, the adversary sequentially presents m jobs. To be strictly better than 2–competitive, any algorithm must schedule each job on a different machine. Then, the adversary presents a final job, which may be scheduled on any machine. Now, the perturbation affects one job which is scheduled alone on a machine, and decreases its processing time to 0. Thus, the computed schedule has an empty machine, and a makespan of 2. The offline optimal perturbed schedule assigns each of the now m jobs on a different machine, and has a makespan of 1. □

We now focus on Graham's algorithm, and analyze the case where the perturbed jobs are scheduled on r out of the m available machines by Graham's algorithm. We have the following theorem:

Theorem 2. *Consider the instances of online scheduling on m identical parallel machines where perturbations may decrease the processing times of some jobs arbitrarily. Restricted to these instances, if Graham's algorithm schedules the perturbed jobs on r machines, Graham's algorithm is $2 + \frac{r-1}{m-r}$-competitive, and this bound is best possible. For $r = 1$, Graham's algorithm is optimal.*

Before proving this theorem, we introduce a compact notation for the makespans to improve readability. We write $\mathcal{L}_{\text{OPT}}$ for $\mathcal{L}(\text{OPT}(J, P), P)$, and $\mathcal{L}^{\downarrow}_{\text{OPT}}$ for $\mathcal{L}(\text{OPT}(J, P^{\downarrow}), P^{\downarrow})$. Similarly for Graham's algorithm, we write $\mathcal{L}^{\downarrow}_{\text{List}}$ instead of $\mathcal{L}(\text{List}(J, P), P^{\downarrow})$. The notation for increases in processing times is obtained accordingly.

Proof. Consider Graham's schedule $\text{List}(J, P)$. We refer to the jobs scheduled on the unaffected machines $M^{=}$ as $J^{=}$. Hence, by definition, these jobs do not change their processing time. To estimate the makespan after the perturbation we analyze the schedule of the unaffected $m - r$ machines $M^{=}$. We distinguish two cases: in the first, the makespan $\mathcal{L}^{\downarrow}_{\text{List}}$ is attained by at least one machine in $M^{=}$, while in the second no machine in $M^{=}$ attains it.

For the first case, we let $\bar{\jmath}_\mu$ be a job attaining the makespan after the perturbation on an unaffected machine $\mu \in M^=$. In this case, the time spent by the machines $M^=$ up to the makespan is given by:

$$\begin{aligned}(m-r)\cdot\mathcal{L}^{\downarrow}_{\text{List}} &= \sum_{j\in J^=} p^{\downarrow}_j + \sum_{i\in M^=} s^{\downarrow}_i \\ &\leq \sum_{j\in J} p^{\downarrow}_j + (m-r-1)p^{\downarrow}_{\bar{\jmath}_\mu} \leq m\mathcal{L}^{\downarrow}_{\text{OPT}} + (m-r-1)\mathcal{L}^{\downarrow}_{\text{OPT}},\end{aligned}$$

since $J^= \subseteq J$ and at most $(m-r-1)$ machines have some idle time, which due to Graham's algorithm is smaller than $p^{\downarrow}_{\bar{\jmath}_\mu}$. Thus, the bound follows: $\mathcal{L}(\text{List}(J,P),P^{\downarrow}) \leq \left(2+\frac{r-1}{m-r}\right)\cdot\mathcal{L}(\text{OPT}(J,P^{\downarrow}),P^{\downarrow})$

For the second case, let $\bar{\jmath}_\mu$ be a job attaining the makespan after the perturbation on an affected machine $\mu \in M^{\neq}$. Because of Graham's algorithm, when $\bar{\jmath}_\mu$ was originally scheduled, μ was a machine with least load, say with load ℓ. Since the processing times on unaffected machines $M^=$ remain unchanged, the latter machines have load at least ℓ after the decreases. Thus, their idle time is $s^{\downarrow}_i \leq p^{\downarrow}_{\bar{\jmath}_\mu}, i \in M^=$, since $\bar{\jmath}_\mu$ attains the makespan. Note that this bound holds both if $\bar{\jmath}_\mu$ is perturbed or it remains unchanged. Thus, the time spent by the machines in $M^=$ up to $\mathcal{L}^{\downarrow}_{\text{List}}$ can be represented as follows:

$$\begin{aligned}(m-r)\cdot\mathcal{L}^{\downarrow}_{\text{List}} &= \sum_{j\in J^=} p^{\downarrow}_j + \sum_{i\in M^=} s^{\downarrow}_i \leq \sum_{j\in J} p^{\downarrow}_j - p^{\downarrow}_{\bar{\jmath}_\mu} + (m-r)\cdot p^{\downarrow}_{\bar{\jmath}_\mu} \\ &\leq m\cdot\mathcal{L}^{\downarrow}_{\text{OPT}} + (m-r-1)\cdot\mathcal{L}^{\downarrow}_{\text{OPT}},\end{aligned}$$

since by the case analysis $\bar{\jmath}_\mu$ is not scheduled on a machine in $M^=$. The bounds lead to the same expression as in the previous case, thus concluding the first part of the proof.

A worst-case instance for $r \leq m-2$ machines has the following structure, illustrated in Figure 1. First, the adversary presents r huge jobs with processing time $C > 1+\frac{r-1}{m-r}$ each. Next, the adversary presents $m-r$ big jobs with processing time one. Then, the adversary presents $(r-1)\cdot(m-r)$ small jobs with processing time $\frac{1}{m-r}$. Finally, a big job with processing time one is presented. Now, the perturbation decreases the processing time of all huge jobs from C to zero and the achieved makespan is $2+\frac{r-1}{m-r}$. The optimal offline algorithm achieves a makespan of 1. Simple examples show worst-case behaviors for $m-1 \leq r \leq m$ according to the stated bound. Note that the analysis is nevertheless not suited for $r = m$. The optimality of Graham's algorithm for $r = 1$ follows from Theorem 1. □

Intuitively, this proof shows that the worst-case scenario happens if the affected machines are blocked with jobs whose processing time decreases to zero. Hence, the adversary and the perturbation force the online algorithm to work with

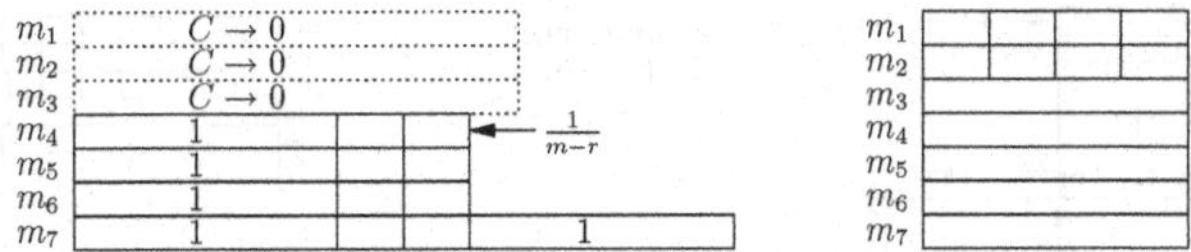

Fig. 1. A worst-case example matching the bound of Theorem 2. Left, Graham's schedule on the original instance; the perturbed jobs have a dotted outline. Right, the optimal schedule of the perturbed instance.

r machine less than initially stated. Surprisingly, this affects the competitive ratio only with an additive term of $\frac{r}{m-r}$.

2.2 Arbitrary Increases in Processing Time

We now consider arbitrary increases in the job's processing times. For clarity, we first analyze the perturbation where r jobs increase in processing time arbitrarily. Then, we give a similar analysis for the case where any number of jobs may be perturbed, and the perturbed jobs are scheduled on r machines of an optimal offline schedule of the perturbed instance.

Theorem 3. *Consider the instances of online scheduling on m identical parallel machines where perturbations may increase the processing times of r jobs arbitrarily. Restricted to these instances, Graham's algorithm has a competitive ratio of $2 + r - \frac{r+1}{m}$, and this bound is best possible for $r \leq m - 2$.*

Proof. Let $J^{\uparrow} \subset J, |J^{\uparrow}| = r$ be the arbitrarily ordered list of jobs whose processing times increase and let $\psi_j = p_j^{\uparrow} - p_j$ be the increase of job $j \in J^{\uparrow}$. Recall that for the analysis we assume the perturbations to occur sequentially. Thus, $\psi_j \leq p_j^{\uparrow} \leq \mathcal{L}_{\text{OPT}}^{\uparrow}, j \in J^{\uparrow}$, and $-\psi_j \leq \delta_j^k \leq \psi_j, k \in M, j \in J^{\uparrow}$. Furthermore, at least one machine $\mu \in M$ attains the makespan before the perturbation, and has idle time $s_\mu = 0$. The idle times of the remaining machines satisfy $s_i \leq \mathcal{L}_{\text{OPT}} \leq \mathcal{L}_{\text{OPT}}^{\uparrow}, i \in M$. Furthermore, the increase of job $j \in J^{\uparrow}$ does not increase the idle time of the machine the job is scheduled on (although it may decrease it). The new makespan of the machines is thus given by:

$$\begin{aligned} m \cdot \mathcal{L}_{\text{List}}^{\uparrow} &= \sum_{j \in J} p_j + \sum_{i \in M} s_i + \sum_{j \in J^{\uparrow}} \psi_j + \sum_{j \in J^{\uparrow}, k \in M} \delta_j^k \\ &= \sum_{j \in J} p_j^{\uparrow} + \sum_{i \in M} s_i + \sum_{j \in J^{\uparrow}, k \in M} \delta_j^k \\ &\leq m \cdot \mathcal{L}_{\text{OPT}}^{\uparrow} + (m-1) \cdot \mathcal{L}_{\text{OPT}}^{\uparrow} + (m-1) \cdot r \cdot \mathcal{L}_{\text{OPT}}^{\uparrow}, \\ \mathcal{L}(\text{List}(J, P), P^{\uparrow}) &\leq \left(2 + r - \frac{r+1}{m}\right) \cdot \mathcal{L}(\text{OPT}(J, P^{\uparrow}), P^{\uparrow}). \end{aligned}$$

To see that the analysis is best possible for $r \leq m - 2$, we give an example, illustrated in Figure 2, where this bound is achieved up to an arbitrarily small

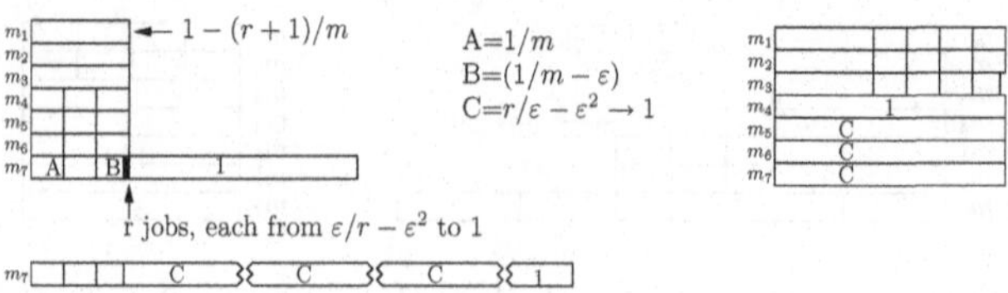

Fig. 2. A worst-case instance for arbitrary increase of r jobs. Left, Graham's schedule before the perturbation, and the affected machine m_7 after the perturbation. Right, the optimal schedule of the perturbed instance.

$\varepsilon \in \mathbb{Q}_0^+, 0 < \varepsilon < \frac{1}{m}$. First, the adversary presents $m - r - 1$ big jobs with processing time $1 - \frac{(r+1)}{m}$. Next, the adversary presents $m \cdot (r+1) \cdot (1 - \frac{r+1}{m}) - 1 = (r+1) \cdot (m - r - 1) - 1$ small jobs with processing time $\frac{1}{m}$, followed by one small job with processing time $\frac{1}{m} - \varepsilon$, for ε arbitrarily small. Now, the adversary presents r tiny jobs with processing time $\frac{\varepsilon}{r} - \varepsilon^2$ and a last job with processing time 1. The perturbation increases the processing times of tiny jobs from $\frac{\varepsilon}{r} - \varepsilon^2$ to 1. The makespan of Graham's schedule is $2 + r - \frac{r+1}{m} - \varepsilon$. On the other hand, the optimal offline algorithm on the perturbed instance has a makespan of 1. □

Theorem 3 can be generalized as follows. Let $J_i^{\uparrow}$ be the arbitrarily ordered list of perturbed jobs which are scheduled on machine $i \in M$ in an optimum offline solution $\mathrm{OPT}(J, P^{\uparrow})$. Let $M^{\neq} = \{i \in M | J_i^{\uparrow} \neq \emptyset\}$ be the set of machines in the considered optimal offline solution which process at least one job with perturbed processing time, and define $r = |M^{\neq}|$. Note that here the perturbed machines are defined with respect to $\mathrm{OPT}(J, P^{\uparrow})$, and not to Graham's schedule.

Theorem 4. *Consider the instances of online scheduling on m identical parallel machines with the following properties: first, the processing times of some jobs may increase arbitrarily; second, the perturbed jobs are scheduled on r machines of an optimal offline solution. For these instances, Graham's algorithm has a competitive ratio of $2 + r - \frac{r+1}{m}$, and this bound is best possible for $r \leq m - 2$.*

The proof follows the lines of the previous one. The additional ingredients are bounding the sum of the variations in processing time of all perturbed jobs within $J_i^{\uparrow}, i \in M$ by $\mathcal{L}_{\mathrm{OPT}}^{\uparrow}$ and noting that for each perturbed job $j \in J$ there is a machine $k \in M$ with $\delta_j^k \leq 0$.

3 Bounded Perturbations

In the following, we bound the perturbation of each job to a constant factor of the job's original processing time. Thus, the perturbed processing times are of the form $\tilde{p} = \alpha p$ for some α.

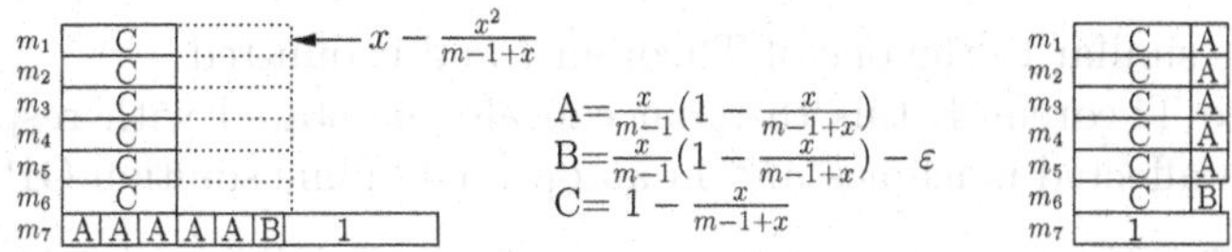

Fig. 3. The example of Theorem 5 for $x = 2$. Left, the Graham's schedule on the original instance. The perturbed jobs are shown as dotted with their original processing time, and with solid lines in their perturbed state. Right, $\mathrm{OPT}(J, P^{\downarrow})$.

3.1 Bounded Decreases in Processing Times

In this section, we analyze the behavior of the makespan given that jobs may decrease by a bounded factor.

Theorem 5. *Consider the instances of online scheduling on m identical parallel machines where the processing times of an arbitrary number of jobs may decrease to a factor at least $\frac{1}{x}$ of their original processing time, for $x > 1, x \in \mathbb{Q}_0^+$. Restricted to these instances, Graham's algorithm has a competitive ratio between $1+x-\frac{x^2}{m-1+x} - \varepsilon$ and $1+x-\frac{x}{m}$, for a small $\varepsilon \in \mathbb{Q}_0^+, 0 < \varepsilon < \frac{x}{m-1}(1-\frac{x}{m-1+x})$.*

Proof. Consider a machine μ attaining the makespan after the perturbation. Let $\bar{\jmath}_\mu$ be the last job scheduled on μ, and let ℓ be the load of μ on the original instance when $\bar{\jmath}_\mu$ was presented. Hence, the total load of μ before the perturbation is $\ell + p_{\bar{\jmath}_\mu}$. Because $\bar{\jmath}_\mu$ was scheduled on μ, all machines have at least load ℓ with the original processing times. Therefore, $m\ell \leq \sum_{j \in J \setminus \{\bar{\jmath}_\mu\}} p_j$. Since at most all jobs are perturbed and decrease to a factor at least $\frac{1}{x}$, $\sum_{j \in J \setminus \{\bar{\jmath}_\mu\}} p_j \leq x \sum_{j \in J \setminus \{\bar{\jmath}_\mu\}} p_j^{\downarrow} \leq xm \cdot \mathcal{L}_{\mathrm{OPT}}^{\downarrow} - x \cdot p_{\bar{\jmath}_\mu}^{\downarrow}$, hence $\ell \leq x\mathcal{L}_{\mathrm{OPT}}^{\downarrow} - \frac{x}{m} p_{\bar{\jmath}_\mu}^{\downarrow}$. Thus:

$$\begin{aligned}\mathcal{L}(\mathrm{List}(J,P),P^{\downarrow}) &\leq \ell + p_{\bar{\jmath}_\mu}^{\downarrow} \leq x \cdot \mathcal{L}_{\mathrm{OPT}}^{\downarrow} - \frac{x}{m} p_{\bar{\jmath}_\mu}^{\downarrow} + p_{\bar{\jmath}_\mu}^{\downarrow} = x \cdot \mathcal{L}_{\mathrm{OPT}}^{\downarrow} + \left(1 - \frac{x}{m}\right) p_{\bar{\jmath}_\mu}^{\downarrow} \\ &\leq \left(1 + x - \frac{x}{m}\right) \mathcal{L}(\mathrm{OPT}(J,P^{\downarrow}),P^{\downarrow}).\end{aligned}$$

An example which comes close to this upper bound is the following sequence of jobs, illustrated in Figure 3 for the case $x = 2$. The adversary presents $m-1$ big jobs with processing time $x - \frac{x^2}{m-1+x}$, $m-2$ small jobs with processing time $\frac{x}{m-1}(1-\frac{x}{m-1+x})$, one small with processing time $\frac{x}{m-1}(1-\frac{x}{m-1+x}) - \varepsilon$, for a small $\varepsilon \in \mathbb{Q}_0^+, 0 < \varepsilon < \frac{x}{m-1}(1-\frac{x}{m-1+x})$, followed by a last job with processing time 1. The perturbation affects the $m-1$ big jobs, decreasing their processing time to $1 - \frac{x}{m-1+x}$. Graham's schedule on the perturbed instance has makespan of $1 + x - \frac{x^2}{m-1+x} - \varepsilon$, whereas $\mathcal{L}(\mathrm{OPT}(J,P^{\downarrow}),P^{\downarrow}) = 1$. □

Theorem 6. *Consider the instances of online scheduling on m identical parallel machines where the perturbations may decrease the processing times of r jobs to a factor at least $\frac{1}{x}$ of their original processing time, for $x \in \mathbb{Q}_0^+, x > 1$. Restricted to these instances, Graham's algorithm has a competitive ratio of $2 + \frac{r \cdot (x-1) - 1}{m}$.*

The proof is similar to the one of Theorem 7 and is omitted.

Similar to Theorem 4, this theorem can also be stated with respect to the number r of affected machines $M^{\neq}$ in an optimal offline solution $\mathrm{OPT}(J, P^{\downarrow})$.

Theorem 7. *Consider the instances for online scheduling on m identical parallel machines with the following two properties: first, the perturbations may decrease the processing times of some jobs to a factor of at least $\frac{1}{x}$ of their original processing time, for $x \in \mathbb{Q}_0^+, x > 1$; second, the perturbed jobs are scheduled on r machines in an optimal offline solution. Restricted to these instances, Graham's algorithm has a competitive ratio of $2 + \frac{r \cdot (x-1) - 1}{m}$.*

Proof. Let $J_i^{\downarrow}$ be the set of perturbed jobs scheduled on machine $i \in M$ in $\mathrm{OPT}(J, P^{\downarrow})$, and let $M^{\neq} = \{i \in M | J_i^{\downarrow} \neq \emptyset\}, r = |M^{\neq}|$. Let $\tilde{J} = \cup_{i \in M^{\neq}} J_i^{\downarrow}$ be the set of all perturbed jobs. Consider a machine μ attaining the makespan after the perturbation, let $\bar{\jmath}_\mu$ be the last job scheduled on μ, and let ℓ be the load of μ in the original instance when $\bar{\jmath}_\mu$ was presented. Therefore, $\mathcal{L}(\mathrm{List}(J, P), P^{\downarrow}) \leq \ell + p_{\bar{\jmath}_\mu}$. Now:

$$\begin{aligned} m\ell &\leq \sum_{j \in J \setminus \{\bar{\jmath}_\mu\}} p_j \leq \sum_{j \in \tilde{J}} p_j + \sum_{j \in J \setminus \tilde{J}} p_j - p_{\bar{\jmath}_\mu} \leq x \sum_{j \in \tilde{J}} p_j^{\downarrow} + \sum_{j \in J \setminus \tilde{J}} p_j^{\downarrow} - p_{\bar{\jmath}_\mu} \\ &\leq \sum_{j \in J} p_j^{\downarrow} + (x-1) \sum_{j \in \tilde{J}} p_j^{\downarrow} - p_{\bar{\jmath}_\mu} = \sum_{j \in J} p_j^{\downarrow} + (x-1) \sum_{i \in M^{\neq}} \sum_{j \in J_i^{\downarrow}} p_j^{\downarrow} - p_{\bar{\jmath}_\mu} \\ &\leq m\mathcal{L}_{\mathrm{OPT}}^{\downarrow} + (x-1) r \mathcal{L}_{\mathrm{OPT}}^{\downarrow} - p_{\bar{\jmath}_\mu} = \mathcal{L}_{\mathrm{OPT}}^{\downarrow} (m + (x-1)r) - p_{\bar{\jmath}_\mu}. \end{aligned}$$

where the last inequality follows because $\sum_{j \in J_i^{\downarrow}} p_j^{\downarrow} \leq \mathcal{L}_{\mathrm{OPT}}^{\downarrow}$. Thus,

$$\mathcal{L}_{\mathrm{List}}^{\downarrow} \leq \mathcal{L}_{\mathrm{OPT}}^{\downarrow} + \frac{(x-1)r}{m} \mathcal{L}_{\mathrm{OPT}}^{\downarrow} + \left(1 - \frac{1}{m}\right) p_{\bar{\jmath}_\mu} \leq \left(2 + \frac{(x-1)r - 1}{m}\right) \mathcal{L}_{\mathrm{OPT}}^{\downarrow}.$$

□

3.2 Bounded Increases in Processing Times

Now, we focus on the case where the perturbation increases the processing times of some jobs to a factor $x \geq 1, x \in \mathbb{Q}_0^+$ of the job's original processing times. Special results hold if the perturbation affects a single job. For this case, we can show that the competitive ratio of Graham's algorithm is $3 - \frac{1}{x} - \frac{\left(2 - \frac{1}{x}\right)}{m}$, and this bound is best possible for $m \geq 3$ and $x \geq \frac{m-1}{m-2}$. Furthermore, for $x = 2$, a very simple example with 3 jobs and 2 machines gives a lower bound of 1.5 for the competitive ratio of any online algorithm.

We do not prove these results here, since our focus is on bounded increases of many jobs.

Theorem 8. *Consider the instances of online scheduling on m identical parallel machines where perturbations may increase the processing times of many jobs*

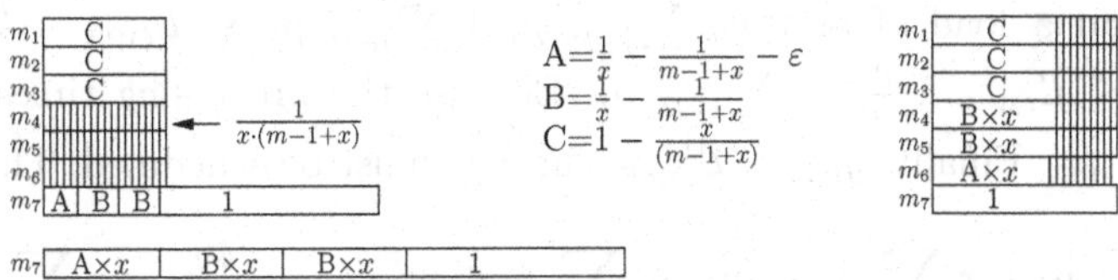

Fig. 4. The example of Theorem 8 for $x = 3$. Left, Graham's schedule on the original instance with the schedule of the affected machine at the bottom. Right, the optimal offline schedule $\mathrm{OPT}(J, P^{\uparrow})$.

to a factor $x \geq 1, x \in \mathbb{Q}_0^+$ of their original processing time. Restricted to these instances, Graham's algorithm is $1+x-\frac{1}{m}$-competitive. For $x \in \mathbb{N}$ and $x \leq m-1$, the competitive ratio of Graham's algorithm is at least $1 + x - \frac{x^2}{m-1+x} - \varepsilon$, for arbitrarily small $\varepsilon \in \mathbb{Q}_0^+, 0 < \varepsilon < \frac{1}{x\cdot(m-1+x)}$.

Proof. We omit the proof of the upper bound, which is similar to proof of Theorem 9. An example achieving a bad competitive ratio for $x \leq m-1, x \in \mathbb{N}$, has the following structure, shown in Figure 4. First, the adversary presents $m-x-1$ big jobs with processing time $1 - \frac{x}{m-1+x}$. Next, the adversary presents the following sequence of jobs: one medium job with processing time $\frac{1}{x} - \frac{1}{m-1+x} - \varepsilon$, with $\varepsilon \in \mathbb{Q}_0^+, 0 < \varepsilon < \frac{1}{x\cdot(m-1+x)}$, and $(m-1)\cdot x$ small jobs with processing time $\frac{1}{x\cdot(m-1+x)}$. The same sequence is repeated $x-1$ times with the difference that the medium jobs have processing time $\frac{1}{x} - \frac{1}{m-1+x}$. At last, the adversary presents a big job with processing time 1. Now, the perturbation affects the medium jobs, which increase to $1 - \frac{x}{m-1+x}$ (except for the first medium job, which is εx smaller than that) and enforce a makespan of $1 + x - \frac{x^2}{m-1+x} - x\varepsilon$. The optimal offline algorithm, on the other hand, achieves a makespan of 1. In this example, it is sufficient to perturb x jobs to get this bad behavior. □

In the following, we consider increases in processing times which may be different for all jobs, but where the impact on the schedule is bounded. To that end, consider Graham's schedule $\mathrm{List}(J, P)$. For each machine $i \in M$, we partition its assigned jobs as follows: let $\bar{\jmath}_i$ be the last job scheduled on i, P_i be the set of jobs, excluding $\bar{\jmath}_i$, which are not perturbed, and $\tilde{P}_i$ the set of jobs, excluding $\bar{\jmath}_i$, which are perturbed.

Theorem 9. *Graham's algorithm for online scheduling on m identical parallel machines, applied to an instance where perturbations may increase the processing times many jobs, and where the perturbed jobs assigned to a machine μ attaining the makespan $\mathcal{L}_{\mathrm{List}}^{\uparrow}$ satisfy $\sum_{j\in\tilde{P}_\mu} p_j^{\uparrow} = x\sum_{j\in\tilde{P}_\mu} p_j$ and $p_{\bar{\jmath}_i}^{\uparrow} = yp_{\bar{\jmath}_i}$, has a competitive ratio of $1 + x - \frac{x}{ym}$, for $x \geq 1, x \in \mathbb{Q}_0^+, y \geq 1, y \in \mathbb{Q}_0^+$.*

Proof. Consider a machine μ attaining the makespan $\mathcal{L}_{\mathrm{List}}^{\uparrow}$ as it was in the original instance. Because of Graham's algorithm, when $\bar{\jmath}_\mu$ was presented, each machine had a load of at least $\sum_{j\in\tilde{P}_\mu} p_j + \sum_{j\in P_\mu} p_j \leq \mathcal{L}_{\text{OPT}} - \frac{p_{\bar{\jmath}_\mu}}{m}$. Thus, $\sum_{j\in\tilde{P}_\mu} p_j \leq \mathcal{L}_{\text{OPT}} - \frac{p_{\bar{\jmath}_\mu}}{m} - \sum_{j\in P_\mu} p_j$. Because the processing times increase, $\mathcal{L}_{\text{OPT}} \leq \mathcal{L}^\uparrow_{\text{OPT}}$. Finally, $yp_{\bar{\jmath}_\mu} \leq \mathcal{L}^\uparrow_{\text{OPT}}$, since $\bar{\jmath}_\mu$ must be scheduled. Thus,

$$\begin{aligned}\mathcal{L}^\uparrow_{\text{List}} &= \sum_{j\in P_\mu} p_j + x\sum_{j\in\tilde{P}_\mu} p_j + yp_{\bar{\jmath}_\mu} \leq \sum_{j\in P_\mu} p_j + x\mathcal{L}_{\text{OPT}} - x\frac{p_{\bar{\jmath}_\mu}}{m} - x\sum_{j\in P_\mu} p_j + yp_{\bar{\jmath}_\mu} \\ &\leq (1-x)\sum_{j\in P_\mu} p_j + x\mathcal{L}_{\text{OPT}} + yp_{\bar{\jmath}_\mu}(1-\frac{x}{ym}) \leq (1+x-\frac{x}{ym})\mathcal{L}^\uparrow_{\text{OPT}}.\end{aligned}$$

Therefore, $\mathcal{L}(\text{List}(J,P),P^\uparrow) \leq \left(1+x-\frac{x}{ym}\right)\mathcal{L}(\text{OPT}(J,P^\uparrow),P^\uparrow)$. □

References

1. Ahuja, R.K., Magnanti, T.L, Orlin, J.B.: Network Flows: Theory, Algorithms and Applications. Prentice-Hall, Englewood Cliffs (1993)
2. Albers, S.: On randomized online scheduling. In: STOC '02: Proceedings of the thiry-fourth annual ACM symposium on Theory of computing, pp. 134–143. ACM Press, New York (2002)
3. Ben-Tal, A., Nemirovski, A.: Robust optimization – methodology and applications. Mathematical Programming 92(3), 453–480 (2002)
4. Chvátal, V.: Linear Programming. Freeman, San Francisco (1983)
5. Fleischer, R., Wahl, M.: Online scheduling revisited. Journal of Scheduling, special issue on approximation algorithms for scheduling algorithms (part 2) 3(6), 343–353 (2000)
6. Graham, R.L.: Bounds for certain multiprocessor anomalies. Bell System Techical Journal 45(9), 1563–1581 (1966)
7. Graham, R.L.: Bounds on multiprocessing timing anomalies. SIAM Journal on Applied Mathematics 17(2), 416–429 (1969)
8. Hall, N.G., Posner, M.E.: Sensitivity analysis for scheduling problems. Journal of Scheduling 7, 49–83 (2004)
9. Kasperski, A., Zieliński, P.: An approximation algorithm for interval data min-max regret combinatorial optimization problems. Inf. Process. Lett. 97(5), 177–180 (2006)
10. Kolen, A., Kan, A., van Hoesel, C., Wagelmans, A.: Sensitivity analysis of list scheduling heuristics. Discrete Applied Mathematics 55, 145–162 (1994)
11. Kouvelis, P., Yu, G.: Robust discrete optimization and its applications. Kluwer Academic Publishers, Dordrecht (1997)
12. Möhring, R.H., Schulz, A.S., Uetz, M.: Approximation in stochastic scheduling: the power of LP-based priority policies. Journal of the ACM 46(6), 924–942 (1999)
13. Montemanni, R., Gambardella, L.M.: An exact algorithm for the robust shortest path problem with interval data. Computers & Operations Research 31, 1667–1680 (2004)
14. Moukrim, A., Sanlaville, E., Guinan, F.: Parallel machine scheduling with uncertain communication delays. RAIRO Oper. Res. 37, 1–16 (2003)

15. Penz, B., Rapine, C., Trystram, D.: Sensitivity analysis of scheduling algorithms. European Journal of Operational Research 134, 606–615 (2001)
16. Pinedo, M.: Scheduling: Theory, Algorithms, and Systems. Prentice-Hall, Englewood Cliffs (2002)
17. Sanlaville, E.: Sensitivity bounds for machine scheduling with uncertain communication delays. Journal of Scheduling 8(5), 461–473 (2005)
18. Scharbrodt, M., Schickinger, T., Steger, A.: A new average case analysis for completion time scheduling. Journal of the ACM 53(1), 121–146 (2006)

Improved Results for a Memory Allocation Problem

Leah Epstein[1] and Rob van Stee[2,*]

[1] Department of Mathematics, University of Haifa, 31905 Haifa, Israel
lea@math.haifa.ac.il

[2] Department of Computer Science, University of Karlsruhe,
D-76128 Karlsruhe, Germany
vanstee@ira.uka.de

Abstract. We consider a memory allocation problem that can be modeled as a version of bin packing where items may be split, but each bin may contain at most two (parts of) items. This problem was recently introduced by Chung et al. [3]. We give a simple 3/2-approximation algorithm for it which is in fact an online algorithm. This algorithm also has good performance for the more general case where each bin may contain at most k parts of items. We show that this general case is also strongly NP-hard. Additionally, we give an efficient 7/5-approximation algorithm.

1 Introduction

A problem that occurs in parallel processing is allocating the available memory to the processors. This needs to be done in such a way that each processor has sufficient memory and not too much memory is being wasted. If processors have memory requirements that vary wildly over time, any memory allocation where a single memory can only be accessed by one processor will be inefficient. A solution to this problem is to allow memory sharing between processors. However, if there is a single shared memory for all the processors, there will be much contention which is also undesirable. It is currently infeasible to build a large, fast shared memory and in practice, such memories are time-multiplexed. For n processors, this increases the effective memory access time by a factor of n.

Chung et al. [3] studied this problem and described the drawbacks of the methods given above. Moreover, they suggested a new architecture where each memory may be accessed by at most *two* processors, avoiding the disadvantages of the two extreme earlier models. They abstract the memory allocation problem as a bin packing problem, where the bins are the memories and the items to be packed represent the memory requirements of the processors. This means that the items may be of any size (in particular, they can be larger than 1, which is the size of a bin), and an item may be split, but each bin may contain at most two parts of items. The authors of [3] give a 3/2-approximation for this problem.

* Research supported by Alexander von Humboldt Foundation.

F. Dehne, J.-R. Sack, and N. Zeh (Eds.): WADS 2007, LNCS 4619, pp. 362–373, 2007.

We continue the study of this problem and also consider a generalized problem where items can still be split arbitrarily, but each bin can contain up to k parts of items, for a given value of $k \geq 2$.

We study approximation algorithms in terms of the *absolute approximation ratio* or the *absolute performance guarantee*. Let $\mathcal{B}(\mathcal{I})$ (or $\mathcal{B}$, if the input $\mathcal{I}$ is clear from the context), be the cost of algorithm $\mathcal{B}$ on the input $\mathcal{I}$. An algorithm $\mathcal{A}$ is an $\mathcal{R}$-approximation (with respect to the absolute approximation ratio) if for every input $\mathcal{I}$, $\mathcal{A}(\mathcal{I}) \leq \mathcal{R} \cdot \text{OPT}(I)$, where OPT is an optimal algorithm for the problem. The absolute approximation ratio of an algorithm is the infimum value of $\mathcal{R}$ such that the algorithm is an $\mathcal{R}$-approximation. The *asymptotic* approximation ratio for an online algorithm $\mathcal{A}$ is defined to be

$$\mathcal{R}_{\mathcal{A}}^{\infty} = \limsup_{n \to \infty} \sup_{\mathcal{I}} \left\{ \frac{\mathcal{A}(\mathcal{I})}{\text{OPT}(\mathcal{I})} \middle| \text{OPT}(\mathcal{I}) = n \right\} .$$

Often bin packing algorithms are studied using this measure. The reason for that is that for most bin packing problems, a simple reduction from the PARTITION problem (see problem SP12 in [6]) shows that no polynomial-time algorithm has an absolute performance guarantee better than $\frac{3}{2}$ unless P=NP. However, since in our problem items can be split, but cannot be packed more than a given number of parts to a bin, this reduction is not valid. In [3], the authors show that the problem they study is NP-hard in the strong sense for $k = 2$. They use a reduction from the 3-PARTITION problem (see problem [SP15] in [6]). Their result does not seem to imply any consequences with respect to hardness of approximation.

Independently of our work and simultaneous with it, Graham and Mao [7] analyzed the asymptotic approximation ratio of several algorithms, giving upper bounds of 1.498 for $k = 2$, $3/2$ for $k = 3$ and $2 - 2/k$ for $k \geq 4$. They also showed an upper bound of $2 - 1/k$ for NEXT FIT and a lower bound of $1 + (k + \frac{1}{k+1})^{-1}$ for online algorithms.

A related, easier problem is known as bin packing with cardinality constraints. In this problem, all items have size at most 1 as in regular bin packing, and the items cannot be split, however there is an upper bound of k on the amount of items that can be packed into a single bin. This problem was studied with respect to the asymptotic approximation ratio. It was introduced and studied in an offline environment as early as 1975 by Krause, Shen and Schwetman [10,11]. They showed that the performance guarantee of the well known FIRST FIT algorithm is at most $2.7 - \frac{12}{5k}$. Additional results were offline approximation algorithms of performance guarantee 2. These results were later improved in two ways. Kellerer and Pferschy [9] designed an improved offline approximation algorithm with performance guarantee 1.5 and finally a PTAS was designed in [2] (for a more general problem).

On the other hand, Babel et al. [1] designed a simple *online* algorithm with asymptotic approximation ratio 2 for any value of k. They also designed improved algorithms for $k = 2, 3$ of asymptotic approximation ratios $1 + \frac{\sqrt{5}}{5} \approx 1.44721$ and 1.8 respectively. The same paper [1] also proved an almost matching lower bound of $\sqrt{2} \approx 1.41421$ for $k = 2$ and mentioned that the lower bounds of [14,13] for the classic problem hold for cardinality constrained bin packing as well. The lower bound of 1.5 given by Yao [14] holds for small values of $k > 2$ and the lower bound of 1.5401 given by Van Vliet [13] holds for sufficiently large k. No other lower bounds are known.

Finally, Epstein [4] gave an optimal online bounded space algorithm (i.e., an algorithm which can have a constant number of active bins at every time) for this problem. Its asymptotic worst-case ratio is an increasing function of k and tends to $1 + h_\infty \approx 2.69103$, where h_∞ is the best possible performance guarantee of an online bounded space algorithm for regular bin packing (without cardinality constraints). Additionally, she improved the online upper bounds for $3 \leq k \leq 6$. In particular, the upper bound for $k = 3$ was improved to $\frac{7}{4}$.

Another related problem was studied recently by Shachnai et al. [12]. They considered an offline bin packing problem where items may be split arbitrarily. However, to make the problem non-trivial, there are some restrictions. In one model, each part of a split item increases by a constant additive factor. Another variant gives an upper bound on the number of split items. They showed that both these problems do not admit a PTAS unless P = NP. They designed a dual PTAS and an AFPTAS for both problems. Their problem is different from our problem since in their case all items have size at most 1. In their case it is possible to exploit the existence of simple structures of optimal solutions, which are more complicated in our case.

Our results. In the current paper, we begin by showing that this problem is NP-hard in the strong sense for any fixed value of k. This generalizes a result from Chung et al. [3]. We also show that the simple NEXT FIT algorithm has an absolute approximation ratio of $2 - 1/k$. Note that Graham and Mao [7] prove only an *asymptotic* upper bound of $2 - 1/k$ for NEXT FIT. Finally, we give an efficient 7/5-approximation algorithm for $k = 2$.

2 NP-Hardness of the Problem (in the strong sense)

Theorem 1. *Packing splittable items with a cardinality constraint of k parts of items per bin is NP-hard in the strong sense for any fixed $k \geq 3$.*

Proof. Given a fixed value of k, we show a reduction from the 3-Partition problem defined as follows (see problem [SP15] in [6]). We are given a set of $3m$ positive numbers $s_1, s_2, \ldots, s_{3m}$ such that $\sum_{j=1}^{3m} s_j = mB$ and each s_i satisfies $\frac{B}{4} < s_i < \frac{B}{2}$. The goal is to find out whether there exists a partition of the numbers into m sets of size 3 such that the sum of elements of each set is exactly B. The 3-Partition problem is known to be NP-hard in the strong sense.

Given such an instance of the 3-Partition problem we define an instance of the splittable item packing with cardinality constraints as follows. There are $m(k-3)$ items, all of size $\frac{3k-1}{3k(k-3)}$ (for $k = 3$, no items are defined at this point). These items are called padding items. In addition, there are $3m$ items, where item j has size $\frac{s_j}{3kB}$ (for $k = 3$ we define the size to be $\frac{s_j}{B}$). These items are called adapted items. The goal is to find a packing with exactly m bins. Since there are mk items, clearly a solution which splits items must use at least $m+1$ bins. Moreover, a solution in m bins contains exactly k items per bin. Since the sum of items is exactly m, all bins in such a solution are completely occupied with respect to size.

If there exists a partition of the numbers into m sets of sum B each, then there is a partition of the adapted items into M sets of sum $\frac{1}{3k}$ each (the sum is 1 for $k = 3$). Each bin is packed with $k - 3$ padding items and one such triple, giving m sets of k items, each set of sum exactly 1.

If there is a packing into exactly m bins, as noted above, no items are split and each bin must contain exactly k items. If $k = 3$, this implies the existence of a partition. Consider the case $k \geq 4$. We first prove that each bin contains exactly $k - 3$ padding items.

If a bin contains at least $k - 2$ padding items, their total size is at least $\frac{(3k-1)(k-2)}{3k(k-3)} = \frac{3k^2-7k+2}{3k^2-9k} = 1 + \frac{2k+2}{3k(k-3)}$. For $k \geq 4$ this is strictly larger than 1 and cannot fit into a bin. If there are at most $k - \ell \leq k - 4$ padding items, then there are ℓ additional items of size at most $\frac{1}{6k}$ ($\ell \geq 4$). The total size is therefore at most $\frac{(3k-1)(k-\ell)}{3k(k-3)} + \frac{\ell}{6k} = \frac{6k^2-2k-5\ell k-\ell}{6k(k-3)}$. This value is maximized for the smallest value of ℓ which is $\ell = 4$. We get the size of at most $\frac{6k^2-22k-4}{6k(k-3)} = 1 - \frac{4(k+1)}{6k(k-3)}$. For $k \geq 4$ this is strictly less than 1, which as noted above does not admit a packing into m bins.

Since each bin contains exactly $k - 3$ padding items, it contains exactly three adapted items, whose total size is exactly $\frac{1}{3k}$. The original sum of such three items is B, we get that a solution in m bins implies a partition. □

3 The NEXT FIT Algorithm

We can define NEXT FIT for the current problem as follows. This is a straightforward generalization of the standard NEXT FIT algorithm. An item is placed (partially) in the current bin if the bin is not full *and* the bin contains less than k item parts so far. If the item does not fit entirely in the current bin, the current bin is filled, closed, and as many new bins are opened as necessary to contain the item.

Note that this is an online algorithm. The absolute approximation ratio of NEXT FIT for the classical bin packing problem is 2, as Johnson [8] showed. Surprisingly, its approximation ratio for our problem tends to this value for large k. The two problems are different, and the two results seem to be unrelated.

We show that the approximation ratio of NEXT FIT is exactly $2-1/k$. Thus, this extremely simple algorithm performs as well as the algorithm from [3] for $k = 2$.

Theorem 2. *The approximation ratio of NEXT FIT is* $2 - 1/k$.

Proof. We first show a lower bound. The instance contains an item of size $Mk-1$ followed by $M(k - 1)k$ items of size ε, where M is large and $\varepsilon = 1/(Mk(k - 1))$.

Then the first item occupies $Mk-1$ bins, and the rest of the items are k per bin, in $M(k-1)$ bins. OPT has Mk bins in total. This proves a lower bound of $(M(2k-1)-1)/(Mk)$, which tends to $2-1/k$ for $M \to \infty$.

Now we show a matching upper bound. We define a *block* as a maximal set of bins which were consecutively filled by NEXT FIT (NF) in which each pair of consecutive bins contains parts of the same item. A block may contain only one bin. Denote the number of blocks by m. Let $u_1, u_2, \ldots, u_m$ be sizes of the blocks $1, \ldots, m$ of NF. In each block, all bins are full except perhaps the last one, which contains k parts of items (except for block m, perhaps). We assign weights to items. Let the size of item i be s_i. Then $w_i = \lceil s_i \rceil / k$. Note that in any packing, there are at least $\lceil s_i \rceil$ parts of item i. Since there can be at most k parts in a bin, this means

$$\text{OPT} \geq \frac{1}{k} \sum_i \lceil s_i \rceil = \sum_i \frac{\lceil s_i \rceil}{k} . \tag{1}$$

This explains our definition of the weights. This generalizes the weight definition from Chung et al. [3].

Consider the last bin from a block $i < m$. Since NF started a new bin after this bin, it contains k parts of items. Thus it contains at least $k-1$ items of weight $1/k$ (the last $k-1$ items are not split by the algorithm). If $u_i = 1$, there are k such items. If $u_i > 1$, consider all items excluding the $k-1$ last items in the last bin. We do not know how many items there are in the first $u_i - 1$ bins (where the last item extends into bin u_i). However, for a fixed size s, the weight of a group of items of total size s is minimized if there is a single item in the group (since we round up the size for each individual item to get the weight). This implies the total weight in a block of u_i bins is at least $u_i/k + (k-1)/k = (u_i + k - 1)/k$.

Now consider block m. If $u_i = 1$, the weight is at least $1/k$ since there is at least one item. Else, as above the weight is at least u_i/k, since the last bin of this block has at least one item or a part of an item.

We have $\text{NF} = \sum u_i$. Therefore

$$\text{OPT} \geq \sum_i w_i \geq \frac{\sum_{i=1}^{m}(u_i + k - 1) - (k-1)}{k} = \frac{\text{NF} + (m-1)(k-1)}{k} . \tag{2}$$

Also by size, $\text{OPT} > \text{NF} - m$ and thus $\text{OPT} \geq \text{NF} - m + 1$. Multiply this inequality by $(k-1)/k$ and add it (2) to get

$$\frac{2k-1}{k} \cdot \text{OPT} \geq \text{NF}\left(\frac{1}{k} + \frac{k-1}{k}\right) + (m-1)\frac{k-1}{k} - (m-1)\frac{k-1}{k} = \text{NF}.$$

We conclude $\text{NF} \leq (2 - 1/k)\text{OPT}$. □

4 The Structure of the Optimal Packing for $k = 2$

Before we begin our analysis, we make some observations regarding the packing of OPT. A packing can be represented by a graph where the items are nodes and edges correspond (one-to-one) to bins. If there is a bin which contains (parts of) two items, there is an edge between these items. A bin with only one item corresponds to a loop on that item. The paper [3] showed that for any given packing, it is possible to modify the packing such that there are no cycles in the associated graph. Thus the graph consists of a forest together with some loops. We start by analyzing the structure of the graph associated with the optimal packing. Items of size at most 1/2 are called *small*.

Lemma 1. *There exists an optimal packing in which all small items are leaves.*

Proof. Consider a small item that has edges to at least two other items. Note that if two small items share an edge, the packing can be changed so that these two items form a separate connected component with a single edge. Thus we may assume that all neighbors are (parts of) medium or large items.

Order the neighbors in some way and consider the first two neighbors. Denote the small item by s and the sizes of its neighboring parts by w_1 and w_2. In bin i, w_i is combined with a part s_i of the small item s ($i = 1, 2$).

We have $s_1 + s_2 \leq 1/2$. If $s_1 \leq w_2$, we can cut off a part of size s_1 from w_2 and put it in bin 1, while putting s_1 in bin 2. This removes neighbor w_1 from the small item s. Otherwise, $w_2 < s_1 \leq 1/2$, which means that we can put s_1 into bin 2 without taking anything out of bin 2: we have $w_2 < 1/2$ and $s_1 + s_2 \leq 1/2$. Again, w_1 is no longer a neighbor of s (or even connected to s).

Thus we can remove one neighbor from s. We can continue in this way until s has only one neighbor left. □

Due to space constraints, we omit the proof of the following lemma.

Lemma 2. *An item of size in $((i-1)/2, i/2]$ has at most i neighbors for all $i \geq 2$.*

5 A 7/5-Approximation for $k = 2$

Let $k = 2$. We call items of size in $(1/2, 1]$ *medium* and remaining items *large*. Our algorithm works as follows. We present it here in a simplified form which ignores the fact that it might run out of small items in the middle of step 2(b) or while packing a large item in step 4. We will show later how to deal with these cases while maintaining an approximation ratio of 7/5. See Figure 1.

We begin by giving an example which shows that this algorithm is not optimal. For some integer N, consider the input which consists of $4N$ small items of size $2/N$, $2N$ medium items of size $1 - 1/N$, $3N$ medium items of size $1 - 2/N$.

ALG packs the items of size $1 - 1/N$ in $4N$ bins, together with $4N$ small items. It needs $3N(1 - 2/N) = 3N - 6$ bins for the remaining medium items. Thus it needs $7N - 6$ bins in total. OPT places $3N$ small items in separate bins (one per bin), and N small items are split into two equal parts. This gives $5N$ bins in which there is exactly enough room to place all the medium items.

Theorem 3. *This algorithm achieves an absolute approximation ratio of* 7/5.

1. Sort the small items in order of increasing size, the medium items in order of decreasing size, and the large items in order of decreasing size.
2. Pack the medium items one by one, as follows, until you run out of medium or small items.
   (a) If the current item fits with the smallest unpacked small item, pack them into a bin.
   (b) Else, pack the current item together with the two *largest* small items in two bins.
3. If no small items remain unpacked, pack remaining medium and large items using Next Fit and halt. Start with the medium items.
4. Pack all remaining small items in separate bins. Pack the large items one by one into these bins using Next Fit (starting with the largest large item and smallest small item).
5. If any bins remain that have only one small item, repack these small items in pairs into bins and halt.
6. Pack remaining large items using Next Fit.

Fig. 1. The approximation algorithm for $k = 2$

The analysis has three cases, depending on whether the algorithm halts in step 3, 5 or 6. The easiest case among these is without a doubt step 5, at least as long as all bins packed in step 5 contain two small items.

5.1 Algorithm Halts in Step 5

Based on inequality (1), we define weights as follows.

Definition 1. *The weight of an item of size* w_i *is* $\lceil w_i \rceil/2$.

In our proofs, we will also use weights of parts of items, based on considering the total weight of an item and the number of its parts. By Definition 1, small and medium items have weight 1/2. Therefore, we have the following bounds on total weight of bins packed in the different steps:

2.(a) $1/2 + 1/2 = 1$

2.(b) We pack three items of weight 1/2 in two bins, or 3/4 weight per bin on average.

4. Consider a large item which is packed in g bins, that is, together with in total g small items. Its size is strictly larger than $\frac{g-1}{2}$ and thus its weight is at least $g/4$. Each small item has a weight of 1/2, so we pack a weight of at $3g/4$ in these g bins.

5. $1/2 + 1/2 = 1$

This immediately proves an upper bound of 4/3 on the absolute approximation ratio. There is, however, one special case: it can happen that one small item remains unpaired in step 5. Since this case requires deeper analysis, we omit it due to space constraints.

5.2 Critical Items

Definition 2. *A* critical *item is a medium item that the algorithm packs in Step 2(b).*

From now on, for the analysis we use a fixed optimal packing, denoted by OPT. We consider the critical items in order of decreasing size. Denote the current item by x. We will consider how OPT packs x and define an **adjusted weight** based on how much space x occupies in the bins of OPT. Denote the adjusted weight of item i by W_i. The adjusted weights will satisfy the following condition:

$$\sum_{i=1}^{n} \frac{\lceil w_i \rceil}{2} \leq \sum_{i=1}^{n} W_i \leq \text{OPT}. \tag{3}$$

Specifically, we will have $W_i \geq \lceil w_i \rceil / 2$ for $i = 1, \dots, n$. Thus the numbers W_i will generate a better lower bound for OPT, that we can use to show a better upper bound for our algorithm. This is the central idea of our analysis. We initialize $W_i = \lceil w_i \rceil / 2$ for $i = 1, \dots, n$. There are four cases.

Case 1. OPT packs x by itself. In this case we give x adjusted weight 1, and so our algorithm packs an adjusted weight of 1 in each of the (two) bins that contain x.

Case 2. OPT packs x with part *of a small item.* Again x and the bins with x get an adjusted weight of 1. This holds because when OPT splits a small item (or a medium item), it is as if it packs two small items, both of weight 1/2. Therefore such an item gets adjusted weight 1. We can transfer the extra 1/2 from the small item to x.

Case 3. OPT combines x with a complete *small item y.* Since our algorithm starts by considering the smallest small items, y must have been packed earlier by our algorithm, i.e. with a larger medium item x' (which is not critical!). If OPT packs x' alone or with part of a small item, it has an adjusted weight of 1 (Cases 1 and 2). Thus the bin with x' has an adjusted weight of 3/2, and we transfer 1/2 to x. If OPT packs x' with a full small item y', then y' is packed with a larger non-critical item x'' by our algorithm, etc. Eventually we find a non-critical medium item x^* which OPT packs alone or with part of a small item, or for which Case 4 holds. The difference between the weight and the adjusted weight of x^* will be transferred to x. Note that the bin in which our algorithm packs x^* has a weight of 1 since x^* is non-critical. All intermediate items $x', x'', \dots$ have weight 1/2 and are non-critical as well, and we change nothing about those items.

Case 4. OPT packs x with a split medium or large item, or splits x itself.
Since there might be several critical items for which Case 4 holds, we need to consider how OPT packs all these items to determine their adjusted weight. We are going to allocate adjusted weights to items according to the following rules:

1. Each part of a small item (in the OPT packing) gets adjusted weight 1/2.
2. A part of a large item which is in a bin by itself gets adjusted weight 1.
3. Other parts of large items get adjusted weight 1/2.

We do not change the weight of non-critical items. The critical items receive an adjusted weight which corresponds to the number of bins that they occupy in the packing of OPT. As noted above, this packing consists of trees and loops. Loops were treated in Case 1. To determine the adjusted weights, we consider the **non-medium** items that are cut into parts by OPT. Each part of such an item is considered to be a single item for this calculation and has adjusted weights as explained above. We then have that the optimal packing consists only of trees with small and medium items, and loops. It can be seen that each part of a non-medium item (for instance, part of a large item) which is in a tree has weight 1/2.

Consider a tree T in the optimal packing. Denote the number of edges (bins) in it by t. Since all items in T are small or medium, there are $t+1$ items (nodes) in T by Lemmas 1 and 2. Any items that are small (or part of a small item) or medium but non-critical have adjusted weight equal to the weight of a regular small or medium item which is 1/2. Denoting the number of critical items in T by c, we find that the $t+1-c$ non-critical items have weight $\frac{t+1-c}{2}$. All items together occupy t bins in the optimal packing. This means we can give the critical items each an adjusted weight of $(t-\frac{t+1-c}{2})/c = \frac{1}{2}+\frac{t-1}{2c}$ while still satisfying (3). This expression is minimized by taking c maximal, $c=t+1$, and is then $t/(t+1)$. We can therefore assign an adjusted weight of $t/(t+1)$ to each critical item in T.

Since the algorithm combines a critical item with two small items of weight (at least) 1/2, it packs a weight of $1+t/(t+1)=\frac{2t+1}{t+1}$ in two bins, or $\frac{2t+1}{2t+2}$ per bin. This ratio is minimized for $t=2$ and is 5/6.

However, let us consider the case $t=2$ in more detail. If the OPT tree with item x (which is now a chain of length 2) consists of three critical items, then the sum of *sizes* of these items is at most 2. Our algorithm packs each of these items with two small items which do not fit with one such item. Let the sizes of the three medium items be m_1, m_2, m_3. Let the two small items packed with m_i be $s_{i,j}$ for $j=1,2$. We have that $m_1+m_2+m_3 \le 2$ but $m_i+s_{i,j}>1$ for $i=1,2,3$ and $j=1,2$. Summing up the last six inequalities and subtracting the one before, we get that the total size of all nine items is at least 4. Thus the area guarantee in these six bins is at least 2/3.

If one of the items in the chain is (a part of) a small or large item, or a medium non-critical item, it has adjusted weight 1/2. This leaves an adjusted weight of 3/4 for the other two items. In this case we pack at least $3/4+1=7/4$ in two bins, or 7/8 per bin. For $t \ge 3$, we also find a minimum ratio of 7/8.

Thus we can divide the bins with critical items into two subtypes: A with an adjusted weight of 5/6 and area 2/3, and B with an adjusted weight of (at least) 7/8 and area 1/2.

5.3 Algorithm Halts in Step 3

We divide the bins that our algorithm generates into types. We have

1. groups of two small items and one medium item in two bins
2. pairs of one small item and one medium item in one bin
3. groups of four or more medium items in three or more bins
4. groups of three medium items in two bins
5. one group of bins with 0 or more medium items and all the large items

Note that bins of type 4 contain a total weight of at least 3/4 (3/2 per two bins), as well as a total size of at least 3/4 (3 items of size more than 1/2 in two bins). Thus, whether we look at sizes or at weights, it is clear that these bins can be ignored if we try to show a ratio larger than 4/3.

Furthermore, in the bins of type 5 we ignore that some of the items may be medium. The bounds that we derive for the total size and weight packed into these bins still hold if some of the items are only medium-sized.

The bins of type 1 contain the critical items. We say the bins with subtype A are of type $1a$, and the bins with subtype B are of type $1b$. Define $x_{1a}, x_{1b}, x_2, x_3, x_4$ as the number of bins with types $1a, 1b, 2, 3$, and 5, respectively.

Consider the bins of type 3. Let k be the number of groups of medium items. Let $t_i \geq 3$ be the number of bins in group $1 \leq i \leq k$. The items in group i have total size more than $t_i - 1/2$, since the last bin contains a complete medium item. The total weight of a group is $\frac{t_i+1}{2}$, since it contains $t_i + 1$ items, each of weight $\frac{1}{2}$. We get that the total size of items in bins of type 3 is at least $\sum_{i=1}^{k}(t_i - \frac{1}{2}) = x_3 - \frac{k}{2}$, and the total weight of these items is $\sum_{i=1}^{k} \frac{t_i+1}{2} = \frac{x_3+k}{2}$.

We find two different lower bounds on OPT.

Adjusted weight:

$$OPT \geq \frac{5}{6}x_{1a} + \frac{7}{8}x_{1b} + x_2 + \frac{x_3}{2} + \frac{k}{2} + \frac{x_5}{2}. \tag{4}$$

Size:

$$OPT \geq \frac{2}{3}x_{1a} + \frac{x_{1b}}{2} + \frac{x_2}{2} + x_3 - \frac{k}{2} + \max(x_5 - 1, 0). \tag{5}$$

Multiplying the first inequality by $\frac{4}{5}$ and the second one by $\frac{3}{5}$ we get

$$\frac{7}{5}OPT \geq \frac{16}{15}x_{1a} + x_{1b} + \frac{11}{10}x_2 + x_3 + \frac{k}{10} + \frac{2}{5}x_5 + \frac{3}{5}\max(x_5 - 1, 0). \tag{6}$$

If $x_5 = 0$ we are done. Else, (5) is strict and we get

$$OPT > \frac{2}{3}x_{1a} + \frac{x_{1b}}{2} + \frac{x_2}{2} + x_3 - \frac{k}{2} + x_5 - 1. \tag{7}$$

This means x_3 and x_5 occur with the same fractions in (4) and (7). Thus we can set $x_3 := x_3 + x_5$ and $x_5 := 0$. Adding (4) and (7) and dividing by 2 gives

$$\text{OPT} > \frac{3}{4}(x_{1a} + x_2 + x_3) + \frac{11}{16}x_{1b} - \frac{1}{2}\,.$$

This implies we are done if $x_{1a} + x_2 + x_3 \geq \frac{3}{4}x_{1b} + 14$. Clearly, this holds if any of x_{1a}, x_2 or x_3 are at least 14. Finally, by (4) we are also done if

$$\frac{5}{6}x_{1a} + \frac{7}{8}x_{1b} + x_2 + \frac{x_3}{2} + \frac{k}{2} \geq \frac{5}{7}(x_{1a} + x_{1b} + x_2 + x_3).$$

This holds if

$$\frac{5}{42}x_{1a} + \frac{9}{56}x_{1b} + \frac{2}{7}x_2 + \frac{k}{2} \geq \frac{3}{14}x_3.$$

Since we may assume $x_3 < 14$, we are in particular done if $x_{1b} \geq 18$ or $k \geq 6$.

This leaves a limited set of options for the values of x_{1a}, x_{1b}, x_2, x_3 and k that need to be checked. It is possible to verify that for almost all combinations, we find OPT $\geq \frac{5}{7}$ALG. One exception is $x_3 = 3$, $k = 1$. However, going back to the original variables, this means $x_3 + x_5 = 3$ and $k = 1$. But x_3 is either 0 or at least 3. If $k = 1$, we must have $x_3 = 3$ and $x_5 = 0$, so we treated this case already. Two other cases require special attention and are described below.

Special cases. Step 2(b) requires two small items. If only one is left at this point, and there is also no remaining medium item with which it could be packed, we redefine it to be a medium item and pack it in step 3. This leads to it being packed in a bin of type 3 (or 4). Note that in this case, this small item and any medium item we tried to pack with it in Step 2 have total size more than 1. Thus if the small item ends up in a group of type 4 (a group of two bins), the total size of the items in these bins (as well as the total weight) is still at least 3/2, and we can ignore these bins in the analysis. Therefore the analysis still holds.

There are two cases where OPT $< \frac{5}{7}$ALG is possible. If $x_2 = 1$ and $x_5 = 2$, a packing into two bins could exist in case there is only one large item. (If the bins counted in x_5 contain two medium items, then we have that the three medium items require (at least) two bins and the small item requires an extra bin.) If such a packing exists, it works as follows: pack first the medium item, then the large item (partially in the second bin), then the small item. If this gives a packing into two bins, this is how our algorithm packs the items. Otherwise we already have an optimal packing.

If $x_{1b} = 4$, $x_2 = 1$ and $x_5 = 5$, it is a simple matter to try all possible packings for the items in 7 bins and check if one is valid. (We can try all possible forests on at most 13 nodes and at most 7 edges.) If there is no packing in 7 bins, then our algorithm maintains the ratio of 7/5. If there is one, we use it.

5.4 Algorithm Halts in Step 6

In this case we have the following bin types.

1. groups of two small items and one medium item in two bins
2. pairs of one small item and one medium item in one bin
3. groups of large items with small items
4. one group of large items

The analysis in this case is similar to the one in the previous case. We omit it due to space constraints.

6 Conclusions

In this paper, we gave the first absolute upper bounds for general k for this problem. Furthermore we provided an efficient algorithm for $k = 2$. An interesting question is whether it is possible to give an efficient algorithm with a better absolute approximation ratio for $k = 2$ or for larger k. In a forthcoming paper [5] we will present approximation schemes for these problems. However, these schemes are less efficient than the algorithms given in this paper already for $\epsilon = 2/5$.

References

1. Babel, L., Chen, B., Kellerer, H., Kotov, V.: Algorithms for on-line bin-packing problems with cardinality constraints. Discrete Applied Mathematics 143(1-3), 238–251 (2004)
2. Caprara, A., Kellerer, H., Pferschy, U.: Approximation schemes for ordered vector packing problems. Naval Research Logistics 92, 58–69 (2003)
3. Chung, F., Graham, R., Mao, J., Varghese, G.: Parallelism versus memory allocation in pipelined router forwarding engines. Theory of Computing Systems 39(6), 829–849 (2006)
4. Epstein, L.: Online bin packing with cardinality constraints. SIAM Journal on Discrete Mathematics 20(4), 1015–1030 (2006)
5. Epstein, L., van Stee, R.: Approximation schemes for packing splittable items with cardinality constraints. Manuscript.
6. Garey, M.R., Johnson, D.S.: Computers and Intractability: A Guide to the theory of NP-Completeness. W. H. Freeman and Company, New York (1979)
7. Graham, R.L., Mao, J.: Parallel resource allocation of splittable items with cardinality constraints. Manuscript.
8. Johnson, D.S.: Fast algorithms for bin packing. Journal of Computer and System Sciences 8(3), 272–314 (1974)
9. Kellerer, H., Pferschy, U.: Cardinality constrained bin-packing problems. Annals of Operations Research 92, 335–348 (1999)
10. Krause, K.L., Shen, V.Y., Schwetman, H.D.: Analysis of several task-scheduling algorithms for a model of multiprogramming computer systems. Journal of the ACM 22(4), 522–550 (1975)
11. Krause, K.L., Shen, V.Y., Schwetman, H.D.: Errata: Analysis of several task-scheduling algorithms for a model of multiprogramming computer systems. Journal of the ACM 24(3), 527 (1977)
12. Shachnai, H., Tamir, T., Yehezkely, O.: Approximation schemes for packing with item fragmentation. In: Erlebach, T., Persinao, G. (eds.) WAOA 2005. LNCS, vol. 3879, pp. 334–347. Springer, Heidelberg (2006)
13. van Vliet, A.: An improved lower bound for online bin packing algorithms. Information Processing Letters 43(5), 277–284 (1992)
14. Yao, A.C.C.: New algorithms for bin packing. Journal of the ACM 27, 207–227 (1980)

Computational and Structural Advantages of Circular Boundary Representation*

O. Aichholzer[1], F. Aurenhammer[1], T. Hackl[1],
B. Jüttler[2], M. Oberneder[2], and Z. Šír[2]

[1] University of Technology Graz, Austria
{oaich,thackl}@ist.tugraz.at, auren@igi.tugraz.at
[2] Johannes Kepler University of Linz, Austria
{bert.juettler,margot.oberneder,zbynek.sir}@jku.at

Abstract. Boundary approximation of planar shapes by circular arcs has quantitive and qualitative advantages compared to using straight-line segments. We demonstrate this by way of three basic and frequent computations on shapes – convex hull, decomposition, and medial axis. In particular, we propose a novel medial axis algorithm that beats existing methods in simplicity and practicality, and at the same time guarantees convergence to the medial axis of the original shape.

1 Introduction

The plain majority of algorithms in computational geometry have been designed for processing *linear* objects, like lines, planes, or polygons. On the one hand, this is certainly due to the fact that many interesting and deep computational and combinatorial questions do arise already for inputs of this simple form. Again, the pragmatic reason is that algorithms for linear objects are usually both easier to develop and simpler to implement. To make things work for nonlinear objects, which arise frequently in practical settings, such objects are usually approximated in a piecewise-linear manner and up to a tolerable error. Existing approaches [10] to directly extending polygonal algorithms to curved objects are rare and, due to their generality, of limited practical use.

In its simplest form, the input object is a single planar shape, $\mathcal{A}$, with curved and connected boundary $\partial\mathcal{A}$. Frequent tasks to be performed on $\mathcal{A}$ – each being prior to a variety of more involved computations – include constructing the convex hull of $\mathcal{A}$, decomposing $\mathcal{A}$ into primitives, and calculating the medial axis of $\mathcal{A}$. These tasks are well investigated in the case of polygonal shapes. In certain situations, however, the number of line segments required for approximating $\partial\mathcal{A}$ with high accuracy may be prohibitively large. Even more seriously, making a piecewise-linear approximation of $\partial\mathcal{A}$ and invoking a polygonal-shape algorithm may generate results that are topologically incorrect; the medial axis is a well-known example.

* Supported by the Austrian FWF JRP 'Industrial Geometry', S9200.

F. Dehne, J.-R. Sack, and N. Zeh (Eds.): WADS 2007, LNCS 4619, pp. 374–385, 2007.

The intention of the present paper is to highlight the use of circular arcs for boundary representation. It is well known that for nonlinear curve segments the approximation order increases in comparison to using straight-line segments. In particular, if a given accuracy ε is achieved by using N line segments, then as few as $n = \Theta(N^{2/3})$ circular arcs can accomplish the same. This has been an issue in approximation theory, but in computational geometry this gain seems to have been less valued than eliminating small factors in the complexity of the subsequently applied algorithm. Boundary approximation by circular arcs may be of advantage also in a qualitative respect. For instance, it avoids the mentioned topological inconsistencies in medial axis computations, and it supports the computation of shape offsets, as the class of shapes bounded by circular arcs is closed under offset operations.

We will show that for the three basic problems mentioned above – convex hull, triangulation, and medial axis – simple and practical, though still efficient algorithms exist that work for circular arc inputs. The first two problems are less demanding; we treat them mainly to point out the respective favorable (in our opinion) approach, whose practicality shall encourage the use of circular arc boundary representation. Nevertheless, substantial differences to the polygonal case occur; see below. For computing the medial axis, we propose a novel and extremely simple algorithm that is based on a known (though less recognized) decomposition lemma. After having computed a purely combinatorial description of the medial axis using tailored shape splitting, its individual parts (conics and line segments, like in the polygonal case) are reassembled *without* the need of merging.

Suitable circular arc approximations of shapes can be found in linear time. In summary, the obtained shape processing algorithms are superior in runtime and output volume to their line segment based counterparts, retain much (if not all) of their simplicity, and are even more natural in some cases.

1.1 Outline and Background

We briefly describe the contributions of this paper and relate them to existing literature.

Section 2 deals with approximating general curves by suitable primitives. This is a topic of importance in geometric modeling and in CAD and NC applications, and many quite recent results are available [24,25,27,33,17,31]. Our aim is to approximate a parametric curve $\boldsymbol{c}(t)$ by circular arcs. We assume that $\boldsymbol{c}(t)$ is piecewise-polynomial of constant degree, and we use biarcs (pairs of smoothly joined circular arcs) [30,25,31] as primitives. A straight-forward bisection algorithm for biarc generation already fits our purposes. It uniquely assigns biarcs to parameter intervals, which facilitates the error evaluation. An approximating spline curve $\boldsymbol{b}$ of size n is computed in $O(n)$ time. It fits the input curve $\boldsymbol{c}(t)$ in slope at biarc endpoints, and can be tuned to match $\boldsymbol{c}(t)$ in curvature at certain points (a fact being important in subsequent medial axis computations). Though not being optimal in the number of arcs, the approximation order of $\boldsymbol{b}$ is still three [24,31]. In contrast, with line segments one cannot exceed order two, and a polyline of size $N = \Theta(n^{3/2})$ is needed to arrive at the same precision.

The remaining sections propose algorithms for *circular arc shapes* $\mathcal{A}$, where the boundary $\partial\mathcal{A}$ of $\mathcal{A}$ is given as a simple curve composed of n circular arcs. Choice is guided by efficiency as well as by reducibility to basic operations that have robust implementations [11]. Due to lack of space, we had to skip two sections from this version of the paper. Let us nevertheless provide here a short description of the material they contain.

The first topic is computing the convex hull of a circular arc shape $\mathcal{A}$. This task is one of the most basic to be performed for a given shape, and has a variety of applications including shape fitting, motion planning, shape separation, and many others. At least four linear-time algorithms have been developed for polygonal shapes [4,16,23,26]. The incremental method by Melkman [26] stands out by its simplicity, and it is this candidate we generalize for circular arc shapes. Compared to the original setting, two difficulties arise. Deciding inclusion for a currently inserted arc in the convex hull constructed so far is no trivial test, and the convex hull cannot be described by a sequence of input vertices of the shape. We show that a runtime of $O(n)$ is still possible. The basic subroutine of the algorithm computes the convex hull of only two circular arcs.

The second topic is shape triangulation, a fundamental building block in algorithms for decomposition, shortest path finding, and visibility – to name a few. Most existing algorithms are meant for polygonal shapes. They partition a given (simple) n-vertex polygon into triangles without introducing Steiner points. Efficient candidates are [14,22,3,18,7] which all show an $O(n \log n)$ runtime. Theoretically more efficient methods do exist, but when aiming at simplicity, choice should be made from the list above. When trying to generalize to shapes $\mathcal{A}$ bounded by circular arcs, we face two problems. First of all, if the use of Steiner points is disallowed, then a partition of $\mathcal{A}$ into primitives bounded by a constant number of circular arcs need not exist. Also, not all triangulation methods are suited to generalization. This applies, for instance, to the extremely simple ear cutting method in [20] which runs in time $O(r \cdot n)$, where r is the number of reflex vertices of $\mathcal{A}$. The triangulation algorithm we propose is closest to Chazelle's [7]. It manages with an (almost) worst-case minimal number of Steiner points on $\partial\mathcal{A}$, runs in $O(n \log n)$ time, and uses a dictionary as its only nontrivial data structure. The produced primitives are arc triangles with at least one straight edge. The most complex geometric operation is intersecting a circle with a line.

Section 3 is devoted to the medial axis, a frequently used structure associated with a given input shape. Its main applications include shape recognition, solid modeling, pocket machining, and others. Interest in mathematical properties of the medial axis for general shapes found renewal in recent years [9,28,5,6,2]. In our case, where the shape $\mathcal{A}$ is simply connected and $\partial\mathcal{A}$ consists of n circular arcs, its medial axis $M(\mathcal{A})$ is known to be a tree composed of $O(n)$ conic edges. Algorithmic work on the medial axis either concentrated on the case where $\mathcal{A}$ is a polygon [21,7,8], or on general sets of curved arcs [33,19,28,1] (and their Voronoi diagram) without, however, exploiting the fact that the input arcs define a simple curve. (The various existing methods for computing digital versions of the medial axis are not considered here.) Though theoretically efficient as $O(n \log n)$ or better, these algorithms suffer from involved merge or insertion steps which, even for straight arcs as input, are difficult to implement. In addition, numerical stability issues arise heavily; intersections of conics have to be determined repeatedly which, when not calculated exactly, are bound to cumulate the error.

We present a simple randomized divide-and-conquer algorithm for computing $M(\mathcal{A})$ that overcomes these drawbacks. In contrast to comparable algorithms, the costly part is delegated to the divide step. The basic subroutine there is an inclusion test for an arc in a circle. The merge step is trivial: it concatenates two medial axes. The expected runtime is bounded by $O(n^{3/2})$, but is provably better for most types of shape. For example, $O(n \log n)$ expected time suffices if the diameter of $M(\mathcal{A})$ is $\Theta(n)$. No nontrivial data structures are used.

To guarantee applicabiliy of our methods to approximating the medial axes of general shapes $\mathcal{B}$, a convergence result is needed. We prove in Section 4 that, for a suitable approximation of $\partial\mathcal{B}$ by biarcs, $M(\mathcal{B})$ is the limit of $M(\mathcal{A})$ when the approximating arc shape $\mathcal{A}$ converges to $\mathcal{B}$. Related results exist, but either presuppose C^2 conditions on $\partial\mathcal{A}$ not attainable by circular arcs [6], or concern subsets of the medial axis [5] that survive after pruning the Voronoi diagram of point samples from $\partial\mathcal{B}$. (As a negative side effect, the medial axis approximation obtained from a point sample is not C^1.) It is well known [2] that medial axis convergence is *not* given for polygonal approximations of $\mathcal{B}$. In conclusion, circular arcs are the simplest possible tool for boundary conversion that guarantees a stable medial axis approximation.

2 Circular Arcs

In order to represent a general shape $\mathcal{A}$ in a form suitable for geometric computations, we discuss methods for approximating $\partial\mathcal{A}$ by circular arcs. We assume that $\partial\mathcal{A}$ is given as a polynomial spline curve of constant degree. Attention is restricted to degree 3, as every free-form curve can efficiently and with any desired precision be converted into cubics [29], and in many applications the input will already be available in this common form [12].

Several approaches to generating circular arc splines exist; see [24] for a review. We consider a simple bisection algorithm consisting of two steps, approximation and error measurement. A geometric primitive b (an arc or a biarc) is fitted to a segment s of the given cubic curve $\mathbf{c}(t)$, and the distance from b to s is computed. The algorithm is relatively easy to implement and still adapts the degrees of freedom to the input data. As a slight disadvantage, the number of primitives (the resulting data volume) is minimal only in the asymptotic sense.

Define the one-sided Hausdorff distance from a primitive b to a segment $s \subseteq \mathbf{c}(t)$ as $\delta(b, s) = \max_{p \in b} \min_{q \in s} ||p - q||$. (We consider b and s as closed sets.) Let ε denote the error tolerance to be met by the algorithm.

Algorithm BISECT(t_0, t_1)

Construct b
Compute $\delta = \delta(b, \mathbf{c}[t_0, t_1])$
If $\delta \le \varepsilon$ then return $\{b\}$
Else return BISECT$(t_0, \frac{t_0+t_1}{2}) \cup$ BISECT$(\frac{t_0+t_1}{2}, t_1)$

Depending on the primitive b used, Algorithm BISECT produces splines of different quality: merely continuous (C^0) circular arc splines, or tangent continuous (C^1) arc splines. When being content with the former type, we can simply choose for b the unique circular arc passing through the three points $\mathbf{c}(t_0)$, $\mathbf{c}(\frac{(t_0+t_1)}{2})$, and $\mathbf{c}(t_1)$. To obtain C^1 arc splines, so-called biarcs [30] are utilized.

A *biarc* b consists of two circular arcs with common unit tangent vector at their joint. Usually, b is described by its source x with associated unit tangent vector v_x, and its target y with unit tangent vector v_y. Given these data, there exists a one-parameter family of interpolating biarcs. All possible joints are located on the circle σ passing through x and y and having the same oriented angles with v_x and v_y. Several ways for choosing the joint m have been proposed; see e.g. [25,31]. For many applications, taking $m = \sigma \cap \mathbf{c}[t_0, t_1]$ is appropriate. To calculate m, a polynomial of degree 4 has to be solved (where a closed-form solution is still available). The output is a C^1 arc spline with all arc endpoints sitting on $\mathbf{c}(t)$.

In view of subsequent stable medial axis computations, the choice of m has to be made more carefully. Define an *apex* of $\mathbf{c}(t)$ as a local curvature maximum. The apices split the curve $\mathbf{c}(t)$ into pieces of monotonic signed curvature, so-called *spirals*. Following [25], we aim at approximating spirals of $\mathbf{c}(t)$ by circular arc spirals. To this end, we split $\mathbf{c}(t)$ at its apices. These points can be found by solving polynomials of degree 5. Now, we exploit that spiral biarcs can be constructed that connect two given points x and y, match unit tangents there, and assume a predefined curvature in one of them. Let k_x and k_y be the curvature of $\mathbf{c}(t)$ at x and y, respectively, and suppose $k_x < k_y$. To match curvature at x, we choose the radius of the first arc, b_1, equal to $r_x = 1/k_x$. The joint m is obtained by intersecting the circle supporting b_1 with the joint circle σ. According to [25], the radii and curvatures satisfy $r_x > r_y > 1/k_y$. When starting the next biarc from y with $r_y = 1/k_y$ (unless y is an apex), monotonicity of signed curvature will be preserved.

Each arc is found in $O(1)$ time, where the constant depends on the degree of the polynomial to be solved. Concerning the error measurement, each produced circular arc b_i has to be matched to its corresponding segment s on the curve $\mathbf{c}$. We then compute an upper bound for the one-sided Hausdorff distance $\delta(b_i, s)$ by substituting the parametric representation of s into the implicit equation of the circle supporting b_i. A simpler upper bound can be calculated (without polynomial solving) by using Bernstein-Bézier representations [13]. In summary, when algorithm BISECT spans a binary recursion tree with n leaves (the returned n primitives), any of the described arc splines can be constructed in $O(n)$ time.

Let us discuss the asymptotic behaviour of the number n for decreasing tolerance ε. For a given curve $\mathbf{c}(t)$ with domain $[t_0, t_1]$, which is assumed to contain neither inflections nor apices, we consider primitives having approximation order k. Adapting the analysis in [24,31], we get $\delta = \Theta(h^k)$ for the one-sided Hausdorff distance δ, provided that $\mathbf{c}(t)$ is approximated with (small) parameter step size h, and that k is considered a constant.

This relation implies a general lower bound. For *any* approximation of $\mathbf{c}(t)$ by n primitives of order k, the largest step size satisfies $\Delta t \geq \frac{t_1 - t_0}{n}$. Thus from $\varepsilon \geq \delta$, which is to be achieved by the approximation, and from $\delta = \Theta((\Delta t)^k)$, we get $n = \Omega(1/\varepsilon^{1/k})$. On the other hand, the smallest step size Δt taken by algorithm BISECT satisfies $\Delta t \leq \frac{t_1 - t_0}{n}$. When doubling the step size we have $\delta = \Theta((2\Delta t)^k)$ but $\varepsilon < \delta$, as the tolerance is not yet achieved. Thus $n = O(1/\varepsilon^{1/k})$.

In conclusion, for sufficiently small tolerance ε, the number n of primitives constructed by algorithm BISECT is asymptotically optimal. This is also true in the general case where $\mathbf{c}(t)$ contains inflections and apices, because the resulting number of spirals of $\mathbf{c}(t)$ is independent of n. In conclusion, to arrive at tolerance ε, Algorithm BISECT needs $n = \Theta(1/\sqrt[3]{\varepsilon})$ cicular arcs (order 3), whereas $N = \Theta(1/\sqrt{\varepsilon})$ line segments (order 2) have to be invested by any polygonal approximation method.

Lemma 1. *Compared to approximating* $\mathbf{c}(t)$ *with a polyline, the data volume drops from* N *to* $\Theta(N^{2/3})$ *when circular arc splines are used.*

It should be observed that, the other way round, when approximating $\mathbf{c}(t)$ with a point sample (as commonly done for medial axis computations [2]), the data volume increases to $\Theta(n^3)$ compared to n circular arcs.

3 Medial Axis

Let $\mathcal{A}$ be the circular arc shape under consideration. (All objects are considered to be closed sets in the sequel). Call a disk $D \subseteq \mathcal{A}$ *maximal* if there exists no disk D' different from D such that $D \subset D'$ and $D' \subseteq \mathcal{A}$. The medial axis, $M(\mathcal{A})$, of $\mathcal{A}$ is defined as the set of all centers of maximal disks.

As the boundary of $\mathcal{A}$ is a connected and simple curve with n circular arcs, $M(\mathcal{A})$ is finite, connected, and cycle-free [9] and thus forms a tree. $M(\mathcal{A})$ can be decomposed into $O(n)$ *edges*, which are maximal pieces of straight lines and (possibly all four types of) conics. Endpoints of edges will be called *vertices* of $M(\mathcal{A})$.

The contribution of this section is a simple and practical randomized algorithm for computing $M(\mathcal{A})$. It works by divide-and-conquer and accepts as input any description of $\partial\mathcal{A}$ by circular arcs and/or line segments. The costly part is delegated to the divide step, which basically consists of inclusion tests for arcs in circles. The merge step is trivial; it just concatenates two partial medial axes. The expected runtime is bounded by $O(n^{3/2})$, and will be proved to be $O(n \text{ polylog } n)$ for several types of shape. A qualitative difference to existing medial axis algorithms is that a *combinatorial* description of $M(\mathcal{A})$ is extracted first, which can then be directly (and robustly) converted into a geometric representation. We base our algorithm on the following simple though elegant decomposition lemma [9].

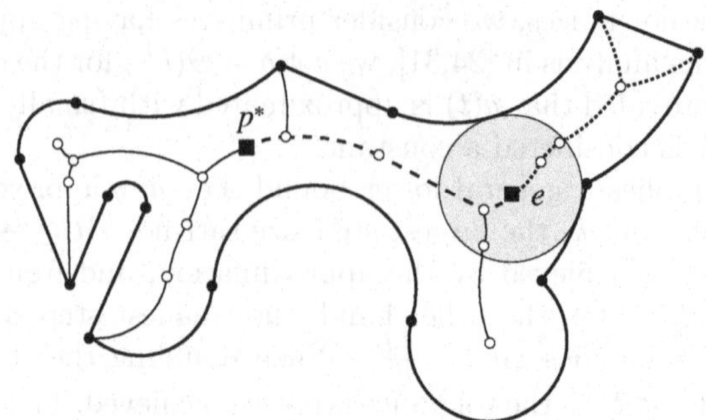

Fig. 1. Walk (dashed) and cut (dotted)

Lemma 2. *Consider any maximal disk D for $\mathcal{A}$. Let $A_1, \ldots, A_t$ be the connected components of $\mathcal{A} \setminus D$, and denote with p the center of D. The following holds.*

$$(1)\ M(\mathcal{A}) = \bigcup_{i=1}^{t} M(A_i \cup D) \qquad (2)\ \{p\} = \bigcap_{i=1}^{t} M(A_i \cup D)$$

In plain words, having at hands some maximal disk one can compute the medial axes for the resulting components recursively, and then glue them together at a single point. However, the desired efficiency of this strategy calls for a balanced decomposition. Its existence is given below.

Lemma 3. *There exists a maximal disk D for $\mathcal{A}$ such that at most $\frac{n}{2}$ arcs from $\partial\mathcal{A}$ are (completely) contained in each component of $\mathcal{A} \setminus D$.*

Proof. Each point $p \in M(\mathcal{A})$ corresponds to a unique maximal disk D_p for $\mathcal{A}$. Let $f(D_p)$ be the number of arcs from $\partial\mathcal{A}$ in the largest component induced by D_p. As long as $f(D_p) > \frac{n}{2}$, the component that realizes $f(D_p)$ is unique, and we can decrease $f(D_p)$ by continuously moving p on $M(\mathcal{A})$ such that D_p enters into this component. This process terminates at some point p^* where $f(D_{p^*}) \leq \frac{n}{2}$. We never move back the way we came, as the component we move out never exceeds a size of $\frac{n}{2}$.

We are left with the algorithmic problem of finding some maximal disk that yields a well-balanced partition. Observe that the optimal point p^* above may be not unique, because the number $f(D_p)$ is invariant under motion of p within the relative interior of any fixed edge $e \subset M(\mathcal{A})$. Let us define $Walk(e)$ as the path length in $M(\mathcal{A})$ from e to p^*. Further, define $Cut(e)$ as the size of the smaller one among the two subtrees which constitute $M(\mathcal{A}) \setminus \{e\}$. See Figure 1. Any tree with small 'cuts' tends to have short 'walks', in the following respect.

Lemma 4. *Let e be an edge of $M(\mathcal{A})$, chosen uniformly at random. Then $E[Walk(e)] = \Theta(E[Cut(e)])$.*

Proof. Orient all the paths in $M(\mathcal{A})$ away from the point p^*. This defines a partial order $\prec$ on the edges of $M(\mathcal{A})$. We have the set equality

$$\bigcup_{e \in M(\mathcal{A})} \{(a,e) \mid a \prec e\} = \bigcup_{e \in M(\mathcal{A})} \{(e,b) \mid b \succ e\}$$

because either set contains each pair of the relation exactly once. The (disjoint) subsets united in the left set, L, represent all the paths in $M(\mathcal{A})$ between its edges e and p^*. Thus we have $E[Walk(e)] = \frac{1}{m} \cdot |L|$, where m is the number of edges of $M(\mathcal{A})$. The (disjoint) subsets united in the right set, R, represent those subtrees defined by the edges e of $M(\mathcal{A})$ which avoid p^*. If we neglect subtrees of sizes larger than $\frac{m}{2}$, then the cardinality of the set drops by a constant factor (of at most 4, if $\prec$ would be a total order, hence less). This implies $\frac{1}{m} \cdot |R| > E[Cut(e)] > \frac{1}{m} \cdot \frac{|R|}{4}$. The lemma now follows from $|R| = |L|$.

Lemma 4 motivates the following disk finding algorithm which combines random cutting with local walking. Its main subroutine, MAX(b), selects for an arc $b \subset \partial\mathcal{A}$ its midpoint x and returns the unique maximal disk for $\mathcal{A}$ with x on its boundary. For the ease of description, we assume that this disk splits $\mathcal{A}$ into exactly two components. Let $c \geq 3$ be a (small) integer constant.

Procedure CUT($\mathcal{A}$)

Put $A' = \mathcal{A}$
Repeat
  Choose a random arc b of $\partial A'$
  Compute D=MAX(b) and let $\mathcal{A}_0$ be the larger component of $\mathcal{A}$ induced by D
  Assign $A' = A' \cap \mathcal{A}_0$
Until $\mathcal{A}_0$ contains less than $n - \frac{n}{c}$ arcs
Report D

Procedure WALK($\mathcal{A}$)

Choose a random arc b of $\partial\mathcal{A}$
Compute D=MAX(b)
Put $\mathcal{A}_0$ =larger component induced by D
While $\mathcal{A}_0$ contains $> n - \frac{n}{c}$ arcs do
  Let b_1 (b_2) be the first (last) complete arc of $\partial\mathcal{A}$ in $\mathcal{A}_0$
  Find D_1=MAX(b_1) and D_2=MAX(b_2)
  Put $\mathcal{A}_0$ = smallest of the respective larger components of $\mathcal{A}$ for D_1 and D_2
  Let $D \in \{D_1, D_2\}$ be the respective disk
Report D

The disk finding algorithm now runs CUT($\mathcal{A}$) and WALK($\mathcal{A}$) in parallel and terminates as soon as the first disk is reported. To analyze its runtime, let us first consider the assignment of arcs on $\partial\mathcal{A}$ to edges of $M(\mathcal{A})$, as done in subroutine MAX. Namely, if MAX(b)=D then arc b is mapped to the edge e that contains the center of D. Observe that either $0, 1$, or 2 arcs are mapped to a fixed edge. Moreover, no two unaddressed edges and no two doubly addressed edges are neighbored. This assignment is sufficiently uniform to convey randomness from arcs to edges. So Lemma 4 applies, and in the worst case of walk length being balanced with cut number, a bound of $O(\sqrt{n})$ on the expected number of loop executions in at least one of CUT($\mathcal{A}$) and WALK($\mathcal{A}$) holds.

The costly part in both procedures is their subroutine MAX, whose expected number of calls obeys the same bound. D=MAX(b) has a simple implementation which runs in $O(n)$ time: We initialize the disk D as the (appropriately oriented) halfplane that supports b at its midpoint x and, for all remaining arcs $b_i \subset \partial\mathcal{A}$ that intersect D, we shrink D so as to touch b_i while still being tangent to b at x. The most complex operation needed for shrinking D is an inclusion test of a point in a circle. In particular, and unlike previous medial axis algorithms, no conics take part in geometric operations.

In summary, the randomized complexity for computing the medial axis is given by $T(n) = T(\frac{1}{c}n) + T((1 - \frac{1}{c})n) + O(n^{3/2}) = O(n^{3/2})$. In many cases, however, will the algorithm perform substantially better. Let d be the graph diameter of $M(\mathcal{A})$. Then the loop in WALK($\mathcal{A}$) is executed less than d times. So, for example, if $d = \Theta(\log n)$ then an overall runtime of $O(n \log^2 n)$ is met. For the other extreme case, $d = \Theta(n)$, our strategy is even faster. With constant probability, an edge on the diameter is chosen, and $\Theta(n)$ such edges e have $Cut(e) = \Theta(n)$. The expected number of loop executions in CUT($\mathcal{A}$) now is only $O(1)$, and an $O(n \log n)$ algorithm results. We conjecture that the latter situation is quite relevant in practice. In many applications, for typical shapes their medial axes will not branch extensively. Even if so, the branching will be independent of n, because each branch will be approximated by a large number of circular arcs in order to achieve the necessary precision.

The output of the algorithm is a list of $O(n)$ points on $M(\mathcal{A})$, namely, the centers of the splitting disks, plus a list of $O(n)$ edges connecting them. Each edge is given implicitly by its defining two arcs on $\partial\mathcal{A}$. To make sure that the reported point list includes all the vertices of $M(\mathcal{A})$, base cases that involve constantly many (pieces of) original arcs from $\partial\mathcal{A}$ have to be solved directly. (The constant equals 2 or 3 if $\partial\mathcal{A}$ is C^1.) Note that the algorithm works exclusively on $\partial\mathcal{A}$ except for a final step, where the conic edges of $M(\mathcal{A})$ are explicitly calculated and reassembled. This gives rise to increased numeric stability in comparison to existing approaches.

Opposed to approximating $\partial\mathcal{A}$ with the same accurracy by a polyline of size N, our circular arc algorithm takes $O(n^{3/2}) = O(N)$ time; see Lemma 1 in Section 2. Thus, even for (probably rare) worst-case inputs, our simple algorithm competes asymptotically well with previous methods. Another advantage over polygonal approximation is described in Section 4.

4 Convergence

A well-known unpleasant phenomenon of the medial axis is its instability under perturbations of the shape boundary. Several papers discussing this issue have been published recently. A result in [6] shows that stability is, in general, not given unless perturbations are C^2. To deal with general shapes, the so-called λ-medial axis has been introduced as a tool in [5]. After drawing a point sample from the shape boundary, the Voronoi diagram of these points is constructed and pruned appropriately. The λ-medial axis converges to the original for vanishing sample distance. Drawbacks are the large sample size for a close (and homotopy-equivalent) approximation, the lack of its C^1 behavior, and the need of computing a general planar Voronoi diagram.

We prove in this section that medial axis convergence under the Hausdorff distance comes as a byproduct of the careful (though, of course, still C^1) biarc boundary conversion described in Section 2. We start with two technical lemmas, whose proofs are omitted due to lack of space.

For some shape $\mathcal{A}$ and a point $p \in M(\mathcal{A})$, let D_p denote the unique maximal disk with center p. Recall that $M(\mathcal{A})$ is defined as the union of the centers of all maximal disks. Define $\xi_p \leq \pi$ as the largest angle at p spanned by two points in the set $D_p \cap \partial\mathcal{A}$. The assertion below does not assume any regularity condition for the shape boundaries.

Lemma 5. *Let $\mathcal{A}$ and $\mathcal{B}$ be two shapes whose (two-sided) Hausdorff distance satisfies $H(\partial\mathcal{A}, \partial\mathcal{B}) = \varepsilon$. Define $k = \frac{4}{1-\cos(\xi_p/2)}$ and let D_p denote any maximal disk for $\mathcal{A}$ whose radius r fulfills $r > k \cdot \varepsilon > 0$. Then there exists a maximal disk D_q for $\mathcal{B}$ such that $\|p - q\| < k \cdot \varepsilon$.*

Define a *leaf* of the medial axis as a vertex with a single incident edge. The following lemma describes the behavior of $M(\mathcal{A})$ in the vicinity of its leaves. Recall that an apex of $\partial\mathcal{A}$ is a point of maximal curvature.

Lemma 6. *For an apex x of $\partial\mathcal{A}$, consider the unique maximal disk D_p that avoids a segment of $\partial\mathcal{A}$ through x with fixed (small) length ℓ. Further, consider the maximal disk D_q osculating at x. If $\partial\mathcal{A}$ is piecewise analytic C^2 in the neighborhood of x then*

$$\|q - p\| \to 0 \text{ as } O(\ell^2).$$

We are now prepared to prove the claimed convergence result. Slightly more general than in Section 2, we assume that $\partial\mathcal{A}$ for the original shape $\mathcal{A}$ is C^2 and piecewise analytic. (These requirements are fulfilled if $\partial\mathcal{A}$ is a cubic spline.) The proof generalizes easily to the case where $\partial\mathcal{A}$ is an arbitrary concatenation of analytic pieces, and thus, in particular, is allowed to contain 'sharp' vertices.

Let $\mathcal{B}_n$ denote some circular arc shape that comes from approximating $\partial\mathcal{A}$ by a suitable biarc spline; see Section 2. For sufficiently large n, each leaf of $M(\mathcal{A})$ is also a leaf of $M(\mathcal{B}_n)$, and all leaves of $M(\mathcal{B}_n)$ are contained in $M(\mathcal{A})$. This is because the spline preserves not only spirals, but also position, normal vector, and curvature at each apex x of $\partial\mathcal{A}$. All leaves are centers of osculating disks at some apex x.

Let us remove from $\partial\mathcal{B}_n$ the containing circular arc b_x for each apex x whose osculating disk is included in $\mathcal{B}_n$ (and hence is maximal for $\mathcal{B}_n$). This decomposes $\partial\mathcal{B}_n$ into components. The lengths of the arcs b_x shrink to zero as $\Omega(n^{-1})$ by construction of $\mathcal{B}_n$, as does their minimum d_n. Apart from disks for leaves, each maximal disk D_p for $\mathcal{B}_n$ has contact to at least two different components. (Otherwise, there would be a supplementary leaf of $M(\mathcal{B}_n)$.) For such a disk D_p, we have the angle inequality $\xi_p \geq \xi_n$, for $\xi_n = 2\arcsin(d_n/2L)$, and L denoting the geometric diameter of $\mathcal{B}_n$. Because $d_n \to 0$ as $\Omega(n^{-1})$ and since L is a constant, we have $1 - \cos(\xi_n/2) = \Omega(n^{-2})$. Moreover, $H(\partial\mathcal{A}, \partial\mathcal{B}_n) \to 0$ as $O(n^{-3})$ by construction. That is, the condition in Lemma 5 holds for almost all maximal disks D_p for $\mathcal{B}_n$ when n is sufficiently large. Consequently, for each point

$p \in M(\mathcal{B}_n)$ there exists a point $q \in M(\mathcal{A})$ such that $\|p - q\| \rightarrow 0$ as $O(n^{-1})$. That is, the one-sided Hausdorff distance $\delta(M(\mathcal{B}_n), M(\mathcal{A}))$ converges at this speed.

The other direction can be proved similarly. For each apex x of $\mathcal{A}$, we define a neighborhood c_x on $\partial\mathcal{A}$ of length $n^{-3/4}$. Removal of the segments c_x leads us to two types of maximal disks D_q for $\mathcal{A}$, depending on whether D_q touches a single segment c_x (q is then close to x), or not. For the latter type, the analysis is the same as above, and shows that q approaches the center of some maximal disk for $\mathcal{B}_n$ at speed $O(n^{-3/2})$. For the former type, due to Lemma 6, the distance $\|q - p\|$ between q and the leaf $p \in M(\mathcal{B}_n)$ associated with c_x behaves as $\Theta(n^{-3/4})^2$, i.e., the same. The one-sided Hausdorff distance $\delta(M(\mathcal{A}), M(\mathcal{B}_n))$ thus converges at that speed.

Note that the global convergence speed of the medial axis with respect to the Hausdorff distance is $\Theta(n^{-1})$, whereas the error of the boundary approximation improves as $\Theta(n^{-3})$. This is due to the behavior of the medial axis close to its leaves. When we restrict ourselves to the λ-medial axis [5] for *any* $\lambda > 0$, then d_n in the formula for ξ_n becomes a constant, and the approximation speed is $\Theta(n^{-3})$ by Lemma 5. This well compares to using a size-n point sample on $\partial\mathcal{A}$ and pruning its Voronoi diagram, as the approximation speed then is only $\Theta(n^{-1})$.

Acknowledgements. Thanks go to Raimund Seidel for discussions on Section 3.

References

1. Alt, H., Cheong, O., Vigneron, A.: The Voronoi diagram of curved objects. Discrete & Computational Geometry 34, 439–453 (2005)
2. Attali, D., Boissonnat, J.-D., Edelsbrunner, H.: Stability and computation of medial axes – a state-of-the-art report. In: Mller, T., Hamann, B., Russell, B. (eds.) Mathematical Foundations of Scientific Visualization, Computer Graphics, and Massive Data Exploration, Springer Series on Mathematics and Visualization (to appear)
3. Avis, D., Toussaint, G.T.: An efficient algorithm for decomposing a polygon into star-shaped polygons. Pattern Recognition 13, 395–398 (1981)
4. Bhattacharya, B.K., El Gindy, H.: A new linear convex hull algorithm for simple polygons. IEEE Trans. Information Theory IT-30, 85–88 (1984)
5. Chazal, F., Lieutier, A.: Stability and homotopy of a subset of the medial axis. In: Proc. 9th ACM Symp. Solid Modeling and Applications, pp. 243–248 (2004)
6. Chazal, F., Soufflet, R.: Stability and finiteness properties of medial axis and skeleton. J. Dynamical and Control Systems 10, 149–170 (2004)
7. Chazelle, B.: A theorem on polygon cutting with applications. In: Proc. 23rd IEEE Symp. FOCS, pp. 339–349 (1982)
8. Chin, F., Snoeyink, J., Wang, C.A.: Finding the medial axis of a simple polygon in linear time. In: Staples, J., Katoh, N., Eades, P., Moffat, A. (eds.) ISAAC 1995. LNCS, vol. 1004, pp. 382–391. Springer, Heidelberg (1995)
9. Choi, H.I., Choi, S.W., Moon, H.P.: Mathematical theory of medial axis transform. Pacific J. Mathematics 181, 57–88 (1997)
10. Dobkin, D.P., Souvaine, D.L.: Computational geometry in a curved world. Algorithmica 5, 421–457 (1990)

11. Emiris, I.Z., Kakargias, A., Pion, S., Teillaud, M., Tsigaridas, E.P.: Towards an open curved kernel. In: Proc. 20th Ann. ACM Symp. Computational Geometry, pp. 438-446 (2004)
12. Farin, G.: Curves and Surfaces for Computer Aided Geometric Design. Academic Press, San Diego (1997)
13. Farin, G., Hoschek, J., Kim, M.-S.: Handbook of Computer Aided Geometric Design. Elsevier, Amsterdam (2002)
14. Garey, M.R., Johnson, D.S., Preparata, F.P., Tarjan, R.E.: Triangulating a simple polygon. Information Processing Letters 7, 175–179 (1978)
15. Graham, R.L.: An efficient algorithm for determining the convex hull of a finite planar set. Information Processing Letters 1, 132–133 (1972)
16. Graham, R.L., Yao, F.F.: Finding the convex hull of a simple polygon. J. Algorithms 4, 324–331 (1984)
17. Held, M., Eibl, J.: Biarc approximation of polygons with asymmetric tolerance bands. Computer-Aided Design 37, 357–371 (2005)
18. Hertel, S., Mehlhorn, K.: Fast triangulation of the plane with respect to simple polygons. Information & Control 64, 52–76 (1985)
19. Klein, R., Mehlhorn, K., Meiser, S.: Randomized incremental construction of abstract Voronoi diagrams. Computational Geometry: Theory and Applications 3, 157–184 (1993)
20. Kong, X., Everett, H., Toussaint, G.T.: The Graham scan triangulates simple polygons. Pattern Recognition Letters 11, 713–716 (1990)
21. Lee, D.T.: Medial axis transformation of a planar shape. IEEE Trans. Pattern Analysis and Machine Intelligence PAMI-4, 363–369 (1982)
22. Lee, D.T., Preparata, F.P.: Location of a point in a planar subdivision and its applications. SIAM J. Computing 6, 594–606 (1977)
23. McCallum, D., Avis, D.: A linear algorithm for finding the convex hull of a simple polygon. Information Processing Letters 9, 201–206 (1979)
24. Meek, D.S., Walton, D.J.: Approximation of a planar cubic Bézier spiral by circular arcs. J. Computational and Applied Mathematics 75, 47–56 (1996)
25. Meek, D.S., Walton, D.J.: Spiral arc spline approximation to a planar spiral. J. Computational and Applied Mathematics 107, 21–30 (1999)
26. Melkman, A.: On-line construction of the convex hull of a simple polygon. Information Processing Letters 25, 11–12 (1987)
27. Ong, C.J., Wong, Y.S., Loh, H.T., Hong, X.G.: An optimization approach for biarc curve fitting of B-spline curves. Computer-Aided Design 28, 951–959 (1996)
28. Ramamurthy, R., Farouki, R.T.: Voronoi diagram and medial axis algorithm for planar domains with curved boundaries I. Theoretical foundations. J. Computational and Applied Mathematics 102, 119–141 (1999)
29. Reif, U.: Uniform B-spline approximation in Sobolev spaces. Numerical Algorithms 15, 1–14 (1997)
30. Sabin, M.A.: The use of circular arcs to form curves interpolated through empirical data points. Rep. VTO/MS/164, British Aircraft Corporation (1976)
31. Šír, Z., Feichtinger, R., Jüttler, B.: Approximating curves and their offsets using biarcs and Pythagorean hodograph quintics. Computer-Aided Design 38, 608–618 (2006)
32. Yang, X.: Efficient circular arc interpolation based on active tolerance control. Computer-Aided Design 34, 1037–1046 (2002)
33. Yap, C.K.: An $O(nlogn)$ algorithm for the Voronoi diagram of a set of simple curve segments. Discrete & Computational Geometry 2, 365–393 (1987)

Alpha-Beta Witness Complexes*

Dominique Attali[1], Herbert Edelsbrunner[2], John Harer[3], and Yuriy Mileyko[4]

[1] LIS-CNRS, Domaine Universitaire, BP 46, 38402 Saint Martin d'Hères, France
[2] Departments of Computer Science and Mathematics, Duke University, Durham, and Geomagic, Research Triangle Park, North Carolina
[3] Department of Mathematics and Center for Computational Science, Engineering, and Medicine, Duke University, Durham, North Carolina
[4] Department of Computer Science, Duke University, Durham, North Carolina

Abstract. Building on the work of Martinetz, Schulten and de Silva, Carlsson, we introduce a 2-parameter family of witness complexes and algorithms for constructing them. This family can be used to determine the gross topology of point cloud data in $\mathbb{R}^d$ or other metric spaces. The 2-parameter family is sensitive to differences in sampling density and thus amenable to detecting patterns within the data set. It also lends itself to theoretical analysis. For example, we can prove that in the limit, when the witnesses cover the entire domain, witness complexes in the family that share the first, scale parameter have the same homotopy type.

1 Introduction

The analysis of large data sets is a paradigm of growing importance in the sciences. Broad advances in technology are leading to ever larger data sets capturing information in unprecedented detail. Examples are micro-arrays that probe gene activity for entire genomes and sensor networks that challenge our ability to integrate time-series of distributed measurements. After distilling such data and giving it a geometric interpretation as a point cloud in possibly high-dimensional ambient space, we are faced with the problem of extracting properties of that cloud, such as its gross topology, various patterns within it, or its geometric shape. We see the study of these point clouds as an extension of the reconstruction of surfaces from point clouds in $\mathbb{R}^3$; see [1].

In this paper we adopt the point of view that the goal is not the reconstruction of a unique shape but rather a hierarchy that captures the data at different scale levels. In this we are inspired by the work on alpha shapes where scale is captured by the radius of the spherical neighborhoods defined around the data points [2]. Our point of departure is in the method of reconstruction. Instead of appealing to the metric of the ambient space we use the data itself to drive the formation of the family of complexes. Specifically, we distinguish data points by the way we use them: the *landmarks* form the vertices of the complexes we build and the

* Research by the authors is partially supported by DARPA under grant HR0011-05-1-0007, by CNRS under grant PICS-3416 and by IST Program of the EU under Contract IST-2002-506766.

F. Dehne, J.-R. Sack, and N. Zeh (Eds.): WADS 2007, LNCS 4619, pp. 386–397, 2007.

witnesses provide support for simplices we add to connect the vertices. This idea can be traced back to the topology adapting networks of Martinetz and Schulten [3], who draw an edge between two landmarks if there is a witness for which they are the two nearest. We may interpret the witness as a proof for the edge to belong to the Delaunay triangulation of the landmark points. Unfortunately, a witness is not proof for its three nearest landmarks forming a triangle in the Delaunay triangulation. The resulting impasse was overcome for ordinary Delaunay triangulations by de Silva [4]. He proved that if for every subset of $p+1$ landmarks there is a witness for which the points in the subset are at least as close as any other landmarks, then this is a proof for the $p+1$ landmarks to form a p-simplex in the Delaunay triangulation. This insight motivated de Silva and Carlsson to introduce a generalization of the Martinetz-Schulten networks to two- and higher-dimensional complexes [5]. They used their new tool to study the picture collection of van Hateren and van der Schaaf [6], also considered by Lee, Pedersen and Mumford [7]. The main insight from their work is that a majority of small pixel subarrays can be parametrized on a (two-dimensional) Klein bottle in 7-dimensional ambient space [8].

If the witness complex is patterned after the Delaunay triangulation, why do we not just construct the latter? There is a variety of reasons, including

- the size of the complex can be controlled by choosing the landmarks while not ignoring the information provided by the possibly many more sample points;
- distances are easier to compute than the primitives required to construct Delaunay triangulations;
- extending the definition of witness complexes to metric spaces different from Euclidean spaces is comparatively straightforward;

all already mentioned in [5]. There are also significant drawbacks, such as the locally imperfect reconstruction caused by the finiteness of the witness set. The main purpose of this paper is to present methods that cope with the mentioned drawback of witness complexes. Our main contributions are theoretical, in understanding the family of witness complexes and its algorithms. Specifically,

(i) we introduce a 2-parameter family that contains prior witness complexes as sub-families;
(ii) we generalize de Silva's result for Delaunay triangulations to witness complexes in the limit;
(iii) we analyze the structure of the family of witness complexes by subdividing its parameter plane;
(iv) we give algorithms to construct this subdivision, compute homology within it, and visualize the result.

Outline. Section 2 presents the complexes after which we model our witness complexes. Section 3 introduces the 2-parameter family of witness complexes. Section 4 studies the family through subdivisions of the parameter plane. Section 5 describes algorithms constructing alpha-beta witness complexes. Section 6 concludes the paper.

2 Complexes

In this section, we introduce the family of complexes that provide the intuition for our witness complexes. The family contains the 1-parameter families of Čech and alpha complexes and uses a second parameter to interpolate between them. We begin with definitions from algebraic topology.

Simplicial Complexes. The geometric notion of a *simplex*, σ, is the convex hull of a collection of affinely independent points in $\mathbb{R}^d$. We say the points *span* the simplex. If there are $p+1$ points in the collection, we call σ a *p-simplex* and $p = \dim \sigma$ its *dimension*. Any subset of the $p+1$ points defines another simplex, $\tau \leq \sigma$, and we call τ a *face* of σ and σ a *coface* of τ. A *simplicial complex* is a finite collection of simplices, K, that is closed under the face relation and satisfies the extra condition that any two of its simplices are either disjoint or their intersection is a face of both. A *subcomplex* is a simplicial complex $K' \subseteq K$. It is *full* if it contains all simplices in K exclusively spanned by vertices in K'. We often favor the abstract view in which a p-simplex is just a collection of $p+1$ points, a face is simply a subset, and a simplicial complex is a finite system of such collections closed under the subset relation. For every finite abstract simplicial complex, there is a large enough finite dimension, d, such that the complex can be realized as a simplicial complex in $\mathbb{R}^d$. For example, d equal to one plus twice the largest dimension of any simplex is always sufficient. The primary use of a simplicial complex is to construct or represent a topological space. Its *underlying space* is the subset of $\mathbb{R}^d$ covered by the simplices, together with the topology inherited from $\mathbb{R}^d$. Finally, K *triangulates* a topological space if its underlying space is homeomorphic to that topological space.

A computationally efficient approach to classifying topological spaces is based on *homology groups* [9]. For a given space, there is one group for each dimension p capturing, in some sense, the holes with p-dimensional boundaries. We use modulo-2 arithmetic and thus get homology groups isomorphic to $\mathbb{Z}/2\mathbb{Z}$ to some non-negative integer power. That power is the *rank* of the group and the *p-th Betti number* of the topological space. The classification of spaces by homology groups is strictly coarser than by homotopy type. It follows that two spaces with the same homotopy type have isomorphic homology groups, of all dimensions. Building a simplicial complex incrementally and writing down the result at every stage, we get a nested sequence of complexes, $\emptyset = K_0 \subset K_1 \subset \ldots \subset K_n = K$, which we refer to as a *filtration* of K. The inclusion $K_i \subset K_j$ induces a homomorphism from the p-th homology group of K_i to the p-th homology group of K_j, for every $p \geq 0$. We refer to the image of the homomorphism as a *persistent homology group* and to its rank as a *persistent Betti number*. For more information on these groups we refer to [10, 11].

Čech and Alpha Complexes. There are but a few complexes that have been used to turn a finite set of points into a multi-scale representation of the space from which the points are sampled. Perhaps the oldest construction is the nerve of a collection of spherical neighborhoods, one about each data point. To formalize this idea, let $L \subseteq \mathbb{R}^d$ be a finite set of points.

Definition. For any real number $\alpha \geq 0$, the *Čech complex* of L, Čech(α), consists of all simplices $\sigma \subseteq L$ for which there exists a point $x \in \mathbb{R}^d$ such that $\|x - k\| \leq \alpha$, for all vertices $k \in \sigma$.

The Nerve Lemma implies that Čech(α) has the same homotopy type as the union of the balls with radius α and centered at points in L [12]. A similar construction requires, in addition, that x be closest to and equally far from the relevant data points [2].

Definition. For any real number $\alpha \geq 0$, the *alpha complex* of L, Alpha(α), consists of all simplices $\sigma \subseteq L$ for which there exists a point $x \in \mathbb{R}^d$ such that $\|x - k\| \leq \alpha$ and $\|x - k\| \leq \|x - \ell\|$, for all $k \in \sigma$ and all $\ell \in L$.

Equivalently, Alpha(α) is the nerve of the collection of balls of radius α, each clipped to within the Voronoi cell of its center. The Nerve Lemma implies that Alpha(α) also has the homotopy type of the union of balls. In summary, Alpha(α) is a subcomplex of Čech(α) and the two have the same homotopy type, for every $\alpha \geq 0$. Alpha complexes are more efficient than Čech complexes but require the evaluation of a more complicated geometric primitive. For $\alpha = \infty$, we have the nerve of the collection of Voronoi cells, also known as the *Delaunay complex* of L, Delaunay = Alpha(∞) [13].

Almost Alpha Complexes. We interpolate between Čech and alpha complexes using a second parameter, β.

Definition. For any real numbers $\alpha, \beta \geq 0$, the *almost alpha complex*, AA(α, β), consists of the simplices $\sigma \subseteq L$ for which there exists a point $x \in \mathbb{R}^d$ such that $\|x - k\| \leq \alpha$ and $\|x - k\|^2 \leq \|x - \ell\|^2 + \beta^2$, for all $k \in \sigma$ and all $\ell \in L$.

As suggested by the name, these complexes are similar to but different from the almost Delaunay complexes introduced in [14]. For $\beta \geq \alpha$, the second constraint is redundant, and for $\beta = 0$, it requires that x be equidistant from all $k \in \sigma$. In other words, AA(α, α) = Čech(α) and AA(α, 0) = Alpha(α).

Let $a_k(\alpha)$ be the closed ball with center k and radius α, and write $a_\sigma(\alpha)$ for the common intersection of the balls $a_k(\alpha)$, for $k \in \sigma$. Similarly, let $b_{k,\ell}(\beta)$ be the closed half-space of points whose square distance to k exceeds that to ℓ by at most β^2, and write $b_{\sigma,v}(\beta)$ for the common intersection of the half-spaces $b_{k,\ell}(\beta)$, for $k \in \sigma$ and $\ell \in v$. Then σ belongs to AA(α, β) iff $\text{region}_\sigma(\alpha, \beta) = a_\sigma(\alpha) \cap b_{\sigma,L}(\beta)$ is non-empty. But this region is the intersection of the regions of its vertices, $\text{region}_\sigma(\alpha, \beta) = \bigcap_{k \in \sigma} \text{region}_k(\alpha, \beta)$. Hence, AA($\alpha, \beta$) is the nerve of the regions of the vertices. Independent of β, the union of these regions is the union of balls of radius α, same as for the Čech and the alpha complexes. Indeed, β only controls the amount of overlap between the regions, which increases with increasing β. Since the regions are convex, the Nerve Lemma implies that the homotopy type of AA(α, β) is the same as that of the union of balls. We summarize,

$$\text{Alpha}(\alpha) \subseteq \text{AA}(\alpha, \beta) \subseteq \text{Čech}(\alpha), \quad (1)$$

$$\text{Alpha}(\alpha) \simeq \text{AA}(\alpha, \beta) \simeq \text{Čech}(\alpha), \quad (2)$$

for all $\alpha, \beta \geq 0$.

3 Alpha-Beta Witness Complexes

The almost alpha complexes have witness versions obtained by collecting all simplices whose regions contain at least one of a finite set of sampled points. This construction is problematic for small values of β, for which the regions of the vertices have only small overlap. Following de Silva [4], we introduce the concept of a weak witness and show that the resulting witness complexes are better approximations of the complexes than the mentioned witness versions.

Weak and Strong Witnesses. The general set-up consists of a finite set $X \subseteq \mathbb{R}^d$ of *witnesses* and another, usually smaller finite set $L \subseteq \mathbb{R}^d$ of *landmarks.* We consider complexes over L consisting of simplices that have the backing of witnesses in X. Specifically, we call $x \in X$ a *weak* (α, β)*-witness* of $\sigma \subseteq L$ if

[I] $\|x - k\| \leq \alpha$, for all $k \in \sigma$, and
[II] $\|x - k\|^2 \leq \|x - \ell\|^2 + \beta^2$, for all $k \in \sigma$ and all $\ell \in L - \sigma$.

Equivalently, x belongs to $a_\sigma(\alpha) \cap b_{\sigma, L-\sigma}(\beta)$. We call a weak (α, β)-witness a *strong* (α, β)*-witness* if the inequality in Condition [II] holds for all $k \in \sigma$ and all $\ell \in L$ or, equivalently, if $x \in a_\sigma(\alpha) \cap b_{\sigma, L}(\beta)$. The difference is in the set of landmarks that compete with the vertices of σ. For a weak witness this set excludes the vertices of σ which therefore do not compete with each other. This subtle difference has important consequences.

Definition. For any real numbers $\alpha, \beta \geq 0$, the *alpha-beta witness complex*, Witness(α, β), consists of the simplices $\sigma \subseteq L$ such that every face $\tau \leq \sigma$ has a weak (α, β)-witness in X.

Condition [II] is redundant unless α exceeds β so we restrict the 2-parameter family to $0 \leq \beta \leq \alpha \leq \infty$. With increasing value of α and, independently, of β, the requirements for being a weak witness get more tolerant, which implies Witness$(\alpha, \beta) \subseteq$ Witness(α', β') whenever $\alpha \leq \alpha'$ and $\beta \leq \beta'$.

Witness Complexes in the Limit. Similar to almost alpha complexes, the alpha-beta witness complexes have a nice geometric interpretation. We describe it in the full version of the paper, where we also show how to extend de Silva's result on Delaunay triangulations to almost alpha complexes. In particular, we prove that the existence of a weak (α, β)-witness for each face implies the existence of a strong (α, β)-witness for the simplex. In other words, if $X = \mathbb{R}^d$ then the alpha-beta witness complex is the same as the almost alpha complex.

Weak Almost Alpha Theorem. If $X = \mathbb{R}^d$ then Witness$(\alpha, \beta) =$ AA(α, β).

For finite sets X, the alpha-beta witness complex can only be smaller than for $X = \mathbb{R}^d$, which implies Witness$(\alpha, \beta) \subseteq$ AA(α, β). This should be contrasted with the fact that a strong witness for a simplex is a weak witness for all faces of the simplex. Hence, the witness version of the almost alpha complex, which collects all simplices with strong (α, β)-witnesses in X, is a subcomplex of Witness(α, β). By (2), the homotopy type of the almost alpha complex does not depend on β. Any variation in the homotopy type of the alpha-beta witness complex for fixed value of α must therefore be attributed to insufficient sampling.

4 2-Parameter Family

In this section, we focus on the family of witness complexes, describing properties in terms of subdivisions of the parameter plane. In this plane of points (α, β) the balls grow from left to right and the Voronoi cells grow from bottom to top. Potentially interesting sub-families arise as horizontal and vertical lines but also as 45-degree lines along which the balls and cells grow at the same rate.

Comparison with Prior Notions. Several versions of witness complexes have been defined in [5]. We compare them with the 2-parameter family, limiting ourselves to Čech-like constructions. We start with the first version introduced by de Silva and Carlsson.

Definition. The *strict witness complex*, W_∞, consists of the simplices $\sigma \subseteq L$ whose faces belong to W_∞ and for which there exists a witness $x \in X$ such that

[S] $\|x - k\| \leq \|x - \ell\|$, for all $k \in \sigma$ and all $\ell \in L - \sigma$.

Condition [S] is the same as Condition [II] for $\beta = 0$. There is no counterpart to [I] but we can make this condition redundant by setting $\alpha = \infty$. In other words, $W_\infty = \text{Witness}(\infty, 0)$ in our family, as indicated in Fig.1. To introduce the other three constructions in [5], let p be the dimension of σ and $\text{dist}_j(x)$ the distance of $x \in X$ from its j-nearest landmark point. Using a non-negative real parameter R, we get three 1-parameter families of witness complexes, each obtained by substituting one of

[**0**] $\|x - k\| \leq R$, for all vertices $k \in \sigma$;
[**1**] $\|x - k\| \leq R + \text{dist}_1(x)$, for all vertices $k \in \sigma$;
[Δ] $\|x - k\| \leq R + \text{dist}_{p+1}(x)$, for all vertices $k \in \sigma$;

for Condition [S] in the definition of W_∞. Following [5], we denote the members of the three families as $W(R, 0)$, $W(R, 1)$, and $W(R, \Delta)$. The members of the first family are the witness versions of the Čech complex, $W(R, 0) = \text{Witness}(R, R)$. For $R = 0$ in the second family, we get a p-simplex σ iff there is a witness in the intersection of the $p+1$ Voronoi cells of its vertices, which happens with probability 0 unless $p = 0$. As R increases, we get more tolerant about the precise location of the witness. Equivalently, we can think of growing the Voronoi cells and adding a simplex whenever we find a witness in the common intersection of the enlarged cells. The effect of increasing R is therefore similar to that of increasing β in Condition [II], although the enlarged cells have different shape. Condition [Δ] is less restrictive than Condition [**1**] so we have $W(R, 1) \subseteq W(R, \Delta)$. We can interpret [$\Delta$] in terms of growing order-$(p+1)$ Voronoi cells. This makes the complexes in the third family rather similar to alpha-beta witness complexes for $\alpha = \infty$, although the geometric details are again different. The growth prescribed by Condition [II] is milder and more controlled than that prescribed by Condition [Δ]. Indeed, we have $\text{Witness}(\infty, R) \subseteq W(R, \Delta)$, for all $R \geq 0$. To see this, consider Conditions [II] and [Δ] for a witness x and a p-simplex σ. If the $p+1$ vertices of σ are the $p+1$ closest landmarks then x and σ satisfy both conditions for all values of β and R. Otherwise, the smallest distance from x to a landmark ℓ not in σ is at most $\text{dist}_{p+1}(x)$. For $\beta = R$, Condition [II] is equivalent to $\|x - k\|^2 \leq R^2 + \|x - \ell\|^2$ for all $\ell \in L - \sigma$. It follows that $\|x - k\|^2 \leq R^2 + \text{dist}^2_{p+1}(x)$ which implies Condition [Δ]. The containment relation cannot be reversed, meaning there is no positive constant c such that $W(R, \Delta)$ is necessarily a subcomplex of $\text{Witness}(\infty, cR)$.

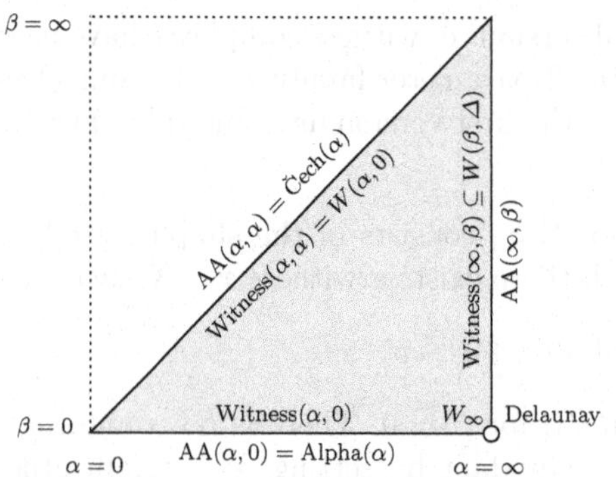

Fig. 1. The parameter plane of alpha-beta witness complexes. We find the Čech and alpha complexes and the witness complexes of de Silva and Carlsson along the edges of the triangle.

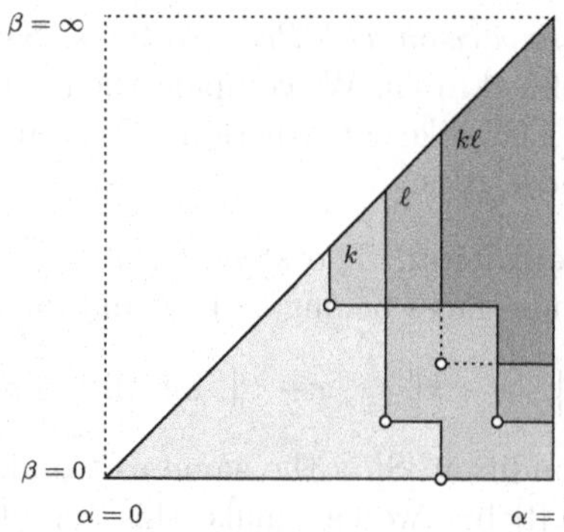

Fig. 2. Since vertices have no proper faces, $Q(k, X)$ and $Q(\ell, X)$ are unions of quadrants. For the edge, $Q(k\ell, X)$ is the portion of its union of quadrants inside $Q(k, X)$ and $Q(\ell, X)$.

Birthline Subdivision. We decompose the parameter plane into maximal regions within which the alpha-beta witness complexes are the same. For this purpose, we introduce two collections of functions, $A_\sigma, B_{\sigma,\upsilon} : \mathbb{R}^d \to \mathbb{R}$, defined by

$$A_\sigma(x) = \max_{k \in \sigma} \|x - k\|^2;$$

$$B_{\sigma,\upsilon}(x) = \max_{k \in \sigma} \|x - k\|^2 - \min_{\ell \in \upsilon} \|x - \ell\|^2.$$

Both are convex. It follows that their sublevel sets are convex regions, namely the intersections of balls and half-spaces used earlier, $A_\sigma^{-1}(-\infty, \alpha^2] = a_\sigma(\alpha)$ and $B_{\sigma,\upsilon}^{-1}(-\infty, \beta^2] = b_{\sigma,\upsilon}(\beta)$. Hence, a point $x \in X$ is a weak (α, β)-witness for σ iff $A_\sigma(x) \leq \alpha^2$ and $B_{\sigma,L-\sigma}(x) \leq \beta^2$. The two conditions are independent implying the set of points (α^2, β^2) whose coordinates satisfy them form an upper right quadrant which we denote $Q(\sigma, x)$. Since σ can have more than one weak witness, we consider the union of quadrants they define, and since we require all faces of σ to have weak witnesses, we take the intersection of these unions,

$$Q(\sigma, X) = \bigcap_{\tau \leq \sigma} \left(\bigcup_{x \in X} Q(\tau, x) \right),$$

calling its boundary the *birthline* of σ. It decomposes the parameter plane into two regions such that σ belongs to Witness(α, β) iff the point (α^2, β^2) lies on or to the upper right of the birthline; see Fig.2.

The birthlines decompose the parameter plane into the *birthline subdivision* consisting of maximal regions within which the alpha-beta witness complexes are the same. Neighboring regions are separated by curves, each belonging to one or more birthlines. Curves meet at common endpoints where birthlines merge or cross. Curves that belong to two or more birthlines are common, even in the generic case. In a typical example, the witness complexes in two neighboring regions differ by a collapse, which consists of all faces of a simplex that are cofaces of a proper face of that simplex. A collapse does not affect the homotopy type of the complex, implying that we get isomorphic homology groups in the two regions, for all dimensions.

5 Algorithms

We focus on algorithms that construct the family rather than individual alpha-beta witness complexes. We begin by constructing the birthline subdivision of the parameter plane, which we use as a representation of the family. We then discuss an algorithm for computing the homology of the complexes in the family. To extract patterns we consider classes that persist while we vary the two parameters.

Constructing Birthlines. Recall that a p-simplex σ and a witness x define a quadrant above and to the right of its corner point. The first coordinate of the corner is $A_\sigma(x) = \max_{k \in \sigma} \|x - k\|^2$. To get the second coordinate, we find the set of $p + 1$ landmarks closest to x and distinguish between two cases. If this set is σ then x is a weak witness of σ for all values of β so the second coordinate of the corner is zero. Else this set contains a closest landmark ℓ not in σ and we get the second coordinate as $B_{\sigma, L-\sigma}(x) = A_\sigma(x) - \|x - \ell\|^2$. Clearly these computations benefit from a data structure that provides fast access to the landmarks near a query point. There are many data structures available for this task and we refer to Indyk [15] for a recent survey of this literature. The union of the quadrants $Q(\sigma, x)$, over all witnesses x, is the lower staircase of their corner points. Constructing this staircase is another classic problem in computational geometry [16]. There are many fast methods including a plane-sweep algorithm that constructs the staircase from left to right. This algorithm is convenient for our purposes since it can be reused to compute the birthline of σ as the upper envelope of the staircases of all faces of σ. Finally, we use the plane-sweep algorithm a third time to convert the collection of birthlines into the birthline subdivision. Alternatively, we can do all three plane-sweeps in one, constructing the birthline subdivision directly from the corner points of the quadrants.

What we described is hardly the most efficient method to construct the birthline subdivision. In particular, we expect that most of the quadrants are redundant. It would be interesting to prove bounds on the output size, the number of edges in the birthline subdivision, and to find an algorithm that avoids looking at redundant quadrants and achieves a running time sensitive to the output size.

Computing Homology. We now describe an algorithm that computes the p-th Betti number for each region in the subdivision. It does this for all values of p. The main idea is to explore the parameter plane in a topological sweep that advances a directed path connecting the start-point, $(0, 0)$, with the end-point, (∞, ∞), while remaining monotonically non-decreasing in both parameters at all times. Initially, the path follows the lower edge of the parameter plane, from $(0, 0)$ to $(\infty, 0)$, and then the right edge, from $(\infty, 0)$ to (∞, ∞). We represent this combinatorially by the sequence of simplices labeling the birthlines the path crosses. If m denotes the number of landmarks, we go from the empty complex at $(0, 0)$ to the m-simplex at (∞, ∞), which implies that the sequence contains all $M = 2^m$ simplices spanned by the landmark points. An elementary move pushes the path locally across a vertex of the subdivision. This corresponds to locally reordering the simplices, which we do one transposition at a time. After processing all transpositions, we arrive at the final path, which follows the diagonal from $(0, 0)$ to (∞, ∞). The purpose of the sweep is to compute the Betti numbers of the regions, which we do using the algorithm in [10] for the initial sequence and the algorithm in [17] to update the information for each transposition. In the worst case, the initialization takes time cubic in M and each transposition takes time linear in M.

The algorithm's biggest impediment is the large size of the complex at (∞, ∞). To make it feasible for landmark sets that are not very small, we choose an upper bound b for β. Shrinking the parameter domain this way seems appropriate since α and β play fundamentally different roles. The first parameter, α, controls the resolution of the reconstruction, allowing small features to form for small α and letting gross features take over for large α. The second parameter, β, controls how tolerantly we interpret witnesses. The strict interpretation at $\beta = 0$ combined with occasional gaps in the distribution of witnesses leads to holes caused by sporadically missing simplices. The findings in [5] suggest that small non-zero values of β suffice to repair these holes. Although our mathematical formulation of tolerance is different from that paper, we expect the same holds for alpha-beta witness complexes.

Persistence. We now address the question of how to read the Betti numbers of the family represented by the birthline subdivision. We are not after finding the "best" complex since we cannot expect that a single complex would contain all interesting patterns in the data. Since these patterns are expressed at different scale levels a simultaneous representation may indeed be impossible. Instead, we are looking for homology classes that persist while α and β vary. Ideally, we would like to define a notion of two-parameter persistence but there are algebraic difficulties [18]. We therefore fall back on the one-parameter notion introduced in [10] which measures the length of the interval in a path along which a homology class persists. Since the scale level is controlled solely by α it makes sense to draw the path horizontally in the parameter plane so that persistence captures scale. In other words, the directed path used in the computation of homology sweeps the parameter plane from bottom to top. More precisely, we gradually increase β from 0 to b and restrict the path to two turns, one at (β^2, β^2) and the other at (∞, β^2), with a horizontal line in between. To simulate monotonicity, which is necessary to reduce the sweep to transpositions, we advance the horizontal line by processing the simultaneous elementary moves from right to left. For each value of β we can visualize the persistence information in a two-dimensional diagram as defined in [19]. Each homology class is represented by a point whose first coordinate marks its birth and whose second coordinate marks its death. Since birth occurs before death this point lies above the diagonal and its vertical distance from the diagonal is its *persistence.*

As proved by Cohen-Steiner et al. [19], small changes in the function cause only small changes in the diagram. In the case at hand, the function is the value of α at which a simplex is added to the witness complex. As β increases the value of α at which the simplex enters stays the same or decreases. The changes correspond to the steps in the birthlines and are therefore not continuous. Most of the time the steps are small but not always. In particular the first step at which a simplex is introduced can be large. Nevertheless it is useful to stack up the persistence diagrams and to describe the evolution of a homology class as a possibly discontinuous curve in three-dimensional space. In a context in which these curves are continuous they have been referred to as *vines* forming a collection called a *vineyard* [17]. The vineyard of the family of alpha-beta complexes enhances the visualization of persistent homology classes by showing how the persistence changes with varying β, the amount of tolerance with which we recognize a witness of a simplex.

6 Questions and Extensions

We conclude this paper with a list of open questions and suggestions for further research motivated by our desire to improve the algorithms.

Can we take advantage of the hole repairing quality of β without paying the high price of exploding numbers of simplices? Evidence in support of this possibility is that an overwhelming majority of changes caused by increasing β are collapses, which preserve the homotopy type. This is consistent with our observation that in the limit, for $X = \mathbb{R}^d$, the homotopy type of Witness(α, β) is independent of β.

Under reasonable assumptions on the distribution of witnesses and landmarks, what is the expected size of the alpha-beta witness complex as a function of α

and β? Similarly, what is the expected number of corners per birthline and what is the expected size of the birthline subdivision?

There are strong parallels between work on witness complexes and on surface and shape reconstruction. Are there versions of witness complexes analogous to the Wrap complex [20], which may be viewed as following Forman's theory of discrete Morse functions [21]? Similarly, are there relaxations of the alpha-beta witness complexes akin to the independent complexes studied in [22]?

Data sets are often contained in subspace of Euclidean space. Recent work in this direction proves that every smoothly embedded compact manifold of dimension 1 or 2 in $\mathbb{R}^d$ has sufficiently fine samplings of landmarks and witnesses such that Witness$(\infty, 0)$ is homeomorphic to the manifold [23]. A counterexample to extending this result to manifolds of dimension 3 or higher is described in [24]. The counterexample is based on slivers, very flat tetrahedra in the Delaunay triangulation, suggesting the use of sliver exudation methods to remedy the situation [25]. It would be interesting to extend these results to samplings of submanifolds in which density variations encode important information about the data.

Acknowledgments

The authors thank David Cohen-Steiner and Dmitriy Morozov for helpful technical discussions.

References

[1] Dey, T.K.: Curve and Surface Reconstruction. Cambridge Univ. Press, England (2007)
[2] Edelsbrunner, H., Mücke, E.P.: Three-dimensional alpha shapes. ACM Trans. Comput. Graphics 13, 43–72 (1994)
[3] Martinetz, T., Schulten, K.: Topology representing networks. Neural Networks 7, 507–522 (1994)
[4] de Silva, V.: A weak definition of Delaunay triangulation. Manuscript, Dept. Mathematics, Pomona College, Claremont, California (2003)
[5] de Silva, V., Carlsson, G.: Topological estimation using witness complexes. In: Proc. Sympos. Point-Based Graphics, pp. 157–166 (2004)
[6] van Hateren, J.H., van der Schaaf, A.: Independent component filters of natural images compared with simple cells in primary visual cortex. Proc. Royal Soc. London B 265, 359–366 (1998)
[7] Lee, A.B., Pedersen, K.S., Mumford, D.: The nonlinear statistics of high-contrast patches in natural images. Intl. J. Comput. Vision 54, 83–103 (2003)
[8] Carlsson, G., Ishkhanov, T., de Silva, V., Zomorodian, A.: On the local behavior of spaces of natural images. Manuscript, Dept. Mathematics, Stanford Univ., Stanford, California (2006)
[9] Munkres, J.R.: Elements of Algebraic Topology. Redwood City, California (1984)
[10] Edelsbrunner, H., Letscher, D., Zomorodian, A.: Topological persistence and simplification. Discrete Comput. Geom. 28, 511–533 (2002)

[11] Zomorodian, A., Carlsson, G.: Computing persistent homology. Discrete Comput. Geom. 33, 249–274 (2005)

[12] Leray, J.: Sur la forme des espaces topologiques et sur les point fixes des représentations. J. Math. Pures Appl. 24, 95–167 (1945)

[13] Delaunay, B.: Sur la sphère vide. Izv. Akad. Nauk SSSR, Otdelenie Matematicheskikh i Estestvennykh Nauk 7, 793–800 (1934)

[14] Bandyopadhyay, D., Snoeyink, J.: Almost-Delaunay simplices: nearest neighbor relations for imprecise points. In: Proc. 15th Ann. ACM-SIAM Sympos. Discrete Alg. pp. 403–412 (2004)

[15] Indyk, P.: Nearest neighbors in high-dimensional space. In: Goodman, J.E., O'Rourke, J. (eds.) CRC Handbook of Discrete and Computational Geometry, 2nd edn. Chapman & Hall/CRC, Baton Rouge, Louisiana, pp. 877–892 (2004)

[16] Kung, H.T., Luccio, F., Preparata, F.P.: On finding the maxima of a set of vectors. J. Assoc. Comput. Mach. 22, 469–476 (1975)

[17] Cohen-Steiner, D., H.E., Morozov, D.: Vines and vineyards by updating persistence in linear time. In: Proc. 22nd Ann. Sympos. Comput. Geom, 119–126 (2006)

[18] Carlsson, G., Zomorodian, A.: The theory of multidimensional persistence. In: Proc. 23rd Ann. Sympos. Comput. Geom. (2007) (to appear)

[19] Cohen-Steiner, D., Edelsbrunner, H., Harer, J.: Stability of persistence diagrams. Discrete Comput. Geom. 37, 103–120 (2007)

[20] Edelsbrunner, H.: Surface reconstruction by wrapping finite sets in space. In: Aronov, B., Basu, S., Pach, J., Sharir, M. (eds.) Discrete and Computational Geometry — The Goodman-Pollack Festschrift, pp. 379–404. Springer, Heidelberg (2003)

[21] Forman, R.: Combinatorial differential topology and geometry. In: Billera, L.J., Björner, A., Greene, C., Simion, R., Stanley, R.P. (eds.) New Perspective in Geometric Combinatorics, vol. 38, pp. 177–206. Cambridge Univ. Press, Cambridge (1999)

[22] Attali, D., Edelsbrunner, H.: Inclusion-exclusion formulas from independent complexes. Discrete Comput. Geom. 37, 59–77 (2007)

[23] Attali, D., Edelsbrunner, H., Mileyko, Y.: Weak witnesses for Delaunay triangulations of submanifolds. In: ACM Sympos. Solid Phys. Modeling (to appear)

[24] Oudot, S.Y.: On the topology of the restricted Delaunay triangulation and witness complexes in higher dimensions. Manuscript, Dept. Comput. Sci., Stanford Univ., Stanford, California (2007)

[25] Boissonnat, J.D., Guibas, L.J., Oudot, S.Y.: Manifold reconstruction in arbitrary dimensions using witness complexes. In: Proc. 23rd Ann. Sympos. Comput. Geom (to appear)

Cauchy's Theorem and Edge Lengths of Convex Polyhedra

Therese Biedl, Anna Lubiw, and Michael Spriggs

University of Waterloo, Waterloo, ON, N2L 3G1, Canada
{alubiw,biedl,mjspriggs}@uwaterloo.edu

Abstract. In this paper we explore, from an algorithmic point of view, the extent to which the facial angles and combinatorial structure of a convex polyhedron determine the polyhedron—in particular the edge lengths and dihedral angles of the polyhedron. Cauchy's rigidity theorem of 1813 states that the dihedral angles are uniquely determined. Finding them is a significant algorithmic problem which we express as a spherical graph drawing problem. Our main result is that the edge lengths, although not uniquely determined, can be found via linear programming. We make use of significant mathematics on convex polyhedra by Stoker, Van Heijenoort, Gale, and Shepherd.

1 Introduction

Cauchy proved that convex polyhedra are rigid in the sense that if the faces were metal plates, and the edges were hinges, no flexing would be possible. Although most presentations (e.g. see [6,1]) don't point this out, Cauchy actually proved a stronger result: that the dihedral angles of a convex polyhedron are completely determined by the facial angles and the combinatorial structure (the list of faces, edges, vertices, and their containments). In other words, Cauchy's proof makes no use of the edge lengths. See [26] or [3] for presentations where this is explicit.

In this paper we examine the relationships among these four attributes of convex polyhedra: the facial angles and combinatorial structure, which we consider as givens; and the dihedral angles and the edge lengths, which we consider as unknowns. Dihedral angles are uniquely determined, by Cauchy's Theorem; edge lengths are not, as the family of boxes demonstrates. This paper addresses the question: Are there efficient algorithms to find dihedral angles/edge lengths for given facial angles and combinatorial structure?

Our main result is an algorithm for the case of edge lengths (see Section 4). We give a linear system expressing the edge lengths in terms of facial angles and combinatorial structure. Correctness depends on a new characterization (in Section 3) of the edge lengths, facial angles, and dihedral angles of convex polyhedra. This is related to recent work on "local" or "easily checkable" conditions for convex polyhedra [17,10,20].

It is a major open question to devise an efficient algorithm to find the dihedral angles corresponding to given facial angles and combinatorial structure—i.e. to

F. Dehne, J.-R. Sack, and N. Zeh (Eds.): WADS 2007, LNCS 4619, pp. 398–409, 2007.

devise an algorithm for Cauchy's theorem. It is usually assumed that edge lengths are given as well. Such an algorithm can be used to find the unique polyhedron formed by "folding up" a surface satisfying Alexandrov's conditions [3]. (See [9] for further explanation.) In 1998 Sabitov ([21], or see the sketch in [9]) gave a finite algorithm for Cauchy's theorem. His algorithm uses the edge lengths heavily and crucially, and involves trying out all roots of the "volume polynomial". The algorithm takes an exponential number of steps, where solving a high-degree polynomial is counted as one step. Recently Bobenko and Izmestiev ([5], see also [19]) gave an algorithm for the more general Alexandrov's problem that, although it is not polynomial time, is effective in practice and has been implemented.

As noted, both these algorithms assume edge lengths are given—an assumption that Cauchy did not make. Our work fills in this gap: it is possible to compute edge lengths from facial angles and combinatorial structure and then proceed with these algorithms.

The present work falls under the general topic of reconstructing polyhedra. For other results and open problems in this area see [14,8,15].

We assume standard definitions about convex polyhedra and graphs, see e.g. [6]. By the "combinatorial structure" of a convex polyhedron we mean the list of faces, edges, and vertices and their containments. Equivalently, the combinatorial structure is a 3-connected planar graph, from which we can uniquely determine a *combinatorial embedding*, i.e. a cyclic list of the edges around each vertex that can be used to draw the graph in the plane without edge crossings.

2 The Spherical Dual

In this section we describe the spherical dual of a convex polyhedron, which we will use in the next section for our characterization of the edge lengths, facial angles, and dihedral angles of convex polyhedra.

In a 1968 paper Stoker [26] gave some generalizations of Cauchy's rigidity theorem. He pointed out explicitly that Cauchy's proof does not use edge lengths, and transformed Cauchy's problem in a way that abolishes edge lengths. This transformation is known as the spherical dual or "Gauss map" (see, e.g. [2]).

Given a convex polyhedron P, map it to the unit sphere as follows. Translate the sphere so that the origin is an interior point. Consider any point p on the unit sphere as a vector. Because the origin is inside the polyhedron, this vector p is the outward normal to a unique supporting plane π_p of the polyhedron. Label the point p on the sphere according to the vertex/edge/face of the polyhedron that lies in the plane π_p. Thus each face f of P, having a single supporting plane, maps to a single point $s(f)$ on the sphere. Each edge e of P has a one-dimensional space of supporting planes and maps to a geodesic arc $s(e)$; if f and g are the two faces on either side of e, $s(e)$ is the shorter arc of the great circle though $s(f)$ and $s(g)$. Each vertex v of P maps to a convex spherical polygon bounded by the arcs corresponding to the edges incident to v. See Figure 1.

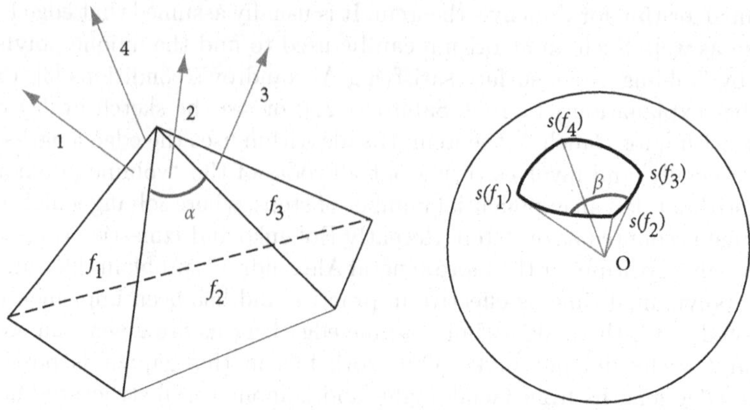

Fig. 1. The Gauss map and the correspondence of facial angles, $\beta = \pi - \alpha$

Let G be the skeleton graph of the polyhedron P. Mapping P to the sphere produces an embedded graph $s(G)$ with vertices $\{s(f) : f \text{ a face of } P\}$, and edges $\{s(e) : e \text{ an edge of } P\}$ and the cyclic order of edges around vertices induced by the order of edges around faces in G. Clearly $s(G)$ is the dual graph of G.

We now consider what the Gauss map does to facial and dihedral angles. Let e and e' be two incident edges of a face f of P. Let α be the facial angle between them. Let β be the angle between $s(e)$ and $s(e')$ at the point $s(f)$. Then, as Stoker [26] shows, β and α are supplementary angles, i.e. $\beta = \pi - \alpha$. Let f and g be two faces of P joined at edge e. Let γ be the dihedral angle between f and g, and let δ be the length of the arc $s(e)$ (measured in radians as an angle from the origin). Then γ and δ are supplementary angles [26].

In our situation, we know the embedded graph G and the facial angles. Applying the Gauss map, we know the embedded graph $s(G)$ and its facial angles. Dihedral angles of the original map to edge lengths in the spherical dual, and Cauchy's theorem says that these edge lengths are unique. Under this transformation the algorithmic form of Cauchy's theorem becomes: Given an embedded 3-connected planar graph and an angle between each consecutive pair of edges incident on a vertex, find a drawing of the graph on the sphere with non-crossing geodesic arcs for the edges, and with the specified angles. Supplements of arc lengths in the drawing provide the dihedral angles for the original problem.

No efficient algorithm is known for this problem, but it is connected to quite a body of work in graph drawing. The problem of drawing a graph in the plane with specified angles was first considered by Vijayan [27] and later proved NP-hard by Garg [12]. Drawing on the sphere might be harder, but all our angles are convex, which should be easier. See also [4] and [23] for the case of triangulated graphs. Spherical drawing of triangulated graphs has been addressed in the graphics community in the context of spherical parameterization, see in particular [24]. A closely related problem is to efficiently represent a 3-connected planar graph as the skeleton graph of a convex polyhedron [7].

3 Conditions for Existence of a Convex Polyhedron

In this section we give conditions on the edge lengths, facial angles, and dihedral angles that are necessary and sufficient for the existence of a convex polyhedron. In some sense this is solved by the "local" or "easily checkable" conditions for convex polyhedra [17,10,20] (see below for more details); however, our goal is to give conditions that separate the role of edge lengths and dihedral angles.

Theorem 1. *A 3-connected planar graph (with its unique combinatorial embedding) and given edge lengths and convex facial and dihedral angles are those of a convex polyhedron iff*

(1) the edges around every face form a simple convex polygon
(2) in the spherical dual, the arcs of the edges around every face form a simple convex spherical polygon

We will prove Theorem 1 using a 1952 result of Van Heijenoort [13] on locally convex manifolds. We first discuss algorithmic ramifications for computing edge lengths and/or dihedral angles. Note that Condition (1) depends on facial angles and edge lengths; Condition (2) depends on facial angles and dihedral angles—recall that dihedral angles correspond to arc lengths in the spherical dual. We were unable to use the conditions to help us find dihedral angles, but we can use them to find edge lengths. We separate Condition (1) into a part (1a) depending only on facial angles, and a part (1b) depending on both facial angles and edge lengths and expressible via linear inequalities. Facial angles determine edge directions. More precisely, if we choose a unit vector in the plane for the direction of one edge of face f, then the facial angles determine unit direction vectors $d(e) \in R^2$ for each edge e as we go around f from the initial edge. Condition (1b) is that the sum of these unit vectors times the appropriate edge lengths is the zero vector, i.e. that the sequence of edges closes up to form a polygon.

What are the conditions on facial angles? It is not sufficient that every facial angle be convex, i.e. in the interval $(0, \pi)$: the polygon in Figure 2, from [17], has only convex angles, but is not a *simple* convex polygon, nor would any other choice of edge lengths make it so. To guarantee a simple convex polygon, we impose Condition (1a): that the facial angles around a face of k edges sum to $\pi(k-2)$. We therefore arrive at the following conditions:

(1a) for each face f of k edges, $\pi(k-2) = \sum\{\alpha : \alpha \text{ a facial angle of } f\}$.
(1b) for each face f, $\sum_{e \in f} l(e)d(e) = (0,0)$ where $l(e)$ is the length of edge e and $d(e)$ is its direction, relative to some initial choice for one edge.

Lemma 1. *Condition (1) is equivalent to Conditions (1a) and (1b).*

Proof. The main part of this is proved by Vijayan [27]. The proof is easy; we outline it for completeness. Clearly (1) implies (1a) and (1b). For the other direction, (1b) implies that we have a sequence of line segments that closes up to form a cycle. We prove by induction on k, the number of segments, that a

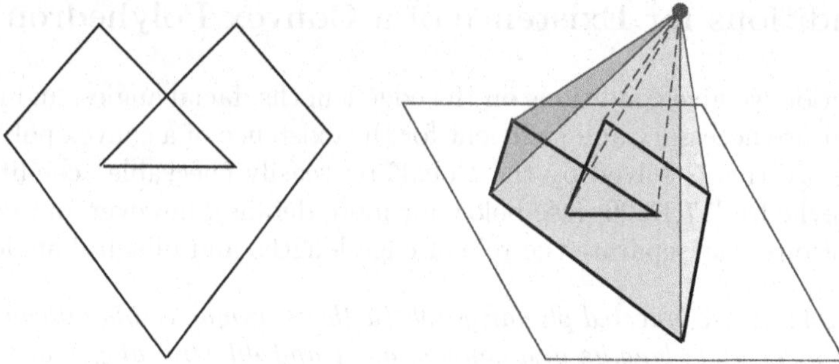

Fig. 2. Convex angles do not always make a simple polygon (left), or a simple polyhedron (right)

simple convex polygon is formed. This is obvious for $k = 3$, and easy for $k = 4$. Consider $k > 4$. If there are two consecutive angles with sum greater than π we can eliminate the edge between them by extending the two neighbouring edges, and apply induction. If all pairs of consecutive angles have sum at most π then $2\sum \alpha \leq k\pi$. Applying condition (1a), $2\pi(k-2) \leq \pi k$, so $2k - 4 \leq k$, so $k \leq 4$.

Translated to the sphere, Condition (2) seems symmetric with Condition (1) except that it involves arc lengths (corresponding to dihedral angles) in place of edge lengths. It seems tantalizing to express Condition (2) using subconditions analogous to those above, and thus obtain an algorithm to find dihedral angles.

Condition (1b)—that following the sequence of edges around a face "closes up" the polygon—can be transferred to the spherical situation, though it is computationally more difficult for the following reason. In the plane the edges around a face provide successive translations, so we get a linear system for the edge lengths; however, on the sphere the edges around a face provide successive rotations, so we get a non-linear system. This is not the main difficulty. In the plane, the remaining condition (1a) did not depend on edge lengths, but on the sphere it does, as we now show. Recall that condition (1a) precluded a polygon "wrapping around" more than once as in Figure 2 (left). The same issue arises on the sphere. As described by Mehlhorn et al. [17], the example of Figure 2 can be extended to three dimensions by adding two vertices, one above the plane and one below, with triangular faces joining each of these vertices to each edge of the polygon. Figure 2 (right) shows the new upper vertex. The resulting object is combinatorially a bipyramid with a 7-sided base; each face is a triangle; and each facial angle and each dihedral angle is in the range $(0, \pi)$. In the spherical dual the face corresponding to the upper vertex is a spherical polygon that wraps around twice and intersects itself.

In the plane, condition (1a) excluded such "wrapping around" by requiring that the sum of facial angles be $\pi(k-2)$ for any face of k edges. On the sphere no condition on facial angles alone will suffice: we give an example (prior to the Gauss map) of two spherical polygons with the same facial angles, exactly one of which is simple. Consider the example of Figure 2 with the upper vertex far away from the plane of the rest of the Figure, and with many acute triangles incident to it. With these same acute angles we can instead make the dihedral angles larger and connect to a simple polygon—see Figure 3.

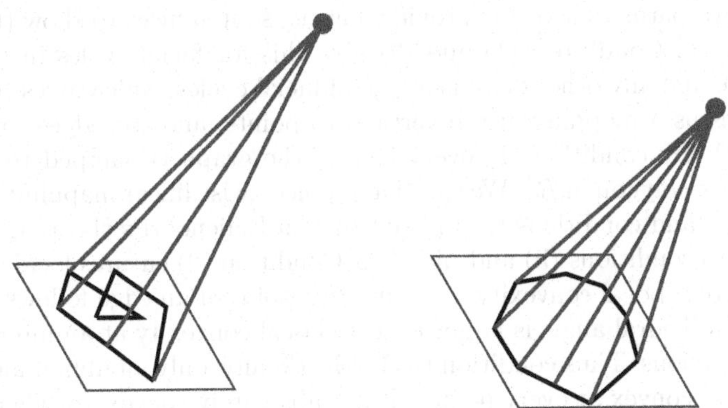

Fig. 3. The same facial angles at a vertex can form a simple or non-simple cone

In the remainder of this section we give a proof of Theorem 1 using the following result of Van Heijenoort. The terms used in the theorem are defined just below. Our situation is more specialized in that we have a piece-wise linear manifold, which, as we shall see, makes the topological conditions straightforward.

Theorem 2. *[13] If a 2-dimensional manifold M is*

(i) mapped into R^3 by a locally topological mapping f
(ii) locally convex under f
(iii) absolutely convex at a point
(iv) complete under f

then $f(M)$ is the boundary of a 3-dimensional convex set.

Van Heijendoort defines the manifold M to be *complete* if "every bounded infinite subset of M has an accumulation point in M". "Bounded" in this case means that the distances are bounded, using the metric induced by the mapping of M into R^3. M is *locally convex* under f if every point p of M has a neighbourhood N s.t. $f(N)$ lies on the boundary of a convex body K. *Absolute convexity* means that, in addition, there is a support plane of K at $f(p)$ that contains no other point of K.

Proof (of Theorem 1). The forward direction is clear. For the other direction, assume conditions (1) and (2) hold. We need to prove Van Heijenoort's conditions. The embedded 3-connected planar graph drawn on the surface of a sphere provides a manifold. We begin by assigning vertex coordinates. Arbitrarily choose coordinates for one vertex v and directions for two consecutive edges incident with that vertex, forming the correct facial angle for face f between them. The plane of face f is now determined. So are the coordinates of the vertices around face f. From these, and the dihedral angles, we get the planes of the faces adjacent to face f. Continuing in this way, we obtain coordinates for all the vertices as we expand outward from the initial choices. We claim that these coordinates are well-defined—i.e. that they are independent of the order in which we expand outward. Two paths to a vertex provide a cycle, so it suffices to show that every cycle closes up. Conditions (1) and (2) give this for facial cycles in the graph and its dual, and any other cycle is a sum of facial cycles, which gives the result.

This gives us a mapping of the vertices to points, and the edges to line segments in R^3. By condition (1) every face of the graph is mapped to a simple planar convex polygon in R^3. We thus have a piece-wise linear mapping of a manifold into R^3, and conditions (i) and (iv) of Van Heijenoort's theorem follow.

We turn to conditions (ii) and (iii). Our Condition (2) ensures local convexity at every vertex. Local convexity at an interior point of an edge follows from the fact that no dihedral angle is larger than π. Local convexity at an interior point of a face is obvious. Thus condition (ii) holds. Finally, only an unbounded object can be locally convex at every point but not absolutely convex anywhere, giving condition (iii). Thus by Van Heijenoort's Theorem we have the boundary of a piece-wise linear 3-dimensional convex set—i.e. a convex polyhedron.

3.1 Background: Local Conditions for Convexity

Although we found Van Heijenoort's conditions most useful, there is more recent, more algorithmic work on conditions for a polyhdron to be convex. In this section we briefly describe such work by Mehlhorn et al. [17], Devillers et al. [10], and Rybnikov [20]. The conditions of Mehlhorn et al. involve checking if a ray from a point that lies on the "inside" of the plane through every face intersects only one face. The conditions of Devillers et al. are that all dihedral angles be convex and that the projection of the *seam* to the x-y plane be a convex polygon. The *seam* consists, roughly speaking, of the edges that are extreme with respect to some plane perpendicular to the x-y plane.

The idea of specializing Van Heijenoort's conditions to piece-wise linear manifolds is due to Rybnikov. In 3-dimensions it is clear that it suffices to check local convexity at vertices. Rybnikov's result [20], which he proves using Van Heijenoort's higher dimensional extension [13], is that to check convexity of piece-wise linear hypersurfaces in n dimensions, it suffices to check local convexity at the $(n-3)$-dimensional faces. Rybnikov gives a convexity-testing algorithm; the main step is to transform the local convexity test at an $(n-3)$-dimensional face to a convexity test for a [possibly self-intersecting] polygon, for which he gives a straight-forward algorithm (Devillers et al. [10] also give an algorithm for this.)

4 Determining Edge Lengths

In this section we consider the following problem: given the combinatorial structure of a convex polyhedron and given the facial angles, find edge lengths for the polyhedron. The edge lengths are not unique, even discounting scaling. For example, a cube can be stretched along any of its three axes. Non-uniqueness is discussed in section 4.4. It turns out to be equivalent to "indecomposability", a notion introduced by Gale [11], and studied by Shephard [25], Meyer [18], and McMullen [16] among others.

We will make use of the conditions for the existence of a convex polyhedron from the previous section, which were expressed in terms of facial angles, dihedral angles, and edge lengths. Recall that the only condition involving edge lengths was Condition (1b); we will express that condition in terms of linear inequalities.

In section 4.2 we consider the version of the problem where the dihedral angles are known, and we apply duality to give a characterization of when a polyhedron exists with given facial and dihedral angles.

4.1 An LP Formulation

Let V, E and F be the vertices, edges and faces of the graph, respectively. For each face f, choose one edge e_0 and choose a unit-length direction vector $d_f(e_0) \in R^2$ for it. Based on this choice, the facial angles determine unit direction vectors $d_f(e)$ for all the edges e in clockwise order around the face f. Note that an edge is in two faces, and may be assigned totally different edge direction vectors in those two faces. The question of whether there exist edge lengths satisfying condition (1b) is equivalent to feasibility of the following linear system in variables $\lambda(e)$, $e \in E$.

$$\begin{aligned} &\forall e \in E \;\; \lambda(e) > 0 \\ &\forall f \in F \;\; \sum_{e \in f} \lambda(e) d_f(e) = (0,0) \end{aligned} \tag{1}$$

Theorem 3. *Suppose a convex polyhedron exists with given facial angles and combinatorial structure. Then its edge lengths satisfy (1) and any solution to (1) gives edge lengths of such a polyhedron.*

The problem of finding edge lengths is thus solvable via linear programming algorithms [22]. Note that we need an algebraic model of computing to go from facial angles to $d(e)$. Linear programming, however, is only solvable in polynomial time in the bit complexity model, so we cannot claim a polynomial time algorithm to find edge lengths. Still, the simplex method should be practical. Note also that solving the linear system says nothing about whether the input facial angles and combinatorial structure are those of a convex polyhedron.

4.2 With Dihedral Angles

The above method computes direction vectors for edges within the plane of each face. If we have dihedral angles, we can compute true 3-D direction vectors for edges. We make an initial choice of coordinates for one vertex, and direction vectors for two edges consecutively incident at the vertex, ensuring that the angle between the two vectors matches the required facial angle. Based on these initial choices, we can compute direction vectors for all edges in 3-D. For edge $e = (u, v) \in E$, let $d(e) \in R^3$ be the direction vector of the edge from u to v. Note that we (arbitrarily) choose an order (u, v) or (v, u) to do this. For face $f \in F$, distinguish $cw(f)$, the edges of face f whose vector $d(e)$ is directed clockwise around f, and $ccw(f)$, the edges of face f whose vector is directed counter-clockwise around f. The linear system becomes:

$$\forall e \in E \;\; \lambda(e) > 0$$

$$\forall f \in F \;\; \sum_{e \in cw(f)} \lambda(e) d(e) \;\; - \sum_{e \in ccw(f)} \lambda(e) d(e) = (0, 0, 0) \tag{2}$$

Theorem 4. *There exists a convex polyhedron with given face and dihedral angles and given combinatorial structure iff conditions (1a) and (2) hold, and the linear system (2) is feasible.*

Our purpose in this section is to give duality conditions for feasibility of (4), but we mention first that it is possible to test conditions (1a) and (2) in polynomial time—see the work referenced in section 3.1.

Duality theory gives a characterization of when the linear system (2) is feasible. The linear system has the form $Ax = b, x > 0$. By Stiemke's Transposition Theorem (see Schrijver [22, p. 95]), there is a solution x iff for any y, $yA \geq 0$ implies $yA = 0$. Translating into our situation, we have a dual variable $\nu(f) \in R^3$ for each face $f \in F$. For edge $e = (u, v)$ let $f_r(e)$ be the face to the right of e and let $f_l(e)$ be the face to the left of e. The dual linear system is:

$$\forall e \in E \;\; d(e) \cdot (\nu(f_r(e)) - \nu(f_l(e))) \geq 0 \tag{3}$$

A change of variables gives more intuitive conditions. For each edge e let $\nu(e) = \nu(f_r(e) - \nu(f_l(e))$. Formula (3) becomes $d(e) \cdot \nu(e) \geq 0$. We can recover the $\nu(f)$ vectors from the $\nu(e)$ vectors so long as the sum of the $\nu(e)$'s is 0 around any dual cycle. Let $\overline{F}$ be the faces of the dual graph. We obtain:

Theorem 5. *Given an embedded 3-connected planar graph with specified facial and dihedral angles s.t. conditions (1a) and (2) hold, either there exists a corresponding convex polyhedron OR there are vectors $\nu(e) \in R^3$, $e \in E$ s.t.*

$$\forall e \in E \;\; d(e) \cdot \nu(e) \geq 0 \tag{4}$$

$$\forall \overline{f} \in \overline{F} \;\; \sum_{e \in \overline{F}} \nu(e) = 0 \tag{5}$$

with strict inequality in (4) for at least one edge e. Furthermore, NOT BOTH the polyhedron and the vectors can exist.

Proof. Straightforward: If there is no convex polyhedron then there are vectors $\nu(f) \in R^3, f \in F$ s.t. (3) holds and with strict inequality for at least one edge e. Performing a change of variables as described above, gives vectors $\nu(e) \in R^3, e \in E$ s.t. (4) and (5) hold, and with strict inequality in (4) for at least one edge.

Conversely, if vectors $\nu(e) \in R^3, e \in E$ exist s.t. (4) and (5) hold, and with strict inequality in (4) for at least one edge, then define vectors $\nu(f) \in R^3$ for each $f \in F$ as follows. Begin by choosing one $f_0 \in F$ and setting $\nu(f_0) = (0, 0, 0)$. Then use the formula $\nu(e) = \nu(f_r(e) - \nu(f_l(e))$ to define $\nu : F \rightarrow R^3$. Note that ν is well-defined by (5). From (4) we obtain (3), so there is no convex polyhedron satisfying the requirements.

4.3 Example

Recall that in Section 3 we gave conditions (1a), (1b) and (2) for the existence of a convex polyhedron with specified combinatorial structure, edge lengths, and facial and dihedral angles. That section contained an example to show that the "convexity" condition (1a) was necessary. In this section we show that condition (1b) is necessary by giving an example where Conditions (1a) and (2) hold but the linear system (1) is not feasible.

The construction starts with an octahedron, which has facial angles of $\frac{\pi}{3} = 60°$ and dihedral angles of $\cos^{-1}(-\frac{1}{3}) \approx 109.47°$. Split one vertex and add a new edge e as shown in Figure 4. The four new facial angles are $\frac{2}{3}\pi = 120°$. All other facial and dihedral angles stay the same. Consistent edge direction vectors exist, and all convexity conditions are satisfied. The linear system (1) is not feasible: in order for edge e to have positive length while maintaining the specified angles, the square visible in Figure 4 as the silhouette of the octahedron must become a rectangle—but this destroys the bottom half of the octahedron.

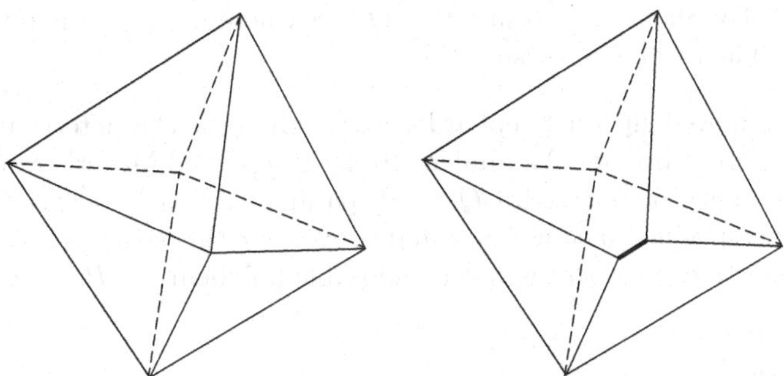

Fig. 4. An octahedron (left) and the addition of one new edge (right) making an example where angle convexity conditions hold, but no feasible edge lengths exist

4.4 Relation to Decomposability of Polyhedra

The current Section 4 has been about the existence of a convex polyhedron with specified combinatorial structure and facial and dihedral angles. There is a considerable body of work on the related uniqueness question: given a convex polyhedron, can we preserve all facial and dihedral angles but alter edge lengths (other than by scaling). In this subsection we briefly summarize this work.

For polytopes P and Q, Gale [11] defined $Q \leq P$ if for every direction u, the extreme set of Q in direction u has dimension less than or equal to the dimension of the extreme set of P in direction u. In particular, this means that any face of Q has a corresponding face of P with the same normal; however, the combinatorial structure may be different in that a face of P may have shrunk to an edge or vertex of Q, and an edge of P may have shrunk to a vertex. Thus this concept seems at first glance to be more general than the uniqueness of edge-lengths question mentioned above. But in fact the notions are equivalent.

Gale [11] defined a convex polyhedron P to be *decomposable* if P can be expressed as a Minkowski sum, $P = R + S$ where neither R nor S is homothetic to (i.e. a scaled translated version of) P. Shephard [25] proved that a convex polyhedron P is decomposable iff there is a convex polyhedron $Q \leq P$ that is not homothetic to P. In fact he proved a stronger thing, that such a Q can be used in a decomposition of P. We will use Shephard's result to relate uniqueness of edge lengths to the relation $\leq$.

Lemma 2. *For convex polyhedron P, the following are equivalent:*

(i) P is decomposable
(ii) there is a convex polyhedron $Q \leq P$ that is not homothetic to P
(iii) there is a convex polyhedron R with the same combinatorial structure as P and the same facial and dihedral angles, but with different non-zero edge lengths (not just re-scaled)

Proof. Equivalence of (i) and (ii) is Shephard's result. Clearly, (iii) implies (ii). Suppose (ii). By Shephard's result $P = Q + S$ where neither Q nor S is homothetic to P. Then $Q + \frac{1}{2}S$ satisfies (iii).

Meyer [18] followed up on Shephard's work, giving a characterization of decomposable polyhedra, and McMullen [16] later reproved Meyer's results—they consider the space of polyhedra $Q \leq P$, parameterizing in terms of the face offsets for the specified face normals and prove that this space is a cone, whose extreme rays correspond to the indecomposable polyhedra $\leq P$.

References

1. Aigner, M., Ziegler, G.M.: Proofs from the Book, 3rd edn. Springer, Heidelberg (2003)
2. Alboul, L., Echeverria, G., Rodgrigues, M.: Discrete curvatures and gauss maps for polyhedral surfaces. In: European Workshop on Computational Geometry (2005) `http://www.win.tue.nl/EWCG2005/Proceedings/18.pdf`
3. Alexandrov, A.D.: Convex Polyhedra. Springer, Heidelberg (2005)
4. Battista, G.D., Vismara, L.: Angles of planar triangular graphs. SIAM J. Discrete Math. 9, 349–359 (1996)
5. Bobenko, A.I., Izmestiev, I.: Alexandrov's theorem, weighted delaunay triangulations, and mixed volumes (2006) `http://www.citebase.org/abstract?id=oai:arXiv.org:math/0609447`

6. Cromwell, P.R.: Polyhedra. Cambridge University Press, Cambridge (1997)
7. Das, G., Goodrich, M.T.: On the complexity of optimization problems for 3-dimensional convex polyhedra and decision trees. Computational Geometry Theory and Applications 8, 123–137 (1997)
8. Demaine, E.D., Erickson, J.: Open problems on polytope reconstruction (July 1999) `http://theory.csail.mit.edu/~edemaine/papers/PolytopeReconstruction/`
9. Demaine, E.D., O'Rourke, J.: Geometric Folding Algorithms: Linkages, Origami, and Polyhedra. Cambridge University Press, Cambridge (2007)
10. Devillers, O., Liotta, G., Preparata, F., Tamassia, R.: Checking the convexity of polytopes and the planarity of subdivisions. Computational Geometry: Theory and Applications 11, 187–208 (1998)
11. Gale, D.: Irreducible convex sets. In: Proc. International Congress of Mathematicians, vol. II, pp. 217–218, Amsterdam, North-Holland (1954)
12. Garg, A.: New results on drawing angle graphs. Computational Geometry Theory and Applications 9, 43–82 (1998)
13. Heijenoort, J.V.: On locally convex manifolds. Communications on Pure and Applied Mathematics 5, 223–242 (1952)
14. Kaibel, V., Pfetsch, M.E.: Some algorithmic problems in polytope theory. Algebra, Geometry, and Software Systems 23–47 (2003)
15. Lucier, B.: Unfolding and reconstructing polyhedra. Master's thesis, David R. Cheriton School of Computer Science, University of Waterloo (2006)
16. McMullen, P.: Representations of polytopes and polyhedral sets. Geometriae Dedicata 2, 83–99 (1973)
17. Mehlhorn, K., Näher, S., Seel, M., Schilz, T., Schirra, S., Uhrig, C.: Checking geometric programs or verification of geometry structures. Computational Geometry: Theory and Applications 12, 85–103 (1999)
18. Meyer, W.: Indecomposable polytopes. Transactions of the American Mathematical Society 190, 77–86 (1974)
19. O'Rourke, J.: Computational geometry column. SIGACT News 38(2) (to appear)
20. Rybnikov, K.A.: Fast verification of convexity of piecewise-linear surfaces. CoRR, cs.CG/0309041 (2003) `http://arxiv.org/abs/cs.CG/0309041`
21. Sabitov, I.K.: The volume as a metric invariant of polyhedra. Discrete and Computational Geometry 20(4), 405–425 (1998)
22. Schrijver, A.: Theory of Linear and Integer Programming. Wiley, Chichester (1986)
23. Sheffer, A., de Sturler, E.: Parameterization of faceted surfaces for meshing using angle based flattening. Engineering with Computers 17(3), 326–337 (2001)
24. Sheffer, A., Gotsman, C., Dyn, N.: Robust spherical parameterization of triangular meshes. Computing 72, 185–193 (2004)
25. Shephard, G.C.: Decomposable convex polyhedra. Mathematika 10, 89–95 (1963)
26. Stoker, J.J.: Geometrical problems concering polyhedra in the large. Communications on Pure and Applied Mathematics 11, 119–168 (1968)
27. Vijayan, G.: Geometry of planar graphs with angles. In: Proc. 2nd Annual ACM Symposium on Computational Geometry, pp. 116–124 (1986)

Fixed-Parameter Tractability for Non-Crossing Spanning Trees

Magnús M. Halldórsson[1,⋆], Christian Knauer[2], Andreas Spillner[3], and Takeshi Tokuyama[4,⋆⋆]

[1] Dept. of Computer Science, Reykjavik University
mmh@ru.is
[2] Institute of Computer Science, Freie Universität Berlin
christian.knauer@inf.fu-berlin.de
[3] Institute of Computer Science, Friedrich-Schiller-Universität Jena
spillner@minet.uni-jena.de
[4] Graduate School of Information Sciences, Tohoku University, Sendai, 980-8579 Japan
tokuyama@dais.is.tohoku.ac.jp

Abstract. We consider the problem of computing non-crossing spanning trees in topological graphs. It is known that it is NP-hard to decide whether a topological graph has a non-crossing spanning tree, and that it is hard to approximate the minimum number of crossings in a spanning tree. We consider the parametric complexities of the problem for the following natural input parameters: the number k of crossing edge pairs, the number μ of crossing edges in the given graph, and the number ι of vertices in the interior of the convex hull of the vertex set. We start with an improved strategy of the simple search-tree method to obtain an $O^*(1.93^k)$ time algorithm. We then give more sophisticated algorithms based on graph separators, with a novel technique to ensure connectivity. The time complexities of our algorithms are $O^*(2^{O(\sqrt{k})})$, $O^*(\mu^{O(\mu^{2/3})})$, and $O^*(2^{O(\sqrt{\iota})})$. By giving a reduction from 3-SAT, we show that the $O^*(2^{\sqrt{k}})$ complexity is hard to improve under a hypothesis of the complexity of 3-SAT.

1 Introduction

A *topological graph* is a graph with an embedding of its edges as curve segments in the plane such that each pair of edge curves intersects at most once. We refer to the embeddings of the vertices also as vertices, and to the geometric curve segments as *curves*. A topological graph is said to be *non-crossing* if none of the edge curves cross. We consider non-crossing subgraph problems that involve finding a non-crossing subgraph satisfying some property: spanning tree, s–t path, and cycle. All of these problems are known to be NP-hard [10,6]. In this article we focus on the non-crossing spanning tree problem (NCST). The corresponding minimization problem may be of interest when focusing on finding structures in the

⋆ Work done at University of Iceland, and supported by the Icelandic Research Fund.
⋆⋆ Supported by the project *New Horizons in Computing*, MEXT Japan.

F. Dehne, J.-R. Sack, and N. Zeh (Eds.): WADS 2007, LNCS 4619, pp. 410–421, 2007.

drawing of an embedded graph. Removing as many edges and crossings as possible makes it easier to recognize the structure of the graph in terms of connectivity.

Let G be a topological graph on n vertices and m edges. A *crossing* is a pair of edges that meet in a non-vertex point, and a *crossing edge* is one that participates in some crossing. A *crossing point* is a non-vertex point that is contained in at least two edge curves. Note that if d edges intersect in a crossing point, they create $\binom{d}{2} = d(d-1)/2$ crossings. Let X be the set of crossings in G, and let E_X be the set of crossing edges. Let $k = |X|$ be the number of crossings and let $\mu = |E_X|$ be the number of crossing edges. Observe that $\mu/2 \leq k \leq \mu(\mu-1)/2$. We assume without loss of generality that the curves intersect only in individual points, not in curve segments. Note that sometimes [8], a topological graph is allowed to have multiple crossings between a pair of edges, and our theory can be easily modified to that definition as long as the number of multiple crossings between each pair of edges is bounded by a constant.

A very naive method for a noncrossing subgraph problem is to exhaustively check the noncrossing properties for all subgraphs with the requested properties. This needs exponential time in the number m of edges of the graph. However, if k is small and the problem is polynomial time solvable without the noncrossing condition (e.g., spanning tree, cycle and s-t path), we have the following better strategy: For every crossing pair of edges, we delete one of the crossing edges to have a non-crossing subgraph. We have 2^k possible combinations of deletions, and it takes polynomial time for each fixed combination to find a spanning tree (for example) in the subgraph if it is connected. We can see that if G has a noncrossing spanning tree, we can find one by the above method. Thus, it is clear that the problem is computed in $O^*(2^k)$ time, where the O^*-notation hides polynomial terms. Recently, Knauer et al. [8] gave algorithms for NCST with improved time complexity of $O^*(1.9999992^k)$. This left the question of how far down the complexity can be brought down.

Our results. We give a number of results that answer many of the open questions about the fixed-parameter tractability of non-crossing subgraph problems.

We first give an improved $O^*(1.928^k)$-time algorithm for NCST. This is based on a compact *kernel* for the problem, and on a new set of reduction rules that takes advantage of limited recurrences for low-degree vertices. This approach actually applies to a generalized problem, involving arbitrary pairwise conflicts on the edges.

One of the main contributions of this paper is an algorithm for NCST with an asymptotic improvement in the time complexity to $2^{O(\sqrt{k})}$ (we ignore polynomial time preprocessing), see Section 3. This is based on finding a cyclic separator in a related planar graph. Thus turns out to be best possible, under the *exponential time hypothesis* that 3-SAT does not have a $2^{o(n)}$-time algorithm (where n is the number of variables), as shown in Section 5.

We also present fixed-parameter algorithms for two further parameters. For the parameter μ, the number of crossing edges, we give a $\mu^{O(\mu^{2/3})}$-time algorithm.

A *geometric graph* is a topological graph whose edges correspond to the straight-line segments that connect their endpoints. For geometric graphs, we use another measure to design a fixed-parameter algorithm. Consider the vertex set of the embedded graph as a set of points in the plane. Then we can refer to the points that lie in the interior of the convex hull of the point set as *inner* points. The number ι of inner points has been used successfully to parameterize some hard geometric problems on points in the plane, including Minimum Weight Triangulation problem (MWT) [4,13]. For this parameter, we give an algorithm that solves NCST for geometric graphs in $O^*(\iota^{O(\sqrt{\iota})})$ time. Note that it is easy to come up with geometric graphs where ι is small but k is large. We also show that it is unlikely that a $2^{o(\sqrt{\iota})}$-time algorithm exists.

2 Improved Search-Tree Algorithm

We first give a simple search-tree method to find a non-crossing spanning tree in a topological graph with k crossings in time 1.9276^k (plus polynomial time preprocessing) if one exists. This improves on the previous bounds of 1.99999^k, as well as on the 1.968^k bound for a Monte-Carlo algorithm [8]. Although it will be greatly improved asymptotically to $2^{O(\sqrt{k})}$ in Section 3, we feel the above result is valuable since the search-tree algorithm is preferable in practice for the range of k that the problem is solvable in real feasible time, and our improved method gives little additional burden to programmers who want to implement a search-tree method. We reduce the original problem to a compact kernel problem, and then introduce some simple rules for a naive search-tree algorithm to obtain the improved time complexity.

Kernel. A kernel is a reduced problem instance, whose solution can be "easily" turned into a solution of the original instance. To form a kernel for NCST we use edge contractions, where contracting the edge uv in a graph G results in the graph where the vertices u and v have been merged into a single vertex that has all the neighbors of that either of its original vertices had.

Edges that cross are said to be *crossing edges*; if they share an endpoint v, we say they are *tangled*, more specifically, they are *tangled at* v.

To form a small kernel, we contract all non-crossing edges of the graph G yielding a new topological graph G'. More precisely, for each connected component of the induced subgraph of G by the non-crossing edges, we select any spanning tree, and contract it. The other edges in the connected component are deleted. It is clear that the kernel is obtained in polynomial (indeed, linear) time.

Note that this does not affect the crossing properties of the crossing edges. However, it can lead to non-tangled pairs to become tangled. A planar subgraph H' of G' maps to a subgraph H of G; adding the contracted edges to H still retains planarity. Hence, there is a bijective mapping between maximal planar subgraphs of G and G'.

Every edge in G' is crossing, thus the number of edges in G' is at most μ. Since the graph G' is necessarily connected and non-acyclic, the number of vertices in G' is at most its number of edges. We further delete all loop edges in G' even if they are crossing. This resolves some crossings, but does not affect the problem solution because of the property of a spanning tree.

Proposition 1. *A kernel for NCST with at most μ edges and vertices can be computed in linear time.*

Search-tree approach. We give reduction rules that result in an efficient search tree for a non-crossing connected spanning subgraph. A non-crossing spanning trees can then be easily found.

In most nodes of the search tree we select an edge e for *branching*: either a solution contains e or it does not. If it contains e it cannot contain the edges C_e crossing e. Hence, we obtain two subproblems: $G - \{e\}$ and $G - C_e$. In either of the subproblems, we eliminate all crossings incident on e, and apply the available contractions. The measure, $T(k)$, of a subproblem is the number of search tree leaves in terms of the number k of crossings. In subproblem $G - C_e$, the number of crossings is reduced by one, for a measure of $T(k-1)$. We want to show that the measure of $G - \{e\}$ is less.

We select branching edges in the following order of preference:

1. If there is an edge crosses two or more edges, then we choose such an edge. Crossing number k is reduced by at least two in $G - \{e\}$ (also in $G - C_e$).
2. For tangled parallel edges, we can pick either of them, yielding the same subproblem, since neither is twice-crossing (otherwise, we should apply the rule 1). This allows us to contract both edges, reducing k by one.
3. Consider a node v of degree ≤ 3. At least one edge e incident on v is not tangled with either of the other incident edges; otherwise, one of them would be twice-crossing. We branch on e and obtain on one branch a degree-2 node. For a degree-2 node with two incident tangled edges, branching on either edges yields the same subproblem after contractions. Otherwise, we branch on one of the incident edges, obtaining on one branch a degree-1 node. A degree-1 node must be connected in a spanning tree, thus only one choice is then possible. Hence, a problem with a node of degree at most 3 has a measure of at most $T(k-1) + T(k-2) + T(k-3)$.
4. Consider a degree-4 node v with an edge untangled at v. Let e be an edge incident on v that is not tangled with the other edges incident on v. When we branch on e, the non-included case leaves us with v being of degree-3. We then apply the degree-3 case above.
5. When none of the above rules apply, we branch on an arbitrary edge.

In each case, except when we reach the last rule, we measure the decrease in the number of crossings. This allows us to bound the size of the search tree.

Let us first consider what happens when we reach the last rule. In that case, all nodes are of degree at least four. Further, only nodes that have two tangled incident edge pairs have degree 4, while the others are of degree at least 5. Thus, each edge that is tangled at node v appears untangled at the other endpoint, since there are no tangled parallel edges and no twice-crossing edges. Thus, no two degree 4 nodes are adjacent to each other. We claim that the number of nodes, n, is at most $9\mu/20$. Let a denote the number of degree 4 nodes, and note that all neighbors of degree-4 nodes are of degree at least 5. Therefore, counting edge incidences, $\mu \geq \frac{4a+5(n-a)}{2} = \frac{5n-a}{2}$. But clearly, $\mu \geq 4a$. Combining the two inequalities, we have that $\mu \geq (20/9)n$. We contract an edge, eliminating a vertex, in each round. Hence, the depth of the recursion is at most $n - 1 \leq (9/20)\mu = (9/10)k$, for a time complexity of $2^{0.9k}$.

Let us now evaluate the effects of the other branching rules. In each rule, we perform one or more branches, yielding a set of subproblems measured in terms of the number of crossings remaining. We express each case as a recurrence relation:

$$T(k) \leq \max \begin{cases} 2T(k-2), & \text{Twice-crossing edge} \\ T(k-1), & \text{Tangled parallel edges} \\ T(k-1) + T(k-2) + T(k-3), & \text{Degree-3 case} \\ T(k-1) + T(k-2) + T(k-3) + T(k-4), & \text{Degree-4 case} \\ 2^{0.9k}. & \text{Dense case} \end{cases}$$

The worst case is the degree-4 case, which yields $T(k) \leq 1.9276^k \approx 2^{0.9468k}$.[1]

Generalized structures. Our arguments do not use planarity in any way, except indirectly as prescribing conflicts between edges. Thus, the approach works more generally for finding spanning forests of graphs with conflicts between edges. More generally, we can formulate the Conflict-Free Spanning Tree (CFST) problem, where we are given a graph G and a conflict graph H defined on the edge set $E(G)$. We are to determine whether there exists a subset of mutually non-conflicting edges forming a spanning tree. In NCST, the conflicts are given by the crossings, and $|E(H)| = k$. For another example, the algorithm can be applied to layouts of graphs on surfaces of higher genus.

Theorem 1. *Given graphs G and H, CFST can be solved in time $O^*(1.9276^{|E(H)|})$.*

3 Separator-Based Algorithm

We describe here our algorithm for the non-crossing spanning tree problem. The approach bears some similarity to the algorithm of Deineko et al [1] for the Hamilton cycle problem in planar graphs. that has complexity $O^*(2^{O(\sqrt{n})})$. Our method is based on a cycle separator theorem of Miller.

Proposition 2. *(Miller [12]) Let G' be an embedded triangulated planar graph on n vertices. Then, there is a linear time algorithm that finds in G' a simple cycle C of at most $\sqrt{8n}$ vertices that partitions $G' - C$ into a vertex set A that lies within the region inside of C, and a vertex set B that lies outside of C, with $|A| \leq 2n/3$ and $|B| \leq 2n/3$.*

Before applying the above theorem, we resolve the multiplicities of the kernel. The *multiplicity* of a crossing is the number of pairs of edges that meet in the same point. Large multiplicity can confuse good algorithms, especially those based on separators, and the same can be said of high-degree vertices.

[1] 1.9276 represents the positive-valued solution of the equation $x^4 = 1 + x + x^2 + x^3$.

Fortunately, we can assume without loss of generality that crossings are of unit multiplicity and vertices of maximum degree 3. We omit details, but the basic idea is to clip edges at high-degree vertices and to replace the clipped stars by binary trees and to wiggle edge curves in order to avoid degenerate crossings. We have the following theorem:

Theorem 2. *Suppose there is an algorithm that solves NCST on degree-3 graphs with unit crossing multiplicity in time $T(k, \mu, n)$. Then, there is an algorithm for NCST for general topological graphs running in time $O(T(k, \mu, n))$.*

Given a topological graph H, we form an associated triangulated plane graph $P = P_H$ as follows. We replace each crossing point of H by a vertex and the curve of each crossing edge by line segments connecting the vertices and the crossing points. Finally, we arbitrarily triangulate the graph. The edges of the resulting graph P are therefore of three kinds: non-crossing edges from H, segments of crossing edges (connected a crossing point to either another crossing point or to an original vertex), and newly introduced "dummy" edges. Observe that the number $n(P_H)$ of vertices in H equals $\mu + k$.

The idea of our algorithm is as follows. In the preprocessing step, we find a kernel, as guaranteed by Proposition 1, and apply the multiplicity reduction of Theorem 2 to ensure each crossing point involves exactly two crossing edges.

The main algorithm finds a cycle separator in the derived plane graph P_H, and solves the two resulting subgraphs of H recursively, under all possible ways of constraining one subsolution to contribute to the connectedness of the whole solution. More precisely, if C is a cycle separator of P_H, we partition its nodes into C_v, a set of vertices of H, and C_c, a set of crossing points in H. The algorithm tries all $2^{|C_c|}$ ways of breaking the crossings of C_c. Consider one such decision vector D, and let D_v be the set of vertices of the chosen crossing edges that are on the inside of the cycle C. Consider now the set $S = C_v \cup D_v$. This set can be topologically arranged on a circle C', such that no edges cross the circle. Let H_A be the subgraph of H induced by vertices on the inside of *or on* the circle C', and H_B the subgraph on the outside of or on C'. Thus, $V(H_A) \cap V(H_B) = S$.

Given H_A and H_B, the algorithm examines all the ways that the vertices of S can be connected inside C' (i.e. within H_B) while maintaining planarity. Namely, if we view S as being an ordered set, we seek, in combinatorial terminology, a *non-crossing partition* of S. A partition of an ordered set is non-crossing if no two blocks "cross" each other, i.e. whenever a and b belong to one block and x and y to another, they are not arranged in the order $axby$. For each non-crossing partition Π, we form a star forest $X = X_\Pi$ with the leaves of each star corresponding to a block of the partition and a new node as the root of the star. Let $H'_B = H_B \cup X$. The algorithm recursively solves H'_B, yielding a non-crossing forest F_B in H_B. By induction, crossing edges in G have either all of its segments in H in F_B or none. The algorithm then recursively solves $H'_A = H_A \cup F_B$, giving a non-crossing spanning tree in G.

Theorem 3. *The algorithm solves NCST in time $2^{O(\sqrt{k})} + O(m)$ and polynomial space.*

Proof. We first show that the correctness of the algorithm. Suppose that the input graph G contains a non-crossing spanning tree T. Let T_A (T_B) be the restriction of T to H_A (H_B). Each tree of the forest T_A contains some nodes of S; for the purpose of the solution of H_B, all that matters is that it connects those vertices together. Thus, if we replace each tree U of T_A by a star with nodes in $S \cap U$ as leaves, the resulting union, joined with T_B, induces a connected tree spanning all the nodes. Hence, by induction, the first recursive call of the algorithm returns a spanning tree of H'_B, whose restriction to H_B is the forest F_B. Now, $F_B \cup T_A$ is connected and spans $F_B \cup H_A$. Hence, the second recursive call will also result in a non-crossing spanning tree T' of $H'_A = H_A \cup F_B$. The nodes of F_B are the nodes of H_B; hence, we have spanned all of G. Thus, the algorithm correctly computes a non-crossing spanning tree. On the other hand, if G does not contain a non-crossing spanning tree, the second recursive call never finds a non-crossing spanning tree.

Next, we analyze the complexity. Let $\nu = n(P_H) = \mu + k$ to be the *measure* of the problem. By Proposition 2, the algorithm finds a cycle separator in P of size at most $z = \sqrt{8\nu}$. We have at most 2^z ways of resolving the crossing edges on the separator. The size of S and the cycle C' is still z. The number of non-crossing partitions of S equals the Catalan number $C_z = \frac{1}{z+1}\binom{2z}{z} < 4^z$.Thus, there are less than 8^z cases considered by the algorithm.

Each case involves two subproblems. The larger of the subproblems is of measure M of at most $2\nu/3 + z$. A more careful analysis actually shows that most of the cases involve smaller subproblems. The measure of the smaller subproblem is at most $(\nu - M) + 2z$. The time complexity for any subproblem, aside from recursive calls, is linear in the size of the graph. Thus, the complexity of the algorithm is bounded by $T(\nu) = O(8^z/z^{3/2}) \cdot (T(2\nu/3 + z) + T(\nu/3 + z)) + O(\nu)$. This leads to $T(\nu) = O(2^{18\sqrt{\nu}})$. Since $\nu = \mu + k \leq 3k$, $T(\nu) = 2^{O(\sqrt{k})}$. QED.

The parameter μ. A straightforward $O^*(2^\mu)$ algorithm for NCST follows by considering all subsets of the set of crossing edges, and $O^*(2^{0.552\mu})$ can be obtained by the search-tree method (omitted in this version). We further give the following asymptotic improvement by combining the search method and the separator-based method:

Theorem 4. *NCST can be solved in $\mu^{O(\mu^{2/3})} + O(m)$ time and polynomial space.*

Proof. We split the computation into two cases, depending on the size of μ relative to $\nu = \mu + k$. Let $R(\mu)$ be the number of subproblems in an instance with μ crossing edges. If $\nu < 2\mu^{4/3}$ (Case 1), then the separator-based algorithm gives $R(\mu) < 2^{c\sqrt{\nu}} < 2^{2c\mu^{2/3}}$. Otherwise (Case 2), $\nu \geq 2\mu^{4/3}$. Then there exists an edge that participates in at least $2\mu^{1/3}$ crossings. We branch on that edge, resulting in two subproblems: one without that edge, and the other without all the edges crossing it. This gives the recurrence $R(\mu) \leq R(\mu-1) + R(\mu - 2\mu^{1/3}) + 1$. The time complexity follows from this recurrence using Case 1 as the induction basis.

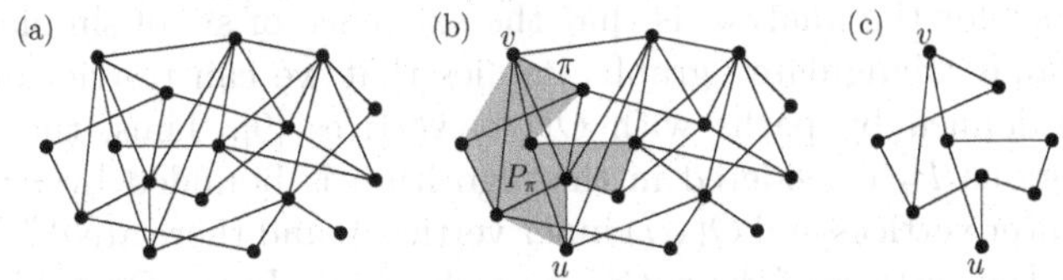

Fig. 1. A geometric graph, the polygon P_π and the subgraph G_π

4 Dynamic Programming Approach for the Parameter ι

A necessary prerequisite to successfully parameterize a problem with the number of inner points is that we can solve the problem in polynomial time for sets of points in convex position. For geometric graphs whose vertices are in convex position, it is easy to see that NCST can be solved using dynamic programming in $O(n^3)$ time. So this parameter could be viewed as a measure that tries to capture for each input its "distance from triviality" [3]. The key observation that provides a unified view on many of the problems mentioned in the introduction is that we can reformulate them as the search for a certain kind of triangulation. For NCST, we are given a geometric graph $G = (V, E)$ and the goal is to find a triangulation $\mathcal{T}$ of V such that the graph formed by those edges of G contained in the triangulation $\mathcal{T}$ is connected. Then, we can easily find a noncrossing spanning tree by using only those edges.

We describe the subproblem considered in our dynamic programming algorithm to solve NCST. The subproblem is defined by a crossing-free path π that starts at an outer vertex u, visits some inner vertices and ends at another outer vertex v. Such a path π splits the convex hull of V into two polygonal regions. Note that π is not necessarily a path in the input graph, but any noncrossing path connecting vertices by line segments is fine; indeed, it is a path in the (unknown) triangulation we are searching for. By P_π we denote the polygonal subregion to the left of π. An example is given in Figure 1(b), where P_π is shaded. The subgraph G_π induced by π consists of all those vertices and edges of G that are contained in P_π. This is illustrated in Figure 1(c).

We now describe what we actually want to compute for each P_π. It is not enough to decide whether or not there is a crossing-free spanning tree in G_π. Intuitively, we need a list of those crossing-free spanning forests of G_π where each tree in the forest shares at least one vertex with the path π. However, it is costly to consider the complete list of such spanning forests. Instead, it suffices to know which vertices on π belong to the same tree in the spanning forest. We can handle this by considering partitions of the set of vertices of the path π. For each such path π we have a collection of subproblems: one for each partition of the vertices of π. For such a subproblem we must decide whether or not there is a spanning forest of G_π such that every tree in the forest has at least one vertex on π and vertices on π in a component of the partition belong to the same tree in the forest.

The key fact for the analysis is that the existence of small simple cycle separators in planar triangulated graphs implies that we can restrict ourselves to subproblems defined by paths with $O(\sqrt{\iota})$ vertices [9]. Thus, the number of polygonal regions P_π considered in the algorithm is bounded by $n^2\iota^{O(\sqrt{\iota})}$ (selecting two outer vertices and $O(\sqrt{\iota})$ inner vertices), and there are $\iota^{O(\sqrt{\iota})}$ possible partitions for the vertices of the path π of each region. In the DP table we record whether there is a triangulation containing a feasible forest for each partition of each such polygonal region. Thus, the table size is $O(n^2\iota^{O(\sqrt{\iota})})$.

It remains to sketch how we process a subproblem in P_π by using information for smaller polygons stored in the dynamic programming table. We check every triangle Δ that is contained in P_π, shares an edge with the path π, and does not contain a vertex of V in its interior. Checking Δ means to decide whether a suitable triangulation for the subproblem containing Δ exists. By removing Δ from P_π we have one or two subpolygons, and this leads us to one or two smaller subproblems. We remark that we discard the choice of Δ if it generates a subpolygon with too many interior points on its boundary. It is routine to see that we can now solve the subproblem for P_π by referring the dynamic programming table. Thus, we have the following theorem:

Theorem 5. *Given a geometric graph G with n vertices we can decide in $O^*(\iota^{O(\sqrt{\iota})})$ time and $O^*(\iota^{O(\sqrt{\iota})})$ space whether or not G admits a crossing-free spanning tree.*

The time and space complexities are $O(n^3\iota^{O(\sqrt{\iota})})$ and $O(n^2\iota^{O(\sqrt{\iota})})$ if we consider polynomial factors of n. We can also compute a crossing-free spanning tree (not only decision) if exists in the same time and space complexities.

5 Hardness Results

We show here that the results of Section 3 are in some sense best possible, assuming the well-known *Exponential time hypothesis*, which is that 3-SAT cannot be solved in sub-exponential time. This hypothesis was formalized by Impagliazzo, Paturi, and Zane [5]. Evidence was given there and in later papers for support of the hypothesis. We are interested in the NCST_κ problem, where we decide whether an input geometric graph $G = (V, E)$ with k crossings has a crossing free spanning tree, and we use $\kappa(G) = \lceil\sqrt{k}\rceil$ as the parameter. We want to relate the question of whether there is an algorithm solving NCST_κ in $O^*(2^{o(\kappa(G))})$ time to an open question concerning the $\mathsf{3SAT}_\nu$ (3-SAT with the parameter ν):

> Instance: Exact 3-SAT formula (CNF formula with exactly three literals per clause) F.
> Parameter: The number $\nu(F)$ of variables occurring in F.
> Problem: Decide whether F is satisfiable.

The exponential time hypothesis is that $\mathsf{3SAT}_\nu$ cannot be solved in time $O^*(2^{o(\nu(F))})$. If we take the closure of $\mathsf{3SAT}_\nu$ under so called *subexponential reduction families (serf)* (cf. [2]) we obtain the class S[1]. Our goal is to show that NCST_κ is S[1]-hard. S[1]-hardness can be also shown for the parameter $\sqrt{\iota}$, but we omit it because of space limitation.

To achieve the S[1]-hardness, it suffices to give a *parameter preserving polynomial time reduction* from $\mathsf{3SAT}_\nu$ to NCST_κ . Such a reduction transforms a given instance F of $\mathsf{3SAT}_\nu$ in polynomial time into a instance G of NCST_κ such that $\kappa(G) \in O(\nu(F))$. We can give such a reduction through some intermediate problems. The first is $\mathsf{3SAT}_\mu$, which has the same instance and problem as $\mathsf{3SAT}_\nu$ but the parameter is the number $\mu(F) = 3m$ where m is the number of clauses of F. $\mathsf{3SAT}_\mu$ is known to be S[1]-complete (cf. [2]).

With every 3-CNF formula F we can associate a bipartite graph $H(F) = ((V, C), E)$. The vertices in V represent the variables occurring in F. The vertices in C represent the clauses of F. A variable is connected to a clause by an edge in E iff the variable occurs in this clause. Lichtenstein [11] gives a polynomial time algorithm that computes for every 3-CNF formula F a 3-CNF formula F' such that (1) formula F is satisfiable iff formula F' is satisfiable, (2) the associated bipartite graph $H(F')$ is planar, and (3) Formula F' has $O((\mu(F))^2)$ clauses.

This immediately gives a parameter preserving polynomial time reduction from $\mathsf{3SAT}_\mu$ to the following $\mathsf{planar\ 3SAT}_{\mu'}$.

> Instance: Exact 3SAT formula F such that the graph $H(F)$ is planar.
> Parameter: $\mu'(F) = \lceil \sqrt{m} \rceil$ where m is the number of clauses of F.
> Problem: Decide whether F is satisfiable.

Moreover, it is shown in [11] that we can restrict to instances F of $\mathsf{planar\ 3SAT}_{\mu'}$ where the bipartite graph $H(F)$ has a drawing satisfies the following conditions: Every vertex of $H(F)$ that represents a variable in F lies on a horizontal line, no edge crosses the horizontal line, and no vertex representing a clause lies on the horizontal line. Hence $\mathsf{planar\ 3SAT}_{\mu'}$ with these properties is S[1]-hard.

Thus, it suffices to give a polynomial time reduction from this restricted version of $\mathsf{planar\ 3SAT}_{\mu'}$ to NCST_κ . We remark that this reduction was also given in [7] in the context of NP-hardness and approximation hardness.

Our reduction maps a given instance F of $\mathsf{planar\ 3SAT}_{\mu'}$ to a geometric graph G_F such that G_F has a crossing-free spanning tree iff F is satisfiable. The overall structure of G_F is indicated in the left picture of Figure 2 for $F = (x_1 \vee x_2 \vee \overline{x}_4) \wedge (x_2 \vee \overline{x}_3 \vee x_4) \wedge (x_1 \vee x_3 \vee x_4)$.

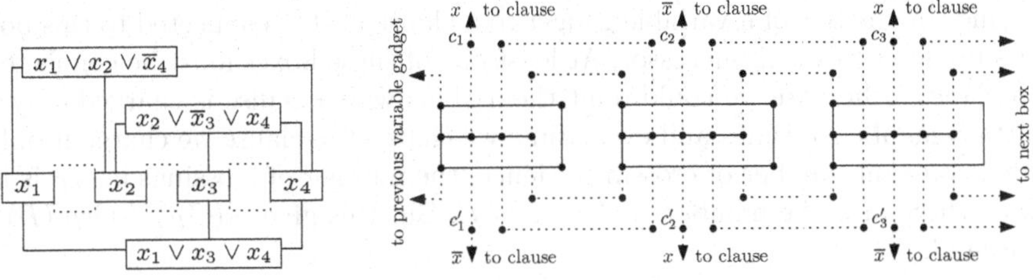

Fig. 2. Overall structure of G_F (left), and a part of a variable gadget (right)

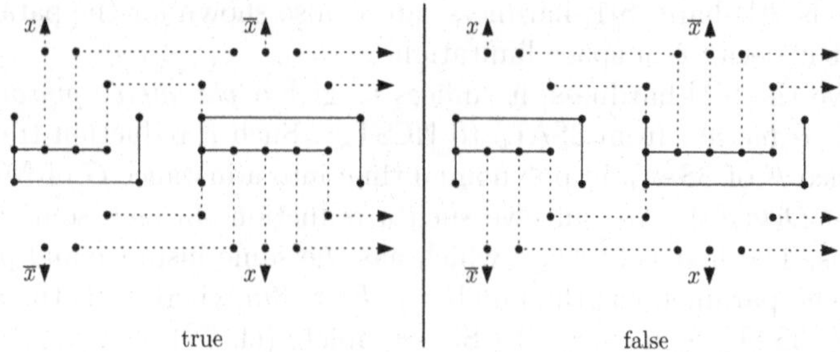

Fig. 3. Spanning trees encoding true and false for a variable

We have a gadget for every variable occurring in F. These gadgets are arranged along a horizontal line ℓ. We further have a gadget for every clause in F which is connected with every variable occurring in the clause. This gadget looks like a three-legged comb.

Now let's have a closer look at the gadgets. The leftmost part of the gadget for a variable x is shown as the right picture in Figure 2. The gadget for x consists of at most twice as many boxes as there are clauses in F that contain x. Three of these boxes are drawn with solid edges in Figure 2. The dotted edges that emanate from the boxes fulfill three tasks. First they connect consecutive boxes within one variable gadget. Second they connect the first and last box of variable gadgets that are consecutive on the line ℓ. Third they connect boxes to clause gadgets. Each dotted edge that connects a variable gadget to a clause gadget is associated with a literal. This literal will be true if the dotted edge is part of the spanning tree of G_F.

The intended way of simulating the truth assignment of the variable x is indicated in Figure 3. The Boolean values of x correspond to the two ways in which a crossing-free spanning tree can be chosen among the edges of the gadget of x. Note that only every other box can be connected to a clause gadget above (below) ℓ. This way we ensure that according to the value of x either only the dotted edges associated to positive literals or only the dotted edges associated to negative literals can connect x to clause gadgets. Not all points of type c_i or c_i' in a variable gadget are used—only those where the variable is in fact connected to a clause gadget in G_F. A clause gadget is just a vertex of degree three connecting to the corresponding literals.

It remains to argue that our reduction is parameter preserving. We charge the crossings in one box of a variable gadget to a clause that is connected to this box or its predecessor or its successor. At least one of these boxes must be connected to a clause, otherwise we could omit them. This way a clause is charged only a constant number of times and every time we charge the clause we charge it only with a constant number of crossings. Hence, the number of crossings in G_F is in $O(m)$ where m is the number of clauses of F. But this gives $\kappa(G_F) \in O(\mu'(F))$, as desired.

6 Concluding Remarks

As we have claimed in the introduction, we can apply our method to several other problems such as non-crossing s–t paths and cycles. We can also deal with the optimization problems, minimizing either the number of components in a non-crossing spanning forest or the number of crossing edges in a spanning tree. These extensions will be given in the full paper.

Acknowledgement. The authors gratefully acknowledge Alexander Wolff for his valuable suggestions and help in organizing this joint research.

References

1. Deineko, V.G., Klinz, B., Woeginger, G.J.: Exact algorithms for the Hamilton cycle problem in planar graphs. Inf. Process. Lett. 34, 269–274 (2006)
2. Flum, J., Grohe, M.: Parameterized Complexity Theory. Springer, Heidelberg (2006)
3. Guo, J., Hüffner, F., Niedermeier, R.: A structural view on parameterizing problems: Distance from triviality. In: Downey, R.G., Fellows, M.R., Dehne, F. (eds.) IWPEC 2004. LNCS, vol. 3162, pp. 162–173. Springer, Heidelberg (2004)
4. Hoffmann, M., Okamoto, Y.: The minimum weight triangulation problem with few inner points. Computational Geometry 34(3), 149–158 (2006)
5. Impagliazzo, R., Paturi, R., Zane, F.: Which problems have strongly exponential complexity. Journal of Comput. Syst. Sci. 63, 512–530 (2001)
6. Jansen, K., Woeginger, G.J.: The complexity of detecting crossingfree configurations in the plane. BIT 33, 580–595 (1993)
7. Knauer, C., Schramm, É., Spillner, A., Wolff, A.: Spanning trees with few crossings in geometric and topological graphs. In: Proc. European Workshop on Computational Geometry, pp. 195–198 (2005)
8. Knauer, C., Schramm, É., Spillner, A., Wolff, A.: Configurations with few crossings in topological graphs. Computational Geometry 37(2), 104–114 (2007)
9. Knauer, C., Spillner, A.: A fixed-parameter algorithm for the minimum weight triangulation problem based on small graph separators. In: Fomin, F.V. (ed.) WG 2006. LNCS, vol. 4271, pp. 49–57. Springer, Heidelberg (2006)
10. Kratochvíl, J., Lubiw, A., Nešetřil, J.: Noncrossing subgraphs in topological layouts. SIAM J. Disc. Math. 4(2), 223–244 (1991)
11. Lichtenstein, D.: Planar formulae and their uses. SIAM J. Computing 11, 329–343 (1982)
12. Miller, G.L.: Finding small simple cycle separators for 2-connected planar graphs. J. Comput. Syst. Sci. 32, 265–279 (1986)
13. Spillner, A.: A faster algorithm for the minimum weight triangulation problem with few inner points. In: Proc. Workshop Algorithms and Complexity in Durham (ACID'05), pp. 135–146. KCL Publications (2005)

Improved Algorithms for the Feedback Vertex Set Problems*,**

Jianer Chen[1], Fedor V. Fomin[2], Yang Liu[1], Songjian Lu[1], and Yngve Villanger[2]

[1] Department of Computer Science, Texas A&M University, College Station, Texas 77843-3112, USA
{chen,yangliu,sjlu}@cs.tamu.edu

[2] Department of Informatics, University of Bergen, N-5020 Bergen, Norway
{fomin,yngvev}@ii.uib.no

Abstract. We present improved parameterized algorithms for the Feedback Vertex Set problem on both unweighted and weighted graphs. Both algorithms run in time $O(5^k kn^2)$. The algorithms construct a feedback vertex set of size bounded by k (in the weighted case this set is of minimum weight among the feedback vertex set of size at most k) in a given graph G, or reports that no such a feedback vertex set exists in G.

1 Introduction

Let G be a graph. A *feedback vertex set* (FVS) F in G is a set of vertices in G whose removal results in an acyclic graph (or equivalently, every cycle in G contains at least one vertex in F). The problem of finding a minimum feedback vertex set in a graph is one of the classical NP-complete problems [13] and has many applications. The history of the problem can be traced back to the early '60s. For several decades, many different algorithmic approaches were tried on this problem, including approximation algorithms, linear programming, local search, polyhedral combinatorics, and probabilistic algorithms (see the survey of Festa et al. [9]). There are also exact algorithms finding a minimum FVS in a graph on n vertices in time $O(1.9053^n)$ [16] and in time $O(1.7548^n)$ [10].

An important application of the FVS problem is *deadlock recovery* in operation systems [18], in which a deadlock is presented by a cycle in a *system resource-allocation graph* G. Therefore, in order to recover from deadlocks, we need to abort a set of processes in the system, i.e., to remove a set of vertices in the graph G, so that all cycles in G are broken. Equivalently, we need to find an FVS in G. The problem also has a version on weighted graphs, where the weight of a vertex can be interpreted as the cost of aborting the corresponding process. In this case, we are looking for an FVS in G whose weight is minimized.

* This work was supported in part by the National Science Foundation of USA under the Grant CCF-0430683.

** This work is supported by the Research Council of Norway.

F. Dehne, J.-R. Sack, and N. Zeh (Eds.): WADS 2007, LNCS 4619, pp. 422–433, 2007.

In a practical system resource-allocation graph G, it can be expected that the size k of the minimum FVS in G, i.e., the number of vertices in the FVS, is fairly small. This motivated the study of *parameterized algorithms* for the FVS problem that find an FVS of k vertices in a graph of n vertices (where the weight of the FVS is minimized, in the case of weighted graphs), and run in time $f(k)n^{O(1)}$ for a fixed function f (thus, the algorithms become practically efficient when the value k is small).

This line of research has received considerable attention, most are on the FVS problem on unweighted graphs. The first group of parameterized algorithms of running time $f(k)n^{O(1)}$ for the FVS problem on unweighted graphs was given by Bodlaender [3] and by Downey and Fellows [7]. Since then a chain of dramatic improvements was obtained by different researchers (see Figure 1 for references.)

Bodlaender, Fellows [3,7]	$O(17k^4!n^{O(1)})$
Downey and Fellows [8]	$O((2k+1)^k n^2)$
Raman et al.[15]	$O(\max\{12^k, (4\log k)^k\}n^{2.376})$
Kanj et al.[12]	$O((2\log k + 2\log\log k + 18)^k n^2)$
Raman et al.[14]	$O((12\log k/\log\log k + 6)^k n^{2.376})$
Guo et al.[11]	$O((37.7)^k n^2)$
Dehne et al.[6]	$O((10.6)^k n^3)$

Fig. 1. The history of parameterized algorithms for the unweighted FVS problem

Our results. In this paper we use the technique of iterative compression that was already applied for several similar problems including the FVS problem [6,11,17]. The novel part of our approach is the new recursive procedure and its analysis which allow us to reduce the running time of the algorithm significantly. We show that the problem of finding an FVS of size k of minimum weight in a weighted graph G of n vertices can be solved in time $O(5^k kn^2)$. This improves and generalizes a long chain of results in parameterized algorithms.

We remark that randomized parameterized algorithms have also been studied in the literature for the FVS problem, for both unweighted and weighted graphs. The best known randomized parameterized algorithms for the FVS problems are due to Becker et al. [2], who developed a randomized algorithm of running time $O(4^k kn^2)$ for the FVS problem on unweighted graphs, and a randomized algorithm of running time $O(6^k kn^2)$ for the FVS problem on weighted graphs. Compared to these results, the running time of our (deterministic) algorithm comes close to that of the best randomized algorithm for the FVS problem on unweighted graphs, while our (deterministic) algorithm has been better than the previous best randomized algorithm for the FVS problem on weighted graphs.

Due to space limitations the proofs of Lemmas 4, 5 and Theorem 2 have been omitted. For a full version see [4].

2 On Feedback Vertex Sets in Unweighted Graphs

In this section, we consider the FVS problem on unweighted graphs.

We start with some terminologies. A *forest* is a graph that contains no cycles. A *tree* is a forest that is connected (therefore, a forest can be equivalently defined as a collection of disjoint trees). Let W be a subset of vertices in a graph G. We will denote by $G[W]$ the subgraph of G that is induced by the vertex set W. A pair (V_1, V_2) of vertex subsets in a graph $G = (V, E)$ is a *forest bipartition* of G if $V_1 \cup V_2 = V$, $V_1 \cap V_2 = \emptyset$, and both induced subgraphs $G[V_1]$ and $G[V_2]$ are forests.

Let G be a graph and let F be a subset of vertices in G. The set F is a *feedback vertex set* (shortly, FVS) of G if $G - F$ is a forest (or equivalently, if every cycle in G contains at least one vertex in F). The *size* of an FVS F is the number of vertices in F.

Our main problem is formally defined as follows.

> FEEDBACK VERTEX SET: given a graph G and an integer k, either find an FVS of size bounded by k in G, or report that no such an FVS exists.

Before we present our algorithm for the FEEDBACK VERTEX SET problem, we first consider a special version of the problem, defined as follows:

> F-BIPARTITION FVS: given a graph G, a forest bipartition (V_1, V_2) of G, and an integer k, either find an FVS of size bounded by k for the graph G *in the subset* V_1, or report that no such an FVS exists.

Note that the main difference between the F-BIPARTITION FVS problem and the original FEEDBACK VERTEX SET problem is that we require that the FVS in the F-BIPARTITION FVS is contained in the given subset V_1.

A *bypass* operation will be used heavily in our process. Let w be a degree-2 vertex with two neighbors u and v in a graph G. We say that a graph G' is obtained from G by *bypassing* the degree-2 vertex w if G' is obtained from G by first removing the vertex w then adding a new edge between u and v.

The algorithm, **Feedback**(G, V_1, V_2, k), for the F-BIPARTITION FVS problem is given in Figure 2. We first discuss the correctness of the algorithm. The correctness of step 1 and step 2 of the algorithm is obvious. Now consider step 3. Let w be a vertex in V_1 that has at least two neighbors in V_2.

If the vertex w has two neighbors in V_2 that belong to the same tree T in the induced subgraph $G[V_2]$, then the tree T plus the vertex w contains at least one cycle. Since we are restricted to find an FVS in the vertex subset V_1, the only way to break the cycles in $T + w$ is to include the vertex w in the objective FVS. Moreover, the objective FVS of size bounded by k exists in G if and only if the remaining graph $G - w$ has an FVS of size bounded by $k - 1$ in the subset $V_1 - w$ (note that $(V_1 - w, V_2)$ is a valid forest bipartition of the graph $G - w$). Therefore, step 3.1 correctly handles this case.

If no two neighbors of the vertex w belong to the same tree in the induced subgraph $G[V_2]$, then the vertex w is either in the objective FVS or not in the

```
Algorithm-1 Feedback(G, V1, V2, k)
Input: G = (V, E) is a graph with a forest bipartition (V1, V2), k is an integer
Output: An FVS F of G such that |F| ≤ k and F ⊆ V1; or report "No" (i.e.,
        no such an FVS)
1.    if (k < 0) or (k = 0 and G is not a forest) then return "No";
2.    if (k ≥ 0) and G is a forest then return ∅;
3.    if a vertex w in V1 has at least two neighbors in V2 then
3.1.      if two of the neighbors of w in V2 belong to the same tree in G[V2]
          then F' = Feedback(G − w, V1 − w, V2, k − 1);
               if F' = "No" then return "No" else return F' + w;
3.2.      else F1 = Feedback(G − w, V1 − w, V2, k − 1);
               F2 = Feedback(G, V1 − w, V2 + w, k);
               if F1 ≠ "No" then return F1 + w
               else if F2 ≠ "No" then return F2
               else return "No";
4.    else pick any vertex w that has degree ≤ 1 in G[V1];
4.1.      if w has degree ≤ 1 in the original graph G
          then return Feedback(G − w, V1 − w, V2, k)
4.2.      else let Gw be the graph obtained from G by bypassing w;
               return Feedback(Gw, V1 − w, V2, k)
```

Fig. 2. Algorithm for unweighted FVS problem

objective FVS. If w is in the objective FVS, then we should be able to find an FVS F_1 in the graph $G - w$ such that $|F_1| \leq k - 1$ and $F_1 \subseteq V_1 - w$ (again note that $(V_1 - w, V_2)$ is a valid forest bipartition of the graph $G - w$). On the other hand, if w is not in the objective FVS, then the objective FVS for G must be contained in the subset $V_1 - w$. Also note that in this case, the subgraph $G[V_2 + w]$ induced by the subset $V_2 + w$ is still a forest since no two neighbors of w in V_2 belong to the same tree in $G[V_2]$. In consequence, $(V_1 - w, V_2 + w)$ still makes a valid forest bipartition for the graph G. Therefore, step 3.2 handles this case correctly.

Now we consider step 4. At this point, every vertex in V_1 has at most one neighbor in V_2. Moreover, since the induced subgraph $G[V_1]$ is a forest, there must be a vertex w in V_1 that has degree bounded by 1 in $G[V_1]$ (note that V_1 cannot be empty at this point since otherwise the algorithm would have stopped at step 2). If the vertex w also has degree bounded by 1 in the original graph G, then removing w does not help breaking any cycles in G. Therefore, the vertex w can be discarded. This case is correctly handled by step 4.1. Otherwise, the vertex w has degree bounded by 1 in the induced subgraph $G[V_1]$ but has degree larger than 1 in the original graph G. Observing also the fact that w has at most one neighbor in V_2, we can easily derive in this case that the degree of w in the original graph G must be 2, and that w has two neighbors u_1 and u_2 such that u_1 is in V_1 and u_2 is in V_2. Therefore, if w is in the objective FVS F, then the set $F' = F - w + u_1$ will also make a valid solution to the given problem instance. Thus, by bypassing the degree-2 vertex w in G, we obtain a graph G_w, with the forest bipartition $(V_1 - w, V_2)$, such that G_w has an FVS of size bounded by k in $V_1 - w$ if and only if the original graph G has an FVS of size bounded by k in V_1. In conclusion, step 4.2 correctly handles this case.

Now we are ready to present the following lemma.

Lemma 1. *The algorithm* **Feedback**(G, V_1, V_2, k) *solves the* F-BIPARTITION FVS *problem correctly. The running time of the algorithm is bounded by $O(2^{k+l}n^2)$, where n is the number of vertices in the graph G, and l is the number of connected components (i.e. trees) in the induced subgraph $G[V_2]$.*

Proof. The correctness of the algorithm has been verified by the above discussion. Now we consider the complexity of the algorithm.

The recursive execution of the algorithm can be described as a search tree $\mathcal{T}$. We first count the number of leaves in the search tree $\mathcal{T}$. Note that only step 3.2 of the algorithm corresponds to branches in the search tree $\mathcal{T}$. Let $T(k, l)$ be the total number of leaves in the search tree $\mathcal{T}$ for the algorithm **Feedback**(G, V_1, V_2, k), where l is the number of trees in the forest $G[V_2]$. Inductively, the number of leaves in the search tree $\mathcal{T}_1$ corresponding to the recursive call **Feedback**$(G - w, V_1 - w, V_2, k - 1)$ is bounded by $T(k - 1, l)$. Moreover, we assume at step 3.2 that w has at least two neighbors in V_2 and that no two neighbors of w in V_2 belong to the same tree in $G[V_2]$. Therefore, the vertex w "merges" at least two trees in $G[V_2]$ into a single tree in $G[V_2 + w]$. Hence, the number of trees in $G[V_2 + w]$ is bounded by $l - 1$. In consequence, the number of leaves in the search tree $\mathcal{T}_2$ corresponding to the recursive call **Feedback**$(G, V_1 - w, V_2 + w, k)$ is bounded by $T(k, l - 1)$. This gives the following recurrence relation:

$$T(k, l) \leq T(k - 1, l) + T(k, l - 1)$$

It is easy to derive from this relation that $T(k, l) = O(2^{k+l})$. Finally, observe that along each root-leaf path in the search tree $\mathcal{T}$, the total number of executions of steps 1, 2, 3.1, and 4 of the algorithm is bounded by $O(n)$ because each of these steps either stops immediately, or reduces the input graph size by at least 1. Moreover, it is also easy to verify that each of these steps takes time $O(n)$.

Therefore, the computation time along each root-leaf path in the search tree $\mathcal{T}$ is bounded by $O(n^2)$. In conclusion, the running time of the algorithm **Feedback**(G, V_1, V_2, k) is bounded by $O(2^{k+l}n^2)$. This completes the proof of the theorem. □

Following the idea of *iterative compression* proposed by Reed et al. [17], we formulate the following problem:

> FVS REDUCTION: given a graph G and an FVS F of size $k + 1$ for G, either construct an FVS of size bounded by k for G, or report that no such an FVS exists.

Lemma 2. *The* FVS REDUCTION *problem can be solved in time $O(5^k n^2)$.*

Proof. We use the algorithm **Feedback** to solve the FVS REDUCTION problem. Let F be the FVS of size $k + 1$ in the graph $G = (V, E)$. Every FVS F' of size bounded by k for G is a union of a subset F_1 of at most $k-j$ vertices in $V-F$ and a subset F_2 of j vertices in F, for some integer j, $0 \le j \le k$. Note that since we assume that no vertex in $F-F_2$ is in the FVS F', the induced subgraph $G[F-F_2]$ must be a forest. Therefore, for each j, $0 \le j \le k$, we enumerate all subsets of j vertices in F. For each such a subset F_2 in F such that $G[F-F_2]$ is a forest, we seek a subset F_1 of at most $k-j$ vertices in $V-F$ such that $F_1 \cup F_2$ makes an FVS for the graph G.

Fix a subset F_2 in F, where $|F_2| = j$. Note that the graph G has an FVS $F_1 \cup F_2$ of size bounded by k, where $F_1 \subseteq V-F$, if and only if the subset F_1 of $V-F$ is an FVS for the graph $G-F_2$ and the size of F_1 is bounded by $k-j$. Therefore, to solve the original problem, we can instead consider how to construct an FVS F_1 for the graph $G-F_2$ such that $|F_1| \le k-j$ and $F_1 \subseteq V-F$.

Since F is an FVS for G, we have that the induced subgraph $G[V-F] = G-F$ is a forest. Moreover, by our assumption, the induced subgraph $G[F-F_2]$ is also a forest. Note that $(V-F) \cup (F-F_2) = V-F_2$, which is the vertex set for the graph $G' = G-F_2$. Therefore, $(V-F, F-F_2)$ is a forest bipartition of the graph G'. Thus, an FVS F_1 for the graph G' such that $|F_1| \le k-j$ and $F_1 \subseteq V-F$ can be constructed by the algorithm **Feedback**$(G', V-F, F-F_2, k-j)$.

Since $|F| = k+1$ and $|F_2| = j$, we have that $|F-F_2| = k+1-j$. Therefore, the forest $G[F-F_2]$ contains at most $k+1-j$ trees. By Lemma 1, the running time of the algorithm **Feedback**$(G', V-F, F-F_2, k-j)$ is bounded by $O(2^{(k-j)+(k+1-j)}n^2) = O(4^{k-j}n^2)$. Now for all integers j, $0 \le j \le k$, we enumerate all subsets F_2 of j vertices in F and apply the algorithm **Feedback**$(G', V-F, F-F_2, k-j)$ for those F_2 such that $G[F-F_2]$ is a forest. As we discussed above, the graph G has an FVS of size bounded by k if and only if for some $F_2 \subseteq F$, the algorithm **Feedback**$(G', V-F, F-F_2, k-j)$ produces an FVS F_1 for the graph G'. The running time of this process is bounded by

$$\sum_{j=0}^{k} \binom{k+1}{j} \cdot O(4^{k-j}n^2) = \sum_{j=0}^{k} \binom{k+1}{k-j+1} O(4^{k-j+1}n^2) = O(5^k n^2).$$

This completes the proof of the lemma. □

Finally, by combining Lemma 2 with iterative compression, we obtain the main result of this section.

Theorem 1. *The* FEEDBACK VERTEX SET *problem is solvable in time* $O(5^k kn^2)$.

Proof. To solve the FEEDBACK VERTEX SET problem, for a given graph $G = (V, E)$, we start by applying Bafna et al.'s 2-approximation algorithm for the MINIMUM FEEDBACK VERTEX SET problem [1]. This algorithm runs in $O(n^2)$ time, and either returns an FVS F' of size bounded by $2k$, or verifies that no FVS of size bounded by k exists. If no FVS is returned, the algorithm is terminated with the result that no FVS of size bounded by k exists. In the case of the opposite result, we use any subset $V' \subseteq F'$ of k vertices, and let $V_0 = V' \cup (V-F')$. Of course, the induced subgraph $G[V_0]$ has an FVS of size k, since $G[V_0 - V']$ is a forest. Let $F' - V_0 = \{v_1, v_2, \ldots, v_{|F'|-k}\}$, and let

$V_i = V_0 \cup \{v_1, \ldots, v_i\}$ for $i = 0, 1, \ldots, |F'| - k$. Inductively, suppose that we have constructed an FVS F_i for the graph $G[V_i]$, where $|F_i| = k$. Then the set $F'_{i+1} = F_i + v_{i+1}$ is obviously an FVS for the graph $G[V_{i+1}]$ and $|F'_{i+1}| = k + 1$.

Now the pair $(G[V_{i+1}], F'_{i+1})$ is an instance for the FVS REDUCTION problem. Therefore, in time $O(5^k n^2)$, we can either construct an FVS F_{i+1} of size k for the graph $G[V_{i+1}]$, or report no such an FVS exists. Note that if the graph $G[V_{i+1}]$ does not have an FVS of size k, then the original graph G cannot have an FVS of size k. In this case, we simply stop and claim the non-existence of an FVS of size k for the original graph G. On the other hand, with an FVS F_{i+1} of size k for the graph $G[V_{i+1}]$, our induction proceeds to the next graph $G[V_{i+1}]$, until we reach the graph $G = G[V_{|F'|-k}]$. Clearly, this process runs in time $O(5^k kn^2)$ since $|F'| - k \leq k$, and solves the FEEDBACK VERTEX SET problem. □

3 On Feedback Vertex Sets in Weighted Graphs

In this section, we discuss the FVS problem on weighted graphs. A weighted graph $G = (V, E)$ is an undirected graph, where each vertex $u \in V$ is assigned a non-negative weight. The weight of a vertex set $A \subseteq V$ is the sum of the vertex weights of all vertices in A. We denote by $|A|$ the size of, i.e., the number of vertices in, the set A . The (parameterized) feedback vertex set problem on weighted graphs is formally defined as follows:

> WEIGHTED-FVS: given a weighted graph G and an integer k, either find an FVS F of minimum weight for G such that $|F| \leq k$, or report that no FVS of size bounded by k exists in G.

Our algorithm for the weighted case has several similarities with the unweighted case, but also has a significant difference. The difference is that the bypass operation for unweighted graphs can no longer be used in the weighted case. Indeed, a degree-2 vertex in a weighted graph may be necessarily included in the objective FVS if its weight is very small.

On the other hand, if two degree-2 vertices v and w are adjacent, then we can always bypass the one with a larger weight. This is because every cycle in the graph either contains both v and w or contains neither of them, so we can always assume that the one with a larger weight is not included in the objective FVS. We call this operation that bypasses the vertex of a larger weight in two adjacent degree-2 vertices in a weighted graph the *restricted bypass* operation.

However, since the restricted bypass operation cannot guarantee to eliminate all degree-2 vertices in a weighted graph, step 4.2 in the algorithm **Feedback** is not always possible. To overcome this difficulty, we introduce a new partition structure of the vertices in a weighted graph.

A triple (V_0, V_1, V_2) is a *independent-forest partition* (briefly, an *IF-partition*) of a graph $G = (V, E)$ if (V_0, V_1, V_2) is a partitioning of V such that (1) $G[V_1]$ and $G[V_2]$ are forests; (2) $G[V_0]$ is an independent set; and (3) every vertex $u \in V_0$ is of degree 2 in G, and all neighbors of u are in V_2.

We consider the following problem on weighted graphs that is similar to the F-BIPARTITION FVS problem on unweighted graphs.

WEIGHTED IF-PARTITION FVS: given a weighted graph G, an IF-partition (V_0, V_1, V_2) of G, and an integer k, either find an FVS F of minimum weight for G that satisfies the conditions $|F| \leq k$ and $F \subseteq V_0 \cup V_1$, or report that no such an FVS exists.

To study WEIGHTED IF-PARTITION FVS, we introduce the following concept.

Definition 1. Let (G, k) be an instance of WEIGHTED IF-PARTITION FVS, where an IF-partition (V_0, V_1, V_2) of G is given. The *deficiency* of (G, k) is defined by

$$\tau(k, V_0, V_1, V_2) = k - (|V_0| - \#c(V_2) + 1)$$

where $\#c(V_2)$ is the number of connected components in the subgraph $G[V_2]$.

Intuitively, the deficiency $\tau(k, V_0, V_1, V_2)$ of the instance (G, k) is the maximum number of vertices in the objective FVS that are in the set V_1 (this will become clearer during our discussion below). Our algorithm for the WEIGHTED IF-PARTITION FVS problem is based on the following observation: once we have correctly determined all vertices in the objective FVS that are in the set V_1, the problem will become solvable in polynomial time, as shown in the following lemma.

Lemma 3. *Let* (G, k) *be an instance of* WEIGHTED IF-PARTITION FVS, *with an IF-partition* (V_0, V_1, V_2) *of* G. *If* $V_1 = \emptyset$ *or* $\tau(k, V_0, V_1, V_2) \leq 0$, *then the solution to the instance* (G, k) *can be constructed in time* $O(n^2)$.

Proof. Construct a new graph $\mathcal{H} = (\mathcal{V}, \mathcal{E})$, where each vertex μ in $\mathcal{V}$ corresponds to a connected component in the induced subgraph $G[V_2]$, and each edge $[\mu, \nu]$ in $\mathcal{E}$ corresponds to a vertex v in the set V_0 such that the two edges incident to v in G are connected to the connected components in $G[V_2]$ that correspond to the two vertices μ and ν, respectively, in $\mathcal{H}$. Equivalently, the graph $\mathcal{H}$ can be obtained from the induced subgraph $G[V_0 \cup V_2]$ by shrinking each connected component in $G[V_2]$ into a single vertex and bypassing each degree-2 vertex in V_0. Moreover, we give each edge in $\mathcal{H}$ a weight that is equal to the weight of the corresponding vertex in V_0. Thus, the graph $\mathcal{H}$ is a graph with edge weights.

First consider the case of $V_1 = \emptyset$. If $k < 0$ then the solution to (G, k) is "No": we cannot remove a negative number of vertices. Assuming $k \geq 0$. Then we need to find a minimum-weight subset of at most k vertices in the set V_0 whose removal from $G = G[V_0 \cup V_2]$ results in an acyclic graph. Note that removing vertices in V_0 in the graph G corresponds to removing edges in the graph $\mathcal{H}$. Therefore, this problem is equivalent to finding a minimum-weight subset of at most k edges in the graph $\mathcal{H}$ whose removal from $\mathcal{H}$ results in an acyclic graph (note that each connected component in $G[V_2]$ is a tree). Let $\mathcal{H}_1$, ..., $\mathcal{H}_s$ be the connected components of the graph $\mathcal{H}$, where for each i, the component $\mathcal{H}_i$ has n_i vertices and m_i edges. In order to get an acyclic graph from $\mathcal{H}$, it is necessary and sufficient to remove $m_i - n_i + 1$ edges from $\mathcal{H}_i$ for each i (i.e., to make each connected component in $\mathcal{H}$ a tree). In consequence, in order to get an acyclic graph from $\mathcal{H}$, it is necessary and sufficient to remove $\sum_i^s(m_i - n_i + 1) = |\mathcal{E}| - |\mathcal{V}| + s$ edges from the graph $\mathcal{H}$.

Correspondingly, in case $V_1 = \emptyset$, a minimum-weight FVS in V_0 for the graph G contains exact $|\mathcal{E}| - |\mathcal{V}| + s$ vertices. Note that $|\mathcal{E}| = |V_0|$, and $|\mathcal{V}|$ is equal to the number $\#c(V_2)$ of connected components in the induced subgraph $G[V_2]$. Thus, every FVS in the graph G contains at least $|V_0| - \#c(V_2) + s$ vertices. Therefore, if $\tau(k, V_0, V_1, V_2) = k - (|V_0| - \#c(V_2) + 1) < 0$, or $\tau(k, V_0, V_1, V_2) = 0$ but $s > 1$, then we have $k < |V_0| - \#c(V_2) + s$. That is, the graph G has no FVS of size bounded by k and the solution to the instance (G, k) is a "No".

The remaining case is that $s = 1$, and $\tau(k, V_0, V_1, V_2) = k - (|V_0| - \#c(V_2) + 1) = 0$. In this case, to find a minimum-weight FVS of k vertices in V_0, we construct a maximum-weight spanning tree in the graph $\mathcal{H}$ (this can be done by a modified minimum spanning tree algorithm of time $O(n^2)$ [5]). The remaining $|\mathcal{E}| - |\mathcal{V}| + 1 = |V_0| - \#c(V_2) + 1 = k$ edges in $\mathcal{H}$ then correspond to k vertices in the set V_0 that make a minimum-weight FVS for the graph G. Summarizing the above discussion, we conclude that if $V_1 = \emptyset$, then the solution to the instance (G, k) can be constructed in time $O(n^2)$.

Now consider the case $\tau(k, V_0, V_1, V_2) \leq 0$. As shown above, even to break all cycles in the induced subgraph $G[V_0 \cup V_2]$ requires removing at least $|V_0| - \#c(V_2) + 1$ vertices in the set V_0. Therefore, if $\tau(k, V_0, V_1, V_2) \leq 0$, then $k \leq (|V_0| - \#c(V_2) + 1)$, and all k vertices in the objective FVS must be in the set V_0 in order to break all cycles in the induced subgraph $G[V_0 \cup V_2]$, and no vertex in the objective FVS can be in the set V_1. Hence, if the induced subgraph $G[V_1 \cup V_2]$ contains a cycle, then the solution to (G, k) is a "No". On the other hand, suppose that $G[V_1 \cup V_2]$ is a forest, then the graph G has another IF-partition (V_0', V_1', V_2'), where $V_0' = V_0$, $V_1' = \emptyset$, and $V_2' = V_1 \cup V_2$. It is easy to verify that in this case the instance (G, k) with the IF-partition (V_0', V_1', V_2') has the same solution set as the same instance with the IF-partition (V_0, V_1, V_2). Now since $V_1' = \emptyset$, by the first part of this lemma, the solution to (G, k) with the IF-partition (V_0', V_1', V_2') can be constructed in time $O(n^2)$. This completes the proof of the lemma. □

Now we are ready for our main algorithm, which is given in Figure 3 and solves the WEIGHTED IF-PARTITION FVS problem. As explained for the unweighted case, vertices of degree less than 2 cannot contribute to the objective FVS, thus can be directly deleted. Moreover, each restricted bypass operation takes time $O(n)$ and eliminates a degree-2 vertex in a pair of adjacent degree-2 vertices. Therefore, we can perform a preprocessing of time $O(n^2)$ and assume that the input graph G of the algorithm contains no vertex of degree less than 2, and no two adjacent degree-2 vertices. Moreover, for each tree in the forest $G[V_1]$, we fix a root so that we can talk about the "lowest leaf" in a tree in $G[V_1]$.

We first discuss the correctness of the algorithm. Step 1 of the algorithm is justified by Lemma 3. Justifications for steps 2, 3.1, and 3.2 are exactly the same as those for the unweighted case. Now consider step 4. When the algorithm reaches step 4, the following conditions hold: (1) every vertex in G has degree at least 2; (2) there are no two adjacent degree-2 vertices in G; (3) $V_1 \neq \emptyset$; and

Algorithm-1 W-Feedback(G, V_0, V_1, V_2, k)
Input: $G = (V, E)$ is a graph with an IF-partition (V_0, V_1, V_2), k is an integer
Output: a minimum-weight FVS F of G such that $|F| \leq k$ and $F \subseteq V_0 \cup V_1$;
    or report "No" (i.e., no such an FVS).
1. **if** $V_1 = \emptyset$ or $\tau(k, V_0, V_1, V_2) \leq 0$ **then** solve the problem in time $O(n^2)$;
2. **if** $(k < 0)$ or ($k = 0$ and G is not a forest) **then** return "No";
3. **if** a vertex w in V_1 is incident to 2 edges whose other ends are in V_2 **then**
3.1 **if** 2 edges incident to w have their other ends in the same tree in $G[V_2]$
    then return w + **W-Feedback**$(G - w, V_0, V_1 - w, V_2, k - 1)$;
3.2 **else** $F_1 = w$ + **W-Feedback**$(G - w, V_0, V_1 - w, V_2, k - 1)$;
    F_2 = **W-Feedback**$(G, V_0, V_1 - w, V_2 + w, k)$;
    return the one of F_1 and F_2 that has a smaller weight;
4. **else** pick a lowest leaf w_1 in any tree T in $G[V_1]$;
4.1 let w be the parent of w_1 in T, and let $w_1, \ldots, w_t$ be the children
    of w in T, where for each i, w_i has a neighbor v_i in V_2;
4.2 **if** w has a neighbor v in V_2 **then**
4.2.1 **if** for some i, v and v_i are in the same tree in $G[V_2]$
4.2.2 **then** $F_1 = w$ + **W-Feedback**$(G - w, V_0, V_1 - w, V_2, k - 1)$;
    $F_2 = w_i$ + **W-Feedback**$(G - w_i, V_0, V_1 - w_i, V_2, k - 1)$;
    return the one of F_1 and F_2 that has a smaller weight;
4.2.3 **else** $F_1 = w$ + **W-Feedback**$(G - w, V_0, V_1 - w, V_2, k - 1)$;
4.2.4 F_2 = **W-Feedback**$(G, V_0 + w_1 + \cdots + w_t, V_1 - w, V_2 + w, k)$;
    return the one of F_1 and F_2 that has a smaller weight;
4.3 **else** $F_1 = w$ + **W-Feedback**$(G - w, V_0, V_1 - w, V_2, k - 1)$;
    F_2 = **W-Feedback**$(G, V_0 + w_1 + \cdots + w_t, V_1 - w, V_2 + w, k)$;
    return the one of F_1 and F_2 that has a smaller weight;

Fig. 3. Algorithm for weighted FVS problem

(4) every vertex in V_1 is incident to at most one edge whose other end is in V_2. Conditions (1) and (2) hold because of our assumption on the input graph G; condition (3) holds because of step 1; and condition (4) holds because of step 3.

By condition (3) and because the induced subgraph $G[V_1]$ is a forest, step 4 can always pick the vertex w_1. By conditions (1) and (4), the vertex w_1 is adjacent to a unique vertex v_1 in V_2. Then by condition (1) again, w_1 must have a parent w in the tree T in $G[V_1]$. In consequence, the vertex w_1 has degree exactly 2 in the graph G. Finally, since w_1 is the lowest leaf in the tree T, all children $w_1, \ldots, w_t$ of w in T are also leaves in T. By conditions (1) and (4) again, each child w_i of w has a unique neighbor v_i in the set V_2, and every child w_i of w has degree exactly 2 in the graph G.

If the vertex w has a (unique) neighbor v in V_2, and the vertices v and v_i for some i belong to the same tree T' in $G[V_2]$, as given in step 4.2.1, then this tree T' plus the edges $[v_i, w_i]$, $[w_i, w]$, and $[w, v]$ must contain a cycle, and in this cycle all vertices are in V_2 except w_i and w. Therefore, to break this cycle using vertices not in V_2, one of the vertices w and w_i must be included in the objective

FVS. Thus, step 4.2.2 correctly handles this case. On the other hand, suppose that the vertex v is in a tree in $G[V_2]$ that does not contain any of the vertices $v_1, \ldots, v_t$. Then we simply branch on the vertex w: step 4.2.3 includes w in the objective FVS, and step 4.2.4 excludes w from the objective FVS, by moving w from V_1 to V_2. Note that in case of step 4.2.4, each of the degree-2 vertices $w_1, \ldots, w_t$ now is incident to two edges whose other ends are in V_2. Thus, we can correctly move these vertices from V_1 to V_0. Moreover, it is easy to verify that the new partition in each of the cases still makes a valid IF-partition of the graph G.

Finally, suppose that the vertex w has no neighbor in V_2. By condition (2) and because the vertex w_1 has degree 2 in G, the vertex w must have at least two children in the tree T (i.e., $t \geq 2$). Again we branch on w by either including or excluding w in the objective FVS, as given by step 4.3.

Since all possible cases are covered in the algorithm, we conclude that when the algorithm **W-Feedback** stops, it must output a correct solution to the given instance (G, k).

Lemma 4. *The algorithm* **W-Feedback**(G, V_0, V_1, V_2, k) *solves the* WEIGHTED IF-BIPARTITION FVS *problem correctly, and runs in time* $O(2^{\tau(k,V_0,V_1,V_2)}n^2)$, *where* n *is the number of vertices in the graph* G.

With Lemma 4, we can now proceed the same way as for the unweighted case to solve the original WEIGHTED-FVS problem. Consider the following weighted version of the FVS REDUCTION problem.

> WEIGHTED FVS REDUCTION: given a weighted graph G and an FVS F of size $k+1$ for G, either construct an FVS F' of minimum weight that satisfies $|F'| \leq k$, or report that no such an FVS exists.

Note that in the definition of WEIGHTED FVS REDUCTION, we do not require that the given FVS F' of size $k+1$ have the minimum weight.

Lemma 5. *The* WEIGHTED FVS REDUCTION *problem is solvable in time* $O(5^k n^2)$.

Now using Theorem 1 and Lemma 5, we obtain the main result of this paper.

Theorem 2. *The* WEIGHTED-FVS *problem is solvable in time* $O(5^k kn^2)$.

Acknowledgment

The authors wish to thank Hans L. Bodlaender for the suggestion of combining iterative compression and an approximation algorithm, to further reduce the polynomial factor in the complexity of the algorithms.

References

1. Bafna, V., Berman, P., Fujito, T.: A 2-approximation algorithm for the undirected feedback vertex set problem. SIAM J. Discrete Math. 12(3), 289–297 (1999)
2. Becker, A., Bar-Yehuda, R., Geiger, D.: Randomized algorithms for the loop cutset problem. J. Artif. Intell. Res (JAIR) 12, 219–234 (2000)

3. Bodlaender, H.L.: On disjoint cycles. Int. J. Found. Comput. Sci. 5(1), 59–68 (1994)
4. Chen, J., Fomin, F.V., Liu, Y., Lu, S., Villanger, Y.: Improved algorithms for the feedback vertex set problems. Technical Report 348, University of Bergen, Norway (2007)
5. Cormen, T.H., Leiserson, C.E., Rivest, R.L., Stein, C.: Introduction to Algorithms, 2nd edn. The MIT Press and McGraw-Hill Book Company (2001)
6. Dehne, F.K.H.A., Fellows, M.R., Langston, M.A., Rosamond, F.A., Stevens, K.: An $o(2^{o(k)}n^3)$ fpt algorithm for the undirected feedback vertex set problem. In: Wang, L. (ed.) COCOON 2005. LNCS, vol. 3595, pp. 859–869. Springer, Heidelberg (2005)
7. Downey, R.G., Fellows, M.R.: Fixed Parameter Tractability and Completeness. In: Complexity Theory: Current Research, pp. 191–225. Cambridge University Press, Cambridge (1992)
8. Downey, R.G., Fellows, M.R.: Parameterized Complexity. Springer, Heidelberg (1999)
9. Festa, P., Pardalos, P.M., Resende, M.G.C.: Feedback set problems. In: Handbook of combinatorial optimization, Supplement vol. A, pp. 209–258. Kluwer Acad. Publ., Dordrecht (1999)
10. Fomin, F.V., Gaspers, S., Pyatkin, A.V.: Finding a minimum feedback vertex set in time $O(1.7548^n)$. In: Bodlaender, H.L., Langston, M.A. (eds.) IWPEC 2006. LNCS, vol. 4169, pp. 184–191. Springer, Heidelberg (2006)
11. Guo, J., Gramm, J., Hüffner, F., Niedermeier, R., Wernicke, S.: Compression-based fixed-parameter algorithms for feedback vertex set and edge bipartization. J. Comput. Syst. Sci. 72(8), 1386–1396 (2006)
12. Kanj, I.A., Pelsmajer, M.J., Schaefer, M.: Parameterized algorithms for feedback vertex set. In: Downey, R.G., Fellows, M.R., Dehne, F. (eds.) IWPEC 2004. LNCS, vol. 3162, pp. 235–247. Springer, Heidelberg (2004)
13. Karp, R.M.: Reducibility among combinatorial problems. In: Complexity of computer computations, pp. 85–103. Plenum Press, New York (1972)
14. Raman, V., Saurabh, S., Subramanian, C.R.: Faster fixed parameter tractable algorithms for finding feedback vertex sets. ACM Trans. Algorithms 2(3), 403–415 (2006)
15. Raman, V., Saurabh, S., Subramanian, C.R.: Faster fixed parameter tractable algorithms for undirected feedback vertex set. In: Bose, P., Morin, P. (eds.) ISAAC 2002. LNCS, vol. 2518, pp. 241–248. Springer, Heidelberg (2002)
16. Razgon, I.: Exact computation of maximum induced forest. In: Arge, L., Freivalds, R. (eds.) SWAT 2006. LNCS, vol. 4059, pp. 160–171. Springer, Heidelberg (2006)
17. Reed, B., Smith, K., Vetta, A.: Finding odd cycle transversals. Oper. Res. Lett. 32(4), 299–301 (2004)
18. Silberschatz, A., Galvin, P.: Operating System Concepts, 4th edn. Addison-Wesley, London (1994)

Kernelization Algorithms for d-Hitting Set Problems*

Faisal N. Abu-Khzam

Division of Computer Science and Mathematics
Lebanese American University
Beirut, Lebanon
faisal.abukhzam@lau.edu.lb
http://www.csm.lau.edu.lb/fabukhzam

Abstract. A kernelization algorithm for the 3-Hitting-Set problem is presented along with a general kernelization for d-Hitting-Set problems. For 3-Hitting-Set, a quadratic kernel is obtained by exploring properties of yes instances and employing what is known as *crown reduction*. Any 3-Hitting-Set instance is reduced into an equivalent instance that contains at most $5k^2 + k$ elements (or vertices). This kernelization is an improvement over previously known methods that guarantee cubic-size kernels. Our method is used also to obtain a quadratic kernel for the Triangle Vertex Deletion problem. For a constant $d \geq 3$, a kernelization of d-Hitting-Set is achieved by a generalization of the 3-Hitting-Set method, and guarantees a kernel whose order does not exceed $(2d-1)k^{d-1} + k$.

1 Introduction

For a given collection C of subsets of a finite set S, a hitting set is a subset of S that has a nonempty intersection with every element of C. Identifying small hitting sets proved to have a wide range of applications in a variety of domains that include Bioinformatics, Computer Networks as well as Software Testing [14,10,9]. Formally, the decision version of the Hitting Set problem can be defined as follows:

Given: A collection C of subsets of a set S and a parameter k.

Question: Does C have a hitting set of size k or less?

Hitting Set is $\mathcal{NP}$-complete, even if the cardinality of every element of C is bounded by 2 [8], in which case it coincides with the Vertex Cover problem. From the viewpoint of parameterized complexity, Hitting Set is $\mathcal{W}[2]$-hard. However, the problem becomes fixed-parameter tractable ($\mathcal{FPT}$) when the cardinality of every element of C is upper-bounded by a fixed number d. In this latter case, the problem is called d-Hitting-Set (henceforth dHS). Many fixed-parameter

* This research has been supported in part by the Lebanese American University under grant URC-2004-c63.

F. Dehne, J.-R. Sack, and N. Zeh (Eds.): WADS 2007, LNCS 4619, pp. 434–445, 2007.

algorithms appeared that address dHS problems. 3HS, in particular, has been a focal point of attention in recent Hitting Set algorithms [12,7]. This paper is concerned mainly with kernelization methods for the parameterized search version of dHS.

Kernelization has become a major pre-processing tool for solving problems that fall in the class $\mathcal{FPT}$. For an arbitrary instance of an $\mathcal{FPT}$ problem, a kernelization procedure is a polynomial-time pre-processing algorithm that could result in one of the following outputs:

- An early detection of failure;
- An early solution;
- An equivalent instance whose size is bounded by a function of the input parameter.

In the third case, the function is often polynomial in the input parameter(s). The reduced instance is called a problem kernel.

$3HS$ has a cubic-size kernel that was achieved by a clever use of the pigeon hole principle [12]. In this paper, we show how to obtain a quadratic kernel for $3HS$. Our method relies on crown reduction, a fairly new method introduced by the authors of [6] and developed further in [1,2]. Our kernelization can be used to obtain a quadratic kernel for the Triangle Vertex Deletion problem. We also show how our method extends to a d-Hitting Set kernelization.

2 Background

We shall treat the input, (S, C), of a hitting set instance as a hypergraph whose vertices and edges are the elements of S and C respectively. This allows us to use a few known graph-theoretic terminologies. In particular, we use $N_H(A)$ to denote the set of neighbors of A in H, where A is a set of vertices and H is a graph or subgraph. We use the term order to denote the number of vertices (or elements of S). An independent set of a hypergraph is a set of vertices that are pairwise non-adjacent, in the sense that no two of them belong to an element of C. Moreover, a hitting set of (S, C) is a vertex cover of the corresponding hypergraph. We shall say that a subset A of S **hits** a subset C' of C if every element of C' has a non-empty intersection with A. For each vertex $x \in S$, we denote by $E(x)$ the set of all edges containing x. For $E \subset C$, $V(E)$ denotes the set of all vertices that are elements of elements of E.

The first cubic-size (and cubic order) kernelization algorithm for 3HS is due to Niedermeier and Rossmanith [12]. Their algorithm starts by employing the following two pre-processing rules, dubbed **domination rules**, that are due to Weihe [16].

- **Vertex Domination Rule:** If $x, y \in S$ are such that $E(x) \subset E(y)$, then delete x.
- **Edge Domination Rule:** If $e_1, e_2 \in C$ are such that $V(\{e_1\}) \subset V(\{e_2\})$ then delete e_2.

For two vertices x, y of S, the co-occurrence of x and y, denoted $co(x, y)$ is the number of edges in $E(x) \cap E(y)$. This function plays an important role in the kernelization algorithms that appeared in [12] and [3]. In particular, the following pair of "successive" reduction rules were used in [12] to obtain the cubic-size kernel.

1. If $x, y \in S$ are such that $co(x, y) > k$, then delete all elements of $E(x) \cup E(y)$ and add $\{x, y\}$ to C.
2. After applying the above rule, if $|E(x)| > k^2$ for some $x \in S$, then add x to the hitting set, decrement k, and delete all elements of $E(x)$.

These two rules are enough to reduce any arbitrary instance into one whose size is bounded by k^3. We refer the reader to [12] for further details.

2.1 Crown Decomposition

Our kernelization will be based on the notion of crown decomposition, a recent $\mathcal{FPT}$ technique that was introduced by Fellows et al. in [6] and was further studied in conjunction with the Vertex Cover problem in [1] and [2]. In short, a crown decomposition (or just crown) of a simple undirected graph G is a triple (H, I, M) such that:

- I is an independent set of G.
- H is the neighborhood of I in G.
- M is a matching in G that satisfies:
  (i) Every edge of M joins a vertex from H to a vertex of I.
  (ii) Every vertex of H is matched, under M, to a vertex of I.

The set H is called the head of a crown (H, I, M) and $|H|$ is the crown width. Figure 1 illustrates the notion of a crown decomposition.

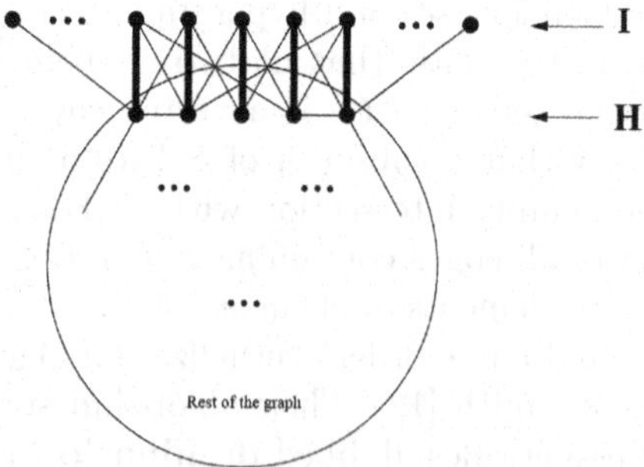

Fig. 1. A crown decomposition of width 5. Bold edges are elements of M.

Deciding whether a graph has a crown decomposition is solvable constructively in polynomial time [2]. The following lemma, adopted from [6], provides a necessary and sufficient condition for a graph to have a crown.

Lemma 1. A graph G has a crown decomposition of positive width if and only if G has an independent set I that satisfies $|N_G(I)| < |I|$.

Once an independent set I that satisfies the condition of Lemma 1 is found in $G = (V, E)$, a crown can be constructed in time $O(|V|^{2.5})$ by a procedure provided also in [6]. We shall refer to this procedure by *Construct_Crown* in this paper. *Construct_Crown* starts by constructing a maximum matching M_1 of the bipartite subgraph $(I \cup N(I), \{e = uv \in E : u \in I\})$, then identifying a crown (H, I', M) such that $H \subset N(I)$, $M \subset M_1$, $I' \subset I$ and (most importantly) $(I - (V(M) \cap I)) \subset I'$.

Remark 1. The condition $(I - (V(M) \cap I)) \subset I'$ plays a very important role in crown reductions, and it is the main reason behind the statement of Lemma 1. In fact, *Construct_Crown* starts by placing all vertices of $(I - (V(M) \cap I))$ (the vertices of I that are not matched under M_1) in I' to make sure that the number of elements of $I - I'$ is equal to the number of edges in M_1 that are not added to M.

2.2 Crown Reduction for Vertex Cover

Let (G, k) be an instance of the Vertex Cover problem and let (H, I, M) be a crown decomposition in G. The crown reduction rule consists of placing all vertices of H in the potential vertex cover of G, thus deleting all elements of $H \cup I$. This rule is sound because of the following observations:

- Every edge of M must be covered by at least one distinct vertex. Therefore, the subgraph G_c induced by $(I \cup H)$ has a vertex cover of size bounded below by $|M|$.
- H is a vertex cover of size $|M|$ of G_c, so H is an optimal cover for the subgraph G_c.
- Using vertices from $H \cup I$ in a global vertex cover is not better than using all vertices of H as the latter ones may also be used to cover other edges of G.

It was shown in [6] that crown decompositions can be used to obtain a linear kernel for Vertex Cover. The reader can verify this easily as follows. Let A be an approximate vertex cover of G. If (G, k) is a yes instance, then the size of A can be guaranteed not to exceed $2k$ [11]. If the complement, $\bar{A}$, of A has more than $|A|$ vertices, then a crown (H, M, I) can be constructed with $H \subset A$, $I \subset \bar{A}$ and $M \subset M_1$ where M_1 is a maximum matching in which every edge joins a vertex from A to one from $\bar{A}$ (see above discussion of *Construct_Crown*). If a solution of size k exists, then $|M_1| \leq k$. Moreover, by Remark 1, $|\bar{A} - I| \leq |M_1 - M| \leq k$. Finally, after applying the crown reduction rule, the resulting vertex set of G is $(A - H) \cup (\bar{A} - I)$.

3 A Quadratic Kernel for 3-Hitting Set

Our kernelization algorithm is based on an extension of the notion of crowns to hypergraphs. A crown in a hypergraph is defined as a tripe (H, I, M) such that:

- I is an independent set.
- $H = \{A \subset S : A \cup \{x\} \in C \text{ for some } x \in I\}$
- M is a matching in the bipartite graph $(I \cup H, E)$ where $E = \{(x, A) : x \in I$ and $A \in H$ and $(\{x\} \cup A) \in C\}$.
- Every element of H is matched under M.

The crown reduction rule for 3HS consists of deleting all elements of I without making any further changes to C and k. The elements of H will automatically become edges of C. The soundness of this rule is based on the following lemma.

Lemma 2. Let (S, C, k) be an instance of 3HS and let (H, I, M) be a crown of (S, C). Then (S, C, k) is equivalent to the instance $(S - I, (C - E(I)) \cup H, k)$.

Proof. Let A be a hitting set of size k of (S, C) and let B be the subset of H that consists of pairs that have non-empty intersection with A. Then every pair $\{y, z\}$ of $H - B$ is a subset of an edge $\{x, y, z\}$ of M, where $x \in A \cap I$ is used to hit $\{x, y, z\}$. Replacing every such x by an element of its matched pair $\{y, z\}$ produces another hitting set whose size is bounded above by $|A|$. (It could be smaller than A when the pairs of $H - B$ have common elements that belong to a solution of (S, C, k)). Q.E.D.

In addition to the use of hypergraph crowns, our kernelization algorithm makes use of the following new reduction rule, which generalizes the high-degree rule used in the Vertex Cover kernelization of [4].

High-degree rule: If $x \in S$ has more than k edges whose pairwise intersection is $\{x\}$, then:

- add x to any potential solution of (S, C, k),
- delete all elements of $E(x)$, and
- decrement k.

We shall see that a crown can be constructed in any 3HS instance as long as its order exceeds $5k^2 + k$. In such cases, Lemma 2 is applied to produce the desired kernel.

Throughout this section, our main focus is on how to obtain an upper bound on the order of our 3HS kernel. It should be clear that exact 3HS algorithms take exponential time. Therefore, we shall omit the details of efficiency analysis whenever it is clear that the run time is polynomial in the input size.

3.1 Weakly Related Edges

Let (S, C, k) be a 3HS instance. Two elements of C are said to be weakly related if they don't share more than one common vertex. A subset W of C is a maximal collection of (pair-wise) weakly related edges if every edge of C is either in W or shares at least two vertices with at least one element of W.

Constructing maximal sets of weakly related edges is simple and can be done by starting with a singleton subset W of C and greedily filling it with elements of C as long as it satisfies the above condition. The construction can be made more efficient by keeping a list of pairs that co-occur in an edge of W. The length of this list does not exceed $3|W|$. We shall assume that edges of size 2 are placed in W before any edge of size 3. Weekly related edges play a central role in our algorithm, mainly due to the following lemma.

Lemma 3. *Let (S, C, k) be a yes instance of 3HS and let W be a collection of weakly related edges of (S, C), then $|W| \leq k^2$.*

Proof. At most k vertices hit all elements of W. By the high-degree rule and the definition of W, each such vertex belongs to at most k edges of W.

3.2 A Crown Kernelization of 3HS

Again, we assume that (S, C, k) is a yes instance of 3HS. Let W be any maximal set of weakly related edges, $I = \{x \in S : x \notin V(W)\}$, and $H = \{\{x, y\} : \{x, y\} \subset e$ for some edge $e \in W\}$. Then we observe the following:

- All elements of C whose cardinality is less than three belong to W.
- I is an independent set.
  To see this, note that any edge e that contains two elements of I qualifies for membership in W. This would violate the assumed maximality of W.
- $V - I$ has at most $2k^2 + k$ elements.
  This is guaranteed by the existence of a solution A of size $\leq k$ that hits all the elements of W. Each element x of A belongs to at most k edges of W. So the restriction of $E(x)$ to W comprises at most $2k$ vertices. In addition to the $\leq k$ vertices of A, there are at most $2k^2$ vertices that belong to $\cup_{x \in A} E(x)$.
- Every edge e of size three that contains an element x of I consists of x and a pair from H.
  Otherwise, e could be added to W, violating W's maximality.
- $|H| \leq 3k^2$.
  Each edge of size three in W gives rise to three (unique) pairs of H.

Consider the simple bipartite graph $G_{I,H} = (H \cup I, E_I)$ where edges of E_I are treated as simple edges connecting an element x of I to an element $\{y, z\}$ of H whenever $\{x, y, z\}$ is in E (or just E_I).

Lemma 4. The graph $G_{I,H}$ has a crown (H', I', M) such that: $H' \subset H$, $I' \subset I$, and $(I - (V(M) \cap I)) \subset I'$.

Proof. If $|I| > H$, then I satisfies the condition of Lemma 1 and the promised crown (H', I', M) can be constructed using *Construct_Crown* (the procedure described in [6]).

By the crown reduction rule, if the original instance (S, C, k) of 3HS has a solution, then some hitting set contains at least one element of every pair of H' and excludes all elements of I'. Therefore, we can delete all I' but we have to keep the elements of H', which replace their corresponding edges in C (those that contain elements of I').

To summarize, our 3HS kernelization algorithm is the following:

Algorithm 3HS-Kernel

Input: Instance (S, C, k) of 3HS.

Output: Either NO if (S, C) has no hitting set of size $\leq k$, or an instance (S', C', k') and a partial solution A such that $k' = k - |A|$ and $|S'| \leq 5k^2 + k$.

Begin

While C contains a singleton edge $\{x\}$
  add x to A, delete $\{x\}$ from C and decrement k.
Apply the high-degree rule
Construct W
If $|W| > k^2$ return NO
Construct the simple bipartite graph $G_{I,H}$
If $|I| > |H|$
  Call *Construct_Crown* on $G_{I,H}$ to obtain a crown (H', I', M).
  Remove all elements of I'
Return the (possibly) new instance

End

Finally, we can state our kernelization theorem.

Theorem 1. *3HS has a kernel of order* $5k^2 + k$.

Proof. Based on the above discussion, the number of elements that remain in I cannot exceed the number of pairs that belong to $H - H'$. The theorem follows from the facts that H has at most $3k^2$ pairs and the vertex set of W has at most $2k^2 + k$ elements.

Note that our 3HS kernelization focused only on reducing the order of the input hypergraph. If the number of edges is not reduced below some large multiple of k^2, then we obtain an instance with high degree vertices. This is a favorable condition for subsequent exact search algorithms, as illustrated in [7] where the algorithm starts by branching at vertices of high degree. The hard instances for the algorithm of [7] are those with small vertex degrees (about 3), which are guaranteed to have quadratic order and size after applying our kernelization.

3.3 Vertex Inducedness and the Triangle Vertex Deletion Problem

A vertex-induced kernel of a graph problem is a reduced instance whose corresponding subgraph is induced. This property is of great importance in applications where edges of a Hitting Set instance model constraints like forbidden correlations or co-existence.

Vertex Inducedness of kernels was studied in [3] where a cubic-size (and cubic-order) vertex-induced kernel of 3HS was obtained. We could force our quadratic kernel to be vertex-induced, simply by making sure that no edge is deleted without deleting at least one of its elements. In fact, the only reduction that deletes edges is the edge domination rule which we may opt not to use (as we already did).

Theorem 2. 3HS has a vertex-induced kernel of order $5k^2 + k$.

An application of Theorem 2 is a quadratic kernel for the Triangle Vertex Deletion Problem (TVD), which takes a simple undirected graph G and a positive integer k as input and asks whether G has a set of k or less vertices whose complement induces a triangle-free subgraph. A kernelization algorithm for TVD consists of the following main steps.

- Enumerate all triangles of the input graph $G = (V, E)$. This take time $O(|\Delta(G)||E|)$.
- Use the list of triangles as the set C of a 3HS instance (V, C, k).
- Apply the 3HS kernelization detailed in this section to produce a new list $C' \subset C$ and a new vertex set $V' \subset V$ such that $|V'| \leq 5k^2 + k$. In this application, we avoid adding sets of size two to C'. Therefore, we modify the last step in the crown reduction by keeping only the matching edges in the crown of the graph $G_{I,H}$. In other words, we keep those elements of I that are matched under M.

The change in the last step does not affect the solution since each vertex x of the independent set I (of the crown) is forced to belong to only one triangle, say T_x: the one that corresponds to a matching edge of the crown. If a solution is later found that contains $x \in I$, then it must be changed by replacing x with any of its two neighbors in T_x. (Operations of this type are common in kernelization algorithms. An example is the vertex folding rule, used in Vertex Cover kernelization [5]).

Note the importance and necessity of vertex inducedness in this case. If, for example, the edge domination rule were applied, we could have a triangle deleted without deleting any of its vertices. This could lead to wrong answers in the subsequent search algorithm. Of course, a subsequent 3HS solver does not produce wrong results as it would make sure that at least one of the three vertices belongs to any solution. However, TVD may have another search algorithm that is to be used after kernelization.

Our argument here has to do with a proper definition of kernelization. A problem kernel is defined as an instance of the same problem. In order to guarantee this in the case of TVD, we have to cease the usage of the edge domination rule.

Another important remark about the TVD kernelization is the fact that post-kernelization exact algorithms would benefit from the presence of high degree vertices that belong to a large number of triangles. Again, this highlights the importance of this particular kernelization, in which the reduction targets the number of vertices only.

4 *d*-HS Kernelization

A kernelization algorithm for dHS can be obtained by a careful generalization of our 3HS approach. We discuss this generalization briefly in this section, keeping our main focus on how to obtain an upper bound on the kernel size. Therefore, we omit the details of efficiency analysis whenever it is clear that the run time is polynomial in the input size. Yet, we assume a dHS instance is represented by a data structure that allows us to perform efficient search, insertion and deletion. Moreover, the reader should keep in mind that d is a small constant, since the parameterized complexity of the problem gets higher as the value of d increases.

Again, we deal with an instance (S, C, k) of dHS. The concept of a subedge plays a central role in this section. As the name suggests, a subedge of (S, C) is a non-empty subset of S that is contained in an element (edge) of C. A j-subedge is a subedge of cardinality j. For an edge $e \in C$, we shall use the expression "e' occurs in e" when e' is a subedge of e.

4.1 The High-Degree Rule for *d*HS

Our high-degree rule for dHS generalizes its 3HS homonym, as well as the first reduction rule used in [12] (discussed in section 2). The rule is stated as follows:

If a subedge $e \subset S$ satisfies:

(i) $|e| < d - 1$, and
(ii) e is the pair-wise intersection of more than k edges.

Then add e to C and delete all elements of C that contain e (as a subset).

The soundness of this rule is obvious. if a subedge e satisfies (i) and (ii) and does not contain any element of some solution, A, then each of the edges that intersect at e has a distinct element of A. This is not possible, unless (S, C, k) is a no instance. Note the implicit application of the edge domination rule in the edge deletion part. Therefore, the resulting instance is not vertex-induced. This did not happen in the 3HS case since e was a singleton, which allowed us to delete e by placing its unique element in a potential solution.

If the high-degree rule does not apply to a dHS instance (S, C, k), then we shall say that (S, C, k) is **reduced**. This could be the result of applying the rule iteratively until it cannot be applied.

4.2 Weakly Related Edges

Two edges are weakly related if their intersection contains at most $d-2$ elements and neither of them is a subset of the other. Again, our kernelization algorithm proceeds by constructing a maximal set, W, of weakly related edges. An iterative greedy construction can be performed by adding edges to W in order of increasing size and keeping track of a sorted list L of all edges of size $d-2$ or less, as well as all subedges of size $d-1$. At each stage of this construction, an edge e of C is selected and checked against the elements of L. If none of the elements of L is a subset of e, then e is added to W and L is updated accordingly. Note that $|L| \leq d|W|$, since each element of W whose size is d has d $(d-1)$-subedges (also note the analogy with the 3HS case).

Let A be a solution of a reduced instance (S, C, k) of dHS, and let W be as described above. We observe that any $(d-2)$-subedge is contained in at most k edges of W. This follows from the high-degree rule since a $(d-2)$-subedge of W is the (pairwise) intersection of all edges (of W) that contain it. To achieve our sought upper bound, we need to generalize this observation. This is possible after applying the following reduction algorithm.

The High Occurrence Rule:

```
For i = d − 2 downto 1 do
    For each i-subedge e of W do
        if e is a subedge of more than k^{d−1−i} edges of W, then
            Add e to W
            Delete (from C and W) all edges containing e
```

Knowing that every $(d-2)$-subedge of W occurs in at most k edges (of W), we prove the soundness of each iteration of the above rule as follows.

Assume that every i-subedge occurs in at most k^{d-1-i} edges of W, and let e be an $(i-1)$-subedge occurring in more than k^{d-i} edges. Denote by W_e the set of edges of W that contain e properly. If e contains an element of A, then it is safe to add e to W (and C) and delete all elements of W_e. Otherwise ($e \cap A$ is empty), each element of W_e must have (in addition to elements of e) at least one element from A. Since $|A| \leq k$, some element x of A must appear in more than k^{d-1-i} edges of W_e. It follows that $e \cup \{x\}$ is an i-subedge occurring in more than k^{d-1-i} edges of W. This is a contradiction.

Lemma 5. Let (S, C, k) be a reduced yes instance of dHS and let W be a maximal set of weakly related edges that result from applying the high-occurrence rule. Then $|W| \leq k^{d-1}$.

Proof. Follows immediately from the high-occurrence rule since any 1-subedge is contained in at most k^{d-2} edges and the elements of A form at most k singleton subedges that occur in every element of W.

Every edge of W has at most d elements. Thus $|V(W)| \leq dk^{d-1}$. However, one can prove a tighter bound just as in the 3HS case. In fact, at most k vertices hit all edges of W, and each hits at most k^{d-2} such edges. In addition to the k hitting elements, each edge has at most $d-1$ non-hitting vertices. It follows that there are $k+(d-1)k^{d-1}$ vertices in $V(W)$.

4.3 Crown Reduction

Let I be the complement of $V(W)$ in S, and let H be the set of all $(d-1)$-subedges of W. Then we observe the following:

- Elements of C whose cardinality is less than d are placed in W.
- I is an independent set.
- Every edge e of size d that contains an element x of I consists of x and a $(d-1)$-subedge H. Otherwise, e could be added to W, violating W's maximality.
- $|V(H)| \leq dk^{d-1}$: each edge of size d of W gives rise to d $(d-1)$-subedges.

Again, we consider the bipartite graph $G_{I,H}$ as in the 3HS case. The rest of the work is identical to the crown construction and reduction that was used in the 3HS case. The number of vertices in the resulting instance is bounded above by $|V(H)|+|V(W)| \leq dk^{d-1}+(d-1)k^{d-1}+k$. To conclude:

Theorem 3. dHS has a kernel of order $(2d-1)k^{d-1}+k$.

5 Concluding Remarks

We presented an algorithm that produces quadratic-order kernels for 3HS. Previously known kernelization strategies could guarantee cubic-order kernels only. An attempt for improving the kernel size appeared also in [13] by local usage of the NT kernelization of Vertex Cover [11]. However, it was observed that such technique does not always produce correct kernels [15].

We generalized our kernelization approach to work for any dHS problem. For $d>3$, our dHS kernel is not necessarily vertex-induced, in the sense that the resulting hyper-subgraph could have vertices that share edges in the original instance. Any vertex-induced version has to cease the use of the edge domination rule, which is also part of the general high-degree rule (when $d>3$). It would thus be challenging, but interesting, to find a general vertex-induced kernel.

Finally, an important feature of our 3HS method is its isolation of a subgraph that is almost simple in the input hypergraph: the subgraph formed by the set W of weakly related edges. We are investigating the potential use of this preprocessing step in improved exact and approximate 3HS algorithms.

Acknowledgment. We would like to thank Daniel Raible for his valuable comments.

References

1. Abu-Khzam, F.N., Collins, R.L., Fellows, M.R., Langston, M.A., Suters, W.H., Symons, C.T.: Kernelization Algorithms for the Vertex Cover Problem: Theory and Experiments. In: Workshop on Algorithm Engineering and Experiments (ALENEX), pp. 62–69 (2004)
2. Abu-Khzam, F.N., Fellows, M.R., Langston, M.A., Suters, W.H.: Crown Structures for Vertex Cover Kernelization. Theory of Computing Systems (TOCS) (accepted for publication) (2006)
3. Abu-Khzam, F.N., Fernau, H.: Kernels: Annotated, Proper and Induced. In: Bodlaender, H.L., Langston, M.A. (eds.) IWPEC 2006. LNCS, vol. 4169, pp. 264–275. Springer, Heidelberg (2006)
4. Buss, J.F., Goldsmith, J.: Nondeterminism within P. SIAM Journal on Computing 22, 560–572 (1993)
5. Chen, J., Kanj, I., Jia, W.: Vertex cover: further observations and further improvements. Journal of Algorithms 41, 280–301 (2001)
6. Chor, B., Fellows, M.R., Juedes, D.: Linear Kernels in Linear Time, or How to Save k Colors in $O(n^2)$ Steps. In: International Workshop on Graph Theoretic Concepts in Computer Science (WG), pp. 257–269 (2004)
7. Fernau, H.: A top-down approach to search-trees: Improved algorithmics for 3-Hitting Set. Electronic Colloquium on Computational Complexity (ECCC), p. 073 (2004)
8. Garey, M.R., Johnson, D.S.: Computers and Intractability. W. H. Freeman, New York (1979)
9. Jones, J.A., Harrold, M.J.: Test-Suite Reduction and Prioritization for Modified Condition/Decision Coverage. IEEE Trans. Software Eng. 29(3), 195–209 (2003)
10. Kuhn, F., von Rickenbach, P., Wattenhofer, R., Welzl, E., Zollinger, A.: Interference in Cellular Networks: The Minimum Membership Set Cover Problem. In: Wang, L. (ed.) COCOON 2005. LNCS, vol. 3595, Springer, Heidelberg (2005)
11. Nemhauser, G.L., Trotter, L.E.: Vertex Packings: Structural Properties and Algorithms. Mathematical Programming 8, 232–248 (1975)
12. Niedermeier, R., Rossmanith, P.: An efficient fixed-parameter algorithm for 3-Hitting Set. Journal of Discrete Algorithms 1, 89–102 (2003)
13. Nishimura, N., Ragde, P., Thilikos, D.M.: Smaller kernels for hitting set problems of constant arity. In: Downey, R.G., Fellows, M.R., Dehne, F. (eds.) IWPEC 2004. LNCS, vol. 3162, pp. 121–126. Springer, Heidelberg (2004)
14. Ruchkys, D., Song, S.: A Parallel Approximation Hitting Set Algorithm for Gene Expression Analysis. In: SBAC-PAD '02: Proceedings of the 14th Symposium on Computer Architecture and High Performance Computing, Washington, DC, USA, p. 75. IEEE Computer Society, Los Alamitos (2002)
15. Thilikos, D.M.: Private communication (2005)
16. Weihe, K.: Covering trains by stations or the power of data reduction. In: Battiti, R., Bertossi, A.A. (eds.) International Conference on Algorithms and Experiments, pp. 1–8 (1998)

Largest Bounding Box, Smallest Diameter, and Related Problems on Imprecise Points*

Maarten Löffler and Marc van Kreveld

Department of Information and Computing Sciences
Utrecht University, The Netherlands
{loffler,marc}@cs.uu.nl

Abstract. Imprecise points are regions in which one point should be placed. We study computing the largest and smallest possible values of various basic geometric measures on sets of imprecise points, such as the diameter, width, closest pair, smallest enclosing circle, and smallest enclosing bounding box. We give efficient algorithms for most of these problems, and identify the hardness of others.

1 Introduction

Given a set of points in the plane, various measures exist that try to capture certain properties of that point set. Examples of such measures include the *diameter*: the largest distance between any pair of points, the smallest distance between any pair of points, the *width*: the smallest distance between two parallel lines with all points between them, the smallest circle containing all points, or the smallest axis-aligned bounding box containing all points. All these measures have been well studied and optimal algorithms to compute them are known.

When dealing with real-world data, however, locations of input points are often not known exactly. If we know for each point that it lies inside some region, but not where in that region, it becomes interesting to compute bounds on the possible values of these basic geometric measures. To make this more precise, we are given a set of regions $\mathcal{L}$ and a measure μ that takes a set of points an gives a real number, an we want to place one point in each region of $\mathcal{L}$ such that the resulting point set maximises or minimises μ.

We study five basic measures both for maximisation and minimisation, when the imprecise points are modelled as squares or discs, possibly overlapping and of different sizes. Some of these problems have already been studied in other contexts, and efficient algorithms or hardness results are known. For most remaining problems, we present efficient algorithms here. Table 1 summarises all previous and new results. For the problem of computing the largest possible width we have not found any satisfying result, although we can prove NP-hardness when the points are modelled as line segments. For the smallest diameter of a set of discs, we have no exact algorithm but a polynomial time approximation scheme.

* This research was partially supported by the Netherlands Organisation for Scientific Research (NWO) through the project GOGO.

F. Dehne, J.-R. Sack, and N. Zeh (Eds.): WADS 2007, LNCS 4619, pp. 446–457, 2007.

Table 1. New and known results

problem	model	largest	smallest
smallest bounding box	squares	$O(n)$	$O(n)$
	discs	$O(n)$	$O(n^2)$
smallest enclosing circle	squares	$O(n)$	$O(n)$ [7]
	discs	$O(n)$	$O(n)$ expected
diameter	squares	$O(n \log n)$	$O(n \log n)$
	discs	$O(n \log n)$	$(1+\varepsilon)$-approx. in $O(n^{c\varepsilon^{-\frac{1}{2}}})$
width	squares		$O(n \log n)$ [13]
	discs		$O(n \log n)$
	line segments	NP-hard	$O(n \log n)$ [13]
closest pair	squares	NP-hard [5]	$O(n \log n)$
	discs	NP-hard [5]	$O(n \log n)$

Related work. Data imprecision in computational geometry is often considered in stochastic or fuzzy models. Recently, interest in exact imprecision models has been rising. *Espilon geometry* [6] is a framework for robust computations on imprecise points, and the *tolerance* [1] of a geometric structure is the largest perturbation of the vertices such that the topology remains the same.

Colley *et al.* [4] compute the smallest area axis-aligned rectangle that intersects a set of convex polygons in $O(n \log n)$ time. Fiala *et al.* [5] consider the problem of finding distant representatives in a collection of subsets of a given space. In particular, they prove that maximizing the smallest distance in a set of n imprecise points, modeled as circles or squares, is NP-hard. Cabello [3] gives approximation algorithms for this case. Jadhav *et al.* [7] consider the *intersection radius* of a set of objects: the smallest circle that intersects them all. Robert and Toussaint [13] develop an algorithm for computing the smallest strip that intersects a set of convex regions, while surveying several facility location problems. Such stabbing and facility location problems are related to the minimisation variants of the problems we study.

Averbakh and Bereg [2] also consider imprecise points. Where we compute the point set for which the smallest enclosing circle is worst, they compute the smallest circle that encloses the worst case point set. They study this problem in a facility location context, and they also consider weighted points and different metrics. Nagai and Tokura [10] compute the union and intersection of all possible convex hulls to obtain bounds on the area. As imprecision regions they use discs or convex polygons, and they give an $O(n \log n)$ time algorithm. They also compute (possibly non-obtainable) lower and upper bounds for the diameter.

In [8], we also study another classical geometric problem, the convex hull, in an imprecise context. Results for the different variants of this problem range from $O(n \log n)$ time to $O(n^{10})$ time or NP-hardness.

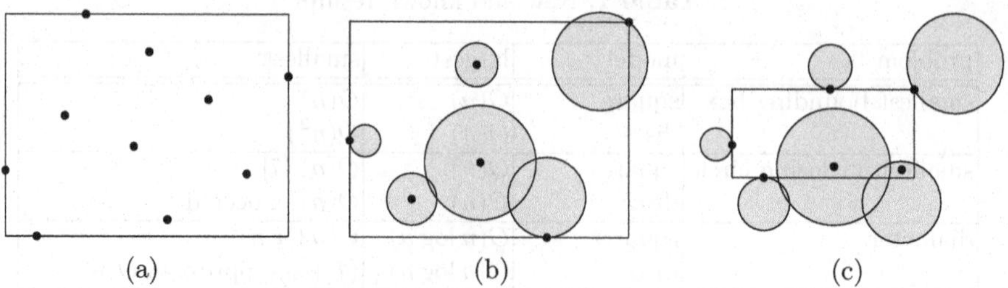

Fig. 1. (a) The axis-aligned bounding box of a set of points in the plane. (b) The largest possible AABB of a set of imprecise points. (c) The smallest possible AABB of a set of imprecise points.

2 Axis-Aligned Bounding Box

We start with a problem that is very simple in the classical case. Given a set of points P, the axis-aligned bounding box (AABB) is the smallest axis-parallel rectangle that contains P, see Figure 1(a). In an imprecise context, we are given a set $\mathcal{L}$ of regions, and we want to place a point in each region such that the bounding box of the resulting point set is as large or as small as possible, see Figures 1(b) and 1(c). We will measure the size of a rectangle by its area, but the algorithms we describe work for the perimeter as well.

2.1 Largest Possible AABB

The largest possible AABB can be computed in linear time for both the square and disc model (or any other constant complexity model). Let $\mathcal{L}' \subset \mathcal{L}$ be the set of the four topmost regions from $\mathcal{L}$ (that is, the four regions with the topmost highest points), the four bottommost, the four leftmost and the four rightmost regions. We can find $\mathcal{L}'$ in linear time.

Lemma 1. *The largest possible AABB of $\mathcal{L}'$ is equal to the largest possible AABB of $\mathcal{L}$.*

Proof. Suppose this is not the case. Then there is a region l in $\mathcal{L} \setminus \mathcal{L}'$ that contributes to the AABB of $\mathcal{L}$, say, to the top boundary. However, there are at least four regions in $\mathcal{L}$ that extend higher than l, of which only three can contribute to another boundary. That means we can place the point of the fourth at its topmost position and get a larger AABB. Contradiction. ⊠

The AABB of $\mathcal{L}'$ can be determined by four points each lying on one of the sides of the bounding box, or by only three or two points when one or two points lie on corners. Since $\mathcal{L}'$ has only constant size, we can try all possibilities and report the largest one.

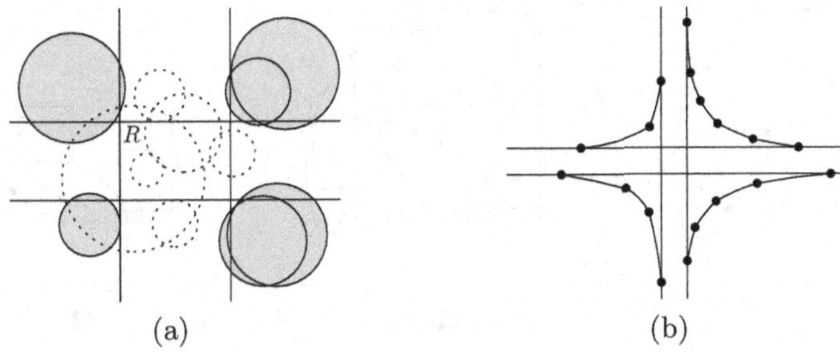

Fig. 2. (a) The discs that intersect the rectangle R (bounded by the extreme lines) are always accounted for. (b) The remaining discs form four chains of circular arcs.

2.2 Smallest Possible AABB

The smallest possible AABB is the smallest rectangle that contains at least one point of each region, so it is actually the smallest rectangle that intersects all regions. Let the *left extreme line* be the leftmost of the lines through the rightmost points of all regions. Similarly we define the *right*, *top* and *bottom* extreme lines. When the left extreme line is to the right of the right extreme line, or the top extreme line is below the bottom extreme line, there exists a zero area solution. Otherwise, they define a rectangle R. For the square model, when there is no zero area solution, R is the smallest possible AABB.

Discs. When the imprecise points are modelled as convex polygons, the problem can be solved in $O(n \log n)$ time [4]. When the points are modelled as discs, we can use a similar transformation to obtain an $O(n^2)$ algorithm. In this case R does not necessarily intersect all discs, see Figure 2(a), but it needs to be enclosed by the smallest AABB, so we only need to consider the discs that do not intersect R. The centre points of these discs lie outside the two strips between the extreme lines; this divides them in four groups. The four corners of the smallest AABB must lie inside or behind all discs of their respective groups. We define a border between valid and invalid corner placements; these borders are convex chains of circular arcs, see Figure 2(b), and can be computed in $O(n \log n)$ time.

Either the top left and bottom right corners of the smallest AABB lie on their respective chains, or the bottom left and top right corners. Suppose the former is the case. We try all possibilities by keeping the top left point on a fixed arc, and moving the bottom right point over the opposite chain. While doing this, we keep track of the projections of those two points on the bottom left and top right chains. When the projection of the top left point is to the left of the projection of the bottom right point, on both other chains, a solution is valid, otherwise it is invalid. In quadratic time we can try all combinations and keep track of the smallest valid solution. There are some subtleties involved with the fact that the top left point is not really fixed; more details are in the full paper.

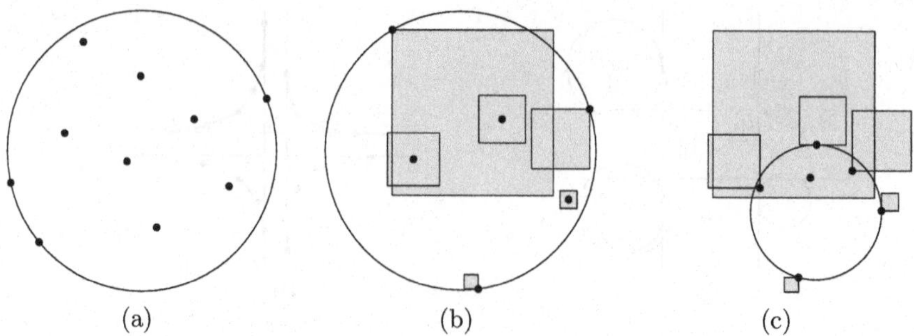

Fig. 3. (a) The smallest enclosing circle of a set of points in the plane. (b) The largest SEC of a set of imprecise points. (c) The smallest SEC of a set of imprecise points.

3 Smallest Enclosing Circle

We proceed with another relatively simple problem. Given a set of points P, the smallest enclosing circle (SEC) is the smallest circle that contains P, see Figure 3(a). When we are given a set $\mathcal{L}$ of imprecise points, we want to place a point in each region such that the SEC of the resulting point set is as large or as small as possible, see Figures 3(b) and 3(c).

3.1 Largest Possible SEC

Squares. The largest smallest enclosing circle of a set of squares (or constant size convex polygons) can be computed by first computing the smallest enclosing circle of the set of corners of all squares using any classical algorithm, e.g. Welzl [14]. If the three points that determine this circle belong to different squares, we are done. Otherwise, there is one square of which multiple corners contribute to the smallest enclosing circle, and we know that this square has to contribute to the optimal solution. So we just try all corners of this square and compute the smallest enclosing circle of this single point and all other corners of the other squares. The two remaining contributing points can again belong to the same square, so we do the same thing once more. We are done in linear time.

Discs. To compute the largest possible smallest enclosing circle of a set of discs, we observe that there are only two possibilities. Either the largest SEC contains all discs, or it does not, see Figure 4. If it does, then the largest SEC is just the smallest circle containing a set of discs, which can be computed in $O(n)$ time [9]. If it does not, there must be one disc D among the input discs that contains all other discs. In this case, the largest SEC is determined by the point p of all other discs closest to D, and the point q on D furthest away from p, see Figure 4(b). This case can clearly be computed in linear time.

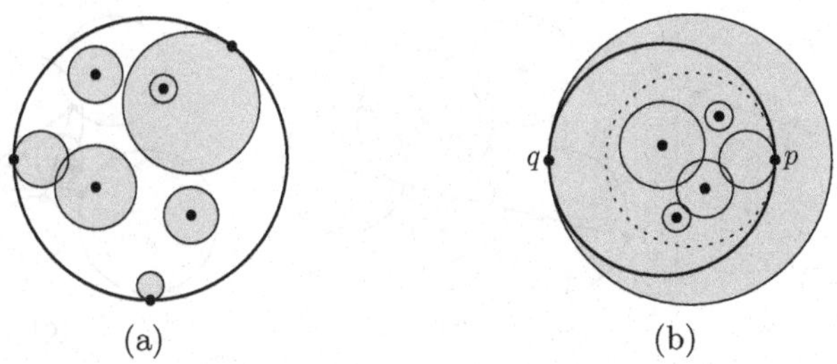

Fig. 4. (a) All discs are completely within the LSEC. (b) One disc contains all others.

3.2 Smallest Possible SEC

The smallest possible SEC for a set of imprecise points is the smallest circle that intersects all regions. This is also called the *intersection radius* of a set of regions. When the regions are modelled as squares (or other convex polygons), it can be computed in linear time [7]. When the points are modelled as discs, we can use an adapted version of the randomised incremental construction algorithm by Welzl [14]. The adaptation is straightforward; details are in the full paper.

4 Closest Pair

The closest pair is the smallest distance between any two points in a given set. Figure 5 shows examples of the precise and imprecise instances of this problem.

Computing the largest possible closest pair is also known as *spreading points*, which is NP-hard for both the square and the circle model [5], but it can be approximated within a constant factor [3]. The smallest possible closest pair is the smallest distance between any pair of regions. We can find this in $O(n \log n)$ time by computing the Voronoi diagram of a set of convex objects [15].

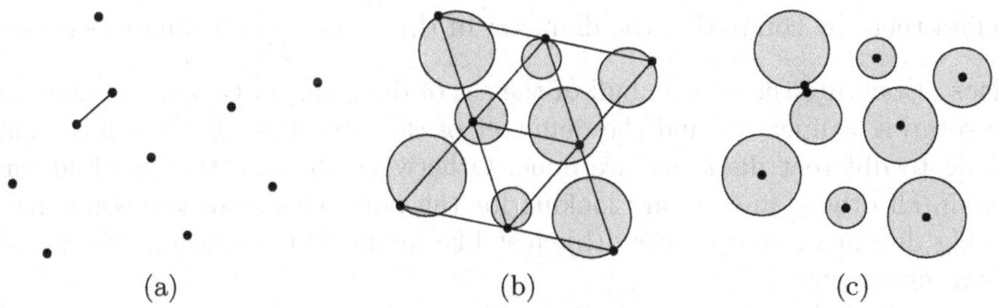

Fig. 5. (a) The closest pair of a set of points in the plane. (b) The largest possible closest pair of a set of imprecise points, reached by many pairs simultaneously. (c) The smallest possible closest pair of a set of imprecise points.

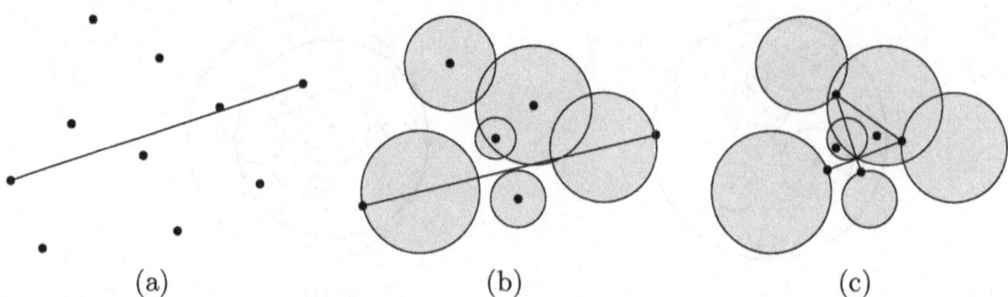

Fig. 6. (a) The diameter of a set of points in the plane. (b) The largest possible diameter of a set of imprecise points. (c) The smallest possible diameter of a set of imprecise points, determined by three pairs simultaneously.

5 Diameter

Given a set of points P, the diameter is the largest distance between any pair of points in P, see Figure 6(a). When the points are imprecise, we are given a set $\mathcal{L}$ of regions, and we want to place a point in each region such that the diameter of that point set is as large or small as possible, see Figures 6(b) and 6(c).

5.1 Largest Possible Diameter

The largest diameter is formed by the pair of points among the input regions that are furthest away from each other, unless they belong to the same region.

Squares. When the points are modelled as squares, the two points forming the largest diameter must be among the corners of the squares. This means we can just compute the diameter of the set of all corners using a conventional diameter algorithm in $O(n \log n)$ time. If the two points found belong to different regions, we are done. Otherwise, they are diagonally opposite corners of one square s, and there are two options. Either the largest diameter is formed by one corner of s and one point among the other corners; we can check all these in $O(n)$ time. Or the largest diameter is formed by two points among the other squares, see Figure 7(a). If that is the case, they must belong to two different squares, and we find them by computing the diameter of the corners of all squares except s.

Discs. Compute the convex hull of the set of discs [12] in $O(n \log n)$ time, and use rotating calipers to find the diameter of this structure. If the points found belong to different discs, we are done. Otherwise, the disc that we find must contain all others, and we are looking for the point closest to the boundary of the big disc again, see Figure 4(b), just like in the SEC problem. We can find this in linear time.

5.2 Smallest Possible Diameter

The smallest possible diameter d can be determined by multiple pairs of points simultaneously, as in Figure 6(c). Moving any of the four points involved would

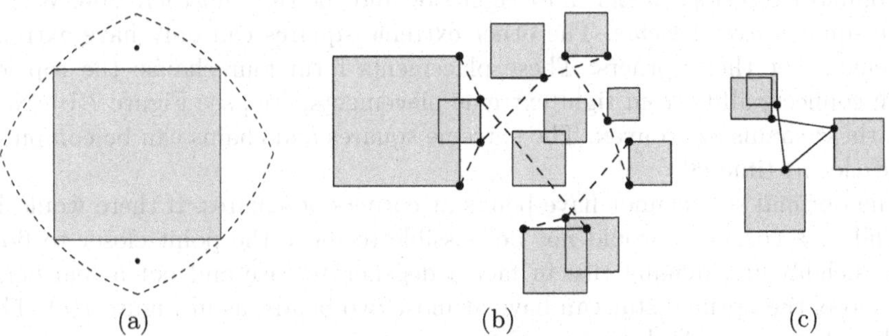

Fig. 7. (a) The largest possible diameter is formed by the two single points, even though the largest diameter among all corners is a diagonal of the square. (b) Four extreme squares and four chains. (c) A star with two bends.

increase the distance between at least one pair of them. In general, this could happen arbitrarily often, see Figure 8(a) where none of the points can be moved without increasing the diameter. Note that these situations are not degenerate. This makes computing the smallest possible diameter a difficult problem.

In the optimal solution, there will be some points that are exactly d away from each other. In general, this will not occur at two different places, unless they depend on each other: there may be a point p that has distance d to two other points, and if we would move it closer to one of the other points, it would move away from the other. If this is the case, we call p a *bend*. Note that the two other points cannot be more than d away from each other, so the angle at p is at most 60°. In the optimal solution, there can be many bends that together form a *star*.

For any subset $\mathcal{L}' \subset \mathcal{L}$, let d' be the value of the smallest possible diameter of $\mathcal{L}'$. We define the *star* of $\mathcal{L}'$ to be the sequence of points that have distance exactly d' from each other in the optimal solution for this subset. This star consists of a startpoint, zero or more bends, and an endpoint, or it is cyclic with only bends. Every connection between two consecutive points in the sequence must intersect all others. Examples of stars are in Figures 6(c), 7(c) and 8(a). We call the star of $\mathcal{L}$ the *optimal star*.

Let $l \in \mathcal{L}$ be a region. We call l an *extreme* region if there exists a line that has l completely on one side, but no other region of $\mathcal{L}$ completely on the same side. We call a point $p \in l$ an *extreme placement* if such a line exists that goes through p. All points of the optimal star must be on extreme placements in extreme regions. Furthermore, if p and q are adjacent points on the optimal star, then no region is entirely on the other side of the line through p perpendicular to $\overline{pq}$.

Squares. When the points are modelled as squares, we can solve the problem in $O(n \log n)$ time. Among the extreme squares, there are only four with infinitely many extreme placements, being the squares with the topmost bottom side, the bottommost top side, the leftmost right side and the rightmost left side. We call these squares *axis-extreme*. The other extreme squares can only have extreme placements at their corners. These placements form four chains: the top left chain connects all bottom right extreme placements, etc., see Figure 7(b). Note that these chains are convex. The extreme squares and chains can be computed in $O(n \log n)$ time [8].

The optimal star cannot have bends at corners of squares. If there would be a bend at a corner, it would not be possible to move the point closer to both of its neighbours, making this in fact a degenerate case and not a real bend. Therefore, the optimal star can have at most two bends, as in Figure 7(c). This implies that we can find it efficiently.

Lemma 2. *We can find the optimal star by computing the star of every set of four extreme squares, of which two are axis-extreme, and reporting the largest among these.*

We can compute all of these stars in $O(n^2)$ time. However, after precomputing the chains we can also find the optimal star in linear time, by using a careful case analysis and using the structure of the chains. For every placement of an axis-extreme point, there is one vertex of a chain that is furthest away from it, and this determines the best possible diameter in this case. As the axis-extreme point moves over its region, this furthest vertex can move only in restricted ways, which saves us a linear factor. The details of this refinement are omitted here, due to lack of space.

We have now computed the value d of the smallest possible diameter. If needed, we can also compute a placement of the points in their regions that realises this diameter. We first compute valid placements of the four axis-extreme points, and then observe that all other points should be placed 'as far inward' as possible.

To compute a valid placement for an axis-extreme point, note the following. If we find a placement such that any star that includes this point has at most length d, then this placement is valid for a global solution of diameter d. This means we can check for all possible stars in which interval the point is allowed to lie if that star must be at most d long, and then place the point somewhere in the intersection of all these intervals. This can trivially be done in quadratic time, but again we can improve that to linear time using the same case analysis as above.

For the rest of the points, if a point is to the left of the lines through the topmost and bottommost axis-extreme points, moving it to the right can only decrease the diameter, and the same for the other directions. Because the regions are squares, every point will end either in a corner of its region or somewhere in the middle of the whole construction.

Discs. When the points are modelled as discs, stars can have up to n bends, see Figure 8(a). This leads to algebraic difficulties: even if we would know the combinatorial structure of the optimal star, computing it exactly would not be possible.

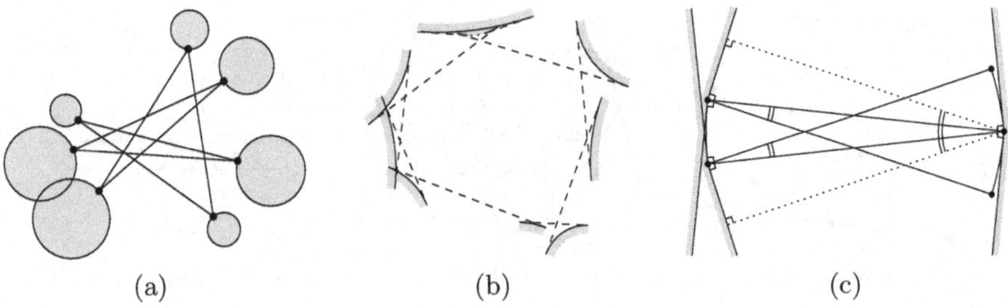

Fig. 8. (a) A cyclic star that visits seven regions. (b) Circular arcs that are extreme in some direction. (c) Three consecutive angles of at most α give a diameter of $\cos\alpha$.

We can make some observations about the combinatorial structure. We can still define the extreme discs, that is, discs that have a tangent line with no other disc completely at the same side. These discs form a chain of arcs with this property, see Figure 8(b). Any bend in the star still needs to be on such an extreme arc. However, it is possible that all discs have extreme arcs.

Since we cannot compute the optimum efficiently, we approximate. A factor $\frac{2}{3}\sqrt{3} \approx 1.15$ approximation can be computed in linear time by computing the smallest smallest enclosing circle of the imprecise points. Details are in the full paper. We can also compute a $(1+\varepsilon)$-approximation of the smallest diameter in $O(n^{c\varepsilon^{-\frac{1}{2}}})$ time, where $c = 1\frac{1}{2}\sqrt{2}\pi \approx 6.66$. The idea of this algorithm is to consider only stars of at most k bends, for k chosen suitably.

Lemma 3. *Suppose the optimal solution is given by a star of at least k bends. Then there exists a star with only one bend that approximates it.*

Proof. Suppose that the optimal diameter is 1. The sum of the angles that the star makes in the bends is at most π. That means that there are three consecutive bends a, b and c somewhere that together make an angle of at most $\alpha = \frac{3\pi}{k}$. Therefore the individual angles are also at most α, see Figure 8(c). The regions of a, b and c are convex and do not intersect the lines perpendicular to $\overline{ab}$ and $\overline{bc}$, otherwise they were not optimal. This means that in the worst case they are arbitrarily close to those lines, and in that case the best possible diameter of a, b and c would be $\cos\alpha$, which is more than $1 - \frac{1}{2}\alpha^2$. Thus, if we take $k = 1\frac{1}{2}\sqrt{2}\pi\varepsilon^{-\frac{1}{2}}$ this gives a $(1-\varepsilon)$-approximation. ⊠

So, if the optimal star has more than k bends, we find an approximation in cubic time. Otherwise, we find the optimum in $O(n^k)$ time.

6 Width

Given a set of points P, the width is the smallest distance between any pair of parallel lines that contains P, see Figure 9(a). Examples of the imprecise case are in Figures 9(b) and 9(c).

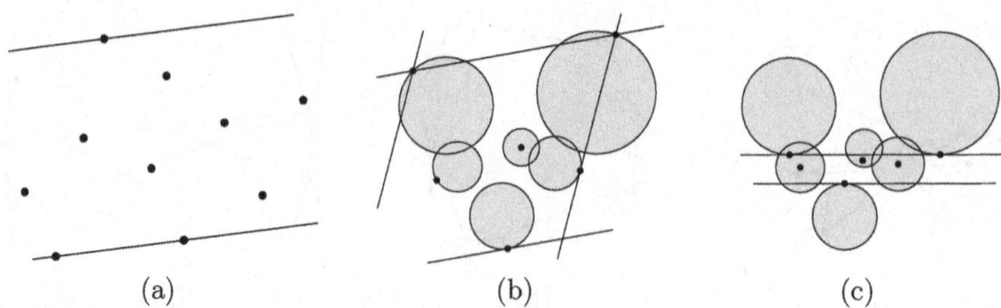

Fig. 9. (a) The width of a set of points in the plane. (b) The largest possible width of a set of imprecise points, reached at two different locations simultaneously. (c) The smallest possible width of a set of imprecise points.

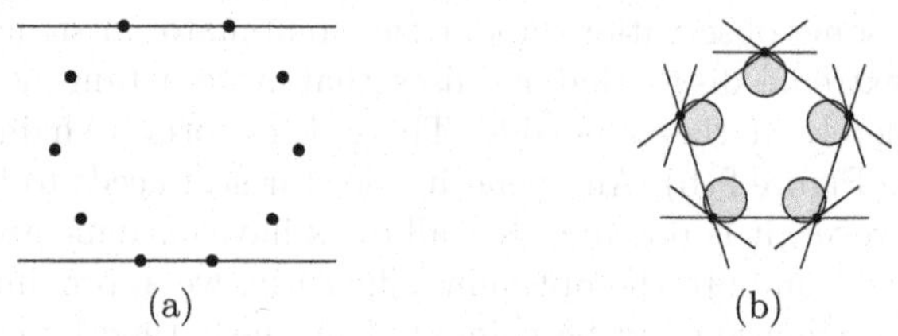

Fig. 10. (a) A set of n points, such that the width changes when any of them is removed. (b) Many triples of points define the width simultaniously.

The problem of computing the largest possible width seems to be hard. When the points are modelled as line segments, it is even NP-hard; the proof is in the full paper. The width of a set of points has the property that there can be points that do not contribute to the width, but are needed to make the width valid, see Figure 10(a). Removing any point here would result in a smaller width. In an imprecise context, this means that the placement of all points is important, and we cannot look only at subsets of the regions. Furthermore, it can happen that many points are simultaneously involved in the optimal width, see Figure 10(b), in a similar way as the stars in the smallest diameter problem.

The smallest width of a set of squares, or any other convex polygonal regions, can be computed in $O(n \log n)$ time [13]. The smallest possible width of a set of discs can be computed by using the structure of Figure 8(b) again. For circles, this can be computed in $O(n \log n)$ time [11]. We use rotating calipers to find the smallest possible width between two extreme lines.

7 Conclusions

We have given a structured overview of the imprecise variants of several basic geometric measures, reusing known results from various other contexts and designing efficient algorithms for the remaining cases. Most problems can be computed efficiently, while some are NP-hard. One remaining open problem is that of computing the largest possible width for the square or disc model. Also, the 3D versions of most problems are still open.

References

1. Abellanas, M., Hurtado, F., Ramos, P.A.: Structural tolerance and Delaunay triangulation. Inf. Proc. Lett. 71, 221–227 (1999)
2. Averbakh, I., Bereg, S.: Facility location problems with uncertainty on the plane. Discrete Optimization 2, 3–34 (2005)
3. Cabello, S.: Approximation algorithms for spreading points. Journal of Algorithms (to appear)
4. Colley, P., Meijer, H., Rappaport, D.: Optimal nearly-similar polygon stabbers of convex polygons. In: Proc. 6th Canad. Conf. Comput. Geom. pp. 269–274 (1994)
5. Fiala, J., Kratochvil, J., Proskurowski, A.: Systems of distant representatives. Discrete Applied Mathematics 145, 306–316 (2005)
6. Guibas, L.J., Salesin, D., Stolfi, J.: Constructing strongly convex approximate hulls with inaccurate primitives. Algorithmica 9, 534–560 (1993)
7. Jadhav, S., Mukhopadhyay, A., Bhattacharya, B.K.: An optimal algorithm for the intersection radius of a set of convex polygons. J. Algorithms 20, 244–267 (1996)
8. Löffler, M., van Kreveld, M.: Largest and smallest tours and convex hulls for imprecise points. In: Arge, L., Freivalds, R. (eds.) SWAT 2006. LNCS, vol. 4059, pp. 375–387. Springer, Heidelberg (2006)
9. Megiddo, N.: On the ball spanned by balls. Discrete Comput. Geom. 4, 605–610 (1989)
10. Nagai, T., Tokura, N.: Tight error bounds of geometric problems on convex objects with imprecise coordinates. In: Proc. Jap. Conf. on Discrete and Comput. Geom. pp. 252–263 (2000)
11. Nagai, T., Yasutome, S., Tokura, N.: Convex hull problem with imprecise input. In: Akiyama, J., Kano, M., Urabe, M. (eds.) Discrete and Computational Geometry. LNCS, vol. 1763, pp. 207–219. Springer, Heidelberg (2000)
12. Rappaport, D.: A convex hull algorithm for discs, and applications. Comput. Geom. Theory Appl. 1(3), 171–181 (1992)
13. Robert, J., Toussaint, G.: Computational geometry and facility location. Technical Report SOCS 90.20, McGill Univ., Montreal, PQ (1990)
14. Welzl, E.: Smallest enclosing discs (balls and ellipsoids). In: Maurer, H.A. (ed.) New Results and New Trends in Computer Science. LNCS, vol. 555, pp. 359–370. Springer, Heidelberg (1991)
15. Yap, C.K.: An $O(n \log n)$ algorithm for the Voronoi diagram of a set of simple curve segments. Discrete Comput. Geom. 2, 365–393 (1987)

Maximizing Maximal Angles for Plane Straight-Line Graphs*

Oswin Aichholzer[1], Thomas Hackl[1], Michael Hoffmann[2], Clemens Huemer[3], Attila Pór[4], Francisco Santos[5], Bettina Speckmann[6], and Birgit Vogtenhuber[1]

[1] Institute for Software Technology, Graz University of Technology
{oaich,thackl,bvogt}@ist.tugraz.at
[2] Institute for Theoretical Computer Science, ETH Zürich
hoffmann@inf.ethz.ch
[3] Departament de Matemática Aplicada II, Universitat Politécnica de Catalunya
clemens.huemer@upc.edu
[4] Dept. of Appl. Mathem. and Inst. for Theoretical Comp. Science, Charles Univ.,
por@kam.mff.cuni.cz
[5] Dept. de Matemáticas, Estadística y Computación, Universidad de Cantabria
francisco.santos@unican.es
[6] Department of Mathematics and Computer Science, TU Eindhoven
speckman@win.tue.nl

Abstract. Let $G = (S, E)$ be a plane straight-line graph on a finite point set $S \subset \mathbb{R}^2$ in general position. The *incident angles* of a point $p \in S$ in G are the angles between any two edges of G that appear consecutively in the circular order of the edges incident to p. A plane straight-line graph is called *φ-open* if each vertex has an incident angle of size at least φ. In this paper we study the following type of question: What is the maximum angle φ such that for any finite set $S \subset \mathbb{R}^2$ of points in general position we can find a graph from a certain class of graphs on S that is φ-open? In particular, we consider the classes of triangulations, spanning trees, and paths on S and give tight bounds in most cases.

1 Introduction

Conditions on angles in plane straight-line graphs have been studied extensively in discrete and computational geometry. It is well known that Delaunay triangulations maximize the minimum angle over all triangulations, and that in a (Euclidean) minimum weight spanning tree each angle is at least $\frac{\pi}{3}$. In this paper we address the fundamental combinatorial question, what is the maximum value

* O.A., T.H., and B.V. were supported by the Austrian FWF Joint Research Project 'Industrial Geometry' S9205-N12. C.H. was partially supported by projects MEC MTM2006-01267 and Gen. Cat. 2005SGR00692. A.P. was partially supported by Hungarian National Foundation Grant T60427. F.S. was partially supported by grant MTM2005-08618-C02-02 of the Spanish Ministry of Education and Science. Preliminary results of this article have been presented in [1].

F. Dehne, J.-R. Sack, and N. Zeh (Eds.): WADS 2007, LNCS 4619, pp. 458–469, 2007.

α such that for each finite point set in general position there exists a (certain type of) plane straight-line graph where each vertex has an incident angle of size at least α. In other words, we consider min – max – min – max problems, where we minimize over all finite point sets S in general position in the plane, the maximum over all plane straight-line graphs G (of the considered type), of the minimum over all $p \in S$, of the maximum angle incident to p in G. We present bounds on α for three classes of graphs: spanning paths, (general and bounded degree) spanning trees, and triangulations. Most of the bounds we give are tight. In order to show that, we describe families of point sets for which no graph from the respective class can achieve a greater incident angle at each vertex.

Background. Our motivation for this research stems from the investigation of "pseudo-triangulations", a straight-line framework which apart from deep combinatorial properties has applications in motion planning, collision detection, ray shooting and visibility; see [3,12,13,15,16] and references therein. Pseudo-triangulations with a minimum number of pseudo-triangles (among all pseudo-triangulations for a given point set) are called *minimum* (or *pointed*) pseudo-triangulations. They can be characterized as plane straight-line graphs where each vertex has an incident angle greater than π. Furthermore, the number of edges in a minimum pseudo-triangulation is maximal, in the sense that the addition of any edge produces an edge-crossing or negates the angle condition.

In comparison to these properties, we consider connected plane straight-line graphs where each vertex has an incident angle α—to be maximized—and the number of edges is minimal (spanning trees) and the vertex degree is bounded (spanning trees of bounded degree and spanning paths). We further show that any planar point set has a triangulation in which each vertex has an incident angle which is at least $\frac{2\pi}{3}$. Observe that perfect matchings can be described as plane straight-line graphs where each vertex has an incident angle of 2π and the number of edges is maximal.

Related Work. There is a vast literature on triangulations that are optimal according to certain criteria, cf. [2]. Similar to Delaunay triangulations which maximize the smallest angle over all triangulations for a point set, farthest point Delaunay triangulations minimize the smallest angle over all triangulations for a convex polygon [9]. If all angles in a triangulation are $\geq \frac{\pi}{6}$ then it contains the relative neighborhood graph as a subgraph [14]. The relative neighborhood graph for a point set connects any pair of points which are mutually closest to each other (among all points from the set). Edelsbrunner et al. [10] showed how to construct a triangulation that minimizes the maximum angle among all triangulations for a set of n points in $O(n^2 \log n)$ time.

In applications where small angles have to be avoided by all means, a Delaunay triangulation may not be sufficient in spite of its optimality because even there arbitrarily small angles can occur. By adding so-called Steiner points one can construct a triangulation on a superset of the original points in which there is some absolute lower bound on the size of the smallest angle [7]. Dai et al. [8] describe several heuristics to construct minimum weight triangulations

(triangulations which minimize the total sum of edge lengths) subject to absolute lower or upper bounds on the occurring angles.

Spanning cycles with angle constraints can be regarded as a variation of the traveling salesman problem. Fekete and Woeginger [11] showed that if the cycle may cross itself then any set of at least five points admits a locally convex tour, that is, a tour in which the angle between any three consecutive points is positive. Arkin et al. [5] consider as a measure for (non-)convexity of a point set S the minimum number of (interior) reflex angles (angles $> \pi$) among all plane spanning cycles for S. Aggarwal et al. [4] prove that finding a spanning cycle for a point set which has minimal total angle cost is NP-hard, where the angle cost is defined as the sum of direction changes at the points. Regarding spanning paths, it has been conjectured that each planar point set admits a spanning path with minimum angle at least $\frac{\pi}{6}$ [11]; recently, a lower bound of $\frac{\pi}{9}$ has been presented [6].

Definitions and Notation. Let $S \subset \mathbb{R}^2$ be a finite set of points in general position, that is, no three points of S are collinear. In this paper we consider plane straight-line graphs $G = (S, E)$ on S. The vertices of G are the points in S, the edges of G are straight-line segments that connect two points in S, and two edges of G do not intersect except possibly at their endpoints. The *incident angles* of a point $p \in S$ in G are the angles between any two edges of G that appear consecutively in the circular order of the edges incident to p. We denote the *maximum incident angle* of p in G with $\mathrm{op}_G(p)$. For a point $p \in S$ of degree at most one we set $\mathrm{op}_G(p) = 2\pi$. We also refer to $\mathrm{op}_G(p)$ as the *openness* of p in G and call $p \in S$ *φ-open* in G for some angle φ if $\mathrm{op}_G(p) \geq \varphi$. Consider for example the graph depicted in Fig. 1. The point p has four incident edges of G and, therefore, four incident angles. Its openness is $\mathrm{op}_G(p) = \alpha$. The point q has only one incident angle and correspondingly $\mathrm{op}_G(q) = 2\pi$.

Similarly we define the *openness* of a plane straight-line graph $G = (S, E)$ as $\mathrm{op}(G) = \min_{p \in S} \mathrm{op}_G(p)$ and call G *φ-open* for some angle φ if $\mathrm{op}(G) \geq \varphi$. In other words, a graph is φ-open if and only if every vertex has an incident angle of size at least φ. The *openness* of a class $\mathcal{G}$ of graphs is the supremum over all angles φ such that for every finite point set $S \subset \mathbb{R}^2$ in general position there exists a φ-open connected plane straight-line graph G on S and G is an embedding of some graph from $\mathcal{G}$. For example, the openness of minimum pseudo-triangulations is π.

Observe that without the general position assumption many of the questions become trivial because for a set of collinear points the non-crossing spanning tree is unique—the path that connects them along the line—and its interior points have no incident angle greater than π.

The convex hull of a point set S is denoted with $CH(S)$. Points of S on $CH(S)$ are called vertices of $CH(S)$. Let a, b, and c be three points in the plane that are not collinear. With $\angle abc$ we denote the counterclockwise angle between the segment (b, a) and the segment (b, c) at b.

Results. In this paper we study the openness of several well-known classes of plane straight-line graphs, such as triangulations (Section 2), (general and

Table 1. Openness of several classes of plane straight-line graphs. All given values except for paths on point sets in general position are tight.

Triangulations	Trees	Trees with maxdeg. 3	Paths (convex sets)	Paths (general)
$\frac{2\pi}{3}$	$\frac{5\pi}{3}$	$\frac{3\pi}{2}$	$\frac{3\pi}{2}$	$\frac{5\pi}{4}$

bounded degree) trees (Section 3), and paths (Section 4). The results are summarized in Table 1 above.

2 Triangulations

Theorem 1. *Every finite point set in general position in the plane has a triangulation that is $\frac{2\pi}{3}$-open and this is the best possible bound.*

Proof. Consider a point set $S \subset \mathbb{R}^2$ in general position. Clearly, $\text{op}_G(p) > \pi$ for every point $p \in \text{CH}(S)$ and every plane straight-line graph G on S. We recursively construct a $\frac{2\pi}{3}$-open triangulation T of S by first triangulating $\text{CH}(S)$; every recursive subproblem consists of a point set with a triangular convex hull.

Let S be a point set with a triangular convex hull and denote the three points of $\text{CH}(S)$ with a, b, and c. If S has no interior points, then we are done. Otherwise, let a', b' and c' be (not necessarily distinct) interior points of S such that the triangles $\Delta a'bc$, $\Delta ab'c$ and $\Delta abc'$ are empty (see Fig. 2). Since the sum of the six exterior angles of the hexagon $ba'cb'ac'$ equals 8π, the sum of the three angels $\angle ac'b$, $\angle ba'c$, and $\angle cb'a$ is at least 2π. In particular, one of them, say $\angle cb'a$, is at least $\frac{2\pi}{3}$. We then recurse on the two subsets of S that have $\Delta b'bc$ and $\Delta b'ab$ as their respective convex hulls.

The upper bound is attained by a set S of n points as depicted in Fig. 3. S consists of a point p and of three sets S_a, S_b, and S_c that each contain $\frac{n-1}{3}$ points. S_a, S_b, and S_c are placed at the vertices of an equilateral triangle Δ and p is placed at the barycenter of Δ. Any triangulation T of S must connect p

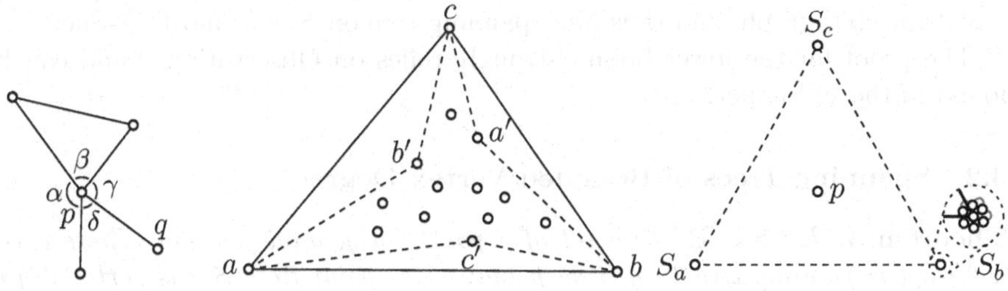

Fig. 1. The incident angles of p

Fig. 2. Constructing a $\frac{2\pi}{3}$-open triangulation

Fig. 3. The openness of triangulations of this point set approaches $\frac{2\pi}{3}$

with at least one point of each of S_a, S_b, and S_c and hence $\mathrm{op}_T(p)$ approaches $\frac{2\pi}{3}$ arbitrarily close. □

3 Spanning Trees

In this section we give tight bounds on the φ-openness of two basic types of spanning trees, namely general spanning trees and spanning trees with bounded vertex degree. Consider a point set $S \subset \mathbb{R}^2$ in general position and let p and q be two arbitrary points of S. Assume w.l.o.g. that p has smaller x-coordinate than q. Let l_p and l_q denote the lines through p and q that are perpendicular to the edge (p,q). We define the *orthogonal slab* of (p,q) to be the open region bounded by l_p and l_q.

Observation 1. *Assume that $r \in S \setminus \{p,q\}$ lies in the orthogonal slab of (p,q) and above (p,q). Then $\angle qpr \leq \frac{\pi}{2}$ and $\angle rqp \leq \frac{\pi}{2}$. A symmetric observation holds if r lies below (p,q).*

Recall that the diameter of a point set is the distance between a pair of points that are furthest away from each other. Let a and b define the diameter of S and assume w.l.o.g. that a has a smaller x-coordinate than b. Clearly, all points in $S \setminus \{a,b\}$ lie in the orthogonal slab of (a,b).

Observation 2. *Assume that $r \in S \setminus \{a,b\}$ lies above a diametrical segment (a,b) for S. Then $\angle arb \geq \frac{\pi}{3}$ and hence at least one of the angles $\angle bar$ and $\angle rba$ is at most $\frac{\pi}{3}$. A symmetric observation holds if r lies below (a,b).*

3.1 General Spanning Trees

Theorem 2. *Every finite point set in general position in the plane has a spanning tree that is $\frac{5\pi}{3}$-open and this is the best possible bound.*

The upper bound is attained by the point set depicted in Fig. 6. Each of the sets $S_i, i \in 1,2,3$ consists of $\frac{n}{3}$ points. If a point $p \in S_1$ is connected to any other point from $S_1 \cup S_2$, then it can only be connected to a point of S_3 forming an angle of at least $\frac{\pi}{3} - \varepsilon$. As the same argument holds for S_2 and S_3, respectively, any connected graph, and thus any spanning tree on S is at most $\frac{5\pi}{3}$-open.

The proof for the lower bound strongly relies on Observation 2 and can be found in the full paper.

3.2 Spanning Trees of Bounded Vertex Degree

Theorem 3. *Let $S \subset \mathbb{R}^2$ be a set of n points in general position. There exists a $\frac{3\pi}{2}$-open spanning tree T of S such that every point from S has vertex degree at most three in T. The angle bound is best possible, even for the much broader class of spanning trees of vertex degree at most $n-2$.*

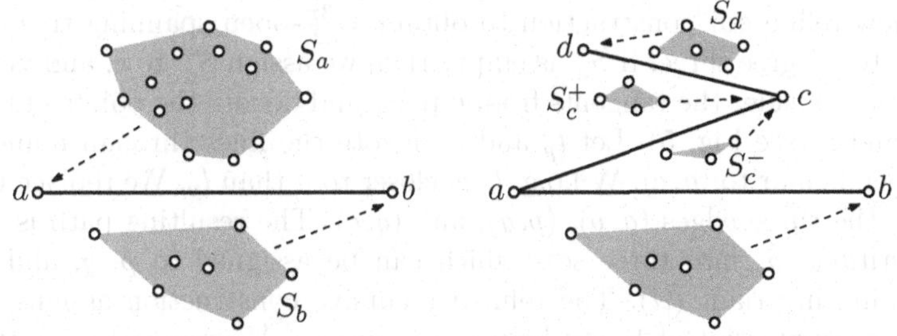

Fig. 4. Constructing a $\frac{3\pi}{2}$-open spanning tree with maximum vertex degree four

Proof. We show in fact that S has a $\frac{3\pi}{2}$-open spanning tree with maximum vertex degree three. To do so, we first describe a recursive construction that results in a $\frac{3\pi}{2}$-open spanning tree with maximum vertex degree four. We then refine our construction to yield a spanning tree of maximum vertex degree three.

Let a and b define the diameter of S. W.l.o.g. a has a smaller x-coordinate than b. The edge (a, b) partitions $S \setminus \{a, b\}$ into two (possibly empty) subsets: the set S_a of the points above (a, b) and the set S_b of the points below (a, b). We assign S_a to a and S_b to b (see Fig 4). Since all points of $S \setminus \{a, b\}$ lie in the orthogonal slab of (a, b) we can connect any point $p \in S_a$ to a and any point of $q \in S_b$ to b and by this obtain a $\frac{3\pi}{2}$-open path $P = \langle p, a, b, q \rangle$. Based on this observation we recursively construct a spanning tree of vertex degree at most four.

If S_a is empty, then we proceed with S_b. If S_a contains only one point p then we connect p to a. Otherwise consider a diametrical segment (c, d) for S_a. W.l.o.g. d has a smaller x-coordinate than c and d lies above (a, c). Either $\angle adc$ or $\angle dca$ must be less than $\frac{\pi}{2}$. W.l.o.g. assume that $\angle dca < \frac{\pi}{2}$. Hence we can connect d via c to a and obtain a $\frac{3\pi}{2}$-open path $P = \langle d, c, a, b \rangle$. The edge (d, c) partitions S_a into two (possibly empty) subsets: the set S_d of the points above (d, c) and the set S_c of the points below (d, c). The set S_c is again partitioned by the edge (a, c) into a set S_c^+ of points that lie above (a, c) and a set S_c^- of points that lie below (a, c). We assign S_d to d and both S_c^+ and S_c^- to c and proceed recursively.

The algorithm maintains the following two invariants: (i) at most two sets are assigned to any point of S, and (ii) if a set S_p is assigned to a point p then p can be connected to any point of S_p and $\text{op}_T(p) \geq \frac{3\pi}{2}$ for any resulting tree T.

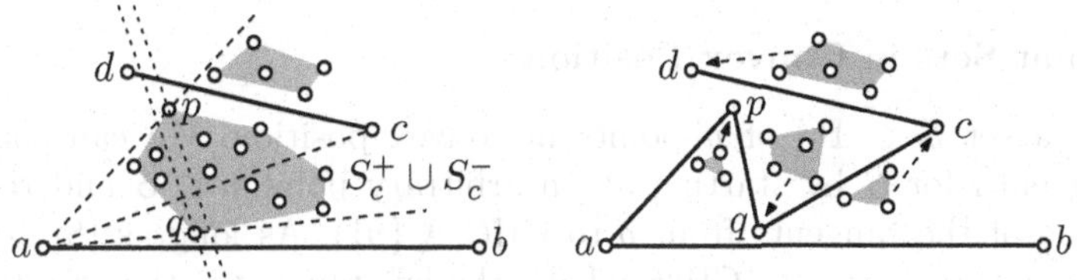

Fig. 5. Constructing a $\frac{3\pi}{2}$-open spanning tree with maximum vertex degree three

We now refine our construction to obtain a $\frac{3\pi}{2}$-open spanning tree of maximum vertex degree three. If S_c^+ is empty then we assign S_c^- to c, and vice versa. Otherwise, consider the tangents from a to S_c and denote the points of tangency with p and q (see Fig. 5). Let l_p and l_q denote the lines through p and q that are perpendicular to (a, c). W.l.o.g. l_q is closer to a than l_p. We replace the edge (a, c) by the three edges (a, p), (p, q), and (q, c). The resulting path is $\frac{3\pi}{2}$-open and partitions S_c into three sets which can be assigned to p, q, and c while maintaining invariant (*ii*). The refined recursive construction assigns at most one set to every point of S and hence constructs a $\frac{3\pi}{2}$-open spanning tree with maximum vertex degree three.

The upper bound is attained by the set S of n points depicted in Fig. 7. S consists of $n-1$ near-collinear points close together and one point p far away. In order to construct any connected graph with maximum degree at most $n-2$, one point of S_1 has to be connected to another point of S_1 and to p. Thus any spanning tree on S with maximum degree at most $n-2$ is at most $\frac{3\pi}{2}$-open. □

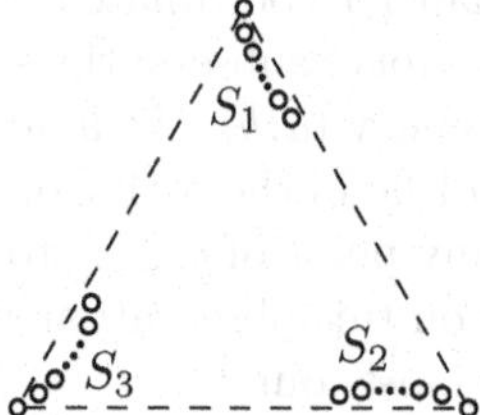

Fig. 6. Every spanning tree is at most $\frac{5\pi}{3}$-open

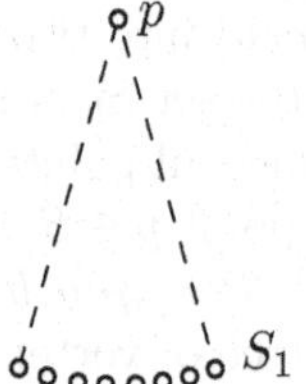

Fig. 7. Every spanning tree with vertex degree at most $n-2$ is at most $\frac{3\pi}{2}$-open

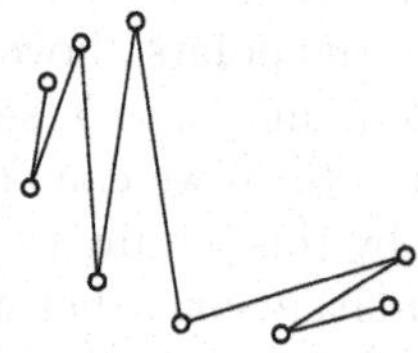

Fig. 8. A zigzag path

4 Spanning Paths

Spanning paths can be regarded as spanning trees with maximum vertex degree two. Therefore, the upper bound construction from Fig. 7 applies to paths as well. We will show below that the resulting bound of $\frac{3\pi}{2}$ is tight for points in convex position, even in a very strong sense: There exists a $\frac{3\pi}{2}$-open spanning path starting from any point.

4.1 Point Sets in Convex Position

Consider a set $S \subset \mathbb{R}^2$ of n points in convex position. We can construct a spanning path for S by starting at an arbitrary point $p \in S$ and recursively taking one of the tangents from p to $\mathrm{CH}(S \setminus \{p\})$. As long as $|S| > 2$, there are two tangents from p to $\mathrm{CH}(S \setminus \{p\})$: the left tangent is the oriented line t_ℓ through p and a point $p_\ell \in S \setminus \{p\}$ (oriented in direction from p to p_ℓ) such that no point from S is to the left of t_ℓ. Similarly, the right tangent is the oriented line t_r through p and a point $p_r \in S \setminus \{p\}$ (oriented in direction from p to p_r)

such that no point from S is to the right of t_r. If we take the left and the right tangent alternately, see Fig. 8, we call the resulting path a *zigzag* path for S.

Theorem 4. *Every finite point set in convex position in the plane admits a spanning path that is $\frac{3\pi}{2}$-open and this is the best possible bound.*

Proof. As a zigzag path is completely determined by one of its endpoints and the direction of the incident edge, there are exactly n zigzag paths for S. (Count directed zigzag paths: There are n choices for the startpoint and two possible directions to continue in each case, that is, $2n$ directed zigzag paths and, therefore, n (undirected) zigzag paths.)

Now consider a point $p \in S$ and sort all other points of S radially around p, starting with one of the neighbors of p along $\mathrm{CH}(S)$. Any angle that occurs at p in some zigzag path for S is spanned by two points that are consecutive in this radial order. Moreover, any such angle occurs in exactly one zigzag path because it determines the zigzag path completely. Since the sum of all these angles at p is less than π, for each point p at most one angle can be $\geq \frac{\pi}{2}$. Furthermore, if p is an endpoint of a diametrical segment for S then all angles at p are $< \frac{\pi}{2}$. Since there is at least one diametrical segment for S, there are at most $n-2$ angles $> \frac{\pi}{2}$ in all zigzag paths together. Thus, there exist at least two spanning zigzag paths that have no angle $> \frac{\pi}{2}$, that is, they are $\frac{3\pi}{2}$-open.

To see that the bound of $\frac{3\pi}{2}$ is tight, consider again the point set shown in Fig. 7. □

A constructive proof for Theorem 4 is given in the full paper. There we also prove the following stronger statement.

Corollary 1. *For any finite set $S \subset \mathbb{R}^2$ of points in convex position and any $p \in S$ there exists a $\frac{3\pi}{2}$-open spanning path for S which has p as an endpoint.*

4.2 General Point Sets

The main result of this section is the following theorem about spanning paths of general point sets.

Theorem 5. *Every finite point set in general position in the plane has a $\frac{5\pi}{4}$-open spanning path.*

Let $S \subset \mathbb{R}^2$ be a set of n points in general position. For a suitable labeling of the points of S we denote a spanning path for (a subset of k points of) S with $\langle p_1, \ldots, p_k \rangle$, where we call p_1 the starting point of the path. Then Theorem 5 is a direct consequence of the following, stronger result.

Theorem 6. *Let S be a finite point set in general position in the plane. Then*

(1) For every vertex q of the convex hull of S, there exists a $\frac{5\pi}{4}$-open spanning path $\langle q, p_1, \ldots, p_k \rangle$ on S starting at q.

(2) *For every edge $\overline{q_1q_2}$ of the convex hull of S there exists a $\frac{5\pi}{4}$-open spanning path starting at either q_1 or q_2 and using the edge $\overline{q_1q_2}$, that is, a spanning path $\langle q_1, q_2, p_1, \ldots, p_k\rangle$ or $\langle q_2, q_1, p_1, \ldots, p_k\rangle$.*

Proof. For each vertex p in a path G the maximum incident angle $\mathrm{op}_G(p)$ is the larger of the two incident angles (except for start- and endpoint of the path). To simplify the case analysis we will consider the smaller angle at each point and prove that we can construct a spanning path such that it is at most $\frac{3\pi}{4}$. We denote with (q, S) a spanning path for S starting at q, and with $(\overline{q_1q_2}, S)$ a spanning path for S starting with the edge connecting q_1 and q_2. The *outer normal cone* of a vertex y of a convex polygon is the region between two half-lines that start at y, are respectively perpendicular to the two edges incident at y, and are both in the exterior of the polygon.

We prove the statements (1) and (2) of Theorem 6 by induction on $|S|$. The base cases $|S| = 3$ are obviously true.

Induction for (1): Let $\mathcal{K} = CH(S \setminus \{q\})$.

Case 1.1. q lies between the outer normal cones of two consecutive vertices y and z of $\mathcal{K}$, where z lies to the right of the ray $\overrightarrow{qy}$.

Induction on $(\overline{yz}, S\setminus\{q\})$ results in a $\frac{5\pi}{4}$-open spanning path $\langle y, z, p_1, \ldots, p_k\rangle$ or $\langle z, y, p_1, \ldots, p_k\rangle$ of $S \setminus \{q\}$. Obviously $\angle qyz \leq \frac{\pi}{2} < \frac{3\pi}{4}$ and $\angle yzq \leq \frac{\pi}{2} < \frac{3\pi}{4}$, and thus we get a $\frac{5\pi}{4}$-open spanning path $\langle q, y, z, p_1, \ldots, p_k\rangle$ or $\langle q, z, y, p_1, \ldots, p_k\rangle$ for S (see Fig. 9).

Case 1.2. q lies in the outer normal cone of a vertex of $\mathcal{K}$.

Let p be that vertex and let y and z be the two vertices of $\mathcal{K}$ adjacent to p, z being to the right of the ray $\overrightarrow{py}$. The three angles $\angle qpz$, $\angle zpy$ and $\angle ypq$ around p obviously add up to 2π. We consider subcases according to which of the three angles is the smallest, the cases of $\angle qpz$ and $\angle ypq$ being symmetric (see Fig. 10).

Case 1.2.1. $\angle zpy$ is the smallest of the three angles.

Then, in particular, $\angle zpy < \frac{3\pi}{4}$. Assume without loss of generality that $\angle qpz$ is smaller than $\angle ypq$ and, in particular, that it is smaller than π. Since q is in the normal cone of p, $\angle qpz$ is at least $\frac{\pi}{2}$, hence $\angle pzq$ is at most $\frac{\pi}{2} < \frac{3\pi}{4}$. Let $S' = S \setminus \{q, z\}$ and consider the path that starts with q and z followed by (p, S'), that is $\langle q, z, p, p_1, \ldots, p_k\rangle$. Note that $\angle zpp_1 \leq \angle zpy$.

Case 1.2.2. $\angle ypq$ is the smallest of the three angles.

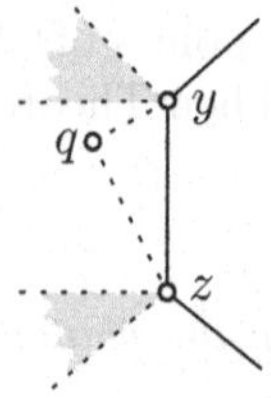

Fig. 9. Case 1.1

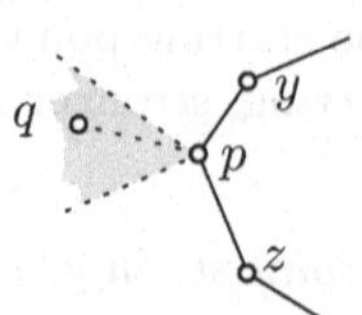

Fig. 10. Case 1.2

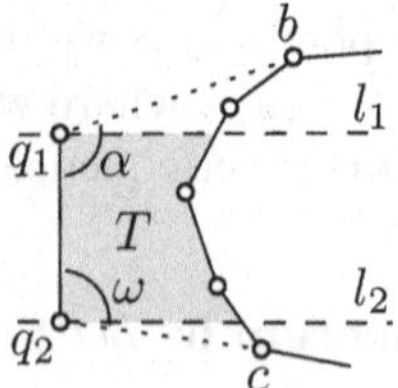

Fig. 11. Case 2

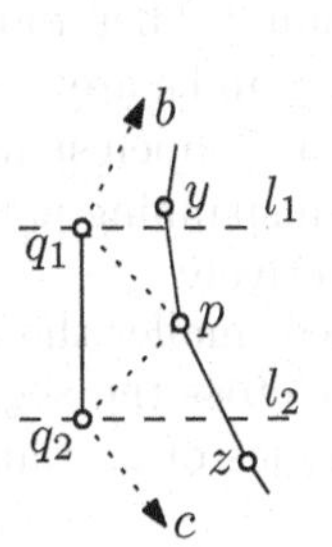

Fig. 12. Case 2.2.1

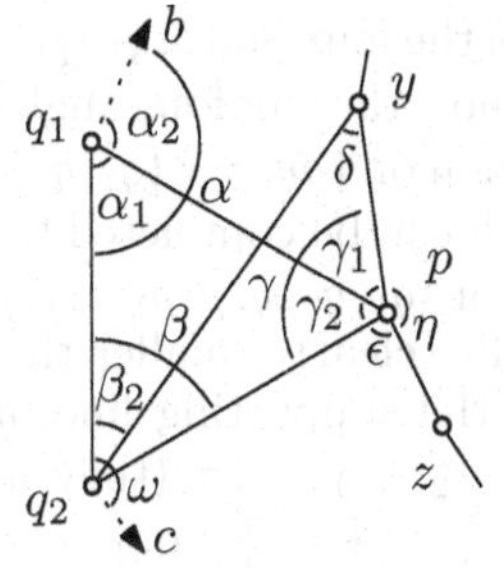

Fig. 13. Case 2.2.1.[1,2]

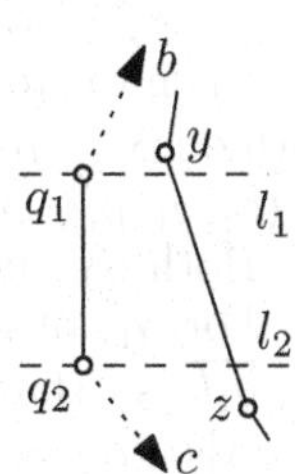

Fig. 14. Case 2.2.2

Then $\angle ypq < \frac{3\pi}{4}$. Moreover, in this case all three angles $\angle qpz$, $\angle ypq$ and $\angle zpy$ are at least $\frac{\pi}{2}$, the first two because q lies in the normal cone of p, the latter because it is is not the smallest of the three angles. We have $\angle qyp < \frac{\pi}{2}$ because this angle lies in the triangle containing $\angle ypq \geq \frac{\pi}{2}$, and $\angle ypq < \frac{3\pi}{4}$ by assumption. We iterate on $(\overline{py}, S \setminus \{q\})$ and get a $\frac{5\pi}{4}$-open spanning path on $S \setminus \{q\}$ by induction, which can be extended to a $\frac{5\pi}{4}$-open spanning path on S, $\langle q, p, y, p_1, \ldots, p_k \rangle$ or $\langle q, y, p, p_1, \ldots, p_k \rangle$, respectively.

Induction for (2): Let b and c be the neighboring vertices of q_1 and q_2 on $CH(S)$, such that $CH(S)$ reads $\ldots, b, q_1, q_2, c, \ldots$ in ccw order (see Fig. 11).

Case 2.1. $\alpha < \frac{3\pi}{4}$ or $\omega < \frac{3\pi}{4}$ (see Fig. 11).
Without loss of generality assume that $\alpha < \frac{3\pi}{4}$. By induction on $(q_1, S \setminus \{q_2\})$ we get a $\frac{5\pi}{4}$-open spanning path $\langle q_1, p_1, \ldots, p_k \rangle$ on $S \setminus \{q_2\}$. As $\angle q_2 q_1 p_1 \leq \alpha < \frac{3\pi}{4}$ we get a $\frac{5\pi}{4}$-open spanning path $\langle q_2, q_1, p_1, \ldots, p_k \rangle$ on S.

Case 2.2. Both α and ω are at least $\frac{3\pi}{4}$.
Let l_1 and l_2 be the lines through q_1 and q_2, respectively, and orthogonal to $\overline{q_1 q_2}$. Further let $\mathcal{K} = CH(S \setminus \{q_1, q_2\})$ and with T we denote the region bounded by $\overline{q_1 q_2}$, l_1, l_2 and the part of $\mathcal{K}$ closer to $\overline{q_1 q_2}$ (see Fig. 11).

Case 2.2.1. At least one vertex p of $\mathcal{K}$ exists in T.
If there exist several vertices of $\mathcal{K}$ in T, then we choose p as the one with smallest distance to $\overline{q_1 q_2}$ (see Fig. 12). Obviously the edges $\overline{q_1 p}$ and $\overline{q_2 p}$ intersect $\mathcal{K}$ only in p and the angles α_1 and β are each at most $\frac{\pi}{2}$ (see Fig. 13).

Case 2.2.1.1. $\gamma_2 > \frac{\pi}{2}$ (see Fig.13).
By induction on $(p, S \setminus \{q_1, q_2\})$ we get a $\frac{5\pi}{4}$-open spanning path $\langle p, p_1, \ldots, p_k \rangle$ for $S \setminus \{q_1, q_2\}$. Moreover the smaller of $\angle q_2 p p_1$ and $\angle p_1 p q_1$ is at most $\frac{2\pi - \frac{\pi}{2}}{2} = \frac{3\pi}{4}$. Thus we get a $\frac{5\pi}{4}$-open spanning path $\langle q_1, q_2, p, p_1, \ldots, p_k \rangle$ or $\langle q_2, q_1, p, p_1, \ldots, p_k \rangle$ for S.

Case 2.2.1.2. $\gamma_2 \leq \frac{\pi}{2}$ (see Fig.13).
Let y and z be vertices of $\mathcal{K}$, with y being the clock-wise neighbor of p and z being the counterclockwise one (b might equal y and c might equal z). At least one of α_1 or β is $\geq \frac{\pi}{4}$. Without loss of generality assume that $\beta \geq \frac{\pi}{4}$, the other case is symmetric. Then q_1, q_2, p, y form a convex four-gon because $\alpha \geq \frac{3\pi}{4}$ and

$\beta \geq \frac{\pi}{4}$ imply that $\angle bpq_2$ in the four-gon b, q_1, q_2, p is less than π. Therefore also $\gamma \leq \angle bpq_2 < \pi$. We will show that all four angles α_1, γ_1, β_2 and δ are at most $\frac{3\pi}{4}$. Then we apply induction on $(\overline{py}, S \setminus \{q_1, q_2\})$ and get a $\frac{5\pi}{4}$-open spanning path on $S \setminus \{q_1, q_2\}$, which can be completed to a $\frac{5\pi}{4}$-open spanning path for S, $\langle q_2, q_1, p, y, p_1, \ldots, p_k\rangle$ or $\langle q_1, q_2, y, p, p_1, \ldots, p_k\rangle$, respectively.

- Both α_1 and $\beta_2 < \beta$ are clearly smaller than $\frac{\pi}{2}$, hence smaller than $\frac{3\pi}{4}$.
- For γ_1, observe that the supporting line of $\overline{yp}$ must cross the segment $\overline{q_1 b}$, so that we have $\alpha_2 + \gamma_1 < \pi$ (they are two angles of a triangle). Also, $\alpha_2 = \alpha - \alpha_1 \geq \frac{3\pi}{4} - \frac{\pi}{2} = \frac{\pi}{4}$, so $\gamma_1 < \frac{3\pi}{4}$.
- Analogously, for δ, observe that the supporting line of $\overline{yp}$ must cross the segment $\overline{q_2 c}$, so that we have $\omega - \beta_2 + \delta < \pi$. Also $\omega - \beta_2 \geq \frac{\pi}{4}$, so $\delta < \frac{3\pi}{4}$.

Case 2.2.2. No vertex of $\mathcal{K}$ exists in T.

Both, l_1 and l_2, intersect the same edge $\overline{yz}$ of $\mathcal{K}$ (in T), with y closer to l_1 than to l_2 (see Fig. 14). We will show that the four angles $\angle yzq_1$, $\angle q_2q_1z$, $\angle yq_2q_1$ and $\angle q_2yz$ are all smaller than $\frac{3\pi}{4}$. Then induction on $(\overline{yz}, S\setminus\{q_1, q_2\})$ yields a path that can be extended to a $\frac{5\pi}{4}$-open path $\langle q_2, q_1, z, y, p_1, \ldots, p_k\rangle$ or $\langle q_1, q_2, y, z, p_1, \ldots, p_k\rangle$. Clearly, the angles $\angle q_2q_1z$ and $\angle yq_2q_1$ are both smaller than $\frac{\pi}{2}$. The sum of $\angle q_2yz + \angle cq_2y$ is smaller than π because the supporting line of $\overline{yz}$ intersects the segment $\overline{q_2c}$. Now, $\angle cq_2y$ is at least $\frac{\pi}{4}$ by the assumption that $\angle cq_2q_1 \geq \frac{3\pi}{4}$. So, $\angle q_2yz < \frac{3\pi}{4}$. The symmetric argument shows that $\angle yzq_1 < \frac{3\pi}{4}$. □

Note that for Theorem 6 it is essential that the predefined starting point of a $\frac{5\pi}{4}$-open path is an extreme point of S, as an equivalent result is in general not true for interior points. As a counter example consider a regular n-gon with an additional point in its center. It is easy to see that for sufficiently large n starting at the central point causes a path to be at most $\pi + \varepsilon$-open for a small constant ε. Similar, non-symmetric examples already exist for $n \geq 6$ points, and analogously, if we require an interior edge to be part of the path, there exist examples bounding the openness by $\frac{4\pi}{3} + \varepsilon$ [17]. Despite these examples we conclude this section with the following conjecture.

Conjecture 1. *Every finite point set in general position in the plane has a $\frac{3\pi}{2}$-open spanning path.*

Acknowledgments. Research on this topic was initiated at the third European Pseudo-Triangulation working week in Berlin, organized by Günter Rote and André Schulz. We thank Sarah Kappes, Hannes Krasser, David Orden, Günter Rote, André Schulz, Ileana Streinu, and Louis Theran for many valuable discussions. We also thank Sonja Čukić and Günter Rote for helpful comments on the manuscript.

References

1. Aichholzer, O., Hackl, T., Hoffmann, M., Huemer, C., Santos, F., Speckmann, B., Vogtenhuber, B.: Maximizing Maximal Angles for Plane Straight Line Graphs - Extended Abstract. In: Abstracts 23rd European Workshop Comput. Geom. 98–101 (2007)

2. Aurenhammer, F., Xu, Y.-F.: Optimal Triangulations. Encyclopedia of Optimization 4, 160–166 (2000)
3. Aichholzer, O., Aurenhammer, F., Krasser, H., Brass, P.: Pseudo-Triangulations from Surfaces and a Novel Type of Edge Flip. SIAM J. Comput. 32(6), 1621–1653 (2003)
4. Aggarwal, A., Coppersmith, D., Khanna, S., Motwani, R., Schieber, B.: The Angular-Metric Traveling Salesman Problem. SIAM J. Comput. 29(3), 697–711 (1999)
5. Arkin, E., Fekete, S., Hurtado, F., Mitchell, J., Noy, M., Sacristán, V., Sethia, S.: On the Reflexivity of Point Sets. In: Discrete and Computational Geometry: The Goodman-Pollack Festschrift, Springer, Heidelberg, Algorithms and Combinatorics 25, 139–156 (2003)
6. Bárány, I., Pór, A., Valtr, P.: Paths with no Small Angles. Manuscript in preparation (2006)
7. Bern, M., Eppstein, D., Gilbert, J.: Provably Good Mesh Generation. J. Comput. Syst. Sci. 48(3), 384–409 (1994)
8. Dai, Y., Katoh, N., Cheng, S.-W.: LMT-Skeleton Heuristics for Several New Classes of Optimal Triangulations. Comput. Geom. Theory Appl. 17(1-2), 51–68 (2000)
9. Eppstein, D.: The Farthest Point Delaunay Triangulation Minimizes Angles. Comput. Geom. Theory Appl. 1(3), 143–148 (1992)
10. Edelsbrunner, H., Tan, T.S., Waupotitsch, R.: An $O(n^2 \log n)$ Time Algorithm for the Minmax Angle Triangulation. SIAM J. Sci. Stat. Comput. 13(4), 994–1008 (1992)
11. Fekete, S.P., Woeginger, G.J.: Angle-Restricted Tours in the Plane. Comput. Geom. Theory Appl. 8(4), 195–218 (1997)
12. Haas, R., Orden, D., Rote, G., Santos, F., Servatius, B., Servatius, H., Souvaine, D., Streinu, I., Whiteley, W.: Planar Minimally Rigid Graphs and Pseudo-Triangulations. Comput. Geom. Theory Appl. 31(1-2), 31–61 (2005)
13. Kirkpatrick, D., Snoeyink, J., Speckmann, B.: Kinetic Collision Detection for Simple Polygons. Internat. J. Comput. Geom. Appl. 12(1-2), 3–27 (2002)
14. Keil, J.M., Vassilev, T.S.: The Relative Neighbourhood Graph is a Part of Every 30deg-Triangulation. Abstracts 21st European Workshop Comput. Geom. 9–12 (2005)
15. Rote, G., Santos, F., Streinu, I.: Pseudo-Triangulations – a Survey. Manuscript (2006) `http://arxiv.org/abs/math/0612672`
16. Streinu, I.: Pseudo-Triangulations, Rigidity and Motion Planning. Discrete Comput. Geom. 34(4), 587–635 (2005)
17. Vogtenhuber, B.: On Plane Straight Line Graphs. Master's Thesis, Graz University of Technology, Graz, Austria (January 2007)

Cuttings for Disks and Axis-Aligned Rectangles

Eynat Rafalin[1], Diane L. Souvaine[2], and Csaba D. Tóth[3]

[1] Google Inc., Mountain View, CA
eynat@google.com
[2] Department of Computer Science, Tufts University, Medford, MA
dls@cs.tufts.edu
[3] Department of Mathematics, MIT, Cambridge, MA
toth@math.mit.edu

Abstract. We present new asymptotically tight bounds on *cuttings*, a fundamental data structure in computational geometry. For n objects in space and a parameter $r \in \mathbb{N}$, an $\frac{1}{r}$*-cutting* is a covering of the space with simplices such that the interior of each simplex intersects at most n/r objects. For n pairwise disjoint disks in $\mathbb{R}^3$ and a parameter $r \in \mathbb{N}$, we construct a $\frac{1}{r}$-cutting of size $O(r^2)$. For n axis-aligned rectangles in $\mathbb{R}^3$, we construct a $\frac{1}{r}$-cutting of size $O(r^{3/2})$.

As an application related to multi-point location in three-space, we present tight bounds on the cost of spanning trees across barriers. Given n points and a finite set of disjoint disk barriers in $\mathbb{R}^3$, the points can be connected with a straight line spanning tree such that every disk cuts at most $O(\sqrt{n})$ edges of the tree. If the barriers are *axis-aligned rectangles*, then there is a straight line spanning tree such that every rectangle cuts $O(n^{1/3})$ edges. Both bounds are the best possible.

1 Introduction

Divide-and-conquer strategies are omnipresent in computer science. In problems involving multivariable reals, one of the most successful methods over the last two decades has been the *partition technique* (in particular, *cuttings*) in computational geometry. They are indispensable for optimal data structures that support range searching, point location, motion planning, among others, and they are also used in currently best combinatorial bounds for hard Erdős-type discrete geometry problems [7,24]. For a set of n objects in $\mathbb{R}^d$ and a parameter $r \in \mathbb{N}$, a $\frac{1}{r}$*-cutting* is a finite collection of simplices that cover $\mathbb{R}^d$ and such that the interior of each simplex intersects at most n/r objects. Even though the definition of cuttings allows overlapping simplices, all cuttings we construct consist of interior-disjoint simplices, which form *subdivisions* of $\mathbb{R}^d$.

Optimal size cuttings are known for hyperplanes and $(d-1)$-dimensional simplices in $\mathbb{R}^d$. In this paper, we present new tight bounds on the minimum size of cuttings for disjoint 2-dimensional objects in three-space. Our main result is an optimal size cutting for disjoint disks in three-space. For brevity, we write *disk* for any planar set of constant description complexity (that is, a set in $\mathbb{R}^3$

F. Dehne, J.-R. Sack, and N. Zeh (Eds.): WADS 2007, LNCS 4619, pp. 470–482, 2007.

that lies in a plane and is described by a constant number of algebraic inequalities of constant degree).

Theorem 1. *For every set of n pairwise disjoint disks in* $\mathbb{R}^3$ *and every r,* $1 \leq r \leq n+1$*, there is a* $\frac{1}{r}$*-cutting of size* $O(r^2)$*. This bound is the best possible.*

Similar bounds were previously known only for disjoint *triangles* in $\mathbb{R}^3$. Note that a subdivision of $\mathbb{R}^3$ into $O(r^2)$ cells, each bounded by a constant number of algebraic surfaces (rather than simplices bounded by hyperplanes), easily follows from known techniques. In some applications, such *pseudo-cuttings* (with curved boundaries) are satisfactory, others require partitions into *convex* simplices (e.g., Theorems 3 and 4 below). The challenge, that we partially resolve in this paper, is to construct optimal size cuttings (with "straight" simplices) for curved objects. We also give a randomized algorithm that computes, for an input of n disjoint disks, an $\frac{1}{r}$-cutting of size $O(r^2)$ in polynomial expected time. It can be derandomized with standard techniques [21].

For disjoint *axis-aligned rectangles*, we can construct substantially smaller cuttings. This is the first sub-quadratic bound for a family of 1- or 2-dimensional objects in $\mathbb{R}^3$.

Theorem 2. *For every set of n pairwise disjoint axis-aligned rectangles in* $\mathbb{R}^3$ *and every r,* $1 \leq r \leq n+1$*, there is an* $\frac{1}{r}$*-cutting of size* $O(r^{3/2})$*. This bound is best possible.*

An application: Spanning trees across barriers. A $(d-1)$-dimensional object b *cuts* a line segment e (equivalently, e *stabs* b) in $\mathbb{R}^d$ if the relative interiors of e and b intersect but the hyperplane spanned by b does not contain e. Chazelle and Welzl [11] showed that a set S of n points (*sites*) in $\mathbb{R}^d$, $d \geq 2$, can be connected by a straight line spanning tree such that every hyperplane cuts at most $O(n^{1-1/d})$ edges of the tree. This bound is tight apart from the constant factor (e.g., for any spanning tree on n points in the $\lfloor n^{1/d} \rfloor \times \ldots \times \lfloor n^{1/d} \rfloor$ integer grid in $\mathbb{R}^d$, there is an axis-aligned hyperplane that cuts $\Omega(n^{1-1/d})$ edges). In particular, for n points in the plane, there is a spanning tree that stabs every line $O(\sqrt{n})$ times. Interestingly, if we replace the lines by a set of disjoint line segments, there is a spanning tree that stabs every segment only a constant number of times. This constant is between 3 and 4 in the worst case by recent results of [2] and [15]. We address analogous problems involving disjoint barriers in $\mathbb{R}^3$, which have applications on multi-point location data structures in three-space [29].

Theorem 3. *Given a set S of n points and a finite set B of pairwise disjoint disk barriers in* $\mathbb{R}^3$*, there is a straight line spanning tree T on the vertex set S such that every barrier in B cuts at most* $O(\sqrt{n})$ *edges of T. There are n points and* $2\sqrt{n}$ *disjoint circular disks in* $\mathbb{R}^3$ *such that for every spanning tree T on S, a disk cuts* $\Omega(\sqrt{n})$ *edges of T on average.*

Theorem 4. *Given a set S of n points and a finite set B of pairwise disjoint axis-aligned rectangles in* $\mathbb{R}^3$*, there is a straight line spanning tree T on the vertex* *set S such that every rectangle in B cuts at most $O(n^{1/3})$ edges of T. There are n points and $O(\sqrt{n})$ disjoint axis-aligned rectangles in $\mathbb{R}^3$ such that for every spanning tree T on S, a rectangle cuts $\Omega(n^{1/3})$ edges of T on average.*

It follows that the total cost of such a spanning tree is bounded by $O(|B| \cdot n^{1/3})$ for disks and by $O(|B| \cdot \sqrt{n})$ for axis-aligned rectangles in the worst case, which are also best possible bounds. Note, however, that for computing a *minimum* cost spanning tree for given sets S and B, one can use a *min-max weight spanning tree* algorithm by Camerini [5]. Compute the number of barriers stabbed by each of the $\binom{n}{2}$ edges of a complete straight line graph on S, which gives an integer weight on each edge. Camerini's algorithm computes a spanning tree, for which our combinatorial bounds apply, in $O(n^2)$ time. A randomized linear-time algorithm of Krager *et al.* [16] can compute the *minimum weight spanning tree* for a given configuration.

Related previous work on cuttings. Cuttings were introduced in the late eighties by Clarkson and Shor [12], and their ideas were later gradually improved and simplified [7,8,22,30]. For an arrangement of hyperplanes in $\mathbb{R}^d$, there is a $\frac{1}{r}$-cutting of size $O(r^d)$ [20]. For disjoint $(d-1)$-dimensional simplices in $\mathbb{R}^d$, there is an $\frac{1}{r}$-cutting of size $O(r^{d-1})$. Pellegrini [28] combined these results and showed that if the arrangement of n $(d-1)$-simplices in $\mathbb{R}^d$ has K vertices, there is an $\frac{1}{r}$-cutting of size $O(r^{d-1} + (K/n^d)r^d)$, which is best possible. Similar arguments show that there is a $\frac{1}{r}$-pseudocutting of size $O(r^d \text{ polylog } r)$ for $(d-1)$-dimensional semi-algebraic surfaces of constant maximal degree [9]; however, pseudo-cuttings typically produce a covering with nonconvex regions. Two essentially different methods have been developed for constructing optimal cuttings: one is based on vertical decompositions and the other on a sparse ε-nets. We briefly compare them with our proof strategies.

One method is a delicate construction by Chazelle and Friedman [10,22] based on the following components: (i) the Clarkson-Shor random sampling technique [12]; (ii) a vertical decomposition algorithm; (iii) and a certain locality property of the decomposition that allows using tail estimates (for more details and examples, see [22]). The bottleneck for this technique is often the *vertical decomposition*, which is the partition of the space into vertical cylinders (often called *trapezoids* in the plane) of bounded complexity such that each cylinder has at most two nonvertical sides. For example, Mulmuley [23] showed that r disjoint triangles in $\mathbb{R}^3$ have a vertical decomposition into $O(r^2)$ trapezoids (see also [4] [1]). A straightforward application of this method, for instance, gives an $\frac{1}{r}$-cutting of size $O(r^2)$ for m disjoint triangles in $\mathbb{R}^3$.

Vertical decompositions exists for several types of objects in space, although tight bounds are known only in very few cases. Chazelle *et al.* [9] showed that there is a vertical decomposition of size $O(r^2)$ for r disjoint disks, and this bound is sharp. An almost sharp upper bound on cuttings for hypersurfaces

[1] De Berg, Guibas, and Halperin [3] extended this result and showed that m not necessarily disjoint triangles in $\mathbb{R}^3$ have a vertical decomposition of size $O(n^{2+\varepsilon} + K)$ for every $\varepsilon > 0$, where and K is the complexity of the arrangement of the triangles.

in four-space was recently proved by Koltun [17,18]. The cells of such vertical decompositions, however, are bounded by semi-algebraic surfaces, and may not be *convex*. This explains why the technique of Chazelle and Friedman [10] cannot produce optimal size $\frac{1}{r}$-cuttings (with "straight" simplices) for disks in $\mathbb{R}^3$.

The other method known for constructing optimal cuttings for hyperplanes in $\mathbb{R}^d$ is due to Chazelle [6] (following earlier work by Agarwal [1]). It is a hierarchical decomposition based on so-called sparse ε-nets (a multi-purpose random sample related to ε-nets—a combinatorial tool introduced by Haussler and Welzl [14]). We adapt this method to disjoint disks in $\mathbb{R}^3$ by combining it with techniques from binary space partitions. Two recent results also followed an argument reminiscent of [6]: a construction of almost optimal pseudocuttings "sensitive" to a collection of algebraic curves (which means that a curve intersects few pseudo-simplices on average [19]); and an algorithm for counting the number of intersecting pairs in a set of triangles in 3-space [13]. Prior to our work, Pellegrini [27] has combined binary space partition methods with sparse ε-nets when he constructed optimal cuttings for triangles in $\mathbb{R}^3$.

Binary space partitions. The *binary space partition* (for short, *BSP*) is a data structure produced by a simple hierarchical partition scheme, called *BSP algorithm*: Given a set B of disjoint $(d-1)$-dimensional objects in the interior of a convex cell σ, $\sigma \subset \mathbb{R}^d$, a BSP algorithm partitions σ along a hyperplane into two convex subcells σ_1 and σ_2 (while fragmenting the input objects as well), and recurses on the objects clipped in σ_1 and σ_2, independently, until the interior of each resulting cell is empty of input objects. The *size* of a BSP is the number of fragments of the input objects; intuitively, it measures the fragmentation caused by the partition. Paterson and Yao [25,26] constructed a BSP of size $O(r^2)$ for r disjoint triangles and a BSP of size $O(r^{3/2})$ for r disjoint axis-aligned rectangles in $\mathbb{R}^3$; they also showed that these bounds are best possible.

Notice that our bounds on the size of $\frac{1}{r}$-cuttings match Paterson and Yao's bounds on the size of the BSPs of r disjoint objects. A BSP naturally constructs a subdivision of the input cell σ into convex subcells. A convex cell in $\mathbb{R}^3$ with k vertices can be partitioned into $O(k)$ simplices, so a BSP leads to a subdivision of σ into simplices whose number is proportional to the *complexity* of the subdivision (which is the total number of faces of dimensions 1, 2, and 3 over all cells). Note, however, that the size of a BSP may be much smaller than the complexity of the resulting subdivision (for example, a BSP of size $O(n^2)$ can produce a subdivision of the space into $O(n^2)$ convex cells of $\Theta(n^3)$ total complexity). For disjoint axis-aligned rectangles, we devise a deterministic BSP-like partition scheme for constructing a $\frac{1}{r}$-cutting of size $O(r^{3/2})$ in the full version of this paper.

2 Preliminaries

Overview of our construction of cuttings. We construct an optimal size cutting for disjoint disks in $\mathbb{R}^3$ by combining several layers of hierarchical space decompositions. We introduce a binary relation between disks (a *avoids* b), and

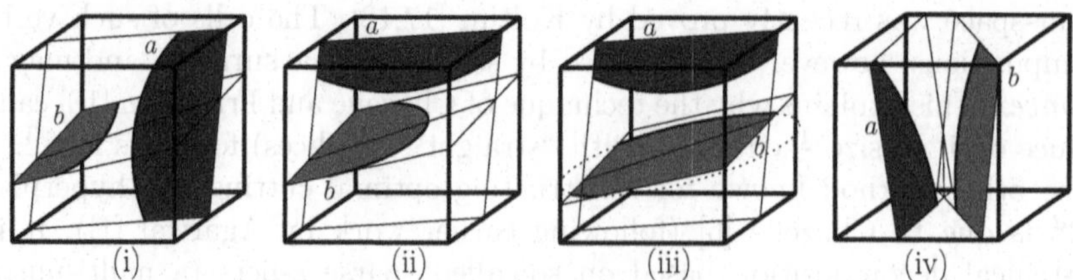

Fig. 1. Disks a and b clipped within a cube σ. The (relative) boundaries ∂a and ∂b clipped within σ are bold. In (i) a and b do not avoid each other; in (ii) b avoids a, but a does not avoid b; in (iii) a and b avoid each other but $\pi_a \cap b \cap \sigma \neq \emptyset$; and in (iv) a and b avoid each other and $\pi_a \cap b \cap \sigma = \pi_b \cap a \cap \sigma = \emptyset$.

call a configuration *sparse* if the disks mutually avoid each other. For k mutually avoiding disks, we subdivide the space into $k+1$ convex polytopes, the interior of each of which is disjoint from the disks. For mutually avoiding disks, we use this algorithm recursively to build an $\frac{1}{r}$-cutting of size $O(r^{1+\delta})$ for every fixed $\delta > 0$. Our main algorithm for arbitrary disjoint disks is reminiscent of a scheme originally introduced by Agarwal and Chazelle [1,6] to construct optimal size cuttings for hyperplanes. We partition $\mathbb{R}^3$ into simplices recursively using planes spanned by so-called *sparse ε-nets* of a constant number of disks. The size of the resulting $\frac{1}{r}$-cutting is bounded by $O(r^2)$, which is shown by a charging scheme. The basis for our charging scheme is that the total number of nonavoiding pairs cannot increase when the problem in split into subproblems (even though a disk may be fragmented into many pieces, which can occur in several subproblems).

Full and sparse configurations. We define a binary relation between disks in $\mathbb{R}^3$. Consider a set B of n disjoint disks in $\mathbb{R}^3$. Let π_b denote the plane containing disk $b \in B$, and let ∂b denote the (relative) boundary of b. For two disks $a, b \in B$, we say that *a avoids b* with respect to a convex cell σ if $\pi_a \cap \partial b \cap \mathrm{int}(\sigma) = \emptyset$ (Fig. 1). This relation is not symmetric: It is possible that a avoids b but b does not avoid a in a cell σ. Note, also, that even if a avoids b in σ, the plane π_a may intersect $b \cap \sigma$ (but then, π_a intersects ∂b outside of σ).

The *multiplicity* of a nonavoiding pair $(a, b) \in B^2$ with respect to σ is the number of intersection points of $\pi_a \cap \partial b$ that lie in $\mathrm{int}(\sigma)$. We measure the complexity of a set B of disjoint disks relative to the interior of a convex cell σ by $\tau(B, \sigma)$, which is the sum of multiplicities of all nonavoiding pairs with respect to σ.

For a convex cell σ, we define $B_\sigma = \{b \in B : b \cap \mathrm{int}(\sigma) \neq \emptyset\}$ to be the set of disks that intersect the interior of σ, and $\hat{B}_\sigma = \{b \cap \sigma : b \cap \sigma \neq \emptyset\}$ to be the portions of the disks of B_σ clipped within σ. Letting $n_\sigma = |B_\sigma|$, it is clear that $\tau(B, \sigma) = O(n_\sigma^2)$, since the multiplicity of every pair is bounded by a constant (depending on their description complexity). We use the following crucial property of the measure τ.

Proposition 1. *For any subdivision Ξ of a convex cell σ into convex subcells, we have* $\sum_{\xi\in\Xi}\tau(B,\xi)\leq\tau(B,\sigma)$.

Proof. Every intersection point $p\in\pi_a\cap\partial b\cap\mathrm{int}(\sigma)$ lies in the interior of at most one subcell $\xi\in\Xi$. Even if $p\in\mathrm{int}(\xi)$ for some $\xi\in\Xi$, it is counted in $\tau(B,\xi)$ only if disk a intersects $\mathrm{int}(\xi)$. □

The technically most difficult part of the proof of Theorem 1 is the construction of a subquadratic $\frac{1}{r}$-cutting for a set of mutually avoiding disks. It uses sparse ε-nets and a randomized incremental subdivision scheme. In particular, it uses two randomized constructions: (1) it chooses a random sample of disks (which is a sparse ε-net) *and* (2) it subdivides a cell incrementally along sample disks in a random order. Our incremental subdivision scheme produces convex cells, but the benefits of ε-nets hold for simplices only: our construction maneuvers between convex and simplicial subdivisions.

ε-nets. We use a few basic facts about ε-nets and range spaces in Sections 3 and 4. (For an in-depth overview, refer to [8] or [22]) A *range space* is a set system (P,Q) on a ground set P and some subsets $Q\subset 2^P$. The *VC-dimension* of a range space is the size of the largest subset $S\subset P$ such that all $2^{|S|}$ subsets of S can be written as $S\cap q$ for some range $q\in Q$. In this paper, we consider range spaces (B,Q) where B is the set of geometric objects in $\mathbb{R}^d$ and every simplex σ defines a range $q_\sigma=\{b\in B: b\cap\mathrm{int}(\sigma)\neq\emptyset\}$. It is not difficult to see that if B is a set of objects of bounded description complexity, then the corresponding range space (B,Q) has bounded VC-dimension [8,22].

Haussler and Welzl [14] introduced the concept of *ε-nets* in range spaces. An ε-net of a finite range space (P,Q), for a constant $\varepsilon>0$, is a set $S\subset P$ such that for every $q\in Q$ with $|P\cap q|\geq\varepsilon\cdot|P|$, we have $S\cap q\neq\emptyset$. For every range space of VC-dimension δ, a Bernoulli sample $S\subset P$, that contains every element $p\in P$ independently at random with probability $(\frac{cd}{\varepsilon}\log\frac{d}{\varepsilon})^{-1}$ for an absolute constant $c>0$, is an ε-net with constant probability [8].

Sparse ε-nets. Chazelle [6] noticed that a random sample drawn from a bounded VC-dimensional range space (P,Q), $|P|=n$, is not only an ε-net, but it may also preserve other properties of P with constant probability. He considered the range space of hyperplanes intersecting a simplex σ in $\mathbb{R}^d$. We state this result in broader terms. Assume that we are given a t-uniform hypergraph G with e hyperedges[2] on the vertex set P. The random sample $S\subset P$ that contains every element $p\in P$ independently at random with probability $p=(\frac{cd}{\varepsilon}\log\frac{d}{\varepsilon})^{-1}$ is expected to span $e\cdot p^t$ hyperedges of G. Chazelle showed that (i) S is an ε-net, (ii) it has size $\Theta(n\cdot p)$, *and* (iii) it spans at most $2e\cdot p^t$ hyperedges of G with constant probability. We use this observation for $t=2$ (that is, when G is a graph defined on B) in Section 4.

[2] A t-uniform hypergraph, $t\in\mathbb{N}$, is a set system on a ground set (*vertex set*) where every set (*hyperedge*) has t elements. A 2-uniform hypergraph, for instance, is a simple graph.

3 Subdivisions for Mutually Avoiding Disks

In this section, we construct in three steps an $O(r^{1+\delta})$ size $\frac{1}{r}$-cutting for mutually avoiding disks for every $\delta > 0$. First, we present a simple algorithm that subdivides a convex cell into $k+1$ convex subcells along k mutually avoiding disks. Second, we extend this scheme to a randomized algorithm that subdivides a simplex into $k+1$ convex subcells along k disks drawn randomly from a set of n mutually avoiding disks, and partitions the n disks into $O(n \log k)$ pieces. Third, we apply this subdivision hierarchically to construct an $\frac{1}{r}$-cutting of size $O(r^{1+\delta})$ for mutually avoiding disks, where $\delta > 0$ is an arbitrary but fixed constant.

Lemma 1. *For k mutually avoiding disks w.r.t. a convex polytope σ in $\mathbb{R}^3$, there is a subdivision of σ into $k+1$ convex cell such that the interior of every cell is disjoint from the disks and every cell is bounded by planes spanned by the disks and sides of σ.*

Extend incrementally every disk b_i, $i = 1, 2, \ldots, k$, to a planar polygon w_i (*wall*) in the plane π_i spanned by b_i so that the (relative) interior of w_i is disjoint from other disks and previous walls; and its boundary ∂w_i lies on the boundary of σ, another disk, or a previous wall. See the full version of the paper for a detailed proof.

Subdivision for a subset of disks. Next we subdivide a simplex σ along the elements of a random sample $S \subset B$ of disks. If we compute a subdivision described in Lemma 1 for cell σ and the k disks in S, then some of the disks $b \in B \setminus S$ may be split into several *fragments*. We can deduce an upper bound on the number of fragments with the following lemma.

Lemma 2. *Given a set B of n mutually avoiding disks w.r.t. a simplex σ and a random sample $S \subset B$ of size k. With probability at least 3/4 there is a subdivision of σ into $k+1$ convex cells such that the interior of every cell is disjoint from every sample disk in S, every cell is bounded by planes spanned by disks of S and sides of σ, and the total number of fragments of disks of B is $O(n \log k)$.*

Proof. Let (B, R) be the range space over B where every planar quadrilateral domain q defines a range $\{b \in B : b \cap q \neq \emptyset\}$, containing all disks in B that intersect q. The range space (B, R) has finite VC-dimension. This implies that with probability at least 3/4, the sample S is an ε-net for (B, R) with some $\varepsilon = \Omega((\log k)/k)$. In what follows, we assume that S is an ε-net for (B, R).

Label the k sample disks of S by a random permutation as $b_1, b_2, \ldots, b_k$; and apply the subdivision algorithm described in Lemma 1. It subdivides σ along polygonal walls, each lying in a plane spanned by a disk of S, into $k+1$ convex cells such that the interior of each cell is disjoint from the sample disks. It remains to bound the expected number of fragments of disks of $B \setminus S$.

We show that the wall erected in plane π_i in step i of the incremental subdivision algorithm is expected to cut at most $O(n/i)$ disks. This implies that the expected number of fragments over k steps is $O(n \log k)$, and so there is a permutation where the number of fragments is $O(n \log k)$.

Consider the set $B_i \subset B$ of disks that intersect the plane π_i spanned by $b_i \in S$. Let $L_i = \{b \cap \pi_i : b \in B_i\}$ be the set of intersection segments of these disks with the polygon $\Delta_i = \sigma \cap \pi_i$. Since b_i avoids every disk in B_i, every segment in L_i connects two points on the boundary of the polygon $\Delta_i = \sigma \cap \pi_i$. We introduce a partial order on the segments of L_i: we say that $\ell_1 \prec \ell_2$ if ℓ_1 separates b_i and ℓ_2 in Δ_i. Note that Δ_i is a triangle or a quadrilateral (since σ is a simplex), and each segment in L connects two sides of Δ_i. Select $O(1/\varepsilon) = O(k/\log k)$ quadrilaterals in π_i, each containing at least εn segments of L_i. Since S is an ε-net for (B, R), every quadrilateral contains a segment $b_j \cap \Delta_i$, $b_j \in S$.

At step i, a disk $b \in B \setminus S$ is split into two fragments if the segment $b \cap \Delta_i$ lies in a quadrilateral that is not separated from b_i by any previous sample disk b_j, $j < i$. Each of the $O(1/\varepsilon) = O(k/\log k)$ quadrilaterals contains a sample disk b_j, $j < i$, with probability at least $(i-1)/k$. The expected number of quadrilaterals not separated from b_i is $O((1/\varepsilon)/i)) = O(k/(i \log k))$. Hence, the expected number of disks fragmented in step i is $O(1/(\varepsilon i)) \cdot O(\varepsilon n) = O(n/i)$. □

Cuttings for mutually avoiding disks in a simplex. We are now ready to prove the main result of this section.

Lemma 3. *For every $\delta > 0$, there is a constant $c(\delta)$ with the following property. Given a set B of n mutually avoiding disks w.r.t. a simplex σ_0 in $\mathbb{R}^3$, there is a subdivision of σ_0 into $c(\delta)n^{1+\delta}$ simplices such that the interior of every simplex is disjoint from any disk of B.*

Proof. We show that the following algorithm computes a required subdivision. The input is a simplex σ_0 and a set B of mutually avoiding disks w.r.t. σ_0. Let $k > 0$ be a constant to be specified later. Put $i := 0$ and $\mathcal{C}_0 := \{\sigma_0\}$. For every $i \in \mathbb{N}$, $\mathcal{C}_i$ will be a subdivision of σ_0 into convex cells (not necessarily simplices). Repeat the following step until all cells in $\mathcal{C}_i$ are disjoint from any disk in B for some $i \in \mathbb{N}$.

We compute a subdivision $\mathcal{C}_{i+1}$, which is a refinement of $\mathcal{C}_i$. Cells of $\mathcal{C}_i$ that are already disjoint from the disks are not refined. Refine every convex cell $\sigma \in \mathcal{C}_i$ where $B_\sigma \neq \emptyset$ as follows. With probability at least 3/4, a random sample S_σ of size $c_1 k \log k$ drawn from B_σ is an $\frac{1}{k}$-net for the range space (B_σ, Q) defined in Section 2, where $c_1 > 0$ is an absolute constant (which depends only on the VC-dimension of the range space). Apply Lemma 2 for the disks B_σ and the *initial simplex* σ_0. With probability at least 3/4, we obtain a subdivision $\mathcal{D}_\sigma$ of σ_0 into at least $|S_\sigma| = c_1 k \log k$ convex cells such that the interior of every cell is disjoint of the sample disks, and the disks of $\hat{B}_\sigma$ are fragmented into at most $c_2 n_\sigma \log k$ pieces, with another absolute constant $c_2 > 0$.

We say that a cell $\sigma' \in \mathcal{D}_0$, $\sigma' \subset \sigma_0$, is *heavy* if it intersects more than $n_\sigma/(c_2 \log k)^\alpha$ disks of $\hat{B}_\sigma$, where α satisfies $(c_2 \log k)^\alpha \leq k$ (or, equivalently, $\alpha \leq \log k/\log(c_2 \log k)$). At most $(c_2 \log k)^{\alpha+1}$ cells of $\mathcal{D}_0$ are heavy, since there are at most $c_2 n_\sigma \log k$ fragments of $\hat{B}_\sigma$ over all cells of $\mathcal{D}_\sigma$. Triangulate every heavy cell $\sigma' \in \mathcal{D}_0$. Since S_σ is an $\frac{1}{k}$-net for B_σ, each simplex in the triangulation of σ' intersect at most $n_\sigma/k \leq n_\sigma/(c_2 \log r)^\alpha$ disks of B_σ. Since every cell of $\mathcal{D}_\sigma$ is bounded by at most $c_1 k \log k + 4$ faces (by Lemma 2), its triangulation consists of at most $4c_1 k \log k$ simplices. (Here we use Euler's polyhedron theorem pertaining to the surface of a polytope in 3-space.) We obtain a subdivision of σ_0 into at most $c_1 k \log k + (c_2 \log k)^{\alpha+1}(4c_1 k \log k) \leq c_1(4c_2^{\alpha+1}+1)k \log^3 k = O(k \log^{\alpha+1} k)$ convex cells, each containing at most $n_\sigma/(c_2 \log k)^\alpha$ fragments of $\hat{B}_\sigma$. By clipping these cells in σ, we have a subdivision $\mathcal{E}_\sigma = \{\sigma \cap \sigma' : \sigma \in \mathcal{D}_\sigma\}$ of σ into $O(k \log^{\alpha+1} k)$ convex cells. This completes the description of the refinement of a cell $\sigma \in \mathcal{C}_i$ with $B_\sigma \neq \emptyset$.

The recursion terminates with a subdivision $\mathcal{C} = \mathcal{C}_i$ of σ_0 into convex cells, each of which is empty of disks. Triangulate each convex cell of $\mathcal{C}$. Return the resulting subdivision $\mathcal{F}$ of σ_0. This completes the description of our algorithm.

Next, we bound the number of simplices in $\mathcal{F}$. We can represent the convex cells in all subdivisions $\mathcal{C}_i$ by a rooted tree: The root corresponds to σ_0; and if our algorithm subdivides a cell $\sigma \in \mathcal{C}_i$, then the children of σ correspond to the cells of $\mathcal{E}_\sigma$. Notice that for every $\sigma \in \mathcal{C}_i$, we have $n_\sigma \leq n/(c_2 \log k)^{i\alpha}$. This implies that the depth of the recursion tree is at most $\log n/\log(c_2 \log k)^\alpha$. Each convex cell in $\mathcal{C}$ is bounded by at most $c_1 k \log k \cdot \log n/\log(c_2 \log k)^\alpha = O(\log n)$ planes. The triangulation of a cell of $\mathcal{C}$ creates at most $O(\log n)$ simplices, and so $\mathcal{F}$ consists of $O(|\mathcal{C}| \log n)$ simplices. It remains to bound the size of $\mathcal{C}$. Since we subdivide cells hierarchically until every cell is empty of disks of B, the size of $\mathcal{C}$ is proportional to the total number of fragments of B produced during the algorithm.

Consider a recursion step. A subdivision along disks of an $\frac{1}{k}$-net S_σ of a cell $\sigma \in \mathcal{C}_i$ increases the number of fragments by a factor of at most $c_2 \log k$ (c.f., Lemma 2). The triangulation of heavy cells also increases the number of fragments: There are up to $(c_2 \log k)^{\alpha+1}$ heavy cells, each heavy cell is triangulated into at most $4c_1 k \log k$ simplices, and each simplex contains at most n_σ/k fragments. This gives an upper bound of

$$(c_2 \log k)^{\alpha+1} \cdot 4c_1 k \log k \cdot \frac{n_\sigma}{k} = n_\sigma \cdot 4c_1 c_2^{\alpha+1} \log^{\alpha+2} k$$

on the number of fragments in the resulting simplices. That is, the refinement of $\mathcal{C}_i$ into $\mathcal{C}_{i+1}$ increases the number of fragments by a factor of at most $c_3 \log^{\alpha+2} k$, where $c_3 > 0$ is an absolute constant. Throughout the algorithm, the total number of fragments produced in $\log n/\log(c_2 \log k)^\alpha$ recursive steps is at most

$$n \cdot (c_3 \log^{\alpha+2} k)^{\frac{\log n}{\log(c_2 \log k)^\alpha}} = n \cdot n^{\frac{\log(c_3 \log^{\alpha+2} k)}{\log(c_1 \log k)^\alpha}} \leq n^{1+\frac{\log c_3 + (\alpha+2)\log\log k}{\log c_1 + \alpha \log\log k}} \leq n^{1+\delta},$$

if α is sufficiently large. We can set α, $\alpha \leq \log k/\log(c_2 \log k)$, to be arbitrarily large, if k is a sufficiently large integer. This completes the proof of Lemma 3. □

Corollary 1. *For every $\delta > 0$, there is a constant $c(\delta)$ with the following property. For every set B of n mutually avoiding disks and every $r \in \mathbb{N}$, there is a $\frac{1}{r}$-cutting of size $c(\delta)r^{1+\delta}$.*

Proof. Perform the recursive algorithm in the proof of Lemma 3 until the interior of every cell intersects at most n/r disks. The result follows by an analogous argument. □

4 Optimal Cuttings for Disjoint Disks

In this section, we prove Theorem 1 and construct an $\frac{1}{r}$-cutting of size $O(r^2)$ for a set B of m mutually disjoint disks in $\mathbb{R}^3$ and an $r \in \mathbb{N}$. We use a hierarchical partition scheme of Agarwal [1] and Chazelle [6], originally designed for hyperplane arrangements.

Let σ_0 be a bounding simplex of all input disks in $\mathbb{R}^3$. We construct a subdivision recursively, in stages. In stage $k \in \mathbb{N}$, we have a subdivision $\mathcal{D}_k$ of a bounding volume into simplices, initially $\mathcal{D}_0 = \{\sigma_0\}$. Recall that in Section 2, we defined $\tau(B,\sigma) = \sum_{a,b\in B} |\pi_a \cap \partial b \cap \text{int}(\sigma)|$, the total number of points in the interior of σ that lie on a boundary of one disk and in the plane spanned by another disk; and it is at most quadratic in the number n_σ of disks that intersect σ, that is, $\tau(B,\sigma) = O(n_\sigma^2)$. For a small constant $c_0 > 0$, to be specified later we distinguish three types of cells:

- $\sigma \in \mathcal{D}_k$ is *dormant* if $n_\sigma < n/r_0^{k+1}$ (ie, it intersects less than n/r_0^{k+1} disks of B);
- $\sigma \in \mathcal{D}_k$ is *full* if $n_\sigma \geq n/r_0^{k+1}$ and $\tau(B,\sigma) \geq c_0 \cdot n_\sigma^2$;
- $\sigma \in \mathcal{D}_k$ is *sparse* if $n_\sigma \geq n/r_0^{k+1}$ and $\tau(B,\sigma) < c_0 \cdot n_\sigma^2$.

Our algorithm proceeds as follows. In stage k, we leave every dormant cell intact and refine the full and sparse cells of $\mathcal{D}_k$ to obtain the next subdivision $\mathcal{D}_{k+1}$. Consider every full or sparse simplex $\sigma \in \mathcal{D}_k$, and define the (undirected) graph G_σ on B_σ that has an edge between two disks $a, b \in B_\sigma$ if they do not avoid each other. By our remarks on sparse ε-nets in Section 2, if we pick every element of B_σ independently at random with probability $p = cdr_0 \log(dr_0)/n_\sigma$, then with positive probability we obtain a $\frac{1}{r_0}$-net $S_\sigma \subset B_\sigma$ of size $\Theta(dr_0 \log(dr_0))$ such that $\tau(S_\sigma,\sigma) = \Theta(p^2\tau(B_\sigma,\sigma))$.

If σ is full, then compute the full arrangement of the planes spanned by S_σ within σ and triangulate the resulting convex cells. We obtain a subdivision of σ into $O(r_0^3 \log^3 r_0)$ simplices. If σ is sparse, we can apply Lemma 2 for B_σ and σ. Indeed, if $c_0 < 1/r_0^2$, then $\tau(S_\sigma,\sigma) = 0$ with constant probability. By Lemma 3, we obtain a subdivision of σ into $O((r_0 \log r_0)^{1+\delta})$ simplices, for some $\delta < 1$. Note that $O((r_0 \log r_0)^{1+\delta})$ is less than $r_0^2/2$ if r_0 is a sufficiently large constant. This is the refinement of σ into simplices in $\mathcal{D}_{k+1}$. Iterate until the resulting subdivision is a $\frac{1}{r_0}$-cutting. The analysis of this algorithm is based on a charging scheme, reminiscent of [6,7,8].

Lemma 4. *$\mathcal{D}_k$ is an $O(r_0^{2k})$ size $1/r_0^k$-cutting for B.*

Proof. Consider the subdivision $\mathcal{D}_k$, for $k \in \mathbb{N}$. It is easy to show (by induction on k) that every simplex $\sigma \in \mathcal{D}_k$ intersects at most n/r_0^k and at least n/r_0^{k+1}

disks in B, and so $\mathcal{D}_k$ is an $1/r_0^k$-cutting for B. It remains to estimate the size of $\mathcal{D}_k$.

First, consider the full simplices of $\mathcal{D}_k$ and those dormant simplices of $\mathcal{D}_k$ produced by subdividing a full simplex of $\mathcal{D}_j$, for some $1 \leq j \leq k$. A full simplex $\sigma \in \mathcal{D}_j$ intersects at least n/r_0^{j+1} disks, and since it is full, we have $\tau(B,\sigma) \geq c_0(n/r_0^{j+1})^2$. Since we initially have $\tau(B,\sigma_0) = O(n^2)$, at stage j at most $n^2/(c_0(n/r_0^{j+1})^2) = r_0^{2(j+1)}/c_0$ full simplices of $\mathcal{D}_j$ are partitioned, and each of them produces $O(r_0^3 \log^3 r_0)$ dormant simplices. In the first k stages, the total number of dormant simplices produced by full simplices is at most $\sum_{j=1}^{k}(r_0^{2(j+1)}/c_0)O(r_0^3 \log^3 r_0) = O(r_0^{2k+5} \log^3 r_0/c_0)$.

Next, consider the sparse simplices of $\mathcal{D}_k$ and those dormant simplices of $\mathcal{D}_k$ produced by subdividing some sparse ones. We charge every sparse simplex to its closest full ancestor or to σ_0 (even if σ_0 is sparse). A full simplex $\sigma \in \mathcal{D}_j$ (and σ_0) may produce $O(r_0^3 \log^3 r_0)$ offsprings, but every sparse simplex produces at most $r_0^2/2$ offsprings. So each sparse simplex $\sigma \in \mathcal{D}_j$ (resp., σ_0) leads to at most $O(r_0^3 \log^3 r_0) \cdot \sum_{i=0}^{\infty}(r_0^2/2)^i = O(r_0^5 \log^3 r_0)$ sparse simplices and dormant simplices produced by sparse ones. Hence $\mathcal{D}_k$ is a $1/r_0^k$-cutting and it consists of $r_0^{2k} \cdot O(r_0^5/c_0 \log^3 r_0)$ simplices. Since r_0 and the constant of proportionality do not depend on k, we can consider $O(r_0^5/c_0 \log^3 r_0)$ as a constant. We conclude that $\mathcal{D}_k$ is a $\frac{1}{r}$-cutting of size $O(r^2)$ for $k = \lceil \log r/\log r_0 \rceil$. We give a matching lower bound in the full version of the paper. □

5 Open Problems

In our proof, we assumed that the disks are pairwise disjoint. For possibly intersecting planar objects in $\mathbb{R}^3$, the smallest $\frac{1}{r}$-cutting may have $\Theta(r^3)$ size, the size of a $\frac{1}{r}$-cutting for a plane arrangement in $\mathbb{R}^3$. However, if just a few pairs of disks intersect, our results are likely to extend in the sense of Pellegrini [28]: we conjecture that for n disks in $\mathbb{R}^3$ and a parameter r, there is an $\frac{1}{r}$-cutting of size $O(r^2 + (K/n^3)r^3)$, where K is the number of vertices of the arrangement of the disks.

We leave higher dimensional generalizations as an open problem: what is the minimum value $f_d(r)$ such that any finite set of disjoint objects in $\mathbb{R}^d$, each having bounded description complexity and lying in a hyperplane, (e.g., disjoint $(d-1)$-balls in $\mathbb{R}^d$) admits a $\frac{1}{r}$-cuttings of size $f_d(r)$?

We do not know if our results hold if we drop the condition that each object in $\mathbb{R}^3$ is planar: is there an $\frac{1}{r}$-cutting of size $O(r^2)$ for any finite set of disjoint objects in $\mathbb{R}^3$, each having constant description complexity and lying in a 2-dimensional algebraic variety (e.g., disjoint spherical caps)? We do not even know if there is an $\frac{1}{r}$-cutting of size $O(r^3)$ for any finite set of 2-dimensional algebraic varieties of constant description complexity (it is known only that a *pseudo-cutting* of size $O(r^3 \text{ polylog } r)$ always exists).

References

1. Agarwal, P.K.: Partitioning arrangements of lines I: an efficient deterministic algorithm. Discrete Comput. Geom. 5, 449–483 (1990)
2. Asano, T., de Berg, M., Cheong, O., Guibas, L.J., Snoeyink, J., Tamaki, H.: Spanning trees crossing few barriers. Discrete and Comput. Geom. 30(4), 591–606 (2003)
3. de Berg, M., Guibas, L.J., Halperin, D.: Vertical decompositions for triangles in 3-space. Discrete Comput. Geom. 15(1), 35–61 (1996)
4. Boissonnat, J-D., Yvinec, M.: Algorithmic geometry. Cambridge Univ. Press, Cambridge (1998)
5. Camerini, P.M.: The min-max spanning tree problem and some extensions. Inf. Process. Lett. 7(1), 10–14 (1978)
6. Chazelle, B.: Cutting hyperplanes for divide-and-conquer. Discrete Comput. Geom. 9, 145–158 (1993)
7. Chazelle, B.: Cuttings. In: Handbook of Data Structures and Applications, pp. 501–511. CRC Press, Boca Raton (2005)
8. Chazelle, B.: The Discrepancy Method. Press, Cambridge (2000)
9. Chazelle, B., Edelsbrunner, H., Guibas, L., Sharir, M.: A singly-exponential stratification scheme for real semi-algebraic varieties. Theoret. Comp. Sci. 84, 77–105 (1991)
10. Chazelle, B., Friedman, J.: A deterministic view of random sampling and its use in geometry. Combinatorica 10(3), 229–249 (1990)
11. Chazelle, B., Welzl, E.: Quasi-optimal range searching in spaces of finite VC-dimension. Discrete Comput. Geom. 4(5), 467–489 (1989)
12. Clarkson, K.L., Shor, P.W.: Application of random sampling in computational geometry, II. Discrete Comput. Geom. 4, 387–421 (1989)
13. Ezra, E., Sharir, M.: Counting and representing intersections among triangles in three dimensions. Comput. Geom. Theory Appl. 32(3), 196–215 (2005)
14. Haussler, D., Welzl, E.: ϵ-nets and simplex range queries. Discrete Comput. Geom. 2, 127–151 (1987)
15. Hoffmann, M., Tóth, Cs.D.: Connecting points in the presence of obstacles in the plane. In: Proc. 14th Canadian Conf. Comput. Geom, pp. 63–67 (2002)
16. Karger, D.R., Klein, P.N., Tarjan, R.E.: A randomized linear-time algorithm to find minimum spanning trees. J. ACM 42, 321–328 (1995)
17. Koltun, V.: Almost tight upper bounds for vertical decompositions in four dimensions. J. ACM 51(5), 699–730 (2004)
18. Koltun, V.: Sharp bounds for vertical decompositions of linear arrangements in four dimensions. Discrete Comput. Geom. 31(3), 435–460 (2004)
19. Koltun, V., Sharir, M.: Curve-sensitive cuttings. SIAM J. Comput. 34(4), 863–878 (2005)
20. Matoušek, J.: Cutting hyperplane arrangements. Discrete Comput. Geom. 6, 385–406 (1991)
21. Matoušek, J.: Derandomization in computational geometry. In: Sack, J.-R., Urrutia, J. (eds.) Handbook of Computational Geometry, pp. 559–595. North-Holland, Amsterdam (2000)
22. Matoušek, J.: Lectures on Discrete Geometry. Springer, Heidelberg (2002)
23. Mulmuley, K.: Hidden surface removal with respect to a moving view point. In: Proc. Sympos. Theory of Computing, pp. 512–522. ACM Press, New York (1991)
24. Pach, J., Sharir, M.: Geometric incidences, Towards a theory of geometric graphs, AMS. Contemp. Math. 342, 185–223 (2004)

25. Paterson, M., Yao, F.F.: Efficient binary space partitions for hidden-surface removal and solid modeling. Discrete Comput. Geom. 5, 485–503 (1990)
26. Paterson, M., Yao, F.F.: Optimal binary space partitions for orthogonal objects. J. Algorithms 13(1), 99–113 (1992)
27. Pellegrini, M.: On point location and motion planning among simplices. SIAM J. Comput. 25(5), 1061–1081 (1996)
28. Pellegrini, M.: On counting pairs of intersecting segments and off-line triangle range searching. Algorithmica 17(4), 380–398 (1997)
29. Sarnak, N., Tarjan, R.E.: Planar point location using persistent search trees. Commun. of the ACM 29(7), 669–679 (1986)
30. Sharir, M.: The Clarkson-Shor technique revisited and extended. Combinatorics Probability and Computation 12(2), 191–201 (2003)

Kernelization and Complexity Results for Connectivity Augmentation Problems

Jiong Guo* and Johannes Uhlmann**

Institut für Informatik, Friedrich-Schiller-Universität Jena,
Ernst-Abbe-Platz 2, D-07743 Jena, Germany
{guo,uhlmann}@minet.uni-jena.de

Abstract. Connectivity augmentation problems ask for adding a set of at most k edges whose insertion makes a given graph satisfy a specified connectivity property, such as bridge-connectivity or biconnectivity. We show that, for bridge-connectivity and biconnectivity, the respective connectivity augmentation problems admit problem kernels with $O(k^2)$ vertices and links. Moreover, we study partial connectivity augmentation problems, naturally generalizing connectivity augmentation problems. Here, we do not require that, after adding the edges, the entire graph should satisfy the connectivity property, but a large subgraph. In this setting, two polynomial-time solvable connectivity augmentation problems behave differently, namely, the *partial* biconnectivity augmentation problem remains polynomial-time solvable whereas the *partial* strong connectivity augmentation problem becomes W[2]-hard with respect to k.

1 Introduction

Connectivity augmentation problems on undirected and directed graphs have as input a graph $G = (V, E)$, a set E' of edges, and a non-negative integer k, and ask for a set E'' of at most k edges from E' such that $(V, E \cup E'')$ satisfies a specified connectivity property. The edges in E' are called the *links*. Eswaran and Tarjan [2] introduced connectivity augmentation problems and described their numerous applications.

We use $G = (V, E)$ and $D = (V, A)$ to denote undirected and directed graphs. A *path* from vertex u_1 to vertex u_l in $G = (V, E)$ (or $D = (V, A)$) is a sequence of edges $\{u_1, u_2\}$, $\{u_2, u_3\}$, ..., $\{u_{l-1}, u_l\}$ (or arcs (u_1, u_2), (u_2, u_3), ..., (u_{l-1}, u_l)). A *cycle* is a path with $u_1 = u_l$.

A vertex u in an undirected graph is called a *cut-vertex* if there are two vertices v, w with $v \neq u$ and $w \neq u$ such that every path from v to w contains u. If an undirected graph G is connected and has no cut-vertex, then G is *biconnected.* A *bridge* in an undirected graph is an edge $\{u, v\}$ such that every path

* Supported by the Deutsche Forschungsgemeinschaft (DFG), Emmy Noether research group PIAF (fixed-parameter algorithms), NI 369/4.

** Partially supported by the Deutsche Forschungsgemeinschaft (DFG), research project OPAL (optimal solutions for hard problems in computational biology), NI 369/2.

F. Dehne, J.-R. Sack, and N. Zeh (Eds.): WADS 2007, LNCS 4619, pp. 483–494, 2007.

between u and v contains $\{u, v\}$. If G is connected and has no bridge, then G is *bridge-connected*. A directed graph $D = (V, A)$ is *strongly connected* if, for all pairs of vertices u and v, there is a path from u to v. The *connected (biconnected, bridge-connected, strongly connected) components* of a graph are its maximal connected (biconnected, bridge-connected, strongly connected) subgraphs. We consider here two connectivity augmentation problems, namely, BRIDGE-CONNECTIVITY AUGMENTATION (BCA) and BICONNECTIVITY AUGMENTATION (BIA), where we are asked to add at most k links to make the given graph bridge-connected or biconnected.

There is a long history of research dealing with BCA and BIA starting in 1976 with the work of Eswaran and Tarjan [2]. They showed that, in the case that E' is *complete*, that is, graph (V, E') is a complete graph, both problems are polynomial-time solvable. In 1981, Frederickson and JáJá [4] proved the NP-completeness of both problems if E' is *incomplete*. Motivated by the NP-completeness, the approximability of the optimization versions of these problems has been extensively studied in the literature. Frederickson and JáJá [4] gave polynomial-time factor-2 approximation algorithms for both problems. For BCA, Nagamochi [8] improved the approximation factor to 1.875. Later, Even et al. [3] presented a factor-1.5 approximation algorithm for BCA. In case of BIA, Khuller and Thurimella [5] improved the running time of the factor-2 approximation algorithm in [4]. In an unpublished manuscript, Kortsarz and Nutov [7] claimed a polynomial-time factor-$\frac{12}{7}$ algorithm for BIA. On the negative site, Kortsarz et al. [6] proved that there exists an $\epsilon > 0$ for which it is NP-hard to approximate BCA and BIA within a factor of $1 + \epsilon$.

Concerning the parameterized complexity of these problems, we are only aware of one result due to Nagamochi [8]. Since the bridge-connected components of a graph form a tree, we may assume that the input graph of a BCA-instance is a tree by contracting these components. Nagamochi [8] showed that BCA is fixed-parameter tractable with respect to the number of leaves ℓ in the given tree. More precisely, there is an algorithm solving this problem in $O(\ell^{\ell+1} \log \ell \cdot (|V| + |E'|))$ time. Since the number of leaves provides a lower bound on the solution size k, BCA is fixed-parameter tractable with respect to k. Nothing has been known concerning the kernelization of these problems. Problem kernelization is one of the most important contributions of fixed-parameter algorithmics to practical computing [1,9]. A *kernelization* is a polynomial-time algorithm that transforms a given instance I with parameter k of a problem P into a new instance I' with parameter $k' \leq k$ of P such that the original instance I is a yes-instance with parameter k iff the new instance I' is a yes-instance with parameter k' and $|I'| \leq g(k)$ for a function g. The instance I' is called the *problem kernel*. We complement the result of Nagamochi with a problem kernel with $O(k^2)$ vertices and links for BCA and BIA.

Furthermore, we study *partial* connectivity augmentation problems, a natural generalization of the connectivity augmentation problems, where we have as input a graph $G = (V, E)$, a set E' of links, and two non-negative integers k, Φ

and ask for a subset E'' of E' with $|E''| \leq k$ such that graph $(V, E \cup E'')$ has a subgraph that contains at least Φ vertices and satisfies the given connectivity property. Clearly, if $\Phi = |V|$, then we have the connectivity augmentation problems. We consider two partial connectivity augmentation problems, PARTIAL BRIDGE-CONNECTIVITY AUGMENTATION and PARTIAL STRONG CONNECTIVITY AUGMENTATION. For both connectivity properties, their corresponding non-partial connectivity augmentation problems are solvable in polynomial-time if E' is complete [2]. We show in Sect. 5 that PARTIAL BRIDGE-CONNECTIVITY AUGMENTATION with a complete link set remains polynomial-time solvable but PARTIAL STRONG CONNECTIVITY AUGMENTATION with a complete link set is W[2]-hard, that is, it is very unlikely that this problem is fixed-parameter tractable with respect to the parameter k.

Most proofs are deferred to a long version of this paper.

2 Preliminaries

Parameterized complexity is a two-dimensional framework for studying the computational complexity of problems [1,9]. One dimension is the input size n (as in classical complexity theory) and the other one the *parameter* k (usually a positive integer). A problem is called *fixed-parameter tractable* (fpt) if it can be solved in $f(k) \cdot n^{O(1)}$ time, where f is a computable function only depending on k. A core tool in the development of fixed-parameter algorithms is polynomial-time preprocessing by *data reduction rules*, often yielding a *kernelization*. Herein, the goal is, given any problem instance I with parameter k, to transform it in polynomial time into a new instance I' with parameter k' such that the size of I' is bounded from above by some function only depending on k, $k' \leq k$, and (I, k) is a yes-instance iff (I', k') is a yes-instance. A data reduction rule is *correct* if the new instance after an application of this rule is a yes-instance iff the original instance is a yes-instance. Throughout this paper, we call a problem instance *reduced* if the corresponding data reduction rules cannot be applied anymore. A formal framework to show *fixed-parameter intractability* was developed by Downey and Fellows [1] who introduced the concept of *parameterized reductions*. A parameterized reduction from a parameterized language L to another parameterized language L' is a function that, given an instance (I, k), computes in $f(k) \cdot n^{O(1)}$ time an instance (I', k') (with k' only depending on k) such that $(I, k) \in L \Leftrightarrow (I', k') \in L'$. The basic complexity class for fixed-parameter intractability is W[1] as there is good reason to believe that W[1]-hard problems are not fixed-parameter tractable [1].

Throughout this paper, we set $n := |V|$ and $m := |E'|$ for a given graph $G = (V, E)$ and a given link set E'. For a graph G, we also use $V(G)$ and $E(G)$ to denote its vertex and edge set, respectively. The *neighborhood* $N(v)$ of a vertex $v \in V$ is the set of vertices that are adjacent to v. The *degree* of a vertex v, denoted by $\deg(v)$, is the size of $N(v)$.

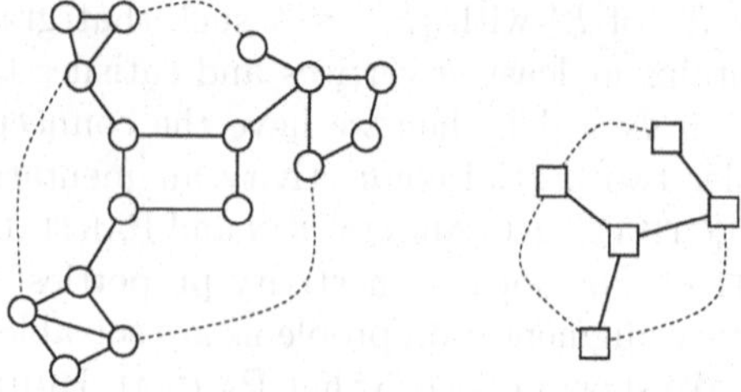

Fig. 1. Example of contracting bridge-connected components of a given graph. The links are drawn as dashed lines.

3 The Bridge-Connectivity Augmentation Problem

The main result of this section is a data reduction for Bridge-Connectivity Augmentation that leads to a quadratic-size problem kernel. Given an instance of BCA, we can assume that the input graph G is a tree [2,4,3]: Each bridge-connected component of $G = (V, E)$ can be contracted into a single vertex by contracting all edges in this component, resulting in a tree. The set of links has to be adapted accordingly. The contraction of the bridge-connected components can be done in $O(|V| + |E|)$ time [11]. See Fig. 1 for an example. Hence, in the following, the input instance is always denoted by T.

In contrast to the tree edges, denoted by $\{u, v\}$, we denote links by (u, v). We use $p_{u,v}$ to denote the uniquely determined path between two vertices u and v in T. In the course of the data reduction process, if a link $(u, v) \in E'$ is added to a solution E'', then the vertices from the path $p_{u,v}$ form a bridge-connected component and we contract all edges in this component, obtaining a tree again. We say a link (u, v) *covers* an edge e if e lies on $p_{u,v}$. For an edge $e \in E$, we use $l(e)$ to denote the set of links covering e. A link $(u, v) \in E'$ is called a *shadow* if there exists a link $(u', v') \in E'$ with $V(p_{u,v}) \subsetneq V(p_{u',v'})$. Let $N_{E'}(u) := \{v \mid (u, v) \in E'\}$. For a vertex $v \in V$ and an edge $e \in E$, we use $T_{v,e}$ to denote the subtree of $(V, E \setminus \{e\})$ that contains v.

The following observation provides the starting point for our kernelization:

Lemma 1 ([2]). *Let $L(T)$ be the set of leaves of the tree T of a BCA-instance. Every solution of this instance contains at least $|L(T)|/2$ many links, that is, $k \geq |L(T)|/2$.*

We can conclude that every yes-instance of BCA contains at most $2k$ leaves and, also, at most $2k - 1$ internal vertices with degree at least three. It remains to upper-bound the number of internal vertices of degree two. If we can bound the maximum length of the paths that consist solely of degree-2 vertices, then we can achieve an upper bound on the number of degree-2 vertices. To this end, we apply four data reduction rules. We begin with three data reduction rules that are also used in [3,8] and whose correctness is easy to verify.

Shadow Deletion: Delete all shadows in E'.
Unit Link: If there is an edge $e \in E$ with $l(e) = \{(u,v)\}$, then contract $p_{u,v}$ and decrease the parameter k by one.
Covered Edge: If $l(e_1) \subseteq l(e_2)$ for two edges $e_1, e_2 \in E$, then contract e_2.

Lemma 2. *The above three rules can be executed in $O(n \cdot m^3 + n^3 \cdot m)$ time.*

Before we present the fourth data reduction rule, we show some structural properties of a BCA-instance that is reduced with respect to the above three rules. In particular, we show that, in a reduced instance, the links over a path consisting solely of degree-2 vertices have some "consecutiveness" property, which provides the basis for the fourth data reduction rule.

Lemma 3. *Let $(T = (V,E), E', k)$ be a reduced instance with respect to the above three rules, let $v \in V$ be a degree-2 vertex in T, and let e, e' be the edges incident to v. Then, there exists at least one link (v,x) in E' with $x \in V(T_{v,e})$ and at least one link (v,y) in E' with $y \in V(T_{v,e'})$.*

Lemma 4. *Let $(T = (V,E), E', k)$ be a reduced instance with respect to the above three rules. Then, for every link $(u,v) \in E'$, it holds that $|E(p_{u,v})| \geq 2$.*

Lemma 5. *Let $(T = (V,E), E', k)$ be a reduced instance with respect to the above three rules. Consider a path $P = \{u, v_1\}, \{v_1, v_2\}, \ldots, \{v_l, w\}$ in T with $deg(v_i) = 2$ for all $1 \leq i \leq l$, $deg(u) \geq 3$, and $deg(w) \geq 3$. Let E'_v denote the set of links with both endpoints in $\{v_1, v_2, \ldots, v_l\}$. If $E'_v \neq \emptyset$, then there exists an integer N with $1 \leq N \leq l-1$ such that $E'_v = \{(v_i, v_{i+N}) \mid 1 \leq i \leq l-N\}$ and there exists no link (x,y) with $x \in V(T_{u,\{u,v_1\}})$ and $y \in V(T_{w,\{v_l,w\}})$.*

The fourth data reduction rule restricts the length of paths that consist solely of degree-2 vertices. By Lemma 5, the links with both endpoints from such a path admit a "consecutiveness" property. By making use of this property, the next data reduction rule replaces a long degree-2 path by a shorter "equivalent" degree-2 path.

Degree-2-Path: Let $(T = (V,E), E', k)$ be a reduced instance with respect to the above three rules. Let $P = \{u, v_1\}, \{v_1, v_2\}, \ldots, \{v_l, w\}$ be a path in T such that $\deg(v_i) = 2$ for all $1 \leq i \leq l$, $\deg(u) \geq 3$, and $\deg(w) \geq 3$ and let E'_v denote the set of links with both endpoints from $v_1, v_2, \ldots, v_l$. If there exists an integer N with $l \geq 2N$ such that $E'_v = \{(v_i, v_{i+N}) \mid 1 \leq i \leq l-N\}$, then proceed as follows: Let $c := \lfloor \frac{l}{N} \rfloor - 1$ and $d := (l \mod N)$. Replace P by a path $P' = \{u, x_1\}, \{x_1, x_2\}, \ldots, \{x_{N+d}, w\}$. Remove all links in E'_v from E' and add the links (x_i, x_{i+N}) for $1 \leq i \leq d$ to E'. Replace every link (v_i, y) with $y \in V(T_{u,\{u,v_1\}})$ by the link (x_i, y) and replace every link (v_i, y) with $y \in V(T_{w,\{v_l,w\}})$ by the link $(x_{N+d-(l-i)}, y)$. Finally, decrease parameter k by c.

See Fig. 2 for an example of the application of the Degree-2-Path rule.

Lemma 6. *The Degree-2-Path rule is correct and can be executed in $O(n^2 + nm)$ time.*

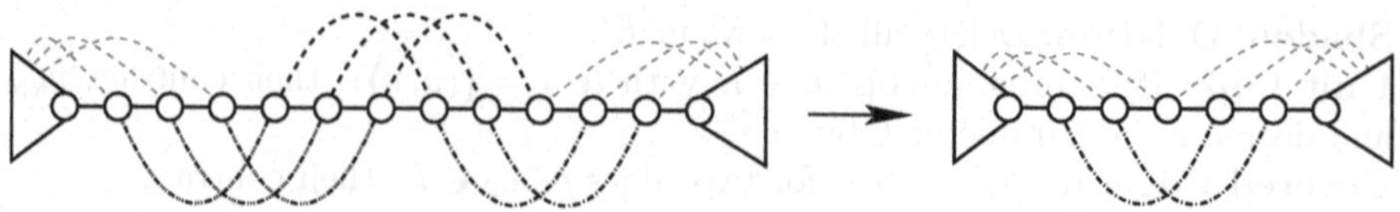

Fig. 2. Example for the Degree-2-Path rule for $N = 3$, $l = 11$, $c = 2$, and $d = 2$

Proof. Let $(T = (V, E), E', k)$ be a BCA-instance reduced with respect to the first three rules and let (T_a, E'_a, k_a) be the resulting instance after one application of the Degree-2-Path rule. Let $P = \{u, v_1\}, \{v_1, v_2\}, \dots, \{v_l, w\}$ be the path for which the conditions of the Degree-2-Path rule are fulfilled and let E'_v be the set of links with both endpoints from $v_1, \dots, v_l$. By Lemma 5, either $E'_v = \emptyset$ or there is an integer N with $E'_v = \{(v_i, v_{N+i}) \mid 1 \leq i \leq l - N\}$ and there is no link between a vertex in $V(T_{u,\{u,v_1\}})$ and a vertex in $V(T_{w,\{v_l,w\}})$. Since the Degree-2-Path rule is applicable to P, we know $E'_v \neq \emptyset$ and $l \geq 2N$. Then, we have $c = \lfloor \frac{l}{N} \rfloor - 1$, $d = (l \mod N)$, and $k_a = k - c$. We show that (T, E', k) is a yes-instance if and only if (T_a, E'_a, k_a) is a yes-instance. The correctness of the Degree-2-Path rule then follows by induction on the number of applications of the rule. Here, we give only the proof of "$\Rightarrow$"-direction. The proofs of "$\Leftarrow$"-direction and the running time are omitted due to lack of space.

"$\Rightarrow$": Let $E'' \subseteq E'$ with $|E''| \leq k$ be a solution for the original instance. We first show some properties of $E'' \cap E'_v$ and then construct a solution E''_a for the new instance with $|E''_a| \leq k_a$ from E''.

We can assume that $E'' \cap E'_v$ contains only pairwise "non-overlapping" links, that is, there are no two links $(v_i, v_{i+N}), (v_j, v_{j+N}) \in E'' \cap E'_v$ with $i < j < i+N$: If there are two links $(v_i, v_{i+N}), (v_j, v_{j+N}) \in E'' \cap E'_v$ with $i < j < i + N$, then we construct another solution from E'' by replacing (v_j, v_{j+N}) by (v_{i+N}, z), where $z = v_{i+2N}$ or z is a vertex from $T_{w,\{v_l,w\}}$. Link (v_{i+N}, z) exists due to Lemma 3. Obviously, this yields a solution of the same size (or even of smaller size if (v_{i+N}, z) is already part of the solution).

Next, we show that $c \leq |E'' \cap E'_v| \leq c + 1$. On the one hand, since every link in E'_v covers exactly N edges and the links in $E'' \cap E'_v$ are pairwise non-overlapping, there can be at most $\lfloor \frac{l-1}{N} \rfloor = \lfloor \frac{(c+1)N+d-1}{N} \rfloor = c + 1$ pairwise non-overlapping links in $E'' \cap E'_v$. On the other hand, all edges of path P which lie between vertex v_N and vertex v_{l-N+1} have to be covered by links in E'_v. This means that E'' has cardinality at least $\lceil \frac{l-2N+1}{N} \rceil = \lceil \frac{(c+1)N+d-2N+1}{N} \rceil = (c - 1) + \lceil \frac{d+1}{N} \rceil = c$.

In the following, let $i_l := \max\{i \mid (y, v_i) \in E'' \wedge y \in V(T_{u,\{u,v_1\}})\}$ and $i_r := \min\{i \mid (v_i, z) \in E'' \wedge z \in V(T_{w,\{v_l,w\}})\}$. We distinguish two cases, namely, $|E'' \cap E'_v| = c$ and $|E'' \cap E'_v| = c + 1$, and construct in both cases a solution for the new instance.

In the first case, we can assume that $E'' \cap E'_v = \{(v_{i_l}, v_{i_l+N}), (v_{i_l+N}, v_{i_l+2N}), \dots, (v_{i_l+(c-1)N}, v_{i_l+cN})\}$. We construct the new solution E''_a from E'' as follows: We replace every link $(y, v_i) \in E''$ by a link (y, x_i). Note that, since the given instance is reduced with respect to the first three rules, the existence of the link (y, v_i) implies that $i < N \leq N + d$ and, thus, the link (y, x_i) exists in E'_a. The links $(v_i, z) \in E''$ with $z \in V(T_{w,\{v_l,w\}})$ are also replaced by links $(x_{N+d-(l-i)}, z)$. With the same reason as above, links (x_i, z) exist in E'_a. Finally, we remove all links in $E'' \cap E'_v$ from E''. The resulting set E''_a is then a solution for the new instance. Since $|E'' \cap E'_v| = c$ in this case, we have $|E''_a| \leq k_a$. Obviously, all edges of T_a that are not between the new vertices $x_1, \ldots, x_{N+d}$ are covered. To show that the edges between the new vertices are also covered, observe that the edges on the path between x_1 and x_{i_l} and the edges on the path between $x_{N+d-(l-i_r)}$ and x_{N+d} are covered by the links (y, x_{i_l}) with $y \in V(T_{u,\{u,v_1\}})$ and $(x_{N+d-(l-i_r)}, z)$ with $z \in V(T_{w,\{v_l,w\}})$ that replace the links (y, v_{i_l}) and (v_{i_r}, z), respectively. Then, it suffices to show that $N + d - (l - i_r) \leq i_l$. Since E'' is a solution of the non-reduced instance, we get $i_l + cN \geq i_r$. This is equivalent to $i_l \geq i_r - cN = N + d - cN - N - d + i_r = N + d - (l - i_r)$.

In the second case, we assume that $E'' \cap E'_v = \{(v_{i_l}, v_{i_l+N}), (v_{i_l+N}, v_{i_l+2N}), \ldots, (v_{i_l+cN}, v_{i_l+(c+1)N})\}$. We construct the solution E''_a for the new instance from E'' as follows: We replace every link $(y, v_i) \in E''$ with $y \in V(T_{u,\{u,v_1\}})$ by a link (y, x_i). The links $(v_i, z) \in E''$ with $z \in V(T_{w,\{v_l,w\}})$ are also replaced by links $(x_{N+d-(l-i)}, z)$. We remove all links in $E'' \cap E'_v$ from E''. Finally, we add link (x_{i_l}, x_{i_l+N}) to E''. With a similar argument, we can show that the resulting set is a solution of the new instance. □

Next, we show that a BCA-instance reduced with respect to the four data reduction rules has $O(k^2)$ vertices and $O(k^2)$ links. The key point in the following is to upper-bound the number of the internal vertices with degree two. Herein, we consider the paths formed by degree-two vertices. The next two lemmas are used to show the upper bounds on the number and the length of such paths. The first one is due to Even et al. [3] and shows that there is no such degree-two vertex path between a leaf and an internal vertex of degree at least three.

Lemma 7 ([3]). *Let (T, E', k) be a BCA-instance to which the Shadow Deletion rule, the Unit Link rule, and the Covered Edge rule cannot be applied. Let v be a leaf of T and u be the parent of v. Then, $deg(u) \geq 3$.*

In the next lemma, we upper-bound the length of a path in a reduced instance that consists of degree-two vertices. Herein, we use $L(T)$ to denote the set of leaves in tree T.

Lemma 8. *Let (T, E', k) be a reduced BCA-instance and let $P = \{u, v_1\}, \{v_1, v_2\}, \ldots, \{v_l, w\}$ be a path in T with $deg(u) \geq 3$, $deg(w) \geq 3$, and $deg(v_i) = 2$ for all $1 \leq i \leq l$. Let $L_u := L(T_{u,\{u,v_1\}})$ and $L_w := L(T_{w,\{v_l,w\}})$. Then,*

(1) there are at most $2|L_u| + 2|L_w|$ links that have exactly one endpoint in $V_P := \{v_1, v_2, \ldots, v_l\}$.
(2) $l \leq 4 \cdot \min(|L_u|, |L_w|)$.

Proof. In the following, we use T_u and T_w to denote $T_{u,\{u,v_1\}}$ and $T_{w,\{v_l,w\}}$, respectively, and consider them as two rooted trees with roots u and w, respectively. For a vertex x in T_u or T_w, we use T_x to denote the subtree of T_u or T_w

rooted at x. Moreover, we use A_u and A_w to denote the set of internal vertices of degree at least 3 in T_u and T_w.

The key for proving the lemma is the following observation: There exist at most $|A_u| + |L_u|$ links in E' with one endpoint in T_u and one endpoint in V_P and there exist at most $|A_w| + |L_w|$ links in E' with one endpoint in T_w and one endpoint in V_P. Here, we prove this observation for T_u. Consider a degree-2 vertex x in T_u. By Lemma 3, there exists a link $(x, y) \in E'$ with $y \in V(T_x)$. The existence of the link (x, y) then excludes any link (a, b) with $a \in V(T_y)$ and $b \in V_P$, since, otherwise, (x, y) would be a shadow and the Shadow Deletion rule would be applied. In this way, every degree-2 vertex x in T_u "blocks" at least one vertex from T_x from building links with the vertices in V_P. Thus, by a simple calculation, we arrive at the $|A_u| + |L_u|$-bound on the number of links between the vertices in T_u and the vertices in V_P.

From the above observation, we know that there are at most $|A_u| + |L_u| + |A_w| + |L_w|$ links with exactly one endpoint in V_P. Since $|A_u| \leq |L_u| - 1$ and $|A_w| \leq |L_w| - 1$, the first part of the lemma follows.

To prove the second part of the lemma, we distinguish two cases. First, suppose that there exists at least one link $(x, y) \in E'$ with $x \in V(T_u)$ and $y \in V(T_w)$. Then, since the instance is reduced with respect to the Shadow Deletion rule, there is no link in E' between two vertices in V_P. By Lemma 3, for every vertex v_i in V_P, there are at least two links $(v_i, a), (v_i, b) \in E'$ with $a \in V(T_u)$ and $b \in V(T_w)$. According to the above observation, there are at most $2|L_u| - 1$ (or $2|L_w| - 1$) links between V_P-vertices and the vertices in $V(T_u)$ (or $V(T_w)$). Therefore, $|V_P| \leq \min(2|L_u| - 1, 2|L_w| - 1)$.

In the second case, there is no link $(x, y) \in E'$ with $x \in V(T_u)$ and $y \in V(T_w)$. Let v_{i_l} be the vertex in V_P such that there exist a link $(v_{i_l}, z) \in E'$ with $z \in V(T_u)$ and, for all $i_l \leq i \leq l$, there is no link $(v_i, z) \in E'$ with $z \in V(T_u)$. From the above observation, we know $i_l \leq 2|L_u| - 1$. By Lemma 3 and the fact that the instance is reduced with respect to the Degree-2-Path rule, for every vertex v_i with $i_l \leq i \leq l$, there is a link $(v_i, v_j) \in E'$ with $1 \leq j \leq i_l$. Since the instance is reduced with respect to the Shadow Deletion rule, there cannot be two vertices from $\{v_{i_l+1}, \ldots, v_l\}$ which form two links with one vertex from $\{v_1, \ldots, v_{i_l}\}$, we can conclude $l \leq 2i_l \leq 4|L_u| - 2$. Obviously, the same argument can also be applied to obtain $l \leq 4|L_w| - 2$. The second part of the lemma follows. □

Now, we prove the size bound of the problem kernel for BCA.

Theorem 1. BRIDGE-CONNECTIVITY AUGMENTATION *admits a problem kernel with $O(k^2)$ vertices and $O(k^2)$ links.*

4 The Biconnectivity Augmentation Problem

In this section, by studying BICONNECTIVITY AUGMENTATION (BIA), we deal with a more general problem setting than in the previous section. Hence, based on the previous section, we extend and refine our kernelization technique introduced there. As shown by Frederickson and JáJá [4], we can assume that the input

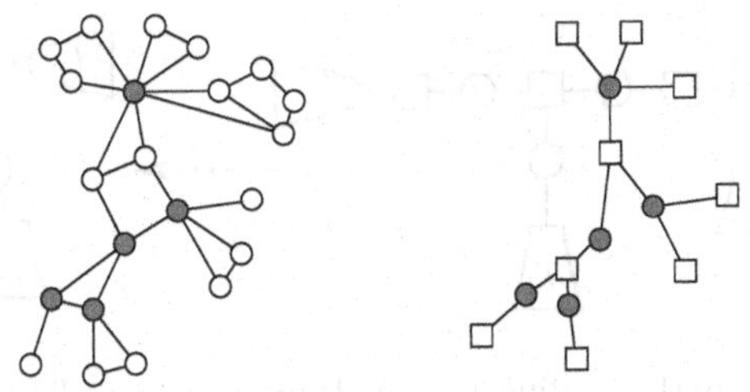

Fig. 3. A graph together with its block tree. The cut-vertices are colored gray and the block-vertices are drawn as rectangles.

graph is a so-called *block tree.* A block tree $T = (V_T, E_T)$ is a tree over the vertex set $V_T := B \cup C$ with $B \cap C = \emptyset$ where the leaves of T form a subset of B and the edges in E_T have one endpoint from B and one endpoint from C.

We can easily compute a block tree from a given undirected and connected graph $G = (V, E)$: Identify B as the set of biconnected components of G and C as the set of cut-vertices of G. Insert an edge between a biconnected component and a cut-vertex into E_T if the cut-vertex belongs to the biconnected component. In the following, the vertices in B are called *block-vertices* and the vertices in C are called *cut-vertices.* See Fig. 3 for an example of a graph and its block tree.

Eswaran and Tarjan [2] gave a lower bound on the size of the solutions of a BIA-instance.

Lemma 9 ([2]). *If (T, E', k) is a yes-instance of BIA, then $k \geq \lceil |L|/2 \rceil$ where L is the set of leaves of T.*

By Lemma 9, the number of leaves and the number of internal vertices of degree at least three of a given block tree can be easily bounded from above by $2k$ and $2k - 1$. Again, we focus on the internal vertices of degree two of T. The decisive difference between a BIA-instance and a BCA-instance lies in the partition of the tree vertices into two subsets, the block-vertices and the cut-vertices. A block-vertex can only have cut-vertices as neighbors and vice versa. In the following, we present first a preprocessing, which ensures that the links in E' are all between block-vertices.

Preprocessing: While there exists a link $(u, v) \in E'$ with $u \in C$ and $v \in B \cup C$, replace (u, v) by the link (w, v) where $w \in B$ is the neighbor of u that lies on the path between u and v in T. Finally, for all $u \in B \cup C$, remove all links (u, u) from E'.

To see the correctness of the preprocessing, the following equivalent formulation of BIA is helpful: Given a block tree $T = (B \cup C, E_T)$, a set of links E', and a non-negative integer k, find a subset E'' of links with $|E''| \leq k$ such that, for every $c \in C$, if c is removed from the graph $(B \cup C, E_T \cup E'')$, then the resulting graph is still connected.

Lemma 10. *The preprocessing is correct and can be executed in $O(n \cdot m)$ time.*

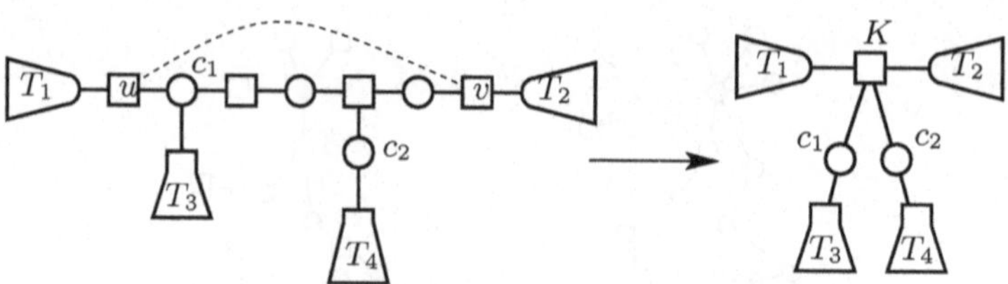

Fig. 4. An illustration of the modification made after adding a link (u, v) to the solution set

Next, we present the data reduction rules for BIA which generalize the data reduction rules in Sect. 3. Herein, if we add a link (u, v) to the solution E'', then, following Rosenthal and Goldner [10], we modify the instance as follows: Let P denote the path in T between u, v, let C be the set of cut-vertices on P that have degree at least three, and let N be the set of cut-vertices which are neighbors of the block-vertices in P and do not lie on P. Replace P by a single block-vertex K. Every link (u, v) with at least one endpoint being in P, say u, is replaced by link (K, v). For every vertex $v \in N$, add edge $\{K, v\}$ and, for every $c \in C$, add edge $\{K, c\}$. An illustration is given in Fig. 4.

The data reduction rules use the following terms and notations: For a vertex u, we use E'_u to denote the links in E' which cover u or have u as one of its endpoints. We call a path between two cut vertices a *degree-2-cut-path* if all vertices on this path are degree-two vertices. A degree-2-cut-path is *maximal* if it is not a proper subpath of another degree-2-cut-path.

Shadow Deletion: Delete all shadows.
Unit Link: If there exists a cut-vertex u with $E'_u = \{(x, y)\}$, then add (x, y) to E'' and decrease the parameter k by one.
Covered Cut-Vertex: If there are two cut-vertices u and v with $E'_u \subseteq E'_v$ and $N(v) = \{w_1, w_2\}$, then add a new block-vertex w and make it adjacent to the vertices in $(N(w_1) \cup N(w_2)) \setminus \{v\}$ and replace every link of the form (w_1, x) or (w_2, x) by a link (w, x). Finally, remove v, w_1, w_2 from T.
Degree-2-Cut-Path: Let (T, E', k) be a BIA-instance to which the first three rules do not apply, let $P = \{c_1, b_1\}, \{b_1, c_2\}, \{c_2, b_2\}, \ldots, \{c_l, b_l\}, \{b_l, c_{l+1}\}$ be a maximal degree-2-cut-path in T with $\{b_1, \ldots, b_l\} \subseteq B$ and $\{c_1, \ldots, c_{l+1}\} \subseteq C$, and E'_b be the set of links with both endpoints from $\{b_1, \ldots, b_l\}$. If there exists an integer N with $2N \leq l$ such that $E'_b = \{(b_i, b_{i+N}) \mid 1 \leq i \leq l - N\}$, then proceed as follows: Let $c := \lfloor \frac{l}{N} \rfloor - 1$ and $d = l \mod N$. Replace P by a path $P' = \{c'_1, b'_1\}, \ldots, \{c'_{N+d}, b'_{N+d}\}, \{b'_{N+d}, c'_{N+d+1}\}$. Remove all links in E'_b from E' and add links (b'_i, b'_{i+N}) for $1 \leq i \leq d$ to E'. Replace every link (b_i, y) with $y \in V(T_{b_1, \{b_1, c_1\}})$ by link (b'_i, y) and replace every link (b_i, y) with $y \in V(T_{b_l, \{c_{l-1}, b_l\}})$ by link $(b'_{N+d-(l-i)}, y)$. Finally, decrease the parameter k by c.

Due to the similarity of the four rules to the ones in Sect. 3, the running time follows from Lemmas 2 and 6.

Lemma 11. *The four data reduction rules are correct and can be executed in $O(n \cdot m^3 + n^3 \cdot m)$ time.*

Theorem 2. BICONNECTIVITY AUGMENTATION *admits a problem kernel with $O(k^2)$ vertices and $O(k^2)$ links.*

5 Partial Augmentation Problems

In this section, we study partial augmentation problems. Here, given a graph, a set of links, and two non-negative integers Φ, k, one asks for a set of at most k links whose insertion in the graph results in a graph that has a subgraph with at least Φ vertices that satisfies a given connectivity property. We show that, in the case that the link set is complete, that is, it contains all possible edges or arcs, PARTIAL BRIDGE-CONNECTIVITY AUGMENTATION (PBCA) is polynomial-time solvable and PARTIAL STRONG CONNECTIVITY AUGMENTATION (PSCA) is W[2]-hard. Note that the non-partial versions of both problems are polynomial-time solvable if the link set is complete.

5.1 Partial Bridge-Connectivity Augmentation

The PARTIAL BRIDGE-CONNECTIVITY AUGMENTATION problem (PBCA) we study here is defined as follows: Given an undirected and connected graph $G = (V, E)$, a set of links E', and two non-negative integers Φ, k, find a set E'' of at most k links such that the graph $(V, E \cup E'')$ contains a bridge-connected component with at least Φ vertices. Note that in the case that E' is incomplete, BRIDGE-CONNECTIVITY AUGMENTATION is NP-complete, which implies that PBCA is also NP-complete in this case. We show here that PBCA becomes polynomial-time solvable if E' is complete. This extends a result by Eswaran and Tarjan [2] saying that BCA is polynomial-time solvable in the case of a complete link set. Our solving strategy consists of two steps: The first step reduces the augmentation problem to a special subtree problem and the second step applies a dynamic programming approach to solve the subtree problem. The special subtree problem, MAXIMUM d-LEAVES SUBTREE (MLST), is defined as follows: Given a tree $T = (V_T, E_T)$, two non-negative integers N, d, and a weight function $w : V_T \to \mathbb{N}$, find a subtree of T with at most d leaves such that the total weight of the vertices in this subtree is at least N.

Theorem 3. MAXIMUM d-LEAVES SUBTREE *can be solved in $O(|V_T| \cdot d^2)$ time.*

Theorem 4. *In the case of a complete link set,* PARTIAL BRIDGE-CONNECTIVITY AUGMENTATION *is solvable in $O(|V| \cdot k^2)$ time.*

5.2 Partial Strong Connectivity Augmentation

Now, we show that PARTIAL STRONG CONNECTIVITY AUGMENTATION (PSCA) is W[2]-hard, which is defined as follows: Given a directed graph $D = (V, A)$, a set of links $A' \subseteq V \times V$, and two non-negative integers Φ, k, find a subset A'' of A'

with $|A''| \leq k$ such that $(V, A \cup A'')$ contains a strongly connected component with at least Φ vertices. We give a parameterized reduction from the W[2]-hard SET COVER problem [1].

Theorem 5. *In both incomplete and complete link set cases,* PARTIAL STRONG CONNECTIVITY AUGMENTATION *is W[2]-hard.*

6 Open Problems

The most interesting open problem is to study the parameterized complexity of the STRONG CONNECTIVITY AUGMENTATION problem, where we are given a directed graph $D = (V, A)$, a set of links $A \subsetneq V \times V$, and a non-negative integer k and ask for a subset A'' of links such that graph $(V, A \cup A'')$ is strongly connected. We conjecture that this problem is fixed-parameter tractable with respect to k. Improving the size bounds of the problem kernels for BRIDGE-CONNECTIVITY AUGMENTATION and BICONNECTIVITY AUGMENTATION to a linear function in k is another interesting open problem. Further opportunities for future work include investigating the approximability and fixed-parameter tractability of the PARTIAL BRIDGE-CONNECTIVITY AUGMENTATION and PARTIAL BICONNECTIVITY AUGMENTATION problems for the case that the link set E' is incomplete.

References

1. Downey, R.G., Fellows, M.R.: Parameterized Complexity. Springer, Heidelberg (1999)
2. Eswaran, K.P., Tarjan, R.E.: Augmentation problems. SIAM Journal on Computing 5(4), 653–665 (1976)
3. Even, G., Feldman, J., Kortsarz, G., Nutov, Z.: A 3/2-approximation algorithm for augmenting the edge-connectivity of a graph from 1 to 2 using a subset of a given edge set. In: Goemans, M.X., Jansen, K., Rolim, J.D.P., Trevisan, L. (eds.) RANDOM 2001 and APPROX 2001. LNCS, vol. 2129, pp. 90–101. Springer, Heidelberg (2001)
4. Frederickson, G.N., JáJá, J.: Approximation algorithms for several graph augmentation problems. SIAM Journal on Computing 10(2), 270–283 (1981)
5. Khuller, S., Thurimella, R.: Approximation algorithms for graph augmentation. Journal of Algorithms 14(2), 214–225 (1993)
6. Kortsarz, G., Krauthgamer, R., Lee, J.R.: Hardness of approximation for vertex-connectivity network design problems. SIAM Journal on Computing 33(3), 704–720 (2004)
7. Kortsarz, G., Nutov, Z.: A 12/7-approximation algorithm for the vertex-connectivity of a graph from 1 to 2, Manuscipt (2002)
8. Nagamochi, H.: An approximation for finding a smallest 2-edge-connected subgraph containing a specified spanning tree. Discrete Applied Mathematics 126, 83–113 (2003)
9. Niedermeier, R.: Invitation to Fixed-Parameter Algorithms. Oxford University Press, Oxford (2006)
10. Rosenthal, A., Goldner, A.: Smallest augmentations to biconnect a graph. SIAM Journal on Computing 6(1), 55–66 (1977)
11. Tarjan, R.E.: Depth-first search and linear graph algorithms. SIAM Journal on Computing 1(2), 146–160 (1972)

An Improved Parameterized Algorithm for the Minimum Node Multiway Cut Problem*

Jianer Chen, Yang Liu, and Songjian Lu

Department of Computer Science
Texas A&M University
College Station, TX 77843, USA
{chen,yangliu,sjlu}@cs.tamu.edu

Abstract. The PARAMETERIZED NODE MULTIWAY CUT problem is for a given graph to find a separator of size bounded by k whose removal separates a collection of terminal sets in the graph. In this paper, we develop an $O(4^k n^{O(1)})$ time algorithm for this problem, significantly improving the previous algorithm of time $O(4^{k^3} n^{O(1)})$ for the problem. Our result also gives the first polynomial time algorithm for the MINIMUM NODE MULTIWAY CUT problem when the separator size is bounded by $O(\log n)$.

1 Introduction

The MULTITERMINAL CUT problem is a well-known problem, and has been extensively studied ([2,9,12]). Applications of this problem are found in distributed computing [13], VLSI [4], computer vision [1], and many other fields. The problem is defined as follows: given an undirected graph $G = (V, E)$ and a set of l vertices $\{t_1, \dots, t_l\}$ in G (the vertices t_i are called *terminals*), find an edge set E' of minimum size in G such that after the deletion of E', no two terminals are in the same connected component. This problem is NP-hard for general graphs for any fixed integer $l \geq 3$, and is also NP-hard for planar graphs when l is not fixed [6].

A generalization of the MULTITERMINAL CUT problem is the MINIMUM NODE MULTIWAY CUT problem, which, for a given graph and a given set of terminals, is to find a vertex set S of minimum size such that after the deletion of S, no two terminals are in the same connected component. The MINIMUM NODE MULTIWAY CUT problem is at least as hard as the MULTITERMINAL CUT problem, since the latter can be reduced to the former in time $O(|V| + |E|)$, if we require that no terminal be in S [5]. Therefore, the MINIMUM NODE MULTIWAY CUT problem is also NP-hard for $l \geq 3$.

When there are only two terminals s and t, the MULTITERMINAL CUT problem and the MINIMUM NODE MULTIWAY CUT problem become the edge version and the vertex version of the MINIMUM s-t CUT problem, respectively. According to the max-flow min-cut theorem [8], the MINIMUM s-t CUT problem, for both

* This work was supported in part by the National Science Foundation under the Grants CCR-0311590 and CCF-0430683.

F. Dehne, J.-R. Sack, and N. Zeh (Eds.): WADS 2007, LNCS 4619, pp. 495–506, 2007.

the edge version and the vertex version, can be solved via algorithms for the MAXIMUM s-t FLOW problem. For an undirected graph G of n vertices and m edges, the MAXIMUM s-t FLOW problem can be solved in time $O(n^{7/6}m^{2/3})$ [10]. In consequence, the MULTITERMINAL CUT problem and the MINIMUM NODE MULTIWAY CUT problem can also be solved in time $O(n^{7/6}m^{2/3})$.

A natural extension of the MINIMUM NODE MULTIWAY CUT problem is to have a collection of terminal sets, instead of a collection of individual terminals. Formally, let $G = (V, E)$ be an undirected graph, and let $\{T_1, \ldots, T_l\}$ be a collection of *terminal sets* where each T_i is a subset of vertices in G. A *separator* S for $\{T_1, T_2, \ldots, T_l\}$ is a subset of vertices in G such that no vertex in S is in any terminal set, and after deleting S from the graph G, no connected component in the resulting graph contains vertices from more than one terminal set.

In certain real world applications, one may expect that the size of the separator be small. For example, suppose that we are given a network (i.e., a graph) $G = (V, E)$ and a collection of network node groups $\{T_1, \ldots, T_l\}$ in G, and we want to monitor the message communication among the node groups. A separator for $\{T_1, \ldots, T_l\}$ in the network G will well serve for this purpose: any communication path between any two node groups must pass through at least one node in the separator. Therefore, if we set up a monitor process in each of the nodes in the separator, then we can monitor all communications among the node groups. Naturally, we may want to limit the cost of this monitoring system by using only a small number of "monitor nodes" in the network G.

This motivates a parameterized version of the MINIMUM NODE MULTIWAY CUT problem, which will be called the PARAMETERIZED NODE MULTIWAY CUT problem and is defined as follows: given an undirected graph $G = (V, E)$, a collection of pairwise disjoint terminal sets $\{T_1, \ldots, T_l\}$ (where each T_i is a subset of vertices in G), and a parameter k, either construct a separator of at most k vertices in G, or report that no such a separator exists. Our goal is, for the PARAMETERIZED NODE MULTIWAY CUT problem, to develop a *fixed-parameter tractable algorithm* [7], i.e., an algorithm whose running time is of the form $f(k)n^c$ with a function f independent of the input size n and a constant c. In particular, when the parameter value k is small, such a fixed-parameter tractable algorithm will be practically effective. In fact, the study of fixed-parameter tractable algorithms for a variety of parameterized problems has drawn considerable attention recently and has direct impact on real word applications where the selected parameter varies in a small range of values [7].

It can be derived from the graph minor theory of Robertson and Seymour [7] that there is a fixed-parameter tractable algorithm for the PARAMETERIZED NODE MULTIWAY CUT problem. However, the proof is not constructive. An explicit constructive algorithm for the problem was given by Marx [11], who developed an algorithm of running time $O(n^5 4^{k^3})$ for the PARAMETERIZED NODE MULTIWAY CUT problem for its original version (i.e., in which each terminal set is restricted to contain a single terminal). To our knowledge, it is the only known fixed-parameter tractable algorithm for the problem.

In this paper, we present an algorithm of running time $O(n^3 lk4^k)$ for the PARAMETERIZED NODE MULTIWAY CUT problem, which significantly improves the algorithm given in [11]. In the real world of computing, this improvement makes it become possible to practically solve the problem for some reasonable values of the parameter k. For example, for the case of $k = 10$, our algorithm has running time $O(n^3 l4^{10})$, which is practically feasible using the currently available computation power. On the other hand, the algorithm in [11] in this case has running time $O(n^5 2^{2000})$, which is totally infeasible from the practical point of view. Theoretically, our result gives the first polynomial time algorithm for the MINIMUM NODE MULTIWAY CUT problem when the size of the optimal separator is of order $O(\log n)$.

2 On Minimum Cuts Between Two Terminal Sets

We start with some terminology.

Let $G = (V, E)$ be a graph and let u and v be two vertices in G. A *path between u and v* is a simple path in G whose two ends are u and v, respectively. For a subset V of vertices in G, we say that there is a *path between a vertex u and V* if there is a vertex v in V such that there is a path between u and v. For two vertex subsets V_1 and V_2, we say that there is a *path between V_1 and V_2* if there exist a vertex u in V_1 and a vertex v in V_2 such that there is a path between u and v. Two paths are *internally disjoint* if there is no vertex that is an internal vertex for both the paths.

Let G be a graph, and let $\{T_1, \ldots, T_l\}$ be a collection of pairwise disjoint terminal sets (each terminal set is a subset of vertices in G). A subset S of vertices in G is a *separator* for $\{T_1, \ldots, T_l\}$ if S contains no vertex in any of the sets T_1, ..., T_l, and if after deleting all vertices in S from G, there is no path between any two different subsets T_i and T_j in the resulting graph. In particular, a separator S for two terminal sets T_1 and T_2 is also called a *cut* between the two sets T_1 and T_2.

Let T be a subset of vertices in the graph $G = (V, E)$. By *merging T (into a single vertex)*, we mean the operation that first deletes all vertices in T then creates a new vertex w adjacent to each v of the vertices in $V - T$ where v is a neighbor of a vertex in T in the original graph G.

Finally, for a subset V' of vertices in the graph G, we will denote by $G(V')$ the subgraph of G that is induced by the vertex subset V'.

We start with an undirected vertex form of Menger's Theorem, whose proof can be found in [3].

Proposition 1. [3] (The Undirected Vertex Form of Menger's Theorem) *Let u and v be two distinct and nonadjacent vertices in a graph G. Then the maximum number of internally disjoint paths between u and v in G is equal to the size of a minimum cut between u and v in G.*

Proposition 1 can be generalized from the case for two vertices to the case of two vertex subsets, as given in the following lemma.

Lemma 1. *Let T_1 and T_2 be two disjoint vertex subsets in a graph G such that no vertex in T_1 is adjacent to a vertex in T_2. Then the maximum number h of internally disjoint paths between T_1 and T_2 in G is equal to the size of a minimum cut between T_1 and T_2 in G. Moreover, for any set π of h internally disjoint paths between T_1 and T_2 in G, every minimum cut between T_1 and T_2 in G contains exact one vertex in each of the paths in π.*

Proof. Let G' be the graph obtained from the graph G by merging the two vertex subsets T_1 and T_2 into two vertices t_1 and t_2, respectively. Note that t_1 and t_2 are not adjacent in G'.

By the definition of the merge operation, it is easy to verify that a vertex subset S is a cut between the vertex subsets T_1 and T_2 in the graph G if and only if S is a cut between the vertices t_1 and t_2 in the graph G'. In particular, the size of a minimum cut between T_1 and T_2 in G is equal to the size of a minimum cut between t_1 and t_2 in G'. Moreover, it is also easy to verify that for any integer h', from a set of h' internally disjoint paths between T_1 and T_2 in G, we can construct a set of h' internally disjoint paths between t_1 and t_2 in G', and *vice versa.* Therefore, the maximum number of internally disjoint paths between T_1 and T_2 in G is equal to the maximum number of internal disjoint paths between t_1 and t_2 in G'. Now the first part of the lemma follows by applying Proposition 1 to the graph G'.

To prove the second part of the lemma, let S be a minimum cut, of size h, between T_1 and T_2 in G, and let π be a set of h internally disjoint paths between T_1 and T_2. The vertex set S must contain at least one vertex from each of the paths in π: otherwise there would be a path between T_1 and T_2 in $G - S$, contradicting the assumption that S is a cut between T_1 and T_2. Moreover, the set S cannot contain more than one vertex in any path in π: otherwise S would not be able to contain at least one vertex for each of the paths in π (note that the paths in π are internally disjoint). □

Lemma 1 provides an efficient algorithm that constructs the maximum number of internally disjoint paths and a minimum-size cut between two given vertex subsets in a graph.

Lemma 2. *Let T_1 and T_2 be two disjoint vertex subsets in a graph $G = (V, E)$ such that no vertex in T_1 is adjacent to a vertex in T_2. Then in time $O((|V| + |E|)k)$, we can decide if the size h of a minimum cut between T_1 and T_2 is bounded by k, and in case $h \leq k$, construct h internally disjoint paths between T_1 and T_2.*

Proof. Let G' be the graph obtained from the graph G by merging the two vertex subsets T_1 and T_2 into two vertices t_1 and t_2, respectively. As discussed in the proof of Lemma 1, it suffices to show how to decide if the size h of a minimum cut between t_1 and t_2 in G' is bounded by k, and in case $h \leq k$, how to construct h internally disjoint paths between t_1 and t_2.

This can be done based on the standard approach to the MAXIMUM t_1-t_2 FLOW problem [3]. For this, we first transform the undirected graph G' into a directed graph by replacing each edge by two reverse arcs. Then we modify the new directed graph by replacing each vertex u (except the vertices t_1 and t_2) by two vertices u_1 and u_2 with an arc from u_1 to u_2, connecting all u's incoming arcs to the vertex u_1 and connecting all u's outgoing arcs to the vertex u_2. Finally we set all edges to have capacity 1. Let the resulting flow graph be G''.

Applying Ford-Fulkerson's standard approach using augmenting paths, in time $O((|V| + |E|)k)$, we can either construct a t_1-t_2 flow of value larger than k in G'', or end up with a maximum t_1-t_2 flow of value h bounded by k. In the former case, we conclude that the size of a minimum cut between t_1 and t_2 in G' is larger than k, which implies that the size of a minimum cut between T_1 and T_2 in G is larger than k. In the latter case, h internally disjoint paths between t_1 and t_2 in G' can be easily constructed from the maximum t_1-t_2 flow of value h in G'', from which h internally disjoint paths between T_1 and T_2 in G can be constructed. □

3 The Main Algorithm

Now we return back to the PARAMETERIZED NODE MULTIWAY CUT problem. Formally, an instance $(G, \{T_1, \ldots, T_l\}, k)$ of the PARAMETERIZED NODE MULTIWAY CUT problem consists of an undirected graph G, a collection $\{T_1, \ldots, T_l\}$ of pairwise disjoint *terminal sets* (each terminal set is a vertex subset in G), and a parameter k. The objective is to either construct a separator of at most k vertices for $\{T_1, \ldots, T_l\}$, or conclude that no such a separator exists.

Before we formally present our algorithm, we give a less formal but intuitive explanation on the basic idea of the algorithm. For each i, $1 \leq i \leq l$, let the size of a minimum cut between T_i and $\bigcup_{j \neq i} T_j$ be m_i. Define $m = \max\{m_i \mid 1 \leq i \leq l\}$. Without loss of generality, assume $m = m_1$.

Pick a vertex u that is not in any terminal set and has a neighbor in T_1. If u also has a neighbor in another terminal set T_i, $i \neq 1$, then we can directly include u in the separator (this is necessary because the separator must separate T_1 and T_i), and recursively find a separator of size $k-1$ in the remaining graph. On the other hand, if u has no neighbor in other terminal sets, then we compute the size m' of a minimum cut between the sets $T_1' = T_1 \cup \{u\}$ and $\bigcup_{i \neq 1} T_i$. It can be proved that we must have $m \leq m'$. Note that by Lemma 2, the values m and m' can be computed in polynomial time.

In the case $m = m'$, we will show that the instance $(G, \{T_1, T_2, \ldots, T_l\}, k)$ has a separator of size bounded by k if and only if the instance $(G, \{T_1', T_2, \ldots, T_l\}, k)$ has a separator of size bounded by k. Then we recursively work on the new instance $(G, \{T_1', T_2, \ldots, T_l\}, k)$. Thus, in the case of $m = m'$, we can reduce the number of vertices that are not in the separator by 1.

On the other hand, suppose $m < m'$. Then we branch on the vertex u in two cases, one includes u in the separator and the other excludes u from the separator. In the case of including the vertex u in the separator, we recursively work on the instance $(G - \{u\}, \{T_1, T_2, \ldots, T_l\}, k - 1)$, in which the parameter value is decreased by 1; and in the case of excluding the vertex u from the separator, we recursively work on the instance $(G, \{T_1', T_2, \ldots, T_l\}, k)$, in which the size of the minimum cut between T_1' and $\bigcup_{i \neq 1} T_i$ is increased by at least 1.

Therefore, for the given instance $(G, \{T_1, T_2, \ldots, T_l\}, k)$, we can either (1) apply a polynomial time process that either decreases the parameter value by 1 or reduces the number of vertices not in the separator by 1, or (2) branch into two cases, of which one decreases the parameter value by 1 and the other increases the value m by at least 1 (see the definition of m given in the second paragraph in this section). Note that all these generated new instances will be "simpler" than the original given instance: (i) reducing the number of vertices not in the separator will narrow down our search space for the separator; (ii) an instance of parameter value bounded by 1 can be solved in polynomial time; and (iii) an instance in which the value m is larger than the parameter value k obviously has no separator of size bounded by k.

To present our formal discussions, we fix an instance $(G, \{T_1, \ldots, T_l\}, k)$ of the PARAMETERIZED NODE MULTIWAY CUT problem, where $G = (V, E)$ is a graph, and $\{T_1, \ldots, T_l\}$ is a collection of terminal sets in G. For each i, $1 \leq i \leq l$, let the size of a minimum cut between T_i and $\bigcup_{j \neq i} T_j$ be m_i. Define $m = \max\{m_i \mid 1 \leq i \leq l\}$, and assume, without loss of generality, that $m = m_1$. Moreover, fix a vertex u that is not in any of the terminal sets but has a neighbor in the terminal set T_1. Let $T_1' = T_1 \cup \{u\}$.

Lemma 3. *Let m' be the size of a minimum cut between the two sets T_1' and $\bigcup_{j \neq 1} T_j$. Then $m' \geq m$.*

Proof. The lemma follows from the observation that every cut between the sets T_1' and $\bigcup_{j \neq 1} T_j$ is also a cut between the sets T_1 and $\bigcup_{j \neq 1} T_j$. □

The following theorem is the most crucial observation for our algorithm.

Theorem 1. *If the minimum cuts between the sets T_1 and $\bigcup_{j \neq 1} T_j$ and the minimum cuts between the sets T_1' and $\bigcup_{j \neq 1} T_j$ have the same size, then the instance $(G, \{T_1, T_2, \ldots, T_l\}, k)$ has a separator of size bounded by k if and only if the instance $(G, \{T_1', T_2, \ldots, T_l\}, k)$ has a separator of size bounded by k.*

Proof. If the instance $(G, \{T_1', T_2, \ldots, T_l\}, k)$ has a separator S of size bounded by k, then it is obvious that S is also a separator of the instance $(G, \{T_1, T_2, \ldots, T_l\}, k)$. Now we consider the other direction. To simplify the discussion, let $T_{other} = \bigcup_{j \neq 1} T_j$.

Suppose that S_m is a minimum cut between T_1' and T_{other}, then S_m is also a cut between T_1 and T_{other}. In fact, by the assumption of the theorem, S_m is also a minimum cut between T_1 and T_{other}. Let $C(T_1)$ be the set of vertices x such that either $x \in T_1$ or there is a path between x and T_1 in the induced subgraph $G(V - S_m)$. In particular, $u \in C(T_1)$. Moreover, let $C(T_{other}) = V - C(T_1) - S_m$.

By Lemma 1, there exist $|S_m|$ internally disjoint paths between T_1 and T_{other}, each contains exactly one vertex in the set S_m. Therefore, each of these $|S_m|$ paths is cut into two subpaths by a vertex in S_m, such that one subpath is in the induced subgraph $G(C(T_1))$ and the another subpath is in the induced subgraph

$G(C(T_{other}))$. From this, we derive that there are $|S_m|$ internally disjoint paths between T_1 and S_m in the induced subgraph $G(C(T_1) \cup S_m)$, each contains a distinct vertex in the set S_m.

Let S_k be a separator of the instance $(G, \{T_1, T_2, \ldots, T_l\}, k)$ of size bounded by k. Define $A = S_k \cap C(T_1)$, $B = S_k \cap S_m$, and $C = S_k \cap C(T_{other})$. Finally, let S'_m be the set of vertices x in S_m such that there is a path between x and T_{other} in the induced subgraph $G(C(T_{other}) \cup S_m - S_k)$ (see Figure 1 for an intuitive illustration of these sets).

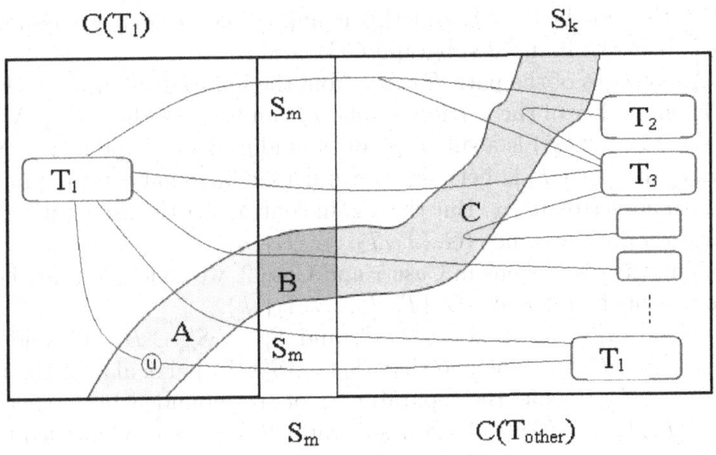

Fig. 1. Decomposition of Separators

We first prove that $|A| \geq |S'_m|$.

From the fact that there are $|S_m|$ internally disjoint paths between T_1 and S_m in the induced subgraph $G(C(T_1) \cup S_m)$ in which each path contains a distinct vertex in the set S_m, we derive that there are $|S'_m|$ internally disjoint paths between T_1 and S'_m in the induced subgraph $G(C(T_1) \cup S'_m)$. If $|A| < |S'_m|$, then there must be a path P_1 between T_1 and a vertex v' in S'_m in the subgraph $G(C(T_1) \cup S'_m - A) = G(C(T_1) \cup S'_m - S_k)$. Moreover, by the definition of the set S'_m, there is also a path P_2 between v' and T_{other} in the induced subgraph $G(C(T_{other}) \cup S_m - S_k)$. The concatenation of the paths P_1 and P_2 would give a path between T_1 and T_{other} in the induced subgraph $G(V - S_k)$, which contradicts the assumption that S_k is a separator of the instance $(G, \{T_1, T_2, \ldots, T_l\}, k)$. Therefore, we must have $|A| \geq |S'_m|$.

Define a set $S'_k = S'_m \cup B \cup C$. We now prove that the set S'_k is a separator of the instance $(G, \{T'_1, T_2, \ldots, T_l\}, k)$. Suppose that the set S'_k is not a separator of the instance $(G, \{T'_1, T_2, \ldots, T_l\}, k)$, then there are two vertices v_1 and v_2 that are in two different terminal sets in $\{T'_1, T_2, \ldots, T_l\}$ and there exists a path P between v_1 and v_2 in the induced subgraph $G(V - S'_k)$. We discuss this in two possible cases.

Case 1: There is a vertex w in the path P such that $w \in C(T_1)$. Because (1) at least one of the vertices v_1 and v_2 is in the set T_{other}, (2) there is a path between T_1 and w in the induced subgraph $G(C(T_1))$, and (3) S_m is a cut between T_1 and T_{other}, we conclude that there must be a vertex $s \in S_m$ that is also on the path P. Without loss of generality, we can suppose that the vertex v_1 is in the set T_{other}, and that the subpath P' of P that begins from v_1 and ends at s has no vertices from $C(T_1)$ – for this we only have to pick the first vertex s in S_m when we traverse on the path P from v_1 to v_2. Then the path P' is in the induced subgraph $G(C(T_{other}) \cup S_m - S'_k)$, which is a subgraph of the induced subgraph $G(C(T_{other}) \cup S_m - S_k)$. Now by the definition of the set S'_m, the vertex s is in the set S'_m, thus in the set S'_k. But this is impossible because we assumed that the path P is in the induced subgraph $G(V - S'_k)$.
Case 2: All vertices of the path P come from the induced subgraph $G(V - S'_k - C(T_1))$. Then neither of the vertices v_1 and v_2 can be from the set T_1. Moreover, since $G(V - S'_k - C(T_1))$ is a subgraph of the induced subgraph $G(V - S_k)$, this implies that the path P is between two different terminal sets in $\{T_2, \dots, T_l\}$ and contains no vertex in S_k. But this again contradicts the assumption that S_k is a separator of the instance $(G, \{T_1, T_2, \dots, T_l\}, k)$.

Combining the discussions in Case 1 and Case 2, we conclude that the set S'_k is a separator of the instance $(G, \{T'_1, T_2, \dots, T_l\}, k)$.

Since $|A| \geq |S'_m|$, $S_k = A \cup B \cup C$, and $S'_k = S'_m \cup B \cup C$, and A does not intersect $B \cup C$, we conclude that $|S_k| \geq |S'_k|$. In particular, if the instance $(G, \{T_1, T_2, \dots, T_l\}, k)$ has the separator S_k of size bounded by k, then the instance $(G, \{T'_1, T_2, \dots, T_l\}, k)$ has the separator S'_k of size also bounded by k. □

The proof of Theorem 1 becomes complicated partially because the vertex u may be included in a separator for the instance $(G, \{T_1, T_2, \dots, T_l\}, k)$. If we restrict that the vertex u is not in the separators for the instance $(G, \{T_1, T_2, \dots, T_l\}, k)$, then a result similar to Theorem 1 can be obtained much more easily, even without the need of the condition that the minimum cuts between T_1 and $\bigcup_{j \neq 1} T_j$ and the minimum cuts between T'_1 and $\bigcup_{j \neq 1} T_j$ have the same size. This is given in the following lemma. This result will also be needed in our algorithm.

Lemma 4. *Let S be a vertex subset in the graph G such that S does not include the vertex u. Then S is a separator for the instance $(G, \{T_1, T_2, \dots, T_l\}, k)$ if and only if S is a separator for the instance $(G, \{T'_1, T_2, \dots, T_l\}, k)$.*

Proof. If S is a separator for the instance $(G, \{T'_1, T_2, \dots, T_l\}, k)$, then as explained in Theorem 1, S is also a separator for the instance $(G, \{T_1, T_2, \dots, T_l\}, k)$.

For the other direction, suppose that the set S is a separator for the instance $(G, \{T_1, T_2, \dots, T_l\}, k)$. We show that S is also a separator for the instance $(G, \{T'_1, T_2, \dots, T_l\}, k)$. Suppose that S is not a separator for the instance $(G, \{T'_1, T_2, \dots, T_l\}, k)$. Then there is a path P in $G - S$ between two different terminal sets in $\{T'_1, T_2, \dots, T_l\}$. Let one of these two terminal sets in $\{T'_1, T_2, \dots, T_l\}$ be T_i, where $i \neq 1$. The path P must contain the vertex u (recall that S does not contain u) – otherwise the path P in $G - S$ would be between two different terminal sets in $\{T_1, T_2, \dots, T_l\}$, contradicting the assumption that

S is a separator for $(G, \{T_1, T_2, \ldots, T_l\}, k)$. However, this would imply that the path from T_1 to u (recall that u has a neighbor in T_1) then following the path P to the terminal set T_i would give a path in $G - S$ between T_1 and T_i, again contradicting the assumption that S is a separator for $(G, \{T_1, T_2, \ldots, T_l\}, k)$. Therefore, S is also a separator for the instance $(G, \{T_1', T_2, \ldots, T_l\}, k)$. □

Now, we are ready to present our algorithm. For an instance $(G, \{T_1, \ldots, T_l\}, k)$ of the PARAMETERIZED NODE MULTIWAY CUT problem, a vertex in the graph G that does not belong to any terminal sets will be called a "non-terminal".

The algorithm is given in Figure 2.

Algorithm $\mathbf{NMC}(G, \{T_1, T_2, \ldots, T_l\}, k)$
input: an instance $(G, \{T_1, T_2, \ldots, T_l\}, k)$ of the PARAMETERIZED NODE MULTIWAY CUT problem
output: a separator of size bounded by k for $(G, \{T_1, T_2, \ldots, T_l\}, k)$, or report "No" (i.e., no such a separator)

1. **If** an edge has its two ends in two different terminal sets **then** return "No";
2. **if** a non-terminal w has two neighbors in two different terminal sets **then** return $w + \mathbf{NMC}(G - w, \{T_1, \ldots, T_l\}, k - 1)$;
3. **for** $i = 1$ **to** l **do**
   let m_i be the size of a minimum cut between T_i and $\bigcup_{j \neq i}^{l} T_j$;
   (suppose $m_1 = \max\{m_i \mid 1 \leq i \leq l\}$);
4. **If** $m_1 > k$ **then** return "No";
5. **else if** $m_1 = 0$ **then** return $\emptyset$;
6. **else** pick a non-terminal u that has a neighbor in T_1; let $T_1' = T_1 + u$;

6.1 **if** the size of a minimum cut between T_1' and $\bigcup_{j \neq 1} T_j$ is equal to m_1 **then** return $\mathbf{NMC}(G, \{T_1', T_2, \ldots, T_l\}, k)$;
6.2 **else** $S = u + \mathbf{NMC}(G - u, \{T_1, T_2, \ldots, T_l\}, k - 1)$;
If S is not "No" **then** return S;
6.3 **else** return $\mathbf{NMC}(G, \{T_1', T_2, \ldots, T_l\}, k)$.

Fig. 2. New algorithm for the k-NODE MULTIWAY CUT problem

Theorem 2. *The algorithm* $\mathbf{NMC}(G, \{T_1, T_2, \ldots, T_l\}, k)$ *in Figure 2 solves the* PARAMETERIZED NODE MULTIWAY CUT *problem in time* $O(n^3 lk4^k)$.

Proof. We first prove the correctness of the algorithm. Let $(G, \{T_1, T_2, \ldots, T_l\}, k)$ be an input to the algorithm, which is an instance of the PARAMETERIZED NODE MULTIWAY CUT problem, where $G = (V, E)$ is a graph, $\{T_1, T_2, \ldots, T_l\}$ is a collection of terminal sets, and k is the upper bound of the size of the separator we are looking for.

If there is an edge whose two ends are in two different terminal sets, then we have no way to separate these two terminal sets since all vertices in a separator are supposed to be non-terminals. Step 1 handles this case correctly.

If a non-terminal w has two neighbors that are in two different terminal sets, then w must be in the separator because otherwise the two terminal sets will not be separated. Thus, we can simply include the vertex w in the separator, and recursively find a separator of size bounded by $k-1$ for the same collection of terminal sets $\{T_1, T_2, \ldots, T_l\}$ in the remaining graph $G-w$. This case is correctly handled by step 2.[1]

Step 3 computes the size of a minimum cut between the sets T_i and $\bigcup_{j \neq i} T_j$ for each i, $1 \leq i \leq l$. By Lemma 2, each m_i can be computed in time $O((|V|+|E|)k)$. Thus, step 3 takes time $O((|V|+|E|)kl)$.

If $m_1 > k$, then the size of a minimum cut between T_1 and $\bigcup_{j \neq 1} T_j$ is larger than k, which implies that even separating the set T_1 from the other sets $\bigcup_{j \neq 1} T_j$ requires more than k vertices. Thus, no separator of size bounded by k can exist for the terminal sets T_1, T_2, ..., T_l. This is handled by step 4.

In step 5 we handle the case $m_1 = 0$. Note that m_1 is the largest m_i we computed in step 3. Thus, $m_1 = 0$ implies that for every i, there is no path in the graph G from the terminal set T_i to any other terminal sets. Therefore, the terminal sets T_1, T_2, ..., T_l have already been separated, and we can simply return an empty set $\emptyset$ as a separator of size 0 (note that because of step 4, here we must have $k \geq 0$).

When the algorithm reaches step 6, the following conditions hold true: (1) no edge has its two ends in two different terminal sets (because of step 1); (2) no non-terminal has two neighbors in two different terminal sets (because of step 2); (3) $0 < m_1 \leq k$ (because of steps 4-5). In particular, by condition (3), there must be a non-terminal u that has a neighbor in T_1.

Let m' be the size of a minimum cut between the sets T_1' and $\bigcup_{j \neq 1} T_j$. If $m' = m_1$, then by Theorem 1, the instance $(G, \{T_1, T_2, \ldots, T_l\}, k)$ has a separator of size bounded by k if and only if the instance $(G, \{T_1', T_2, \ldots, T_l\}, k)$ has a separator of size bounded by k. In particular, as shown in the proof of Theorem 1, a separator of size bounded by k for the instance $(G, \{T_1', T_2, \ldots, T_l\}, k)$ is actually also a separator for the instance $(G, \{T_1, T_2, \ldots, T_l\}, k)$. Therefore, in this case, we can recursively work on the instance $(G, \{T_1', T_2, \ldots, T_l\}, k)$, as given in step 6.1. On the other hand, if $m' \neq m$ ($m' > m$), then we simply branch on the vertex u in two cases: (1) one case includes u in the separator and recursively works on the remaining graph for a separator of size bounded by $k-1$, as given by step 6.2; and (2) the other case excludes u from the separator thus looks for a separator that does not include u and is of size bounded by k for the instance $(G, \{T_1, T_2, \ldots, T_l\}, k)$. By Lemma 4, the second case is equivalent to finding a separator of size bounded by k for the instance $(G, \{T_1', T_2, \ldots, T_l\}, k)$. This case is thus handled by step 6.3.

This completes the proof of the correctness of the algorithm. Now we analyze the complexity of the algorithm.

---

[1] To simplify the expression, we suppose that "No" plus any vertex set gives a "No". Therefore, step 2 will return a "No" if **NMC**$(G-w, \{T_1, \ldots, T_l\}, k-1)$ returns a "No".

The recursive execution of the algorithm can be described as a search tree $\mathcal{T}$. We first count the number of leaves in the search tree $\mathcal{T}$. Note that only steps 6.2-6.3 of the algorithm correspond to branches in the search tree $\mathcal{T}$. Let $D(k, m)$ be the total number of leaves in the search tree $\mathcal{T}$ for the algorithm $\mathbf{NMC}(G, \{T_1, T_2, \ldots, T_l\}, k)$, where $m = \max\{m_i \mid 1 \leq i \leq l\}$, and m_i is the size of a minimum cut between the sets T_i and $\bigcup_{j \neq i} T_j$. Then steps 6.2-6.3 induce the following recurrence relation:

$$D(k, m) \leq D(k-1, m'') + D(k, m''') \tag{1}$$

where $m'' = \max\{m''_i \mid 1 \leq i \leq l\}$, and m''_i is the size of a minimum cut between T_i and $\bigcup_{j \neq i} T_j$ *in the graph* $G - u$ as given in step 6.2, and m''' is similarly defined based on the instance $(G, \{T'_1, T_2, \ldots, T_l\}, k)$ as given in step 6.3. Note that $m - 1 \leq m'' \leq m$ because removing the vertex u from G cannot increase the size of a minimum cut between two sets, and can decrease the size of a minimum cut between the two sets by at most 1. Moreover, by Lemma 3 and because of step 6.1, the size m' of a minimum cut between T'_1 and $\bigcup_{j \neq 1} T_j$ in step 6.3 is at least $m + 1$. By the definition of m''', we have $m''' \geq m' \geq m + 1$. Summarizing these, we have

$$m - 1 \leq m'' \leq m \quad \text{and} \quad m''' \geq m + 1 \tag{2}$$

Introduce a new function D' such that $D(k, m) = D'(2k - m)$, and let $t = 2k - m$. Then by Inequalities (1) and (2), the branch in step 6.2-6.3 in the algorithm becomes

$$D'(t) \leq D'(t_1) + D'(t_2)$$

where when $t = 2k - m$ then $t_1 = 2(k-1) - m'' \leq t - 1$, and $t_2 = 2k - m''' \leq t - 1$. Our initial instance starts with $t = 2k - m \leq 2k$. In the case $t = 2k - m = 1$, because we also have the conditions $k \geq m > 0$, we can derive $m = 1$ and $k = 1$, in this case the algorithm can solve the instance without further branching. Therefore, we have $D'(1) = 1$. Combining all these, we derive

$$D(k, m) = D'(2k - m) \leq 2^{2k} = 4^k,$$

and the search tree $\mathcal{T}$ has at most 4^k leaves.

Finally, it is easy to verify that along each root-leaf path in the search tree $\mathcal{T}$, the running time of the algorithm is bounded by $O(n^3 lk)$, where n is the number of vertices in the graph. In conclusion, the running time of the algorithm $\mathbf{NMC}(G, \{T_1, T_2, \ldots, T_l\}, k)$ is bounded by $O(n^3 lk 4^k)$. □

4 Conclusion

We developed an algorithm of running time $O(n^3 lk 4^k)$ for a more general PARAMETERIZED NODE MULTIWAY CUT problem. Our algorithm finds a separator that has no vertices in any terminal set. We call such a separator a *restricted separator.* If a separator is allowed to include vertices from terminal sets, the separator is called an *unrestricted separator.* It can be verified easily that the instance $(G, \{T_1, \ldots, T_l\}, k)$ has an unrestricted separator of size k if and only if the instance $(G', \{\{x_1\}, \ldots, \{x_l\}\}, k)$ has a restricted separator of size k, where the graph G' is obtained by adding l vertices $x_1, \ldots, x_l$ to the graph G and connecting x_i to each vertex in T_i for all $1 \leq i \leq l$. Therefore, our algorithm can also be used to find an unrestricted separator for a given instance.

One related problem is the PARAMETERIZED NODE MULTICUT problem [11] where we look for a separator of size k to separate each of the l given pairs of terminals. When both k and l are used as parameters, based on the techniques developed in the current paper, the fixed parameter tractable algorithm presented in [11] for the PARAMETERIZED NODE MULTICUT problem can be improved. On the other hand, if only k is used as the parameter, or if the graph G is a directed graph (or even just a directed acyclic graph), it is currently unknown whether the PARAMETERIZED NODE MULTICUT problem has fixed parameter tractable algorithms, which seem very interesting topics for further research.

References

1. Boykov, Y., Veksler, O., Zabih, R.: Markov random fields with efficient approximations. In: Proc. IEEE Conference on Computer Vision and Pattern Recognition, pp. 648–655 (1998)
2. Calinescu, G., Karloff, H., Rabani, Y.: An improved approximation algorithm for multiway cut. Journal of Computer and System Science 60, 564–574 (2000)
3. Chartrand, G., Lesniak, L.: Graphs & Digraphs, 2nd edn. The Wadsworth & Brooks/Cole Mathematics Series (1986)
4. Cong, J., Labio, W., Shivakumar, N.: Multi-way VLSI circuit partitioning based on dual net representation. In: Proc. IEEE International Conference on Computer-Aided Design, pp. 56–62 (1994)
5. Cunningham, W.: The optimal multiterminal cut problem. DIMACS Series in Discrete Mathematics and Theoretical Computer Science 5 (1991)
6. Dahlhaus, E., Johnson, D., Papadimitriou, C., Seymour, P., Yannakakis, M.: The complexity of multiterminal cuts. SIAM J. Comput. 23, 864–894 (1994)
7. Downey, R., Fellows, M.: Parameterized Complexity. In: monograph in computer science, Springer, Heidelberg (1999)
8. Ford Jr., L., Fulkerson, D.: Flows in Networks. Princeton University Press, Princeton (1962)
9. Karger, D., Klein, P., Stein, C., Thorup, M., Young, N.: Rounding algorithms for a geometric embedding of minimum multiway cut. In: Proc. on 31th Annual ACM Syposium on Theory of Computing, pp. 668–678 (1999)
10. Karger, D., Levine, M.: Finding maximum flows in undirected graphs seems easier than bipartite matching. In: Proc. on 30th Annual ACM Syposium on Theory of Computing, pp. 69–78 (1998)
11. Marx, D.: Parameterized graph separation problems. Theoretical Computer Science 351, 394–406 (2006)
12. Naor, J., Zosin, L.: A 2-approximation algorithm for the directed multiway cut problem. SIAM J. Comput. 31, 477–482 (2001)
13. Stone, H.: Multiprocessor scheduling with the aid of network flow algorithms. IEEE Transactions on Software Engineering 3, 85–93 (1977)

Branch and Recharge: Exact Algorithms for Generalized Domination*

Fedor V. Fomin[1], Petr A. Golovach[1], Jan Kratochvíl[2], Dieter Kratsch[3], and Mathieu Liedloff[3]

[1] Department of Informatics, University of Bergen, 5020 Bergen, Norway
{fomin, petrg}@ii.uib.no

[2] Department of Applied Mathematics, and Institute for Theoretical Computer Science, Charles University, Malostranské nám. 25, 118 00 Praha 1, Czech Republic
honza@kam.ms.mff.cuni.cz

[3] LITA, Université Paul Verlaine - Metz, 57045 Metz Cedex 01, France
{kratsch, liedloff}@univ-metz.fr

Abstract. Let σ and ϱ be two sets of nonnegative integers. A vertex subset $S \subseteq V$ of an undirected graph $G = (V, E)$ is called a (σ, ϱ)-dominating set of G if $|N(v) \cap S| \in \sigma$ for all $v \in S$ and $|N(v) \cap S| \in \varrho$ for all $v \in V \setminus S$. This notion introduced by Telle generalizes many domination-type graph invariants. For many particular choices of σ and ϱ it is NP-complete to decide whether an input graph has a (σ, ϱ)-dominating set.

We show a general algorithm enumerating all (σ, ϱ)-dominating sets of an input graph G in time $O^*(c^n)$ for some $c < 2$ using only polynomial space, if σ is successor-free, i.e., it does not contain two consecutive integers, and either both σ and ϱ are finite, or one of them is finite and $\sigma \cap \varrho = \emptyset$. Thus in this case one can find maximum and minimum (σ, ϱ)-dominating sets in time $o(2^n)$. Our algorithm straightforwardly implies a non trivial upper bound c^n with $c < 2$ for the number of (σ, ϱ)-dominating sets in an n-vertex graph under the above conditions on σ and ϱ.

Finally, we also present algorithms to find maximum and minimum $(\{p\}, \{q\})$-dominating sets and to count the $(\{p\}, \{q\})$-dominating sets of a graph in time $O^*(2^{n/2})$.

1 Introduction

We consider finite undirected graphs without loops or multiple edges. Thus a graph is a pair $G = (V, E)$ where V is the (finite) set of vertices and E the set of edges. The size of G is the number of vertices, and throughout the paper we reserve $n = |V|$ to denote this quantity. We call two vertices u, v *adjacent* if they form an edge, i.e., if $uv \in E$. The *open neighborhood* of a vertex $u \in V$ is the set of the vertices adjacent to it, denoted by $N(u) = \{x : xu \in E\}$, and the *closed neighborhood* is defined as $N[u] = N(u) \cup \{u\}$. A set of vertices $S \subseteq V$ is *dominating* if every vertex of G is either in S or adjacent to a vertex in S, i.e., if

* Part of the research was done when some of the authors were visiting DIMATIA, Prague, in April 2006.

F. Dehne, J.-R. Sack, and N. Zeh (Eds.): WADS 2007, LNCS 4619, pp. 507–518, 2007.

$V = \bigcup_{u \in S} N[u]$. Finding a dominating set of the smallest possible size belongs to the most important optimization problems on graphs. In some sense the problem is harder than typical graph invariants such as clique or independent set – the problem is NP-hard even for chordal graphs (cf. [11]), and the parameterized version is W[2]-complete [5]. Many generalizations have been studied, such as independent dominating set, connected dominating set, efficient dominating set, etc. (cf. [11]).

In [15], Telle introduced the following framework of domination-type graph invariants (see also [10,12]). Let σ and ϱ be two sets of non negative integers. A vertex subset $S \subseteq V$ of an undirected graph $G = (V, E)$ is called a (σ, ϱ)-*dominating set* of G if $|N(v) \cap S| \in \sigma$ for all $v \in S$ and $|N(v) \cap S| \in \varrho$ for all $v \in V \setminus S$. The following table shows a sample of previously defined and studied graph invariants which can be expressed in this framework ($\mathbb{N}$ is the set of positive integers, $\mathbb{N}_0$ is the set of nonnegative integers).

σ	ϱ	(σ, ϱ)-dominating set	σ	ϱ	(σ, ϱ)-dominating set
$\mathbb{N}_0$	$\mathbb{N}$	dominating set	$\{0\}$	$\mathbb{N}_0$	independent set
$\mathbb{N}_0$	$\{1\}$	efficient dominating set	$\{0\}$	$\{1\}$	1-perfect code
$\{0\}$	$\{0,1\}$	strong stable set	$\{0\}$	$\mathbb{N}$	independent dominating set
$\{1\}$	$\{1\}$	total perfect dominating set	$\mathbb{N}$	$\mathbb{N}$	total dominating set
$\{1\}$	$\mathbb{N}_0$	induced matching	$\{r\}$	$\mathbb{N}_0$	r-regular induced subgraph

We are interested in the computational complexity of decision, search and counting problems related to (σ, ϱ)-domination. Explicitly, we consider the following problems parameterized by σ and ϱ.

$\exists(\sigma, \varrho)$-DS: Does an input graph G contain a (σ, ϱ)-dominating set?
ENUM-(σ, ϱ)-DS: Given a graph G, list all (σ, ϱ)-dominating sets of G.
#-(σ, ϱ)-DS: Given a graph G, determine the number of (σ, ϱ)-dominating sets of G.
MAX-(σ, ϱ)-DS: Given a graph G, find a (σ, ϱ)-dominating set of maximum size.
MIN-(σ, ϱ)-DS: Given a graph G, find a (σ, ϱ)-dominating set of minimum size.

Obviously, the enumeration problem ENUM-(σ, ϱ)-DS is the most difficult one, since as soon as we have all (σ, ϱ)-dominating sets in a list, we can quickly see if the list is nonempty (and hence answer the $\exists(\sigma, \varrho)$-DS problem), we can also compare the sizes of the listed sets to answer the minimization and maximization questions, and we can quickly count the number of listed sets. However, maybe slightly surprisingly, the existence problem is NP-complete for many parameter pairs σ and ϱ, including some of those listed in Table 1 (1-perfect code and total perfect dominating set). In fact, Telle [15] proves that $\exists(\sigma, \varrho)$-DS is NP-complete for every two finite nonempty sets σ, ϱ such that $0 \notin \varrho$.

There are several ways how to deal with NP-hard problems. One may look for heuristics or approximation algorithms, or aim at speeding up the (exponential) running time of exact algorithms. The latter approach is also the goal of the present paper. We present a $O^*(c^n)$-time[1] algorithm for the ENUM-(σ, ϱ)-DS

[1] As has recently become standard, we write $f(n) = O^*(g(n))$ if $f(n) \leq p(n) \cdot g(n)$ for some polynomial $p(n)$.

problem, where the constant $c < 2$ depends on σ and ϱ, for a fairly wide class of parameter sets σ and ϱ.

The main contribution of the paper is three-fold. First it is the general algorithm for a large class of domination-type problems. Secondly, the technique used in the running time analysis is a combination of a standard Branching algorithm with Recharging (motivated by recharging techniques used in many proofs of especially graph coloring theorems). We call the new technique Branch & Recharge and we hope it can be found useful for other problems as well. Thirdly, an upper bound on the running time of an enumeration algorithm immediately implies an upper bound on the number of enumerated objects. Thus our Branch & Recharge algorithm has a combinatorial corollary that every graph with n vertices contains at most $O^*(c^n)$ (σ, ϱ)-dominating sets (under the same assumptions on σ and ϱ).

This situation is interesting in its own. For several fast exact algorithms, the running time analysis is based on combinatorial theorems bounding the number of certain objects. For example a number of coloring algorithms are based on bounds on the number of certain maximal independent sets and bipartite subgraphs in a graph [2,3,6] and the algorithm for domatic number in [1] is based on a bound for the number of minimal dominating sets.

From the other side, the time analysis of a branching algorithm often yields the proof of combinatorial upper bounds. The most famous combinatorial result of this type is the well-known Moon-Moser theorem stating that the maximum number of maximal cliques (resp. maximal independent sets) of an n-vertex graph is $3^{n/3}$ [13] (while its original proof is combinatorial it can be easily turned into a branching algorithm enumerating all maximal independent sets). Techniques inspired by the analysis of exact algorithms were later used to obtain the bounds on the number of minimal dominating sets, minimal feedback vertex sets, and maximal r-regular subgraphs among others [7,8,9]. In general, exact algorithms and their analysis seem to be useful tools to obtain such combinatorial results (up to a polynomial factor).

Finally, we use a classical technique for the design of exact algorithms (see e.g. [14,16]). In his seminal survey paper [16] on exact algorithms Woeginger describes how to design algorithms using this paradigm to solve the subset sum problem and the binary knapsack problem in time $O^*(2^{n/2})$ (instead of time $O^*(2^n)$ by exhaustive search). The basic idea is a clever use of sorting and searching, and thus we call it Sort & Search. We establish $O^*(2^{n/2})$ time algorithms for the $\exists(\sigma, \varrho)$-DS, Min-(σ, ϱ)-DS, Max-(σ, ϱ)-DS and the #-(σ, ϱ)-DS problem in the case that σ and ϱ are singletons.

2 Preliminaries and the Main Theorem

We call a set of integers *successor-free* if it contains no pair of consecutive integers. In the rest of the paper we use the notation $p = \max \sigma$ and $q = \max \varrho$ (with $p = \infty$ if σ is infinite, and $q = \infty$ if ϱ is infinite). We denote by $\mathbb{N}$ the set of positive integers and by $\mathbb{N}_0$ the set of nonnegative integers.

A graph on n vertices may contain as many as 2^n (σ, ϱ)-dominating sets, e.g., if $0 \in \sigma \cap \varrho$, then the edgeless graph does. Another less trivial example is $\sigma = \varrho = \{0, 1, \ldots, d\}$, since then any set of vertices in a graph of maximum degree d is (σ, ϱ)-dominating. Therefore, if we aim at enumeration algorithms significantly faster than $\Theta(2^n)$, some restrictions must be imposed on the parameter sets σ and ϱ. The crucial condition required by our algorithm is the successor-freeness of σ. However, simple examples show that this itself is not enough. E.g., if σ is the set of even integers and ϱ the set of odd integers, then the complete graph $G = K_n$ contains 2^{n-1} (σ, ϱ)-dominating sets (every odd subset of vertices induces one), and yet σ and ϱ are successor-free and disjoint (but both are infinite). Similarly, for $\sigma = \{0\}$ and $\varrho = \mathbb{N}_0$ (σ successor-free and finite, but σ and ϱ are not disjoint), the star $K_{1,n-1}$ contains $2^{n-1}+1$ (σ, ϱ)-dominating sets.

Another obvious observation concerns disconnected graphs. The number of (σ, ϱ)-dominating sets in a graph is equal to the product of the numbers of (σ, ϱ)-dominating sets in its connected components. Hence it would suffice to consider connected graphs. However, the analysis of our algorithm also works for isolate-free input graphs (i.e., graphs without isolated vertices), which is more interesting for the main combinatorial result of our paper:

Theorem 1. *If σ is successor-free and either both σ and ϱ are finite, or one of them is finite and $\sigma \cap \varrho = \emptyset$, then every isolate-free graph contains at most c^n (σ, ϱ)-dominating sets, where $c = c_{\sigma,\varrho} < 2$ is a constant depending on σ and ϱ. Moreover, all the (σ, ϱ)-dominating sets can be enumerated in time $O^*(c^n)$ (where c is the same constant).*

The theorem is proved algorithmically in the next section. As already mentioned in the Introduction, the algorithm solves not only the enumeration problem Enum(σ, ϱ)-DS in time $O^*(c^n)$, but also the $\exists(\sigma, \varrho)$-DS, Max-(σ, ϱ)-DS, Min-(σ, ϱ)-DS and #-(σ, ϱ)-DS problems (only the polynomial factor in the star-notation is different). It is worth noting that the constant c depends only on p and q.

Recent results of Gupta et al. [9] give enumeration algorithms and lower and upper bounds for the number of maximal r-regular subgraphs in a given graph. Induced r-regular subgraphs are $(\{r\}, \mathbb{N}_0)$-dominating sets, and thus this result is related to our work. As can be expected for particular (σ, ϱ), their bounds are better than our general bounds.

3 The Branch and Recharge Algorithm

Throughout the rest of the paper we assume that σ and ϱ satisfy the assumptions of Theorem 1. The algorithm is based on the Branch & Reduce paradigm and consists of only one simple branching rule. To guarantee a running time faster than 2^n, the branching is combined with a recharging mechanism. The key idea of the algorithm is to guarantee that in any branching step on a chosen vertex v the weight of the input graph decreases by at least 1 when v is discarded, and it decreases by at least $1 + \epsilon$ when v is selected into the so far generated candidate for a (σ, ϱ)-dominating set. Here $\epsilon > 1$ is a constant dependent on σ and ϱ (in fact, on p and q only). The usage of the branching vector $(1, 1+\epsilon)$ with $\epsilon > 0$ immediately implies that the running time is faster than 2^n.

Initially every vertex $v \in V$ of the isolate-free input graph $G = (V, E)$ is assigned weight $w(v) = 1$. The total weight of the graph is $w(G) = \sum_{v \in V} w(v) = n$. The algorithm recursively builds candidate sets S for (σ, ϱ)-dominating sets in G. It recursively calls procedure `SigmaRho` which consists of three subroutines: `Forcing` (which identifies vertices that must or must not be placed in S), `Recharge` (which prepares the grounds for the next subroutine by sending charges of ϵ along certain edges), and `Branch` (the core of the algorithm, as it is responsible both for the exponential running time of the algorithm and for the base of the exponential function). All these subroutines and the procedure work with the same graph G and leave it unchanged, and with a global variable L which is the list of candidate sets S. Their parameters are S, $\overline{S}$ (containing the vertices discarded from the candidate set), the weight function w and an auxiliary directed graph H which is an orientation of a spanning subgraph of G (H is tracking the recharging moves). Moreover, `Forcing`, `Recharge` and `Branch` are called on a particular free vertex v. At every stage of the algorithm a vertex is called *free* if it does not belong to $S \cup \overline{S}$. Free vertices keep positive weights, allocated vertices get weight zero. Once a vertex is allocated in S (we say it is *selected*) or in $\overline{S}$ (we say it is *discarded*) it never changes its status during further calls. We will now describe the details in pseudocode (the global variable L and the input graph G are not listed in the preamble). The choice of ϵ is

$$\epsilon = \begin{cases} \frac{1}{1+\max(p,q)} & \text{if both } \sigma \text{ and } \varrho \text{ are finite} \\ \frac{1}{1+\min(p,q)} & \text{if at least one of } \sigma, \varrho \text{ is finite and } \sigma \cap \varrho = \emptyset. \end{cases}$$

```
Procedure SigmaRho(S, S̄, w, H)
  if there is no free vertex then  L := L ∪ {S}
  else
      let v be the last free vertex in the BFS ordering of V
      if v = v1 then
          /* v1 is the first vertex in the BFS ordering computed in the
             preprocessing of the Algorithm Main-EnumSigmaRho (see
             below)                                                    */
          L := L ∪ {S, S ∪ {v}}
      else
          Forcing(v, S, S̄, w, H)
          if Forcing halted then  Halt
          if v is still free then
              Recharge(v, S, S̄, w, H)
              Branch(v, S, S̄, w, H)
          else  SigmaRho(S, S̄, w, H)
```

```
Subroutine Recharge(v, S, S̄, w, H)
  if w(v) < 1 then
    let {w_1, ..., w_t} = {x : vx ∈ E(H)}
    for i := 1 to t do let u_i be another free neighbor (in G) of w_i
    for i := 1 to t do
      w(u_i) := w(u_i) − ε
      E(H) := (E(H) ∪ {u_i w_i}) \ {v w_i}
    w(v) := 1
```

Comment: Note that $w_1, \ldots, w_t$ are distinct vertices, while $u_1, \ldots, u_t$ need not be. If some u is the chosen free neighbor of several, say k, vertices from $w_1, \ldots, w_t$, then its weight drops by $k\epsilon$ and also k edges starting in u are added to H. Lemma 4 shows that w_i always has another free neighbor in G.

```
Subroutine Branch(v, S, S̄, w, H)
  1. S' := S; S̄' := S̄ ∪ {v}; w' := w; w'(v) := 0; H' := H;
  SigmaRho(S', S̄', w', H')
  2. let u be a free neighbor of v
     S := S ∪ {v}; w(v) := 0; w(u) := w(u) − ε; E(H) := E(H) ∪ {uv};
     SigmaRho(S, S̄, w, H)
```

The last subroutine depends on σ and ϱ and the choice of ϵ. If both sets are finite and $\epsilon(1 + \max(p, q)) = 1$, we define `Forcing` as the subroutine `Forcing-a` (see below). If $\sigma \cap \varrho = \emptyset$ and $\epsilon(1+\min(p, q)) = 1$ we define the `Forcing` subroutine as `Forcing-b` (note that at least one of σ, ϱ must be finite in this case).

```
Subroutine Forcing-a(v, S, S̄, w, H)
  if ∃x ∈ S s.t. v is its unique free neighbor then
    case
      |N(x) ∩ S| ∈ σ then S̄ := S̄ ∪ {v}, w(v) := 0
      |N(x) ∩ S| + 1 ∈ σ then S := S ∪ {v}, w(v) := 0
      {|N(x) ∩ S|, |N(x) ∩ S| + 1} ∩ σ = ∅ then Halt
  if ∃x s.t. |N(x) ∩ S| > max{p, q} then Halt
```

Having described the recursive procedure and its subroutines, the entire algorithm named `Main-EnumSigmaRho` (see below) can be formalized as one call of the recursive procedure (and necessary preprocessing and final check of the items in the candidate list).

The correctness of the algorithm follows from the following lemmas, and from the fact that it branches on each vertex whose membership to S or $\overline{S}$ is not forced.

Lemma 1 (Loop invariant). *At the time of each call of the* `SigmaRho` *procedure, the following invariants are fulfilled:*

1. *if x is free then $w(x) = 1 - d\epsilon$, where d is the outdegree of x in H,*
2. *H is the disjoint union of out-oriented stars, and $xy \in E(H)$ implies that $y \in S$,*
3. *$w(x) = 0$ for every $x \in S \cup \overline{S}$,*
4. *$w(x) \geq 0$ for every free vertex x, and moreover $w(x) > 0$ after the execution of the* `Forcing` *subroutine.*

```
Subroutine Forcing-b(v, S, S̄, w, H)
  while (∃x s.t. x is free and |N(x) ∩ S| > min{p,q}) or
  (∃y ∈ S with a unique free neighbor z) or (∃ a free vertex u with no free
  neighbor) do
    let x or y, z or u be such vertices
    case
      |N(x) ∩ S| > max{p,q} then Halt
      |N(x) ∩ S| > p then S̄ := S̄ ∪ {x}; w(x) := 0
      |N(x) ∩ S| > q then S := S ∪ {x}; w(x) := 0
      |N(y) ∩ S| ∈ σ then S̄ := S̄ ∪ {z}; w(z) := 0
      |N(y) ∩ S| + 1 ∈ σ then S := S ∪ {z}; w(z) := 0
      {|N(y) ∩ S|, |N(y) ∩ S| + 1} ∩ σ = ∅ then Halt
      |N(u) ∩ S| ∈ σ then S := S ∪ {u}; w(u) := 0
      |N(u) ∩ S| ∈ ϱ then S̄ := S̄ ∪ {u}; w(u) := 0
      |N(u) ∩ S| ∉ σ ∪ ϱ then Halt
```

```
Algorithm Main-EnumSigmaRho(G)
  Preprocessing : Choose an arbitrary vertex v1 and order the vertex set of G
  in a BFS ordering starting in v1
  Initialization : L := ∅; S := ∅; S̄ := ∅; H := (V(G), ∅)
  for v ∈ V(G) do w(v) := 1
  SigmaRho(S, S̄, w, H)
  for S ∈ L do
    if S is not a (σ,ϱ)-dominating set in G then L := L \ {S}
  output(L)
```

Proof 1. The weight of free vertices. At the beginning the weight of every vertex is 1, and also H is edgeless, so the outdegree of every vertex is 0 (in H). The invariant follows by induction on the number of recursive calls. The weights of free vertices are changed in the `Recharge` and `Branch` subroutines, and in each case the multiple of ϵ that is subtracted from (or added to) the weight of the vertex is the same as the number of oriented edges starting in the vertex that are added to (deleted from, respectively) H.

2. The shape of H. At the beginning, H is edgeless. It gets modified in the `Recharge` and `Branch` subroutines. When recharging, edges vw_i are replaced by u_iw_i, and when branching, the edge uv is added. In each case the endpoint is a vertex allocated in S and every vertex from S is the endvertex of at most one arc of H. (In the case of branching this follows from the fact that $w(v) = 1$ before `Branch` is called on v, and hence has neither outgoing nor incoming arcs.)

3. The weight of a vertex allocated to S or $\overline{S}$ becomes 0 at the time of allocation.

4. Weights of free vertices are nonnegative. This is guaranteed by the `Forcing` subroutines. We distinguish two cases.

4a. `Forcing-a`: A free vertex, say x, would have weight 0 or less only if it had outdegree $t > \max(p,q)$ in H. But then x must have t neighbors in S. Certainly

at the beginning of the first call of `SigmaRho`, no such vertex exists. The number of S-neighbors may get raised during the first part of the `Forcing-a` subroutine, but that is immediately discovered by the second part of the subroutine and the execution is halted. The only other possibility is during the second part of the `Branch` subroutine, when v is selected into S. In that case the weight of u, the free neighbor of v, was positive before `Branch` was called and becomes 0 for just a short moment – the `Forcing-a` subroutine of the subsequent call of `SigmaRho` discovers that u has too many neighbors in S and halts the execution (unless this call is one of the final ones and we are in a leaf of the search tree).
4b. `Forcing-b`: A free vertex, say x, would have weight 0 or less only if it had outdegree $t > \min(p, q)$ in H, and hence at least as many neighbors in S. Again such a vertex is discovered in the `Forcing-b` subroutine and allocated in S or $\overline{S}$ (or the execution halts). After that the number of S-neighbors of a free vertex may get raised only in the second part of the `Branch` subroutine. In that case the weight of such free vertex, called u in the subroutine, drops only by one ϵ, and since it was positive before, it becomes 0, and only for a short while. The `Forcing-b` subroutine of the subsequent call of `SigmaRho` discovers that u has too many neighbors in S and allocates u (or halts or contributes to the list L of candidate sets when we are in a leaf of the search tree). □

Lemma 2 (Halting). *If* `Forcing` *halted with current values* $S, \overline{S}$*, then* G *contains no* (σ, ϱ)*-dominating set* M *such that* $S \subseteq M \subseteq V \setminus \overline{S}$.

Proof a) `Forcing-a`: If `Forcing-a` halts because some x has more than $\max(p, q)$ neighbors in S, then such an S cannot be a subset of any (σ, ϱ)-dominating set M. Indeed, if $x \in M$ then $|N(x) \cap M| \geq |N(x) \cap S| > p = \max \sigma$ and $|N(x) \cap M|$ cannot be in σ, as well as $|N(x) \cap M| \geq |N(x) \cap S| > q = \max \varrho$ and $|N(x) \cap M|$ cannot be in ϱ if $x \notin M$. If `Forcing-a` halts because some $x \in S$ has a unique free neighbor, but neither $|N(x) \cap S|$ nor $|N(x) \cap S| + 1$ are in σ, then no M containing S is a (σ, ϱ)-dominating set since $|N(x) \cap M|$ equals $|N(x) \cap S|$ or $|N(x) \cap S| + 1$, depending on whether $v \in M$ or not.

b) `Forcing-b`: The first two reasons why `Forcing-b` may halt are the same as above. If the subroutine halts because for some free vertex x, $|N(x) \cap S|$ is neither in σ nor in ϱ, then no superset M of S can be a (σ, ϱ)-dominating set since x can be neither in M nor outside it. □

Lemma 3 (Necessity). *If at some stage, with current values of* $S, \overline{S}$, `Forcing` *wants to place* x *in* S *(resp. in* $\overline{S}$*), then for every* (σ, ϱ)*-dominating set in* G *such that* $S \subseteq M \subseteq V \setminus \overline{S}$*, it holds that* $x \in M$ *(resp.* $x \notin M$*).*

Proof Again the two cases are distinguished. In both it is assumed that M is a (σ, ϱ)-dominating set such that $S \subseteq M \subseteq V \setminus \overline{S}$.

a) `Forcing-a`: Suppose v is the unique neighbor of $x \in S$ and $|N(x) \cap S| \in \sigma$. Then $|N(x) \cap S| + 1 \notin \sigma$ because σ is successor-free. Thus v cannot be in M, since then $|N(x) \cap M| = |N(x) \cap S| + 1 \notin \sigma$. Similarly, $|N(x) \cap S| + 1 \in \sigma$ implies $|N(x) \cap S| \notin \sigma$ and v must be in M, since it is the only possible additional M-neighbor of x.

b) `Forcing-b`: If z is the only free neighbor of a vertex $y \in S$, the argument is as above. If x is a free vertex such that $|N(x) \cap S| > p$ (and $\leq q$ since the subroutine did not halt in the previous step), x cannot be in M because then $|N(x) \cap M| \geq |N(x) \cap S| > p = \max \sigma$ and $|N(x) \cap M|$ could not be in σ, while if $|N(x) \cap S| > q$, then x cannot be outside M because then $|N(x) \cap M| \geq |N(x) \cap S| > q = \max \varrho$ and $|N(x) \cap M|$ could not be in ϱ. Finally, if u is a free vertex with no free neighbor, then $|N(u) \cap M| = |N(u) \cap S|$ and the membership of u in M is uniquely determined since $\sigma \cap \varrho = \emptyset$. □

Lemma 4 (Correctness). *The subroutines* `Recharge` *and* `Branch` *can always be executed.*

Proof The `Forcing` subroutine guarantees that no S-neighbor of v has v as its only free neighbor. In the `Recharge` subroutine, vw_i is an arc of H and hence $w_i \in S$ for every $i = 1, \ldots, t$. But then each w_i has another free neighbor and `Recharge` does not get stuck.

For the `Branch` subroutine, we distinguish the two cases again.
`Forcing-a`: We note that vertices of G get allocated into S or $\overline{S}$ only when we attempt to branch on them (in the preceding `Forcing-a` subroutine, or in `Branch` itself). Therefore when we consider v to be the last free vertex in the BFS ordering of the vertex set of G, either $v = v_1$ is the root (and then we do not bother checking anything and just add both S and $S \cup \{v\}$ to the candidate list L) or v has a predecessor u in the BFS tree of G. This u comes earlier in the BFS ordering of G, hence was not attempted to branch on yet, and hence is free at the time when v is processed.

`Forcing-b`: If v has no free neighbor, then v (renamed as u) gets allocated in the preceding run of the `Forcing-b` subroutine or that subroutine is halted, but in neither case v remains free for branching. □

Analysis of the running time. The weight of an instance $(G, w, S, \overline{S}, H)$ is $w(G) = \sum_{v \in V} w(v)$. In each branching on a vertex v the measure of the input decreases by 1 when discarding v, and it decreases by $1 + \epsilon$ when selecting v. In the standard terminology of branching algorithms this implies that the branching vector is $(1, 1 + \epsilon)$. The running time of each execution of `SigmaRho` (without recursive calls) is polynomial, and so the total running time is $O^*(T)$ where T is the number of leaves of the search tree. Note that each (σ, ϱ)-dominating set corresponds to one leaf of the search tree.

Let $T[k]$ be the maximum number of leaves of the search tree that any execution of our algorithm may generate on a problem instance of weight k. Due to the branching vector we obtain: $T[k] \leq T[k-1] + T[k-1-\epsilon]$. Thus the number of (σ, ϱ)-dominating sets (which is bounded from above by $T[n]$) in an isolate-free graph on n vertices is at most c^n, and the running time of our algorithm that enumerates all of them is $O^*(c^n)$, where c is the largest real root of the characteristic polynomial $x^{1+\epsilon} - x^{\epsilon} - 1 = 0$.

The table shows the base of the exponential function bounding the running time of our algorithm for some particular values of $\varphi = \frac{1}{\epsilon} - 1$. Note that $\varphi =$

$\max(p,q)$ if both σ and ϱ are finite, and $\varphi = \min(p,q)$ if at least one of them is finite and $\sigma \cap \varrho = \emptyset$.

φ	0	1	2	3	4	5	6	7	8	100
c	1.6181	1.7549	1.8192	1.8567	1.8813	1.8987	1.9116	1.9216	1.9296	1.9932

As can be expected, c converges to 2 when φ converges to infinity (and ϵ converges to 0). This is easily seen from the characteristic polynomial, which tends to $x - 2 = 0$.

4 Lower Bounds

The combinatorial consequence of our algorithm shows that (under certain assumptions on σ and ϱ) every isolate-free graph on n vertices contains at most $2^{n(1-\delta)}$ (σ,ϱ)-dominating sets, for some $\delta > 0$.

Taking $\sigma = \{0\}$ and $\varrho = \mathbb{N}$ we obtain the Independent Dominating Set problem and the (σ,ϱ)-dominating sets are precisely the maximal independent sets. Hence our theorem implies that the maximum number of maximal independent sets is upper bounded by 1.6181^n; while Moon and Moser [13] tells us that the correct value is 1.4423^n.

While the upper bound of Moon and Moser is tight, others like the one by Fomin et al. for the maximum number of minimal dominating sets [8] might not be tight. Likewise, our upper bounds established by a general approach are unlikely to be tight for all particular values of (σ,ϱ). Thus it is natural to look after lower bounds.

Let σ be the set of all even integers from the interval $[0, r-1]$, and ϱ be the set of all odd integers from this interval, where $r \geq 2$ is a positive integer. Consider $G = sK_r$, the disjoint union of s copies of the complete graph K_r. Clearly, this graph G has $2^{(r-1)s} = 2^{\frac{r-1}{r}n}$ (σ,ϱ)-dominating sets.

Since both σ and ϱ are finite, and $\sigma \cap \varrho = \emptyset$, we can use our algorithm in both variants. The next table compares the bases of the exponential upper bounds given by our algorithm (with both variants of $\epsilon = \frac{1}{1+\max(p,q)}$ and $\epsilon = \frac{1}{1+\min(p,q)}$, distinguished as c_{max} and c_{min}, respectively) with the base $a_r = 2^{\frac{r-1}{r}}$ of the exponential function giving the lower bound arising from the example above.

r	2	3	4	5	6	7	8	9	101
c_{max}	1.7549	1.8192	1.8567	1.8813	1.8987	1.9116	1.9216	1.9296	1.9932
c_{min}	1.6181	1.7549	1.8192	1.8567	1.8813	1.8987	1.9116	1.9216	1.9932
a_r	1.4142	1.5874	1.6817	1.7411	1.7817	1.8114	1.8340	1.8517	1.9863

5 A Sort and Search Approach

We use a classical method based on sorting and searching to design exact algorithms. Woeginger's survey [16] and a paper from 1981 by Schroeppel and Shamir [14] show how to use such a paradigm to establish moderately exponential-time algorithms solving some NP-hard problems.

We study the use of Sort & Search for constructing algorithms for $\exists(\sigma,\varrho)$-DS defined in the Introduction. The task is to decide whether the given graph has a (σ,ϱ)-dominating set. Those problems are polynomial-time solvable or NP-complete depending on (σ,ϱ). For more details, we refer to [15].

The NP-complete $\exists(\{0\},\{1\})$-DS problem is called PERFECT CODE and it can be solved in time $O(1.1730^n)$ [4]. The algorithm is based on solving the exact satisfiability problem (called XSAT) and the approach has been generalized to the so-called X_iSAT problem. Those algorithms use Sort & Search and their running time is $O^*(2^{n/2})$.

Our use of Sort & Search was inspired by the aforementioned algorithms.

Theorem 2 *The $\exists(\{p\},\{q\})$-DS problem can be solved in time $O^*(2^{n/2})$.*

Proof Let $p,q \in \mathbb{N}_0$. Let $G=(V,E)$ be the input graph and let $k=\lfloor n/2 \rfloor$. The algorithm partitions the set of vertices into $V_1=\{v_1,v_2,\dots v_k\}$ and $V_2=\{v_{k+1},\dots,v_n\}$. Then for each subset $S_1 \subseteq V_1$, it computes the vector $\boldsymbol{s_1}=(x_1,\dots,x_k,x_{k+1},\dots,x_n)$ where

$$x_i = \begin{cases} p-|N(v_i)\cap S_1| & \text{if } 1\le i\le k \text{ and } v_i \in S_1 \\ q-|N(v_i)\cap S_1| & \text{if } 1\le i\le k \text{ and } v_i \notin S_1 \\ |N(v_i)\cap S_1| & \text{if } k+1\le i\le n \end{cases}$$

and for each subset $S_2 \subseteq V_2$, it computes the corresponding vector $\boldsymbol{s_2}=(x_1,\dots,x_k,x_{k+1},\dots,x_n)$ where

$$x_i = \begin{cases} |N(v_i)\cap S_2| & \text{if } 1\le i\le k \\ p-|N(v_i)\cap S_2| & \text{if } k+1\le i\le n \text{ and } v_i \in S_2 \\ q-|N(v_i)\cap S_2| & \text{if } k+1\le i\le n \text{ and } v_i \notin S_2. \end{cases}$$

After the computation of these 2^{k+1} vectors, the algorithm sorts those corresponding to V_2 in lexicographic order. Then, for each vector $\boldsymbol{s_1}$ (corresponding to an $S_1 \subseteq V_1$) using binary search it tests whether there exists a vector $\boldsymbol{s_2}$ (corresponding to an $S_2 \subseteq V_2$), such that $\boldsymbol{s_2}=\boldsymbol{s_1}$. Note that the choice of the vectors garantees that $\boldsymbol{s_2}=\boldsymbol{s_1}$ iff $S_1 \cup S_2$ is a $(\{p\},\{q\})$-dominating set. Such fixed vector $\boldsymbol{s_1}$ can be found in time $n\log 2^{n/2}$ in the lexicographic order of the vectors of V_2. Thus the overall running time is $O^*(2^{n/2})$. □

Corollary 1 MAX-$(\{p\},\{q\})$-DS, MIN-$(\{p\},\{q\})$-DS *and* #-$(\{q\},\{q\})$-DS *can also be solved in time $O^*(2^{n/2})$.*

Proof The algorithm of the previous theorem only needs to be modified as follows: Instead of sorting all vectors corresponding to V_2 in lexicographic order, multiple copies are removed and each vector is stored with an entry indicating its number of occurrences. Furthermore for minimization (resp. maximization), with each vector we store an $S_i \subseteq V_i$ of minimum (resp. maximum) cardinality that generates this vector. □

Finally, let us mention that the approach can be extended to certain infinite σ and ϱ. Let $m \ge 2$ be a fixed integer and $k \in \{0,1,\dots,m-1\}$. We denote by $k+m\mathbb{N}_0$ the set $\{m\cdot\ell+k:\ \ell\in\mathbb{N}_0\}$.

Theorem 3 *Let $m \geq 2$ and $p, q \in \mathbb{N}_0$. The problems* $\exists(p + m\mathbb{N}_0, q + m\mathbb{N}_0)$-DS, MIN-$(p+m\mathbb{N}_0, q+m\mathbb{N}_0)$-DS, MAX-$(p+m\mathbb{N}_0, q+m\mathbb{N}_0)$-DS *and* #-$(p+m\mathbb{N}_0, q+m\mathbb{N}_0)$-DS *can be solved in time* $O^*(2^{n/2})$.

Proof The corresponding algorithms in Theorem 2 and Corollary 1 are to be modified such that all components of vectors are taken modulo m and the addition and/or subtraction of vector components is taken modulo m. □

References

1. Björklund, A., Husfeldt, T.: Inclusion-exclusion algorithms for counting set partitions. In: Proceedings of FOCS, pp. 575–582 (2006)
2. Byskov, J.M.: Enumerating maximal independent sets with applications to graph colouring. Operations Research Letters 32, 547–556 (2004)
3. Byskov, J.M., Madsen, B.A., Skjernaa, B.: On the number of maximal bipartite subgraphs of a graph. Journal of Graph Theory 48, 127–132 (2005)
4. Dahlöf, V., Jonsson, P., Beigel, R.: Algorithms for four variants of the exact satisfiability problem. Theoretical Computer Science 320, 373–394 (2004)
5. Downey, R.G., Fellows, M.R.: Parameterized complexity. Springer, Heidelberg (1999)
6. Eppstein, D.: Small maximal independent sets and faster exact graph coloring. Journal of Graph Algorithms and Applications 7, 131–140 (2003)
7. Fomin, F.V., Gaspers, S., Pyatkin, A.V.: Finding a minimum feedback vertex set in time $O(1.7548^n)$. In: Bodlaender, H.L., Langston, M.A. (eds.) IWPEC 2006. LNCS, vol. 4169, pp. 184–191. Springer, Heidelberg (2006)
8. Fomin, F.V., Grandoni, F., Pyatkin, A.V., Stepanov, A.A.: Bounding the Number of Minimal Dominating Sets: a Measure and Conquer Approach. In: Deng, X., Du, D.-Z. (eds.) ISAAC 2005. LNCS, vol. 3827, pp. 573–582. Springer, Heidelberg (2005)
9. Gupta, S., Raman, V., Saurabh, S.: Fast exponential algorithms for Maximum r-regular induced subgraph problems. In: Arun-Kumar, S., Garg, N. (eds.) FSTTCS 2006. LNCS, vol. 4337, pp. 139–151. Springer, Heidelberg (2006)
10. Halldorsson, M.M., Kratochvil, J., Telle, J.A.: Independent sets with domination constraints. Discrete Applied Mathematics 99, 39–54 (2000)
11. Haynes, T.W., Hedetniemi, S.T., Slater, P.J.: Fundamentals of Domination in Graphs. Marcel Dekker, New York (1998)
12. Heggernes, P., Telle, J.A.: Partitioning graphs into generalized dominating sets. Nordic Journal of Computing 5, 128–142 (1998)
13. Moon, J.W., Moser, L.: On cliques in graphs. Israel Journal of Mathematics 5, 23–28 (1965)
14. Schroeppel, R., Shamir, A.: A $T = O(2^{n/2})$, $S = O(2^{n/4})$ algorithm for certain NP-complete problems. SIAM Journal on Computing 3, 456–464 (1981)
15. Telle, J.A.: Complexity of domination-type problems in graphs. Nordic Journal of Computing 1, 157–171 (1994)
16. Woeginger, G.J.: Exact algorithms for NP-hard problems: A survey. In: Jünger, M., Reinelt, G., Rinaldi, G. (eds.) Combinatorial Optimization - Eureka, You Shrink! LNCS, vol. 2570, pp. 185–207. Springer, Heidelberg (2003)

On Computing the Centroid of the Vertices of an Arrangement and Related Problems

Deepak Ajwani, Saurabh Ray, Raimund Seidel, and Hans Raj Tiwary*

Max-Planck-Institut für Informatik, Saarbrücken, Germany
Universität des Saarlandes, Saarbrücken, Germany

Abstract. We consider the problem of computing the centroid of all the vertices in a non-degenerate arrangement of n lines. The trivial approach requires the enumeration of all $\binom{n}{2}$ vertices. We present an $\mathcal{O}(n \log^2 n)$ algorithm for computing this centroid. For arrangements of n segments we give an $\mathcal{O}(n^{\frac{4}{3}+\epsilon})$ algorithm for computing the centroid of its vertices. For the special case that all the segments of the arrangement are chords of a simply connected planar region we achieve an $\mathcal{O}(n \log^5 n)$ time bound. Our bounds also generalize to certain natural weighted versions of those problems.

1 Introduction

An arrangement of n lines in the plane has up to $\binom{n}{2}$ vertices. However, these vertices are implicitly specified by only $2n$ real numbers. Thus it is not necessarily surprising that some functions of this vertex set can be computed in subquadratic time: E.g. the vertex with k-th smallest x-coordinate can be computed in $\mathcal{O}(n \log n)$ time [4]. It is an outstanding open problem in computational geometry whether in subquadratic time the true number of vertices can be computed (in other words, whether in subquadratic time degeneracy can be determined).

In this paper we study the problem of computing the centroid of the vertices in an arrangement of lines (and also of line segments). In contrast to the problems mentioned above the centroid function is not combinatorial in the sense that it does not produce an integer value but it produces real values.

We first show that the centroid of intersection points n lines in the plane can be computed in $\mathcal{O}(n \log^2 n)$ time. Using this result and employing a segment query data structure, we show that the centroid of the intersection points of n line segments in the plane can be computed in $\mathcal{O}(n^{4/3+\delta})$ time ($\delta > 0$ arbitrarily small). This should be compared with the complexity of the best known algorithm for counting the number of intersections in the plane by Chazelle [3], which is $\Theta(n^{4/3}(\log n)^{1/3})$. In case the segments have a restricted structure in that they all are chords of a simply connected region, we can do better: we show a bound of $\mathcal{O}(n \log^5 n)$ time. We finally show that all the mentioned bounds continue to hold for a natural weighted generalization of the centroid problem:

* The last author was supported by Graduiertenkolleg der FR Informatik, Universität des Saarlandes while working on this problem.

F. Dehne, J.-R. Sack, and N. Zeh (Eds.): WADS 2007, LNCS 4619, pp. 519–528, 2007.

endow each line (or segment) with a real weight and define the weight of an intersection point to be the sum of the weights of the involved lines.

The main computational ingredient in our approach besides the usual computational geometry machinery is the Fast Fourier Transform.

Our approach does in no way solve the above-mentioned degeneracy problem. However, in the conclusion we offer a brief discussion why algebraic methods as used in this paper may be a viable approach towards a subquadratic solution the degeneracy problem For the sake of presentation we assume non-degeneracy. Degeneracy in the form of non-intersecting lines can easily be taken care of explicitly. Degeneracy in the form of of concurrent lines is ignored in the sense that if an arrangement vertex v is incident to k lines then it is counted $\binom{k}{2}$ times, once for each pair of lines intersecting in v.

The problem of computing the centroid of the vertices of an arrangement is admittedly somewhat academic. For readers with a strong need for applications here is a conceivable scenario where our results would be relevant. Consider the deployment of wireless devices on road-crossings in a city for the purpose of traffic monitoring (finding traffic rule violations or updating the people about overcrowded crossings or traffic jams). These devices need to continously transmit the data to a central base station. An important cost criterion here is the power consumed by these devices. The power needed by a device is proportional to the square (assuming free space) of the distance to which it needs to transmit the data. Thus the location of the central base station should be such that it minimizes the sum of the squares of the distances to the road crossings. This location is realized by the centroid of the crossings. Thus our results apply if all the roads in the city are straight and all intersections are crossings of exactly two roads.

If you assume that the power consumption of the wireless device at an intersection is proportional to the number of cars going by and the average number w_i of cars going along road i per unit of time is independent of the position along the road, then the weighted versions of our centroid problems apply.

Our results heavily rely on the following two facts from computational algebra (see Chapter 1 of [9], [5]).

Fact 1 (Fast polynomial multiplication). *The product of two univariate polynomials over the reals of maximal degree n can be computed in time $O(n \log n)$.*

Fact 2 (Fast multiple evaluation). *Let $p(x)$ be polynomial over the reals of degree at most n and let A be a set of n real number.*

The set $\{(a, p(a)|a \in A\}$ can be computed in $O(n \log^2 n)$ time.

2 Computing the Centroid of the Intersection Points of n Lines

We are given a set L of n lines $l_i : y = m_i x - c_i$ (for $1 \leq i \leq n$) in general position (no three of them intersect at the same point and no two of them are parallel). Let (X_{ij}, Y_{ij}) represent the intersection point of lines l_i and l_j. We want to compute the centroid (X_L, Y_L) of the intersection points (X_{ij}, Y_{ij}). By the definition of centroid,

$$X_L = \binom{n}{2}^{-1} \sum_{\substack{i,j \in [1 \ldots n] \\ i<j}} X_{ij}, \qquad Y_L = \binom{n}{2}^{-1} \sum_{\substack{i,j \in [1 \ldots n] \\ i<j}} Y_{ij}$$

Consider a query line $l \;:\; y = \mu x - \gamma$. We would like to compute the sum of the x-coordinates of the intersection points of l with each of the lines in L. This is given by

$$F_L(\mu, \gamma) = \sum_{1 \le i \le n} \frac{c_i - \gamma}{m_i - \mu}$$

This function can be represented as:

$$F_L(\mu, \gamma) = \frac{P_L(\mu) - \gamma Q_L(\mu)}{S_L(\mu)} \qquad (*)$$

where P_L, Q_L and S_L are single variable polynomials of degree at most n.

We assume that the query line is not parallel to any of the lines in L. We do, however, allow it to be identical to one of the lines in L in which case we want to compute the sum of the x coordinates of the intersection of l with the other lines in L. If l is not identical to any of the lines in L, then $F_L(\mu, \gamma)$ as defined above is well defined. If l is identical to one of the lines (l_j) in L, then $F(\mu, \gamma)$ is of the form $\frac{0}{0}$ and thus F cannot be evaluated using (*). We therefore evaluate F at $\mu = m_j, \gamma = c_j$ applying de l'Hôpital's rule, which yeilds

$$F_L(m_j, c_j) = \frac{P'_L(m_j) - c_j Q'_L(m_j)}{S'_L(m_j)},$$

where P'_L, Q'_L, and S'_L denote the derivatives of P_L, Q_L, and L_L with respect to μ. We will associate the polynomials P_L, Q_L and S_L defined above with the set L of lines.

Lemma 1. *Given a set of n lines L, the associated polynomials P_L, Q_L and S_L can be computed in $\mathcal{O}(n \log^2 n)$ time.*

Proof. We arbitrarily color half the lines blue and the rest red. Let R and B be the set of red and blue lines respectively. We then recursively compute P_R, Q_R and S_R for the red lines and P_B, Q_B and S_B for the blue lines. Then,

$$\begin{aligned} F_L(\mu, \gamma) &= \frac{P_R(\mu) - \gamma Q_R(\mu)}{S_R(\mu)} + \frac{P_B(\mu) - \gamma Q_B(\mu)}{S_B(\mu)} \\ &= \frac{(P_R(\mu) \cdot S_B(\mu) + P_B(\mu) \cdot S_R(\mu)) - \gamma(Q_R(\mu) \cdot S_B(\mu) + Q_B(\mu) \cdot S_R(\mu))}{S_R(\mu) \cdot S_B(\mu)} \end{aligned}$$

Therefore,

$P_L(\mu) = P_R(\mu) \cdot S_B(\mu) + P_B(\mu) \cdot S_R(\mu)$
$Q_L(\mu) = Q_R(\mu) \cdot S_B(\mu) + Q_B(\mu) \cdot S_R(\mu)$
$S_L(\mu) = S_R(\mu) \cdot S_B(\mu)$

Since two polynomials of degree at most n can be multiplied in $\mathcal{O}(n \log n)$ time using FFT (Fact 1), P_L, Q_L and S_L can be computed in $\mathcal{O}(n \log n)$ time from P_R, Q_R, S_R, P_B, Q_B and S_B. Therefore P_L, Q_L and S_L can be computed in $\mathcal{O}(n \log^2 n)$ time (we get one log factor due to recursion).

Theorem 3. *Given a set R of n red lines and a set B of n blue lines such that no red line is parallel (or identical) to any blue line, we can compute the centroid (X_{RB}, Y_{RB}) of red-blue intersection points in $\mathcal{O}(n \log^2 n)$ time.*

Proof. We shall treat each of the red lines as a query line and compute the sum of the x-coordinates of its intersection with the blue lines. We then add the sums for all the red lines and the divide by the number of red-blue intersections (n^2) to get the x-coordinate of the centroid of the red-blue intersections. Since none of the red lines are identical to any of the blue lines, the x-coordinate X_L of the centroid of red-blue intersections is

$$X_L = n^{-2} \sum_{l_i \in R} F_B(m_i, c_i) = n^{-2} \sum_{l_i \in R} \frac{P_B(m_i) - c_i Q_B(m_i)}{S_B(m_i)}$$

Since the polynomials P_B, Q_B and S_B can be computed in $\mathcal{O}(n \log^2 n)$ time using Lemma 1 and also they can be evaluated at the n m_i's corresponding to the n red lines in $\mathcal{O}(n \log^2 n)$ time using FFT (See Fact 2), the overall time for the computation of X_L is $\mathcal{O}(n \log^2 n)$. The y-coordinate of the centroid of red-blue intersections can be computed similarly.

Corollary 1. *Given a set R of r red lines and a set B of b blue lines, the centroid (X_{RB}, Y_{RB}) of the red-blue intersections can be computed in $\mathcal{O}((r + b) \log^3 (r + b))$ time.*

Theorem 4. *Given a set L of n lines, we can compute the centroid (X_L, Y_L) of their intersections in $\mathcal{O}(n \log^2 n)$ time.*

Proof. In this case, we treat each of the lines as a query line and compute the sum of the x-coordinates of its intersections with the lines in L. We know that each of these query lines is identical to exactly one line in L (i.e. the line itself). Therefore, the x-coordinate of the centroid of the intersections of the lines is given by,

$$X_L = n^{-2} \sum_{l_i \in L} F_L(m_i, c_i) = \sum_{l_i \in L} \frac{P'_L(m_i) - c_i Q'_L(m_i)}{S'_L(m_i)}$$

Since P_L, Q_L and S_L can be computed in $\mathcal{O}(n \log^2 n)$ time, P'_L, Q'_L and S'_L can also be computed in $\mathcal{O}(n \log^2 n)$ time. Also, P'_L, Q'_L and S'_L can be evaluated at the n m_i's corresponding to the n lines in $\mathcal{O}(n \log^2 n)$ time. Therefore, X_L can be computed in $\mathcal{O}(n \log^2 n)$ time. Similarly, Y_L can be computed in $\mathcal{O}(n \log^2 n)$ time.

If each of the lines $l_i \in L$ of n lines is endowed with a weight w_i, we define the centroid of weighted lines to be

$$(X_L, Y_L) = \left(\sum_{\substack{l_i, l_j \in L \\ i<j}} X_{ij}(w_i + w_j), \sum_{\substack{l_i, l_j \in L \\ i<j}} Y_{ij}(w_i + w_j) \right)$$

where (X_{ij}, Y_{ij}) is the intersection point of l_i and l_j.

We proceed exactly as in the unweighted case by considering a query line $l : y = \mu x - \gamma$ with weight ω. Then, the weighted sum of the x-coordinates of the intersections of l with the lines in L is given by

$$G_L(\mu, \gamma, \omega) = \sum_{1 \leq i \leq n} \frac{c_i - \gamma}{m_i - \mu} \cdot (\omega + w_i)$$

This function can again be represented as:

$$G_L(\mu, \gamma, \omega) = \omega \cdot \frac{P_L(\mu) - \gamma Q_L(\mu)}{S_L(\mu)} + \frac{U_L(\mu) - \gamma V_L(\mu)}{S_L(\mu)}$$

where P_L, Q_L and S_L are as before and U_L and V_L are some other polynomials of degree at most n in μ.

We can now apply the same techniques as for the unweighted case to obtain the following:

Lemma 2. *Given a set R of n red weighted lines and a set B of n blue weighted lines and their associated polynomials the centroid of the red-blue intersections can be computed in $\mathcal{O}(n \log^2 n)$.*

Theorem 5. *Given a set L of n weighted lines the centroid (X_L, Y_L) of their intersections can be computed in $\mathcal{O}(n \log^2 n)$ time.*

Corollary 2. *Given a set R of r weighted red lines and a set B of b weighted blue lines, the centroid (X_{RB}, Y_{RB}) of the red-blue intersections can be computed in $\mathcal{O}((r+b) \log^2 (r+b))$ time.*

3 Centroid of Line Segment Intersections

Theorem 6. *Given a set R of r red segments and a set B of b blue segments, if R can be preprocessed into a data structure of size $\mathcal{O}(s(r))$ in time $\mathcal{O}(p(r))$ so that all segments intersecting a query segment $t \in B$ can be reported as the union of $\mathcal{O}(u(r))$ "canonical" prestored subsets in $\mathcal{O}(q(r))$ time, then we can form sets $R_1, R_2, \cdots, R_k$ of red segments and sets $B_1, B_2, \cdots, B_k$ of blue segments in $\mathcal{O}(p(r) + b\, q(r))$ time such that*

1. *For* $1 \leq i \leq k$, *all segments in* R_i *intersect all segments in* B_i
2. *If a red segment* ρ *intersects a blue segment* β, *there is a unique* i *such that* $\rho \in R_i$ *and* $\beta \in B_i$
3. $\sum_{i=1}^{n} |R_i| = \mathcal{O}(s(r))$, $\sum_{i=1}^{n} |B_i| = \mathcal{O}(b\ u(r))$

Proof. We use a method used by Agarwal and Varadarajan in [2]. We construct the data structure for the red segments. The canonical prestored subsets produced by the data structure form our sets R_i. With each set R_i we associate a bucket which is initially empty. We query this data structure for each blue segment $\beta \in B$ one by one. The output of the query is given as the disjoint union of $\mathcal{O}(u(r))$ canonical subsets and we put the segment β into the buckets associated with each of those subsets. The set of segments in the bucket associated with R_i forms the set B_i. It is clear that each segment in R_i intersects each segment in B_i since we put exactly those segments in B_i which intersect all segments in R_i. Also, since the output to each query is given as a disjoint union of canonical subsets, whenever a red segment ρ and a blue segment β intersect, there is a unique i such that $\rho \in R_i$ and $\beta \in B_i$. Since the size of the data structure is $\mathcal{O}(s(r))$, $\sum_{i=1}^{n} |R_i| = \mathcal{O}(s(r))$ and since each query returns $\mathcal{O}(u(r))$ canonical subsets, each blue segment is contained in at most $\mathcal{O}(u(r))$ buckets and therefore $\sum_{i=1}^{n} |B_i| = \mathcal{O}(b\ u(r))$. The time required to construct the data structure is $\mathcal{O}(p(r))$ and the time for the b queries is $\mathcal{O}(b\ q(r))$. So, the total time complexity is $\mathcal{O}(p(r) + b\ q(r))$.

Theorem 7. *Given a set* R *of* r *(weighted) red segments and a set* B *of* b *(weighted) blue segments, if* R *can be preprocessed into a data structure of size* $\mathcal{O}(s(r))$ *in time* $\mathcal{O}(p(r))$ *so that all segments intersecting a query segment* $t \in B$ *can be reported as the union of* $\mathcal{O}(u(r))$ *"canonical" prestored subsets in* $\mathcal{O}(q(r))$ *time, then we can compute the number of the red-blue intersections and their centroid in* $\mathcal{O}\left(p(r) + b\ q(r) + (s(r) + b\ u(r)) \log^2 (r+b)\right)$ *time.*

Proof. We use Theorem 6 to form the sets R_i and B_i. Since all segments in R_i intersect all segments in B_i, the total number of intersections between the segments in R_i and the segments in B_i is $m_i = |R_i||B_i|$ and the centroid of those intersections (X_i, Y_i) can be computed in $\mathcal{O}(n_i \log^2 n_i)$ time (Corollaries 1 and 2) where $n_i = |R_i| + |B_i|$. So, the total number of intersections is $m = \sum_i m_i$ and the centroid of all red-blue intersections is $m^{-1} \sum_i m_i X_i$ and can be computed in

$$\mathcal{O}\left(p(r) + b\ q(r) + \sum_i n_i \log^2 n_i\right) = \mathcal{O}\left(p(r) + b\ q(r) + \log^2 (r+b)(\sum_i n_i)\right) =$$

$\mathcal{O}\left(p(r) + b\ q(r) + (s(r) + b\ u(r)) \log^2 (r+b)\right)$ time.

Theorem 8. *If a set S of n (weighted) segments in the plane can be preprocessed into a data structure of size $\mathcal{O}(s(n))$ in time $\mathcal{O}(p(n))$ so that all segments intersecting a query segment $t \in S$ can be reported as the union of $O(u(n))$ "canonical" prestored subsets in $\mathcal{O}(q(n))$ time, then we can compute the number of segment intersections and their centroid in $\mathcal{O}((p(n) + n\, q(n)) \log n + (s(n) + n\, u(n)) \log^4 n)$ time.*

Proof. We color half the segments red and the rest blue and then use Theorem 7 with $r = b = n/2$ to compute the number m_{RB} of red-blue intersections and their centroid (X_{RB}, Y_{RB}). We recursively compute the number m_R of red-red intersections and their centroid (X_R, Y_R) and also the number m_B of blue-blue intersections and their centroid (X_B, Y_B). The total number of intersections is $m = m_{RB} + m_R + m_B$ and their centroid is $(m^{-1}(m_{RB}X_{RB} + m_RX_R + m_BX_B), (m^{-1}(m_{RB}Y_{RB} + m_RY_R + m_BY_B))$. The time required for the computation is $\mathcal{O}((p(n) + n\, q(n)) \log n + (s(n) + n\, u(n)) \log^3 n)$ (we get an extra $\log n$ factor due to the recursion).

Corollary 3. *The centroid of the intersections of arbitrary (weighted) segments in the plane can be computed in $\mathcal{O}(n^{4/3+\epsilon})$ time.*

Proof. Agarwal and Sharir [1] have shown that given a collection S of segments in the plane and a parameter $n^{1+\epsilon} \leq s \leq n^{2+\epsilon}$ we can preprocess S into a data structure of size s, in time $\mathcal{O}(s^{1+\epsilon})$ so that we can report all k segments of S intersecting a query segment in time $\mathcal{O}(n^{1+\epsilon}/s^{\frac{1}{2}} + k)$. Furthermore, the output to such a query is given as the disjoint union of $\mathcal{O}(n^{1+\epsilon}/s^{\frac{1}{2}})$ "canonical" prestored subsets. We put $s = n^{4/3}$ and apply Theorem 8 and the result follows.

Since the time complexity of the best known algorithm even for computing the number of intersections among n line segments in the plane is $\Theta(n^{4/3}(\log n)^{1/3})$ [3], it is unlikely that this can be improved.

4 Intersection of Lines Inside a Polygon

If the configuration of the given set of segments allows a better query structure for segment intersections, we can do better. This for example is the case when the segments are chords of a simple region.

A region $D \subset \Re^2$ is called simple if any line intersects it at most c(a constant) times and the intersection of any set of n lines and the boundary of D can be ordered along the boundary in $\mathcal{O}(n \log n)$ time [8]. A chord of the region is a segment joining two boundary points and lying completely in the interior of the region. Let $c_1, c_2, \cdots, c_n$ be chords of a simple region R. We represent a chord c_i with the pair (l_i, r_i) where l_i and r_i are indices of the endpoints of c_i in the sorted order and $l_i < r_i$. Now given a query chord c, we can do binary searching of its endpoints in the sorted order of endpoints to compute their ranks l_c and r_c in $\mathcal{O}(\log n)$ time. Denote by $I(c)$ the set of chords intersecting c. Observe that $I(c) = I_1(c) \cup I_2(c)$ where $I_1(c) = \{c_i | l_i < l_c \;\&\; l_c < r_i < r_c\}$ and

$I_2(c) = \{c_i | l_c < l_i < r_c \ \& \ r_i > r_c\}$. If we represent c_i by the point (l_i, r_i) in the plane, then I_1 and I_2 can be computed using orthogonal range queries of the types $[-\infty, l_c] \times [l_c, r_c]$ and $[l_c, r_c] \times [r_c, \infty]$. We can use range trees [6], which can be built in $\mathcal{O}(n \log n)$ time and $\mathcal{O}(n \log n)$ space, to answer such queries in $\mathcal{O}(\log^2 n)$ time where the answer to each query is the union of $\mathcal{O}(\log^2 n)$ pairwise disjoint canonical subsets. We can therefore apply Theorem 8 (with $p(n) = \mathcal{O}(n \log n), s(n) = \mathcal{O}(n \log n), u(n) = \mathcal{O}(\log^2 n), q(n) = \mathcal{O}(\log^2 n)$) to show that the centroid of the intersections of the chords can be computed in $\mathcal{O}(n \log^5 n)$ time.

Corollary 4. *The centroid of the intersection points of n (weighted) lines lying inside a given simple region D can be computed in $\mathcal{O}(n \log^5 n)$ time.*

Proof. Since each line intersects D at most c times, the problem can be reduced to computing the centroid of the intersections of at most $\lfloor c/2 \rfloor n$ chords of D.

5 Computing Higher Moments

The x and y coordinates of the centroid of the intersection points are the averages of the coorresponding coordinates of the intersection points. We are now interested in the averages of the higher powers of the x and y coordinates.

As in Section 2, consider a set L of n lines and a query line $l \ : \ y = \mu x - \gamma$. We want to compute the averages of the kth powers, $X_L^{(k)}$ and $Y_L^{(k)}$, of the x and y-coordinates of the intersection points of the line l with the lines in L. The sum of the the kth powers of the x-coordinates of these intersection points is

$$F_L^{(k)}(\mu, \gamma) = \sum_{1 \le i \le n} \left(\frac{c_i - \gamma}{m_i - \mu} \right)^k$$

This function can be represented as:

$$F_L^{(k)}(\mu, \gamma) = \frac{\sum_{0 \le j \le k} \gamma^j P_L^{(j)}(\mu)}{S_L(\mu)}$$

where the $P_L^{(j)}$'s and S_L are polynomials of degree at most kn in μ.

We can therefore proceed as before and prove that computing the averages of the kth powers of the coordinates requires at most k^2 times the time required to compute the centroid (We get one k since there are $\mathcal{O}(k)$ polynomials to deal with and another k since the polynomials are of degree kn).

6 Conclusion and Outlook

In this paper we describe subquadratic algorithms for computing certain functions on the set of all intersecting pairs of lines (or segments) in an arrangement. Which functions do admit such subquadratic computations? The most outstanding question of this sort akss whether two distinct intersecting pairs of lines define the same intersection point, in other word, do some three lines intersect in a point. This is known as the line arrangement degeneracy problem. At this point this problem seems out of reach. However, we would like to point out the related problem of 3-SUM may be amenable to an algebraic approach as used in this paper.

The 3-SUM problem [7] asks whether for three given sets A, B, and C of n real numbers each, we have $(A + B) \cap C = \emptyset$, where $A + B$ is the Minkowski sum $\{a + b | a \in A, b \in B\}$. No subquadratic time solutions to this problem are known, except for the case where the three sets consist of integers in the range of 0 to K. In that case it suffices to compare C with the support of the product polynomial $(\sum_{a \in A} x^a) \cdot (\sum_{b \in B} x^b)$. This can be done in $O(K \log K)$ time.

Let us consider the simpler question whether we have $A \cap B = \emptyset$. There is an obvious $O(n \log n)$ solution via sorting and merging. This solution is combinatorial in that it relies on order comparisons $<, =, >$. It may be somewhat surprising that there is a also subquadratic solution relying solely on equality comparisons: Consider the polynomial $p_A(x) = \prod_{a \in A}(x - a)$. We have $A \cap B \neq \emptyset$ iff $p_A(b) = 0$ for some $b \in B$. The polynomial p_A can be computed in $O(n \log^2 n)$ time using divide-and-conquer and FFT-based polynomial multiplication, and within the same bound p_A can be evaluated for all n elements of B. Thus $A \cap B = \emptyset$ can be decided in $O(n \log^2 n)$ time without using order comparisons.

Algebraically maybe more succinct is the formulation $A \cap B = \emptyset$ iff $\gcd(p_A(x), p_B(x)) = 1$, where of course $p_B(x) = \prod_{b \in B}(x - b)$. This also leads to an $O(n \log^2 n)$ time solution since the gcd of two polynomials of degree n can be computed within this time (See Section 2.4, Chapter 2 of [9]).

The 3-SUM problem allows the following algebraic formulation:

$$(A + B) \cap C = \emptyset \text{ iff} \gcd(\text{ resultant}(p_A(x), p_B(z - x), x) \ , \ p_C(z)) = 1 \, .$$

Here $p_C(z) = \prod_{c \in C}(z - c)$. Note that $\text{resultant}(p_A(x), p_B(z - x), x)$ is nothing but $p_{A+B}(z)$, i.e. the polynomial whose roots are exactly the elements of $A + B$. Current techniques of computational algebra do not seem to allow the evaluation of $\gcd(\text{ resultant}(p_A(x), p_B(z - x), x) \ , \ p_C(z))$ in subquadratic time. However, there may be a better chance of success along such an algebraic approach than along a traditional combinatorial approach.

References

1. Agarwal, P.K., Sharir, M.: Applications of a new space-partitioning technique. Discrete Comput. Geom. 9(1), 11–38 (1993)
2. Agarwal, P.K., Varadarajan, K.R.: Efficient algorithms for approximating polygonal chains. Discrete Computational Geometry 23, 273–291 (2000)
3. Chazelle, B.: Cutting hyperplanes for divide-and-conquer. Discrete Computational Geometry 9(2), 145–158 (1993)
4. Cole, R., Salowe, J., Steiger, W., Szemerédi, E.: Optimal slope selection. SIAM J. Computing 18, 792–810 (1989)

5. Cormen, T.H., Leiserson, C.E., Rivest, R.L., Stein, C.: Introduction to Algorithms, ch. 32, 2nd edn. MIT Press, Cambridge (2001)
6. de Berg, M., van Kreveld, M., Overmars, M., Schwarzkopf, O.: Computational Geometry: Algorithms and Applications, ch. 5. Springer, Heidelberg (1997)
7. Gajentaan, A., Overmars, M.H.: On a class of $\o(n^2)$ problems in computational geometry. Comput. Geom. Theory Appl. 5(3), 165–185 (1995)
8. Langerman, S., Steiger, W.: Ham-sandwich cuts and other tasks in arrangements, Technical report (2001)
9. Yap, C.K.: Fundamental problems of algorithmic algebra. Oxford University Press, New York (2000)

Optimal Algorithms for the Weighted p-Center Problems on the Real Line for Small p

Binay Bhattacharya* and Qiaosheng Shi

School of Computing Science, Simon Fraser University,
Burnaby B.C., Canada V5A 1S6
{binay,qshi1}@cs.sfu.ca

Abstract. An optimal linear time algorithm for the unweighted p-center problems in trees has been known since 1991 [4]. No such worst-case linear time result is known for the weighted version of the p-center problems, even for a path graph. In this paper, for fixed p, we propose two linear-time algorithms for the weighted p-center problem for points on the real line, thereby partially resolving a long-standing open problem. One of our approaches generalizes the trimming technique of Megiddo [10], and the other one is based on the parametric pruning technique, introduced here. The proposed solutions make use of the solutions of another variant of the center problem called the *conditional center location problem* [13].

1 Introduction

In this paper we study the *p-center problem* for points on the real line L is studied when p is a fixed constant. The input is a set V of n points lying on L where each point $v \in V$ is associated with a non-negative weight $w(v)$. V is known as the *demand set.* Let $q(u)$ denote the *coordinate* of a point u on L, and let $d(u,v)$ denote the *distance* between the points u and v, i.e., $d(u,v) = |q(u) - q(v)|$. Let $S(X,V)$ denote the *service cost* of a set $X = \{\alpha_1, \ldots, \alpha_p\}$ on L to the demand set V, that is,

$$S(X,V) = \max_{v \in V} \{w(v) \cdot d(X,v)\}, \text{ where } d(X,v) = \min_{j=1,\ldots,p} d(\alpha_j, v).$$

The p-center problem is to determine a set X of p points on L such that $S(X,V)$ is minimized. When all the weights $w(v)$ are the same, the problem is known as the *unweighted p-center problem.* When the p centers are restricted to be points in V, the problem is known as the *discrete p-center problem.*

As the solution to this problem may be not unique, we settle for computing one optimal solution. The point in V with the smallest (resp. largest) coordinate is labeled v_1 (resp. v_n). It is easy to see that in any optimal solution $X^* = \{\alpha_1, \ldots, \alpha_p\}$, $q(v_1) \le q(\alpha_i) \le q(v_n), 1 \le i \le p$.

The p-center problem and many of its variants on general graphs have been shown to be NP-hard [9,12]. Frederickson [4] presented an $O(n)$ algorithm for

* Research was partially supported by MITACS and NSERC.

F. Dehne, J.-R. Sack, and N. Zeh (Eds.): WADS 2007, LNCS 4619, pp. 529–540, 2007.

locating p facilities in a vertex-unweighted tree. However, no optimal algorithm for the weighted version of the p-center problem is yet known, even for a fixed value of p. Megiddo [10] proposed a linear-time algorithm for the 1-center problem in trees. Very recently it was shown that the weighted 2-center problem in trees can also be solved in linear time [1]. Also recently, by using the Helly property and the generalized linear programming technique, Halman [5,6] proposed randomized linear time algorithms for the weighted p-center problem and the discrete unweighted p-center problem for points on the real line for any fixed p. The above result of Halman is also true for any path graph, an easier case. It is worth mentioning that Hassin and Tamir [7] devised algorithms for several coverage location models on the real line.

Efficient algorithms for the weighted p-center problems in trees are known. Megiddo and Tamir [12] provided an $O(n \log^2 n \log\log n)$ procedure to solve the weighted p-center problem in trees, which can be improved to $O(n \log^2 n)$ by applying a result of Cole [3]. This result is true for any arbitrary value of p. Improved results are known for small p (viz. $p = o(\log n)$). Jeger and Kariv [8] showed that the weighted p-center problem in trees can be solved in $O(pn \log n)$ time. The discrete weighted p-center problem in trees is also solvable in $O(n \log^2 n)$ time for any p [11].

Our problem of computing the weighted p-center problem for points on the real line is similar to the weighted p-center problem in a path graph. The main difference is that the path topology provides the ordering of these n points, whereas no ordering information of the demand points on L is available. The main results of this paper are worst-case linear-time algorithms for the weighted p-center problem of points on the real line when p is fixed. Megiddo [10] used a trimming technique to solve the weighted 1-center problem in linear time. The problem of generalizing the trimming approach to solve the p-center problem for $p > 1$ was open for a long time. In [1], the interactions between the two centers and split edge guide the trimming of the underlying tree. In this paper, we generalize the approaches proposed in [1,10], which are then used to solve the weighted p-center problem, p fixed, for points on the real line L in optimal linear time.

The paper is organized as follows. Our main idea is discussed in Sect. 2 where we discuss Megiddo's trimming method and the proposed parametric-pruning method for the 1-center problem in trees. In Sect. 3, we present the main result of this paper – two optimal linear time algorithms to solve the weighted p-center problem for points on the real line when p is fixed. Section 4 presents a brief summary.

2 Main Idea of Our Algorithm

In many practical situations there may already exist some facilities in the network, and the problem is to find locations for a specified number of new facilities. These types of problems are referred to as *conditional location problems* [13]. In this paper we define the conditional k-center problem for points on L to be the problem of locating k new centers on L, where a facility is already located at v_1 or at v_n, or where two facilities are already located at v_1 and v_n.

The proposed linear-time algorithm to solve the p-center problem on L uses the fact that there exist linear-time algorithms for any k-center problem and any conditional k-center problem for $k < p$. Our objective is to solve the p-center problem incrementally using the solutions of lower-order center problems.

This requires more notations. A link connecting two consecutive points u, v in V is called *an edge* $e : uv$. Let E denote the set of edges constructed from points in V. Clearly, $|E| = n - 1$. Given two real values x_1, x_2 ($x_1 \le x_2$), $[x_1, x_2]$ denotes the interval on L from x_1 to x_2 (including x_1 and x_2); x_1 (resp. x_2) is called the *left* (resp. *right*) *endpoint* of this interval.

Two subsets, V_1 and V_2, of V are *disjoint* if the coordinate of any point in one subset is smaller than the coordinates of all the points of the other subset. Let $X = \{\alpha_1, \cdots, \alpha_p\}$ be a set of p centers on L. A demand point v is called a *dominating demand point* of the center set X if $w(v) \cdot d(X, v) = S(X, V)$. Observe that, in an optimal solution, its center set always has an equal or smaller service cost to the dominating demand points of X.

Let $V_i \subseteq V$ denote the set of demand points closest to a particular center $\alpha_i \in X, i = 1, \cdots, p$:

$$V_i = \{v \in V : d(\alpha_i, v) = \min_{j=1,\ldots,p} d(\alpha_j, v)\}.$$

Clearly, these subsets are mutually disjoint. Let v_i' be the leftmost point and v_i'' be the rightmost point in V_i, that is, $q(v_i') \le q(v) \le q(v_i''), v \in V_i$. The edges of E whose endpoints belong to different subsets $V_i, 1 \le i \le p$, are called *split edges*. Thus, locating p centers on L is equivalent to finding a set of $p - 1$ split edges which define p regions such that the maximum service cost of the 1-centers of these regions is equal to the optimal p-center cost of V on L. Clearly, there is at least one subset V_i in any optimal solution that contains at least n/p demand points in V.

Discarding one demand point v is called a *safe operation for a center* α, if v is served by α, and v is not the weighted farthest demand point to α. Similarly, discarding one demand point is called a *safe operation for an interval* $[x, y]$, if it is a safe operation for any center located in $[x, y]$. Our main idea here is to locate one subset of demand points that are served by the same center in some optimal solution, and then safely pruning a fraction of demand points in this subset will result in a smaller-size similar problem whose optimal solution is the same as that of the original problem.

2.1 The Weighted 1-Center Problem in Trees

For a tree graph T, let $V(T), E(T), A(T)$ denote the vertex set, the edge set and the continuum set of points on the edges of T, respectively. The length of the simple path between two points $x, y \in A(T)$ is denoted by $d_T(x, y)$. Let $\delta_{T'}(v)$ be the degree of vertex v in subtree T' of T. A vertex v is called the *anchor vertex* of T' with respect to T if $v \in V(T'), \delta_{T'}(v) = 1$ and $\delta_{T'}(u) = \delta_T(u), u \in V(T') \backslash \{v\}$. In Fig. 1, T' is a subtree (bold part) and o is the anchor vertex of subtree T' with respect to T.

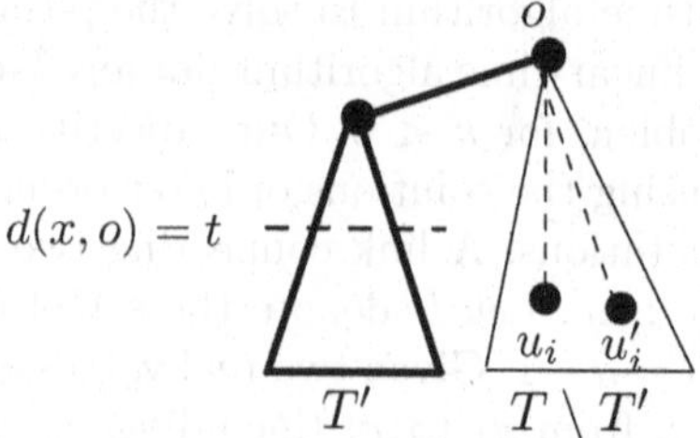

Fig. 1. Megiddo's algorithm for the weighted 1-center problem in trees

The *centroid* vertex of a subtree T, which can be found in linear time, is a vertex $u \in V(T)$, such that each subtree with the removal of u has a size of $|V(T)|/2$ at most.

The weighted 1-center problem in a tree T is to locate a point in $A(T)$ such that its maximum weighted distance to the vertices of T (here the demand set is $V(T)$) is minimized. Megiddo [10] designed a trimming (also known as prune-and-search) algorithm for this problem, which is carried out in two phases. The first phase is to locate a subtree T', anchored at o, which contains an optimal weighted 1-center (refer to Fig. 1). This is determined by computing the weighted distances of the vertices of T to the centroid vertex o. It is easy to see that the optimal 1-center provides service to all the demand points outside T' (i.e., $T \setminus T'$) through the vertex o. Therefore, the topology of the subtree $T \setminus T'$ is not important. For each vertex in $T \setminus T'$, we only need to keep its distance information to o. We call a subtree of T a *big component* if it contains a constant fraction of the vertices of T. Clearly the subtree $T \setminus T'$ is a big component, since $|V(T \setminus T')| \geq n/2$.

In the second phase, the following *key question* is answered: determine whether there is an optimal 1-center in T' within the distance t to o. An appropriate value of t is determined in the following way. We arbitrarily pair the vertices in $T \setminus T'$. Let $(u_1, u'_1), (u_2, u'_2), \ldots, (u_l, u'_l)$ be the pairs where $w(u_i) \geq w(u'_i)$, and if $w(u_i) = w(u'_i)$ then $d(u_i, o) \geq d(u'_i, o)$. For every such pair $(u_i, u'_i), 1 \leq i \leq l$ let $t_i = [w(u'_i)d_T(u'_i, o) - w(u_i)d_T(u_i, o)]/(w(u_i) - w(u'_i))$ if $w(u_i) \neq w(u'_i)$, and let $t_i = 0$ otherwise. Note that when $t_i \leq 0$, the dominating demand point in $T \setminus T'$ of the 1-center in T' cannot be u'_i, and therefore u'_i can be discarded. Without any loss of generality we assume that $t_i > 0$ for all i. Thus u_i and u'_i have the same weighted distance to a point with the distance t_i to o outside $T \setminus T'$. Let t be the median of these positive values. Once the answer to the key question is known, approximately $1/4$ of the vertices in $T \setminus T'$ cannot determine an optimal solution, and therefore can be safely discarded. The algorithm performs $O(\log n)$ such iterations, and each iteration takes linear time, linear in the size of the current tree. Thus the weighted 1-center problem in trees can be solved in linear time.

Below we introduce another approach to perform the pruning step. The method is called parametric pruning, which is a generalization of the Megiddo's trimming technique [10].

For every pair $(u_i, u'_i), 1 \leq i \leq l$, $w(u_i) \geq w(u'_i)$, let $c_i = w(u_i) \cdot (d_T(u_i, o) + t_i) = w(u'_i) \cdot (d_T(u'_i, o) + t_i)$, where t_i is the positive weighted equidistant point of u_i, u'_i as described above and c_i is called the *switch service cost* of this pair. If the optimal service cost is larger (resp. smaller) than c_i, then u_i (resp. u'_i) dominates vertex u'_i (u_i). Let c be the median of the switch service costs $c_i, i = 1, 2, \ldots, l$. We can find either $c^* > c$ or $c^* \leq c$ after solving the following feasibility decision problem: does there exist a point $x \in A(T)$ such that $S(x, V(T)) \leq c$? Here c^* denotes the optimal service cost. We know that the feasibility decision problem in a tree can be solved in linear time [11]. Therefore, the pruning of the vertices in $T \setminus T'$ can also be performed using this parametric-pruning method. Our approach is general in the sense that $T \setminus T'$ does not have to be connected. When $T \setminus T'$ is not connected, Megiddo's trimming technique [10] does not work.

3 The Weighted p-Center Problem on L, for a Fixed p

In this section, we first describe our algorithm for the conditional 1-center problem. Then we present our algorithm for the weighted p-center problem of V on L. The algorithm for the weighted p-center problem uses the solutions of weighted k-center and conditional k-center problems on L for all $k < p$.

3.1 The Conditional 1-Center Problems on L

In this section, we first describe a linear-time algorithm for the conditional 1-center problem for points on L with one existing facility located at v_n. Then we describe its extension to the conditional 1-center problem for points on L with any fixed number of existing facilities.

We divide V into two disjoint subsets of almost equal size: V_1 and V_2 ($|V_1| = \lceil n/2 \rceil$). Suppose that V_1 is served by the new facility and V_2 is served by v_n. We first solve the weighted 1-center problem for the demand points of V_1 only, and letting z be the optimal cost. We consider the following situations.

(a) If $z \geq S(v_n, V_2)$, then V_2 is served by v_n in an optimal solution. Let u be a demand point in V_2 such that $w(u) \cdot d(u, v_n) = S(v_n, V_2)$. Clearly, all the demand points of $V_2 \setminus \{u, v_n\}$ can be safely discarded.

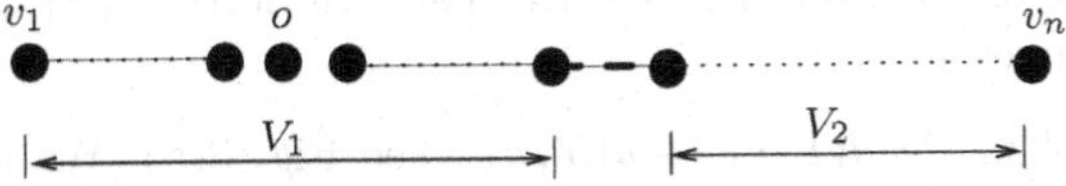

Fig. 2. The conditional 1-center problem with one existing facility v_n

(b) Suppose that $z < S(v_n, V_2)$. In this case V_1 is served by the new facility in some optimal solution. Refer to Fig. 2. Let o be a median point of V_1. Based on the location of dominating demand points of center set $\{o, v_n\}$, we can determine the relative location of the new facility with respect to o in some optimal solution. That is, if these dominating demand points of $\{o, v_n\}$ lie in $[q(o), q(v_n)]$, then in an optimal solution the new facility lies within $[q(o), q(v_n)]$, otherwise it lies within $[q(v_1), q(o)]$. Note that if the dominating points of $\{o, v_n\}$ lie in both $[q(o), q(v_n)]$ and $[q(v_1), q(o)]$, the o is the optimal conditional 1-center with an existing facility located at v_n.

Without loss of generality, assume that all the dominating demand points of $\{o, v_n\}$ lie in $[q(o), q(v_n)]$. Then, arbitrarily pair the demand points in $[q(v_1), q(o)]$ and compute the appropriate value of t by using Megiddo's method (as in the second phase described in Sect. 2.1). Let u be the point on L with coordinate $q(o) + t$. We can determine that in some optimal solution either the new facility lies in $[q(o), q(u)]$ or the new facility lies in $[q(u), q(v_n)]$ by just computing the dominating demand points of $\{u, v_n\}$ (similar idea is used above). Approximately $1/4$ of the demand points in $[q(v_1), q(o)]$ can be safely discarded.

Therefore, we can prune approximately $n/16$ of the demand points of V from further consideration. The algorithm performs $O(\log n)$ such iterations, and each iteration takes linear time, linear in the current size of the demand set. Thus the conditional 1-center problem with one existing facility at v_n can be solved in linear time.

In fact, a linear-time solution for the conditional 1-center problem with any fixed number of existing facilities can be similarly developed, described as follows.

Let F be the set of a fixed number of existing facilities. First of all, it is not difficult to design a linear-time computation step to find the distance $d(F, v)$ for each vertex $v \in V$. After this preprocessing step, it is safe to discard the points of F.

Let o' be a median point of V. Based on the location of the dominating demand points of center set $\{o'\} \cup F$, we can determine the relative location of another new facility with respect to o' in some optimal solution. Without loss of generality, assume that all the dominating demand points of $\{o'\} \cup F$ lie in $[q(o'), q(v_n)]$. Then, arbitrarily pair the demand points in $[q(v_1), q(o')]$. For each such pair (v_i, v_j), the new facility x serves them through o', (see Fig. 3).

Below we show that at most two intersections exist between the two service cost functions $S(\{x\} \cup F, v_i)$ and $S(\{x\} \cup F, v_j)$. If $d(o', v) \geq d(F, v)$ for a demand point in $[q(v_1), q(o')]$, then v will always be served by some existing facility. Without loss of generality, assume that $d(o', v_i) < d(F, v_i), d(o', v_j) < d(F, v_j)$, and $w(v_i) \geq w(v_j)$. Figure 3 shows the service cost functions $S(\{x\} \cup F, v_i), S(\{x\} \cup F, v_j)$ with a change of $d(x, o')$.

- $d(F, v_i) - d(o', v_i) \geq d(F, v_j) - d(o', v_j)$: see Fig. 3(a). We fix the function $S(\{x\} \cup F, v_i)$ and use the dashed lines to represent all possibilities of the function $S(\{x\} \cup F, v_j)$. There is at most one intersection between the service cost functions $S(\{x\} \cup F, v_i), S(\{x\} \cup F, v_j)$.
- $d(F, v_i) - d(o', v_i) < d(F, v_j) - d(o', v_j)$: see Fig. 3(b). We fix the function $S(\{x\} \cup F, v_j)$ and use the dashed lines to represent all possibilities of the function $S(\{x\} \cup F, v_i)$. In this case, it is possible to have at most two intersections.

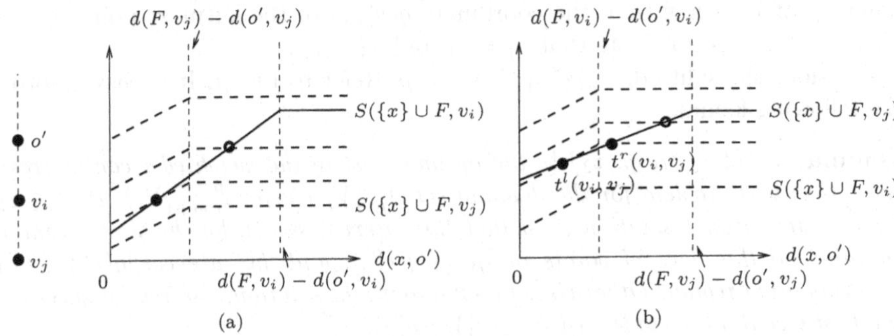

Fig. 3. The number of intersections between the two service cost functions $S(\{x\} \cup F, v_i)$ and $S(\{x\} \cup F, v_j)$ in the conditional model: (a) $d(F, v_i) - d(o', v_i) \geq d(F, v_j) - d(o', v_j)$; (b) $d(F, v_i) - d(o', v_i) < d(F, v_j) - d(o', v_j)$

When there are two intersections for (v_i, v_j), we call the distance $d(x, o')$ with smaller value *left switch distance*, denoted by $t^l(v_i, v_j)$, and call the other one *right switch distance*, denoted by $t^r(v_i, v_j)$.

For those pairs having only one switch distance, we can locate the non-dominating vertices easily by checking the key question with the median switch distance, similar to Megiddo's trimming approach [10]. Suppose that (v_1, v_1'), (v_2, v_2'), $\cdots$, (v_k, v_l') are the pairs of demand points with two switch distances, where $w(v_i) \geq w(v_i'), 1 \leq i \leq l$. Select one value t^l (resp. t^r) such that one third of left (resp. right) switch distances $t^l(v_i, v_i') > t^l$ (resp. $t^r(v_i, v_i') \leq c^r$) and the remaining ones are no greater than (resp. larger than) it. After answering the key questions with the distances t^l and t^r, we can determine at least $\lfloor \frac{l}{3} \rfloor$ non-dominating demand points for an optimal facility [14].

In this way, we can prune approximately $\lfloor \frac{n}{12} \rfloor$ of the demand points of V from further consideration, since the number of pairs of demand points in $[q(v_1), q(o')]$ is at least $\lfloor \frac{n}{4} \rfloor$. Therefore, we have the following theorem.

Theorem 1. *The weighted conditional* 1*-center problem of* V *where any fixed number of existing facilities on the real line* L *can be solved in linear time.*

3.2 The Weighted p-Center Problem on L (fixed p)

Suppose that the weighted k-center problem and conditional k-center problem of V on L can be solved in linear time, for all k less than p. We show that for fixed p the weighted p-center problem of V can also be solved in linear time.

First we determine one region which contains at least $\lfloor n/p \rfloor$ demand points of V served by the same center in some optimal solution. Such a region is called a *big region*.

Since the ith largest element of a set can be found in linear time [2], it costs $O(n \log p)$ time to divide V into p mutually disjoint subsets: $\{V_1, \ldots, V_p\}$ where $\lfloor n/p \rfloor \leq |V_i| < \lceil n/p \rceil, 1 \leq i \leq p$, and the coordinate $q(v_i'')$ of the rightmost point v_i'' of V_i is less than the coordinate $q(v_{i+1}')$ of the leftmost point v_{i+1}' of $V_{i+1}, i = 1, \ldots, p-1$. Note that $v_1' = v_1$ and $v_p'' = v_n$.

Consider the split edge $v_i''v_{i+1}'$, $1 \le i < p$. Refer to Fig. 4. It is easy to prove the following lemma.

Lemma 1. *Let z_1 (resp. z_2) be the optimal cost of the weighted i-center (resp. $(p-i)$-center) problem for the demand set $\cup_{j=1}^{i} V_j$ (resp. $\cup_{j=i+1}^{p} V_j$). If $z_1 \le z_2$, then, $\exists$ an optimal solution such that the interval served by the first i centers contains all the demand points in $[q(v_1'), q(v_i'')]$, and the interval $[q(v_1'), q(v_i'')]$ contains a big region. Otherwise, in some optimal solution, the region served by the first i centers is contained in $[q(v_1'), q(v_i'')]$.*

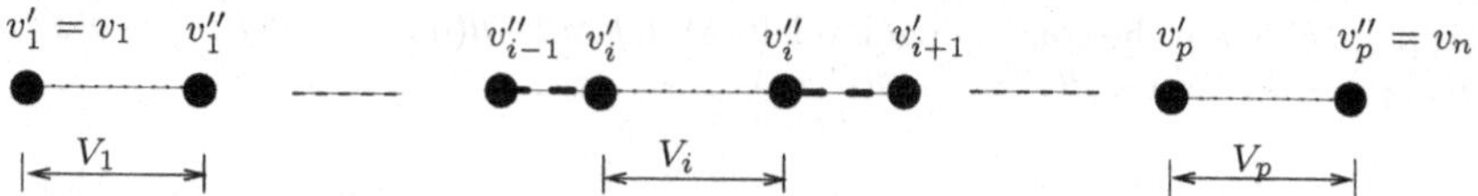

Fig. 4. Locating p centers of V on the real line

As a consequence of Lemma 1, one of the following cases must be true for the p-center problem on L.

Case 1. The optimal cost of the 1-center problem with the demand set V_1 is less than the optimal cost of the $(p-1)$-center problem with the demand set $V \setminus V_1$. In this case, the subset V_1 is served by the same center in some optimal solution.

Case 2. The optimal cost of the 1-center problem with the demand set V_p is less than the optimal cost of the $(p-1)$-center problem with the demand set $V \setminus V_p$. In this case, the subset V_p is served by the same center in some optimal solution.

Case 3. There exists an i, $2 \le i \le p-1$, such that the optimal cost of the i-center problem with the demand set $\cup_{j=1}^{j=i} V_j$ is no more than the optimal cost of the $(p-i)$-center problem with the demand set $\cup_{j=i+1}^{j=p} V_j$, and the optimal cost of the $(i-1)$-center problem with the demand set $\cup_{j=1}^{j=i-1} V_j$ is larger than the optimal cost of the $(p-i+1)$-center problem with the demand set $\cup_{j=i}^{j=p} V_j$. In this case, there exists some optimal solution where V_i is served by the same center. The reason is as follows.

In some optimal solution, the first i centers serve all the demand points lying within $[q(v_1'), q(v_i'')]$, since the cost of an optimal i-center solution of $\cup_{j=1}^{j=i} V_j$ is less than the cost of an optimal $(p-i)$-center solution of $\cup_{j=i+1}^{j=p} V_j$ (Lemma 1). Similarly, all the demand points served by the first $i-1$ centers lie within $[v_1', v_{i-1}'']$, since the cost of an optimal $(i-1)$-center solution of $\cup_{j=1}^{j=i-1} V_j$ is larger than the cost of an optimal $(p-i+1)$-center solution of $\cup_{j=i}^{j=p} V_j$ (Lemma 1).

Thus we can find one big region after considering each split-edge $v_i''v_{i+1}'$, $i = 1, \ldots, p-1$. Note that we need to solve two weighted center problems for each split-edge $v_i''v_{i+1}'$: one i-center problem on a demand set $\cup_{j=1}^{j=i} V_j$ and one $(p-i)$-center problem on a demand set $\cup_{j=i+1}^{j=p} V_j$. The total time to solve all these center problems is linear when p is fixed. Therefore, we have the following result.

Lemma 2. *It takes a linear time to locate one big region served by the same center, in some optimal solution to the weighted p-center problem of V on L when p is a fixed number.*

Actually, one big region can be located using a binary search instead of a linear search on the edges $v_i''v_{i+1}'$, $1 \le i \le p-1$. Let V_l be the big region thus computed.

Next, we show a method to identify approximately 1/8 of demand points of V_l that are not dominating, and hence they can be discarded. This method is very similar to the method described in Sect. 3.1. Let o be the median point of V_l which can be found in $O(n/p)$ time. Consider a facility (center) located at o to serve the points of V_l. Let V_l^1 (resp. V_l^2) be the subset of demand points of V_l lying to the left (resp. right) of o.

We now consider Case 1 described above, i.e., when $l = 1$. The arguments for Case 2 (i.e., when $l = p$) are similar. Refer to Fig. 5(a). For Case 1, the conditional $(p-1)$-center problem of the points of $V \setminus V_1$ with one existing center located at o is first solved. All the points of $V \setminus V_1$ lie to the right of o. Let z be the optimal cost of the conditional $(p-1)$-center problem with one existing center located at o.

- If $S(o, V_1^1) < \max\{S(o, V_1^2), z\}$ then in some optimal solution, the center serving V_1 lies to the right of o.
- If $S(o, V_1^1) = \max\{S(o, V_1^2), z\}$ then o is the center serving V_1 in some optimal solution, and the optimal cost is $S(o, V_1^1)$.
- If $S(o, V_1^1) > \max\{S(o, V_1^2), z\}$ then in some optimal solution, the center serving V_1 lies to the left of o.

In Case 3 (refer to Fig. 5(b)), where $2 \le l \le p-1$, we solve the $(l-1)$-center problem of $\cup_{j=1}^{j=l-1} V_j$ with one existing center located at o, and also solve the $(p-l)$-center problem of $\cup_{j=l+1}^{j=p} V_j$ with one existing center located at o. Let z_1 be the optimal cost of the former problem and z_2 be the optimal cost of the latter one. Like Case 1, after comparing the values of $\max\{z_1, S(o, V_l^1)\}$ and $\max\{z_2, S(o, V_l^2)\}$, either we obtain the optimal cost of the p-center problem of V; we find that in some optimal solution the center serving V_l lies to the left of o; or we find that the center serving V_l lies to the right of o.

Now, Megiddo's trimming method can be applied to prune approximately 1/8 of demand points of V_l, i.e., $\lfloor \frac{n}{8p} \rfloor$ demand points of V. The process is repeated with the reduced set of demand points. Thus, for fixed p, the algorithm performs $O(\log n)$ such iterations, and each iteration takes linear time, linear in the size of the current demand set.

Therefore, we have the following theorem.

Theorem 2. *For any fixed p, the weighted p-center problem of V on the real line can be solved in linear time.*

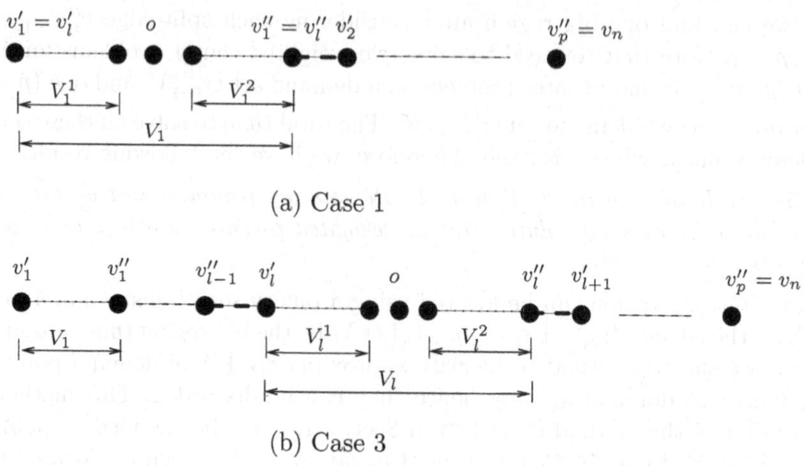

Fig. 5. Case 1 and Case 3

3.3 The Conditional p-Center Problems on L (fixed p)

In this section, we describe a linear-time algorithm for the conditional p-center problem on L with one existing facility located at v_n. The method for this problem is an extension of the steps described in Sect. 3.1 and Sect. 3.2.

We divide V into $p+1$ mutually disjoint subsets of almost equal size. One big region of size no less than $\lfloor n/(p+1) \rfloor$ can be located in linear time, by using the steps similar to the ones described in Sect. 3.2. The only difference is that the subproblems might be conditional center problems here. Still, Megiddo's method is applicable to prune approximately 1/8 of the demand points lying in the big region.

Its extension to the conditional p-center problem with a fixed number of existing facilities is not difficult to obtain.

Theorem 3. *For any fixed p, the weighted conditional p-center problem of V with any fixed number of existing facilities located on the real line L can be solved in linear time.*

3.4 The Second Approach Using Parametric Pruning

In this section, we show that the weighted p-center problem of V on L can also be solved in linear time using the parametric pruning technique introduced in Sect. 2.1.

A value $r \geq 0$ is *feasible* if there exists a set of at most p points $X = \{\alpha_1, \ldots, \alpha_p\}$ on L such that $S(X, V) \leq r$. Clearly, the optimal cost is the minimum value of r that is feasible. Next, we show that a feasibility test can be done in linear time.

Each demand point v in V is bundled with a *center region* $[q(v) - \frac{r}{w(v)}, q(v) + \frac{r}{w(v)}]$ which contains all the points of V with all the weighted distance to v of no more than r. A given value r is feasible if there exists a solution (a set of p points on L) such that at least one center in this solution lies in the center region of each demand point.

Initially, $X = \emptyset$. Among all these center regions, we find a region whose right endpoint has the smallest coordinate. Let it be a center (insert it into X) and remove all demand points whose center regions contain it. Repeat this process on the remaining demand points. Finally, if $|X| \leq p$, then r is feasible, and otherwise r is infeasible. This process takes $O(pn)$ time.

The main difference between this approach and the one in Sect. 3.2 can be described as follows. Recall that when we consider $v_i''v_{i+1}'$ as a split-edge in the first phase to locate a big region, one i-center subproblem and another $(p-i)$-center subproblem need to be solved. In our second approach, only one center subproblem with a smaller demand set needs to be solved, i.e., if $i \leq (p-i)$ then the i-center subproblem is solved. Let z be its optimal cost. Instead of solving another center problem, we can test the feasibility of z. We can have the same effect, as shown in Lemma 3, which is similar to Lemma 1. Assume that $i \leq (p-i)$.

Lemma 3. *Let z be the optimal cost of the weighted i-center problem of the set $\cup_{j=1}^{i} V_j$. If z is feasible for V, then $\exists$ an optimal solution such that the region served by the first i centers contains all the demand points in $[q(v_1'), q(v_i'')]$, and the interval $[q(v_1'), q(v_i'')]$ contains a big region. Otherwise, in some optimal solution, the region served by the first i centers is contained in $[q(v_1'), q(v_i'')]$.*

Similarly, in the second phase, in order to prune demand points in the big region located in the first phase, we only solve one conditional center problem with a smaller demand set and then test the feasibility of its optimal cost.

It is worth noting that, in practice, the second approach is faster than the method in Sect. 3.2. The reason is that the time complexity of our method for the weighted p-center problem is exponential in p, whereas the feasibility test can be done in $O(pn)$ time.

4 Conclusion

In this paper, two methods are proposed which solve, for fixed p, the weighted p-center problem of points on the real line in linear time. The first proposed method is a generalization of Megiddo's trimming method [10]. The second method is based on the parametric-pruning technique. Both of them can be applied to achieve linear time solutions for the weighted p-center and conditional p-center problems on the real line when p is a fixed number. We believe that the the main idea presented in this paper can be generalized to solve the weighted p-center problem in trees.

Unfortunately, the time complexity of our method is exponential in p. One challenging task is to design an $O(f(p) \cdot n)$-time algorithm for the weighted p-center problem on the real line where $f(p)$ is a polynomial in p.

References

1. Ben-Moshe, B., Bhattacharya, B., Shi, Q.: An optimal algorithm for the continuous/discrete weighted 2-center problem in trees. In: Correa, J.R., Hevia, A., Kiwi, M. (eds.) LATIN 2006. LNCS, vol. 3887, pp. 166–177. Springer, Heidelberg (2006)
2. Blum, M., Floyd, R.W., Rivest, R.L., Tarjan, R.E.: Time bounds for selection. J. Comput. Sys. Sci. 7, 448–461 (1972)
3. Cole, R.: Slowing down sorting networks to obtain faster sorting algorithms. J. ACM 34, 200–208 (1987)
4. Frederickson, G.N.: Parametric search and locating supply centers in trees. In: Dehne, F., Sack, J.-R., Santoro, N. (eds.) WADS 1991. LNCS, vol. 519, pp. 299–319. Springer, Heidelberg (1991)
5. Halman, N.: Discrete and lexicographic Helly theorems and their relations to LP-type problems. Ph.D. Thesis, Tel. Aviv. Univ. (2004)
6. Halman, N.: On the power of discrete and lexicographic Helly-type theorems. in 45th FOCS, pp. 463–472 (2004)
7. Hassin, R., Tamir, A.: Improved complexity bounds for location problems on the real line. Oper. Res. Let. 10, 395–402 (1992)
8. Jeger, M., Kariv, O.: Algorithms for finding p-centers on a weighted tree (for relatively small p). Networks 15, 381–389 (1985)
9. Kariv, O., Hakimi, S.L.: An algorithmic approach to network location problems, Part I. The p-centers. SIAM J. Appl. Math. 37, 513–538 (1979)
10. Megiddo, N.: Linear-time algorithms for linear programming in R^3 and related problems. SIAM J. Comput. 12(4), 759–776 (1983)
11. Megiddo, N., Tamir, A., Zemel, E., Chandrasekaran, R.: An $O(n \log^2 n)$ algorithm for the kth longest path in a tree with applications to location problems. SIAM J. Comput. 10(2), 328–337 (1981)
12. Megiddo, N., Tamir, A.: New results on the complexity of p-center problems. SIAM J. Comput. 12(4), 751–758 (1983)
13. Minieka, E.: Conditional centers and medians on a graph. Networks 10, 265–272 (1980)
14. Zemel, E.: An $O(n)$ algorithm for the linear multiple choice knapsack problem and related problems. Info. Proc. Let. 18, 123–128 (1984)

Faster Approximation of Distances in Graphs

Piotr Berman and Shiva Prasad Kasiviswanathan

Computer Science and Engineering, Pennsylvania State University
{berman,kasivisw}@cse.psu.edu

Abstract. Let $G = (V, E)$ be a weighted undirected graph on n vertices and m edges, and let d_G be its shortest path metric. We present two simple deterministic algorithms for approximating all-pairs shortest paths in G. Our first algorithm runs in $\tilde{O}(n^2)$ time, and for any $u, v \in V$ reports distance no greater than $2d_G(u, v) + h(u, v)$. Here, $h(u, v)$ is the largest edge weight on a shortest path between u and v. The previous algorithm, due to Baswana and Kavitha that achieved the same result was randomized. Our second algorithm for the all-pairs shortest path problem uses Boolean matrix multiplications and for any $u, v \in V$ reports distance no greater than $(1 + \epsilon)d_G(u, v) + 2h(u, v)$. The currently best known algorithm for Boolean matrix multiplication yields an $O(n^{2.24+o(1)}\epsilon^{-3}\log(n\epsilon^{-1}))$ time bound for this algorithm. The previously best known result of Elkin with a similar multiplicative factor had a much bigger additive error term.

We also consider approximating the diameter and the radius of a graph. For the problem of estimating the radius, we present an almost 3/2-approximation algorithm which runs in $\tilde{O}(m\sqrt{n} + n^2)$ time. Aingworth, Chekuri, Indyk, and Motwani used a similar approach and obtained analogous results for the diameter approximation problem. Additionally, we show that if the graph has a small separator decomposition a 3/2-approximation of both the diameter and the radius can be obtained more efficiently.

1 Introduction

Consider the all-pairs shortest path (henceforth, referred to as APSP) problem. Given a graph[1] $G = (V, E)$ on $|V| = n$ and $|E| = m$ edges, the goal is to compute the distances between all pairs of vertices. The currently best known upper bound on the worst case complexity is $O(n^3 \log^3 \log n / \log^2 n)$ due to a recent paper by Chan [1]. For the simpler case of unweighted graphs, using fast matrix multiplication, Galil and Margalit [2,3], and Seidel [4] have obtained algorithms that run in $\tilde{O}(n^\omega)$[2] time, where ω denotes the exponent of (square) matrix multiplication algorithm used. The currently best known matrix multiplication algorithm of Coppersmith and Winograd [5] results in $\omega < 2.376$.

We say that an algorithm is a (a, b)-*approximation* of the APSP problem if for any pair of vertices $(u, v) \in V \times V$, the estimate $\delta(u, v)$ produced by the algorithm

[1] Throughout the paper graphs are undirected unless mentioned otherwise.

[2] The notation $\tilde{O}(f) \equiv O(f \text{ poly} \log f)$.

F. Dehne, J.-R. Sack, and N. Zeh (Eds.): WADS 2007, LNCS 4619, pp. 541–552, 2007.

satisfies: $d_G(u, v) \leq \delta(u, v) \leq a d_G(u, v) + b \cdot h(u, v)$. Here d_G is the shortest path metric induced by the connected graph G on its vertices and $h(u, v)$ is the weight of the heaviest edge on a shortest path from u to v. The multiplicative term a is referred to as the *stretch factor* and $b \cdot h(u, v)$ denotes the *additive error*. Note that for unweighted graphs the additive error is just b.

Over the last decade many algorithms have been designed for this problem that achieve sub-cubic time and/or sub-quadratic space. Here we state a few of the relevant results. For a more comprehensive overview of results refer to the survey by Zwick [6].

For unweighted graphs, Aingworth *et al.* [7] used an ingenious method to obtain an $\tilde{O}(n^{5/2})$ time $(1, 2)$-approximation algorithm. Dor *et al.* designed an algorithm which for every even $t > 2$ runs in $\tilde{O}(\min\{n^{2-\frac{2}{t+2}} m^{\frac{2}{t+2}}, n^{2+\frac{2}{3t-2}}\})$ time and is a $(1, t)$-approximation. For weighted graphs, Cohen and Zwick [8] building on the results of Dor *et al.* [9] provided fast algorithms with stretch factor 2, 7/3, and 3. In a recent improvement, Baswana and Kavitha [10] provided faster algorithm for the same stretch factors. They present algorithms that run in expected $\tilde{O}(\sqrt{m} n^{3/2})$ time and expected $\tilde{O}(n^{7/3})$ time for stretch factors of 2 and 7/3 respectively. Furthermore, they designed an expected $\tilde{O}(n^2)$ time $(2, 1)$-approximation algorithm. Also on weighted graphs, Elkin [11] presented an $O(mn^{\rho} + n^{2+\zeta})$ time algorithm that for any $u, v \in V$ reports distance bounded by $(1 + \epsilon) d_G(u, v) + W \cdot \beta(\zeta, \rho, \epsilon)$. Here, W is the ratio between the heaviest and lightest edge in the graph. The constant β depends on ζ as $(1/\zeta)^{\log 1/\zeta}$, depends inverse exponentially on ρ, and depends inverse polynomially on ϵ.

Diameter and radius are two important parameters of a graph. The eccentricity of a vertex is defined as the maximum distance between the vertex and any other vertex. The maximum eccentricity is the graph diameter and the minimum eccentricity is the graph radius. Both diameter and radius can be found by solving the APSP problem. Recently, Chan [12] has shown that diameter of unweighted directed graphs can be obtained in expected $O(mn \log^2 \log n / \log n + n^2 \log n / \log \log n)$ time. For general graphs however it is not clear whether these parameters can be obtained faster than obtaining the whole distance matrix.

On the approximation front, it is easy to estimate both the diameter and radius within a ratio 2 by performing a single-source shortest path from any vertex in the graph. No better result was known until Aingworth *et al.* [7] designed a 3/2-approximation algorithm for the diameter running in $\tilde{O}(m\sqrt{n} + n^2)$ time. Also recently, Boitmanis *et al.* [13] gave algorithms for approximating the diameter and the radius in $\tilde{O}(m\sqrt{n})$ time. The results are produced within an additive error of $O(\sqrt{n})$.

The situation seems no better even if we restrict our attention to the family of separable graphs (i.e., graphs with a small sized vertex separator). The family of separable graphs contains planar graphs, graphs with no fixed minor, k-overlap graphs, bounded tree-width graphs. Even for the generally well-studied planar graphs the only known result seems to be that Eppstein [14], who has shown that if the planar graph has a constant bound on diameter, then the exact diameter can be found in linear time.

In the following we summarize our contributions and put them into context. All missing proofs and details can be found in the full version [15].

1.1 Our Contributions

We design two algorithms for the problem of APSP for weighted undirected graphs. Our first algorithm runs in $\tilde{O}(n^2)$ time and is a $(2,1)$-approximation. Earlier, Thorup and Zwick [16] have shown than for any $t < 3$, a data structure that answers t-approximate distance query in constant time must occupy $\Theta(n^2)$ space. This automatically implies a lower bound of $\Omega(n^2)$ on the space, therefore on time complexity of any $(2,0)$-approximation algorithm. Compared to the $(2,1)$-approximation algorithm of Baswana and Kavitha [10], our algorithm has the advantage of being simpler and deterministic (albeit at a cost of logarithmic factor in the running time).

We extend this result by showing that by relying on fast Boolean matrix multiplication a better approximation could be achieved at the expense of a small increase in the running time. More specifically, for any $\epsilon > 0$ we provide a $(1+\epsilon, 2)$-approximation algorithm. Using the currently best known matrix multiplication algorithms yields an $O(n^{2.24+o(1)}\epsilon^{-3}\log(n\epsilon^{-1}))$ time bound for this algorithm. Moreover, since it is already known that distinguishing between distances 2 and 4 in unweighted graphs is as hard as Boolean matrix multiplication [9], we can't hope to obtain a similar running time for (say) a $(1+\epsilon, 2-3\epsilon)$-approximation algorithm without improving the current Boolean matrix multiplication bounds. As discussed earlier, Elkin [11] has another type of two-parameter approximation, as well as time/quality trade-off. However, it appears that his algorithm is faster only when the approximation quality is inferior. This is because his additive error term is,

$$W \cdot \left(\frac{8c_0}{\epsilon\zeta(\rho - \zeta/2)}\right)^{\lceil \log_{1-(\rho-\zeta/2)} \zeta/2 \rceil + 1} \lceil \log_{1-(\rho-\zeta/2)} \zeta/2 \rceil^{\lceil \log_{1-(\rho-\zeta/2)} \zeta/2 \rceil},$$

for a constant c_0.

We then turn our attention toward approximating the diameter and the radius of a graph. We first show that a variant of the algorithm proposed by Aingworth *et al.* [7] for approximating the diameter can be used for approximating the radius of unweighted undirected graphs. The algorithm gives an almost 3/2-approximation of the radius in $\tilde{O}(m\sqrt{n} + n^2)$ time.

We improve these results for the class of weighted separable graphs. We show if every subgraph of size k has a k^μ-separator, then a 3/2-approximation of the diameter can be achieved in $\tilde{O}(n^{1+\mu} + n^{3\mu})$ time. This result also extends to the case of directed graphs. We also present an algorithm that achieves almost 3/2-approximation of the radius in $\tilde{O}(n^{1+\mu} + n^{3\mu})$ time. As a consequence of these results, the fact that planar graphs have $O(\sqrt{n})$ separator [17], and the fact that single-source shortest path on planar graphs can be done in $O(n)$ time [18], we obtain $O(n^{3/2})$ time algorithms for 3/2-approximation of both the diameter and the radius of positive weighted planar graphs.

2 Preliminaries

For a weighted connected graph $G = (V, E)$ we use the following notation. We use $w_G : E \to \mathbb{R}^+$ to denote the weight function. For a set of vertices $U \subseteq V$, we define $d_G(u, U)$ as $\min_{v \in U} d_G(u, v)$ and $c_G(u, U)$ to be a (any, if more than one) vertex $w \in U$ with $d_G(u, w) = d_G(u, U)$. Similarly we define $f_G(u, U)$ to be a vertex $w \in U$ with $d_G(u, w) = \max_{v \in U} d_G(u, v)$.

We use $N_G(U)$ to denote the neighborhood of U in G. For solving the diameter and radius problems, we also define $min\_ecc(U, G) = \min_{u \in U} d_G(u, f_G(u, V))$ and $max\_ecc(U, G) = \max_{u \in U} d_G(u, f_G(u, V))$.

For a graph G, the center of the graph $cen(G)$ is a vertex of the graph with eccentricity equal to the graph radius. We use $rad(G)$ to denote the radius of G, and $dia(G)$ to denote the diameter of G. Note that $max\_ecc(V, G) = dia(G)$ and $min\_ecc(V, G) = rad(G)$.

In shortest path algorithms, we use a symmetric $n \times n$ distance matrix $\{\delta(u, v)\}_{u,v}$ to hold the currently best upper bound on distance between all pairs of vertices in G. We use *dijkstra*$((V, F), \delta, u)$ to denote an invocation of Dijkstra's single-source shortest path from vertex u on the graph (V, F). Every invocation of the algorithm updates the row and column entries of u in the distance matrix δ, provided the distance found during this run is smaller than previous estimates. Initially $\delta(u, v) = 1$, if $(u, v) \in E$ and ∞ otherwise. We omit the distance matrix argument from *dijkstra*() when not required. We use *bfs*$((V, F), u)$ to denote an invocation of breadth-first search from u on the graph (V, F).

A subset of vertices $S \subseteq V$ of a graph (V, E) is a λ-separator ($\lambda < 1$) if the largest connected component in $V \setminus S$ has at most $\lambda|V|$ vertices. A $[\lambda, \mu]$-separator decomposition of G is a recursive decomposition of G using separators, where subgraphs of size k have λ-separators of size $O(k^\mu)$ for $\mu \in (0, 1)$. Studied in this framework, the planar separator theorem due to Lipton and Tarajn [17] is a $[2/3, 1/2]$-separator decomposition. Henceforth, we call a graph *separable* if it admits a $[\lambda, \mu]$-separator decomposition.

2.1 Estimating Distances Using Dominating Sets

A set of vertices D is said to dominate a set of vertices U if every vertex U has a neighbor in D. The use of dominating sets for solving shortest path problems was first employed by Aingworth *et al.* [7]. The idea is based on the simple observation that there is a small set of vertices that dominates all the high degree vertices of a graph. Therefore, paths going to high degree vertices can be efficiently approximated by taking a small detour through the dominating set.

Cohen and Zwick [8] extended this result to the weighted case. For an input s, they have shown that a dominating set of size $O((n \log n)/s)$ can be constructed such that if $u \in V$ has degree at least s in G, then there is an edge $(u, v) \in E$ with $v \in D$, and (u, v) is one of the s lightest edges incident on u. We use $rank_u(u, v)$ and $rank_v(u, v)$ to denote the index of (u, v) in the sorted adjacency list of u

and v respectively. The following observation based on greedy approximation algorithm for the set cover problem is central to our results.

Lemma 1. *(Cohen and Zwick [8]) Let $G = (V, E)$ be a weighted undirected graph with n vertices and m edges. Let $1 \leq s \leq n$. A dominating set D of size $O((n \log n)/s)$ that dominate all vertices of degree at least s in the graph can be found in $O(m + n)$ time. Furthermore, if $u \in V$ is of degree at least s in G then there is an edge (u, v) with $v \in D$ such that $rank_u(u, v) \leq s$.*

We use an algorithm (details omitted) based on Lemma 1, called $dom(G, s)$. The algorithm receives $G = (V, E)$ and a degree threshold s as inputs, and outputs a set of vertices $D \subseteq V$ satisfying the properties of Lemma 1.

3 Approximation Algorithms for the APSP Problem

The idea behind the preprocessing step (function **preprocess**, Fig. 1) is to split the vertices into classes based on degree. The ith-class contains vertices with degrees between $n/2^i$ to $n/2^{i+1}$. We use Lemma 1 to find a dominating set D_i for the vertices of the ith-class. For a vertex u, $E_{|u}$ represents the set of all edges incident on u in G. The final step involves invoking Dijkstra from all the vertices in the dominating set D_i.

Function preprocess$(G = (V, E), k)$
for $i \leftarrow 0$ to k do $s_i \leftarrow n/2^i$
for $i \leftarrow 1$ to k do $E_i \leftarrow \{(u, v) \in E \mid rank_u(u, v) < s_{i-1}$ or $rank_v(u, v) < s_{i-1}\}$
for $i \leftarrow 1$ to k do $D_i \leftarrow dom(G, s_i)$
for $i \leftarrow 1$ to k do
$\forall u \in D_i$ call $dijkstra((V, E_i \cup E_{\vert u}), \delta, u)$

Fig. 1. Preprocessing function for approximating APSP

The dominating set D_i has a size at most $\min\{(n \log n)/s_i, n\}$ and the graph on which we run Dijkstra from the vertices in D_i has $O(ns_{i-1})$ edges. Therefore, the total time for the preprocessing step is $\tilde{O}(n^2)$.

3.1 (2, 1)-Approximation Algorithm

The algorithm $\mathbf{apasp}_{(2,1)}$ uses the distance matrix from the function **preprocess** to estimate the distances between every pair of vertices. In the final stages when the dominating set grows to linear size, the function **preprocess** just finds all-pairs shortest path in a graph with linear number of edges.

Theorem 1. *The algorithm $\mathbf{apasp}_{(2,1)}$ runs in $O(n^2 \log^2 n)$ time, where n is the number of vertices in the input graph $G = (V, E)$ and for every $u, v \in V$ we have $d_G(u, v) \leq \delta(u, v) \leq 2d_G(u, v) + h(u, v)$.*

```
Algorithm apasp(2,1)(G, δ)
call preprocess(G, ⌈log n⌉)
for every u, v ∈ V do
  for i ← 1 to ⌈log n⌉ do
    u' ← c_G(u, D_i) and v' ← c_G(v, D_i)
    δ(u, v) ← min{δ(u, v), δ(u, u') + δ(u', v), δ(v, v') + δ(v', u)}
```

Fig. 2. (2, 1)-approximation algorithm for the APSP problem

3.2 $(1 + \epsilon, 2)$-Approximation Algorithm

We now describe a simple algorithm that uses fast algorithms for rectangular matrix multiplication of Boolean matrices to obtain a $(1 + \epsilon, 2)$-approximation for APSP. Let W be largest-edge weight in the graph G, after the edge weights are scaled so that the smallest non-zero edge weight in 1, i.e, ratio of heaviest to lightest edge is W. For the sake of simplicity, we will first describe a $(1 + \epsilon, 2)$-approximation algorithm with a running time of $O(n^{2.24+o(1)}\epsilon^{-2} \log(nW\epsilon^{-1}))$. We will later use this algorithm as a sub-routine in the main algorithm.

Preliminary Algorithm: Let j be an integer with $0 \leq j \leq \lceil \log_{1+\epsilon} nW \rceil$. Now with a dominating set D_i, define Boolean matrices of dimensions $n \times |D_i|$ as

$$B_{i,j}[u, v] = 1 \text{ iff } (1 + \epsilon)^j \leq \delta(u, v) < (1 + \epsilon)^{j+1} \text{ for } u \in V \text{ and } v \in D_i.$$

We can ignore all empty matrices, i.e., which don't have at least a 1. For a matrix M, its transpose is denoted by M^T.

Theorem 2. *If Boolean matrix multiplication of $n \times \ell$ by $\ell \times n$ matrices can be performed in $n^{2-\alpha\beta+o(1)}\ell^\beta$ time for constants α, β, then for any $\epsilon > 0$, the algorithm* **apasp**$_{(1+\epsilon,2)}$ *runs in $\tilde{O}(n^{(2+3\beta-\alpha\beta)/(1+\beta)+o(1)} \log^2_{1+\epsilon}(nW))$ time, where n*

```
Algorithm apasp(1+ε,2)(G, δ)
p ← ⌈((1 + αβ) log n)/(1 + β)⌉
call preprocess(G, p)
construct matrices B_{i,j} for integers i, j with i ∈ [1, p] and j ∈ [0, ⌈log_{1+ε} nW⌉]
for i ← 1 to p do
  for j ← 0 to ⌈log_{1+ε} nW⌉ do
    for k ← j + 1 to ⌈log_{1+ε} nW⌉ do
      A_{i,j,k} ← Boolean matrix product of B_{i,j} and B^T_{i,k}
      for every u, v ∈ V do
        δ(u, v) ← { min{⌊(1 + ε)^{j+1} + (1 + ε)^{k+1}⌋, δ(u, v)}  if A_{i,j,k}[u, v] = 1,
                  { δ(u, v)                                       otherwise
Ê ← {(u, v) ∈ E | rank_u(u, v) < 2^p or rank_v(u, v) < 2^p}
for every u ∈ V call dijkstra((V, Ê), δ, u)
```

Fig. 3. $(1 + \epsilon, 2)$-approximation algorithm for the APSP problem

is the number of vertices in the input graph $G = (V, E)$. Also for every $u, v \in V$ we have $d_G(u, v) \leq \delta(u, v) \leq (1 + \epsilon)d_G(u, v) + 2h(u, v)$.

Remark: The dependence on ϵ in the running time can be reduced to $\epsilon^{-1} \ln \epsilon^{-1} \log_{1+\epsilon}(nW)$. The trick is to perform for every index $r \in \{1, 2, \ldots, \lceil \log_{1+\epsilon}(nW) \rceil\}$ a "Boolean OR" of matrices $B_{i,s}$ ($s \leq r - \epsilon^{-1} \ln \epsilon^{-1}$). Then we replace many matrix multiplications of form $B_{i,r} \times B_{i,s}$ by a single matrix multiplication of $B_{i,r}$ with the matrix constructed from the OR operation. The approximation ratio remains practically unchanged.

Let $\omega(1, x, 1)$ be the infimum over all exponents ω' for which $n \times n^x$ by $n^x \times n$ Boolean matrices can be multiplied in $O(n^{\omega'})$ time. Let $\omega = \omega(1, 1, 1)$. The currently best known algorithm for rectangular matrix multiplication from Coppersmith [19] and Huang and Pan [20] provide

$$\omega(1, x, 1) \leq \begin{cases} 2 & \text{if } 0 \leq x < \alpha, \\ 2 + \beta(x - \alpha) & \text{otherwise.} \end{cases}$$

This implies that $n \times \ell$ by $\ell \times n$ Boolean matrices can be multiplied in $n^{2-\alpha\beta+o(1)}\ell^{\beta}$ time. The constant α is defined as the supremum over all constants x for which $\omega(1, x, 1) = 2$. Currently, $\alpha > 0.294$, $\beta = \frac{\omega-2}{1-\alpha}$, $\omega < 2.376$. We immediately get the following corollary from the discussion above.

Corollary 1. *There exists an implementation of the algorithm* **apasp**$_{(1+\epsilon,2)}$ *that runs in $O(n^{2.24+o(1)}\epsilon^{-2} \log(nW\epsilon^{-1}))$ time.*

Main Algorithm: Let APSP(Λ) be a (auxiliary) problem in which APSP the ratio of the heaviest to lightest edge is bounded by Λ. For an instance of APSP(Λ) with n vertices (using **apasp**$_{(1+\epsilon,2)}$) we can compute a $(1 + \epsilon, 2)$-approximation in $O(n^2\phi(n, \epsilon, \Lambda))$ time, where $\phi(n, \epsilon, \Lambda) = n^{0.24+o(1)}\epsilon^{-2} \log(n\Lambda\epsilon^{-1})$. We will be needing the fact that ϕ is a function growing in n. We now describe an $O(n^2\phi(n, \epsilon, \Lambda)\epsilon^{-1} \ln n)$ time algorithm with a $((1 + \epsilon)^2, 2(1 + \epsilon))$-approximation.

In our method, given an input graph G with n vertices, we produce a set of instances of APSP($\epsilon^{-1}n$), say $G_1, \ldots, G_d$ with numbers of vertices $n_1, \ldots, n_d$ such that $n_i \leq n$, and $S_G = \sum_{i=1}^{d} n_i^2 \leq n^2\epsilon^{-1} \ln n$. The time needed to produce these instances and to combine their results will be $O(S_G)$, and the time needed to approximately solve these instances will be $O(\sum_i n_i^2\phi(n, \epsilon, \Lambda)) \leq S_G \cdot \phi(n, \epsilon, \Lambda)$. We define instances of APSP($\epsilon^{-1}n$) for distances in the range $(1+\epsilon)^k$ to $(1+\epsilon)^{k+1}$ (k integer) as follows. We assume w.l.o.g. that the minimum edge weight in G is 1, hence we consider only $k \geq 0$.

① remove edges with cost larger than $(1 + \epsilon)^{k+1}$;
② make a separate instance for each connected component;
③ coalesce vertices that are connected by edges shorter than $((1 + \epsilon)^k\epsilon)/n$ into super-vertices. If a vertex u is not coalesced with any other vertex, we view $\{u\}$ as a super-vertex of this instance;
④ eliminate instances with one vertex only.

Estimating S_G: We first decompose S_G into the sum of contribution of super-vertices: (a) in an instance G_l with n_l super-vertices, each super-vertex contributes n_l to n_l^2, (b) if super-vertex $\mathbf{u}$ in G_l is a set of g_l vertices, we decompose the contribution of $\mathbf{u}$ into g_l equal parts, n_l/g_l for each vertex in $\mathbf{u}$.

We say a vertex u is contained in an instance if there exists a super-vertex of the instance containing u. Now among the instances made for the distance range $[(1+\epsilon)^i, (1+\epsilon)^{i+1}]$, we use $G_i(u)$ to denote the instance containing vertex u (even though we may create many instances for a distance range, only one of them will contain u). We use $g_i(u)$ to denote the number of elements of the super-vertex of $G_i(u)$ that contains u. We denote the number of super-vertices of $G_i(u)$ as $n_i(u)$. The contribution of u to S_G at $G_i(u)$ is $\kappa_i(u) = n_i(u)/g_i(u)$. We want to show that the sum of all $\kappa_i(u)$'s is bounded by $n\epsilon^{-1} \ln n$. Note that for some values of i the instance $G_i(u)$ is not created, e.g., because of ④.

Let $N = \lceil \epsilon^{-1} \ln n \rceil$. The desired inequality holds if for every $j < N$ and every $u \in V$ we have the sum of all $\kappa_{j+iN}(u)$'s is bounded by n. This will bound the sum of all contributions to $n \times n \times N$.

Let $\mathbf{u}'$ be the super-vertex of u in the instance $G_{i+N}(u)$. Consider the instance $G_i(u)$. The key observation is that the union of the set of all super-vertices in the instance $G_i(u)$ is $\mathbf{u}'$. Therefore, $g_{i+N}(u) \geq n_i(u)$ and thus $n_i(u)/g_i(u) \leq g_{i+N}(u)/g_i(u)$. Note that $g_{i+N}(u) \geq g_i(u)$, and if $g_{i+N}(u) = g_i(u)$ then the instance $G_{i+N}(u)$ has one vertex, so it is not created, and thus there is no contribution to S_G.

Therefore, there exists an increasing sub-sequence $1 \leq \bar{g}_1 \leq \bar{g}_2 \ldots \leq \bar{g}_t \leq n$ of $g_i(u)$'s such that the sum of contributions of u for the distance ranges of the form $[(1+\epsilon)^{j+iN}, (1+\epsilon)^{j+iN+1}]$ is at most $\bar{g}_2/\bar{g}_1 + \bar{g}_3/\bar{g}_4 + \ldots \bar{g}_t/\bar{g}_{t-1}$. We can find the largest possible sum of this form as a function of n, say $F(n)$. By induction we show, $F(n) = n$. By considering every possible $\hat{g}$ for $\bar{g}_{t-1}$, we have $F(n) = \max_{\hat{g}}\{n/\hat{g} + F(\hat{g})\}$. It is easy to see that for $\hat{g} > 1$ we have $n/\hat{g} < n - \hat{g}$, so the sum is maximal if it consists of one term only, i.e., $n/1$.

Construction of the instances: The construction uses disjoint sets data-structure and is omitted in this extended abstract (refer [15] for the details). The time for the constructing all the instances is $O(S_G)$.

Combining the results: If we create an instance for distance $[(1+\epsilon)^l, (1+\epsilon)^{l+1}]$, and in that instance we compute, for some u, v a distance approximation larger than $(1+\epsilon+2)(1+\epsilon)^{l+1}$, then we know that the true distance between u and v is above $(1+\epsilon)^{l+1}$, and thus it will be properly estimated in another instance. Similar reasoning applies if the computed distance is smaller than $(1+\epsilon)^l$. Therefore, when we scan the array of results for such an instance, we perform updates only for pairs of super-vertices that have computed distances in the range $(1+\epsilon)^l$ to $(3+\epsilon)(1+\epsilon)^{l+1}$. Given such a pair of super-vertices, say $\mathbf{u}$,$\mathbf{v}$ with computed distance $L' \in [(1+\epsilon)^l, (3+\epsilon)(1+\epsilon)^{l+1}]$, for each $u \in \mathbf{u}$ and $v \in \mathbf{v}$ we update (unless a smaller estimate was already present) the distance from u to v with $L' + (1+\epsilon)^l\epsilon$. The term $(1+\epsilon)^l\epsilon$ is needed to correct for the possible effect of collapsing edges of length less than $((1+\epsilon)^l\epsilon)/n$.

As a result, the time needed to combine the result is the time needed to read the result matrices, which equals to S_G, plus the number of updates in the matrix of final results, and we perform at most $\epsilon^{-1}\ln((1+\epsilon)(3+\epsilon))$ updates for each entry. Since the above discussion holds for any $\epsilon > 0$, we have,

Theorem 3. *Let G be a weighted undirected graph on n vertices. For any $\epsilon > 0$, there exists an $O(n^{2.24+o(1)}\epsilon^{-3}\log(n\epsilon^{-1}))$ time algorithm that for every $u, v \in V$ produces an estimate $\delta(u,v)$ with $d_G(u,v) \leq \delta(u,v) \leq (1+\epsilon)d_G(u,v) + 2h(u,v)$.*

4 Approximating the Radius of Graphs

Aingworth *et al.* [7] presented a 3/2-approximation algorithm for estimating the diameter of weighted directed graphs. In this section we extend their algorithm and show that it can be used to obtain an almost 3/2-approximation of the radius of unweighted undirected graphs. The algorithm is presented in Fig. 4. We assume that $rad(G) \geq 2$, as $rad(G) = 1$ can be easily handled separately.

A s-partial breadth-first search is obtained by performing the breadth-first search from a vertex to the point where exactly s vertices (excluding the starting vertex) have been visited. A s-partial breadth-first from u on graph (V,F) is denoted by s-*bfs*$((V,F),u)$. Let $PBFS(u)$ denote the set of vertices which are visited by an invocation of $\sqrt{n\log n}$-*bfs*(G,u). The size of the dominating set D constructed in the algorithm $\mathbf{rad_{3/2}}$ is $O(\sqrt{n\log n})$ (follows from Lemma 1). The following theorem completes the analysis of the algorithm.

Algorithm $\mathbf{rad_{3/2}}(G,\delta)$
for every $u \in V$ call $\sqrt{n\log n}$-*bfs*(G,u)
$w \leftarrow$ the vertex having the maximum depth partial breadth-first search tree
for every $u \in PBFS(w)$ call *bfs*(G,u)
$\widehat{G} \leftarrow G$ with additional edges of the form (u,v) where either $u \in PBFS(v)$ or $v \in PBFS(u)$
$D \leftarrow dom(\widehat{G}, \sqrt{n\log n})$
for every $u \in D$ call *bfs*(G,u)
output $\min\{min\_ecc(D,G), min\_ecc(PBFS(w),G)\}$

Fig. 4. Almost 3/2-approximation of the radius

Theorem 4. *The algorithm $\mathbf{rad_{3/2}}$ runs in $O(m\sqrt{n}\log n + n^2\log n)$, where n is the number of vertices and m is the number of edges in the input graph G, and gives an estimate of the radius that is at most $\lceil\frac{3}{2}rad(G)\rceil$.*

5 Approximating Diameter, Radius of Separable Graphs

In this section we present algorithms for faster estimation of diameter and radius of graphs having a $[\lambda,\mu]$-separator decomposition, where the decomposition is either provided as part of the input or is quickly obtainable. For most of the well-known separable graphs, the latter condition holds true. We start by proving a general statement about the maximum number of edges that a separable graph can have. Earlier known results had a weaker upper bound of $O(n + n^{2\mu})$ on the number of edges (see for example Cohen [21]).

Lemma 2. *Let G be a graph with n vertices and a $[\lambda, \mu]$-separator decomposition. Then number of edges in G is $O(n)$.*

We use a rooted binary tree T_G to represent a separator decomposition of G (as in [21]). To avoid ambiguities, we refer to the vertices of a graph as *vertices* and vertices of a separator tree as *nodes*. Let $root(T_G)$ be the root node of T_G. Each node $t \in T_G$ is labeled by two subsets of vertices $V(t) \subseteq V$ and $S(t) \subseteq V(t)$. Let $G(t) = (V(t), E(t))$ denote the subgraph induced by $V(t)$. Then $S(t)$ is the separator in $G(t)$. Then $V(root(T_G)) = V$ and $S(root(T_G))$ is a separator in G. For any $t \in T_G$, the labels of its children t_1, t_2 are defined as follows: Let $V_1 \subset V(t)$ and $V_2 \subset V(t)$ be the components separated by $S(t)$ in $G(t)$. Then $V(t_1) = V_1 \cup (S(t) \cap N_G(V_1))$, $V(t_2) = V_2 \cup (S(t) \cap N_G(V_2))$.

We associate *boundary vertices*, $B(t)$ with each node t. The boundary of the $root(T_G)$ is $\emptyset$. The boundary of every other node t is defined as $B(t) = S(p(t)) \cup B(p(t)) \cap V(t)$, where $p(t)$ is the parent of t in T_G. We now describe the preprocessing stage for the algorithms.

Function sep-preprocess(G, t)
create a weighted graph $H(t) = (V(t), E(t) \cup B(t) \times B(t))$, where for an edge $(a, b) \in B(t) \times B(t)$, $w_{H(t)}(a, b) = d_G(a, b)$ and $w_{H(t)}(e) = w_G(e)$ for $e \in E(t)$
for every $u \in S(t)$ call *dijkstra*$(H(t), u)$
create a related graph $\widehat{H}(t)$ (from $H(t)$): merge all vertices of $S(t)$ into a single vertex ϑ and keep all edges in $H(t)$, including parallel edges
call *dijkstra*$(\widehat{H}(t), \vartheta)$ to determine $f_{H(t)}(\vartheta, V(t) \setminus S(t))$

Fig. 5. Preprocessing function for diameter, radius approximation of separable graphs

The algorithm **sep-preprocess** (Fig. 5) does Dijkstra from the vertices in $S(t)$ on a weighted graph $H(t)$. The following lemma shows that the graph $H(t)$ preserves the shortest distance between every pair of vertices from $V(t)$.

Lemma 3. *For any $t \in T_G$ and $u, v \in V(t)$, $d_{H(t)}(u, v) = d_G(u, v)$.*

3/2-approximation of the Diameter: We present an algorithm for weighted undirected graphs. The extension to the directed case is omitted in this extended abstract (see [15] for details). Note that a simple consequence of Lemma 2 is that the diameter approximation algorithm of Aingworth *et al.* [7] runs in $\tilde{O}(n^2)$ time on separable graphs.

We assume that graph is strongly connected. The algorithm **sep-dia$_{3/2}$** (Fig. 6) operates on all nodes in the separator decomposition tree (T_G). For every $t \in T_G$, the function **sep-preprocess** is invoked. One can inductively see

```
Algorithm sep-dia_{3/2}(G, T_G) (G is a separable graph)
for every t ∈ T_G do
    call sep-preprocess(G, t)
    max_1(t) ← max_ecc(S(t), H(t))
    max_2(t) ← max_ecc(f_{H(t)}(ϑ, V(t) \ S(t)), H(t))
    max(t) ← max{max_1(t), max_2(t)}
output max{max(t) | t ∈ T_G}
```

Fig. 6. 3/2-approximation of the diameter of separable graphs

that weights for constructing the graph $H(t)$ is available. For $t = root(T_G)$ it is true. Now consider an edge (v_1, v_2) in $H(t)$. If either of v_1 or v_2 is in $B(p(t))$, inductively we know that weights are available. Otherwise, both v_1 and v_2 are in $S(p(t))$ and the weights are again available (refer Lemma 3).

```
Algorithm sep-rad_{3/2}(G, T_G) (G is a separable graph)
t ← root(T_G)
𝒮 ← ∅
while V(t) ≠ ∅
    call sep-preprocess(G, t)
    𝒮 ← 𝒮 ∪ S(t)
    choose i such that f_{H(t)}(ϑ, V(t) \ S(t)) ∈ V(t_i)
    t ← t_i
for every u ∈ 𝒮 call dijkstra(G, u)
output min_ecc(𝒮, G)
```

Fig. 7. Almost 3/2-approximation of the radius of separable graphs

Theorem 5. *Let G be a weighted undirected separable graph. The algorithm* **sep-dia$_{3/2}$** *runs in $O(n^{1+\mu} \log n + n^{3\mu} \log n)$ time, where n is the number of vertices in G, and gives an estimate of the diameter which is at least $\frac{2}{3}dia(G)$.*

3/2-approximation of the Radius: The algorithm **sep-rad$_{3/2}$** (Fig. 7) follows one path down the separator decomposition tree. For every node t in the path, the function **sep-preprocess** is invoked. As with **sep-dia$_{3/2}$**, we can inductively show that the weights needed for construction of the graphs $H(t)$ are available.

Theorem 6. *Let G be a weighted undirected separable graph. The algorithm* **sep-rad$_{3/2}$** *runs in $O(n^{1+\mu} \log n + n^{3\mu} \log n)$ time, where n is the number of vertices in G, and gives an estimate of the radius which is at most $\lceil \frac{3}{2}rad(G) \rceil$.*

Acknowledgement

The authors would like to thank Martin Fürer for many stimulating discussions. We would also like to thank Surender Baswana for pointing us to references [10] and [11], and Timothy Chan for providing us a preliminary copy of [1].

References

1. Chan, T.M.: More algorithms for all-pairs shortest paths in weighted graphs. In: STOC '07, ACM (to appear)
2. Galil, Z., Margalit, O.: All pairs shortest distances for graphs with small integer length edges. Information and Computation 134(2), 103–139 (1997)
3. Galil, Z., Margalit, O.: All pairs shortest paths for graphs with small integer length edges. JCSS 54(2), 243–254 (1997)
4. Seidel, R.: On the all-pairs-shortest-path problem in unweighted undirected graphs. JCSS 51 (1995)
5. Coppersmith, D., Winograd, S.: Matrix multiplication via arithmetical progressions. Journal of Symbolic Computation 9, 251–280 (1990)
6. Zwick, U.: Exact and approximate distances in graphs - A survey. In: Meyer auf der Heide, F. (ed.) ESA 2001. LNCS, vol. 2161, pp. 33–48. Springer, Heidelberg (2001)
7. Aingworth, D., Chekuri, C., Indyk, P., Motwani, R.: Fast estimation of diameter and shortest paths (without matrix multiplication). SIAM Journal on Computing 28(4), 1167–1181 (1999)
8. Cohen, E., Zwick, U.: All-pairs small-stretch paths. Journal of Algorithms 38(2), 335–353 (2001)
9. Dor, D., Halperin, S., Zwick, U.: All-pairs almost shortest paths. SIAM Journal on Computing 29(5), 1740–1759 (2000)
10. Baswana, S., Kavitha, T.: Faster algorithms for approximate distance oracles and all-pairs small stretch paths. In: FOCS '06, IEEE, pp. 591–602 (2006)
11. Elkin, M.: Computing almost shortest paths. ACM Transactions on Algorithms 1(2), 283–323 (2005)
12. Chan, T.M.: All-pairs shortest paths for unweighted undirected graphs in o(mn) time. In: SODA '06, ACM, pp. 514–523 (2006)
13. Boitmanis, K., Freivalds, K., Ledins, P., Opmanis, R.: Fast and simple approximation of the diameter and radius of a graph. In: Àlvarez, C., Serna, M. (eds.) WEA 2006. LNCS, vol. 4007, pp. 98–108. Springer, Heidelberg (2006)
14. Eppstein, D.: Subgraph isomorphism in planar graphs and related problems. Journal of Graph Algorithms and Applications 3(3) (1999)
15. Berman, P., Kasiviswanathan, S.P.: Faster approximation of distances in graphs (2007) Available at `http://www.cse.psu.edu/~kasivisw/fadig.pdf`
16. Thorup, M., Zwick, U.: Approximate distance oracles. Journal of ACM 52(1), 1–24 (2005)
17. Lipton, R.J., Tarjan, R.E.: A separator theorem for planar graphs. SIAM Journal of Applied Mathematics 36, 177–189 (1979)
18. Henzinger, M.R., Klein, P.N., Rao, S., Subramanian, S.: Faster shortest-path algorithms for planar graphs. JCSS 55(1), 3–23 (1997)
19. Coppersmith, D.: Rectangular matrix multiplication revisited. Journal of Complexity 13(1), 42–49 (1997)
20. Huang, X., Pan, V.Y.: Fast rectangular matrix multiplication and applications. Journal of Complexity 14(2), 257–299 (1998)
21. Cohen, E.: Efficient parallel shortest-paths in digraphs with a separator decomposition. Journal of Algorithms 21(2), 331–357 (1996)

Approximate Shortest Paths Guided by a Small Index*

Jörg Derungs, Riko Jacob, and Peter Widmayer

Institute of Theoretical Computer Science, ETH Zurich, Switzerland
ETH Zentrum, CH-8092 Zürich
{joerg.derungs, riko.jacob, peter.widmayer}@inf.ethz.ch

Abstract. Distance oracles and graph spanners are excerpts of a graph that allow to compute approximate shortest paths. Here, we consider the situation where it is possible to access the original graph in addition to the graph excerpt while computing paths. This allows for asymptotically much smaller excerpts than distance oracles or spanners. The quality of an algorithm in this setting is measured by the size of the excerpt (in bits), by how much of the original graph is accessed (in number of edges), and the stretch of the computed path (as the ratio between the length of the path and the distance between its end points). Because these three objectives are conflicting goals, we are interested in a good trade-off. We measure the number of accesses to the graph relative to the number of edges in the computed path.

We present a parametrized construction that, for constant stretches, achieves excerpt sizes and number of accessed edges that are both sublinear in the number of graph vertices. We also show that within these limits, a stretch smaller than 5 cannot be guaranteed.

1 Introduction

We study the problem of answering approximate shortest path queries on an edge weighted, undirected graph $G = (V, E)$ (called the *base graph* from now on) with n vertices and m edges, where the vertices are labeled with unique bit strings of length at most $\lceil \ell \log n \rceil$ for an $\ell \geq 1$. For practical purposes we also assume that all edge weights are *ℓ-limited precision rational numbers* whose numerators and denominators can be stored with $\lceil \ell \log n \rceil$ bits. In our specific setting, we are allowed to preprocess the (very large) base graph and store information about it (viewed as a bit string) in a memory of severely limited size; we call this an *excerpt* of the graph, and the number of bits the excerpt *size*. A path query is specified by the labels of both end vertices; the answer to the query is a path in G, returned as a sequence of vertex labels. No information can be passed from one query to the next. For answering a path query, we may read from the excerpt at no cost. In addition we may access the graph by *probing* a vertex, which *reports*

* This work has been partially supported under the EU programme COST 295 (DYNAMO) and by the Swiss SBF under contract number C05.0047.

F. Dehne, J.-R. Sack, and N. Zeh (Eds.): WADS 2007, LNCS 4619, pp. 553–564, 2007.

a single edge incident to this vertex at unit cost. In a sequence of probes of the same vertex, the incident edges are returned in order of increasing weights, with equal weight edges in the same order within any probe sequence. A vertex can be *reset* at any time to its *start state*, with the effect that probing starts again at the adjacent edge with the smallest weight. When a path query starts to be processed, all vertices are in their start states. The total cost of accessing the graph for a path query (the total *probe cost*) is the number of probes for that query. Because paths with more edges tend to need more probes, we define the efficiency of probing as the ratio of the probe cost against the number of edges in the returned path, the *probe factor*.

We are interested in the trade-off between excerpt size, probe factor, and approximation ratio of the length of the returned path against the shortest path length in the base graph (also known as the *stretch*). While we are interested in the asymptotic limits of the trade-off in arbitrarily large graphs, we still give concrete values for the trade-off parameters and their limits. Our focus is on showing that graph excerpts and path finding algorithms with the desired properties exist. Still, our proofs are constructive, and the presented and implied algorithms all run in polynomial time.

We came across problems of this nature in the study of public key authentication, but we believe our problem is of interest in other domains as well, such as for external memory and caching. In public key authentication, the certificate of a public key can be retrieved from the organization that issued that certificate. These certificates can be interpreted as the edges of a huge implicit graph. An important and potentially time-consuming part of asserting that a key belongs to an alleged owner translates into finding a path in this graph.

Short paths by means of small excerpts have attracted attention earlier: Graph spanners [16] and approximate distance oracles [20] solve the special case where (expressed in our setting) the allowed probe cost for computing a path is zero, that is, the path has to be computed from the excerpt alone, without access to the base graph. Both the graph spanner and the approximate distance oracle allow for a trade-off between the two conflicting goals of keeping excerpts as well as stretch small. For graphs on n vertices, $O(n \log n)$ bits are sufficient to store a $\log n$-spanner. On the other hand, any excerpt from which paths (of arbitrary stretch) need to be computed without further information (such as base graph accesses) requires at least $\Omega(n \log n)$ bits, as shown in Section 3.1. We are therefore interested in trade-offs where the excerpt size is $o(n \log n)$ bits. Without a limit on the probe factor we can easily restrict the excerpt size. Thus, in our trade-off we limit the probe factor to $o(n)$. This does not even allow the path finding algorithm to exlore the whole neighbourhood of the returned path in dense graphs.

To keep our modelling reasonably simple, we do not restrict the path finding algorithm to be time or space efficient. This strengthens the impossibility results of Section 3. Our construction based on the index graph is not exploiting this loophole, it only computes shortest paths on the index graph and on explored subgraphs with r edges. In particular, for the choice of parameters that achieves sublinear index size and sublinear probe factor, this computation time (per edge in the reported graph) and computation space remain sublinear in the number of nodes of the base graph.

1.1 Results

In Section 2, we define a path finding algorithm and a corresponding excerpt, the *index graph.* We give two trade-off parameters for the construction of the index graph. The first, r, limits the probe factor to $O(r)$, the second, σ, limits the stretch to $O(\sigma)$. With these parameters we get an index graph with $\tilde{n} \leq \min\left(\frac{n \log n}{\sqrt{r}}, n\right)$ nodes and $\tilde{m} \leq \tilde{n}^{1+2/\sigma}$ edges, which can be stored with $O(\tilde{m}\ell \log n)$ bits. As mentioned above, we are interested in trade-offs with both excerpt size and probe factor sublinear in the number of vertices. And indeed, for every $\varepsilon, 0 < \varepsilon < \frac{1}{2}$, we can set $r \geq n^{2\varepsilon}$ and $\sigma \geq 2\left(\frac{1-\varepsilon}{\varepsilon}\right)$ to achieve sublinear size and probe factor and still guarantee a constant stretch.

Our construction can be seen as a natural extension of graph spanners and distance oracles. For probe factor 0, it results in a σ-spanner of the base graph.

Beside the index graph construction, we also provide bounds on the trade-off that can be achieved in different situations. Specifically, we show that for a stretch smaller than 5 (Section 3.3), or if the order in which the edges are accessed is independent of their weights (Section 3.2), the probe factor cannot be smaller than $\frac{n}{8}$, even if the excerpt size is in the same order of magnitude as a graph spanner, namely $\frac{n \log n}{8}$. Thus, the desired trade-off with sublinear size and sublinear probe factor cannot be achieved in these cases.

1.2 Related Work

The problem of finding shortest or short paths in a graph received a lot of attention in the last decades, see for example the recent survey by Zwick [21]. The role of limited amount of working storage space has been investigated, among others, for approximate distance oracles, external memory data structures, and the streaming model, as detailed below.

The approximate distance oracle [20] chooses a $k \geq 1$, uses a data structure of size $O(kn^{1+1/k})$ and returns an approximate shortest path in constant time per path edge. The returned path is at most $2k-1$ times larger than the true distance. The construction of the data structure takes $O(kmn^{1/k})$ expected time, which was recently improved to $\tilde{O}(n^2)$ time [4]. Note that the data structures in approximate distance oracles are always of size $\Omega(n \log n)$ bits, whereas our focus are excerpts that use asymptotically less bits than that.

Another possibility is the relabeling of the entire graph in a preprocessing step [7]. There, the two labels of the source and the sink are sufficient to answer (approximate) distance queries. In this setting, the total space usage of the labels is necessarily $\Omega(n\sqrt{m})$ for some graphs, and is always super-linear in the number of vertices.

Along a different line of thought, external memory algorithms and data structures have been designed for shortest path computations, among other graph problems. See [14] for a recent overview and [15] for one of the latest shortest path algorithms. For external memory algorithms, the assumption is that main memory is limited, and external memory accesses to storage blocks are costly. Distance oracles reduce the required size of main memory, but still they are only suited for what is known as semi-external memory graph algorithms, namely situations in which the memory is big enough to store some information about every node of the graph. In this respect, our approach extends the possibilities since it uses even less space. In contrast to our model, the general external memory model has the focus on algorithms for very limited size memory and arbitrarily large graphs, and assumes that the external memory is organized in blocks (cache-lines). Our setting can be understood as a situation where the algorithm is not able to change the representation of the graph in external memory but can extract a small graph excerpt in an expensive preprocessing phase. Then, a probe corresponds to an access to external memory that retrieves the next block of edges. Hence, our results can be understood in the external memory set-up, but are certainly different in focus from the well established external memory algorithms and data structures.

One of the most recent areas with a strong emphasis on storage efficiency is the streaming model of computation, where only little information can be maintained while a long stream of data passes by. While this area started out with simple statistical and aggregate questions, it has matured to include general algorithmic problems, such as computing some graph property when the huge graph is given as a sequence of edges in arbitrary order. Interestingly, for many of these properties, such as connectedness, which is at the basis of short path computations, at least $\Omega(n)$ bits need to be stored if streams can only be read [11]. The *W-Stream* model [8], where intermediate streams are written in one pass to be read in the next pass, allows for a trade-off between memory and number of passes for single source shortest path. Such an intermediate stream can be viewed as an excerpt of the graph. Nevertheless, one cannot apply these concepts in our setting, since in the first pass the whole graph is read, the size of the intermediate streams may be in the order of the number of graph edges, and the content of the intermediate streams depends on the query and not only on the graph.

Exact and approximate shortest path acceleration [9] can be regarded as closely related to our problem. Shortest path acceleration tries to answer several path queries on the same graph efficiently. To that end, the graph is preprocessed and some data structure is stored additionally to the graph. This preprocessing should also be completed with low running time. In contrast to shortest path acceleration, we try to limit the number of graph edges that need to be examined to answer path queries, and we do not, at this point, consider the cost or running time for preprocessing. On the other hand, the data structure generated in the preprocessing of the shortest path acceleration may be as large as the graph itself, whereas in our model the size of this data structure should be really small, i.e., of size $o(n)$. In particular, the techniques used for shortest path acceleration are not directly applicable to our model. For instance, Goldberg and

Harrelson [13] use a vertex labelling in the preprocessing to be used for triangulation during path queries. This is not feasible in our model since the vertex labels require $\Omega(n \log n)$ space. Sanders and Schultes [19] present a different approach, so called highway hierarchies. While the space needed to store the additional data is considerably smaller than the one needed by vertex labelling, the authors observe that the space is "a small constant factor of the input size" [19].

2 Index Graph and Path Finding Algorithm

In this section, we present an algorithmic solution for the problem of finding approximate shortest paths guided by a small index, or graph excerpt. We describe mainly the mathematical structure of our excerpt. The algorithm to construct it follows directly from this structure. First we describe roughly the path finding algorithm, and then the undirected weighted *index graph*, the graph excerpt that supports such a path finding algorithm. Details on the path finding algorithm and the index graph are given in Section 2.1 and 2.2, respectively.

Note that because we are interested in trade-offs with sublinear excerpt size, not all vertices are represented in the index graph. Thus, the path finding algorithm computes a path between two vertices s and t in the following way. First, it explores the vicinities of s and t in the base graph to find two *index nodes*, i.e., graph vertices represented in the index graph, that are closest to s and t, respectively. Then it computes a preliminary path in the index graph between the two index nodes. This preliminary path is a path with gaps. Finally, the path finding algorithm closes the gaps by exploring the base graph and thus obtains a path between s and t in the base graph.

The two trade-off parameters, r and σ, are not explicitly used in the path finding algorithm. Instead, the structure of the index graph guarantees that the probe cost for finding an index node in the vicinity of any vertex, and for closing a gap in the preliminary path, are bounded by r resp. $2r$. The index graph also limits the stretch of the obtained path to $9\sigma + 2$ (Lemma 9 and Lemma 3).

The main ingredient of constructing the index graph is the selection of the index nodes. As indicated above, we want enough index nodes such that we can find a shortest path from any vertex in the graph to an index node while accessing at most r edges. On the other hand, the limit of the excerpt size imposes a limit on the number of index nodes. To specify a "good" set of index nodes, we introduce the relation $\text{near}_r(u, v)$ over pairs of vertices that is parametrized with the probe cost limit r. We postpone the technical definition of this r-near relation. For now, it is sufficient that if $\text{near}_r(u, v)$, then with probe cost at most r we can find all shortest paths from u to v and to all vertices closer to u than v. Because every vertex is r-near to at least $\sqrt{r}$ other vertices (Lemma 5, page 560), we can compute a set of at most $\frac{n \log n}{\sqrt{r}}$ index nodes in polynomial time (Lemma 7) such that every vertex is r-near an index node. In a similar way as the *neighbourhood* of a vertex v is the set of vertices to which there is an edge from v, we call the set of vertices to which v is r-near the *vicinity* of v; the vertex v is the *center* of the vicinity of v. Note that every vertex is in its own vicinity.

The r-near relation is not only useful to select the index nodes, but also to construct the *index edges*, which represent paths in the base graph. We connect two index nodes v_1 and v_2 if there is an edge $\{u_1, p_2\}$ in the base graph such that a vertex p_1 is r-near v_1 and u_1, and p_2 is r-near v_2. Exploring from the vertices p_1 and p_2, the path finding algorithm can close the gap between v_1 and u_1 and between v_2 and p_2 with probe cost at most $2r$. The weight of the edge is the length of the path from v_1 to v_2 that results from closing the gaps, and the edge is annotated with the vertices p_1, u_1, and p_2. If the path from v_1 to v_2 has less than 6 edges, we annotate the index edge with the whole path. Note that p_1 may be the same vertex as u_1. In particular, with an edge $\{v_1, v_2\}$ in the base graph, $p_1 = u_1 = v_1$ and $p_2 = v_2$.

The limit on the probe cost is not the only feature based on these index edges. Lemma 9 shows that the distance between two index nodes in the *full index graph*, i.e., the graph induced by all index edges, is at most 3 times their distance in the base graph.

Even if we eliminate multiple edges between index nodes, the full index graph may still be arbitrarily dense, causing a large excerpt. Therefore, our *index graph* $I_r^\sigma(G)$ of a base graph G is a greedy σ-spanner [16] of the full index graph.

With Definition 4 (page 560) of the r-near relation, which is based on the exploration of a shortest path tree, we can give the following upper bounds for the size of the index graph and the performance of the path finding algorithm:

Theorem 1. *For all graphs $G = (V, E)$ on n vertices with unique $\lceil \ell \log n \rceil$ bit vertex labels and ℓ-limited precision rational edge weights, for all integer values $r \le n$ and $\sigma, 1 \le \sigma \le \log n$, the index graph $I_r^\sigma(G)$ and the corresponding path finding algorithm $\mathcal{A}$ with input $(s, t, I_r^\sigma(G))$ that reports a path $p(s,t)$ from any vertex s to any vertex t in G have the following properties:*

1. *the index graph $I_r^\sigma(G)$ has $\tilde{n} \le \min\left(\frac{n \log n}{\sqrt{r}}, n\right)$ nodes and $\tilde{m} \le \tilde{n}^{1+2/\sigma}$ edges and can be stored with $O(\tilde{m}\ell \log n)$ bits*
2. *the length of the path $p(s,t)$ reported by the algorithm $\mathcal{A}$ is less than $(9\sigma + 2) \cdot \operatorname{dist}(s,t)$*
3. *the probe cost of the algorithm $\mathcal{A}$ to compute path $p(s,t)$ is at most $2r$ if $p(s,t)$ has less than 6 edges; the probe factor is at most $r/2$ if $p(s,t)$ has at least 6 edges.*

Note that for graphs with positive integer edge weights, the probe factor can also be limited relative to the length of the returned path.

The remainder of this section gives the details of the path finding algorithm and the index graph construction, and defines the r-near relation. Theorem 1 follows directly from these details.

2.1 Path Finding Algorithm

In this section, we formalize the details of the path finding algorithm and present the lemmas used to prove Points 2 and 3 of Theorem 1.

To find a path from a vertex s to a vertex t, the algorithm starts exploring from s and t until an index node $\tilde{s}$ and $\tilde{t}$ is reached, respectively. Note that if $\text{near}_r(s,t)$ or $\text{near}_r(t,s)$, then the algorithm has already found a shortest path from s to t. Next, the algorithm computes a shortest path from $\tilde{s}$ to $\tilde{t}$ in the index graph. For every index edge on that path, the corresponding path in the base graph is computed by exploring from the annotated vertices p_1 and p_2, if it cannot be read directly from the index edge.

We use the term *expanding an index edge* for computing a path in the base graph that connects the two index nodes incident to the index edge. Lemma 2, the limit for the probe factor, is based on the observation that if an index edge is expanded, the corresponding part of the returned path has at least 6 edges.

Lemma 2 (without proof). *The probe factor of the path finding algorithm to compute a path p between two graph vertices is at most $r/2$ if p has at least* 6 *edges; the total probe cost is at most $2r$ if p has less than* 6 *edges.*

The following lemma is based on the observation that if one end point of the path query is closer to the other end point than to an index node, the path finding algorithm will find and return a shortest path.

Lemma 3 (without proof). *If the distance between two index nodes in the index graph $I_r^\sigma(G)$ is at most k times their distance in the base graph, then the length of any path returned by the path finding algorithm is at most $3k+2$ times the distance between the two end points.*

It remains to define an r-near relation that allows the path finding algorithm to find shortest paths from v to all vertices in the vicinity of v, while accessing at most r graph edges. The properties of r-near suggest that we define r-near based on the shortest path tree, which can be computed with Dijkstra's shortest path algorithm [10]. Note that the shortest path algorithm operates on directed graphs. This serves our purpose, since any edge $\{u,v\}$ is reported both by a probe of u and by a probe of v.

However, the shortest path algorithm in its classical formulation examines all arcs of one vertex at once. In a high degree graph, this can induce probe costs higher than any $r \in o(n)$. Therefore, we modify Dijkstra's algorithm slightly to evaluate edges lazily, that is, to only read those arcs that are really needed to determine the next vertex to be added to the shortest path tree. Similar methods to speed up Dijksta's shortest path algorithm by limiting the number of edges in the priority queue have been proposed before, see [1] for an example. The main goal in our case, however, is to limit the number of examined edges. For every vertex u we only need one out-going arc $(u \to v)$ in the priority queue at a time. Remember that we access the edges incident to one vertex in the order of increasing weights. Thus, examining and possibly adding an arc $(u \to w)$ with greater or equal weight can be postponed until $(u \to v)$ is removed from the queue. Specifically, when a vertex u (including the root) is added to the shortest path tree, we add only the first arc reported by a probe of u to the priority queue. Whenever an arc $(u \to v)$ is removed from the queue, we add the next

Algorithm 1. Shortest path with lazy edge evaluation

```
procedure EXPLORE(vertex source, vertex dest)
    add first outgoing arc of source to priority queue
    while dest not in shortest path tree do
        (u → v) ← arc on top of priority queue
        remove (u → v) from priority queue
        if v not in shortest path tree then
            add v to shortest path tree with edge {u, v}
            quit if v = dest
            add first outgoing edge of v to priority queue
        end if
        add next outgoing edge of u to priority queue
    end while
end procedure
```

arc reported by a probe of u. See Algorithm 1 for a formal description of the modified shortest path algorithm.

We want the vertices to be added to the shortest path tree in a deterministic order, for situations in which many vertices have the same distance to the root of the tree, and we want the root to be r-near some but not all of them. For simplicity, we use the label of the arc's target vertex as a tie-breaker, even though using a deterministic priority queue would suffice.

From here on, `Explore(u,v)` refers to the modified shortest path algorithm with source u and destination v. The definition of r-near naturally follows from `Explore`.

Definition 4. *A vertex u is r-near another vertex v, $\text{near}_r(u,v)$, if and only if* `Explore(u,v)` *adds at most r edges to the priority queue.*

Based on this definition, we can bound the size of the vicinity of any vertex.

Lemma 5. *A vertex is r-near at least $\lceil\sqrt{r}\rceil$ vertices, including itself.*

Proof. Assume that `Explore(u,v)` is about to add v as the $(k+1)$st node to the shortest path tree. At that point, at most k^2 arcs can have been added to the priority queue: $k(k-1)$ arcs between tree nodes, and k arcs from tree nodes to vertices that are not in the tree yet (including v). For $k = \lceil\sqrt{r}\rceil - 1 < \sqrt{r}$ we get $k^2 < r$ and therefore $\text{near}_r(u,v)$. □

Lemma 6. *A vertex is r-near at most $r+1$ vertices, including itself.*

Proof. For every edge that is added to the priority queue in `Explore`, at most one vertex can be added to the shortest path tree. □

2.2 Index Graph

In this section, we describe the details of the index graph construction, and present the lemmas used to prove Point 1 of Theorem 1.

As outlined at the beginning of Section 2, we first select the index nodes, then compute the index edges, and lastly build the index graph $I_r^\sigma(G)$ as a greedy σ-spanner [16] of the full index graph induced by the index edges.

Lemma 7. *In every graph G on n vertices there is a set $\tilde{V}$ of at most $\min\left(\frac{n\log n}{\sqrt{r}}, n\right)$ vertices such that every vertex in G is r-near a vertex in $\tilde{V}$. The set $\tilde{V}$ can be computed in $O(rn\log n)$ time.*

Proof. The size limit n is trivial because the set $\tilde{V}$ cannot contain more vertices than the graph G, and every vertex is r-near itself.

The set $\tilde{V}$ of vertices can be interpreted as a Set Cover [12] solution. The graph vertices form the universe of the Set Cover instance, and for every vertex v there is a set containing all vertices that are r-near to v. This Set Cover instance has n elements and n sets. Every element is in at least $\sqrt{r}$ sets because every vertex is r-near at least $\sqrt{r}$ vertices (Lemma 5). Thus, with Lemma 8, Greedy Set Cover [6] computes a set $\tilde{V}$ with at most $(\log_2(\sqrt{r})+1)\,\frac{n}{\sqrt{r}} \leq \frac{n\log n}{\sqrt{r}}$ vertices.

The sets of the Set Cover instance can be computed in $O(r\log r)$ per vertex using `Explore`. The upper bound of $r+1$ for the sets (Lemma 6) limits the running time of Greedy Set Cover to $O(rn\log n)$. □

Lemma 8 (Alon, Spencer [2]). *Given a universe U with at most n elements and at most n subsets of U. If every element is in at least k sets, then the number of sets selected by the Greedy Set Cover algorithm [6] is at most $(\log_2(k)+1)\,\frac{n}{k}$.*

This concludes the selection of the index nodes. The structure of the index edges is already described at the beginning of Section 2. The index edges can be computed efficiently by considering every vertex as point p_1 and creating an index edge for every edge $\{u_1, p_2\}$ with $\text{near}_r(p_1, u_1)$. Lemma 9 limits the distances in the full index graph.

Lemma 9. *The distance between any two index nodes in the* full index graph *induced by the set of all index edges is at most three times their distance in the base graph.*

Proof. Let p be a shortest path in the base graph from index node $\tilde{s}$ to index node $\tilde{t}$. We use a sequence of vertices p_i on the path p, with increasing distance from $\tilde{s}$, to split p into several parts, such that for each part there is an index edge. These index edges will form a path from $\tilde{s}$ to $\tilde{t}$.

For every p_i, let $\tilde{v}_i$ be the closest index node and d_i be the distance from p_i to $\tilde{v}_i$. Note that p_i is r-near $\tilde{v}_i$. As p_1 we take any vertex on p for which $\tilde{v}_1 = \tilde{s}$. The last vertex p_k of the sequence has $\tilde{v}_k = \tilde{t}$. There must be such a p_k because $d_i \leq \text{dist}(p_i, \tilde{t})$ for every p_i.

For a given vertex p_i, we set p_{i+1} as the first vertex on path p with $\text{dist}(p_i, p_{i+1}) \geq d_i$. Since the left neighbour u_i of p_{i+1} on the shortest path is r-near p_i, there is an index edge $\{\tilde{v}_i, \tilde{v}_{i+1}\}$ with weight at most $2\,\text{dist}(p_i, p_{i+1}) + d_{i+1}$. These index edges form a path from $\tilde{s}$ to $\tilde{t}$. The weight of this path is at most $\sum_{i<k}(2\,\text{dist}(p_i, p_{i+1}) + d_{i+1}) \leq d_k + \sum_{i<k} 3\,\text{dist}(p_i, p_{i+1}) \leq 3\,\text{dist}(\tilde{s}, \tilde{t})$.

Note that it does not matter if two index nodes $\tilde{v_i}$ and $\tilde{v_j}$ are actually the same vertex, since this only shortens the length of the path in the full index graph. □

To compute the index graph $I_r^\sigma(G)$ as a σ-spanner of the full index graph, we use the *Greedy σ-Spanner* algorithm [3]. With this algorithm, we can compute the index graph from the index edges in polynomial time. The Greedy σ-Spanner also guarantees an upper bound on the edges in the index graph:

Lemma 10 (Regev [18], combined with Bollobás [5]). *The Greedy σ-Spanner of every graph on $\tilde{n}$ vertices has at most $\tilde{n}^{1+2/\sigma}$ edges.*

3 Impossibility Results for Trade-Offs

In this section, we show various impossibility results for trade-offs between excerpt size, stretch, and probe factor.

Throughout this section, we allow all operations on vertex labels. This is in contrast to the path finding algorithm and the index graph construction presented in Section 2, where we only use comparisons. Additionally, for all impossibility results presented in this section, we restrict the vertex labels to integers from 1 to n on graphs with n vertices. There is also no assumption on the form or content of the graph excerpt, except that it can be stored as a bit string.

We use the same technique to prove all three limits presented in this section. For each limit, we present a family of graphs such that for every pair of graphs in the family, there is at least one pair of vertices for which the connecting paths that may be returned by the path finding algorithm differ. Additionally, the probe cost for distinguishing two graphs of the family is above the allowed limit of the probe cost (which is zero in Section 3.1).

3.1 Lower Bound for Spanners and Oracles

In this section we show that if the path finding algorithm is not allowed to access the base graph, then the size of the excerpt needs to be in $\Omega(n \log n)$.

Lemma 11. *There is no path finding algorithm $\mathcal{A}(s, t, \{0,1\}^L)$, with $L \leq \frac{n \log n}{2}$, that returns a path from vertex s to vertex t for all graphs G on n vertices with integer labels 1 to n.*

Proof. Consider the family of trees with unique vertex labels 1 to n. There are $n^{n-2} > 2^{\frac{n \log n}{2}}$ different trees on the vertices with unique labels 1 through n [17]. Therefore, with $L \leq \frac{n \log n}{2}$, at least two different trees T and T' are represented by the same bit string of length L. However, there is one pair of vertices s and t such that the path from s to t is different in the two trees. Otherwise, all edges in T must also be in T', and the two trees are identical. Hence, it is not possible for the algorithm $\mathcal{A}$ to decide which path to return without accessing the base graph. □

3.2 Edge Orders

In this section we demonstrate that the order in which the edges are reported is important for achieving good trade-offs between excerpt size, number of accessed edges and stretch. More precisely, we show that if the order in which the edges adjacent to one vertex are reported is based on a criterion that is independent of the edge weights, then the stretch cannot be limited to a value independent of the base graph's edge weights, even if the excerpt size is in the same order of magnitude as a $\log n$-spanner and the probe factor is linear in the number of vertices. We denote the combination of the orders in which the edges adjacent to the individual vertices are reported as an *edge order*.

Lemma 12. *For every edge order of the complete graph G_n on n vertices and every positive integer k, there is a family of weight functions $\mathcal{W} : V \times V \to \mathbb{R}^+$ such that there is no path finding algorithm $\mathcal{A}(s, t, \{0,1\}^L)$, with $L \leq \frac{n \log n}{8}$, that returns a path $p(s,t)$ with stretch k and probe factor $\frac{n}{8}$ for every vertex pain s, t and every weight function in $\mathcal{W}$.*

Proof Idea: We divide the vertices into two sets A and B, with $|A| = n/4$. Which vertices are in A depends on the edge order. The set B forms a complete subgraph with all edge weights 1. Every vertex in A has exactly one edge to a vertex in B with weight 1, all other edges have weight $2k+1$. Which edge has weight 1 depends on the specific weight function. The distance between a vertex a in A and a vertex b in B is thus at most 2, and any path from a to b with stretch k must start with the one edge adjacent to a that has weight 1. However, there are more than $2^{\frac{n \log n}{8}}$ such weight functions for which the weights of the first $n/8$ edges adjacent to every vertex do not differ, which makes it impossible to distinguish any two graphs with probe factor $n/8$. □

3.3 Stretch Limit

In this section we argue for finding only approximate shortest paths instead of true shortest paths, although the base graph can be accessed while computing a path. More precisely, we show that even with excerpts whose size is in the same order of magnitude as a $\log n$-spanner, and with a linear probe factor, no stretch smaller than 5 can be achieved.

Lemma 13. *For every $\varepsilon > 0$ there is a family $\mathcal{G}_n^\varepsilon$ of graphs on n vertices with integer labels 1 to n, such that there is no path finding algorithm $\mathcal{A}(s, t, \{0,1\}^L)$, with $L \leq \frac{n \log n}{8}$, that returns a path from s to t with stretch $5 - \varepsilon$ and probe factor $n/8$ for every vertex pair s, t in every graph $G \in \mathcal{G}_n^\varepsilon$.*

Proof Idea: We divide the graph vertices into four equally sized sets A, B, C, and D. Sets A and B form a complete bipartite graph with edge weights $1 - \frac{\varepsilon}{8}$, as do sets C and D. Additionally, there is a perfect matching with edge weights 1 between the vertices of A and the vertices of D. The difference between the graphs in $\mathcal{G}_n$ lies only in the matching. The only path between two vertices $a \in A$

and $d \in D$ with stretch $5 - \varepsilon$ is $\{a, d\}$ if that edge exists in the graph. However, there are more than $2^{\frac{n \log n}{8}}$ such matchings, and the probe cost for finding out if $\{a, d\}$ is in the graph is larger than $\frac{n}{4}$. □

References

1. Aleksandrov, L., Maheshwari, A., Sack, J.-R.: Determining approximate shortest paths on weighted polyhedral surfaces. Journal of the ACM 52(1), 25–53 (2005)
2. Alon, N., Spencer, J.: The Probabilistic Method. John Wiley, New York (1992)
3. Althöfer, I., Das, G., Dobkin, D.P., Joseph, D.: Generating sparse spanners for weighted graphs. In: Gilbert, J.R., Karlsson, R. (eds.) SWAT 1990. LNCS, vol. 447, pp. 26–37. Springer, Heidelberg (1990)
4. Baswana, S., Sen, S.: Approximate distance oracles for unweighted graphs in Õ (n^2) time. In: SODA '04, pp. 271–280 (2004)
5. Bollobás, B.: Extremal Graph Theory. Academic Press, San Diego (1978)
6. Chvátal, V.: A greedy heuristic for the set-covering problem. Mathematics of Operations Research 4, 233–235 (1979)
7. Cohen, E., Halperin, E., Kaplan, H., Zwick, U.: Reachability and distance queries via 2-hop labels. SIAM Journal on Computing 32(5), 1338–1355 (2003)
8. Demetrescu, C., Finocchi, I., Ribichini, A.: Trading off space for passes in graph streaming problems. In: SODA '06, pp. 714–723 (2006)
9. Demetrescu, C., Goldberg, A., Johnson, D. (eds.): 9th DIMACS Challenge on Shortest Paths, available at `http://www.dis.uniroma1.it/~challenge9` (to appear)
10. Dijkstra, E.W.: A note on two problems in connection with graphs. Numerische Mathematik 1, 269–271 (1959)
11. Feigenbaum, J., Kannan, S., McGregor, A., Suri, S., Zhang, J.: Graph distances in the streaming model: the value of space. In: SODA '05, pp. 745–754 (2005)
12. Garey, M.R., Johnson, D.S.: Computer and Intractability: A Guide to the Theory of NP-Completeness. W. H. Freeman (1979)
13. Goldberg, A.V., Harrelson, C.: Computing the shortest path: A* search meets graph theory. In: SODA '05, vol. 16, pp. 156–165 (2005)
14. Katriel, I., Meyer, U.: Elementary graph algorithms in external memory. In: Algorithms for Memory Hierarchies, pp. 62–84 (2002)
15. Meyer, U., Zeh, N.: I/O-efficient undirected shortest paths. In: Di Battista, G., Zwick, U. (eds.) ESA 2003. LNCS, vol. 2832, pp. 434–445. Springer, Heidelberg (2003)
16. Peleg, D., Schäffer, A.A.: Graph spanners. Journal of Graph Theory 13(1), 99–116 (1989)
17. Prüfer, H.: Neuer Beweis eines Satzes über Permutationen. Arch. Math. Phys. 27, 742–744 (1918)
18. Regev, H.: The weight of the greedy graph spanner. Technical Report CS95-22, Weizmann Institute Of Science (July 1995)
19. Sanders, P., Schultes, D.: Highway hierarchies hasten exact shortest path queries. In: Brodal, G.S., Leonardi, S. (eds.) ESA 2005. LNCS, vol. 3669, pp. 568–579. Springer, Heidelberg (2005)
20. Thorup, M., Zwick, U.: Approximate distance oracles. Journal of the ACM 52(1), 1–24 (2005)
21. Zwick, U.: Exact and approximate distances in graphs - a survey. In: Meyer auf der Heide, F. (ed.) ESA 2001. LNCS, vol. 2161, pp. 33–48. Springer, Heidelberg (2001)

Initializing Sensor Networks of Non-uniform Density in the Weak Sensor Model

Martín Farach-Colton and Miguel A. Mosteiro

Department of Computer Science, Rutgers University, Piscataway, NJ 08854, USA
{farach,mosteiro}@cs.rutgers.edu

Abstract. Assumptions about node density in the Sensor Networks literature are frequently too strong or too weak. Neither absolutely arbitrary nor uniform deployment seem feasible in most of the intended applications of sensor nodes. We present a Weak Sensor Model-compatible distributed protocol for hop-optimal network initialization, under the assumption that the maximum density of nodes is some value Δ known by all of the nodes. In order to prove lower bounds, we observe that all nodes must communicate with some other node in order to join the network, and we call the problem of achieving such a communication the *Group Therapy Problem*. We show lower bounds for the Group Therapy Problem in Radio Networks of maximum density Δ, regardless of the use of randomization, and a stronger lower bound for the important class of randomized fair protocols. We also show that even when nodes are distributed uniformly, the same lower bound holds, even in expectation and even for the simpler problem of Clear Transmission.

1 Introduction

Although some papers analyze problems in Radio Networks under the assumption of an arbitrary distribution of nodes, in most applications the layout of nodes is not the result of an uncontrolled random experiment in which the probability of some highly undesirable outcome is positive. On the other hand, a uniform distribution of nodes in the plane, as is customary to assume in the Sensor Networks literature, may be difficult or impossible to achieve in settings where the environment is hostile or remote. Furthermore, for any reasonable model of non-uniform distribution of nodes chosen, a minimum density of nodes has to be ensured in order to guarantee connectivity, and a non-trivial maximum density of nodes can indeed be guaranteed. An example of a feasible model for the distribution of nodes that reflects the random nature of the deployment, yet excludes highly unlikely pathological cases is a multiple bivariate normal distribution.

In this paper, we do not limit ourselves to any particular distribution but we define bounds on the density of nodes for any reasonable model. More specifically, we define a distribution of nodes as a *Smooth Distribution* if the maximum density of nodes in any one-hop neighborhood is some value $\Delta \leq n$, and for any constant $\alpha > 0$, in any disc of radius αr there exists a constant $\beta > 0$ such

F. Dehne, J.-R. Sack, and N. Zeh (Eds.): WADS 2007, LNCS 4619, pp. 565–576, 2007.

that the number of nodes is at least $\beta \log n$[1]. The rationale behind the choice of the lower bound is that, when the nodes are deployed uniformly at random, with enough density to achieve connectivity, a logarithmic density is guaranteed w.h.p.[2] [5]. Given that deterministic deployment is not possible in hostile or remote environments, we assume the random deployment to be a process with the goal of achieving uniform distribution but where nodes are dropped in excess in some areas.

The Problem. We study the network initialization problem with smooth distribution of nodes under the restrictions of the Weak Sensor Model [5], a harsh and comprehensive model that summarizes the literature on sensor node restrictions. The initialization of a Sensor Network is the problem of self-organizing from scratch as a radio-communication network called a sensor network. Even though communication among sensor nodes is through radio broadcast, it is useful to set up explicit links between nodes in order to establish routing paths and prevent flooding. In order to prove lower bounds, we observe that to join the network at least one transmission of each node has to be received by some other node. In this model, collision detection is not available and a transmission in a given time slot is successful if only if exactly one node transmits in the one-hop neigborhood of the receiver. Therefore, achieving non-colliding transmissions fast without knowledge of the topology is not trivial. The problem of achieving non-colliding transmissions has been well studied within other problems. For some problems in multihop networks, such as Sensor Network initialization, the messages transmitted not necessarily must reach all nodes in the network. If in a given time slot a node transmits and all nodes in a two-hop neighborhood do not transmit, we say that a *Clear Transmission* has been produced and all nodes in the one-hop neighborhood of the transmitter have received the Clear Transmission. If all nodes in the network have to either produce or receive a Clear Transmission the problem is called the *Clear Transmission Problem* [6]. For settings where all nodes in the network have to receive all the messages to be transmitted, the various problems studied differ in the number of nodes that have messages. When some arbitrary number k of nodes have a message the problem is known as *k-selection* [10]. If $k = 1$ the problem is called *Broadcast* [2,12], and if $k = n$, it is called *Gossiping* [4,13]. As explained before, for the purpose of proving lower bounds for Sensor Network initialization, we take as a lower bound the problem of transmitting so that at least one transmission of each node has to be received by some other node. Given that to solve the problem all the participants have to be *heard*, we term this problem the *Group Therapy Problem.*[3]

1 Througout this paper, log means $\log_2$ unless otherwise stated.

2 Define *with high probability* to mean with probability at least $1 - O(n^{-\Omega(1)})$.

3 In fact, were it not for the extensive literature on gossiping, we would reverse these terms. After all, it is hardly the point of gossiping to tell everyone your news. Conversely, in group therapy one expects to be heard by all. Nonetheless, we use the current notation for consistency.

Previous work. The literature on Sensor Networks is vast and includes both theoretical and empirical research work. Many of the solutions proposed do not sufficiently handle all the aspects of the problem. The protocol in [21] builds a flat topology, but it is assumed that there are enough channels to accomodate each link among neighbors which is not possible under the Weak Sensor Model. The protocol in [3] builds a network where every node has at most k neighboring nodes. However, the number of available radii of transmission is a function of n, and the protocol relies in distance estimation hardware. The protocol in [22] is an energy efficient topology control scheme. Unfortunately, global synchronization is necessary and an underlying contention resolution mechanism is assumed. In all these protocols the memory size is assumed to be in $\omega(1)$. Recently, a $O(\log^2 n)$ protocol that builds a network under the Weak Sensor Model was presented [5]. This protocol builds a hop-optimal network in settings where the nodes are deployed uniformly at random. This was the first protocol for network formation that is implementable in sensor nodes even theoretically. More general information about sensor networks can be obtained from the surveys [1, 9, 19, 20, 23].

Regarding lower bounds, Kushilevitz and Mansour [12] proved the first lower bound of $\Omega(\log n)$ on the expectation of the running time of any randomized algorithm for clear transmissions in radio networks. A lower bound of $\Omega(\log n \log(1/\epsilon)/(\log\log n + \log\log(1/\epsilon)))$ for achieving a Clear Transmission with probability $1 - \epsilon$ in a one-hop globally-synchronized Radio Network was proved in [8]. Recently, this lower bound was improved to $\Omega(\log n \log(1/\epsilon))$ in [6] closing the gap for this problem. A simple application of this result gives a lower bound of $\Omega(\log \Delta \log(1/\epsilon))$ steps in order to solve the Clear Transmission problem with probability at least $1 - \epsilon$ in Sensor Networks where the maximum density is Δ.

Related work. An important building block of the initialization protocol presented in this paper, which dominates the overall running time, is an algorithm that guarantees that in a network of size n, each node produces a Clear Transmission within $O(\Delta \log n)$ steps w.h.p. A $O(h + \log n \log\log n)$ adaptive algorithm to solve the problem of realizing arbitrary *h-relations* w.h.p. was presented in [7]. In an h-relation, each processor is the source as well as the destination of at most h messages. Also, for the k-selection problem, Martel's [14] randomized adaptive protocol operates in expected time $O(k + \log n)$. These algorithms could seemingly be used as our building block, thus speeding up the overall running time. However, they rely on collision detection and global synchronization.

Our results. We use two different problems to prove our lower bounds, both necessary conditions to solve the Sensor Network initialization problem. We observe that in order to initialize the network the Group Therapy problem must be solved. $\Omega(\Delta)$ steps are required to solve this problem regardless of the use of randomization. Combining this observation and our previous results [6], a lower bound of $\Omega(\Delta + \log \Delta \log(1/\epsilon))$ for solving the problem with probability $1 - \epsilon$ is obtained. Restricting ourselves to the important class of *fair* protocols, i.e., protocols where the probability of transmission of every node is the same in the same execution step, we show here a lower bound of $\Omega(\Delta(\log(1/\epsilon) + \log \Delta))$, which we also show to be tight. The analysis for fair protocols is relevant given that, to the best of our knowledge, asymptotically faster adaptive protocols for the Group Therapy problem are not known. Finally, we also show that even if the nodes are distributed uniformly, and thus $\Delta = \Theta(\log n)$, a seemingly simpler problem such as Clear Transmission, takes $\Omega(\log^2 n)$ *expected* time using fair protocols. As for upper bounds, we present a distributed protocol to self-organize a Sensor Network where the maximum density in any one hop neighborhood is an arbitrary value Δ. More specifically, we show that, upon waking up, a node joins the network w.h.p. within $O(\Delta \log n)$ time steps. A bottleneck of this algorithm is a local-spanner construction where *each* node produces a Clear Transmission within $O(\Delta \log n)$ steps w.h.p. using a fair protocol. Given that if *every* node produces a Clear Transmission, the simpler Group Therapy problem is also solved, this protocol matches the lower bound showed here. We leave open the question of how to use adaptive algorithms that work under these conditions to speed up the overall running time.

Roadmap. In Section 2 we define the models used througout the paper and we prove our lower bounds in Section 3. Upper bounds are showed in Sections 4 and 5. We include some acknowledgements in Section 5.1.

2 The Model

In Sensor Networks nodes are expected to be deployed at random in large quantities over an area of interest. Therefore, an abstraction frequently used in Sensor Networks to model the reachability of nodes is the *Random Geometric Graph Model* (RGG), where the nodes are assumed to be distributed uniformly at random in the plane. In this paper, we relax the assumption about the distribution and we use the *Geometric Graph Model* where the nodes are deployed in $\mathbb{R}^2$, and a pair of nodes is connected if and only if they are at an Euclidean distance of at most a parameter r.

Regarding models of sensor node constraints, Bar-Yehuda, Goldreich and Itai [2] detailed a formal model of a radio network that specifies many of the important restrictions on sensor nodes, including limits on contention resolution, but they make no mention of computational limits such as small memory. After this model was introduced, some papers [11, 18] have added more restrictions, although often such restrictions are implicit in the text or algorithms rather than fully specified. In this paper, we use the Weak Sensor Model [5], a harsh and comprehensive model that summarizes the literature on sensor node restrictions taken the most restrictive choices when possible. In this model, the communication among neighboring nodes is through broadcast on a *shared channel.* A node receives a message if and only if exactly one of its neighbors transmits. There is *no collision detection* mechanism available and the channel is assumed to have only two states: single transmission and silence/collision. Sensors nodes have *non-simultaneous reception and transmission.* Time is assumed to be slotted and all nodes have the same clock frequency, but no global synchronizing mechanism is assumed. Nodes are woken up by an adversary perhaps at different times. Sensor nodes may store only a constant number of $O(\log n)$ bit words. We assume that sensor nodes can adjust their power of transmission to only a *constant* number of levels. Other restrictions include: limited life cycle, short transmission range, only one shared channel of communication, lack of position information and unreliability.

3 Lower Bounds

To solve any problem in a communication network at least one successful transmission is necessary. However, in multihop networks, one non-colliding transmission is not enough to solve most of the problems. For the Radio Network initialization problem we observe that, in order to join the network, for every node at least one transmission has to be received by at least one neighboring node. More precisely, let V be the set of nodes in the network and $N(v) \subseteq V$ denote the set of nodes adjacent to $v \in V$. Then, for all $v \in V$ there is at least one time slot in which there exists a node $u \in N(v)$ such that exactly one node in $N(u)$ transmits and this node is v. We term this problem the *Group Therapy Problem*. In order to provide stronger lower bounds, we relax the Weak Sensor Model to a minimum set of restrictions, namely, low-information channel contention, local synchronism and adversarial node wake-up schedule. We refer to this model as the Radio Network model. We begin observing that a lower bound for the Group Therapy problem, regardless of the use of randomization, is $\Omega(\Delta)$, a claim that we formalize in the following theorem.

Theorem 1. *In order to solve the Group Therapy problem in a multihop Radio Network where the maximum density in any one-hop neighborhood is Δ, any algorithm requires $\Omega(\Delta)$ time-slots.*

Proof. Exploiting the assumption of an adversarial wake-up schedule, let us assume the existence of an adversary that, at a given time, wakes up only a subset of Δ neighboring nodes, i.e., a set of Δ nodes whose connectivity graph is a clique. We call them *active* nodes. Such a subset of nodes exists since the maximum density is Δ. Upon waking up, the active nodes start the execution of the protocol. All the other nodes remain non-active and do not participate in the protocol. In this setting, in order to solve the problem, every node has to achieve a non-colliding transmission in a different time slot, therefore the claim follows.

Combining Theorem 1 with our lower bound [6] of $\Omega(\log n \log(1/\epsilon))$ time steps to solve the Clear Transmission problem with probability $1-\epsilon$ in a one-hop Radio Network of n nodes , the following lower bound for the Group Therapy problem in Radio Networks is obtained.

Corollary 1. *In order to solve the Group Therapy problem with probability $1-\epsilon$ in a multihop Radio Network where the maximum density in any one-hop neighborhood is Δ, any randomized algorithm requires $\Omega(\Delta + \log \Delta \log(1/\epsilon))$ time-slots.*

We consider now lower bounds for *fair* protocols, i.e., protocols where the probability of transmission of every node in the same time step is the same. The analysis for fair protocols is relevant given that, to the best of our knowledge, asymptotically faster adaptive protocols for the Group Therapy problem are not known. We prove this lower bound under the assumption of the existence of a weak adversary that, at a given time, wakes up some subset of nodes of size $\{2^i | 0 \leq i \leq \log \Delta\}$. We call them *active* nodes. Upon waking up, the active nodes start the execution of the protocol. All the other nodes remain non-active and do not participate in the protocol.

We define a *randomized fair protocol* to be a sequence $p_1, p_2, \ldots$ where each node transmits with probability p_ℓ in the ℓth time step after waking up. Given our adversary, this means that all active nodes transmit with the same probability as each other in each time slot. We further assume that all $p_\ell \in \{2^{-j} | 1 \leq j \leq \log \Delta\}$. If this assumption is not true of a particular algorithm A, we can always produce an algorithm A' from A by replacing one attempt in A by a constant number of attempts in A' where the probabilities of transmission in A' have been rounded off to the closest power of $1/2$.

Let p_{ij} denote the probability that a given node fails to achieve a non-colliding transmission when 2^i active nodes transmit with probability 2^{-j}. Then, we know that $p_{ij} = 1 - (1/2^j)(1 - 1/2^j)^{2^i - 1}$. Let t_j be the number of time-slots that nodes are transmitting with probability 2^{-j}. Then, the total probability of failure for any number of active nodes 2^i, needs to be bounded by $2^i \prod_j p_{ij}^{t_j} \leq \epsilon$, or taking logarithms $\sum_j t_j \ln p_{ij} \leq \ln(\epsilon) - \ln 2^i$.

A lower bound can be obtained by minimizing the total number of time-slots needed to satisfy the previous constraints. Here, we reuse our proof technique from [6], i.e., we formulate the problem as a linear program and use a feasible solution of the dual formulation as a lower bound. However, due to the differences between the problems to be solved, the slack variables of the dual need to be more carefully defined, which we do below. The function to be minimized together with the constraints can be formulated as the following *primal* linear program which yields the corresponding *dual*.

Minimize $\mathbf{1}^T\mathbf{t}$,	Maximize $\boldsymbol{\epsilon}^T\mathbf{u}$,	where:
subject to:	subject to:	$\mathbf{t} \triangleq [t_j]$,
$\mathbf{P}\mathbf{t} \geq \boldsymbol{\epsilon}$	$\mathbf{P}^T\mathbf{u} \leq \mathbf{1}$	$\boldsymbol{\epsilon} \triangleq [-\ln(\epsilon) + \ln 2^i]$,
$\mathbf{t} \geq \mathbf{0}$,	$\mathbf{u} \geq \mathbf{0}$,	$\mathbf{P} \triangleq [-\ln(p_{ij})]$.

The primal linear program has a finite minimum solution, and hence its dual has a finite maximum solution. The value of the objective function for every feasible solution of the dual is a lower bound on the minimum value of the objective function for the primal. Thus any feasible solution for the dual will give the lower bound sought. We first define the slack variables as $u_i = 2^i(1 - 1/\sqrt{e})^2$, and show that these values satisfy the constraints of the dual.

Lemma 1. *For any* $1 \leq j \leq \log \Delta$, $\sum_{i=0}^{\log \Delta}(-\ln p_{ij})u_i \leq 1$.

Proof. We want to prove that

$$\sum_{i=0}^{\log \Delta} \left(-\ln\left(1 - \frac{1}{2^j}\left(1 - \frac{1}{2^j}\right)^{2^i-1}\right)\right) 2^i \left(1 - \frac{1}{\sqrt{e}}\right)^2 \leq 1$$

Using that for $0 < x < 1$, $e^{-x/(1-x)} \leq 1 - x \leq e^{-x}$ [15, §2.68] and maximizing for $j = 1$ it is enough to prove $\sqrt{e}(1 - 1/\sqrt{e})^2 \sum_i (2^i/\sqrt{e}^{2^i}) \leq 1$. Diferentiating the arithmetic-geometric series and replacing, the claim follows.

Now, we use the value of the objective function for this feasible solution to show our lower bound.

Theorem 2. *In order to solve the Group Therapy problem with probability* $1-\epsilon$ *in a multihop Radio Network where the maximum density in any one-hop neighborhood is* Δ*, any fair randomized algorithm requires* $\Omega(\Delta(\log(1/\epsilon) + \log \Delta))$ *time-slots.*

Proof. From lemma 1, we know that $u_i = 2^i(1 - 1/\sqrt{e})^2$ satisfies the constraints of the dual LP, replacing

$$\begin{aligned} \epsilon^T \mathbf{u} &= \sum_{i=0}^{\log \Delta} \left(\ln \frac{1}{\epsilon} + \ln 2^i\right) 2^i \left(1 - \frac{1}{\sqrt{e}}\right)^2 \\ &\in \Omega(\Delta(\log(1/\epsilon) + \log \Delta)). \end{aligned}$$

As proved in [5], when the nodes are distributed uniformly, the density of nodes in any disc of radius $\Theta(r)$ is in $\Theta(\log n)$. Therefore, a simple application of Theorem 2 gives a lower bound of $\Omega(\log n(\log(1/\epsilon) + \log \log n))$ for the Group Therapy problem within uniform density settings. However, it can be proved that to solve even a seemingly simpler problem such as Clear Transmission in a Radio Network with uniformly distributed nodes, it takes $\Omega(\log^2 n)$ *expected* time, which we do as follows.

The topology of active nodes chosen by the adversary for this proof consists of a set of disjoint pairs of cliques connected by a single node. One clique of the pair has node density in $\Theta(1)$, the other in $\Theta(\log n)$ and the intermediate node connects to all nodes in both cliques . We call this construction a *clique-pair*. In order to be disjoint, nodes are woken up so the resulting clique-pairs are separated by a distance of r, the maximum range of transmission of any node.

We first give the intuition of why this structure gives a good lower bound on the number of time steps needed to solve the Clear Transmission problem. Recall that in a multi-hop setting a transmission is a Clear Transmission if no node within two hops of the transmitter transmits in the same time slot. To solve the Clear Transmission problem every node has to receive or produce a Clear Transmission. Hence, in order to solve this problem, a necessary condition is that each node in the low-density clique either receives or produces a Clear Transmission. That means that there must exist at least one time slot in which exactly one node in the clique pair transmits, that node being the intermediate node or a node in the low-density clique.

Given the different densities and that the protocol is fair, when the sum of probabilities of transmission in the low density clique reaches a constant, and therefore the probability of having a succesful transmission in that clique is constant, the sum of probabilities of transmission in the 2-hop neighboring high density clique is asimptotically more than a constant and the probability of silence is low. On the other hand, when the sum of probabilities of transmission in the whole clique-pair reaches a constant, and the probability of having a non-colliding transmission is high, the probability that the transmitting node is in the low-density clique or it is the intermediate node is low. Then, the probability that nodes in the low density clique produce or receive a Clear Transmission fast is low.

Lemma 2. *Given a Radio Network with nodes deployed as a connected RGG, the total number of clique-pairs activated by the adversary is in $\Theta(n/\log n)$ w.h.p.*

Proof. It follows from the $\Theta(\log n)$ density bound in any disk of radius $\Theta(r)$ proved in [5].

Theorem 3. *Every fair randomized algorithm takes $\Omega(\log^2 n)$ expected time in order to solve the Clear Transmission problem in a multi-hop Radio Network where nodes are deployed uniformly at random.*

Proof. The proof is based on minimizing the probability of failing to achieve a Clear Transmission in a low density clique. The details are omitted in this extended abstract for brevity.

4 An Optimal Upper Bound for the Group Therapy Problem

A common observation in the literature is that a fair protocol, i.e., a protocol where all nodes are assumed to use the same probability of transmission in the same time slot, has a higher probability of achieving a non-colliding transmission when the probability and the inverse of the number of active nodes agree up to a constant factor and this probability is lower otherwise. Therefore, a main challenge for any protocol is to estimate the density accurately and fast. However, as we show in this section, if all nodes have to achieve successful transmissions by means of a fair protocol it is enough to know the maximum density to achieve a running time of $O(\Delta \log n)$ w.h.p. In achieving a Clear Transmission for all nodes, the Group Therapy problem is also solved. Thus, given the lower bound of Theorem 2, it is optimal for the latter problem. We leave open the question of whether it can be done faster or not using adaptive algorithms. The algorithm is simple to describe, for a network where the maximum density of nodes in any disk of radius r is Δ, every node repeatedly transmits with probability $1/\Delta$.

Theorem 4. *Given a multihop Radio Network where the maximum density in any one-hop neighborhood is Δ, using the protocol described above every node achieves a Clear Transmission within $O(\Delta \log n)$ time steps w.h.p.*

Proof. For a given node, consider a circle of radius $2r$ centered on it. This circle can be completely covered by a constant number, say β_1, of circles of radius r. In each of these circles there are at most Δ nodes, since Δ is the maximum number of nodes in any one-hop neighborhood and all nodes within a circle of radius r are connected. Therefore, $\beta_1\Delta$ is an upper bound of the number of nodes in the 2-hop neighborhood of any node. Hence, the probability of *some* node not achieving a Clear Transmission after $\beta_2\Delta \log n$ steps, where β_2 is a constant is

$$Pr(fail) \leq n \left(1 - \frac{1}{\Delta}\left(1 - \frac{1}{\Delta}\right)^{\beta_1\Delta}\right)^{\beta_2\Delta \log n}$$
$$\in O(n^{-\gamma}), \text{ for some constants } \beta_2, \gamma > 0.$$

Where we used that for all $n \geq 1$ and $|x| \leq n$, $e^x(1-x^2/n) \leq (1+x/n)^n \leq e^x$ [16].

5 Non-uniform Density Network Initialization

As proved in [5], under the Weak Sensor Model, an optimal network should have low hop-stretch while maintaining links to a constant number of neighbors due to memory constraints. The hop-stretch is the maximum, among all pairs of nodes, of the ratio between the minimum number of hops in a path connecting two nodes, and the optimal number of hops given by the Euclidean distance and the maximum range. In the same paper, it was presented a distributed protocol that builds from scratch a network with such a topology, under the assumption that nodes are deployed uniformly at random sufficiently densely to ensure connectivity w.h.p. We show here that even if the density of nodes is not uniform, as long as it is *Smooth* as defined in Section 1, a network with such a topology can be obtained fast using the same general technique adequately implemented for this setting.

To model the reachability of nodes we use the *Geometric Graph Model* or $\mathcal{G}_{n,r,\ell}$, where n nodes are deployed in a space of size $[0, \ell]^2$, and a pair of nodes is connected if and only if they are at an Euclidean distance of at most r. An instance of $\mathcal{G}_{n,r,\ell}$ is called a *Geometric Graph* (GG) and noted $G(n, r, \ell)$. Given that the network we aim to obtain has to have low hop-stretch and constant number of neighbors, the graph that models its topology has to have constant degree and asymptotically optimal path length in terms of number of edges. We call such a graph a *Constant-degree Hop-optimal Spanning Graph* (CHSG).

In [5] was proved the existence of a CHSG subgraph of any connected RGG by means of a dissection technique called bin-covering [17]. Further, in the same paper was given a *Disk Covering Scheme* that produces such a subgraph. Given the smooth distribution assumed here, the minimum density of nodes in any disc of radius $\Theta(r)$ is $\Omega(\log n)$. Therefore, the same results apply to our setting, i.e., given a GG with smooth distribution of nodes, the Disk Covering Scheme produces a CHSG. The Disk Covering Scheme has four phases, namely, small disk layout, bridge interconnection, disks expansion and local spanner construction. The first three phases of the Disk Covering Scheme can be implemented as detailed in the journal version of [5]. Given that in our setting the maximum density is Δ, the probabilities of transmission and counters used need to be changed appropriately. We omit the details in this extended abstract for brevity. The last phase of spanner construction for smooth distributions is detailed in the following section.

5.1 Spanner Construction

The last phase of the Disk Covering Scheme is the construction of a constant-degree spanner within each expanded disk. As shown in [5], the diameter of the spanner must be logarithmic in order to achieve asymptotically optimal hop-stretch. In this paper, we consider Sensor Networks where the maximum density of nodes in any one-hop neighborhood is an arbitrary value Δ bounded only by n. Thus, a straightforward solution such as a linked list can not be used. Instead, we simply use a balanced binary tree. To build such a tree, we locally rank the nodes according with their unique ID and the ID of the bridge node that covers them. Once unique consecutive labels given by the rank within the disk are assigned to all nodes, each node can easily compute to which nodes is connected within the tree.

We give here a description of the distributed algorithm and we omit the details in this extended abstract for brevity. The spanner construction algorithm consists of three phases. First, every node broadcasts its ID keeping track of the ID of its predecessor among the nodes covered by the same bridge for $\Theta(\Delta \log n)$ steps. As we prove in Theorem 4, at this point all nodes have achieved a Clear Transmission w.h.p. so, all nodes have received a transmission from their local predecessor. To obtain their local rank, nodes enumerates themselves one by one in a second phase as follows. Upon receiving the rank i of its predecessor, a node defines its rank as $i+1$ and broadcasts it with constant probability for $\Theta(\log n)$ steps. As shown in lemma 3, there will be at least one transmission without collision w.h.p. The first node in this ordering does not have any predecessor so it starts this phase of the algorithm with rank 1. At this point, all nodes know their local rank and it only remains to connect them as a balanced tree. A final phase broadcasting the rank where node i connects to nodes $\lfloor i \rfloor$, $2i$ and $2i+1$, achieves this. The root of such a tree is therefore the node with the smallest local rank which connects to the bridge.

In order to avoid conflicts with nodes waking up while building the spanner, the Sensor Network initialization protocol has to include an initial waiting phase of $\beta\Delta \log n$ time steps, for some constant β. Nodes can be covered by more than one bridge but, given the geometric restrictions, every node is covered by a constant number of them. Messages and bookkeeping must be replicated for each covering bridge as needed. Nodes running other phases may introduce interference but as long as the sum of their probabilities of transmission is a constant, the analysis can be done as if each phase runs in a different channel in the presence of a source of noise of constant probability of transmission, which we fold into the constants included in the analysis. The details follow.

Lemma 3. *Any node running the second phase of the spanner construction algorithm described above, achieves a transmission without collision within $O(\log n)$ steps w.h.p.*

Proof. Let $\beta_1 \in O(1)$ denote the maximum number of interfering neighbors also running the second phase of the spanner construction algorithm. Let $1/\beta_3 \in O(1)$ be the probability of transmission used by a node running such phase. Let $Pr[\text{fail}]$ denote the probability that any node fails to transmit without collision after $\beta_2 \log n$ steps for some constant β_2. Using the union bound and for some constants $\beta_1, \beta_2, \beta_3$

$$Pr[\text{fail}] \leq n \left(1 - \frac{1}{\beta_3}\left(1 - \frac{1}{\beta_3}\right)^{\beta_1}\right)^{\beta_2 \log n}$$
$$\in O(n^{-\gamma}), \text{ for some constant } \gamma > 0$$

Theorem 5. *Any node running the spanner algorithm joins the spanner within $O(\Delta \log n)$ steps w.h.p.*

Proof. The first and third phase take $O(\Delta \log n)$ time by definition of the algorithm. In the second phase, each of the at most Δ nodes in turn transmit for $O(\log n)$ steps. Hence, the overall running time of the algorithm is $O(\Delta \log n)$. As shown in Theorem 4, every node achieves at least one non-colliding transmission within $O(\Delta \log n)$ steps w.h.p. therefore the claim follows.

Acknowledgements

We would like to thank Rohan Fernandes for helpful discussions. This research was supported in part by DIMACS, Center for Discrete Mathematics & Theoretical Computer Science, grant numbered NSF CCR 00-87022.

References

1. Akyildiz, I.F., Su, W., Sankarasubramaniam, Y., Cyirci, E.: Wireless sensor networks: A survey. Computer Networks 38(4), 393–422 (2002)
2. Bar-Yehuda, R., Goldreich, O., Itai, A.: On the time-complexity of broadcast in multi-hop radio networks: An exponential gap between determinism and randomization. Journal of Computer and System Sciences 45, 104–126 (1992)
3. Blough, D.M., Leoncini, M., Resta, G., Santi, P.: The k-neigh protocol for symmetric topology control in ad hoc networks. In: Proc. of the 4th ACM international symposium on Mobile ad hoc networking and computing, pp. 141–152 (2003)
4. Chlebus, B., Gąsieniec, L., Lingas, A., Pagourtzis, A.: Oblivious gossiping in ad-hoc radio networks. In: Proc. of 5th International Workshop on Discrete Algorithms and Methods for Mobile Computing and Communications, pp. 44–51 (2001)

5. Farach-Colton, M., Fernandes, R.J., Mosteiro, M.A.: Bootstrapping a hop-optimal network in the weak sensor model. In: Brodal, G.S., Leonardi, S. (eds.) ESA 2005. LNCS, vol. 3669, pp. 827–838. Springer, Heidelberg (2005)
6. Farach-Colton, M., Fernandes, R.J., Mosteiro, M.A.: Lower bounds for clear transmissions in radio networks. In: Correa, J.R., Hevia, A., Kiwi, M. (eds.) LATIN 2006. LNCS, vol. 3887, pp. 447–454. Springer, Heidelberg (2006)
7. Gerèb-Graus, M., Tsantilas, T.: Efficient optical communication in parallel computers. In: 4th Annual ACM Symposium on Parallel Algorithms and Architectures, pp. 41–48 (1992)
8. Jurdziński, T., Stachowiak, G.: Probabilistic algorithms for the wakeup problem in single-hop radio networks. Theory of Computing Systems 38(3), 347–367 (2005)
9. Karl, H., Willig, A.: A short survey of wireless networks. Technical Report TKN-03-018, Technical University Berlin (October 2003)
10. Kowalski, D.R.: On selection problem in radio networks. In: Proceedings 24th Annual ACM Symposium on Principles of Distributed Computing, pp. 158–166 (2005)
11. Kumar, V.S.A., Marathe, M.V., Parthasarathy, S., Srinivasan, A.: End-to-end packet-scheduling in wireless ad-hoc networks. In: Proc. of the 15th Annual ACM-SIAM Symposium on Discrete Algorithms, pp. 1021–1030 (2004)
12. Kushilevitz, E., Mansour, Y.: An $\Omega(D\log(N/D))$ lower bound for broadcast in radio networks. SIAM Journal on Computing 27(3), 702–712 (1998)
13. Liu, D., Prabhakaran, M.: On randomized broadcasting and gossiping in radio networks. In: Ibarra, O.H., Zhang, L. (eds.) COCOON 2002. LNCS, vol. 2387, pp. 340–349. Springer, Heidelberg (2002)
14. Martel, C.U.: Maximum finding on a multiple access broadcast network. Inf. Process. Lett. 52, 7–13 (1994)
15. Mitrinović, D.S.: Elementary Inequalities. P. Noordhoff Ltd. - Groningen (1964)
16. Motwani, R., Raghavan, P.: Randomized Algorithms. Cambridge University Press, Cambridge (1995)
17. Muthukrishnan, S., Pandurangan, G.: The bin-covering technique for thresholding random geometric graph properties. In: Proc. of the 16th Annual ACM-SIAM Symposium on Discrete Algorithms, pp. 989–998 (2005)
18. Nakano, K., Olariu, S.: Energy-efficient initialization protocols for radio networks with no collision detection. In: ICPP, pp. 263–270 (2000)
19. Ponduru, V.A.S., Bharathidasan, A.: Sensor networks: An overview. Technical report, University of California, Davis (2003)
20. Rentala, P., Musumuri, R., Saxena, U., Gandham, S.: Survey on sensor networks. `http://citeseer.nj.nec.com/479874.html`
21. Sohrabi, K., Gao, J., Ailawadhi, V., Pottie, G.J.: Protocols for self-organization of a wireless sensor network. Personal Communications, IEEE 7(5), 16–27 (2000)
22. Song, W., Wang, Y., Li, X., Frieder, O.: Localized algorithms for energy efficient topology in wireless ad hoc networks. In: Proc. of the 5th ACM international symposium on Mobile ad hoc networking and computing, pp. 98–108 (2004)
23. Younis, O., Krunz, M., Ramasubramanian, S.: Node clustering in wireless sensor networks: Recent developments and deployment challenges. IEEE Network Magazine 20, 20–25 (2006)

Computing Best Coverage Path in the Presence of Obstacles in a Sensor Field

Senjuti Basu Roy, Gautam Das, and Sajal Das

Department of Computer Sc and Engineering.
University of Texas At Arlington,
Arlington, TX-76019
{roy,gdas,das}@cse.uta.edu

Abstract. We study the presence of obstacles in computing $BCP(s,t)$ (*Best Coverage Path* between two points s and t) in a 2D field under surveillance by sensors. Consider a set of m line segment obstacles and n point sensors on the plane. For any path between s to t, p is the *least protected point* along the path such that the Euclidean distance between p and its closest sensor is maximum. This distance (the path's *cover* value) is minimum for a $BCP(s,t)$. We present two algorithmic results. For *opaque obstacles*, i.e., which obstruct paths and block sensing capabilities of sensors, computation of $BCP(s,t)$ takes $O((m^2n^2 + n^4)\log(mn + n^2))$ time and $O(m^2n^2 + n^4)$ space. For *transparent obstacles*, i.e., which only obstruct paths, but allows sensing, computation of $BCP(s,t)$ takes $O(nm^2 + n^3)$ time and $O(m^2 + n^2)$ space. We believe, this is one of the first efforts to study the presence of obstacles in coverage problems in sensor networks.

1 Introduction

In this paper, we study a specific class of problems that arises in sensor networks, and propose novel solutions based on techniques from computational geometry. Given a 2D field with obstacles under surveillance by a set of sensors, we are required to compute a *Best Coverage Path* (BCP) between two given points that avoids the obstacles. Informally, such a path should stay as close as possible to the sensors, so that an agent following that path would be most "protected" by the sensors. This problem is also related to the classical *art gallery* and *illumination research* type of problems that has been long studied in computational geometry [14,15]. However, there are significant differences between the problem we consider and these other works, which we elaborate later in the paper. Moreover, to the best of our knowledge, ours is one of the first efforts to study the presence of obstacles in coverage problems in sensor networks.

More formally, let $S = \{S_1, \ldots, S_n\}$ be a set of n homogeneous wireless point sensors deployed in a 2D sensor field Ω. Each sensor node has the capability to sense data (such as temperature, light, pressure and so on) in its vicinity (defined by its *sensing radius*). For the purpose of this paper, assume that these sensors are *guards* that can protect any object within their sensing radius, except

F. Dehne, J.-R. Sack, and N. Zeh (Eds.): WADS 2007, LNCS 4619, pp. 577–588, 2007.

that the level of protection decreases as the distance between the sensor and the object increases. Let $P(s,t)$ be any path between a given source point s and a destination point t. The *least protected point* p along $P(s,t)$ is that point such that the Euclidean distance between p and its closest sensor S_i is maximum. This distance between p and S_i is known as the *cover* value of the path $P(s,t)$. A *Best Coverage Path* between s and t, $BCP(s,t)$, is a path that has the minimum cover value.

A BCP is also known as a *maximal support path* (MSP). In recent years there have been several efforts to design efficient algorithms to compute them [5,9,10]. However, one notable limitation of these works is that they have not considered the presence of *obstacles* in the sensor field, i.e., objects that obstruct paths and/or block the line of sight of sensors. In fact, most papers that deal with coverage problems in sensor networks have not attempted to consider the presence of obstacles. This is surprising since obstacles are especially realistic in a deployment of sensors in unmanned terrain (e.g., buildings and trees, uneven surfaces and elevations in hilly terrains, and so on).

In this paper we study the presence of obstacles in computing best coverage paths. However we do not consider the length of the path (i.e., minimizing the path length) in the computation of a BCP. More formally, assume that in addition to the n sensors, there are also m line segment obstacles $O = \{O_1, \ldots, O_m\}$ placed in the sensor field. Line segments are fundamental building blocks for obstacles, as more complex obstacles (e.g., polygonal obstacles) can be modeled via compositions of line segments. We consider two types of obstacles: (a) *opaque obstacles* which obstruct paths as well as block the line of sight of sensors, and (b) *transparent obstacles* which obstruct paths, but allow sensors to "see" through them. Examples of the former may include buildings - they force agents to take detours around them as well as prevent certain types of sensors (such as cameras) from seeing through them - while examples of the latter may include lakes - agents only have to take detours around them but cameras can see across to the other side. When obstacles are opaque, we refer to the best coverage path problem as the *$BCP(s,t)$ Problem for Opaque Obstacles*, whereas for transparent obstacles, we refer to the problem as the *$BCP(s,t)$ Problem for Transparent Obstacles*. Figure 1 is an example of a $BCP(s,t)$ amidst two sensors and four opaque obstacles, whereas Figure 2 shows a $BCP(s,t)$ in the same sensor field but assumes the obstacles are transparent.[1]

To compute $BCP(s,t)$ without obstacles, existing approaches [5,9,10] leverage the fact that the *Delaunay triangulation* of the set of sensors - i.e., the dual of the *Voronoi diagram* - contains $BCP(s,t)$. Furthermore, [9] shows that sparse subgraphs of the Delaunay triangulation, such as *Gabriel Graphs* and even *Relative Neighborhood Graphs* contain $BCP(s,t)$. However, such methods do not easily extend to the case of obstacles. Moreover, as should be clear from Figure 1, the *visibility graph* [2] is also not applicable to the $BCP(s,t)$ problem for opaque

[1] We assume infinite sensing capabilities for our sensors; although sensing intensity decreases with the increased distance. Generalizing our algorithms for a finite sensing radius is straightforward and omitted from this paper.

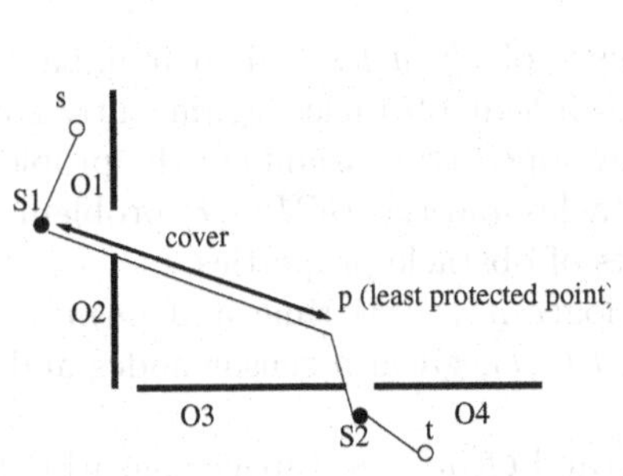

Fig. 1. A $BCP(s, t)$ for Opaque Obstacles

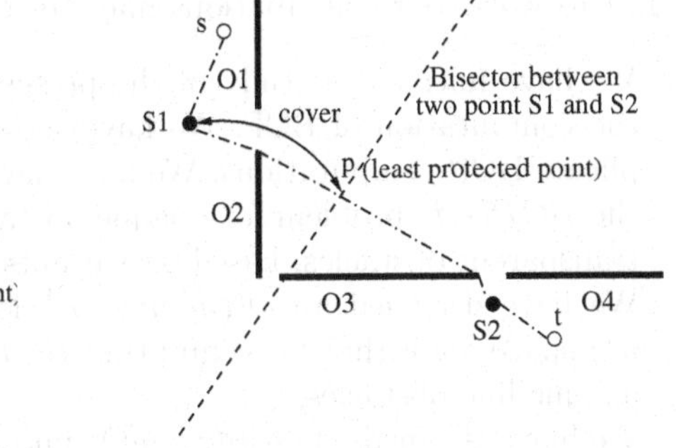

Fig. 2. A $BCP(s, t)$ for Transparent Obstacles

obstacles, as the best coverage paths in this case need not follow edges of the visibility graph. In fact, to solve the $BCP(s, t)$ problem for opaque obstacles, we have developed an algorithm that takes quartic-time, based on constructing a specialized dual of the *Constrained and Weighted Voronoi Diagram* (henceforth known as the *CW-Voronoi diagram*) [13] of a set of point sites in the presence of obstacles. This type of Voronoi diagram is a generalization of *Peeper's Voronoi Diagram* [6] that involves only two obstacles. These two Voronoi diagrams are different from other Voronoi diagrams involving obstacles studied in papers such as [3,1,11] - since the latter requires that every obstacle endpoint be a site whereas in the former the Voronoi sites are distinct from the obstacle endpoints. Unlike standard Voronoi diagrams for point sets without obstacles, the Voronoi regions in a Peeper's Voronoi diagram site may be disconnected [6]. However, for the $BCP(s, t)$ problem with transparent obstacles, we have shown that a best coverage path is contained in the visibility graph (with suitably defined edge weights), which enables us to develop a more efficient algorithm for it.

As mentioned earlier, the $BCP(s, t)$ problems are also related to art gallery problems [14,15], which are concerned with the placement of guards in regions to monitor certain objects in the presence of obstacles. An early result in art gallery research, due to V. Chvátal, asserts that $\lfloor \frac{n}{3} \rfloor$ guards are occasionally necessary and always sufficient to guard an art gallery represented by a simple polygon of n vertices [14,15]. Since then, numerous variations of the art gallery problems have been studied, including mobile guards, guards with limited visibility or mobility, guarding of rectilinear polygon and so on. The main difference between the the art gallery problems and the BCP problems is that the former problems (such as Watchman Route, Robber Route [14,15] and so on) attempt to determine paths that minimize total Euclidean distances under certain constraints, whereas the metric to be minimized in the latter problems (the *cover* of the path) is sufficiently different from Euclidean distance, thus requiring different approaches.[2] A comprehensive survey of art gallery research is presented in [15].

[2] It is easy to prove that the *cover* is a metric, and holds all metric properties. We omit the proof from this version of the paper.

In this work, our contributions may be summarized as follows:

- We have initiated a study of the presence of *obstacles* and their impact in the computation of BCP. We have shown that obstacles significantly complicate BCP computations. We have developed two variants of the problem, the $BCP(s,t)$ problem for opaque obstacles and the $BCP(s,t)$ problem for transparent obstacles, based on variants of obstacle properties.
- We have designed an $O((m^2n^2 + n^4)\log(mn + n^2))$ time and $O(m^2n^2 + n^4)$ space algorithm for computing $BCP(s,t)$, given n sensor nodes and m opaque line obstacles.
- We have designed an $O(nm^2 + n^3)$ time and $O(m^2 + n^2)$ space algorithm for computing $BCP(s,t)$, given n sensor nodes and m transparent line obstacles.

The rest of the paper is organized as follows: In Section 2 we describe our algorithm for the $BCP(s,t)$ problem for opaque obstacles and its correctness. In Section 3, we propose an algorithm for the $BCP(s,t)$ problem for transparent obstacles. Running time analysis and proof of correctness are also presented. Finally, we conclude in Section 4, and give future research directions.

2 The $BCP(s,t)$ Problem for Opaque Obstacles

In this section we study the $BCP(s,t)$ problem for opaque obstacles, defined in Section 1.

We first discuss why the presence of obstacles make the best coverage problem difficult. As discussed earlier, the visibility graph, a standard data structure used for numerous proximity problems in the presence of obstacles, does not necessarily contain a BCP (recall Figure 1 which clearly shows that a BCP from s to t, shown as a solid line path, is not contained in the constructed visibility graph of the 2 sensors and 4 opaque obstacles). Besides the shortcomings of the visibility graph, the existing solutions for the best coverage path problem without obstacles [5,9,10] depended on structures such as the Delaunay triangulation, Gabriel graph, relative neighborhood graph and so on. These structures have no easy generalizations to the case of obstacles.

Instead, we leverage the constrained and weighted Voronoi diagram (CW-Voronoi diagram)[13] for our purposes. The outline of our approach is as follows. We construct the CW-Voronoi diagram of n sensor sites in presence of m line obstacles (where each of the sites have the same weight, i.e., 1). Next we construct a specific weighted *dual* graph of this Voronoi diagram, such that the best coverage path is guaranteed to be contained in this dual graph. We assign edge weights to each of the constructed edges in the dual graph, where the weight of each edge is the distance from its least protected point to its nearest sensor. The dual creation and edge weight assignment is described in Section 2.1. Finally, using the Bellman-Ford algorithm [4], we compute a path between point s and t in the constructed weighted dual graph, whose largest edge is smaller than the largest edge of any other path in the graph.

2.1 CW-Voronoi Diagram and Its Dual

Here we first discuss the CW-Voronoi diagram [13]. As an example, consider Figure 3 which shows the CW-Voronoi diagram where S_1, S_2 are two sensors and O_1, O_2, O_3, O_4 are 4 opaque obstacles. The filled areas are dark regions which cannot be sensed by either of the two sensors S_1, S_2. The remaining cells of the CW-Voronoi diagram are labeled by the sensors to which they are closest. Note that a Voronoi cell of a point (a sensor in our case) may consists of several disjoint subcells (as shown in Figure 3 for S_1 and S_2). The vertex set of this CW-Voronoi diagram consists of (a) the set of obstacle endpoints, (b) the intersection of bisectors between sensors and (c) the intersection of extended visibility lines from sensors passing through the obstacle endpoints. Every Voronoi edge is a section of the bisector of two sensors (e.g., (x, y) in Figure 3), or a section of a visibility line determined by a sensor and an endpoint of an obstacle (e.g., (g, b) in Figure 3), or a section of an obstacle (e.g., (g, h) in Figure 3).

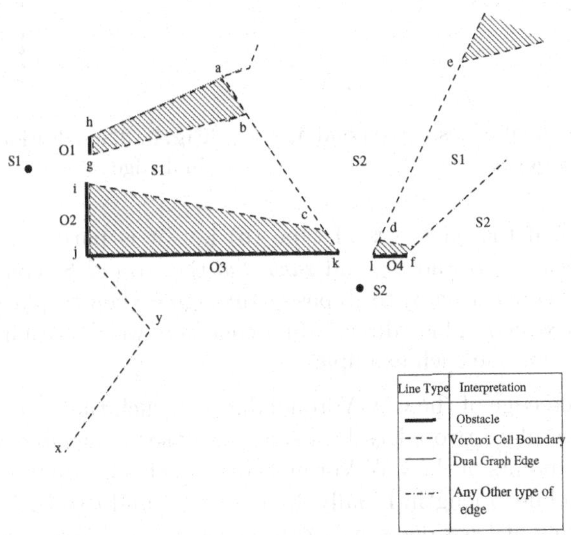

Fig. 3. CW-Voronoi diagram of sensors and obstacles

We are now ready to define a specific *dual* of the CW-Voronoi diagram that will be useful in computing the best coverage path. The dual of the CW-Voronoi diagram is a weighted graph. The vertices are the union of (a) the set of sensors, (b) the vertices of the CW-Voronoi diagram, and (c) the points s and t. We next define the edges of the dual graph.

Consider any edge $e = (u, v)$ of the CW-Voronoi diagram. The edge is one of three types: (a) it is part of an obstacle, (b) it is part of a perpendicular bisector between two sensors, or (c) it is part of an extension of a visibility line from a sensor that passes through an obstacle endpoint. For example, in Figure 3, edge (g, h) is of type(a), edge (x, y) is of type (b), and edge (b, c) is of type (c).

Let $C_1(e)$, $C_2(e)$ be two adjacent CW-Voronoi cells on either side of the edge. Assume that neither of the cells are dark, and let S_1 and S_2 be the labels on these cells, i.e., the two sensors to which the cells are respectively closest. Note that if e is of type (c) then e and one of the two sensors are collinear (e.g., in Figure 3 the edge (b, c) is collinear with sensor S_2).

For each such edge $e = (u, v)$ of the CW-Voronoi diagram, we add edges to the dual graph as follows:

- We add four dual edges (u, S_1), (u, S_2),(v, S_1), and (v, S_2) (see Figure 4). Each dual edge is assigned a weight equal to the Euclidean distance between its endpoints.

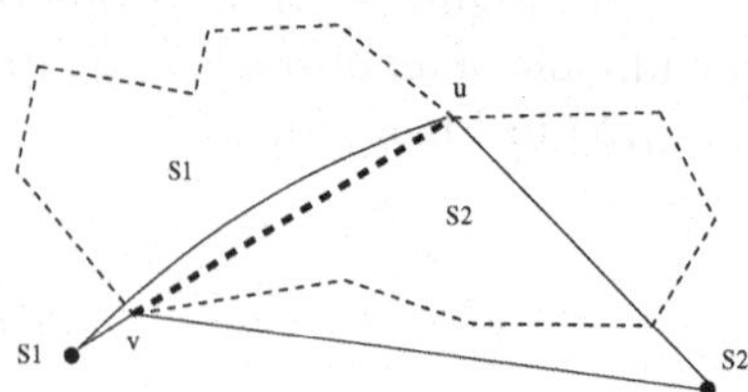

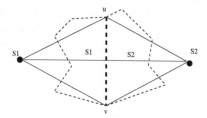

Fig. 4. The four dual edges corresponding to the CW-Voronoi edge $e(u, v)$

Fig. 5. The additional fifth "direct" dual edge

- In addition, if the edge e is of type (b) i.e., it is part of a perpendicular bisector between S_1 and S_2, and *such that* the S_1 and S_2 can see each other and the line connecting S_1 to S_2 passes through e, then we place an additional dual edge (S_1, S_2). This "direct" dual edge gets weight equal to $||S_1S_2||/2$. Figure 5 shows one such example.

Next, for each edge of the CW-Voronoi diagram such that one of the adjacent cells is dark, we place two edges between the sensor associated with the other cell and the endpoints of the CW-Voronoi edge. Each such dual edge gets weight equal to its Euclidean length. Finally, we connect s and t to their closest visible sensors (assuming at least one sensor can see them), and assign weights of these dual edges as their Euclidean distances. This concludes the construction of the dual graph.

Lemma 1. *The dual graph has $O(m^2n^2 + n^4)$ number of vertices and edges.*

Proof. The CW-Voronoi diagram has $O(m^2n^2+n^4)$ number of vertices and edges [13]. As per the dual definition, the dual also has the same number of vertices. Since each CW-Voronoi edge contributes a constant number of edges to the dual, the number of edges in the dual is also $O(m^2n^2 + n^4)$.

We note that way the dual graph has been constructed, at least one endpoint of each edge is a sensor. Thus there cannot be two or more consecutive vertices along any path that are not sensors. We shall next show that there exists a $BCP(s, t)$ that has such a property, hence it can be searched for within this dual graph.

2.2 $BCP(s,t)$ Algorithm for Opaque Obstacles

The algorithm to compute $BCP(s,t)$ for opaque obstacles is as follows:

Algorithm 1. Calculate $BCP(S,O,s,t)$ for opaque obstacles

1: Using the technique of [13], construct the CW-Voronoi diagram of all n sensors and m obstacles (assign each sensor a weight of 1).
2: Construct the dual of the CW-Voronoi diagram as described in Section 2.1.
3: Run *Bellman-Ford* algorithm on this constructed dual graph starting at point s and ending at point t, which computes the *Best Coverage Path* between s and t.
4: The value of $cover = \max(weight(e_1), weight(e_2), \ldots, weight(e_r))$ in the constructed path, where $e_1, e_2, \ldots\ldots, e_r$ are the edges in a best coverage path, $BCP(s,t)$.

Proof of Correctness

Theorem 1. *A $BCP(s,t)$ path for opaque obstacles is contained within the constructed dual graph.*

Proof. The overall idea of the proof is to show that a best coverage path that lies outside the dual graph can be transformed into one that uses only the edges of the dual graph, without increasing the cover value. Consider Figure 6, which shows a best coverage path that does not use the edges of the dual graph. Let us decompose this path into pieces such that each piece lies wholly within a cell of the Voronoi diagram. Consider one such piece within a cell labeled S_i. Let the piece start at a point p and end at a point q, where both p and q are along the cell's boundary. It is easy to see that each such piece can be replaced by the two line segments (p, S_i) and (S_i, q) *without* increasing the cover of the path. Thus, any best coverage path can be transformed into one having linear segments that goes from cell boundary to sensor to cell boundary and so on.

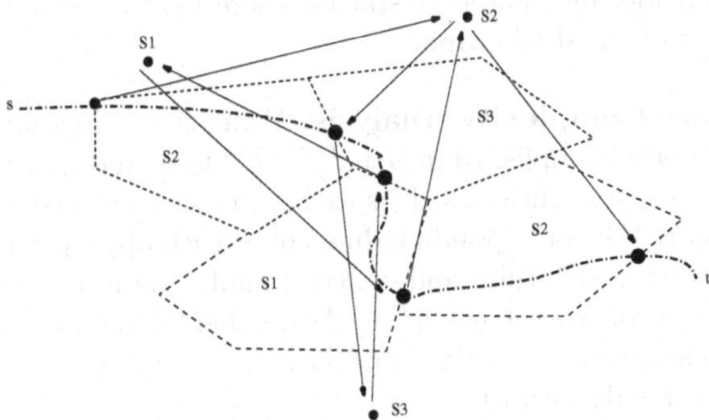

Fig. 6. Transforming a best coverage path

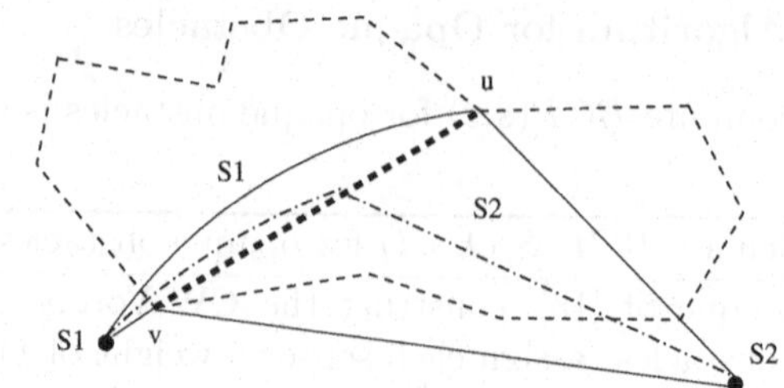

Fig. 7. Moving a $BCP(s,t)$ vertex to a Voronoi vertex

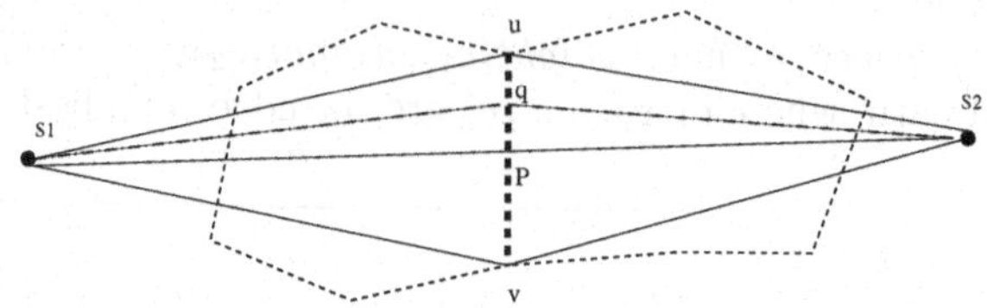

Fig. 8. Moving a $BCP(s,t)$ vertex to the middle of the dual edge

We next show that the points along the cell boundaries of this transformed path can be aligned with Voronoi vertices. To prove this, assume that one such point is not a Voronoi vertex, i.e, it is a point along an edge of a cell boundary. Consider Figure 7, which shows a portion of a $BCP(s,t)$ (transformed as discussed above) that goes from sensor S_1 to a point along the Voronoi edge (u,v) and then to sensor S_2. It should be clear that if we replace this portion by two edges of the dual graph, (S_1,v) and (v,S_2), we will achieve an alternate path whose overall cover value will be the same or less.

One final case needs to be discussed. If a $BCP(s,t)$ passes through a point along a Voronoi edge that is a perpendicular bisector between two sensors S_1 and S_2, such that there is a "direct" dual edge between S_1 and S_2, then the portion of $BCP(s,t)$ from S_1 to S_2 can be replaced by this direct edge without increasing the cover value of the path. This situation is shown in Figure 8.

Thus we conclude, for opaque obstacles, there exists a $BCP(s,t)$ that only follows the edges of the dual graph.

Time and Space Complexity Analysis. Using the techniques in [13], Step 1 of algorithm can be accomplished in $O(m^2n^2+n^4)$ time and space. Likewise, constructing the dual is straightforward, as we have to scan each edge of the Voronoi diagram and insert the corresponding dual edges with appropriate weights. This also takes $O(m^2n^2+n^4)$ time and space. Finally, running the Bellman-Ford Algorithm on a graph with $O(m^2n^2+n^4)$ number of vertices and edges takes $O((m^2n^2+n^4)\log(mn+n^2))$ time and $O(m^2n^2+n^4)$ space (i.e., the overall running time of the algorithm).

3 $BCP(s,t)$ Problem for Transparent Obstacles

In this section, we study the $BCP(s,t)$ problem for transparent obstacles. We note that for transparent obstacles, computing $BCP(s,t)$ is an easier problem than that for opaque obstacles. In contrast to a BCP for opaque obstacles, in the case of transparent obstacles we can show that the visibility graph contains a BCP. However, unlike traditional visibility graph, the edge weights of this graph are more complex than standard Euclidean distances. Consequently, the running time of the algorithm is dominated by the edge weight assignment task.

3.1 $BCP(s,t)$ Algorithm for Transparent Obstacles

The outline of the algorithm is as follows. We create the visibility graph of n sensors and m line segment obstacles, as well as the points s and t. The graph has $O(m^2 + n^2)$ edges. We then assign weights to each of the edges of the constructed visibility graph in a specific manner. Figure 9 explains the edge weight assignment process. A normal Voronoi diagram of the n sensor points (i.e., ignoring obstacles) is first overlayed on top of the constructed visibility graph. Consider the example in Figure 9, where the visibility edge (S_1, S_2) has to be assigned a weight. The edge (S_1, S_2) passes through the Voronoi cells of sensors S_1, S_2, S_3 and S_4. Let us partition (S_1, S_2) into segments that lie wholly within these Voronoi cells. For each segment lying inside the Voronoi cell of a particular sensor, we find out the least protected point. This has to be either of the two endpoints of this segment - e.g., (S_1, S_2) intersects the Voronoi cell of S_4 at the points a and b, so either a or b is the least protected point for this segment, and the cover value of the segment is $\max\{||aS_4||, ||bS_4||\}$. We compute the cover value of all such segments that belong to (S_1, S_2), and the maximum of these values is the cover value of (S_1, S_2), which gets assigned as the weight of (S_1, S_2). Once this weighted visibility graph has been constructed, we run the Bellman-Ford algorithm [4] to compute the best coverage path between s and t. The algorithm is formally described below.

Algorithm 2. Calculate $BCP(S, O, s, t)$ for Transparent Obstacles

---

1: Construct the visibility graph of n sensor nodes, m line obstacles, point s and t.
2: Overlay a (normal) Voronoi diagram of the n sensor nodes on top of the created visibility graph.
3: Assign weight of each edge $e = (u, v) \in$ visibility graph using the procedure described in Section 3.1.
4: Run the Bellman-Ford algorithm on this weighted visibility graph starting at point s and ending at point t to compute a $BCP(s,t)$.
5: $cover = \max(weight(e_1), weight(e_2), \ldots, weight(e_r))$, where $e_1, e_2, \ldots, e_r$ are the edges of $BCP(s,t)$.

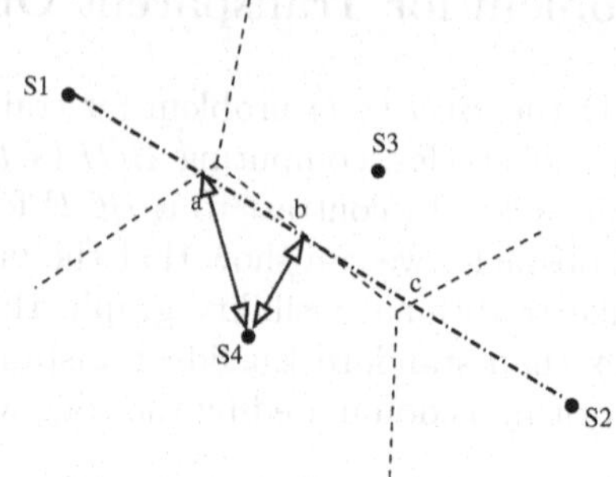

Fig. 9. Weight Assignment of a visibility edge S_1S_2

Next, we validate our proposed solution analytically.

Proof of Correctness

Theorem 2. *A $BCP(s,t)$ for transparent obstacles is contained within the constructed visibility graph.*

Proof. As proved in Theorem 1, the idea here is to show that a BCP that lies outside the visibility graph can be transformed into one that only uses the visibility edges. A BCP that does not follow the visibility edges means that the path makes some bend either:

- *Type 1:* Inside a Voronoi cell, or
- *Type 2:* At a Voronoi bisector, or
- *Type 3:* At a Voronoi vertex

We describe the transformations necessary for bends of the Type 1. Consider Figure 10(a) showing a best coverage path from s to t, that crosses through the Voronoi cell of sensor S_i but it does not follow visibility graph edges (line obstacles are solid thick lines, and BCP is shown as a wriggly path).

Consider two points a and b along this path inside the cell. The cover value of the portion of the path from a to b is at least $\max\{||S_ia||, ||S_i, b||\}$. Thus, if we replace the portion from a to b by the straight line segment (a, b), it is easy to see that the cover value of the transformed path will not have increased. Applying this "tightening" operations to a BCP shall eliminate all bends of Type 1. The resulting portion of a best coverage path within the Voronoi cell for S_i will be eventually transformed to as shown in Figure 10(b). As can be seen, other than the two points on the boundary of the cell (i.e., vertices of bend Type 2 or 3), the rest of the vertices of the path in the interior of the cell will be obstacle endpoints.

If we apply the above transformation to all Voronoi cells, we can eliminate all vertices of bend Type 1 from the path. The elimination of vertices of bend Type 2 and Type 3 follow similar arguments and we omit discussing them in this paper. Thus, we conclude that the visibility graph contains a best coverage path.

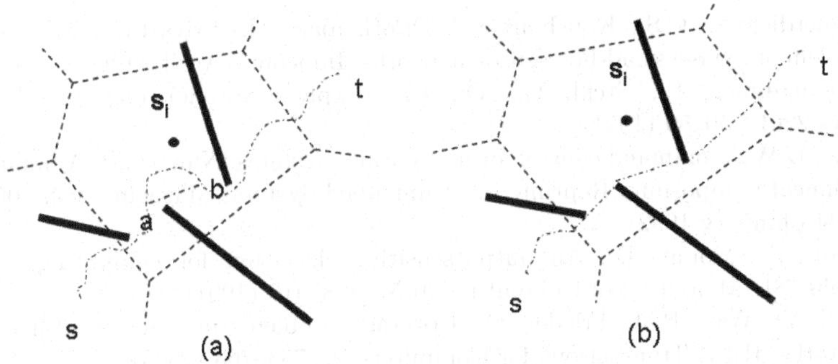

Fig. 10. Applying Type 1 Transformation Inside Voronoi Cells

Time and Space Complexity Analysis. By using the existing algorithm [8] to compute visibility graph, we incur a running time of $O((m+n)\log(m+n)+x)$ and a space requirement of $O(x)$, where x is the number of edges in visibility graph. Assigning weight to each visibility edge takes $O(n)$ time. The running time of this algorithm is dominated by the weight assignment step of all $O(m^2+n^2)$ visibility edges which will take $O(nm^2+n^3)$ time. Bellman-Ford algorithm will take $O((m+n)\log(m+n))$ time to run. The overall running time of this algorithm is $O(nm^2+n^3)$ and the space requirement is $O(m^2+n^2)$.

4 Conclusion

In this work, we have initiated the study of the computation of Best Coverage Paths in a sensor field in the presence of *obstacles*. We have shown that obstacles make the problem significantly difficult, and existing tools and techniques need to be substantially extended to solve the problem. We propose two algorithms to compute a $BCP(s,t)$ in the presence of m line segment obstacles. As future work, we plan to investigate practical techniques such as approximation algorithms and heuristics for solving these problems efficiently. In addition, we are interested to consider an alternative problem which finds out the set of sensors that can be reached in the plane given a source point s and a cover value c (techniques such as parametric search may be useful). We also plan to investigate other types of coverage problems in sensor networks in the presence of obstacles.

References

1. Lingas, A.: Voronoi diagrams with barriers and their applications. Inform. Process. Letters 32, 191–198 (1989)
2. Berg, M., Kreveld, M., Overmars, M., Schwarzkopf, O.: Computational Geometry: Algorithms and Applications. Springer, Heidelberg (1997)
3. Chew, L.P.: Constrained delaunay triangulations. In: Proceedings of the third annual symposium on Computational geometry, pp. 215–222 (1987)
4. Cormen, T.J., Leiserson, C.E., Rivest, R.L.: Introduction to Algorithms. MIT Press and McGraw-Hill (1990)

5. Meguerdichian, C.S., Koushanfar, F., Potkonjak, M., Srivastava, M.: Coverage problems in wireless ad-hoc sensor networks. Infocom. (April 2001)
6. Aurenhammer, F., Stckl, G.: On the peeper's voronoi diagram. SIGACT News 22(4), 50–59 (1991)
7. Gage, D.W.: Command control for many-robot systems. Nineteenth Annual AUVS Technical Symposium, Reprinted in Unmanned Systems Magazine. vol. 10(4), pp. 28–34 (January 1992)
8. Ghosh, S., Mount, D.: An outputsensitive algorithm for computing visibility graphs. SIAM Journal of Computing 20(5), 888–910 (1991)
9. Li, X.-Y., Wan, P.-J., Frieder, O.: Coverage problems in wireless ad-hoc sensor networks. IEEE Transactions for Computers 52, 753–763 (2003)
10. Megerian, S., Koshanfar, F., Potonjak, M., Srivastava, M.B.: Worst and best-case coverage in sensor networks. IEEE Transaction for Mobile Computing 4, 84–92 (2005)
11. Seidel, R.: Constrained delaunay triangulations and Voronoi diagrams with obstacles. Rep. 260, IIG-TU Graz, Austria, pp. 178–191 (1988)
12. Meguerdichian, S., Koushanfar, F., Qu, G., Potkonjak, M.: Exposure in wireless ad hoc sensor networks. In: Procs. of 7th Annual International Conference on Mobile Computing and Networking (MobiCom '01), pp. 139–150 (July 2001)
13. Wang, C.A., Tsin, Y.H.: Finding constrained and weighted voronoi diagrams in the plane. Computational Geometry: Theory and Applications 10(2), 89–104 (1998)
14. Urrutia, J.: Art gallery and illumination problems. In: Sack, J.R., Urrutia, J. (eds.) Handbook on Computational Geometry, pp. 973–1026. Elsevier Science Publishers, Amsterdam (2000)
15. Art gallery theorems and algorithms. Oxford University Press, Inc., Oxford, J. O'Rourke (1987)

35/44-Approximation for Asymmetric Maximum TSP with Triangle Inequality (Extended Abstract)

Łukasz Kowalik and Marcin Mucha*

Institute of Informatics, Warsaw University, Warsaw, Poland
{kowalik,mucha}@mimuw.edu.pl

Abstract. We describe a new approximation algorithm for the asymmetric maximum traveling salesman problem (ATSP) with triangle inequality. Our algorithm achieves approximation factor 35/44 which improves on the previous 31/40 factor of Bläser, Ram and Sviridenko [2].

1 Introduction

The Traveling Salesman Problem and its variants are among the most intensively researched problems in computer science and arise in a variety of applications. In its classical version, given a set of vertices V and a symmetric weight function $w : V^2 \to \mathbb{R}$ one has to find a Hamiltonian cycle of minimum weight. This problem is probably the most widely known example of an inapproximable NP-hard problem. However, there is a lot of research on approximation of several natural variants of TSP. These variants are still NP-hard, but allow approximation. One of the most important problems in this category is the maximization version (maxTSP for short), where w is assumed to have only nonnegative values (otherwise minTSP would reduce to it). There are several variants of maxTSP, e.g. the weight function can be symmetric or asymmetric, it can satisfy the triangle inequality or not, etc. (For some results on maxTSP variants see e.g. [3,4,6,8]).

In this paper, we are concerned with the variant, where the weight function is asymmetric (in other words, the graph is directed) and satisfies the triangle inequality. This variant is often called *the semimetric maxTSP*.

The first approximation algorithm for this problem was proposed by Kostochka and Serdyukov [9] in 1985 and had approximation ratio of $\frac{3}{4}$. Quite recently, Kaplan, Lewenstein, Shafrir and Sviridenko [5] provided a very general and powerful framework for approximating asymmetric TSP variants and gave improved approximation ratios for 3 different problems: $\frac{4}{3}\log_3 n$ for semimetric minTSP, $\frac{10}{13}$ for semimetric maxTSP and $\frac{2}{3}$ for asymmetric maxTSP. Using a different approach, Bläser et. al obtained a $\frac{31}{40}$-approximation algorithm for semimetric maxTSP.

* Part of this work was done while both authors were staying at the Max Planck Institute in Saarbruecken, Germany. This research is partially supported by a grant from the Polish Ministry of Science and Higher Education, project N206 005 32/0807.

F. Dehne, J.-R. Sack, and N. Zeh (Eds.): WADS 2007, LNCS 4619, pp. 589–600, 2007.

We show that in the case of semimetric maxTSP the ideas of Kaplan et al. can be combined with a new patching procedure yielding a $\frac{35}{44}$-approximation.

Overview of the paper. The semimetric max-TSP approximation algorithm of Kaplan et al. combines two ideas: Kostochka and Serdyukov's "patching" algorithm for the same problem and a new framework based on pairs of cycle covers. In Section 2 we briefly review both ideas and the way they can be combined. In Section 3 we introduce a new patching procedure based on Kaplan et al.'s framework. This immediately leads to a relatively simple $\frac{11}{14}$-approximation for semimetric maxTSP. In Section 4 we describe a more elaborate patching method which improves the approximation ratio to $\frac{35}{44}$ by lowerbounding the weight of almost every edge used to form a Hamiltonian cycle.

2 Preliminaries

Throughout the remainder of this paper we assume all graphs to be directed and weighted with a nonnegative weight function w satisfying the triangle inequality.

2.1 Kostochka and Serdyukov's Algorithm

Many approximation algorithms for TSP problems begin with finding a minimum (maximum) cycle cover and then patch it to a Hamiltionian cycle. The following theorem shows how this is done in Kostochka and Serdyukov's algorithm.

Theorem 1. *Let $\mathcal{C} = \{C_1, \ldots, C_k\}$ be a cycle cover in a directed weighted graph G with edge weights satisfying the triangle inequality. Let m_i be the number of edges in C_i and let $w_i = w(C_i)$ be the weight of C_i. Given the cycle cover $\mathcal{C}$, we can find in polynomial time a Hamiltonian cycle of weight $\sum_{i=2}^{k}\left(1 - \frac{1}{2m_i}\right) w_i$.*

A slightly weaker version of the above theorem is due to Kostochka and Serdyukov [7]. The version in this paper is taken from Kaplan et al. [5].

Maximum weight cycle cover (possibly containing 2-cycles) can be found in polynomial time. Such cover has weight at least as large as the maximum weight Hamiltonian cycle. From Theorem 1 it follows that

Theorem 2. *There exists a $\frac{3}{4}$-approximation algorithm for semimetric maxTSP.*

2.2 The Algorithm of Kaplan et al.

The 2-cycles are the obvious bottleneck of the above approach. If we could find, in polynomial time, a maximum weight cycle cover with no 2-cycles, we would get a 5/6-approximation algorithm. Unfortunately, finding such a cover is an NP-hard problem (see e.g. [1]). Kaplan et al. [5] proposed the following alternative approach.

Theorem 3. *Let $G = (V, E)$ be a directed weighted graph. We can find in polynomial time a pair of cycle covers $\mathcal{C}_1, \mathcal{C}_2$ such that (i) $\mathcal{C}_1$ and $\mathcal{C}_2$ share no 2-cycles, (ii) total weight $w(\mathcal{C}_1) + w(\mathcal{C}_2)$ of the two covers is at least $2OPT$, where OPT is the weight of the maximum weight Hamiltonian cycle in G.*

We will call such pairs of cycle covers *nice pairs of cycle covers.*

Observation 1 (Kaplan et al.). *In the above theorem, we can assume that the graph consisting of all the 2-cycles of $\mathcal{C}_1$ and $\mathcal{C}_2$ does not contain oppositely oriented cycles. For if it does contain such cycles, say C and its opposite $\hat{C}$, we can remove all the 2-cycles forming C and $\hat{C}$ from $\mathcal{C}_1$ and $\mathcal{C}_2$ and instead add C to $\mathcal{C}_1$ and $\hat{C}$ to $\mathcal{C}_2$.*

Theorem 4. *There exists a $\frac{10}{13}$-approximation algorithm for semimetric maxTSP.*

The proof of the above theorem can be found in [5]. Since our approach extends that of Kaplan et al., we include it here for completeness. Let us first introduce a few definitions. A *bipath* is a pair of oppositely oriented paths, i.e. a path and its opposite. As a special case, a *biedge* is a single edge together with its opposite edge. A *bicycle* is a pair of oppositely oriented cycles. Finally, a *Hamiltonian bicycle* is a pair of oppositely oriented Hamiltonian cycles.

Proof (of Theorem 4). Let $\mathcal{C}_1$, $\mathcal{C}_2$ be a nice pair of cycle covers. Applying Theorem 1 to $\mathcal{C}_1$ and $\mathcal{C}_2$, we get two Hamiltonian cycles H_1, H_2 with total weight $w(H_1) + w(H_2) \geq \frac{3}{4}W_2 + \frac{5}{6}W_{3+}$, where W_2 is the total weight of 2-cycles in $\mathcal{C}_1$ and $\mathcal{C}_2$ and W_{3+} is the total weight of all the other cycles.

Another way to construct a Hamiltonian cycle using $\mathcal{C}_1$ and $\mathcal{C}_2$ is to consider the graph H consisting of all the 2-cycles of $\mathcal{C}_1$ and $\mathcal{C}_2$. It follows from Observation 1 that H is a sum of disjoint bipaths. We can patch these bipaths arbitrarily to get a Hamiltonian bicycle $\hat{H}$ of weight $w(\hat{H}) \geq W_2$.

Picking the heaviest cycle out of H_1, H_2 and the two cycles of $\hat{H}$ gives a Hamiltonian cycle of weight at least $\frac{1}{2}\max\left\{\frac{3}{4}W_2 + \frac{5}{6}W_{3+}, W_2\right\}$. Since $W_2 + W_{3+} \geq 2\mathrm{OPT}$, easy calculation (or solving a corresponding linear program) shows that the weight of this heaviest cycle is at least $\frac{10}{13}\mathrm{OPT}$. □

3 Spanning Bitrees and 11/14-Approximation

Kaplan et al.'s algorithm (see Theorem 4) balances two solutions. The first one is based on Kostochka and Serdyukov's algorithm and the second one on Kaplan et al.'s approach of constructing a nice pair of cycle covers. However, from these cycle covers they pick only the 2-cycles. The basic idea of our approach is to partially incorporate longer cycles into this second solution by constructing additional bipaths and/or extending existing ones.

Remark 1. Cycles of length > 2 do not contain pairs of opposite edges. Hence, not all the new bipath edges will belong to some cycle.

Let P be a family of disjoint bipaths. We say that set of biedges S is *allowed* w.r.t. P, if S is disjoint from P and the edge sum of P and S is a family of disjoint bipaths (e.g. adding S does not create a bicycle in P). A biedge e is *allowed* w.r.t P if $\{e\}$ is allowed w.r.t. P, otherwise e is *forbidden.*

The following is the skeleton of the algorithm, that we will develop in the remainder of the paper:

Algorithm 3.1. MAIN ALGORITHM

```
Let C_1, C_2 be a nice pair of cycle covers
Let P be the family of bipaths constructed in Kaplan et al.'s Algorithm
Mark all 2-cycles as processed
for all unprocessed cycles C in C_1 and C_2 do
    use C to construct a heavy set S of biedges, allowed w.r.t. P
    P := P ∪ S
    mark C as processed
arbitrarily patch P to a Hamiltonian bicycle
```

Let the degree $\deg_P(v)$ of a vertex v in a family P of bipaths be the number of biedges in P incident with v (and not the number of edges). In the above algorithm S will always be chosen in such a way that the following is satisfied:

Invariant 1. *For any vertex v, $\deg_P(v)$ is not greater than the number of processed cycles containing v.*

How do we construct a heavy set of biedges S using a cycle C? In this section, S will contain only a single biedge e with both ends in C. When choosing $S = \{e\}$, we could pick e to be any of the biedges allowed w.r.t. P. However, we want e to have a large weight.

Let *bitree* be a connected set of biedges with no bicycles. Let C be a cycle and let the vertices of C be numbered $1, \ldots, k$ along the cycle. A bitree T is *plane* w.r.t. C if T does not contain two biedges u_1u_2, v_1v_2 such that $u_1 < v_1 < u_2 < v_2$ (intuitively, this means that if we make a planar drawing of C, we can complete it to a planar drawing of $C \cup T$). We say that T is a *plane spanning bitree* of C if T is plane w.r.t. C and connects all vertices of C. Plane spanning bitrees are interesting because they have large weight.[1]

Lemma 1. *Let T be a plane spanning bitree of a cycle C. Then $w(T) \geq w(C)$.*

Proof. The proof relies on the triangle inequality. The weight of every edge of C is upperbounded by the weight of a certain path in T. Figure 1 shows how this is done. The solid paths incident to a region marked with number i upperbound the weight of the cycle edge i. □

[1] All the plane spanning bitrees we use in this paper are in fact bipaths. We believe, however, that the more general setting might be beneficial in attempts to improve the results of this paper.

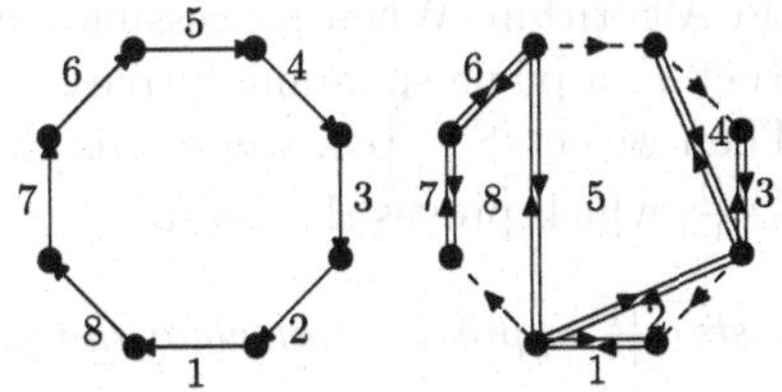

Fig. 1. The proof idea of Lemma 1

Observation 2. *Consider an execution of the Main Algorithm and let C be an unprocessed cycle. If P satisfies Invariant 1, then the set of biedges that have both endpoints in C and are forbidden w.r.t P forms a matching.*

Lemma 2. *Consider an execution of the Main Algorithm, let C be an unprocessed cycle, and let P satisfy Invariant 1. Then, there exists T, a plane spanning bitree w.r.t. C (in fact, a bipath), whose all biedges are allowed w.r.t P.*

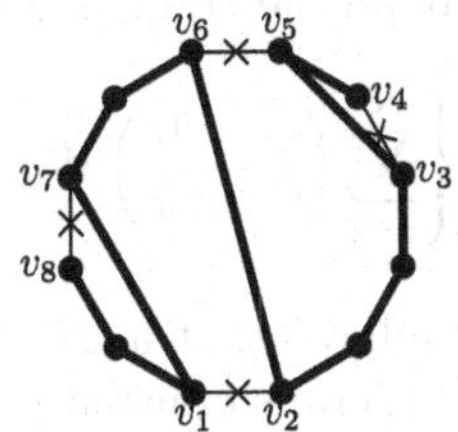

Fig. 2. Finding a plane bipath avoiding forbidden edges

Proof. The path T is constructed as follows. First, for each edge (u, v) of cycle C put biedge uv in T whenever it is allowed. Note that at this point T already contains all vertices of C (because forbidden biedges with endvertices on C form a matching). Let k be the number of forbidden biedges corresponding to edges in $E(C)$. If $k = 0$ we remove any biedge from T and we are done. Otherwise enumerate the endvertices of the k biedges on C from v_1 to v_{2k} along the cycle C. Finally, for every $i = 1, \ldots, k-1$ add edge $v_i v_{2k-i}$ to T. (See Fig. 2). All these edges are allowed since their endvertices are endvertices of distinct forbidden edges and forbidden edges with ends on C form a matching. Also, T forms a path, since all its vertices are of degree 2 except for v_k and v_{2k}, which are of degree 1. Finally, path T is plane: the only edges that may cross are chords of C, however, for any pair of such distinct chords $v_i v_{2k-i}$, $v_j v_{2k-j}$ either $i < j < 2k-j < 2k-i$ or $j < i < 2k-i < 2k-j$. This proves the claim. □

Theorem 5. *Let C_1 and C_2 be a nice pair of cycle covers of G. Then, there exists a Hamiltonian bicycle in G with weight at least $\sum_{i=2}^{\infty} \frac{W_k}{k-1}$, where W_k is the total weight of k-cycles in C_1 and C_2.*

Proof. We use the Main Algorithm. When processing a cycle C of length k, we use Lemma 2 to construct T, a plane spanning bitree w.r.t C, whose all biedges are allowed w.r.t. P. Then we set $S = \{e\}$, where e is the heaviest biedge of T. By Lemma 1 $w(e) \geq \frac{w(C)}{k-1}$, which proves the claim. □

Theorem 6. *There exists a $\frac{11}{14}$-approximation algorithm for semimetric maxTSP.*

Proof. As in the proof of Theorem 4 we construct a nice pair of cycle covers $\mathcal{C}_1$, $\mathcal{C}_2$ and use Theorem 1 to get Hamiltonian cycles H_1, H_2 with total weight

$$w(H_1) + w(H_2) \geq \sum_{i=2}^{\infty} \left(1 - \frac{1}{2k}\right) W_k.$$

Next, by Theorem 5 to get two more Hamiltonian cycles H_3, H_4 with total weight

$$w(H_3) + w(H_4) \geq \sum_{i=2}^{\infty} \frac{1}{k-1} W_k.$$

Picking the heaviest cycle out of all the H_i gives a Hamiltonian cycle H of weight

$$w(H) \geq \frac{1}{2} \max \left\{ \sum_{i=2}^{\infty} \left(1 - \frac{1}{2k}\right) W_k, \sum_{i=2}^{\infty} \frac{1}{k-1} W_k \right\}.$$

From $\sum_{i=2}^{\infty} W_k \geq 2\text{OPT}$, it follows that $w(H) \geq \frac{11}{14}\text{OPT}$. This can be proved by solving a corresponding LP (details omitted in this extended abstract). □

4 Making Ends Meet and 35/44-Approximation

In this section we introduce two improvements. First, we will add more than one biedge to the family P of bipaths, while processing a single cycle C. This is possible if C is long enough. Moreover, recall that in the last step of the algorithm from the previous section we construct a Hamiltonian cycle by patching the bipaths with arbitrary edges. The endvertices of these edges could belong to distinct cycles and we do not lowerbound their weight in any way. The second improvement we are going to present here is to partially incorporate the patching process into the main algorithm in order to be able to lowerbound this weight. We use this approach for processing short cycles.

4.1 Long Cycles

Lemma 3. *Let P be a family of disjoint bipaths satisfying Invariant 1 and let C be an unprocessed cycle of length at least 5. Then there exists an allowed family of biedges S, such that (i) after processing C, the family $P \cup S$ satisfies Invariant 1, (ii) $w(S) \geq \frac{1}{4}w(C)$, (iii) if $|C| \leq 7$ then $w(S) \geq \frac{1}{3}w(C)$, (iv) if $|C| = 5$ then $w(S) \geq \frac{1}{2}w(C)$.*

Proof. In order to keep Invariant 1 satisfied, we make S a set of vertex-disjoint allowed biedges with endvertices in C. Let Q be the plane bipath spanning C with no forbidden biedges, which exists by Lemma 2. We color the edges of Q with two colors: a and b, so that incident biedges get distinct colors. Adding all biedges of one color, say a, to P may create one or more bicycles (note that such a bicycle contains at least two biedges from Q). For each such bicycle we pick one biedge from Q and we recolor it to a new color c. Similarly, we recolor some biedges from b to d.

It is clear that each of the four color classes is an allowed family of biedges. Let S be the heaviest of these four sets. Clearly $w(S) \geq \frac{1}{4}w(Q)$. Since $w(Q) \geq w(C)$ by Lemma 1, we get (ii).

Now, let $|C| \leq 7$. Again, we find the bipath Q and we 2-color it. Suppose that adding all the biedges of color a to P gives a bicycle. Since there are at most 3 biedges colored a and any bicycle contains at least 2 such biedges, we can only get one such bicycle. Similarly, at most one bicycle is formed by P and biedges colored b. Suppose that both bicycles exist (the remaining cases are trivial). We need to recolor one (colored) biedge from each cycle to a new color, so that the recolored edges are not adjacent.

Let us start at one end of Q and go along Q until we encounter a colored cycle biedge. Assume w.l.o.g. that its color is a. Then, we can recolor both this biedge and the furthest cycle biedge colored b to a new color c. Clearly, each of the three color classes is an allowed family of biedges. Again, we let S be the heaviest of them, obtaining $w(S) \geq \frac{1}{3}w(C)$.

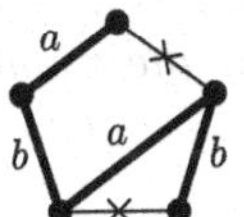

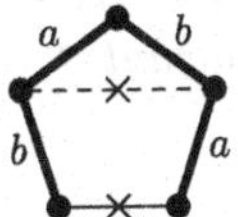

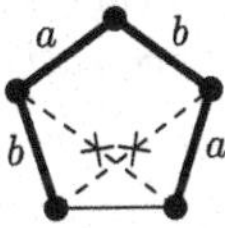

Fig. 3. Coloring a bipath spanning a 5-cycle. Crossed out edges are forbidden.

Finally, consider the case of $|C| = 5$. W.l.o.g. we can assume that there are two forbidden biedges with endvertices on C (if not, we can just "forbid" additional biedges). Figure 3 shows all three possible configurations of these biedges together with our choice of the bipath Q in each case. As before, we 2-color Q, and then set S to be the heavier of the two color classes. This gives $w(S) \geq \frac{1}{2}w(C)$. Observe that in each case both color classes contain a biedge with an endvertex not adjacent to a forbidden biedge. Such a biedge cannot be a part of a bicycle in $P \cup S$, so S is allowed. □

4.2 Short Cycles

To get the approximation ratio better than $\frac{11}{14}$ we need to extract more weight from the 3- and 4-cycles when constructing the bipaths in the Main Algorithm. Unfortunately, it turns out that it is impossible to take more than one edge from *each* such cycle. Note however, that when only a single biedge is put into P when processing a cycle C, at least one vertex v of C becomes a *loose end*, i.e. $\deg_P(v)$ is smaller than the number of processed cycles containing v.

Remark 2. If $\deg_P(v) = 0$ and both cycles containing v have already been processed, we consider v to be *two* loose ends.

We can link loose ends from distinct cycles without violating Invariant 1. Surprisingly, it is possible to lowerbound the weight of such links. First let us see how loose ends are created.

Lemma 4. *Let P be a family of disjoint bipaths satisfying Invariant 1 and let C be an unprocessed k-cycle. Then there exists an allowed family of biedges S such that (i) $w(S) \geq \frac{1}{k-1}w(C)$, (ii) after processing C family $P \cup S$ satisfies Invariant 1, and (iii) the number of loose ends increases by $k-2$.*

Proof. We use the approach described in the previous section, i.e. $S = \{e\}$ where e is the heaviest biedge of the plane spanning bipath of C. All the vertices of C except for the two endvertices of e become loose ends. □

The following two lemmas show how loose ends can be used to extract more weight from 3-cycles and 4-cycles.

Lemma 5. *Let P be a family of disjoint bipaths satisfying Invariant 1 with at least 2 loose ends and let C be an unprocessed 3-cycle. Then there exists an allowed family of biedges S such that (i) $w(S) \geq \frac{3}{4}w(C)$, (ii) after processing C, the family $P \cup S$ satisfies Invariant 1, and (iii) the number of loose ends decreases by 1.*

Proof. Our plan here is to make S contain one biedge with both endvertices in C and one biedge linking the remaining vertex of C with one of the loose ends. This obviously satisfies *(ii)* and *(iii)*. We only need to guarantee that S is allowed and that it has weight at least $\frac{3}{4}w(C)$. We consider one of the following two cases, depending on whether or not there exists a loose end v that is not connected to C with a bipath in P (this bipath might have length 0 in which case one of the vertices of C is a loose end).

Case 1. There exists such v. Let a, b, c be the vertices of C and suppose $Q = abc$ is a plane spanning bipath of C with no forbidden edges. Consider two possibilities for S: $S_1 = \{ab, cv\}$ (ab and cv denote biedges here) and $S_2 = \{bc, av\}$. Both are allowed. For example, if we add S_1 to P, cv lies on a bipath (not a bicycle) because v is not connected with C in P, and ab by itself cannot form a bicycle because it is allowed as a biedge of Q. Similar argument works for S_2. We also have

$$\begin{aligned} w(S_1) + w(S_2) &= w(ab) + w(bc) + w(cv) + w(va) \geq w(ab) + w(bc) + w(ca) \geq \\ &\geq \tfrac{1}{2}[(w(ab) + w(bc)) + (w(bc) + w(ca)) + (w(ca) + w(ab))] \geq \tfrac{3}{2}w(C), \end{aligned}$$

where the second inequality follows from the triangle inequality and the last inequality follows from Lemma 1. Taking S to be the heavier of S_1 and S_2 we get the required lower bound of $\frac{3}{4}w(C)$.

Case 2. Such v does not exists, so we have two loose ends u, v connected to two different vertices of C, say u connected to a, and v connected to b. Let c be the remaining vertex of C. Notice that all biedges of C are allowed. For if any of them, call it xy, were not allowed, then x and y would be connected with a bipath in P, and that cannot happen, since we know that either the bipath starting in x or the bipath starting in y ends in a loose end.

Consider the two solutions defined in the previous case: $S_1 = \{ab, cv\}$ and $S_2 = \{bc, av\}$. They are both allowed. For example, adding S_1 to P forms a bipath $\ldots cv \ldots ba \ldots u$ ending in a loose end u, so no bicycles are formed. Similar argument works for S_2. The weight argument is the same as in Case 1. □

Lemma 6. *Let P be a family of disjoint bipaths satisfying Invariant 1 with at least 2 loose ends and let C be an unprocessed 4-cycle. Then there exists an allowed family of biedges S such that (i) $w(S) \geq \frac{1}{2}w(C)$, (ii) after processing C, the family $P \cup S$ satisfies Invariant 1, and (iii) the number of loose ends does not change.*

Proof. Our plan is to make S contain two biedges with both endvertices on C or one biedge with both endvertices on C and one biedge linking a vertex of C with one of the loose ends. This satisfies *(ii)* and *(iii)* and again we only need to guarantee that S is allowed and that it has weight at least $\frac{1}{2}w(C)$. We consider the same two cases as in the previous lemma.

Case 1. There exists a loose end v not connected to C in P.
Let $C = abcd$ and let Q be a plane spanning bipath of C with no forbidden edges. We consider all solutions of the following form: a biedge of Q and a biedge connecting one of the remaining vertices of C and v. There a six such solutions since Q has 3 edges and there are always 2 remaining vertices. All these solutions are allowed. That is because the bipath edge is allowed by itself, and the linking edge cannot form a cycle in P since v is not connected with C in P.

Let us now bound the total weight of these six solutions. Consider a pair of solutions corresponding to a single biedge of Q, say xy. The total weight of these two solutions is $2w(xy) + w(vz) + w(vw) \geq 2w(xy) + w(zw)$ (by triangle inequality), where z, w are the two remaining vertices. So we get twice the weight of the bipath biedge and the weight of the complementary biedge. Now, notice that for any plane spanning bipath Q of a 4-cycle, the complementary biedges of biedges of Q also form a plane spanning bipath. It follows from Lemma 1 that the total weight of all six solutions is at least $3w(C)$. Taking S to be the heaviest of the six solutions gives the required lower bound of $\frac{1}{2}w(C)$.

Case 2. Such v does not exists, so we have two loose ends u, v connected to two different vertices of C. Let $C = abcd$. We have two cases.

Case 2a. v and u are connected to two successive cycle vertices, say u is connected to a and v is connected to b. Consider two solutions: $S_1 = \{da, bc\}$ and

$S_2 = \{ab, cv\}$ (here cv is a dummy biedge, added only to keep the number of loose ends constant for simplicity). Both solutions are allowed, because if we add any of them to P, each of the added biedges lies on a bipath ending in a loose end. Also $w(S_1) + w(S_2) \geq w(C)$ by Lemma 1, because $\{da, bc, ab\}$ is a plane spanning bitree of C.

Case 2b. v and u are connected to opposite cycle vertices, say u is connected to a and v is connected to c. Consider two solutions: $S_1 = \{ab, cd\}$ and $S_2 = \{ad, bc\}$. The rest of the argument is the same as in the previous Case 2a. □

For technical reasons, that will become clear in the proof of Theorem 7, the very last cycle needs to be processed even more effectively. This is possible, because when processing the last cycle we can make P a Hamiltonian bicycle. To deal with this special case we use the following lemmas (we defer their proofs to the full version of the paper).

Lemma 7. *Let P be a family of disjoint bipaths satisfying Invariant 1 with exactly 1 loose end. Assume that all cycles have been processed except for one 3-cycle C. Then there exists an allowed family of biedges S such that (i) $P \cup S$ is a Hamiltonian bicycle, (ii) $w(S) \geq \frac{3}{4}w(C)$.*

Lemma 8. *Let P be a family of disjoint bipaths satisfying Invariant 1 with exactly 2 loose ends. Assume that all cycles have been processed except for one 4-cycle C. Then there exists an allowed family of biedges S such that (i) $P \cup S$ is a Hamiltonian bicycle, (ii) $w(S) \geq \frac{2}{3}w(C)$.*

Lemma 9. *Let P be a family of disjoint bipaths satisfying Invariant 1 with no loose ends. Assume that all cycles have been processed except for one 4-cycle C. Then there is an allowed family of biedges S such that (i) $P \cup S$ is a Hamiltonian bicycle, (ii) $w(S) \geq \frac{1}{2}w(C)$.*

4.3 Putting It All Together

Theorem 7. *Let C_1 and C_2 be a nice pair of cycle covers of G. Then, there exists a Hamiltonian bicycle in G with weight at least $W_2 + \frac{5}{8}W_3 + \frac{1}{2}W_4 + \frac{1}{2}W_5 + \frac{1}{3}W_6 + \frac{1}{3}W_7 + \frac{1}{4}W_{8+}$, where W_k is the total weight of k-cycles in C_1 and C_2 and W_{8+} is the total weight of cycles of length at least 8 in C_1 and C_2.*

Proof. We use the Main Algorithm and process all the long (i.e. of length at least 5) cycles before the 3- and 4-cycles. Long cycles are processed using Lemma 3. As a result we get a family P of bipaths satisfying Invariant 1 and such that $w(P) \geq W_2 + \frac{1}{2}W_5 + \frac{1}{3}W_6 + \frac{1}{3}W_7 + \frac{1}{4}W_{8+}$. Depending on the number of loose ends in P, we continue in one of the following ways.

Case 1. There are at least 2 loose ends. Then we first process 4-cycles, in any order, using Lemma 6 for each cycle. Note that $w(P)$ increases by at least $\frac{1}{2}W_4$ during this phase. Next we process 3-cycles in order of decreasing weight. The first 3-cycle A is processed using Lemma 5. As a result the number of loose ends drops by 1 and $W(P)$ increases by $\frac{3}{4}w(A)$. Then we process the second 3-cycle B using Lemma 4. We get one loose end and $W(P)$ increases by $\frac{1}{2}w(B)$. We process all the 3-cycles in this way, alternating between Lemmas 5 and 4. Clearly that overall $W(P)$ increases by at least $\frac{5}{8}W_3$, hence after patching P to a Hamiltonian bicycle we get its total weight as claimed.

Case 2. There are no loose ends. Note that, when a cycle C is processed, the number of loose ends increases by $|C| - 2|S|$. Hence, at any time, the parity of the number of loose ends equals the parity of the sum of lengths of the processed cycles. It follows that if there are no loose ends then the sum of lengths of the processed cycles is even. On the other hand, the sum of lengths of all cycles in C_1 and C_2 is $2n$, hence also the sum of lengths of the unprocessed cycles is even. It implies that the number of 3-cycles is even. Now we will consider subcases regarding the number of 3-cycles and 4-cycles.

Case 2a. There are at least two 4-cycles. Then we start by processing the lightest 4-cycle using Lemma 4. This gives us 2 loose ends. Next, all 3-cycles and all but one remaining 4-cycles are processed using the algorithm from Case 1. Again, since the number of 3-cycles is even, we still have 2 loose ends when this phase is finished. It follows that the remaining 4-cycle can be processed using Lemma 8. We see that in total $w(P)$ increases by $\frac{1}{3}$ of the weight of the lightest 4-cycle, $\frac{2}{3}$ of the weight of some other 4-cycle, $\frac{1}{2}$ of the weight of all the other 4-cycles and by $\frac{5}{8}W_3$, which is at least $\frac{5}{8}W_3 + \frac{1}{2}W_4$, as required.

Case 2b. There are at least four 3-cycles. Then we start by processing the two lightest 3-cycles using Lemma 4. This gives us 2 loose ends and $w(P)$ increases by $\frac{1}{2}$ of the weight of these 3-cycles. Next, all 4-cycles and all but two remaining 3-cycles are processed using the algorithm from Case 1. This increases $w(P)$ by $\frac{5}{8}$ of the weight of the triangles processed in this phase and by $\frac{1}{2}W_4$. Note that since the number of 3-cycles is even, we still have 2 loose ends after this phase. The two remaining 3-cycles are processed using Lemma 5 and Lemma 7, respectively. Then $w(P)$ increases by $\frac{3}{4}$ of their weight. During the processing of all short cycles $w(P)$ increases by at least $\frac{5}{8}W_3 + \frac{1}{2}W_4$, as required.

Case 2c. There are two 3-cycles and one 4-cycle. Then we consider two methods of processing these cycles and we choose the more profitable one. Method 1: process the 3-cycles using Lemma 4 and obtaining 2 loose ends, then process the 4-cycle using Lemma 8. In this case $w(P)$ increases by $\frac{1}{2}W_3 + \frac{2}{3}W_4$. Method 2: process the 4-cycle using Lemma 4 and obtaining 2 loose ends, then process the 3-cycles using Lemma 5 for the first one and Lemma 7 for the second one. In this case $w(P)$ increases by $\frac{3}{4}W_3 + \frac{1}{3}W_4$. Clearly the better method gives us $\max\{\frac{1}{2}W_3 + \frac{2}{3}W_4, \frac{3}{4}W_3 + \frac{1}{3}W_4\} \geq \frac{5}{8}W_3 + \frac{1}{2}W_4$, as required .

Case 2d. There are no 3-cycles and there is one 4-cycle. We use Lemma 9.

Case 2e. There are two 3-cycles and no 4-cycles. We process the lighter 3-cycle A using Lemma 4 which gives us 1 loose end. Then the second 3-cycle B can be processed using Lemma 7. This increases $w(P)$ by at least $\frac{1}{2}w(A) + \frac{3}{4}w(B) \geq \frac{5}{8}W_3$ as required.

Case 3. There is exactly one loose end. By the parity argument from Case 2., the number of 3-cycles is odd. We can treat the single loose end as an imaginary 3-cycle I of weight 0. This way the number of 3-cycles becomes even and we again arrive at Case 2. Note that in the algorithms from subcases 2a, 2b and 2e the imaginary triangle would be processed using Lemma 4. If we just do nothing while processing I we get the same effect: $w(P)$ grows by $\frac{1}{2}w(I) = 0$ and we get an additional loose end. Case 2d does not apply since we do have 3-cycles. The only case left is a counterpart of Case 2c: there is one 3-cycle and one 4-cycle. Similarly to Case 2c we consider 2 methods and we choose the more profitable one. Method 1 is: process the 3-cycle using Lemma 4 obtaining the second loose end and then process the 4-cycle using Lemma 8. Method 2 is: process the 4-cycle using Lemma 4 obtaining two more loose ends and then process the 3-cycle using Lemma 5. Performing the same calculations as in Case 2c, we see that $w(P)$ increases by at least $\frac{5}{8}W_3 + \frac{1}{2}W_4$, as required. □

Theorem 8. *There exists a $\frac{35}{44}$-approximation algorithm for semimetric maxTSP.*

Proof. Similarly to the algorithm in Theorem 6, our algorithm chooses the heaviest of the four Hamiltonian cycles: two constructed by Kostochka and Serdukov's algorithm and the two cycles of the bicycle from Theorem 7. Again, by simple LP reasoning, one can show that the resulting cycle has weight $\geq \frac{35}{44}$OPT. □

Acknowledgments. The authors would like to thank Kasia Paluch for many helpful discussions.

References

1. Bläser, M., Manthey, B.: Two approximation algorithms for 3-cycle covers. In: Jansen, K., Leonardi, S., Vazirani, V.V. (eds.) APPROX 2002. LNCS, vol. 2462, pp. 40–50. Springer, Heidelberg (2002)
2. Bläser, M., Ram, S., Sviridenko, M.: Improved approximation algorithms for metric maximum ATSP and maximum 3-cycle cover problems. In: WADS'05, pp. 350–359 (2005)
3. Hassin, R., Rubinstein, S.: Better approximations for max TSP. Inf. Process. Lett. 75(4), 181–186 (2000)
4. Hassin, R., Rubinstein, S.: A 7/8-approximation algorithm for metric Max TSP. Inf. Process. Lett. 81(5), 247–251 (2002)
5. Kaplan, H., Lewenstein, M., Shafrir, N., Sviridenko, M.: Approximation algorithms for asymmetric TSP by decomposing directed regular multigraphs. J. ACM 52(4), 602–626 (2005)
6. Kosaraju, S.R., Park, J.K., Stein, C.: Long tours and short superstrings (preliminary version). In: FOCS'94, pp. 166–177 (1994)
7. Kostochka, A.V., Serdyukov, A.I.: Polynomial algorithms with the estimates 3/4 and 5/6 for the traveling salesman problem of the maximum (in Russian). Upravlyaemye Sistemy 26, 55–59 (1985)
8. Lewenstein, M., Sviridenko, M.: A 5/8 approximation algorithm for the maximum asymmetric TSP. SIAM J. Discrete Math. 17(2), 237–248 (2003)
9. Serdyukov, A.I.: The traveling salesman problem of the maximum (in Russian). Upravlyaemye Sistemy 25, 80–86 (1984)

On Euclidean Vehicle Routing with Allocation

Jan Remy, Reto Spöhel*, and Andreas Weißl

Institute of Theoretical Computer Science, ETH Zurich, 8093 Zurich, Switzerland
{jremy,rspoehel,aweissl}@inf.ethz.ch

Abstract. The (Euclidean) VEHICLE ROUTING ALLOCATION PROBLEM (VRAP) is a generalization of Euclidean TSP. We do not require that all points lie on the salesman tour. However, points that do not lie on the tour are allocated, i.e., they are directly connected to the nearest tour point, paying a higher (per-unit) cost. More formally, the input is a set of points $P \subset \mathbb{R}^d$ and functions $\alpha : P \to [0, \infty)$ and $\beta : P \to [1, \infty)$. We wish to compute a subset $T \subseteq P$ and a salesman tour π through T such that the total length of the tour plus the total allocation cost is minimum. The allocation cost for a single point $p \in P \setminus T$ is $\alpha(p) + \beta(p) \cdot d(p, q)$, where $q \in T$ is the nearest point on the tour. We give a PTAS with complexity $\mathcal{O}\left(n \log^{d+3} n\right)$ for this problem. Moreover, we propose a $\mathcal{O}(n \text{ polylog}(n))$-time PTAS for the Steiner variant of this problem. This dramatically improves a recent result of Armon *et al.* [2].

1 Introduction

Let $P \subset \mathbb{R}^2$ denote a set of points in the plane, and let penalty functions $\alpha : P \to [0, \infty)$ and $\beta : P \to [1, \infty)$ be given. A solution to the (Euclidean) VEHICLE ROUTING ALLOCATION PROBLEM (VRAP) is a subset of *tour points* $T \subseteq P$ and a tour π through T. Each *allocation point* $p \in A := P \setminus T$ is allocated to the nearest tour point $q \in T$ at a cost of $\alpha(p) + \beta(p) \cdot d(p, q)$. We wish to minimize the length of the tour plus the total allocation cost, i.e., we minimize

$$\mathrm{val}(T, \pi) = \sum_{\{p,q\} \in \pi} d(p, q) + \sum_{p \in A} \left(\alpha(p) + \beta(p) \min_{q \in T} d(p, q) \right) .$$

Throughout, let $T^* \subseteq P$ and π^* denote an optimal choice for T and π, i.e., $\mathrm{val}(T^*, \pi^*)$ is minimum.

VRAP is motivated by vehicle routing. For instance, each point represents a bank and we wish to transport cash to the banks using an armored vehicle. The vehicle can visit each bank (which would be a shortest salesman tour), but it might be cheaper to visit only some of the banks while the staff of the other banks have to pick up the cash at the visited banks. Although the total distance is smaller, this way of cash transportation is more risky and needs additional insurance (which can be modeled using the functions α and β).

* The author was supported by SNF grant 200021-108158.

F. Dehne, J.-R. Sack, and N. Zeh (Eds.): WADS 2007, LNCS 4619, pp. 601–612, 2007.

Observe that VRAP becomes the well-known Euclidean traveling salesman problem (TSP) if we have $\beta(p) > 2$ for all $p \in P$, since by the triangle inequality it is always cheaper to include a given point on the tour than to allocate it. As VRAP includes TSP as a special case, we know that VRAP is $\mathcal{NP}$-hard, even in the strong sense (cf. [7]). With $\mathcal{NP}$-hardness at hand approximation algorithms are of interest. A *polynomial time approximation scheme* (PTAS) is an algorithm that for any fixed $\varepsilon > 0$ approximates the optimum within a factor of $(1+\varepsilon)$ in time polynomial in n. Note that the complexity of a PTAS might be exponential in $1/\varepsilon$.

In this paper we show that VRAP admits a PTAS. As our main result, we propose a randomized nearly-linear time approximation scheme.

Theorem 1. *There is a randomized PTAS for* VRAP *with time complexity* $\mathcal{O}\left(n \log^5 n\right)$.

Moreover, we consider the problem Steiner VRAP where we are allowed to include additional points on the tour and allocate to these in order to further reduce the cost. A solution (T, S, π) to Steiner VRAP consists of point sets $T \subseteq P$ and $S \subset \mathbb{R}^2$, and a salesman tour π through $T \dot\cup S$. With $A := P \setminus T$ as before, we wish to find T^*, S^* and π^* minimizing

$$\mathrm{val}^\bullet(T, S, \pi) = \sum_{\{p,q\} \in \pi} d(p,q) + \sum_{p \in A} \left(\alpha(p) + \beta(p) \min_{q \in T \dot\cup S} d(p,q) \right).$$

Theorem 2. *There is a randomized PTAS for* Steiner VRAP *with time complexity* $\mathcal{O}\left(n \log^{\mathcal{O}(1/\varepsilon)} n\right)$.

Theorem 2 improves a recent result of Armon *et al.* [2], where the authors give a randomized PTAS with complexity $\mathcal{O}\left(n^{\mathcal{O}(1/\varepsilon)}\right)$ for a problem called Purchase Cooperative TSP, which is the special case of Steiner VRAP where $\alpha(p) = 0$ and $\beta(p) = 1$ for all points $p \in P$. Their algorithm seems to extend to Steiner VRAP as long as $\beta(p) = \beta(q)$ for all $p, q \in P$.

Both our algorithms extend to the case when $P \subset \mathbb{R}^d$ for some fixed dimension d. The running times increase to $\mathcal{O}(n \log^{d+3} n)$, respectively $\mathcal{O}(n \log^{\xi(d,\varepsilon)} n)$, where $\xi(d,\varepsilon) = \mathcal{O}(\sqrt{d}/\varepsilon)^{d-1}$. Moreover, both algorithms can be derandomized, increasing these complexities by a factor of $\mathcal{O}\left(n^d\right)$. Lastly, if $\beta(p) = \beta(q)$ for all $p, q \in P$, the running time of our PTAS for VRAP can be reduced by a factor of $\mathcal{O}(\log n)$, yielding a complexity of $\mathcal{O}\left(n \log^4 n\right)$ for the two-dimensional case.

Our Methods. Essentially, we prove Theorem 1 by combining the *adaptive dissection* technique due to Kolliopoulos and Rao [9] with dynamic programming on *r-vapid graphs*, as proposed by Rao and Smith [11].

The adaptive dissection technique is used for estimating allocation costs. Its main advantage over the well-known quad tree based methods introduced by Arora [3] is that it allows us to work with only a *constant* (instead of logarithmic) number of portals per rectangle. This improvement is achieved by two key ideas: On the one hand, the location of the tour points is guessed by dynamic programming, and if their bounding box is small, the *zoom tree* – which replaces the quad tree – zooms directly in the 'region of interest', potentially skipping many levels in between. On the other hand, in the resulting nearly-optimal portal-respecting solution, a point is not necessarily allocated to its nearest point, but possibly to a different nearby tour point. This added flexibility turns out to be of advantage. It is worth pointing out that – in contrast to Arora's technique – in the adaptive dissection framework it is necessary to allocate many points simultaneously to the same tour point, since allocating them individually would be too time-consuming.

To estimate the tour length, we transfer ideas presented in [11] for Euclidean TSP from the quad tree setting to the zoom tree setting. To compute a Euclidean spanner quickly, we use the algorithm by Gudmundsson *et al.* [8].

In order to prove Theorem 2, we make use of a relatively simple geometric observation and employ standard quad tree techniques developed in [3] and [4].

Related Work. It is well-known that Euclidean TSP admits a PTAS [3,10], even one with complexity $\mathcal{O}(n \log n)$ [11]. VRAP was introduced in 1996 by Beasley and Nascimento [5] as a network problem, i.e., instead of points in the plane we have a weighted graph as input. In more recent literature, the network version of the problem is usually called MEDIAN CYCLE or RING STAR. References can be found in [13]. Applications of VRAP in bookmobile routing [6] and grass-mower scheduling [12] have been reported.

As mentioned above, a related problem called PURCHASE COOPERATIVE TSP was recently studied by Armon *et al.* [2] in both the network and Euclidean setting. Using methods of Arora [3], they proposed a PTAS with complexity $\mathcal{O}\left(n^{\mathcal{O}(1/\varepsilon)}\right)$ for this problem. In addition, they studied several variants of the problem. As those variants are quite different from VRAP and STEINER VRAP, we refer to [2] for details.

Organization of the Paper. After giving some preliminary remarks in Section 2, we introduce the concepts of *zoom trees* and *portal-respecting distances* in Section 3, following Kolliopoulos and Rao [9]. In Section 4, we adapt the notion of *r-vapid graphs* due to Rao and Smith [11] to our purposes. In Section 5 we describe and analyze our PTAS for VRAP, and in Section 6 we outline how to improve the PTAS for STEINER VRAP proposed by Armon *et al.* [2]. In closing, we discuss the generalization to higher dimensions and explain how our algorithms can be derandomized in Section 7.

2 Preliminaries

We start with a simple but (throughout this paper) important observation: every input point p with $\beta(p) > 2$ is in every optimal solution a tour point due to the triangle inequality. With this fact at hand, one can show by standard techniques that it suffices to consider instances in which the input points have odd integral coordinates and the sidelength of the bounding box is a power of 2 and order of n/ε. We therefore assume throughout this paper that $P \subseteq \{1, 3, \ldots, L-1\}^2$, where $L = \mathcal{O}(n/\varepsilon)$ is a power of two.

3 Zoom Trees and Portal-Respecting Allocations

In this section we mainly simplify concepts appearing in [9], defining a certain distance measure between an allocated point p and the set of tour points T. This distance measure is defined with respect to a dissection tree, called a *zoom tree*, which adapts to a given solution to VRAP. The main result of this section is that, in expectation, this distance measure approximates the real allocation costs quite closely.

For fixed $a, b \in \{0, 2, ..., L-2\}$, let $G_{a,b}(i)$ denote a grid of granularity 2^i with origin (a, b), i.e., the vertical and horizontal grid lines have coordinates $\{a + j2^i : j \in \mathbb{Z}\}$ and $\{b + j2^i : j \in \mathbb{Z}\}$, respectively. Let i_0 denote the smallest integer such that $L = 2^{i_0} \geq 40n/\varepsilon$, and let $\mathfrak{Q}_0 := (a, b) + [-L, L]^2$ denote the square of sidelength $2L$ with center (a, b). Note that $P \subset \mathfrak{Q}_0$. Throughout this paper, we denote for any rectangle $\mathfrak{R} \subset \mathbb{R}^2$ by $|\mathfrak{R}|$ its *sidelength*, that is, the length of the longer sides of $\mathfrak{R}$.

For $1 \leq i \leq i_0$, a rectangle $\mathfrak{R}$ is said to be *i-allowable* if and only if it satisfies the following properties.

- $\mathfrak{R}$ lies in $\mathfrak{Q}_0$ and is bounded by lines of $G_{a,b}(i)$.
- If $i \geq 2$ then $7 \cdot 2^i \leq |\mathfrak{R}| < 7 \cdot 2^{i+1}$.
  If $i = 1$ then $|\mathfrak{R}| < 7 \cdot 2^{i+1} = 28$.

We say that i is the *level* of $\mathfrak{R}$. Note that $|\mathfrak{R}| = \Theta(2^i)$. $\mathfrak{R}$ is said to be *allowable* if there exists an i, $1 \leq i \leq i_0$, such that $\mathfrak{R}$ is i-allowable.

It is easily seen that the aspect ratio of an allowable rectangle is bounded by 14, and that the (non-empty) intersection of two allowable rectangles is an allowable rectangle. Moreover, we have the following Lemma, which will be useful when arguing about running times.

Lemma 1

(i) There are $\mathcal{O}(n \log n)$ allowable rectangles that contain at least one point of P.
(ii) There are $\mathcal{O}(n \log^2 n)$ pairs of allowable rectangles $(\mathfrak{R}', \mathfrak{R})$ such that $\mathfrak{R}'$ contains at least one point of P and $\mathfrak{R}' \subset \mathfrak{R}$.

Next, we introduce a dissection tree that adapts to a given solution to VRAP. The idea is to subdivide $\mathfrak{Q}_0$ recursively by alternately splitting the current rectangle and zooming into the 'area of interest'. We call such a subdivision a *zoom tree*. In principle, a zoom tree $ZT_{a,b}$ is defined with respect to a, b and any fixed subset $T \subseteq P$. However, in this section, as well as in the other analytical parts of this paper, we only consider the zoom tree corresponding to the set $T^* \subseteq P$ of tour points in the optimal solution (of course, the actual algorithm does not know this set in advance and will have to guess T^*, and therefore also the structure of the zoom tree considered here). The root of $ZT_{a,b}$ is $\mathfrak{Q}_0$, and the nodes of $ZT_{a,b}$ are the allowable rectangles recursively obtained from the following parent-child relations. For every rectangle in $ZT_{a,b}$ we either say that it is *split* or that it is *zoomed*, depending on how it is obtained from its parent.

If an i-allowable rectangle $\mathfrak{R} \in ZT_{a,b}$ is *zoomed*, we obtain its two children $\mathfrak{R}'$ and $\mathfrak{R}''$ by cutting $\mathfrak{R}$ parallel to its shorter side along the line of $G_{a,b}(i)$ that minimizes $|\text{area}(\mathfrak{R}') - \text{area}(\mathfrak{R}'')\,|$ (that is, we aim to nearly bisect $\mathfrak{R}$). If this cut is not unique we prefer the leftmost (bottommost) one. We call the line C along which we split $\mathfrak{R}$ the *cutting line*. It is easily seen that the two rectangles we obtain are j-allowable for some $j \in \{i-2, i-1, i\}$. $\mathfrak{R}'$ and $\mathfrak{R}''$ are split rectangles (as they are obtained by splitting $\mathfrak{R}$).

A *split* rectangle $\mathfrak{R}$ has only one child $\mathfrak{R}'$, which is constructed as follows: consider the minimal rectangle B containing all points in $\mathfrak{R} \cap T^*$ (one can show that a split rectangle always contains a tour point). Choose $\widetilde{\mathfrak{R}}$ as the allowable rectangle with smallest circumference such that $d(\partial\widetilde{\mathfrak{R}}, \partial B) \geq |B|/4$. If this does not uniquely define $\widetilde{\mathfrak{R}}$, choose the left- and bottommost candidate.

Let $\mathfrak{R}' := \mathfrak{R} \cap \widetilde{\mathfrak{R}}$. As the intersection of two allowable rectangles, the resulting rectangle $\mathfrak{R}'$ is allowable. $\mathfrak{R}'$ is a zoomed rectangle.

We stop the subdivision process at $\mathfrak{R}$ if either $\mathfrak{R}$ is 1-allowable or if $\mathfrak{R}$ contains at most one point of P. Such rectangles become *leaves* of $ZT_{a,b}$. We define $\mathfrak{Q}_0$ to be split, such that the first dissection step is a zoom step. For an allocation point $p \in P \setminus T^*$ and a tourpoint $q \in T^*$, we say that the rectangle $\mathfrak{R} \in ZT_{a,b}$ *separates* p and q if and only if it is the rectangle in $ZT_{a,b}$ closest to the root such that

- either $\mathfrak{R}$ is split and $p \in \mathfrak{R}$ and $q \notin \mathfrak{R}$,
- or $\mathfrak{R}$ is zoomed and $p \notin \mathfrak{R}$ and $q \in \mathfrak{R}$ (*sic!*).

One easily checks that this uniquely defines $\mathfrak{R}$.

In the sequel, we will introduce the concept of *portal-respecting allocations.* For a given (allowable) rectangle $\mathfrak{R}$, we place a point on each corner and $m-1$ equidistant points subdividing each side. We call these points *portals* and denote by $G_{\text{alloc}} = G_{\text{alloc}}(\mathfrak{R})$ the set of portals on $\partial\mathfrak{R}$. The *portal-respecting distance* $d_{\mathfrak{R}}(p,q)$ between $p \in \mathfrak{R}$ and $q \notin \mathfrak{R}$ is defined as

$$d_{\mathfrak{R}}(p,q) := \min_{g \in G_{\text{alloc}}(\mathfrak{R})} d(p,g) + d(g,q).$$

In other words, we detour the line segment pq over the nearest portal on $\partial\mathfrak{R}$.

It is easily checked that we have the following bound on the difference between the Euclidean distance and the portal-respecting distance with respect to some rectangle $\mathfrak{R}$.

Lemma 2. *For any rectangle $\mathfrak{R}$ and points $p \in \mathfrak{R}$ and $q \notin \mathfrak{R}$, we have*

$$d_{\mathfrak{R}}(p,q) - d(p,q) \leq \frac{|\mathfrak{R}|}{m}.$$

The *portal-respecting distance* $d_{ZT_{a,b}}(p,q)$ between $p \in P \setminus T^*$ and $q \in T^*$ is the portal-respecting distance w.r.t. the rectangle $\mathfrak{R} \in ZT_{a,b}$ that separates p and q. The main result of this section is the next lemma, which appears in similar form already in [9]. It asserts that a *constant* number m of portals per rectangle suffices to guarantee that, in expectation, the portal-respecting distances are good estimates for the real distances. Here we denote by $d_{ZT_{a,b}}(p,T^*)$ the infimum over all portal respecting distances $d_{ZT_{a,b}}(p,q)$, $q \in T^*$.

Lemma 3. *For given $\varepsilon > 0$, there exists $m = m(\varepsilon)$ such that for every allocation point $p \in P \setminus T^*$ and for a and b uniformly at random from $\{0,2,\dots,L-2\}$, we have*

$$\mathbb{E}\left[d_{ZT_{a,b}}(p,T^*)\right] \leq (1+\varepsilon)d(p,T^*).$$

In fact, the proof of Lemma 3 yields a slightly more general statement, which essentially asserts that the lemma still holds if an arbitrary constant factor is inserted in Lemma 2. This is needed because our algorithm introduces several other errors of order $\mathcal{O}\left(|\mathfrak{R}|/m\right)$ when estimating allocation costs.

4 VRAP on Straight-Line Graphs

The results in this section extend work by Rao and Smith [11] to the adaptive dissection setting. Using an algorithm due to Gudmundsson *et al.* [8], we can quickly compute a straight-line graph $\mathcal{S}'$ on a superset of P which has few 'relevant' crossings with the allowable rectangles introduced in the previous section, and such that there exists an expected nearly-optimal tour through $\mathcal{S}'$. This will allow us to quickly find such a tour by dynamic programming.

In the sequel it is of advantage to look at a solution (T,π) from a slightly different viewpoint. Recall that $\mathfrak{Q}_0$ has sidelength $2L = \mathcal{O}(n)$ and its center at (a,b), where $a,b \in \{0,2,\dots,L-2\}$ uniformly at random. For any connected straight-line graph (SLG) $\mathcal{G}$ on a vertex set $P' \supseteq P$, we denote the induced shortest path metric by $d_{\mathcal{G}}(\cdot,\cdot)$. For a given solution (T,π), let

$$\mathrm{val}_{\mathcal{G}}(T,\pi) = \sum_{\{p,q\}\in\pi} d_{\mathcal{G}}(p,q) + \sum_{p\in A}\left(\alpha(p)+\beta(p)\min_{q\in T} d(p,q)\right). \tag{1}$$

Note that only the length of the tour is measured in the shortest path metric. Every solution to VRAP gives rise to a closed walk $\mathcal{W} = \mathcal{W}(T,\pi)$ formed by the shortest paths between subsequent tour points. In principle this walk may include non-tourpoints, but in order to minimize (1) it is always better to 'pick up' such points and include them in T. Thus a solution minimizing (1) can be described as a walk $\mathcal{W}$ through $\mathcal{G}$, where by definition the tourpoints are exactly the points $T_{\mathcal{W}} := T \cap \mathcal{W}$ on the walk, and the remaining points $A_{\mathcal{W}} := P \setminus T_{\mathcal{W}}$ are allocated. Denoting the entire length of the walk $\mathcal{W}$ by $\ell(\mathcal{W})$ we can rewrite (1) as

$$\mathrm{val}_{\mathcal{G}}(\mathcal{W}) = \ell(\mathcal{W}) + \sum_{p\in A_{\mathcal{W}}}\left(\alpha(p)+\beta(p)\min_{q\in T_{\mathcal{W}}} d(p,q)\right). \tag{2}$$

Note that an edge contributes s times to $\ell(\mathcal{W})$ if we traverse it s times on the walk. We denote the optimization problem (2) by VRAP($\mathcal{G}$) in the following.

We say that a crossing of an edge e of a SLG $\mathcal{G}$ and a rectangle $\mathfrak{R}$ is *relevant* if e intersects $\partial\mathfrak{R}$ and exactly one endpoint of e is within $\mathfrak{R}$. The graph $\mathcal{G}$ is said to be *r-sparse* if any allowable rectangle $\mathfrak{R}$ has at most r relevant crossings with edges of $\mathcal{G}$. Note that it depends on the choice of a and b whether a fixed SLG $\mathcal{G}$ is r-sparse or not. Adapting concepts of Rao and Smith [11] and using an algorithm due to Gudmundsson *et al.* [8], one obtains the following Lemma.

Lemma 4. *Let T^* denote the set of tour points of the optimal solution to VRAP. For given $\varepsilon > 0$, there exists $r = r(\varepsilon)$ such that for all choices of $a, b \in \{0, 2, \dots, L-2\}$, one can compute in $\mathcal{O}(n \log^2 n)$ time a point set S and an r-sparse SLG $\mathcal{S}'$ on the point set $P' = P \,\dot\cup\, S$ satisfying the following: If a and b are chosen uniformly at random, the shortest walk $\mathcal{W}^*$ on $\mathcal{S}'$ visiting all points of T^* has expected length*

$$\mathbb{E}\left[\ell\left(\mathcal{W}^*\right)\right] \leq \sum_{\{p,q\}\in\pi^*} d(p,q) + \varepsilon \operatorname{val}\left(T^*, \pi^*\right).$$

Moreover, $\mathcal{W}^*$ uses no edge of $\mathcal{S}'$ more than twice.

Since $T_{\mathcal{W}^*} \supseteq T^*$, the allocation costs in $\operatorname{val}_{\mathcal{S}'}(\mathcal{W}^*)$ do not exceed those in $\operatorname{val}(T^*, \pi^*)$, and thus Lemma 4 immediately implies that

$$\mathbb{E}\left[\operatorname{val}_{\mathcal{S}'}\left(\mathcal{W}^*\right)\right] \leq (1+\varepsilon) \operatorname{val}\left(T^*, \pi^*\right).$$

Together with $\operatorname{val}(T^*, \pi^*) \leq \operatorname{val}_{\mathcal{S}'}(\mathcal{W}^*)$, it follows that $\mathcal{W}^*$ induces an expected nearly-optimal solution to VRAP.

5 A PTAS for VRAP

We now introduce a PTAS for VRAP. Lemma 4 plays a crucial role in our approach. In principle, our PTAS chooses a and b at random, computes $\mathcal{S}'$ and then tries to find an optimal solution $\mathcal{W}_0^*$ to VRAP($\mathcal{S}'$) by dynamic programming, guessing $T_{\mathcal{W}_0^*}$ and the corresponding zoom tree $ZT_{a,b}$ in the process. By Lemma 4 we know that this approach should yield an expected nearly-optimal solution to the original problem VRAP. (Note that $\mathcal{W}_0^*$ does not necessarily equal $\mathcal{W}^*$, as in Lemma 4 we only minimized the tour length and ignored allocation costs.)

This approach needs several extra twists to achieve the desired running time. Most notably, we estimate the allocation costs by the portal-respecting distances. This means that we will not necessarily find $\mathcal{W}_0^*$, but an optimal solution to a slightly modified problem. However, since by Lemma 3 the portal-respecting distances are good estimates for the real distances in expectation, we can keep the expected total error caused by this small.

Moreover, it would be too time-consuming to allocate all points individually, as the same point is considered in many different steps of the dynamic program.

We overcome this difficulty by partitioning the points that need to be allocated in a given step into classes and assigning all points of a class to the same tour point. The errors introduced by this are in the same order of magnitude as the errors inherent to the idea of portal-respecting allocations.

Lastly, to avoid costly shortest path computations we shortcut between vertices of $\mathcal{S}'$ whenever this does not spoil the sparseness properties of $\mathcal{S}'$ which are crucial to our algorithm. One can think of these shortcuts as additional edges that are added to $\mathcal{S}'$. In fact, this causes no problems, as it only decreases the length of the tour the algorithm will output, and does not change the allocation costs.

5.1 Dynamic Programming

Throughout this section, consider $m = m(\varepsilon)$ and $r = r(\varepsilon)$ fixed according to Lemma 3 and Lemma 4. For any allowable rectangle $\mathfrak{R}$ containing at least one point from P, let $G_{\text{alloc}} = G_{\text{alloc}}(\mathfrak{R})$ denote the set of the $4m$ portals on $\partial\mathfrak{R}$ as in Section 3, and let E_{cross} denote the set of edges of $\mathcal{S}'$ crossing the boundary of $\mathfrak{R}$ such that one endpoint is contained in $\mathfrak{R}$. As $\mathcal{S}'$ is r-sparse, we have $|E_{\text{cross}}| \leq r$. A *configuration* $\mathcal{C}$ for $\mathfrak{R}$ is given by

1. a collection S_{con} of pairs from E_{cross}, where each element appears at most twice,
2. functions $\zeta_{\text{in}} : G_{\text{alloc}} \to \{1, ..., 2m, \infty\}$ and $\zeta_{\text{out}} : G_{\text{alloc}} \to \{1, ..., 7m, \infty\}$, and
3. a bit $\sigma \in \{\text{split}, \text{zoomed}\}$.

Since m and r are constant, the total number of configurations for $\mathfrak{R}$ is bounded by a constant depending only on the desired approximation ratio ε.

A configuration $\mathcal{C}$ describes a subproblem, i.e., a local problem for $\mathfrak{R}$ which is interpreted as follows. Firstly, the pairs in S_{con} determine which of the edges in E_{cross} must be connected by walks inside $\mathfrak{R}$. As the walk $\mathcal{W}^*$ we are after might visit edges of $\mathcal{S}'$ twice (cf. Lemma 4), we allow S_{con} to contain duplicates. Secondly, the functions ζ_{in} and ζ_{out} describe the distance from a given portal to the next point on the tour inside resp. outside of $\mathfrak{R}$. More precisely, for every $g \in G_{\text{alloc}}$, the distance from g to the next point on the salesman tour inside $\mathfrak{R}$ is within distance $(|\mathfrak{R}|/m)\zeta_{\text{in}}(g)$. Analogously, the distance to the next point on the salesman tour outside of $\mathfrak{R}$ is encoded by $\zeta_{\text{out}}(g)$. One can show that it suffices to encode these distances up to $7|\mathfrak{R}|$ only.

We ask for a best possible local solution for a given configuration $\mathcal{C}$. More precisely, we try to minimize the length of all tour edges which lie *completely* inside $\mathfrak{R}$ plus the full allocation costs for *all* non-tour points in $\mathfrak{R}$ (regardless of whether they are allocated to a tour point inside or outside $\mathfrak{R}$), subject to the constraints and guarantees given by $\mathcal{C}$. As we do not know the zoom tree corresponding to the optimal solution in advance, we cannot proceed by a top-down divide and conquer approach along the zoom tree. Instead, we proceed bottom-up by dynamic programming, in a much larger structure which can be seen as the union of the zoom trees for all possible choices of $T \subseteq P$. By dynamic programming, we calculate close upper bounds $T[\mathfrak{R}, \mathcal{C}]$ for the optimal solutions to these local optimization problems. The bit σ indicates whether we look at $\mathfrak{R}$ as a split or as a zoomed rectangle in a zoom tree, and thus how our dynamic program calculates $T[\mathfrak{R}, \mathcal{C}]$ from the values previously found for smaller rectangles.

It is crucial that we consider only the $\mathcal{O}(n \log n)$ allowable rectangles $\mathfrak{R}$ containing at least one point of P. We topologically sort those rectangles w.r.t. the partial order given by normal set inclusion, and process them in this order, going through all possible configurations for each rectangle.

Let $\mathfrak{R}$ and $\mathcal{C}$ denote the rectangle and the configuration we currently consider. We now distinguish two cases: if $\mathfrak{R}$ is 1-allowable or $|P \cap \mathfrak{R}| = 1$ (case A), we find $T[\mathfrak{R}, \mathcal{C}]$ by exhaustive search. Otherwise, we calculate $T[\mathfrak{R}, \mathcal{C}]$ by dynamic programming from previously found values in one of two possible ways (case B), as specified by σ.

Case A. $\mathfrak{R}$ is 1-allowable or $|P \cap \mathfrak{R}| = 1$. Note that this means that $\mathfrak{R}$ is a leaf in any zoom tree it is contained in. Also note that an 1-allowable rectangle contains at most $13^2 = 169$ input points from P. This allows us to proceed in brute force fashion.

Let $\mathfrak{R}$ and $\mathcal{C}$ be given. First we choose from $P \cap \mathfrak{R}$ the points that lie on the salesman paths inside $\mathfrak{R}$. Let $T_0 \subseteq P \cap \mathfrak{R}$ denote this point set. The points in $A_0 := (P \cap \mathfrak{R}) \setminus T_0$ need to be allocated to some tour point either inside or outside of $\mathfrak{R}$. We have $\mathcal{O}(1)$ choices of T_0. For every such choice, we check whether it satisfies the restrictions given by ζ_{in} and compute optimal walks visiting exactly the points in T_0 subject to the constraints given by S_{con}. We do not require these walks to use edges of S', but calculate them on the complete graph induced by the (at most 169) points in T_0 and the (at most r) endpoints of the edges from S_{con} (if these points are not adjacent in S', we add the needed edges to S' as shortcuts). This can be done in constant time.

Moreover, we estimate for each $p \in A_0$ its allocation cost, using the function ζ_{out} for allocations to the outside of $\mathfrak{R}$. As we have $4m$ portals and at most 169 points, this can be done in time $\mathcal{O}(1)$.

The total cost for this choice of T_0 is the total length of all edges on the salesman paths that are entirely in $\mathfrak{R}$, plus the total allocation cost for all points in A_0. We identify the choice of T_0 minimizing this cost and store the corresponding value in $T[\mathfrak{R}, \mathcal{C}]$. Thus we can compute $T[\mathfrak{R}, \mathcal{C}]$ in $\mathcal{O}(1)$ time.

Case B. Otherwise. In this case we in particular have no upper bound on the number of input points inside $\mathfrak{R}$. Thus it is not longer possible to compute $T[\mathfrak{R}, \mathcal{C}]$ using brute force search. As we process the rectangles in ascending order, we may assume that we already have the values $T[\mathfrak{R}', \mathcal{C}']$ for all allowable rectangles $\mathfrak{R}' \subset \mathfrak{R}$ and for all configurations $\mathcal{C}'$ to $\mathfrak{R}'$.

Case B1. $\sigma =$ zoomed. We split $\mathfrak{R}$ into two allowable rectangles $\mathfrak{R}'$ and $\mathfrak{R}''$ according to the properties of zoom trees (cf. Section 3).

We enumerate all choices $(\mathcal{C}', \mathcal{C}'')$ of pairs of configurations for $\mathfrak{R}'$ and $\mathfrak{R}''$ with $\sigma' = \sigma'' =$ split, checking for each choice whether it is consistent in itself and compatible with $\mathcal{C}$.

For all pairs of configurations which remain, we compute $T[\mathfrak{R}', \mathcal{C}'] + T[\mathfrak{R}'', \mathcal{C}'']$ and add the total length of the edges in $S'_{\text{con}} \cap S''_{\text{con}}$, which are exactly the tour edges with one endpoint in $\mathfrak{R}'$ and the other in $\mathfrak{R}''$. We choose the pair $(\mathcal{C}', \mathcal{C}'')$ which minimizes this sum and write its value to $T[\mathfrak{R}, \mathcal{C}]$. Note that all computations can be accomplished in $\mathcal{O}(1)$ time.

Case B2. $\sigma =$ split. We enumerate all allowable rectangles $\mathfrak{R}' \subset \mathfrak{R}$ containing at least one point from P and all points $p \in P \cap \mathfrak{R}$ with $\beta(p) > 2$. For any such rectangle $\mathfrak{R}'$, we consider all configurations $\mathcal{C}'$ with $\sigma' =$ zoomed that are compatible with $\mathcal{C}$. For a given choice of $\mathfrak{R}'$ and $\mathcal{C}'$, the total cost for $\mathfrak{R}$ is $T[\mathfrak{R}', \mathcal{C}']$ plus the additional tour costs plus the additional allocation cost. Recall that by definition of the zoom step, there are no tour points in $\mathfrak{R} \setminus \mathfrak{R}'$.

To compute the additional tour cost, we proceed similarly as in case A. Computing the additional allocation cost is somewhat tricky. We need to allocate all input points in $\mathfrak{R} \setminus \mathfrak{R}'$. As we do not have a bound on the number of such points, doing this for each point separately would be prohibitively expensive. However, we can quickly approximate these allocation costs close enough for our purposes. This is achieved by subdividing $\mathfrak{R} \setminus \mathfrak{R}'$ into $\mathcal{O}(\log n)$ cells and assigning all points in a cell to the same portal. By orthogonal semigroup range searching (see [1] for references), we compute for each cell $\mathfrak{C}$ the values $\alpha(\mathfrak{C}) := \sum_{p \in P \cap \mathfrak{C}} \alpha(p)$ and $\beta(\mathfrak{C}) := \sum_{p \in P \cap \mathfrak{C}} \beta(p)$ in time $\mathcal{O}(\log^2 n)$. The resulting total allocation cost is then computed easily.

Lemma 5. *One can preprocess P in $\mathcal{O}(n \log n)$ time such that one can calculate in $\mathcal{O}(\log^3 n)$ time for any pair of allowable rectangles $\mathfrak{R}' \subset \mathfrak{R}$ an upper bound on the (portal-respecting) allocation cost which is tight up to a relative error of $\mathcal{O}(1/m)$ and an absolute error of $\mathcal{O}(k(|\mathfrak{R}|/m) + k'(|\mathfrak{R}'|/m))$, where k and k' denote the number of points that are allocated to a portal of G_{alloc}, respectively G'_{alloc}.*

We choose the configuration $\mathcal{C}'$ which minimizes the accumulated cost and store the total cost in $T[\mathfrak{R}, \mathcal{C}]$.

At the end of the dynamic program, we obtain a value $T[\mathfrak{Q}_0, \bigcirc] =: T[\mathfrak{Q}_0]$ for $\mathfrak{Q}_0 = (a,b) + [-L, L]^2$ and the configuration $\bigcirc$ with $S_{\text{con}} = \emptyset$, $\zeta_{\text{in}}(g) = \zeta_{\text{out}}(g) = \infty$ for all portals $g \in G_{\text{alloc}}(\mathfrak{Q}_0)$, and $\sigma =$ split.

5.2 Analysis

The complexity of our dynamic program is easily checked as follows. Preprocessing the points as required by Lemma 5 and computing the r-sparse SLG guaranteed by Lemma 4 takes time $\mathcal{O}(n \log^2 n)$. In the dynamic program, by Lemma 1(i), cases A and B1 apply $\mathcal{O}(n \log n)$ times and can be computed in time $\mathcal{O}(1)$. By Lemma 1(ii), case B2 applies to $\mathcal{O}(n \log^2 n)$ pairs of rectangles, and can be computed in time $\mathcal{O}(\log^3 n)$ due to Lemma 5. This yields an overall complexity of $\mathcal{O}(n \log^5 n)$.

It remains to argue why our algorithm produces a nearly-optimal solution. Due to space restrictions, we only outline the important points. Near-optimality of the tour found by the algorithm immediately follows from Lemma 4. There are three sources of error in the calculation of allocation costs. The first one is the error inherent to the concept of portal-respecting allocations (cf. Lemma 2), the second one stems from the inaccuracy in the encoding of distances, and the third one is introduced by Lemma 5. In the calculation of the distance of a given point p to a (nearby) tourpoint $q \in T^*$, these errors sum to an absolute error of $\mathcal{O}(|\mathfrak{R}|/m)$ for the rectangle $\mathfrak{R}$ separating p and q, and a relative error of $\mathcal{O}(1/m)$ which is easily bounded by ε choosing m large enough. By Lemma 3, it follows that the absolute errors result in an expected relative error of ε. Adapting ε, we obtain by linearity of expectation that

$$\mathbb{E}\left[T[\mathfrak{Q}_0]\right] \leq (1+\varepsilon)\text{val}\left(T^*, \pi^*\right).$$

Removing all points which are not in P from the walk found by the dynamic program, we obtain a solution (T, π) to VRAP with $\text{val}(T, \pi) \leq T[\mathfrak{Q}_0]$, and the claim follows.

6 Steiner VRAP

The PTAS for Purchase Cooperative TSP proposed in [2] proceeds by dynamic programming in a shifted quadtree $QT_{a,b}$, quite similar to Arora's $\mathcal{O}(n \log^{\mathcal{O}(1/\varepsilon)} n)$ PTAS for TSP [3]. Here, a configuration of a given square $\mathfrak{Q} \in QT_{a,b}$ is defined by specifying for each of $\mathcal{O}(\log n/\varepsilon)$ portals whether a tour- and/or an allocation edge runs through it. This results in $\mathcal{O}\left(n^{\mathcal{O}(1/\varepsilon)}\right)$ possible configurations for $\mathfrak{Q}$ and the same overall complexity for the algorithm. The key observation that allows us to improve on this is the following:

Lemma 6. *Let (T, S, π) denote a solution to* Steiner VRAP *crossing a fixed line segment L of length x five or more times. Then there exists a solution (T, S', π') with $S \subseteq S'$ crossing L not more than four times satisfying* $\text{val}^\bullet(T, S', \pi') \leq \text{val}^\bullet(T, S, \pi) + c \cdot x$ *for some constant c.*

Lemma 6 is an extension of Lemma 3 in [3]. With Lemma 6 at hand, one can show as in [3] that there exists an expected $(1+\varepsilon)$-approximation crossing each square of the shifted quadtree $QT_{a,b}$ only $r = \mathcal{O}(1/\varepsilon)$ times. This makes it possible to bound the number of configurations per square at $\mathcal{O}(\log n^{\mathcal{O}(1/\varepsilon)})$. Combining techniques presented in [3] and [4], one obtains a randomized PTAS for Steiner VRAP with complexity $\mathcal{O}(n \log^{\mathcal{O}(1/\varepsilon)} n)$.

7 Concluding Remarks

It is easily checked that our PTAS for VRAP extends to higher dimensions with minimal modifications. The running time increases to $\mathcal{O}(n \log^{d+3} n)$, as the range searching in Lemma 5 takes time $\mathcal{O}(\log^d n)$ per cell in dimension d. This can be reduced by a factor of $\mathcal{O}(\log n)$ if $\beta(p) = \beta(q)$ for all $p, q \in P$, since then it suffices to count the number of points in a given cell (see [1]).

Our PTAS for Steiner VRAP extends to higher dimensions analogously to Arora's PTAS for TSP [3]. In particular, Lemma 6 extends similarly to Lemma 3 in [3] to any dimension d, yielding a complexity of $\mathcal{O}(n \log^{\xi(d,\varepsilon)} n)$ with $\xi(d, \varepsilon) = \mathcal{O}(\sqrt{d}/\varepsilon)^{d-1}$.

Lastly, our algorithms can be trivially derandomized by enumerating all $\mathcal{O}(n^d)$ choices for the initial random shift of the zoom, respectively quad tree.

References

1. Agarwal, P.K., Erickson, J.: Geometric range searching and its relatives. In: Chazelle, B., Goodman, J.E., Pollack, R. (eds.) Advances in Discrete and Computational Geometry, Contemporary Mathematics, vol. 223, pp. 1–56. American Mathematical Socienty, Providence, RI (1999)
2. Armon, A., Avidor, A., Schwartz, O.: Cooperative TSP. In: Proceedings of the 14th Annual European Symposium on Algorithms, pp. 40–51 (2006)
3. Arora, S.: Polynomial time approximation schemes for Euclidean traveling salesman and other geometric problems. Journal of the ACM 45(5), 753–782 (1998)
4. Arora, S., Raghavan, P., Rao, S.: Approximation schemes for Euclidean k-medians and related problems. In: Proceedings of the 30th Annual ACM Symposium on Theory of Computing, pp. 106–113 (1998)
5. Beasley, J.E., Nascimento, E.: The vehicle routing allocation problem: a unifying framework. TOP 4, 65–86 (1996)
6. Foulds, L.R., Wallace, S.W., Wilson, J., Sagvolden, L.: Bookmobile routing and scheduling in Buskerud County, Norway. In: Proceeding of the 36th Annual Conference of the Operational Research Society of New Zealand, pp. 67–77 (2001)
7. Garey, M., Graham, R., Johnson, D.: Some NP-complete geometric problems. In: Proceedings of the 8th Annual ACM Symposium on Theory of Computing, pp. 10–22 (1976)
8. Gudmundsson, J., Levcopoulos, C., Narasimhan, G.: Improved greedy algorithms for constructing sparse geometric spanners. In: Proceedings of the 7th Scandinavian Workshop on Algorithm Theory, pp. 163–174 (2000)
9. Kolliopoulos, S., Rao, S.: A nearly linear-time approximation scheme for the euclidean k-median problem. In: Proceedings of the 7th Annual European Symposium on Algorithms, pp. 378–389 (1999)
10. Mitchell, J.S.B.: Guillotine subdivisions approximate polygonal subdivisions: A simple polynomial-time approximation scheme for geometric TSP, k-MST and related problems. SIAM Journal on Computing 28(4), 1298–1309 (1999)
11. Rao, S.B., Smith, W.D.: Approximating geometrical graphs via 'spanners' and 'banyans'. In: Proceedings of the 30th Annual ACM Symposium on Theory of Computing, pp. 106–113. ACM Press, New York (1998)
12. Tipping, J.: Scheduling and routing grass mowers around Christchurch. In: Proceeding of the 37th Annual Conference of the Operational Research Society of New Zealand (2002)
13. Vogt, L., Poojari, C.A., Beasley, J.E.: A tabu search algorithm for the single vehicle routing allocation problem. Journal of the Operational Research Society (to appear)

Optimal Lightweight Construction of Suffix Arrays for Constant Alphabets

Ge Nong[1,*] and Sen Zhang[2,**]

[1] Computer Science Department,
Sun Yat-Sen University, GuangZhou 510275, PRC
issng@mail.sysu.edu.cn

[2] Department of Mathematics, Computer Science and Statistics,
SUNY College at Oneonta, NY 07104, U.S.A.
zhangs@oneonta.edu

Abstract. This article presents our divide-and-conquer optimal algorithms for lightweight suffix array construction for constant alphabets. These algorithms can efficiently compute the suffix array of a size-n text T with an alphabet Σ using $O(n \log \Sigma)$ time and $(\ell(T) + |\Sigma|\lceil \log n \rceil + O(1))$-bit *working space* (excluding the space for the output suffix array), where Σ is an integer or constant alphabet, and $\ell(T)$ is the length of T measured in bits. For popular applications in practice with $n \leq 2^{32}$ and $|\Sigma| \leq 256$, these results translate into $O(n)$ time and a total space of $5n + O(1)$ bytes, which are the optimal time and space complexities for lightweight suffix array construction.

1 Introduction

The suffix array is a fundamental index data structure used in a broad range of applications such as compression, string matching and computational biology [1]. For a n-character text T, its suffix array $SA(T)$ is an array of pointers for all suffixes in T sorted lexicographically, which requires $O(n\lceil \log n \rceil)$-bit space. The concept of suffix array was initially proposed by Manber and Myers in 1990 [2,3]. Since then, suffix arrays have been employed widely for data indexing, retrieving, storing and processing. For example, the Burrows-Wheeler transform [4] for building efficient compression solutions can be quickly computed by fast suffixes sorting based on suffix array construction. In many cases where suffix arrays are applied, constructing the suffix arrays generally constitutes the basis for subsequent tasks. Recently, it has been observed that the construction of suffix arrays is needed for large-scale applications where the input texts are huge with over billions characters (e.g., biology genome database) [5,6,7,8,9], which

* Supported in part by the GuangDong Natural Science Foundation research grant number 06023193.
** Supported in part by the SUNY College Oneonta W.B. Ford Grants.

F. Dehne, J.-R. Sack, and N. Zeh (Eds.): WADS 2007, LNCS 4619, pp. 613–624, 2007.

motivated the currently intensive research on time and space efficient suffix array construction algorithms (SACAs).

The suffix array is usually known as a space efficient alternative to the suffix tree. In general, applications built on suffix trees can run times faster than that using suffix arrays; however, the storage of a suffix tree also consumes a memory space times more than its suffix array counterpart. To maximize the benefit from using suffix arrays, it is highly desirable to further reduce the working space required for constructing a suffix array, where the term of *working space* in this context doesn't include the space for the output suffix array [1]. In 2002, Manzini and Ferragina [11] initially raised the problem of lightweight suffix array construction, which was informally termed as quickly constructing the suffix array of a size-n text using $5n+O(1)$ bytes, for $n \leq 2^{32}$ and the alphabet $|\Sigma| \leq 256$. One year later, Burkhardt and Kärkkäinen [10] presented their algorithm based on the concept of "difference covers", with $O(n \log n)$ worst-case runtime and using a working space as the input plus $O(n/\sqrt{\log n})$. In the same year, Hon and Sadakane et al. [7] showed that the suffix arrays can be constructed in optimal $O(n \log n)$ time and $O(n \log n)$-bit space for texts with integer alphabets, or $O(n)$ time and $O(n)$-bit working space for texts with constant alphabets. Later, Na [12] proposed an alphabet-independent linear $O(n)$ time algorithm for constructing suffix arrays using $O(n \log |\Sigma| \log^{\alpha}_{|\Sigma|} n)$-bit working space, where $\alpha = \log_3 2$. Very recently, Puglisi and Smyth et al. [1] conducted a thorough survey on SACAs including [11,10,7,13,14,15,16,17,18,19,20,12], and concluded that to devise an optimal SACA which is *fast*, *lightweight* and *linear* in the worst case still remains as a challenge.

What of our particular interest here is the optimal lightweight suffix array construction for texts with constant alphabets, which we term as to construct suffix arrays with the known optimal time complexities, and meanwhile, using a space as small as possible. Specifically, we present here our novel solution with a set of practical algorithms for optimal lightweight suffix array construction for constant alphabets, which is optimal in the sense that it can compute the suffix array of a size-n text of $n \log |\Sigma|$-bit using $O(n \log |\Sigma|)$ time and $(n \log |\Sigma| + |\Sigma| \log n + O(1))$-bit working space (excluding the space for the output suffix array), where Σ is an integer or constant alphabet. For popular applications in practice with $n \leq 2^{32}$ and $|\Sigma| \leq 256$, these results translate into $O(n)$ time and a total space of $5n + O(1)$ bytes, which are the optimal time and space complexities for *lightweight* suffix array construction termed by Manzini and Ferragina [11]. We intentionally don't use the big-O notations for n and $|\Sigma|$ in the space complexity formulas, in order to show the accurate space requirement which is a main concern for lightweight suffix array construction.

The rest of this article is organized as following. Section 2 introduces the preliminaries including some general notations and assumptions. Our solution for optimal lightweight suffix array construction is presented in Section 3. Section 4 summarizes the main results. Finally, Section 5 gives the conclusion.

[1] In the literature for suffix array construction algorithms, the working space may exclude both the input text and the output suffix array, for example, in [10].

2 Preliminaries

2.1 Notations

Let $T = t_0t_1t_2 \dots t_{n-2}\$$ be the input text with n characters arranged as an array, where the characters are in an alphabet Σ. Two kinds of alphabets are considered for Σ here: (1) an integer alphabet with characters in the range of $[0, n^{O(1)})$, and (2) a constant alphabet of size $O(1)$. Without loss of generality, the characters of T, from left to right, are indexed starting from 0. The last character \$ is called the *sentinel*, which is unique in T, not in Σ and lexicographically smaller than any character in Σ.

For a size-m text $X = x_0x_1x_2 \dots x_{m-2}\$$, we define some notations as below:

- $S(X, i)$: the suffix in X starting at x_i and running to the sentinel \$, i.e. $S(X, i) = x_ix_{i+1} \dots x_{m-2}\$$, where $i \in [0, m)$.
- $SA(X)$: the suffix array of X, which is the pointer array for all the m suffixes in X sorted in their lexicographically order, i.e., each item in $SA(X)$ contains an unique pointer to a suffix in X. Without loss of generality, the sorted order is assumed to be ascending.
- $ISA(X)$: the inverse suffix array of X, defined as $SA[ISA[i]] = i$, for $i \in [0, m)$.

Let $\prec$ and $\succ$ be the lexicographical preceding and succeeding operators, respectively, we define $\tau(T, i)$ to be the LS-type function for T, given as:

$$\tau(T, i) = \begin{cases} 0, & for\ S(T, i) \succ S(T, i+1)\ and\ i \in [0, n); \\ 1, & for\ (S(T, i) \prec S(T, i+1)\ and\ i \in [0, n-1))\ or\ i = n-1. \end{cases}$$

For denotation simplicity, a character $T[i]$ is said to be type-L or type-S for $\tau(T, i) = 0$ or 1, respectively. Moreover, a suffix $S(T, i)$ is said to be a type-L or type-S suffix if $T[i]$ is type-L or type-S, respectively. Let B be the size-n array $[0..n-1]$ of integers allocated for storing the output suffix array $SA(T)$, in which each item is $\lceil \log n \rceil$-bit. Next, $B[i]$ and $T[i]$ are said to be a pair of siblings, and an item $B[i]$ is said to be type-L or type-S if its sibling $T[i]$ is type-L or type-S, respectively. Further, let B_S and B_L be two sets consisting of the last n_0 type-S and type-L items in B, respectively, where $n_0 = |T_0| \leq \lceil n/2 \rceil$ and T_0 will be defined in Section 3.1. To denote a substring $x_ix_{i+1} \dots x_j$ in a text X, a simpler form of $X[i, \dots, j]$ could be used. In addition, we use $\ell(v_0, \dots, v_k)$ to denote the total length of all objects $v_0, \dots, v_k$ measured in bits.

2.2 Assumptions

Unless otherwise specified, in this article, we have the following general assumptions:

- The *working space* of a computation doesn't include the space for the output. In other words, the working space equals to the total space minus the space for the output. Without explicit specification, the term *space* implies the total space.

– The space is provided in a unit-cost RAM with word size $O(\log W)$-bit, where $n \leq W$. Following this assumption, a standard arithmetic or bitwise boolean operation on word-sized operands costs $O(1)$ time.

3 Solution

Our divide-and-conquer solution for computing $SA(T)$ for T is presented in this section. To solve the problem at hand, we first reduce the problem, next compute the suffix array for the reduced problem and then based on which, to derive the suffix array for the original problem.

3.1 Reducing the Problem

First of all, we introduce some basic definitions for reducing the problem. A S-string in T is: (1) $t_i \dots t_j$ $(0 \leq i \leq j < n)$ for both t_i and t_j are type-S; or (2) the sentinel \$ itself. In addition, the rank of t_i is defined as the number of characters less than t_i in T; all the ranks of T starts from 0. Further, let $Z_k(T, i)$ denote a substring in T consisting of the k $(k > 0)$ consecutive S-strings starting at a type-S character $T[i]$, in which fewer S-strings are possible when the sentinel \$ is included. Any k consecutive S-strings in T is called a Z_k-string.

Let T_0 be the text consists of the ranks (also known as the lexicographical names) of all S-strings in T, and let $n_0 = |T_0|$. The Corollary 2 in [16] says that the sorted order of all suffixes in T_0 determines the sorted order of all type-S suffixes in T. Further, let % denote the modulo operator and T_1 be the text consisting of the ranks for triples in $\{T_0[i, i+1, i+2] : i\%3 \neq 2, i \in [0, n_0)\}$, where the ones for $\{i\%3 = 0\}$ are arranged in one consecutive block following by another block consisting of those for $\{i\%3 = 1\}$, and let $n_1 = |T_1|$. Similarly, let T_2 be the text consisting of the ranks for triples in $\{T_0[i, i+1, i+2] : i \bmod 3 = 2, i \in [0, n_0)\}$; and let $n_2 = |T_2|$. Without loss of generality, we assume that there are less [2] type-S characters than the type-L characters in T and n is even for presentation simplicity, which leads to $n_1 \leq \lceil 2n_0/3 \rceil \leq \lceil n/3 \rceil$. Although T_1 can be computed by first computing T_0 from T and then computing T_1 from T_0, doing in this way is too space consuming for our purpose. In the following, we design a space efficient algorithm for directly computing T_1 from T.

To present the algorithm, we continue to introduce some more definitions. Let P_1 be the index array for all ith Z_3-strings of T satisfying $i\%3 \neq 2$, i.e., $P_1[i]$ gives the pointers to all these ith Z_3-strings in T; specifically, in P_1, the ones for $\{i\%3 = 0\}$ and $\{i\%3 = 1\}$ are arranged in two consecutive blocks, respectively. Next, let P_1' be the result array of sorting all elements in P_1 in the lexicographically ascending order of their corresponding Z_3-strings in T, where ties between any two Z_3-strings with different lengths are broken by giving the

[2] In case that the type-S characters are more, the same discussion can be conducted symmetrically on the type-L characters, see [16] for details.

shorter a higher priority [3]. Further, let H_1 be an array for recording the lengths of all Z_3-strings in P_1. From the definitions of P_1, P_1' and H_1, each of them is an array with n_1 items in the range of $[0, n)$, which means that each item can be encoded in $\lceil \log n \rceil$-bit.

Because $n_1 \leq \lceil 2n_0/3 \rceil \leq \lceil n/3 \rceil$, instead of constructing $SA(T)$ directly, we can compute $SA(T_1)$ and then derive $SA(T)$ from $SA(T_1)$ using the algorithms developed below. The whole procedure is started by computing T_1 from T. In brief, computing T_1 from T consists of three steps: (1) compute P_1 from T; (2) compute P_1' from P_1 and T; and (3) compute T_1 from P_1' and T. We further to go through them one-by-one.

First, we have the below lemma for determining the type of a character in T.

Lemma 1. *For $i \in [0, n-1)$, we have (1) $\tau(T, i) = 0$ if $T[i] > T[i+1]$; (2) $\tau(T, i) = 1$ if $T[i] < T[i+1]$; and (3) $\tau(T, i) = \tau(T, i+1)$ if $T[i] = T[i+1]$.*

Proof. The correctness of (1) and (2) is obvious from the definition of $\tau(T, i)$. For (3), if $\tau(T, i+1) = 0$ or 1, we have $S(T, i+1) \succ S(T, i+2)$ or $S(T, i+1) \prec S(T, i+2)$, respectively. Given $T[i] = T[i+1]$, this yields $S(T, i) \succ S(T, i+1)$ or $S(T, i) \prec S(T, i+1)$, respectively, i.e. $\tau(T, i) = \tau(T, i+1)$.

We proceed to compute P_1 and then sort P_1 to obtain P_1'.

Lemma 2. *Given T, P_1 can be computed using $O(n)$ time and a working space of $\ell(T) + O(1)$ bits.*

Proof. This can be done by simply traversing T once from right to left, to record in P_1 the positions of all type-S characters in T. At each step, the type of the current character is derived from that of its immediately succeeding character in $O(1)$ time, using Lemma 1.

Lemma 3. *(time bottleneck) Given T, P_1 and H_1, P_1' can be computed using $O(\ell(T))$ time and a total space of at most $\ell(B, T) + O(1)$ bits.*

Proof. Omitted due to the space limit.

For computing T_1 from P_1', all the Z_3-strings in T need to be sorted. For which, we have the below lemma for retrieving a S-string from its head in T.

Lemma 4. *Given $T[i]$ is type-S, starting from $T[i]$, we can find the first type-S item $T[j]$ succeeding to $T[i]$ by traversing up to the first type-L item $T[k]$ succeeding to $T[j]$, where $i < j < k$.*

Proof. If $T[i] = T[i+1]$, we know that $T[i+1]$ must be type-S and $j = i+1$. Suppose $T[i] < T[i+1]$, starting from $T[i+1]$, we traverse forward to the first $T[k]$ ($k > i+1$) satisfying $T[k-1] < T[k]$. At this moment, we know that $T[k-1]$ must be type-S. From $T[k]$, we traverse backward to the first $T[j]$ satisfying $T[j-1] > T[j]$ ($j < k$).

[3] The correctness of this tie-breaking scheme is supported by the Lemma 2 in [16], which states that if $T[i] = T[j]$, $T[i]$ is type-L and $T[j]$ is type-S, then $S(T, i) \prec S(T, j)$.

Without the space constraint, T_1 can be easily computed from P_1' and T using an auxiliary array as large as B. We design a space efficient algorithm COMPUTE-T1 to avoid using such a large auxiliary array, from which we have the below result, which proof is omitted due to the space limit.

Lemma 5. *Given P_1' and T, T_1 can be computed using $O(n)$ time and a total space of $\ell(B,T)+O(1)$ bits.*

Hence, we have the following result on the complexities for computing T_1 from T.

Lemma 6. *Given T, T_1 can be computed using $O(\ell(T))$ time and a total space of $\ell(T)+3\ell(T_1)+O(1)$ bits.*

Proof. The time and space complexities for the three steps for computing T_1 from T are given by Lemma 2, 3 and 5, respectively. Both the maximum time and the maximum space are observed for computing T_1 from P_1', H_1 and T, which dominate the total time and space complexities.

Now, we have successfully reduced the problem size from n to $\lceil n/3 \rceil$. We proceed to solve the reduced problem, i.e., to compute $SA(T_1)$ from T_1.

3.2 Solving the Reduced Problems

We design in Fig. 1 the algorithm LIGHTWEIGHT-KS-SORT for computing $SA(T_1)$ from T_1 in $O(n)$ time and $\ell(B)$ space, which is a lightweight alternative to the traditional KS algorithm and described as follows:

- Let U, V and X be 3 arrays $[0, m-1]$ of integers, where each item is $\lceil \log m \rceil$-bit. Moreover, suppose the buffers for U, V are allocated consecutively. Let SA_x denote $SA(X)$.
- Let X_1 and X_2 denote the texts consisting of the lexicographical names for triples in $\{X[i,i+1,i+2] : i\%3 \neq 2, i \in [0,m)\}$ and $\{X[i,i+1,i+2] : i\%3 = 2, i \in [0,m)\}$, respectively (in X_1, the names for triples with $\{i\%3 = 0\}$ are in a consecutive block and those for $\{i\%3 = 1\}$ are in another.), and $X_0 = X_1 \oplus X_2$, where "$\oplus$" denotes the text concatenating operator. Moreover, let $m_1 = |X_1|$ and $m_2 = |X_2|$. We first to compute X_1 and X_2 from X into V, using bucket sorting with U as the counter array and V as the bucket array. Then, X_1 and X_2 are copied to X for later use. This step can be done in $O(m)$ time.
- (Now, there are two copies of X_1, one in the last m_1 items of V, and another in X.) Let V_1 denote the size-m_1 array immediately right to the X_1 in V, and U_1 be the size-m_1 array immediately right to V_1. We make the function call LIGHTWEIGHT-KS-SORT(X_1, U_1, V_1) to recursively compute SA_{x1}, where SA_{x1} denotes $SA(X_1)$. This step can be done in $O(m)$ time.
- Provided with SA_{x1} in V and the X_2's copy in X, we use the induced sorting method in the KS algorithm to compute SA_{x2} from SA_{x1} and X_2, where SA_{x2} denotes $SA(X_2)$. This step can be done in $O(m)$ time. As the result, SA_{x1} and SA_{x2} are stored in V.

- Let SA_{x0} denote $SA(X_0)$. Now, we are going to merge SA_{x1} and SA_{x2} to produce SA_{x0}. From SA_{x1} and SA_{x2} in V, we compute ISA_{x1} and ISA_{x2} into U, where ISA_{x1} and ISA_{x2} are the inverse SA_{x1} and SA_{x2}, defined as $SA_{x1}[ISA_{x1}[i]] = i$ ($i \in [0, m_1)$) and $SA_{x2}[ISA_{x2}[j]] = j$ ($j \in [0, m_2)$). This can be done in $O(m)$ time.
- Let Y_1 and Y_2 be defined as $SA_{x1}[i] = SA_{x0}[Y_1[i]]$ and $SA_{x2}[j] = SA_{x0}[Y_2[j]]$, respectively, for $i \in [0, m_1)$ and $j \in [0, m_2)$. Y_1 and Y_2 are called the position index arrays for SA_{x1} and SA_{x2}, which give the position indices for all items in SA_{x1} and SA_{x2} in the merged suffix array SA_{x0}. Computing Y_1 and Y_2 can be done by traversing SA_{x1} and SA_{x2} once and using the buffers for SA_{x1} and SA_{x2} only, i.e., V is updated with Y_1 and Y_2. This can be done in $O(m)$ time.
- Now, U contains ISA_{x1} and ISA_{x2}, and V contains Y_1 and Y_2. To compute ISA_{x0}, which is the inverse SA_{x0}, we simply traverse U once to set $U[i] = V[U[i]]$, for each $i \in [0, m)$. From the definitions of ISA_{x1}, ISA_{x2}, Y_1 and Y_2, it is trivially to see that U contains ISA_{x0} now. This can be done in $O(m)$ time.
- Given ISA_{x0} in U, we compute SA_{x0} into V by traversing U once to set $V[U[i]] = i$, for each $i \in [0, m)$. This can be done in $O(m)$ time.
- Now, because the original positions of the elements of X_1 and X_2 in X are interleaved (every two elements of X_1 followed by an element of X_2), we traverse SA_{x_0} once with a complexity of $O(m)$ to compute SA_x as

$$SA_x[i] = \begin{cases} 3SA_{x_0}[i], \ for \ SA_{x_0}[i] \in [0, \lceil m_1/2 \rceil); \\ 3(SA_{x_0}[i] - \lceil m_1/2 \rceil) + 1, \ for \ SA_{x_0}[i] \in [\lceil m_1/2 \rceil, m_1); \\ 3(SA_{x_0}[i] - m_1) + 2, \ for \ SA_{x_0}[i] \in [m_1, m). \end{cases} \quad (1)$$

- Finally, we copy SA_x in V to X for returning the result.

Lemma 7. *Given a size-n text X with each character encoded in $\lceil \log n \rceil$-bit, $SA(X)$ can be computed in $O(n)$ time and using a total space of $3\ell(X)+O(1)$ bits.*

Proof. In the LIGHTWEIGHT-KS-SORT algorithm, The time is governed by the recurrence $T(n) = T(\lceil 2n/3 \rceil) + O(n)$ and $T(n) = O(1)$ for $n < 3$, which leads to $T(n) = O(n)$. At each iteration, when making the recursion call in line ??, only X is occupied, and U and V are available for use as the buffer space for the recursion. Hence, (omitting the space used for the recursion stack which is $O(\log n)$-bit and commonly neglected in the literature for suffix array construction algorithms) a total space of $\ell(X, U, V) = 3\ell(X) + O(1)$ bits is sufficient for the recurrence.

Recalling that each element of T_1 is $\lceil \log n_1 \rceil$-bit and $n_1 \leq \lceil n/3 \rceil$, Lemma 7 immediately suggests the following result.

Corollary 1. *Given T_1, $SA(T_1)$ can be computed using $O(n)$ time and a total space of $3\ell(T_1) + O(1)$ bits.*

Provided that $SA(T_1)$ is know, it is only a routine job for us to induce $SA(T_2)$ from $SA(T_1)$ using the KS skew algorithm [15], in two steps as follows:

- Compute T_2 from T, using the method for computing T_1. This step can be done using $O(\ell(T))$ time and a total space of $\ell(T) + 3\ell(T_2) + O(1)$ bits (see Lemma 6).
- Compute $SA(T_2)$ from $SA(T_1)$ and T_2, using the KS skew algorithm. This can be done using $O(n)$ time and a total space of $\ell(T_1) + 3\ell(T_2) + O(1)$ bits.

LIGHTWEIGHT-KS-SORT(X, U, V)

▷ Input: X—array $[0..m-1]$ of integer, each item is $\lceil \log m \rceil$-bit.

▷ Output: SA_x—the suffix array of X, returned in the buffer of X.

▷ U, V: array $[0..m-1]$ of integer, each item is $\lceil \log m \rceil$-bit.

▷ Assumption: The buffers for U and V are allocated consecutively .

if $|X| < 3$

  then Directly compute SA_x from X and store the result in X.

    return

Compute X_1 and X_2 from X into V. ▷ X_1 is stored in the last m_1 items of V.

Copy X_1 and X_2 from V to X. ▷ Save X_1 and X_2 for computing SA_{x2} later.

$m_1 \leftarrow |X_1|$; $m_2 \leftarrow |X_2|$

Let U_1 and V_1 be the two arrays $[0..m_1-1]$ immediately right to the X_1 in V.

LIGHTWEIGHT-KS-SORT(X_1, U_1, V_1) ▷ Compute SA_{x1} into V_1.

Induced sorting SA_{x2} from SA_{x1} into U_1.

Traverse SA_{x1} and SA_{x2} in V once to compute ISA_{x1} and ISA_{x2} into U.
▷ Now, U contains $[ISA_{x1}, ISA_{x2}]$; V contains $[SA_{x1}, SA_{x2}]$; X contains $[X_1, X_2]$.

Traverse SA_{x1} and SA_{x2} in V once to update V with the position index arrays Y_1 and Y_2.

For each $i \in [0, m)$, $U[i] \leftarrow V[U[i]]$. ▷ Compute ISA_{x0}.

For each $i \in [0, m)$, $V[U[i]] \leftarrow i$. ▷ Compute SA_{x0}.

For each $i \in [0, m)$, compute SA_x from SA_{x0} by Eq.(1).

Copy V to X. ▷ Return the result in X.

return

Fig. 1. The linear lightweight KS sorting algorithm

Recalling that $n_1 \leq n/3$, $n_2 \leq n/6$, and B, T_1 and T_2 have the same item's size of $\lceil \log n \rceil$-bit, the maximum space of the above two-step procedure is upper bounded by $\ell(T, B) + O(1)$ bits. The complexities for inducing $SA(T_2)$ from $SA(T_1)$ is concluded as follows.

Lemma 8. *Given $SA(T_1)$ and T, $SA(T_2)$ can be computed using $O(\ell(T))$ time and a total space of $\ell(T, B) + O(1)$ bits.*

3.3 Inducing the Final Result

Having solved $SA(T_1)$ and $SA(T_2)$, we proceed to merge $SA(T_1)$ and $SA(T_2)$ into $SA(T_0)$ in $O(n)$ time and $(\ell(T, B) + O(1))$-bit space. If we use the skew method in the LIGHTWEIGHT-KS-SORT algorithm, a total space of $3\ell(T_1, T_2) + O(1) =$

$1.5\ell(B) + O(1)$ bits is required, which is too space consuming. Notice that in the algorithm LIGHTWEIGHT-KS-SORT, X is used only for rank comparisons. Given T, $SA(T_1)$ and SA_2, we design a more space efficient algorithm for this job, which performs rank comparisons using Z_3-strings in T. This algorithm is described below:

- Let SA'_1 and SA'_2 be defined as $SA'_1 = \{P_1[SA(T_1)[i]] : i \in [0, n_1)\}$ and $SA'_2 = \{P_2[SA(T_2)[i]] : i \in [0, n_2)]$, respectively. An item $B[i]$ is said to in a set X if $B[i]$ is allocated for storing an item in X. Further, supposed that SA'_1 and SA'_2 are initially stored in the first n_0 (recalling that $n_0 = n_1 + n_2$) items of B, and SA'_1 is left to SA'_2.
- First, we move SA'_1 and SA'_2 into the type-L items in B_L by traversing B once from right to left. At this step, we record in h the position of the first $B[i]$ in SA'_2. This can be done in $O(n)$ time.
- Next, ISA'_1 and ISA'_2 are computed into the type-S items in B_S, where ISA'_1 is defined as for each $B[i]$ in SA'_1, $ISA'_1[B[i]] = i$, and similarly, ISA'_2 is defined as for each $B[i]$ in SA'_2, $ISA'_2[B[i]] = i$, where $i \in [0, n)$.
- Now, B_L contains SA'_1 and SA'_2, and B_S contains ISA'_1 and ISA'_2. We proceed to merge-sort SA'_1 and SA'_2. We first traverse B_L to compute the position index of each item in SA'_1 and SA'_2 in the merged set SA_0, where SA_0 denotes $SA(T_0)$ and the index starts from 0. To determine the order between an item $B[u]$ in SA'_1 and an item $B[v]$ in SA'_2, let $i = B[u]$ and $j = B[v]$, we compare $Z_3(T, i)$ and $Z_3(T, j)$ (these two strings can be retrieved utilizing Lemma 4). If they are different, the order is immediately determined; or else we continue to compare as follows [4]:
  - Suppose $Z_3(T, i)$ is the kth Z_3-string in T, $k \in [0, n_0)$, we say $Z_3(T, i)$ is a residue-0 or residue-1 string if $k \mod 3 = 0$ or 1, respectively.
  - In case that $Z_3(T, i)$ is a residue-0 string, compare $Z_3(T, i')$ with $Z_3(T, j')$, where $Z_3(T, i')$ and $Z_3(T, j')$ are the first S-strings succeeding to $Z_3(T, i)$ and $Z_3(T, j)$, respectively. The order of $Z_3(T, i')$ and $Z_3(T, j')$ is determined by $B[i']$ and $B[j']$, for ISA'_1 and ISA'_2 are stored in B_S and both $B[i']$ and $B[j']$ are in B_S.
  - In case that $Z_3(T, i)$ is a residue-1 string, compare $Z_3(T, i'')$ with $Z_3(T, j'')$, where $Z_3(T, i'')$ and $Z_3(T, j'')$ are the 2nd S-strings succeeding to $Z_3(T, i)$ and $Z_3(T, j)$, respectively. Similar to the previous case, the order of $Z_3(T, i'')$ and $Z_3(T, j'')$ is determined by $B[i'']$ and $B[j'']$, for both $B[i'']$ and $B[j'']$ are in B_S.

  Retrieving the strings, once more, can be done utilizing Lemma 4. Checking if $Z_3(T, i)$ is a residue-0 or residue-1 string can be done in $O(1)$ time by simply comparing $B[i]$ with h, for all items of SA'_1 are stored before the item $B[h]$. Because each S-string in T can be visited at most four times (1 due to locating the terminating character of the S-string preceding to it and 3 due to S-string comparisons) for merging SA_1 and SA_2, this step is done in $O(n)$ time.

[4] The correctness of this comparison scheme can be trivially seen from the KS skew algorithm in [15].

- Let ISA_0 denote the inverse $SA(T_0)$. ISA_0 is computed by first traversing B once from right to left to set each type-S item $B[i]$ with $B[B[i]]$; and then traverse B once more from right to left to move all the items in B_S into the last n_0 items of B. This can be done in $O(n)$ time.
- Given ISA_0 (in the last n_0 items of B), SA_0 can be easily computed into the left half of B in $O(n)$ time.

As a summary for the above algorithm, we have the following lemma.

Lemma 9. *Given T, $SA(T_1)$ and $SA(T_2)$, $SA(T_0)$ can be computed using $O(n)$ time and a total space of $\ell(B,T)+O(1)$ bits.*

Given SA_0 and T, we can compute $SA(T)$ in two steps, as described below:

- Let P_0 be the S-string array for T, which is the index array for all S-strings in T. In addition, let SA_S denote the suffix array for all type-S suffixes in T, defined as $SA_S = \{SA(T)[i] : T[SA(T)[i]] \; is \; type-S, i \in [0,n)\}$. Given SA_0 (stored in the left half of B), SA_S can be computed in two steps: (1) compute P_0 into the right half of B by traversing B once from right to left; and (2) compute SA_S by traversing SA_0 once to set $SA_0[i] = P_0[SA_0[i]]$. Now, SA_S is contained in the first n_0 items of B. This step can be done in $O(n)$ time.
- Use the KA skew algorithm [16] to induce $SA(T)$ from SA_S and T, which requires a bucket counter array [5] of $|\Sigma|\lceil\log n\rceil$-bit in addition to B and T, and is done in $O(n)$ time. The maximum total space for the whole procedure of computing $SA(T)$ from T is due to this step.

Hence, the complexities for computing $SA(T)$ from $SA(T_0)$ are concluded as follows.

Lemma 10. *(space bottleneck) Given T and $SA(T_0)$, $SA(T)$ can be computed using $O(n)$ time and a total space of $\ell(B,T)+|\Sigma|\lceil\log n\rceil+O(1)$ bits.*

4 Main Results

Theorem 1. *For a size-n text T with an integer or constant alphabet Σ, $SA(T)$ can be computed using $O(n\log|\Sigma|)$ time and a total space of $\ell(T,B)+|\Sigma|\lceil\log n\rceil+O(1)$ bits.*

Proof. The whole procedure for computing $SA(T)$ from T consists of the following steps in sequence:

1. Compute T_1 from T and $SA(T_1)$ from T_1, see Lemma 6 and Corollary1.
2. Compute $SA(T_2)$ from $SA(T_1)$ and T, see Lemma 8.
3. Merge $SA(T_1)$ and $SA(T_2)$ into $SA(T_0)$, see Lemma 9.
4. Induce $SA(T)$ from $SA(T_0)$, see Lemma 10.

[5] Two bucket counter arrays can be used for higher speed when $|\Sigma|$ is not large, e.g., $O(1)$.

The bucket sorting function for computing P_1' from P_1 and T in Step 1, as well as that for computing P_2' from P_2 and T in Step 2, dominate the whole procedure's time complexity, which is $O(\ell(T)) = O(n \log |\Sigma|)$ from Lemma 3.

The total space consists of 3 parts: (1) the array T for the input text; (2) the array B for the output suffix array; and (3) the $|\Sigma|\lceil \log n \rceil$-bit bucket counter array for inducing $SA(T)$ from $SA(T_0)$.

From Theorem 1, we have the following two results for texts with integer and constant alphabets, respectively.

Corollary 2. *For a size-n text T with an integer alphabet Σ, where $|\Sigma| \leq n$, $SA(T)$ can be computed using $O(n \log n)$ time and a total space of at most $3n\lceil \log n \rceil + O(1)$ bits.*

Corollary 3. *(optimal lightweight) For a size-n text T with a constant alphabet Σ, where $n \leq 2^{32}$ and $|\Sigma| \leq 256$, $SA(T)$ can be computed using $O(n)$ time and a total space of $5n + O(1)$ bytes.*

5 Conclusion

A divide-and-conquer solution with a set of practical algorithms has been developed in this work for optimal lightweight suffix array construction. The crucial task for developing this solution, as we have shown, is how to reduce the problem size to be small enough so that the reduced problem can be efficiently computed and meanwhile, the final suffix array can be augmented from the reduced one time and space efficiently. Once the problem has been reduced, a number of traditional suffix array construction algorithms are allowed to be further exploited to solve the reduced problem. This makes the solution flexible to be further improved for better average performance.

Acknowledgment

The authors wish to thank the anonymous reviewers for their constructive suggestions and insightful comments.

References

1. Puglisi, S.J., Smyth, W.F., Turpin, A.: A taxonomy of suffix array construction algorithms. ACM Computing Surveys 2006 (to Appear)
2. Manber, U., Myers, G.: Suffix arrays: A new method for on-line string searches. In: Proceedings of the first ACM-SIAM Symposium on Discrete Algorithms, pp. 319–327 (1990)
3. Manber, U., Myers, G.: Suffix arrays: A new method for on-line string searches. SIAM Journal on Computing 22(5), 935–948 (1993)
4. Burrows, M., Wheeler, D.J.: A block-sorting lossless data compression algorithm. Technical Report SRC Research Report 124, Digital Systems Research Center, California, USA (1994)

5. Manzini, G., Ferragina, P.: Engineering a lightweight suffix array construction algorithm. Algorithmica 40(1), 33–50 (2004)
6. Grossi, R., Vitter, J.S.: Compressed suffix arrays and suffix trees with applications to text indexing and string matching. In: Proceedings of the 32nd Annual ACM Symposium on Theory of Computing (STOC'00), pp. 397–406 (2000)
7. Hon, W.K., Sadakane, K., Sung, W.K.: Breaking a time-and-space barrier for constructing full-text indices. In: Proceedings of FOCS'03, pp. 251–260 (2003)
8. Lam, T.W., Sadakane, K., Sung, W.K., Yiu, S.M.: A space and time efficent algorithm for constructing compressed suffix arrays. In: Proceedings of International Conference on Computing and Combinatorics, pp. 401–410 (2002)
9. Kurtz, S.: Reducing the space requirement of suffix trees. Software Practice and Experience 29, 1149–1171 (1999)
10. Burkhardt, S., Kärkkäinen, J.: Fast lightweight suffix array construction and checking. In: Baeza-Yates, R.A., Chávez, E., Crochemore, M. (eds.) CPM 2003. LNCS, vol. 2676, pp. 55–69. Springer, Heidelberg (2003)
11. Manzini, G., Ferragina, P.: Engineering a lightweight suffix array construction algorithm. In: Möhring, R.H., Raman, R. (eds.) ESA 2002. LNCS, vol. 2461, pp. 698–710. Springer, Heidelberg (2002)
12. Na, J.C.: Linear-time construction of compressed suffix arrays using O(nlog n)-bit working space for large alphabets. In: Apostolico, A., Crochemore, M., Park, K. (eds.) CPM 2005. LNCS, vol. 3537, pp. 57–67. Springer, Heidelberg (2005)
13. Larsson, N.J., Sadakane, K.: Faster suffix sorting. Technical Report LU-CS-TR:99-214, LUNDFD6/(NFCS-3140)/1–20/(1999), Department of Computer Science, Lund University, Sweden (1999)
14. Maniscalco, M.A., Puglisi, S.J.: Faster lightweight suffix array construction. In: Proceedings of 17th Australasian Workshop on Combinatorial Algorithms (AWOCA'06), pp. 16–29 (2006)
15. Kärkkäinen, J., Sanders, P., Burkhardt, S.: Linear work suffix array construction. Journal of the ACM (6), 918–936 (2006)
16. Ko, P., Aluru, S.: Space efficient linear time construction of suffix arrays. In: Baeza-Yates, R.A., Chávez, E., Crochemore, M. (eds.) CPM 2003. LNCS, vol. 2676, pp. 200–210. Springer, Heidelberg (2003)
17. Kim, D.K., Jo, J., Park, H.: A fast algorithm for constructing suffix arrays for fixed-size alphabets. In: Ribeiro, C.C., Martins, S.L. (eds.) WEA 2004. LNCS, vol. 3059, pp. 301–314. Springer, Heidelberg (2004)
18. Itoh, H., Tanaka, H.: An efficient method for in memory construction of suffix arrays. In: Proceedings of String Processing and Information Retrieval Symposium (1999)
19. Seward, J.: On the performance of BWT sorting algorithms. In: Proceedings DCC 2000 Data Compression Conference, Snowbird, UT, USA, pp. 173–82 (2000)
20. Schürmann, K.B., Stoye, J.: An incomplex algorithm for fast suffix array construction. In: Proceedings of 7th Workshop on Algorithm Engineering and Experiments (ALENEX/ANALCO 2005), pp. 77–85 (2005)

Range Non-overlapping Indexing and Successive List Indexing

Orgad Keller, Tsvi Kopelowitz, and Moshe Lewenstein*

Department of Computer Science, Bar-Ilan University, Ramat-Gan 52900, Israel
{kellero, kopelot, moshe}@cs.biu.ac.il

Abstract. We present two natural variants of the indexing problem:

In the *range non-overlapping indexing* problem, we preprocess a given text to answer queries in which we are given a pattern, and wish to find a maximal-length sequence of occurrences of the pattern in the text, such that the occurrences *do not overlap* with one another. While efficiently solving this problem, our algorithm even enables us to efficiently perform so in *substrings* of the text, denoted by given start and end locations. The methods we supply thus generalize the *string statistics* problem [4,5], in which we are asked to report merely the *number* of non-overlapping occurrences in the *entire* text, by reporting the occurrences themselves, even only for substrings of the text.

In the related *successive list indexing* problem, during query-time we are given a pattern and a list of locations in the preprocessed text. We then wish to find a list of occurrences of the pattern, such that the ith occurrence is the leftmost occurrence of the pattern which starts to the right of the ith location given by the input list.

Both problems are solved by using tools from computational geometry, specifically a variation of the *range searching for minimum* problem of Lenhof and Smid [12], here considered over a grid, in what appears to be the first utilization of range searching for minimum in an indexing-related context.

1 Introduction

Given a text string $T = t_1 \dots t_n$ and a pattern string $P = p_1 \dots p_m$, in the *pattern matching* problem [11] we wish to report all the occurrences of P in T. Its online counterpart, the *indexing* problem, is one of the most important paradigms in searching: the idea is to preprocess a text and construct a mechanism that will later provide answers to queries of the form "does a pattern P occur in the text" in time proportional to the length of the *pattern* rather than the text. In addition, if we want to return the occurrences themselves, the time will be proportional to the length of the pattern and the number of actual occurrences.

The *suffix tree* [15,14,7,13] has proven to be an invaluable data structure for indexing, using $O(n)$ space, where n is the text length. Algorithms for the

* This work was supported by a German-Israel Foundation (G.I.F.) young scientists program research grant.

F. Dehne, J.-R. Sack, and N. Zeh (Eds.): WADS 2007, LNCS 4619, pp. 625–636, 2007.

construction of a suffix tree enable $O(n)$ preprocess time when $|\Sigma|$ is constant (where Σ is the alphabet set), and $O(n \log \min(n, |\Sigma|))$ time when $|\Sigma|$ is not. In fact, the suffix tree can be constructed in linear time even for alphabets drawn from a polynomially-sized range, see [7].

The size of the alphabet also affects the query time of the suffix tree: given a pattern P of length m, we can find the set of all occurrences of P in T in $O(m \log \min(n, |\Sigma|) + tocc)$ time for unbounded alphabets, where $tocc$ is the actual number of occurrences of P in T, or accordingly, $O(m + tocc)$ time for constant-sized alphabets.

While the search for P yields an unsorted set of occurrences in T, some may overlap others: a specific location i in the text might participate in several different occurrences of P in T. However, sometimes only *non-overlapping* occurrences are of importance. Such requirement is of interest in fields such as pattern recognition, computational linguistics, speech processing, bio-molecular sequence analysis, code optimization and data compression [4]. For instance, we might want to compress a text by replacing each non-overlapping occurrence of a substring of it with a pointer to a single copy of the substring.

In the *string statistics* problem [4,5], we are interested in finding the maximal *number* of non-overlapping occurrences of P in the entire text T. The solutions proposed in [4] and [5] use properties of periodicity in the text and pattern. However, the methods described there do not report the *actual occurrences* of the pattern. In this paper, we present a solution that returns the *maximal* (sorted by location) sequence of non-overlapping occurrences of P in T[1]. Furthermore, we generalize it such that it can return the maximal sequence of non-overlapping occurrences of P in some substring of T, denoted by start and end locations given alongside the pattern at query time.

In addition, we provide a solution to another problem that incorporates indexing with added location constraints: in the *successive list indexing* problem, we are given a list $L = \langle i_1, \ldots, i_\ell \rangle$ of locations in T together with P, and we wish to find the sequence of occurrences of P in T where the jth occurrence returned is the leftmost occurrence of P in T that occurs after the i_jth location (if such exists). Other kinds of proximity-related indexing variants (for instance, finding the single occurrence of the pattern that overlaps the i_jth location) can be solved by using exactly the same method. We also note that the definition of the matching can be generalized to pattern matching with errors and such [3], but we leave the discussion for the full version of this paper, and assume for the rest of the paper the common matching definition.

Solutions to both problems rely heavily on tools taken from the *computational geometry* area. In the *range searching* problem (see survey in [1]), which is common to this field, we are given a set S of n geometric objects (e.g. points) in a d-dimensional space, which we store in some data structure. When a query object Q (e.g. a hyper-rectangle $[a_1, b_1] \times \cdots \times [a_d, b_d]$) is given, we wish to

[1] Note that we discuss the indexing variant of this problem. If one would like to solve non-overlapping pattern matching, then one could use the simple greedy method discussed in Sect. 5.

return the result of some sort of query on a subset of the points, usually the subset $S \cap Q$. A popular variant of range searching is *range reporting*, in which we are asked to report all the points which are included in the query range $Q = [a_1, b_1] \times \cdots \times [a_d, b_d]$, i.e. the set $S \cap Q$ itself (see [2]).

While range reporting has been used before in several indexing-related papers (e.g. [9,3,8]), to the best of our knowledge, this is the first indexing-related work using a variant of *range searching for minimum* of Lenhof and Smid [12], itself a generalization of a problem presented by Gabow et al. [10]. In Lenhof and Smid's variant, we are given a set of n d-dimensional points, and query them with ranges of type $[a_1, b_1] \times \cdots \times [a_{d-1}, b_{d-1}] \times [a_d, \infty]$, wishing to find a single point in range with minimal dth coordinate. When $d = 2$, they obtain the following bounds: $O(n \log n \log \log n)$ expected preprocessing time, $O(n \log n)$ space, and $O(\log n)$ query time.

We modify the solution from [12] to work on a 2-*dimensional grid*, which suits our purposes. We find it more appropriate to call this variant the *range successor query on a grid* problem. We obtain the following bounds: $O(n \log n \log \log n)$ expected preprocessing time, $O(n \log n)$ space, and $O(\log \log n)$ worst-case query time.

The rest of this paper is organized as follows: in Sect. 2 we provide some formal definitions of our problems. In Sect. 3 we supply an outline of the method we use for the range successor query on a grid problem. In Sect. 4 we solve the successive list indexing problem. In Sect. 5 we finally solve the range non-overlapping indexing problem, and in Sect. 6 we present some concluding remarks.

1.1 Notations

For two integers $i \leq j$, denote by $[i, j]$ the set $\{i, \ldots, j\}$. For an integer u, denote by $[u]$ the set $[0, u-1]$.

Given a string S, denote by $|S|$ the length of S. An integer i is a *location* in S if $i = 1, \ldots, |S|$. Given a string $T = t_1 \ldots t_n$ (i.e. $|T| = n$, hereafter the *text*), a *suffix* of T is a string of the form $t_i \ldots t_n$, for some location i. Given another string $P = p_1 \cdots p_m$ (hereafter the *pattern*), a location i in T is an *occurrence* of P in T if $t_i \ldots t_{i+m-1} = p_1 \cdots p_m = P$. Two occurrences i, j of P in T are said to be *non-overlapping* if $|j - i| \geq m$. The *suffix tree* of T is essentially a compressed trie of the suffixes of T, used as a data structure to efficiently find the occurrences of P in T.

2 Problem Definitions

The *successive list indexing* problem is defined as follows:

Input: a text $T = t_1 \ldots t_n$ over alphabet Σ.
The text will be preprocessed to answer the following:
Query: a pattern $P = p_1 \cdots p_m$ over Σ, and a list $L = \langle i_1, \ldots, i_\ell \rangle$ of locations in T.
Output: the ℓ-length list of occurrences of P in T where the jth occurrence is the leftmost (i.e. minimal) occurrence of P in T that appears after the i_jth location (if such exists).

A simpler version of this problem is the *successive indexing* problem that is defined as follows:

Input: a text $T = t_1 \dots t_n$ over alphabet Σ.
The text will be preprocessed to answer the following:
Query: a pattern $P = p_1 \dots p_m$ over Σ, and a location i in T.
Output: an occurrence $i' \geq i$ of P (i.e. $t_{i'} \dots t_{i'+m-1} = P$) in T for which i' is minimal (if such exists).

The *range non-overlapping indexing* problem is defined as follows:

Input: a text $T = t_1 \dots t_n$ over alphabet Σ.
The text will be preprocessed to answer the following:
Query: a pattern $P = p_1 \dots p_m$ over Σ, and two locations $i \leq j$ in T.
Output: an ascending sequence $L = \langle i_1, \dots, i_k \rangle$ of non-overlapping occurrences of P in T for which $i \leq i_1$ and $i_k \leq j$ (alternatively we can say we require L to be a subsequence of the sorted set $[i, j]$) and k is maximal. Formally, we require that for each $j = 1, \dots, k$, $t_{i_j} \dots t_{i_j+m-1} = P$ and that for each $j = 1, \dots, k-1$, $i_{j+1} - i_j \geq m$.

The (two-dimensional) *range successor query on a grid* problem is defined as follows:

Input: a set $S = \{(x_1, y_1), \dots, (x_n, y_n)\}$ of n points on an $[n] \times [n]$ grid.
Given this input, we will efficiently preprocess it to answer the following queries:
Query: a triplet (x', x'', y).
Output: a specific point $(x_i, y_i) \in S \cap ([x', x''] \times [y, n-1])$ whose y-coordinate (i.e. y_i) is minimal. In other words, (x_i, y_i) is the point with minimal value y_i corresponding to the following conditions:

1. $y_i \geq y$.
2. $x' \leq x_i \leq x''$.

3 Range Successor Query on a Grid

Both solutions for the successive list indexing and range non-overlapping indexing rely heavily on an efficient solution to the range successor query on a grid problem. As mentioned before, this problem, in its version where the points' coordinates are not on a grid (meaning, they are not necessarily integers and are not drawn from a restricted universe $[u]$), and for which the points can also be of dimension greater than 2, was solved by Lenhof and Smid [12]. In their definition, given n points in a d-dimensional space, the query object is a d-dimensional range $[a_1, b_1] \times \cdots \times [a_{d-1}, b_{d-1}] \times [a_d, \infty]$ in which we wish to find the point having the minimal dth coordinate. They issued the problem with the name "range searching for minimum", which was used prior by Gabow et al. [10] to indicate the more particular problem in which the query object is of the form $[a_1, b_1] \times \cdots \times [a_{d-1}, b_{d-1}] \times [-\infty, \infty]$. Again, the goal there is to find the point in the range having the minimal dth coordinate. As Lenhof and Smid's problem is actually the problem of finding the successor of a value, with added range restrictions, we find it more appropriate to name it (in our context) the *range successor query on a grid* problem.

In the solution presented in [12], they used a rank space reduction in order to reduce the given point set in $\mathbb{R}^d$ to a point set on an $[n]^d$ grid. As a result, the query time for the two-dimensional case suffered from an additive $O(\log n)$ time. However, when we solve the problem on an $[n] \times [n]$ grid, we do not need the rank space reduction. Unfortunately, in [12] there is no complete analysis of the query time in absence of the rank space reduction. It can be shown that the query time in such a case is worst-case $O(\log \log n)$. We leave the full details for the full version, as it requires a complete description of the solution presented in [12].

We thus obtain the following:

Theorem 1. *The range successor query on an $[n] \times [n]$ grid problem can be solved with $O(n \log n)$ space and $O(\log \log n)$ query time, using $O(n \log n \log \log n)$ expected preprocess time.*

4 Successive List Indexing

We now present a solution for the successive indexing problem which applies a reduction to the range successor query problem. Later, we will explain how to generalize the solution for solving the successive list indexing problem.

Let $T = t_1 \dots t_n$ be a text over alphabet Σ. When given a pattern $P = p_1 \dots p_m$ over Σ and a location i in T, we wish to find the leftmost occurrence of P in T that still starts to the right of i. Formally, we wish to find the minimal $i' \geq i$ such that $t_{i'} \dots t_{i'+m-1} = P$, if such exists.

We first construct the suffix tree of T, denoted ST(T). In order to prevent the effect of unbounded alphabets on the suffix tree, we can present hashing, as depicted in the following:

Theorem 2. *There exists a* randomized *suffix tree, which can be constructed in* expected *$O(n)$ time (where n is the length of the text), and in which queries can be made in worst-case $O(m + tocc)$ time (where m is the length of the pattern, and tocc is the actual number of occurrences of the pattern in text), for* general *alphabets.*

Proof. Note the construction and query times for constant size alphabets are $O(n)$ and $O(m+tocc)$ respectively. In addition, note that the number of children of any node in the suffix tree is bounded by both $n+1$ and $|\Sigma|$. Hence, we obtain a $\min(n+1, |\Sigma|)$ bound on the number of children of a given node. Thus, if for every node in the suffix tree we maintain pointers to its children in a balanced search tree, the multiplicative $O(\log \min(n, |\Sigma|))$ factor comes from the need to search or to insert elements to balanced search trees. Substituting this balanced search tree with a dynamic hash table (e.g. of Dietzfelbinger et al. [6], supporting worst-case $O(1)$ query time and amortized expected $O(1)$ insertion time), using as before the symbols of the alphabets associated with the edges as keys, would eliminate that factor, thus giving us an *expected* $O(n)$ construction time, and worst-case $O(m + tocc)$ query time (worst-case, since in query-time we do not modify the tree and therefore do not insert elements to those hash tables). □

Note that besides the obvious disadvantage of introducing randomness, another disadvantage of the randomized suffix tree is the fact that now, given a node, the order of its children cannot be efficiently derived from the structure used to hold the pointers to them. As this order eventually determines the order of the leaves of the suffix tree, which is crucial to us during preprocess, suffix trees built throughout this paper will hold the pointers to the children of a given node in both a balanced search tree and a hash table. During query time, since the aforementioned order is of no importance to us, we will use the hash tables option to efficiently navigate through the tree.

4.1 Algorithm Outline

In the suffix tree, each leaf l is associated with a suffix of T and is therefore marked with an integer $y(l)$ which is the start location of that suffix. Assume we go over the leaves of $\mathrm{ST}(T)$ in a left-to-right manner linking them to create a linked list (by using a depth first search). Note that now if we traverse the list, we actually traverse the leaves according to the lexicographical order of the suffixes they are associated with. For a leaf l, let $x(l)$ be the position of l in the linked list. It immediately follows that $x(l)$ is the lexicographical rank of the suffix associated with l. Equivalently: If we lexicographically sort all suffixes of T in an ascending order, then the $x(l)$th suffix is the one associated with l. Setting $x(l)$ for each l can be done by going over the list, marking each leaf l with its position in the list.

When given a pattern $P = p_1 \dots p_m$, we can find all the occurrences of P in T, by traversing $\mathrm{ST}(T)$ from the root downwards according to the symbols in P, until we either conclude that P does not occur in T (in the case we got 'stuck' in the tree, figuratively speaking: this is the case where the next symbol of the pattern cannot be found in our current location in the tree), or that we conclude the traversal at a node v in $\mathrm{ST}(T)$. In the latter, all the leaves in the subtree rooted at v correspond to occurrences of P in T. Denote the subtree rooted at v as T_v. Hence the set $L' = \{y(l) \mid l \text{ is a leaf in } T_v\}$ is the set of all occurrences of P in T.

Note that for the node v mentioned above, the leaves of T_v appear consecutively in the linked list of leaves. Furthermore, since for each leaf l, $x(l)$ is its position in the list, the leaves of T_v form a range $[x(l'_v), x(l''_v)]$ (where l'_v and l''_v are the leftmost and rightmost leaves in T_v, respectively). It immediately follows that for a leaf l, l is a leaf in T_v iff $x(l) \in [x(l'_v), x(l''_v)]$. In other words: $x(l) \in [x(l'_v), x(l''_v)]$ iff P appears in T at location $y(l)$.

Consider the leaf f for which $x(f) \in [x(l'_v), x(l''_v)]$ and $y(f)$ is minimal such that $y(f) \geq i$. By the problem definition, $y(f)$ is exactly what we need to find and return. Now consider the set $\{(x(l), y(l)) \mid l \text{ is a leaf in } \mathrm{ST}(T)\}$. Clearly, this is a set of $n+1$ points on an $[n+1] \times [n+1]$ grid. Since the point $(x(f), y(f))$

Algorithm 1. Successive indexing preprocess

Input: a text $T = t_1 \dots t_n$.
construct $\mathrm{ST}(T)$; `/* assume the field` $y(l)$ `is set for any leaf` l `by the suffix tree algorithm */`
traverse $\mathrm{ST}(T)$ and set the field $x(l)$ for each leaf l;
traverse $\mathrm{ST}(T)$ *using DFS* **:**
  foreach *node* u **do**
    store the values $x(l'_u)$ and $x(l''_u)$ in u; `/*` l'_u `and` l''_u `are the leftmost and rightmost leaves of` T_u`, respectively */`
preprocess the set $\{(x(l), y(l)) \mid l \text{ is a leaf in } \mathrm{ST}(T)\}$ for range successor queries on an $[n+1] \times [n+1]$ grid;

Algorithm 2. Successive indexing query

Input: a pattern $P = p_1 \dots p_m$ and an integer $1 \le i \le n$.
traverse $\mathrm{ST}(T)$ *starting from the root, according to the symbols in* P **:**
  if *stuck* **then return** "no occurrences" **else** let v be the node we reached (if we stopped at a node) or the node immediately below the edge we are at (if we stopped on an edge);
if *the range successor query for* $(x(l'_v), x(l''_v), i)$ *yields a result* (x', y') **then**
  return y';
else return "no occurrence";

is exactly the y-axis successor of i in the range $[x(l'_v), x(l''_v)]$, we can find and return $y(f)$ by using a single range successor query.

The algorithm for the successive indexing problem thus immediately follows, and is presented as Algorithms 1 (preprocess) and 2 (query).

4.2 Analysis

We have obtained the following:

Theorem 3. *The successive indexing problem can by solved with* $O(n \log n)$ *storage and* $O(m + \log\log n)$ *query time, using* $O(n \log n \log\log n)$ *expected preprocess time.*

Proof. The correctness of the proposed algorithm follows immediately from the discussion above. Note that for the values $x(l), y(l)$ for each leaf l in $\mathrm{ST}(T)$, it holds that $x(l), y(l) \in [n+1]$. The space used is therefore:

1. $O(n)$ for the suffix tree itself.
2. $O(n \log n)$ for the data structure supporting range successor queries.

We conclude we use overall $O(n \log n)$ storage space.

The query time consists of:

1. $O(m)$ in order to find node v.
2. $O(\log\log n)$ time for the single range successor query.

Summing up, the query time is worst-case $O(m + \log\log n)$.

The preprocess time consists of:

1. $O(n \log \min(n, |\Sigma|))$ in order to construct the suffix tree with both a balanced search tree and a hash table in each node.
2. $O(n)$ for each traversal on the suffix tree.
3. $O(n \log n \log\log n)$ expected time for preprocessing in order to answer future range successor queries.

We conclude we use overall expected $O(n \log n \log\log n)$ time for preprocess. □

4.3 Solving the Successive List Indexing Problem

Note that after answering the query for some P and i, if we wish to answer this query for the same pattern P and a different location j, we can immediately perform the range successor query and return the result, since we have already found and thus know the x-axis range associated with P. Thus, we have obtained the following:

Theorem 4. *The successive list indexing problem can be solved with $O(n \log n)$ storage and $O(m+\ell \log\log n)$ query time, using $O(n \log n \log\log n)$ expected preprocess time.*

Proof. Given an ℓ-length list L of locations in T, we can use the solution for the successive indexing problem, but repeat the range successor query for every location given in the queried list. □

5 Range Non-overlapping Indexing

We now present a solution for the range non-overlapping indexing problem.

Let $T = t_1 \dots t_n$ be a text over alphabet Σ. When given a pattern $P = p_1 \cdots p_m$ over Σ, and two locations $i \le j$ in T, denote by L' the ascending sequence of locations in T where an occurrence of P in T starts. Denote by L'' the sequence of locations in T which (1) correspond to occurrences of P in T, and (2) are in the range $[i, j]$. Clearly, L'' is a subsequence of L'. We wish to find a subsequence $L = \langle i_1, \dots, i_k \rangle$ of L'' which corresponds to non-overlapping occurrences of P in T in the range $[i, j]$, with maximal k. Notice that L is a subsequence of L'' which is a subsequence of L'. Formally, we require that:

1. $i \le i_1$.
2. $i_k \le j$.
3. For each $d = 1, \dots, k$, $t_{i_d} \cdots t_{i_d+m-1} = P$.
4. For each $d = 1, \dots, k-1$, $i_{d+1} - i_d \ge m$.

Consider the following greedy method for constructing L: we go over $t_i \dots t_{j+m-1}$ by using one of the linear pattern matching algorithms (for instance [11]) which scan the text and return occurrences of P in $t_i \dots t_{j+m-1}$ in an ascending order of positions. We choose the first occurrence the algorithm has outputted to be the first element in L. Note that the occurrence we have just chosen is the leftmost occurrence of P in $t_i \dots t_{j+m-1}$. We then proceed to choose every first occurrence outputted by the algorithm that does not overlap with the last occurrence we have chosen for L. It is easy to see that for the resulting sequence $L = \langle i_1, \dots, i_k \rangle$, it holds that for each $d = 2, \dots, k$, i_d is minimal such that $t_{i_d} \dots t_{i_d+m-1} = P$ and $i_d - i_{d-1} \geq m$.

Lemma 1. *The sequence L is a maximal-length sequence of non-overlapping occurrences of P in $t_i \dots t_{j+m-1}$.*

Proof. Recall that $|L| = k$ and assume by contradiction that L is not maximal, i.e. there is a sequence of non-overlapping occurrences of P in $t_i \dots t_{j+m-1}$, denoted H, such that $|H| \geq k+1$. For an integer d, denote by H_d the dth element of H, if such exists. Since the greedy method always chooses the leftmost non-overlapping occurrence to be included, we can say that for each $d = 1, \dots, k$, $i_d \leq H_d$. In particular, $i_k \leq H_k$. Since H is a sequence of non-overlapping occurrences, we notice that H_{k+1} does not overlap with H_k, and because $i_k \leq H_k$, it follows that H_{k+1} does not overlap with i_k as well. We conclude that the greedy method should have appended the occurrence H_{k+1} to L, which contradicts the fact that i_k is the last element of L. □

Denote $tocc = |L'|$, $k'' = |L''|$ and recall that $|L| = k$. The following lemma tells the relation between the three:

Lemma 2. *$tocc \geq k''$ and $k'' \leq m \cdot k$. Furthermore, there exists a text and a pattern for which $k'' = \Theta(m \cdot k)$.*

Proof. $tocc \geq k''$ since L'' is a subsequence of L'.

L is the maximal set of non-overlapping locations. For two consecutive elements i_d, i_{d+1} in L, if there exists an occurrence of P in T at location e such that $i_d < e < i_{d+1}$, then the occurrence at e certainly overlaps with the occurrence at i_d, otherwise the greedy method would have chosen e instead of i_{d+1}. Therefore, we charge every such occurrence e to i_d, in order to refrain from counting e twice (one time for i_d, and possibly another time for i_{d+1} if it also overlaps with it). Since for each i_d there are $m-1$ locations $i_d + \ell$ ($\ell = 1, \dots, m-1$) for which if P appears at, it would overlap with the occurrence at i_d, the lemma follows.

Finally, if the text is $T = a^n$ (the symbol a repeated n times), the pattern is $P = a^m$, and the query range is $[1, n]$, it is easy to see that $k'' = \Theta(m \cdot k)$ □

Assume we first index T by constructing the suffix tree of T, denoted ST(T). As described in Sect. 4, the suffix tree of T enables us to find all the occurrences of P in T. Therefore, a naive approach for solving the problem will be, when given P, to simply find all occurrences of P in T by using this method, sort them (thus obtaining the sequence L'), choose only those which are in the range

$[i, j]$, and then iterate through them, each time outputting the first location not overlapping with the last location outputted. However, it is clear the the time for such a method will be dependant on *tocc*, which can, as lemma 2 suggests, be $\Theta(m \cdot k)$, which is too large.

Another (slightly better) approach would be to transform the leaves of ST(T) to points on a grid as described in Sect. 4. Assume we have found the node v which is described there (by using the methods described there). Recall that $x(l) \in [x(l'_v), x(l''_v)]$ iff P appears in T at location $y(l)$. We can therefore recover the exact set of occurrences of P in T that are in the range $[i, j]$ by conducting a range reporting query (i.e. searching and reporting all the points in range) of the range $[x(l'_v), x(l''_v)] \times [i, j]$ (latest results of range reporting due to Alstrup et al. [2]). Again, we can sort them (thus obtaining L'') and then iterate through them, each time outputting the first location not overlapping with the last location outputted. However, it is clear that now we still have a dependency on k'', which could be $\Theta(m \cdot k)$. Note that the method of representing the lexicographic order of the suffixes of a text, and the locations in the text, as two axes of a grid, used in this paper, was used before by Ferragina [8]. The goal of [8] required performing range reporting queries on a grid, while we use range successor queries on a grid.

We resort therefore to using a similar method to that which was used to solve the successive indexing problem in Sect. 4: consider the leaf f for which $x(f) \in [x(l'_v), x(l''_v)]$, $y(f) \geq i$ and $y(f)$ is minimal. If $y(f) \leq j$, then $y(f)$ is the leftmost occurrence of P in $t_i \dots t_{j+m-1}$, so according to the greedy scheme, we can include it in L. It is clear that $y(f)$ is exactly the occurrence of P in T successive to i, subject to the requirement that $y(f) \leq j$. Suppose such f exists and therefore we included $y(f)$ in L. We now want to choose the leftmost occurrence of P in T in the range $[i, j]$ not overlapping with the occurrence we have just chosen. In other words: we wish to find a leaf l for which $x(l) \in [x(l'_v), x(l''_v)]$ and $y(l)$ is minimal such that $y(f) + m \leq y(l) \leq j$. Luckily, this is exactly the occurrence of P in T successive to $y(f) + m$, adding the constraint that the occurrence is less than or equal to j. Therefore, this can be solved also by querying for the y-axis successor of $y(f)+m$ in the x-axis range $[x(l'_v), x(l''_v)]$. We can repeat this process in order to obtain the sequence L as it was defined by the greedy method.

The algorithm for the range non-overlapping indexing problem immediately follows, and is described as Algorithms 3 (preprocess) and 4 (query).

5.1 Analysis

Theorem 5. *The range non-overlapping indexing problem can by solved with $O(n \log n)$ storage and $O(m + k \log \log n)$ query time (where k is the maximal number of non-overlapping occurrences of P in T, that are in the range $[i, j]$), using $O(n \log n \log \log n)$ expected preprocess time.*

Algorithm 3. Range non-overlapping indexing preprocess

Input: a text $T = t_1 \dots t_n$
construct ST(T); **/* assume the field $y(l)$ is set for any leaf l by the suffix tree algorithm */**
traverse ST(T) and set the field $x(l)$ for each leaf l;
traverse ST(T) *using DFS* **:**
  foreach *node u* **do**
    store the values $x(l'_u)$ and $x(l''_u)$ in u; **/* l'_u and l''_u are the leftmost and rightmost leaves of T_u, respectively */**
preprocess the set $\{(x(l), y(l)) \mid l \text{ is a leaf in ST}(T)\}$ for range successor queries on an $[n+1] \times [n+1]$ grid;

Algorithm 4. Range non-overlapping indexing query

Input: a pattern $P = p_1 \dots p_m$, and two integers $1 \leq i \leq j \leq n$
let L be the empty sequence;
traverse ST(T) *starting from the root, according to the symbols in P* **:**
  if *'stuck'* **then return** "no occurrences";
  else let v be the node we reached (if we stopped at a node) or the node immediately below the edge we are at (if we stopped on an edge);
$y \leftarrow i$;
while *the range successor query for* $(x(l'_v), x(l''_v), y)$ *yields a result* (x', y'), *for which* $y' \leq j$ **do**
  append y' to L;
  $y \leftarrow y' + m$;
return L;

Proof. The preprocess phase is identical to the one for the successive indexing problem and therefore the space and preprocess time analysis is omitted.

The query time consists of:

1. $O(m)$ in order to find node v.
2. $O(\log \log n)$ time for a successor query to find each element of L, therefore overall $O(k \log \log n)$ for all k non-overlapping occurrences.

We conclude we use overall worst-case $O(m + k \log \log n)$ time. □

6 Conclusions

We have presented solutions for the successive list indexing problem, and the range non-overlapping indexing problems, by using a tool from computational geometry — range successor queries on a grid, which, to our best knowledge, has not been used before in this context. It is conceivable that more indexing problems can be solved by using the tool of range successor queries on a grid.

References

1. Agarwal, P., Erickson, J.: Geometric range searching and its relatives (1999)
2. Alstrup, S., Brodal, G.S., Rauhe, T.: New data structures for orthogonal range searching. In: IEEE Symposium on Foundations of Computer Science, pp. 198–207 (2000)
3. Amir, A., Keselman, D., Landau, G.M., Lewenstein, M., Lewenstein, N., Rodeh, M.: Text indexing and dictionary matching with one error. J. Algorithms 37(2), 309–325 (2000)
4. Apostolico, A., Preparata, F.P.: Data structures and algorithms for the string statistics problem. Algorithmica 15(5), 481–494 (1996)
5. Brodal, G.S., Lyngsø, R.B., Östlin, A., Pedersen, C.N.S.: Solving the string statistics problem in time O(n log n). In: Widmayer, P., Triguero, F., Morales, R., Hennessy, M., Eidenbenz, S., Conejo, R. (eds.) ICALP 2002. LNCS, vol. 2380, pp. 728–739. Springer, Heidelberg (2002)
6. Dietzfelbinger, M., Karlin, A.R., Mehlhorn, K., auf der Heide, F.M., Rohnert, H., Tarjan, R.E.: Dynamic perfect hashing: Upper and lower bounds. vol. 23, pp. 738–761, Philadelphia, PA, USA, Society for Industrial and Applied Mathematics (1994)
7. Farach, M.: Optimal suffix tree construction with large alphabets. In: FOCS '97: Proceedings of the 38th Annual Symposium on Foundations of Computer Science (FOCS '97), Washington, DC, USA, p. 137. IEEE Computer Society Press, Los Alamitos (1997)
8. Ferragina, P.: Dynamic text indexing under string updates. J. Algorithms 22(2), 296–328 (1997)
9. Ferragina, P., Muthukrishnan, S., de Berg, M.: Multi-method dispatching: A geometric approach with applications to string matching problems. In: STOC, pp. 483–491 (1999)
10. Gabow, H.N., Bentley, J.L., Tarjan, R.E.: Scaling and related techniques for geometry problems. In: STOC '84: Proceedings of the sixteenth annual ACM symposium on Theory of computing, pp. 135–143. ACM Press, New York (1984)
11. Knuth, D., Morris, J.H., Pratt, V.: Fast pattern matching in strings. SIAM Journal on Computing 6(2), 323–350 (1977)
12. Lenhof, H.-P., Smid, M.: Using persistent data structures for adding range restrictions to searching problems. RAIRO Theoretical Informatics and Applications 28, 25–49 (1994)
13. McCreight, E.M.: A space-economical suffix tree construction algorithm. J. ACM 23(2), 262–272 (1976)
14. Ukkonen, E.: On-line construction of suffix trees. Algorithmica 14(3), 249–260 (1995)
15. Weiner, P.: Linear pattern matching algorithms. In: 14th Annual Symposium on Switching and Automata Theory, pp. 1–11. IEEE, New York (1973)

Space-Efficient Straggler Identification in Round-Trip Data Streams Via Newton's Identities and Invertible Bloom Filters

David Eppstein and Michael T. Goodrich

Dept. of Computer Science, Univ. of California, Irvine, 92697

Abstract. We study the *straggler identification* problem, in which an algorithm must determine the identities of the remaining members of a set after it has had a large number of insertion and deletion operations performed on it, and now has relatively few remaining members.

1 Introduction

Imagine a security guard, who we'll call Bob, working at a large office building. Every day, Bob comes to work before anyone else, unlocks the front doors, and then staffs the front desk. After unlocking the building, Bob's job is to check in each of a set of n workers when he or she enters the building and check each worker out again when he or she leaves. Most workers leave the building by 6pm, when Bob's shift ends. But, at the end of Bob's shift, there may be a small number, at most $d << n$, of *stragglers*, who linger in the building working overtime. Before Bob can leave for home, he must tell the night guard the ID numbers of all the stragglers. The challenge is that Bob has only a small clipboard of size $o(n)$ to use as a "scratch space" for recording information as workers come and go. That is, Bob does not have enough room on his clipboard to write down all ID numbers of the workers as they arrive and check them off again as they leave. Of course, he also has to deal with the fact that some of the n workers may not come to work at all on any given day. The question we address in this paper is, "How can Bob, the security guard, check workers in and out so as to identify all d stragglers at the end of his shift, using a scratch space of size only $o(n)$?"

Formally, suppose we are given a universe $U = \{x_1, x_2, \ldots, x_n\}$ of unique identifiers, each representable with $O(\log n)$ bits. Given an upper bound parameter $d << n$, the *straggler identification problem* is to design a data structure that uses only $o(n)$ bits and efficiently supports the following operations on an initially-empty subset S of U:

- **Insert** x_i: Add the identifier x_i to S.
- **Delete** x_i: Remove the identifier x_i from S.
- **ListStragglers**: Test whether $|S| \leq d$, and if so, list all the elements of S.

F. Dehne, J.-R. Sack, and N. Zeh (Eds.): WADS 2007, LNCS 4619, pp. 637–648, 2007.

We assume, without loss of generality, that d is small enough so that $d\log(n/d)$ is $o(n)$, since we need $\Omega(d\log(n/d))$ bits just to produce the answer to an **List-Stragglers** query, and if d is close to n we might as well just store all the elements of S explicitly. That is, we are interested in an implicit representation of S, which can be used to list the contents of S when $|S| \leq d$, but makes no such guarantees when $|S| > d$.

In addition to our motivating example of Bob, the security guard (which also applies to other in-and-out physical environments, like amusement parks), the straggler identification problem has the following potential applications:

- In a high bandwidth multicast data stream, a server sends packets to many different clients, which send acknowledgments back to the server identifying each packet that was successfully received. The server then needs to identify and re-send the packets to clients that did not successfully receive them. This round-trip data stream application is an instance of the straggler identification problem, since we expect most of the packets to be sent successfully and we would like to minimize the space needed per client at the server for unacknowledged packet identification.
- In heterogeneous Grid computations, a supervisor sends independent tasks out to Grid participants, who, under normal conditions, perform these tasks and return the results to the supervisor. There may be a few participants, however, who crash, are disconnected from the network, or otherwise fail to perform their tasks. The supervisor would like to identity the tasks without responses, so that they can be sent to other participants for completion.

Our Results. In this paper, we study the straggler identification problem, showing that it can be solved with small space and fast update times. We provide a deterministic solution, which uses $O(d\log n)$ bits to represent the dynamic set S of $O(\log n)$-bit identifiers. Our solution is based on a novel application of Newton's identities and allows for insertions and deletions to be done in $O(d\log^{O(1)} n)$ time. It allows the **ListStragglers** operation to be done in time polynomial in d and $\log n$. This solution does not allow (false) **Delete** x operations that have no matching **Insert** x operations, however. Interestingly, we show that no deterministic algorithm can guarantee correctness in such scenarios, so this drawback should come as no surprise. Nevertheless, we provide a simple randomized solution to the straggler identification problem that uses $O(d\log n\log(1/\epsilon))$ bits and tolerates false deletions, where $\epsilon > 0$ is a user-defined error probability bound. This solution is based on a novel extension to the counting Bloom filter [3, 14], which itself is a dynamic, cardinality-based extension to the well-known Bloom filter data structure [1] (see also [5]). We refer to our extension as the *invertible Bloom filter*, because, unlike the standard Bloom filter and its counting extension—which provide a degree of data privacy protection—the invertible Bloom filter allows for the efficient enumeration of its contents if the number of items it stores is not too large. This might seem like a violation of the spirit of a Bloom filter, which was invented specifically to avoid the space needed for content enumeration. Nevertheless, the invertible Bloom filter is useful for straggler identification, because it can at one time represent, with small space, a multiset that is too large to enumerate, and later, after a series of deletions have been performed, provide for the efficient listing of the remaining elements.

Related Prior Work. Our work is most closely related to the "deterministic k-set structure" of Ganguly and Majumder [16]. This structure solves the straggler detection problem, allowing items to have multiplicity greater than one but disallowing false deletions. This solution, like our deterministic algorithm, is based on finite fields; however the most space-efficient version of their solution uses roughly twice as many bits as ours, and their decoding times are slower: ignoring logarithmic factors, $O(d^3)$ or $O(d^4)$ time, compared to $O(d^2)$ for ours. An additional technical difference is that, for the algorithm of Ganguly and Majumder, the parameter k (analogous to our d) measures the number of distinct stragglers, while for us it measures the total number of stragglers. Independently of our work, Ganguly and Majumder added to the submitted journal version of their paper a lower bound similar to ours proving the impossibility of straggler detection with false deletions (Ganguly, personal communication). Our deterministic solution is also related to work on set reconciliation in communication complexity [23].

Some additional existing work can be adapted to solve the straggler identification problem. For example, Cormode and Muthukrishnan [8] study the problem of identifying the d highest-cardinality members of a dynamic multiset. Their solution can be applied to the straggler identification problem, since whenever there are d or fewer elements in the set, then all elements are of relatively high cardinality. Their result is a randomized data structure that uses $O(d\log^2 n\log(1/\epsilon))$ bits to perform updates in $O(\log^2 n\log(1/\epsilon))$ time and can be adapted to answer **ListStragglers** queries in $O(d\log^2 n\log(1/\epsilon))$ time (in terms of their bit complexities), where $\epsilon > 0$ is a user-defined parameter bounding the probability of a wrong answer.

Also relevant is prior work on combinatorial group testing (CGT), e.g., see [7, 10,11,12,13,15,18,22], and multiple access channels (MAC), e.g., see [6,17,19,20, 21,25,26,28]. In combinatorial group testing, there are d "defective" items in a set U of n objects, for which we are allowed to perform *tests*, which involve forming a subset $T \subseteq U$ and asking if there are any defective items in T. In the standard CGT problem, the outcome is binary—either T contains defective items or it does not. The objective is to identify all d defective items. The CGT algorithms that are most relevant to straggler identification are *nonadaptive*, in that they must ask all of their tests, $T_1, T_2, \ldots, T_m$, in advance. Such an algorithm can be converted to solve the straggler identification problem by creating a counter t_i for each test T_i. On an insertion of x, we would increment each t_i such that $x \in T_i$. Likewise, on a deletion of x, we would decrement each t_i such that $x \in T_i$. The tests with non-zero counters would be exactly those containing our objects of interest, and the nonadaptive CGT algorithm could then be used to identify them. Unfortunately, these algorithms don't translate into efficient straggler-identification methods, as the best known nonadaptive CGT algorithms (e.g., see [11,12]) use $O(d^2 \log n)$ tests, which would translate into a straggler solution needing $O(d^2 \log^2 n)$ bits.

The MAC problem is similar to the CGT problem, except that the items of interest are no longer "defective"—they are d devices, out of a set U, wishing to broadcast a message on a common channel. In this case a "test" is a time slice where members of a subset $T \subseteq U$ can broadcast. Such an event has a three-way outcome, in that there can be 0 devices that use this time slice, 1 device that uses it (in which case it is identified and taken out of the set of potential broadcasters), or there can be 2 or more who attempt to use the channel, in which case none succeed (but all the potential broadcasters learn that T contains at least two broadcasters). Unfortunately, traditional MAC algorithms are adaptive, so do not immediately translate into straggler identification algorithms.

Nevertheless, we can extend the MAC approach further [17,25,26,28], so that each test T returns the actual number of items of interest that are in T. This extension gives rise to a *quantitative* version of CGT (e.g., see [11], Sec. 10.5). Unfortunately, previous approaches to the quantitative CGT problem are either non-constructive [25], adaptive [17,25,26,28], or limited to small values of d. We know of no nonadaptive quantitative CGT algorithms for $d \geq 3$, and the ones for $d = 2$ don't translate into efficient solutions to the straggler identification problem (e.g., see [11], Sec. 11.2).

2 Straggler Detection Via Symmetric Polynomials

We now describe a deterministic algorithm for straggler detection using near-optimal memory. The algorithm is algebraic in nature: it stores as its snapshot of the data stream a collection of *power sums* in a finite field, $GF[p^e]$. The decoding algorithm for this information uses Newton's identities to convert these power sums into the coefficients of a polynomial that has the stragglers as its roots, and finds the roots of this polynomial.

As is standard for this sort of computation, we represent values in $GF[p^e]$ as univariate polynomials of degree at most $e-1$, with coefficients that are integers modulo p; the $GF[p^e]$ arithmetic operations are the standard polynomial arithmetic, modulo a *primitive polynomial* of degree e. Therefore, values in the field $GF[p^e]$ may be represented in space $O(e \log p)$ each. Addition and subtraction of values in $GF[p^e]$ may be performed using modulo-p operations independently over each coefficient, while multiplication of values in $GF[p^e]$ may be performed using a convolution-based polynomial multiplication algorithm, together with reduction modulo the primitive polynomial. Our algorithms also involve division by integers in the range $[2, p-1]$, which may again be done independently on each coefficient. Therefore, each field operation may be performed in bit complexity $\tilde{O}(e \log p)$, where $\tilde{O}(x)$ is a convenient shorthand for $O(x \log^{O(1)} x)$.

Theorem 1. *There is a deterministic streaming straggler detection algorithm using $O(d \log n)$ bits of storage, such that **Insert** and **Delete** operations can be performed in bit complexity $\tilde{O}(d \log n)$, and such that **ListStragglers** operations can be performed in bit complexity $\tilde{O}(d \log^3 n + d^2 \log n + d^{3/2} \log^2 n \min(d, \log n))$.*

Proof. We let p be a prime number, larger than d, and choose e such that $p^e > n$. We perform all operations of the algorithm in the field $GF[p^e]$, and interpret all identifiers in the straggler detection problem as values in this field. The number of bits needed to represent a single value in $GF[p^e]$ is $O(\log n)$, and, with this choice of p and e, each arithmetic operation in the field may be performed in bit complexity $\tilde{O}(\log n)$.

Define the power sums

$$s_k(S) = \sum_{x_i \in S} x_i^k$$

(where x_i and s_k belong to $GF[p^e]$, except for s_0 which we store as a $\log n$ bit integer). Our streaming algorithm stores $s_k(S)$ for $0 \leq k \leq d$. As $s_0(S)$ is the number of stragglers, we can easily compare the number of stragglers to d.

To update the power sums after an insertion of a value x_i, we simply add x_i^k to each power sum s_k; this requires $O(d)$ arithmetic operations in $GF[p^e]$. Similarly, to delete x_i, we subtract x_i^k from each power sum s_k.

At any point in the algorithm, we may define a polynomial in $GF[p^e][x]$,

$$P(x) = \prod_{x_i \in S} (x - x_i) = \sum_{k=0}^{|S|} (-1)^k \sigma_k x^{|S|-k},$$

where σ_k is the kth *elementary symmetric function* of S (the sum of the products of all k-tuples of members of S). These coefficients can be related to the power sums by *Newton's identities* (e.g. see [9]):

$$s_k - k(-1)^k a_k = -\sum_{i=1}^{k-1} (-1)^i \sigma_i s_{k-i}.$$

That is,

$$\begin{aligned}
s_1 - \sigma_1 &= 0 \\
s_2 + 2\sigma_2 &= \sigma_1 s_1 \\
s_3 - 3\sigma_3 &= \sigma_1 s_2 - \sigma_2 s_1 \\
s_4 + 4\sigma_4 &= \sigma_1 s_3 - \sigma_2 s_2 + \sigma_3 s_1 \\
s_5 - 5\sigma_5 &= \sigma_1 s_4 - \sigma_2 s_3 + \sigma_3 s_2 - \sigma_4 s_1,
\end{aligned}$$

and so on. These equations hold over any field, and in particular over $GF[p^e]$. By using these identities, we may calculate the coefficients of P in sequence from the power sums and the earlier coefficients, using $O(d^2)$ arithmetic operations to compute all coefficients. Note that these calculations involve divisions by the numbers 2, 3, 4, ..., d, but all such divisions are possible modulo p. Thus, this stage of the **ListStragglers** operation takes bit complexity $\tilde{O}(d^2 \log n)$.

Finally, to determine the list of stragglers, we find the roots of the polynomial $P(x)$ that has been determined as above. The deterministic root-finding algorithm of Shoup [27] solves this problem in $\tilde{O}(d \log^2 n + d^{3/2} \log n \min(d, \log n))$ field operations; thus, the overall bit complexity bound is as stated. □

We note that a factor of $d^{1/2}$ in Shoup's algorithm [27] occurs only when p has an unexpectedly long repeated subsequence in its sequence of quadratic characters. It seems likely that a more careful choice of p can eliminate this factor, simplifying the time bound for the **ListStragglers** operation to $\tilde{O}(d \log^3 n + d^2 \log n)$. If this is possible, it would be an improvement when d lies in the range of values from $\log^{2/3} n$ to $\log^2 n$.

For $d \leq 4$, the root finding algorithm may be replaced by the usual formulae for solving low degree polynomials in closed form.

3 Impossibility Results for False Deletions

So far, we have assumed that an element deletion can occur only if a corresponding insertion has already occurred. That is, the only anomalous data patterns that might occur are insertions that are not followed by a subsequent deletion. What can we say about more general update sequences in which insertion-deletion pairs may occur out of order, multiple times, or with a deletion that does not match an insertion? We would like to have a streaming data structure that handles these more general event streams and allows us to detect small numbers of anomalies in our insertion-deletion sequences.

Formally, define a *signed multiset* over a set S to be a map f from S to the integers, where $f(x)$ is the number of occurrences of x in the multiset. To insert x into a signed multiset, increase $f(x)$ by one, while to delete x, decrease $f(x)$ by one. Thus, any sequence of insertions and deletions, no matter how ordered, produces a well-defined signed multiset. We wish to find a streaming algorithm that can determine whether all but a small number of elements in the signed multiset have nonzero values of $f(x)$ and identify those elements. But, as we show, for a natural and general class of streaming algorithms, even if restricted to signed multisets in which each x has $f(x) \in \{-1, 0, 1\}$, we cannot distinguish the empty multiset (in which all $f(x)$ are zero) from some nonempty multiset. Therefore, it is impossible for a deterministic streaming algorithm to determine whether a multiset has few nonzeros.

The signed multisets form a commutative group, which we will represent using additive notation: $(f + g)(x) = f(x) + g(x)$. Call this group M. Define a *unit multiset* to be a signed multiset in which all values $f(x)$ are in $\{-1, 0, 1\}$; the unit multisets form a subset of M, but not a subgroup.

Suppose a streaming algorithm maintains information about a signed multiset, subject to insertion and deletion operations. We say that the algorithm is *uniquely represented* if the state of the algorithm at any time depends only on the multiset at that time and not on the ordering of the insertions and deletions by which the multiset was created. That is, there must exist a map u from M to states of the algorithm.

Define a binary operation $+$ on states of a uniquely represented multiset streaming algorithm, as follows. If a and b are states, let A and B be signed multisets such that $u(A) = a$ and $u(B) = b$, and let $a + b = u(A + B)$.

Lemma 1. *If a streaming algorithm is uniquely represented, and $u(P) = u(Q)$, then $u(P + R) = u(Q + R)$.*

Proof. Let s be a sequence of updates that forms R. Then s transforms $u(P)$ to $u(P + R)$ and $u(Q)$ to $U(Q + R)$. Since $u(P) = u(Q)$, $u(P + R)$ must equal $u(Q + R)$. □

Lemma 2. *The operation defined above is well-defined independently of how the representative multisets A and B are chosen, the states of the streaming algorithm form a commutative group under this operation, and u is a group homomorphism.*

Proof. Independence from the choice of representation is Lemma 1. Associativity and commutativity follow from the associativity and commutativity of the corresponding group operation on M. By Lemma 1, $u(A) + u(-A) = u(0)$ and $u(A) + u(0) = u(A)$, so $u(0)$ satisfies the axioms of a group identity; therefore, we have defined a commutative group. That u is a homomorphism follows from the way we have defined our group operations as the images by u of group operations in M. □

Theorem 2. *Any uniquely represented multiset streaming algorithm for a multiset on n items, with fewer than n bits of storage, will be unable to distinguish between the empty set and some nonempty unit multiset.*

Proof. Suppose there are $k < n$ bits of storage, so 2^k possible states. By the pigeonhole principle, two different sets A and B, when interpreted as multisets and mapped to states, map to the same state $u(A) = u(B)$. Then by Lemma 2, $u(A - B) = u(\emptyset)$. $A - B$ is a nonempty unit multiset that cannot be distinguished from the empty set. □

By applying similar ideas, we can prove a similar impossibility result without assumption about the nature of the streaming algorithm.

Theorem 3. *No deterministic streaming algorithm with fewer than n bits of storage can distinguish a stream of matched pairs of insert and delete operations over a set of n items from a stream of insert and delete operations that are not matched in pairs.*

Proof. Suppose that we have a deterministic streaming data structure with $k < n$ bits of storage. For any set A, let $f(A)$ denote the state of the data structure on a stream that starts with an empty set and inserts the items in A in some canonical order. By the pigeonhole principle there exist two sets A and B such that $A \neq B$ but such that $f(A) = f(B)$. Let s_{PQ} ($P, Q \in \{A, B\}$) be the operation stream formed by inserting the items in set P followed by deleting the items in set Q. Then the streaming algorithm must have the same state after stream s_{AA} as it does after stream s_{BA}, but s_{AA} consists of matched insert-delete pairs while s_{BA} does not. □

4 Invertible Bloom Filters

The standard Bloom filter [1] is a randomized data structure for approximately representing a set S subject to insertion operations and membership queries. Given a parameter d on the expected size of S and an error parameter $\epsilon > 0$, it consists of a hash table B containing $m = O(d \log(1/\epsilon))$ single-bit cells (which we denote as a "bit" field), which are initially all 0's, together with $k = \Theta(\log(1/\epsilon))$ random hash functions $\{h_1, \ldots, h_k\}$ that map elements of S to integers in the range $[0, m-1]$. Performing an insert of element x amounts to setting each $B[h_i(x)]$.bit to 1, for $i = 1, \ldots, k$. Likewise, testing for membership of x in S amounts to testing that there is no $i \in \{1, \ldots, k\}$ such that $B[h_i(x)]$.bit $= 0$. Setting the constants appropriately, one can make the probability of returning a false positive to a membership query (that is, an element not in S identified as belonging to S) to be less than ϵ (e.g., see [4]).

The counting Bloom filter [3,14] extends the standard Bloom filter by replacing each "bit" cell of B with a counter cell, "count" (initialized to 0 for each cell). An insertion of item x amounts to incrementing each $B[h_i(x)]$.count by 1, for $i = 1, \ldots, k$. Such a structure also supports the deletion of an item x, by decrementing each cell $B[h_i(x)]$.count by 1, for $i = 1, \ldots, k$. Answering a membership query is similar to that for the standard Bloom filter, amounting to testing that there is no $i \in \{1, \ldots, k\}$ such that $B[h_i(x)]$.count $= 0$.

The *invertible Bloom filter* extends the counting Bloom filter, in several ways, and allows us to solve the straggler identification problem even in the presence of false deletions. It requires that we use three additional random hash functions, f_1, f_2, and g, in addition to the k hash functions, $h_1, \ldots, h_k$, used for B above. The functions, f_1 and f_2 map integers in $[0, n]$ to integers in $[0, m]$. The function g maps integers in $[0, n]$ to integers in $[0, n^2]$. In addition, we add two more fields to each Bloom filter cell, $B[i]$:

- An "idSum" field, which stores the sum of all the elements, x in S, for x's that map to the cell $B[i]$. Note that if $B[i]$ stores m copies of a value x (and no other values), then $B[i]$.idSum $= mx$.
- A "hashSum" field, which stores the sum of all the hash values, $g(x)$, for x's that map to the cell $B[i]$. Note that if $B[i]$ stores m copies of a value x (and no other values), then $B[i]$.hashSum $= mg(x)$.

Moreover, we create a second Bloom filter, C, which has the same number of (count, idSum, and hashSum) fields as B, but uses only the functions f_1 and f_2 to map elements of S to its cells. That is, C is a secondary augmented counting Bloom filter with the same number of cells as B, but with only two random hash functions, f_1 and f_2, to use for mapping purposes. Intuitively, C will serve as a fallback Bloom filter for "catching" elements that are difficult to recover using B alone. Finally, in addition to these fields, we maintain a global count variable, initially 0. Each of our count fields is a signed counter, which (in the case of false deletions) may go negative.

Since all n ID's in U can be represented with $O(\log n)$ bits, their sum can also be represented with $O(\log n)$ bits. Thus, the space needed for B and C is $O(m \log n) = O(d \log n \log(1/\epsilon))$.

We process updates for the invertible Bloom filter as follows.

Insert x:
  increment `count`
  for $i = 1, \ldots, k$ **do**
    increment $B[h_i(x)]$.`count`
    add x to $B[h_i(x)]$.`idSum`
    add $g(x)$ to $B[h_i(x)]$.`hashSum`
  for $i = 1, 2$ **do**
    increment $C[f_i(x)]$.`count`
    add x to $C[f_i(x)]$.`idSum`
    add $g(x)$ to $C[f_i(x)]$.`hashSum`

Delete x:
  decrement `count`
  for $i = 1, \ldots, k$ **do**
    decrement $B[h_i(x)]$.`count`
    subtract x from $B[h_i(x)]$.`idSum`
    subtract $g(x)$ from $B[h_i(x)]$.`hashSum`
  for $i = 1, 2$ **do**
    decrement $C[f_i(x)]$.`count`
    subtract x from $C[f_i(x)]$.`idSum`
    subtract $g(x)$ from $C[f_i(x)]$.`hashSum`

That is, to insert x, we go to each cell that x maps to and increment its `count` field, add x to its `idSum` field, and add $g(x)$ to its `hashSum` field. Thus, the methods for element insertion is fairly straightforward. Deletion is similarly easy, in that we simply decrement counts and subtract out the appropriate summands to reverse the insertion operation.

Our method for performing the **ListStragglers** operation is a bit more involved, however. The basic idea is that some cells of B are likely to be *pure*, that is, to have values that have been affected by only a single item. If we can find a pure cell, we can recover the identity of its item by dividing its `idSum` by its `count`. Once a single item and its count are known, we can remove that item from the data structure and continue until all items have been found.

The difficulty with this approach is in finding the pure cells. Because of the possibility of multiple insertions and false deletions, we cannot simply test whether `count` is one: some pure cells may have larger counts (i.e., have multiple copies of the same value), and some impure cells may have a count equal to one (e.g., because of two insertions of a value x followed by a false deletion of a value y that collides with x at this cell). Instead, to test whether a cell is pure, we use its `hashSum`: in a pure cell, the `hashSum` should equal the `count` times the hash of the item's identifier, while in a cell that is not pure it is very unlikely that the `hashSum`, `idSum`, and `count` fields will match up in this way.

The following pseudo-code expresses the decoding algorithm outlined above.

ListStragglers:
while $\exists i$, s. t. $g(B[i].\texttt{idSum}/B[i].\texttt{count}) = B[i].\texttt{hashSum}/B[i].\texttt{count}$ **do**
  if $B[i].\texttt{count} > 0$ **then** {this is a good element}
    Push $x = B[i].\texttt{idSum}/B[i].\texttt{count}$ onto an output stack O.
    Delete all $B[i].\texttt{count}$ copies of x from B and C (using a method similar to **Delete** x above)
  else {this is a false delete}
    Back out all $-B[i].\texttt{count}$ falsely-removed copies of x from B and C (using a method similar to **Insert** x above)
if $\texttt{count} = 0$ **then**
  Output the elements in the output stack and insert each element back into B and C.
else {we have mutually-conflicting elements in B}
  Repeat the above while loop, but do the tests using C instead of B.
  Output the elements in the output stack, O, and insert each element back into B and C.

There is a slight chance that this algorithm fails. For example, we could have two or more items colliding in a cell of B, but we could nevertheless have the condition, $g(B[i].\texttt{idSum}/B[i].\texttt{count}) = B[i].\texttt{hashSum}/B[i].\texttt{count}$, satisfied (and similarly for C in the second while loop). Fortunately, since g is a random function from $[0, n]$ to $[0, n^2]$, such an event occurs with probability at most $1/n^2$; hence, over the entire algorithm we can assume, with high probability, that it never occurs (since $d << n$). More troubling is the possibility that, even after using the fallback array, C, to find and enumerate elements in the invertible Bloom filter (in the second while loop), we might still have some mutually-conflicting elements in C.

Lemma 3. *If the number of elements in S, which were inserted but not deleted, plus the number of false elements negatively indicated in S, corresponding to items deleted but not inserted, is at most d, then the first while loop will remove all but ϵd such elements from S with probability $1 - \epsilon/2$, for $\epsilon < 1/4$.*

Proof. Omitted due to space limitations. □

Let us assume, therefore, that at most ϵd elements (true and/or false) remain in S after the first while loop. Let us suppose further that each is mapped to two distinct cells in C (the probability there is any such self-collision among the remaining elements in C is at most $\epsilon d/4dk \leq \epsilon/4$). We can envision each cell in C as forming a vertex in a graph, and each selected pair of cells as forming an edge in the graph; thus our data can be modeled as a random multigraph with $x \leq \epsilon d$ edges and $y = 4dk \geq 8d$ vertices. Thus, it is a very sparse graph. Let $c = y/x \geq 8/\epsilon$.

Two types of bad event could prevent us from decoding the data remaining in C after the first loop. First, two items could map to the same pair of cells, so that our multigraph is not a simple graph. There are $x(x-1)/2$ pairs of items, and each two items collide with probability $2/(y(y-1))$, so the expected number of collisions of this type is $x(x-1)/(y(y-1))$, roughly $1/c^2$. Second, the graph may be simple but may contain a cycle. As shown by Pittel [2, Exercise 8, p. 122], the expected number of vertices in cyclic components of a random graph of this size is bounded by $\sum_{k=3}^{\infty} kc^-k = O(1/c^3)$. Therefore, the expected number of events of either type, and the probability that there exists an event of either type, is $O(1/c^2)$. Choosing $c = O(\sqrt{1/\epsilon})$ is sufficient to show that we will fail in the second while loop with probability at most $\epsilon/4$.

Theorem 4. *If the number of elements in S, which were inserted but not deleted, plus the number of false elements negatively indicated in S, which correspond to items deleted but not inserted, is at most d, then the above algorithm correctly answers a **ListStragglers** query with probability at least $1-\epsilon$, where $\epsilon < 1/4$.*

Acknowledgments

We would like to thank Dan Hirschberg for several helpful discussions. We are also grateful to an anonymous reviewer for suggesting the multicast application.

References

1. Bloom, B.H.: Space/time trade-offs in hash coding with allowable errors. Commun. ACM 13, 422–426 (1970)
2. Bollobás, B.: Random Graphs. Academic Press, New York (1985)
3. Bonomi, F., Mitzenmacher, M., Panigrahy, R., Singh, S., Varghese, G.: An improved construction for counting Bloom filters. In: Azar, Y., Erlebach, T. (eds.) ESA 2006. LNCS, vol. 4168, pp. 684–695. Springer, Heidelberg (2006)
4. Bose, P., Guo, H., Kranakis, E., Maheshwari, A., Morin, P., Morrison, J., Smid, M., Tang, Y.: On the false-positive rate of Bloom filters. Report, School of Comp. Sci. Carleton Univ. (2007) `http://cg.scs.carleton.ca/~morin/publications/ds/bloom-submitted.pdf`
5. Broder, A., Mitzenmacher, M.: Network applications of Bloom filters: A survey. Internet Mathematics 1(4), 485–509 (2005)
6. Capetanakis, J.I.: Tree algorithms for packet broadcast channels. IEEE Trans. Inf. Theory IT-25(5), 505–515 (1979)
7. Colbourn, Dinitz, Stinson.: Applications of combinatorial designs to communications, cryptography, and networking. In: Walker. (ed.) Surveys in Combinatorics, 1993. London Mathematical Society Lecture Note Series, vol. 187, Cambridge University Press, Cambridge (1999)
8. Cormode, G., Muthukrishnan, S.: What's hot and what's not: tracking most frequent items dynamically. ACM Trans. Database Syst. 30(1), 249–278 (2005)
9. Cox, D., Little, J., O'Shea, D.: Ideals, Varieties, and Algorithms: An Introduction to Computational Algebraic Geometry and Commutative Algebra. Springer, Heidelberg (1992)
10. DeBonis, A., Gasieniec, L., Vaccaro, U.: Generalized framework for selectors with applications in optimal group testing. In: Baeten, J.C.M., Lenstra, J.K., Parrow, J., Woeginger, G.J. (eds.) ICALP 2003. LNCS, vol. 2719, pp. 81–96. Springer, Heidelberg (2003)

11. Du, D.-Z., Hwang, F.K.: Combinatorial Group Testing and Its Applications, 2nd edn. World Scientific, Singapore (2000)
12. Du, D.-Z., Hwang, F.K.: Pooling Designs and Nonadaptive Group Testing. World Scientific, Singapore (2006)
13. Eppstein, D., Goodrich, M.T., Hirschberg, D.S.: Improved combinatorial group testing for real-world problem sizes. In: Dehne, F., López-Ortiz, A., Sack, J.-R. (eds.) WADS 2005. LNCS, vol. 3608, Springer, Heidelberg (2005)
14. Fan, L., Cao, P., Almeida, J., Broder, A.Z.: Summary cache: a scalable wide-area web cache sharing protocol. IEEE/ACM Trans. Networking 8(3), 281–293 (2000)
15. Farach, M., Kannan, S., Knill, E., Muthukrishnan, S.: Group testing problems with sequences in experimental molecular biology. In: SEQUENCES, p. 357. IEEE Press, New York (1997)
16. Ganguly, S., Majumder, A.: Deterministic k-set structure. In: Proc. 25th ACM SIGMOD Symp. Principles of Database Systems, pp. 280–289 (2006)
17. Georgiadis, L., Papantoni-Kazakos, P.: A collision resolution protocol for random access channels with energy detectors. IEEE Trans. on Communications COM-30(11), 2413–2420 (1982)
18. Goodrich, M.T., Hirschberg, D.S.: Efficient parallel algorithms for dead sensor diagnosis and multiple access channels. In: 18th ACM Symp. on Parallelism in Algorithms and Architectures (SPAA), pp. 118–127. ACM Press, New York (2006)
19. Greenberg, A.G., Ladner, R.E.: Estimating the multiplicities of conflicts in multiple access channels. In: Proc. 24th Annual Symp. on Foundations of Computer Science (FOCS'83), pp. 383–392. IEEE Computer Society Press, Los Alamitos (1983)
20. Greenberg, A.G., Winograd, S.: A lower bound on the time needed in the worst case to resolve conflicts deterministically in multiple access channels. J. ACM 32(3), 589–596 (1985)
21. Hofri, M.: Stack algorithms for collision-detecting channels and their analysis: A limited survey. In: Balakrishnan, A.V., Thoma, M. (eds.) Proc. Inf. Sem. Modelling and Performance Evaluation Methodology, Lecture Notes in Control and Info. Sci., vol. 60, pp. 71–85 (1984)
22. Hwang, F.K., Sós, V.T.: Non-adaptive hypergeometric group testing. Studia Scient. Math. Hungarica 22, 257–263 (1987)
23. Minsky, Y., Trachtenberg, A., Zippel, R.: Set reconciliation with nearly optimal communication complexity. IEEE Trans. Information Theory 49(9), 2213–2218 (2003)
24. Motwani, R., Raghavan, P.: Randomized Algorithms. Cambridge University Press, Cambridge (1995)
25. Pippenger, N.: Bounds on the performance of protocols for a multiple-access broadcast channel. IEEE Trans. on Information Theory IT-27(2), 145–151 (1981)
26. Ruszinkó, M., Vanroose, P.: A code construction approaching capacity 1 for random access with multiplicity feedback. Report, Fakultät für Mathematik der Universität Bielefeld, Report no. 94-025 (1994) `http://www.math.uni-bielefeld.de/sfb343/preprints/abstracts/apr94025.ps.gz`
27. Shoup, V.: A fast deterministic algorithm for factoring polynomials over finite fields of small characteristic. In: Proc. Int. Symp. Symbolic and Algebraic Computation, pp. 14–21. ACM Press, New York (1991)
28. Tsybakov, B.S.: Resolution of a conflict of known multiplicity. Problems of Information Transmission 16(2), 134–144 (1980)

Dynamic TCP Acknowledgment with Sliding Window

Hisashi Koga

Graduate School of Information Systems, University of Electro-Communications, 1-5-1 Chofugaoka Chofu-si, Tokyo 182–8585, Japan

Abstract. *The dynamic TCP acknowledgement problem* formulated by Dooly et al. has been intensively studied in the area of competitive analysis. However, their framework does not consider the *sliding window* that restricts the maximum number of packets that the sender can inject into the network without an acknowledgement in TCP protocol. This paper proposes a new problem in which the sliding window is realistically integrated. We study how the ability of on-line algorithms change, depending on whether the receiver knows the window size. We show that a deterministic on-line algorithm extended from the optimal on-line algorithm for Dooly's framework achieves the best competitive ratio of 2, if the window size is given. By contrast, if the window size is not given, the lower bound of the competitive ratio for an algorithm class which contains the algorithm by Dooly et al. depends on the peak packet rate from the sender and the window size. Significantly, our problem models the situation in which an on-line algorithm involuntarily transforms the input and processes the modified input without noticing the transformation.

1 Introduction

TCP (Transport Control Protocol) is the most used transport protocol in the Internet. Thus, there is a strong need to grasp the behavior of TCP protocol both from theoretical and experimental sides. Among previous works that analyzed TCP theoretically, Dooly et al. [1] focused on the mechanism of TCP acknowledgement. When a sender S sends packets to a receiver R using TCP protocol, a packet arriving at R must be acknowledged by R in order to notify S that the transmission was successful. However, each packet need not be acknowledged individually. Instead, most TCP implementations adopt "delayed ACK" which admits the receiver to acknowledge multiple packets with a single acknowledgement by postponing the acknowledgement. The delayed ACK mechanism contributes to reducing the overhead of the acknowledgements by decreasing their frequency. On the other hand, it has the risk to add excessive latency to the TCP connection. Dooly et al. [1] formulated this trade-off as *the dynamic TCP acknowledgement problem.*

In the dynamic TCP acknowledgement problem, a sequence of n packets $\sigma = (p_1, p_2, \cdots, p_n)$ reach R in order. The arrival time of p_i is denoted by a_i. If $i_i < i_2$, $a_{i_1} \leq a_{i_2}$. An acknowledgment algorithm in R receives the arrival

F. Dehne, J.-R. Sack, and N. Zeh (Eds.): WADS 2007, LNCS 4619, pp. 649–660, 2007.

time sequence $(a_1, a_2, \cdots, a_n)$ as the input and divides σ into m subsequnces $\sigma_1, \sigma_2, \ldots, \sigma_m$, where each subsequence end corresponds to a single acknowledgment. All the packets contained in σ_j $(1 \le j \le m)$ are acknowledged together by the j-th acknowledgement at time t_j. To assure that all the packets should be acknowledged, $m \ge 1$ and $t_m \ge a_n$. In case a packet p is not acknowledged immediately, an extra latency arises. The purpose of the dynamic TCP acknowledgement problem is to minimize the sum of the cost for generating acknowledgements and the cost for the latency of acknowledgements by choosing the acknowledgement time sequence $(t_1, t_2, \cdots, t_m)$ adequately. Ordinarily, an acknowledgment time is decided in an on-line fashion without knowing the future packet arrivals. Dooly et al. propose two kinds of cost functions to be minimized. The first cost function f_{sum} combines the number of acknowledgements with the sum of delays for all the packets. f_{sum} is described as $m + \sum_{j=1}^{m} \sum_{p_i \in \sigma_j} |t_j - a_i|$. The second cost function f_{max} combines the number of acknowledgements with the sum of the maximum delays of a packet in each subsequence σ_j and is described as $m + \sum_{j=1}^{m} \max_{p_i \in \sigma_j} |t_j - a_i|$.

Dooly et al. evaluated on-line algorithms with competitive analysis [2] which compares an on-line algorithm with the optimal off-line algorithm OPT that knows σ in advance. Let $C_A(\sigma)$ be the cost of an algorithm A in processing σ. An on-line algorithm A is called c-competitive if $C_A(\sigma) \le c \cdot C_{OPT}(\sigma) + \beta$ for any σ, where β is a constant independent of σ. For each of f_{sum} and f_{max}, they presented a 2-competitive deterministic optimal on-line algorithm. The best on-line algorithm for f_{max} utilizes a timer: When p_i reaches R at time a_i, the timer is set to $a_i + 1$. If $a_{i+1} > a_i + 1$, the timer expires and the acknowledgement is performed at $a_i + 1$. Otherwise, the timer is updated to $a_{i+1} + 1$ at a_{i+1}. In this paper, this algorithm is referred to as `WAIT(1)`, because it always waits for 1 unit time after the last packet arrival before an acknowledgement. This paper deals with f_{max} only and we abbreviate the dynamic TCP acknowledgement problem with f_{max} simply as DTCP.

We claim that DTCP abstracts the mechanism of TCP acknowledgement only partially, because it misses the concept of *sliding window* that plays a crucial role for congestion control in TCP. The sliding window functions in a TCP sender S and restricts the maximum number of packets that S can transmit without notified acknowledgements. See Fig. 1. The sliding window is depicted as a rectangle and divides the packet sequence into three subsequences. The leftmost subsequence presents the packets which have been already acknowledged. Packets on the right of the left end of the sliding window are still unacknowledged. Among them, S can inject the packets in the sliding window into the network without receiving a new acknowledgement. In other words, the width of the sliding window defines the maximum number of packets that S can inject into the network without a new acknowledgement. Every time S receives an acknowledgement, the sliding window slides rightward. The sliding window forces S to stop sending packets when an acknowledgement has not been returned from R for long. The width of the sliding window is termed *window size*.

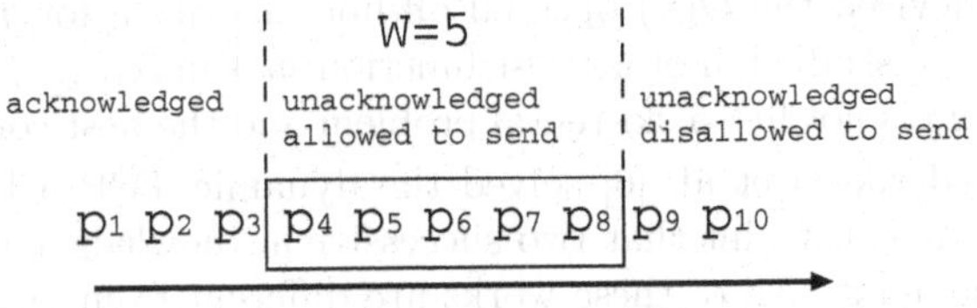

Fig. 1. Sliding Window

Our primary contribution is to propose a new realistic problem which integrates the sliding window into DTCP. Because, in the standard TCP, R operates without knowing if S stops sending packets due to the sliding window, we examine how the power of on-line algorithms depends on whether R recognizes that S is kept waiting by the sliding window. This paper assumes that the window size is a constant integer $W(\geq 1)$. Namely, a packet p_i can never reach R before p_{i-W} is acknowledged by R for $i > W$. Under this simplification, R has only to know the value of W so as to judge whether S is waiting or not. Our new problem is named as DTCPSW (DTCP with Sliding Window). Interestingly, in DTCPSW, a packet arrival time reflects the advance of the sliding window which results from how the acknowledgment algorithm in R acknowledged in the past. Therefore, an on-line algorithm involuntarily changes the original input (the time when a packet becomes ready at S) and processes the modified input (the time when it arrives at R) without noticing that it is changed from the original input. As far as we know, there is no previous work that addresses this intractability in the research area of competitive analysis.

Section 2 defines DTCPSW formally. Section 3 presents the optimal off-line algorithm for DTCPSW. Section 4 discusses the case in which R knows the window size. For this case, we construct a 2-competitive deterministic optimal on-line algorithm by extending `WAIT(1)`. Thus, a result comparable to DTCP is obtained. Section 5 treats the case in which R does not know the window size. Here, we pick up an algorithm class `WAIT`(α) that generalizes `WAIT(1)` such that α is arbitrary positive real. We prove that no on-line algorithm in `WAIT`(α) is better than $\frac{T}{W+\lfloor \frac{T}{W} \rfloor -1}$-competitive, even if the number of packets S wishes to send per unit time never goes beyond T. Thus, Sect. 4 and Sect. 5 contrast, which reveals the importance of the agreement between the sender and the receiver in communications. In addition, since the receiver does not grasp the window size in the real TCP protocol, we suppose that Sect. 5 outputs a more realistic result than other previous theoretical researches. We also show that one instance of `WAIT`(α) becomes $(\lceil \frac{T}{W} \rceil + 2)$-competitive.

1.1 Related Works

The dynamic TCP acknowledgement problem has been intensively studied in terms of competitive analysis. Karlin et al. [3] studied randomized on-line algorithms for f_{sum} and developed an $\frac{e}{e-1}$-competitive randomized on-line algorithm.

Noga et al. [4] devised the $O(n)$ optimal off-line algorithm for f_{sum}. Recently, Albers and Bals [5] studied another cost function $m + \max_{1 \le i \le m} \max_{p_i \in \sigma_j} |t_j - a_i|$. Their problem is not like a ski-rental problem and the best competitive ratio becomes $\frac{\pi^2}{6}$. Frederiksen et al. [6] solved the dynamic TCP acknowledgement problem under the constraint that two successive acknowledgement times must be one unit time apart. All of these works are different from our paper in that they do not take the sliding window into account.

2 Problem Statement

In DTCPSW we are given a sequence of packets $\sigma = (p_1, p_2, \cdots, p_n)$ that S shall send to R in order. DTCPSW differs from the original DTCP in that the next two sorts of times are associated with each packet p_i.

- Ready time r_i: the time when S prepares the transmission of p_i. p_i is called *ready at time* t if $t \ge r_i$.
- Arrival time a_i^A: the time when p_i arrives at R and get eligible to the acknowledgment algorithm A in R. The superscript A indicates that the arrival time is influenced by the action of A as explained below.

S can send p_i at r_i unless kept waiting by the sliding window. On the other hand, the sliding window permits S to send p_i only after A acknowledges p_{i-W} if $i > W$, where W is a constant integer. Let $\text{ack}^A(p)$ be the time when A acknowledges a packet p. In DTCPSW, by assuming that the propagation delay between S and R equals 0, a_i^A is described by (1). Here, when $i \le 0$, $\text{ack}^A(p_i)$ is defined to be $-\infty$. Note that $a_i^A = r_i$ for any i in the original DTCP.

$$a_i^A = \max\{r_i, \text{ack}^A(p_{i-W})\}. \tag{1}$$

If A acknowledges p_{i-W} with the l-th acknowledgement at time t_l, (1) may be written as $a_i^A = \max\{r_i, t_l\}$. At t_l, immediately after A's l-th acknowledgement, a group of packets postponed by the sliding window are passed to A. We allow A to acknowledge them instantly by the $(l+1)$-th acknowledgement. In this case, $t_l = t_{l+1}$. Thus, A can make multiple acknowledgements at a given time. At most W packets are eligible to A simultaneously because of the sliding window. Hence, for kW packets that has the same ready time, A may acknowledge k times at a given time.

Importantly, whereas the ready time sequence $(r_1, r_2, \cdots, r_n)$ is the inherent input that has nothing to do with A, A affects the arrival time sequence $(a_1^A, a_2^A, \cdots, a_n^A)$. Thus, we denote the ready time sequence $(r_1, r_2, \cdots, r_n)$ by σ with equating it with the input packet sequence $(p_1, p_2, \cdots, p_n)$. A encounters the arrival time sequence only and serves it without knowing σ. In the subsequence the name of the acknowledgment algorithm is omitted from variables expressing arrival times, when clear from the context. In DTCPSW, the latency of a packet p is naturally defined as the length of the time period between its ready time and $\text{ack}^A(p)$. Again, A divides σ into m subsequences $\sigma_1, \sigma_2, \cdots \sigma_m$ whose ends correspond to acknowledgments. The purpose of DTCPSW is to minimize the cost function f_{max} shown in (2).

$$f_{max} = m + \sum_{j=1}^{m} \max_{p_i \in \sigma_j} |t_j - r_i|. \tag{2}$$

A cannot know σ even after it finishes the processing of the arrival time sequence, as long as A does not know W exactly. When $W = 1$, a trivial on-line algorithm that acknowledges every packet instantly becomes the optimal for DTCPSW. Hence, we assume $W \geq 2$ in DTCPSW.

DTCPSW opens up a new vista on the research of competitive analysis, because it models the situation in which an on-line algorithm unconsciously changes the original input sequence which is the ready time sequence in our case. It is not rare that an on-line algorithm does not figure out the original input sequence after changing it involuntarily. For example, consider a scenario in which there exist a couple of mice in some house and the inhabitant sees one of them by accident. If he chooses to get rid of the mouse on the spot, the number of mice cannot increase any more. On the contrary, if he lets the mouse escape, he will meet a lot of mice (i.e., the changed input sequence) in future, but he will never become aware that there were only two mice at the beginning.

3 Optimal Off-Line Algorithm

This section presents the optimal off-line algorithm OPT for DTCPSW. OPT knows σ and W in advance. We start with the necessary condition of OPT.

Lemma 1. *Let B be an off-line algorithm. If S has ever defer sending a packet to R because of the sliding window in B's running, B is not the optimal.*

Proof. Let p_i be the first packet which S defers sending to R. Namely, $r_k = a_k$ for $1 \leq k \leq i-1$ and $r_i < a_i$. p_i becomes eligible to R at a_i just after some acknowledgement by B. Consider the subsequence of σ that ends with this acknowledgement. Let this subsequence be σ_j. Since $r_{i-1} \leq r_i$, we have $a_{i-1} \leq r_i < a_i$. Therefore, if B acknowledged at a_{i-1} instead of a_i, the latency cost for σ_j would have been reduced by $a_i - a_{i-1}$. Since this modification does not influence the latency for other subsequences, B cannot be the optimal. □

From Lemma 1, OPT makes an acknowledgement, whenever the number of unacknowledged eligible packets increases to W. Therefore, OPT acknowledges at least once for every W packets. More importantly, $\forall i, a_i = r_i$ for OPT.

When an acknowledgement algorithm A guarantees $a_i = r_i$ for any i, we can proceed the analysis of A by dividing the total cost $C_A(\sigma)$ into $C_A(p_i)$ that is the cost incurred for each packet p_i in the next manner. Dooly et al. [1] exploited the similar technique for DTCP.

- When A acknowledges at a_i, $C_A(p_i) = 1$ which corresponds to the cost for a single acknowledgement.

for $1 \le k \le q$

$$\mathrm{OPTCOST}[k] = \begin{cases} r'_k - r'_1 + 1. & \text{if } k \le W \\ \min_{1 \le l \le W} \mathrm{OPTCOST}[k-l] + r'_k - r'_{k-l+1} + 1. & \text{if } k > W \end{cases}$$

$$\mathrm{PREVACK}[k] = \begin{cases} 0 & \text{if } k \le W \\ \mathrm{argmin}_l\ \mathrm{OPTCOST}[k-l] + r'_k - r'_{k-l+1} + 1. & \text{if } k > W \end{cases}$$

Fig. 2. Update of OPTCOST and PREVACK

- Suppose p_i is not the final packet of σ. When A does not acknowledge at a_i, $C_A(p_i) = r_{i+1} - r_i$ which corresponds to the latency cost.

Note that any algorithm that acknowledges at a halfway time between a_i and a_{i+1} cannot be optimal. It is easy to verify that $C_A(\sigma) = \sum_{i=1}^{n} C_A(p_i)$. Similarly, for a subsequence τ of σ, we define $C_A(\tau)$ as $\sum_{p_i \in \tau} C_A(p_i)$.

Lemma 2. *If* $r_{i+1} - r_i > 1$. *OPT acknowledges at* r_i.

Proof. Recall that $\forall i, a_i = r_i$. If OPT does not acknowledge at r_i, $C_{OPT}(p_i) = r_{i+1} - r_i$. Else if OPT acknowledges at a_i, $C_{OPT}(p_i) = 1$. Thus, if $r_{i+1} - r_i > 1$, OPT must acknowledge at r_i to achieve the optimality. □

Algorithm OPT: OPT first scans $\sigma = (r_1, r_2, r_3, \cdots r_n)$ from its head. When $r_{i+1} - r_i > 1$, OPT puts an acknowledgement at r_i. As the result, σ is cut into subsequences. The distance between any two adjacent ready times is at most 1 in each subsequence.

Consider one of these subsequences, say $\sigma' = (r'_1, r'_2, \cdots, r'_q)$. OPT computes where to put acknowledgements in serving σ' with dynamic programming. OPT prepares two arrays OPTCOST and PREVACK. OPTCOST$[k]$ ($1 \le k \le q$) stores the minimum cost to serve the prefix of σ', i.e., $(r'_1, r'_2, \cdots, r'_k)$. PREVACK$[k]$ remembers the location of the second to last acknowledgment in the optimal solution for $(r'_1, r'_2, \cdots, r'_k)$. OPTCOST and PREVACK are updated according to the procedure in Fig. 2. When $k \le W$, the optimal solution for the prefix of σ' has only to acknowledge once at the end. When $k > W$, it must decide where to put the second-to-last acknowledgement. In the end, OPTCOST$[q]$ holds the cost for processing σ' optimally. The times for acknowledgements are acquired by tracing back the array PREVACK.

Since it takes an $O(W)$ time to perform the min operation which is invoked for each packet, the computational complexity of OPT becomes $O(Wn)$.

4 Known Window Size

This section studies the case when on-line algorithms know W beforehand. Notably our algorithm `WAITSW(1)` extended from `WAIT(1)` accomplishes the same competitive ratio in DTCPSW as the competitive ratio of `WAIT(1)` in DTCP.

Algorithm WAITSW(1): At a_i $(i \geq 1)$, the timer is set to $a_i + 1$. In case $a_{i+1} > a_i + 1$, the timer expires and `WAITSW(1)` acknowledges at $a_i + 1$. Otherwise the timer is updated at a_{i+1} to $a_{i+1} + 1$. In addition, instantly the number of unacknowledged eligible packets is increases to W, they are immediately acknowledged.

Since the sliding window never obstructs S from sending packets, $a_i = r_i$ for any i when `WAITSW(1)` runs.

Theorem 1. `WAITSW(1)` *is 2-competitive.*

Proof. In both of OPT and `WAITSW(1)`, we have $a_i = r_i$ for any i. Furthermore, both algorithms make an acknowledgement before a_{i+1} when $a_{i+1} > a_i + 1$. Hence, we may proceed the analysis by decomposing σ into subsequences and treating them separately. In each subsequence, any two adjacent ready times are distant by less than 1.

Hence, we only consider ready time sequences $\sigma = (r_1, r_2, \cdots, r_n)$ for which $r_{i+1} \leq r_i + 1$ for any $i \leq n - 1$ hereafter. Since $a_i = r_i$ for any i, $a_{i+1} \leq a_i + 1$ for $i \leq n - 1$. Therefore, `WAITSW(1)` acknowledges when W unacknowledged eligible packets are accumulated at R except the last acknowledgement. Suppose that `WAITSW(1)` divides σ into subsequences $\sigma_1, \sigma_2, \cdots, \sigma_m$. Given a subsequence $\sigma_j = (a_{i+1}, a_{i+2}, \ldots a_{i+l})$, we need to consider two cases.
Case 1: Suppose that $j = m$. In this case, `WAITSW(1)` acknowledges at $a_n + 1$ if $|\sigma_m| < W$ and at a_n if $|\sigma_m| = W$. Thus, we have $C_{\texttt{WAITSW(1)}}(\sigma_m) \leq 1 + (a_n + 1 - a_{i+1}) = a_n - a_{i+1} + 2$. On the other hand, $C_{OPT}(\sigma_m) = C_{OPT}(\sigma_m \backslash p_n) + 1 \geq \sum_{k=i+1}^{n-1} \min\{1, a_{k+1} - a_k\} + 1 = a_n - a_{i+1} + 1$ because $a_{k+1} - a_k \leq 1$. Thus, $\frac{C_{\texttt{WAITSW(1)}}(\sigma_m)}{C_{OPT}(\sigma_m)} \leq \frac{a_n - a_{i+1} + 2}{a_n - a_{i+1} + 1} \leq \frac{2}{1} = 2$.
Case 2: Suppose that $j < m$. In this case, σ_j contains just W packets so that $l = W$. Thus $C_{\texttt{WAITSW(1)}}(\sigma_j) = a_{i+W} - a_{i+1} + 1$ because `WAITSW(1)` acknowledges at a_{i+W}. Since σ_j consists of W packets, $C_{OPT}(\sigma_j) \geq 1$. It also holds that $C_{OPT}(\sigma_j) \geq \sum_{k=i+1}^{i+W} \min\{1, a_{k+1} - a_k\} \geq a_{i+W} - a_{i+1}$. Thus, $C_{\texttt{WAITSW(1)}}(\sigma_j) = 2 * \frac{a_{i+W} - a_{i+1} + 1}{2} \leq 2 \max\{a_{i+W} - a_{i+1}, 1\} \leq 2C_{OPT}(\sigma_j)$. □

The proof of the lower bound for DTCP in [1] utilizes an arrival time sequence such that at most two packet arrivals appear in a single subsequence in the runnings of on-line algorithms and OPT both. Hence, this lower bound is also valid for DTCPSW when $W \geq 2$. The next theorem is obtained from Theorem 23 in [1]. From this theorem, `WAITSW(1)` turns out to be the optimal.

Theorem 2. *Let A be any deterministic on-line algorithm for DTCPSW that knows W. Then, there exists a ready time sequence σ s.t. $C_A(\sigma) \geq 2C_{OPT}(\sigma) - \epsilon$, where ϵ can be made arbitrarily small relatively to $C_{OPT}(\sigma)$.*

5 Unknown Window Size

Throughout this section, the window size is unknown to on-line algorithms. We focus on a class of on-line algorithms `WAIT(`α`)` that generalizes `WAIT(1)`.

Algorithm WAIT(α)**:** Let α is an arbitrary positive real value. At a_i ($i \geq 1$), the timer is set to $a_i + \alpha$. In case $a_{i+1} > a_i + \alpha$, WAIT(α) acknowledges at $a_i + \alpha$. Otherwise the timer is updated at a_{i+1} to $a_{i+1} + \alpha$.

We show that the competitive ratio of WAIT(α) depends on the peak packet rate from S whose formal definition is given in Definition 1.

Definition 1 (peak packet rate). *For a TCP connection, if at most T ready times lie in the time interval $[t, t+1)$ for any time t, the peak packet rate from the TCP sender is said to be T.*

The time interval in this definition is half-open. This admits at most $(t_2-t_1)T+T$ ready times to appear in the close time interval $[t_1, t_2]$ for t_1 and t_2 such that $t_2 - t_1 \geq 1$, Throughout this section, we assume that the peak packet rate from the TCP sender is T.

5.1 Lower Bound

The lower bound on the competitiveness of WAIT(α) depends on T and W.

Theorem 3. *For $\forall\alpha$, WAIT(α) is worse than $\frac{T}{W+\lfloor\frac{T}{W}\rfloor-1}$-competitive if $T > W$.*

Proof. Let $k = \lfloor\frac{T}{W}\rfloor$. Since $T > W$, $k \geq 1$. First consider a ready time sequence σ' such that kW ready times appear every unit time until time $I-1$ where I is some integer. Namely, $r_{(i-1)kW+1} = r_{(i-1)kW+2} = \cdots r_{ikW} = i-1$ for $1 \leq i \leq I$. OPT acknowledges these kW packets immediately by acknowledging k times at each time t for $0 \leq t \leq I-1$. Since OPT incurs no delay, $C_{OPT}(\sigma') = kI$.

WAIT(α) waits for α for each clump of W packets before acknowledging them. For the first kW packets whose ready times are 0, $p_{(j-1)W+1}, p_{(j-1)W+2}, \cdots p_{jW}$ for $1 \leq j \leq k$ are acknowledged at time $j\alpha$. Thus, the whole latency costs for the first kW packets becomes $\sum_{i=1}^{k}(j\alpha - 0) = \frac{k(k+1)}{2}\alpha$.

The next W packets become eligible at $a_{kW+1} = \max\{r_{kW+1}, a_{kW-W+1}\} = \max\{1, k\alpha\}$. Therefore, they are acknowledged at time $\max\{1, k\alpha\} + \alpha$. In this way, the whole latency costs for packets from p_{kW+1} to p_{2kW} whose ready times are 1 becomes $\sum_{j=1}^{k}(\max\{1, k\alpha\} + j\alpha - 1) = k\max\{0, k\alpha - 1\} + \frac{k(k+1)}{2}\alpha$.

In general, the whole latency costs for the kW packets from $p_{(i-1)kW+1}$ to p_{ikW} becomes

$$\sum_{j=1}^{k}(\max\{0, (i-1)(k\alpha-1)\} + j\alpha) = k\max\{0, (i-1)(k\alpha-1)\} + \frac{k(k+1)}{2}\alpha. \quad (3)$$

By summing up the right-hand term of (3) for $1 \leq i \leq I$ and adding the acknowledgement cost of kI,

$$C_{\mathtt{WAIT}(\alpha)}(\sigma') = kI + k\max\{0, \frac{I(I-1)}{2}(k\alpha - 1)\} + \frac{Ik(k+1)}{2}\alpha. \quad (4)$$

The competitive ratio $C_{\texttt{WAIT}(\alpha)}(\sigma')/C_{OPT}(\sigma')$ is shown in (5).

$$1 + \max\{0, \frac{(I-1)(k\alpha-1)}{2}\} + \frac{(k+1)}{2}\alpha. \quad (5)$$

Since I may be chosen to be arbitrary large, it has to hold that $\alpha \le \frac{1}{k}$ in order for WAIT(α) to be competitive.

Next, consider another ready time sequence σ" whose length equals W such that $r_1 = a_1 = 0$ and r_{i+1} is the time immediately after WAIT(α) acknowledges the i-th packet at $a_i + \alpha$. Therefore, $r_i = (i-1)\alpha$. Since $W < T$, σ" satisfies the condition that the peak packet rate is T. OPT serves σ" by acknowledging only once at r_W. Since the latency of OPT equals $(W-1)\alpha$, $C_{OPT}(\sigma") = 1+(W-1)\alpha$. WAIT($\alpha$) must pay a latency cost of α and one acknowledgement cost for each packet. Hence, $C_{\texttt{WAIT}(\sigma)}(\sigma") = W(1+\alpha)$. The cost ratio between WAIT(α) and OPT becomes $\frac{W(1+\alpha)}{1+(W-1)\alpha}$. Since $\frac{W(1+\alpha)}{1+(W-1)\alpha}$ decreases with respect to α for $W \ge 2$ and we have $\alpha \le \frac{1}{k}$, $\frac{W(1+\alpha)}{1+(W-1)\alpha}$ becomes the minimum when $\alpha = \frac{1}{k}$ in the next way: $\frac{W(k+1)}{W+k-1} = \frac{W(\lfloor \frac{T}{W} \rfloor+1)}{W+\lfloor \frac{T}{W} \rfloor-1} \ge \frac{T}{W+\lfloor \frac{T}{W} \rfloor-1}$. □

Intuitively, this proof is interpreted as follows. Because the arrival time a_i may fall behind the ready time r_i due to the sliding window when W is unknown, an on-line acknowledgement algorithm cannot wait for so long so as to be competitive. However this impatience becomes harmful when S is not kept waiting by the sliding window. The peak packet rate T determines the maximum possible value of $a_i - r_i$, that is the maximum extent that the ready time sequence is transformed. Thus, the competitive ratio depends on T in DTCPSW.

When $T \le W$, the lower bound of competitiveness becomes 2 for $T \ge 2$.

5.2 Upper Bound

Theorem 4. WAIT($\frac{1}{\lceil \frac{T}{W} \rceil+1}$) *is* $(\lceil \frac{T}{W} \rceil + 2)$*-competitive.*

This theorem states that one instance from WAIT(α) attains the nearly tight competitive ratio, compared with the lower bound in Theorem 3.

Before the proof, we mention two crucial properties of WAIT($\frac{1}{\lceil \frac{T}{W} \rceil+1}$).

Lemma 3. *When the peak packet rate is T, the number of unacknowledged ready packets is always at most $T + W$.*

An unacknowledged ready packet may be either eligible or not eligible to the acknowledgment algorithm. In general, if the number of unacknowledged ready packets is greater than W, only the first W of them are eligible.

Proof. The proof utilizes contradiction. Assume that the number of unacknowledged ready packets reaches $T + W + 1$. Let t_2 be the first time when this event takes place and t_1 be the last time before t_2 such that the number of unacknowledged ready packets increased from $W - 1$ to W. Thus, $T + W + 1 - W = T + 1$

packets become ready in the time interval $[t_1,t_2]$. Since the peak packet rate is T, $t_2 - t_1 \geq 1$.

Meanwhile, W unacknowledged packets are always accumulated at R in $[t_1, t_2]$. Therefore, WAIT($\frac{1}{\lceil \frac{T}{W} \rceil+1}$) decreases the number of unacknowledged ready packets by W every $\frac{1}{\lceil \frac{T}{W} \rceil+1}$ time with an acknowledgment, starting from the time $t_1 + \frac{1}{\lceil \frac{T}{W} \rceil+1}$. The total number of unacknowledged ready packets that are acknowledged in $[t_1,t_2]$ grows

$$\lfloor (t_2 - t_1)(\lceil \frac{T}{W} \rceil + 1) \rfloor W > ((t_2 - t_1)(\lceil \frac{T}{W} \rceil + 1) - 1)W \geq (t_2 - t_1)\lceil \frac{T}{W} \rceil W$$
$$\geq (t_2 - t_1)T.$$

After all, $(t_2-t_1)T+(T+W+1-W) = (t_2-t_1)T+T+1$ packets become ready in $[t_1, t_2]$, which contradicts with the fact that the peak packet rate is T. □

Lemma 4. *If* $r_{i+1} - r_i > 1$, $a_{i+1} = r_{i+1}$.

Proof. At r_i, there are at most $T + W$ unacknowledged ready packets from Lemma 3. Since WAIT($\frac{1}{\lceil \frac{T}{W} \rceil+1}$) decreases these $T + W$ ready unacknowledged packets by W every $\frac{1}{\lceil \frac{T}{W} \rceil+1}$ time, all of the $T + W$ packets are acknowledged before $r_i+(\lceil \frac{T+W}{W} \rceil \times \frac{1}{\lceil \frac{T}{W} \rceil+1}) = r_i+(\lceil \frac{T}{W} \rceil+1)\frac{1}{\lceil \frac{T}{W} \rceil+1} = r_i+1 < r_{i+1}$. Hence p_{i+1} is sent at r_{i+1} without impeded by the sliding window. Thus, $a_{i+1} = r_{i+1}$. □

From Lemma 2, it also holds that $a_{i+1} = r_{i+1}$ if $r_{i+1} - r_i > 1$ for OPT. Hence, we can proceed the analysis by decomposing σ into subsequences in which any two adjacent ready times are at most one unit time apart. Thus, consider $\sigma = (r_1, r_2, \cdots, r_n)$ s.t. $r_{i+1} - r_i \leq 1$ for $1 \leq i \leq n-1$ in the subsequence.

Proof of Theorem 4: Suppose that WAIT($\frac{1}{\lceil \frac{T}{W} \rceil+1}$) acknowledges a subsequence of σ denoted by $\tau = (r_{i+1}, r_{i+2}, \cdots, r_{i+k})$ with a single acknowledgement. Because of the sliding window, $k \leq W$. The cost incurred by WAIT($\frac{1}{\lceil \frac{T}{W} \rceil+1}$) to process τ is obtained as (6), since WAIT($\frac{1}{\lceil \frac{T}{W} \rceil+1}$) acknowledges at time $a_{i+k} + \frac{1}{\lceil \frac{T}{W} \rceil+1}$.

$$C_{\texttt{WAIT}(\frac{1}{\lceil \frac{T}{W} \rceil+1})}(\tau) = a_{i+k} + \frac{1}{\lceil \frac{T}{W} \rceil + 1} - r_{i+1} + 1. \tag{6}$$

Since $k \leq W$, OPT acknowledges at most twice in serving τ. Suppose p_{i+k} is not p_n, the last packet of σ. Let $d_1 = \max_{1 \leq j \leq k}\{r_{i+j+1} - r_{i+j}\}$. Similarly, let d_2 be the length of the second maximum interval between two adjacent ready times in τ. Obviously, $d_1 \leq 1$ and $d_2 \leq 1$. If OPT acknowledges two times in serving τ, $C_{OPT}(\tau) \geq r_{i+k+1} - r_{i+1} - d_1 - d_2 + 2 \geq r_{i+k+1} - r_{i+1} - d_1 + 1$. If OPT acknowledges only once, $C_{OPT}(\tau) \geq r_{i+k+1} - r_{i+1} - d_1 + 1 \geq r_{i+k+1} - r_{i+1}$. If OPT does not acknowledge at all, $C_{OPT}(\tau) \geq r_{i+k+1} - r_{i+1}$. We need to consider two cases.

(Case I) $a_{i+k} = r_{i+k}$: There are two cases depending on whether $k = W$.

If $k = W$, OPT acknowledges at least once. Therefore,

$$\frac{C_{\mathtt{WAIT}(\frac{1}{\lceil \frac{T}{W} \rceil +1})}(\tau)}{C_{OPT}(\tau)} \leq \frac{r_{i+k} + \frac{1}{\lceil \frac{T}{W} \rceil +1} - r_{i+1} + 1}{r_{i+k+1} - r_{i+1} - d_1 + 1} \leq 1 + \frac{\frac{1}{\lceil \frac{T}{W} \rceil +1} + d_1}{r_{i+k+1} - r_{i+1} - d_1 + 1}$$
$$\leq 1 + (\frac{1}{\lceil \frac{T}{W} \rceil + 1} + d_1) \leq 2 + \frac{1}{\lceil \frac{T}{W} \rceil + 1}. \quad (7)$$

Even if $p_{i+k} = p_n$, (7) is obtained because $C_{OPT}(\tau) \geq r_{i+k} - r_{i+1} - d_1 + 1$.

If $k < W$, since OPT may not acknowledge at all, $C_{OPT}(\tau) \geq r_{i+k+1} - r_{i+1}$. Because $a_{i+k} = r_{i+k}$ and $\mathtt{WAIT}(\frac{1}{\lceil \frac{T}{W} \rceil +1})$ acknowledges strictly less than W packets, we have $r_{i+k+1} - r_{i+k} \geq \frac{1}{\lceil \frac{T}{W} \rceil +1}$. Hence, $C_{OPT}(\tau) \geq r_{i+k} - r_{i+1} + \frac{1}{\lceil \frac{T}{W} \rceil +1}$. Thus,

$$\frac{C_{\mathtt{WAIT}(\frac{1}{\lceil \frac{T}{W} \rceil +1})}(\tau)}{C_{OPT}(\tau)} \leq \frac{r_{i+k} + \frac{1}{\lceil \frac{T}{W} \rceil +1} - r_{i+1} + 1}{r_{i+k} - r_{i+1} + \frac{1}{\lceil \frac{T}{W} \rceil +1}} \leq \frac{\frac{1}{\lceil \frac{T}{W} \rceil +1} + 1}{\frac{1}{\lceil \frac{T}{W} \rceil +1}} = \lceil \frac{T}{W} \rceil + 2. \quad (8)$$

In case $p_{i+k} = p_n$, $C_{OPT}(\tau) \geq r_{i+k} - r_{i+1} + 1$ and (8) still holds.

(Case II) $a_{i+k} > r_{i+k}$: We first explain how this case occurs and mention that multiple subsequences categorized into (Case II) may appear in succession.

The first subsequence $\tau_1 = (p_{i+1}, p_{i+2}, \cdots p_{i+k_1})$ categorized into (Case II) is constructed as follows: p_i is the W-th packets in the subsequence previous to τ_1. As the previous subsequence is categorized into (Case I), $a_i = r_i$. Because of the sliding window, the packets next to p_i are not eligible until the previous subsequence is acknowledged at $a_i + \frac{1}{\lceil \frac{T}{W} \rceil +1}$. Here, a_{i+k_1} differs from r_{i+k_1}, if and only if $r_{i+k_1} < a_i + \frac{1}{\lceil \frac{T}{W} \rceil}$. If $k_1 < W$, the subsequence next to τ_1 does not belong to (Case II). However, if $k_1 = W$, the second subsequence τ_2 classified into (Case II) emerges if and only if $r_{i+W+k_2} < (a_i + \frac{1}{\lceil \frac{T}{W} \rceil +1}) + \frac{1}{\lceil \frac{T}{W} \rceil +1}$ where k_2 is the length of τ_2. Note that $(a_i + \frac{1}{\lceil \frac{T}{W} \rceil +1}) + \frac{1}{\lceil \frac{T}{W} \rceil +1}$ presents the time when the packets in τ_1 are acknowledged. If $k_2 = W$, the third subsequence τ_3 classified into (Case II) may appear after τ_2 in the same way. Figure 3 illustrates an example of (Case II) where $\frac{1}{\lceil \frac{T}{W} \rceil +1}$ is denoted by α. Here, τ_4 consists of less than W packets.

Let $\tau_1, \tau_2, \cdots, \tau_X$ be the series of subsequences categorized into (Case II) ($X \geq 1$). All of $\tau_1, \tau_2, \cdots, \tau_{X-1}$ contain exactly W packets and the last τ_X consists of $k_X (\leq W)$ packets. As τ_x for $1 \leq x \leq X-1$ is acknowledged at $a_i + (x+1)\frac{1}{\lceil \frac{T}{W} \rceil +1} = r_i + (x+1)\frac{1}{\lceil \frac{T}{W} \rceil +1}$ and the ready time of the first packet in τ_x is greater than r_i, the latency cost for τ_x is less than $\frac{x+1}{\lceil \frac{T}{W} \rceil +1}$. Adding the cost for a single acknowledgement, $C_{\mathtt{WAIT}(\frac{1}{\lceil \frac{T}{W} \rceil +1})}(\tau_x) \leq 1 + \frac{x+1}{\lceil \frac{T}{W} \rceil +1}$ for $1 \leq x \leq X - 1$. For τ_X, $C_{\mathtt{WAIT}(\frac{1}{\lceil \frac{T}{W} \rceil +1})}(\tau_X) \leq 1 + r_i + \frac{X+1}{\lceil \frac{T}{W} \rceil +1} - r_{i+(X-1)W+1}$, because $p_{i+(X-1)W+1}$ is the first packet in τ_X. In addition, $X \leq \lceil \frac{T}{W} \rceil$. Otherwise, since $X - 1 \geq \lceil \frac{T}{W} \rceil$, $\lceil \frac{T}{W} \rceil W + 1 \geq T + 1$ ready times from r_i to $r_{i+\lceil \frac{T}{W} \rceil W}$ lie in $[r_i, r_i + \frac{\lceil \frac{T}{W} \rceil}{\lceil \frac{T}{W} \rceil +1}]$, contradicting with the fact that the peak rate is T.

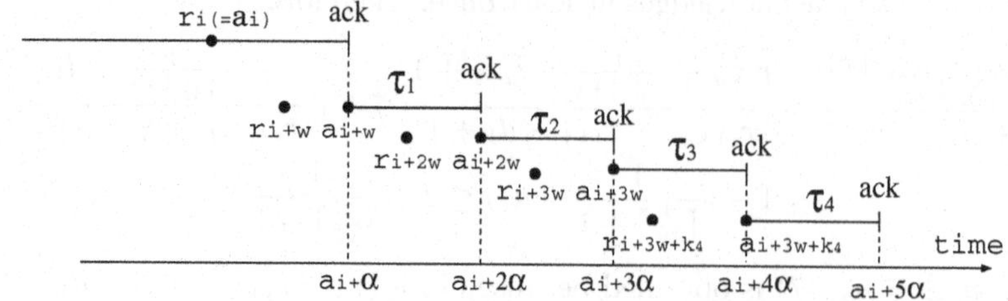

Fig. 3. Example of (Case II)

For $1 \leq x \leq X-1$, $C_{OPT}(\tau_x) \geq 1$ since τ_x contains W packets. Furthermore, for $x = X$, $C_{OPT}(\tau_X) \geq 1$ if OPT acknowledges in serving τ_X. In this case

$$\frac{C_{\mathtt{WAIT}(\frac{1}{\lceil\frac{T}{W}\rceil+1})}(\tau_x)}{C_{OPT}(\tau_x)} \leq C_{\mathtt{WAIT}(\frac{1}{\lceil\frac{T}{W}\rceil+1})}(\tau_x) \leq 1 + \frac{X+1}{\lceil\frac{T}{W}\rceil+1} \leq 2. \tag{9}$$

Unless OPT acknowledges in serving τ_X, $C_{OPT}(\tau_X) \geq r_{i+(X-1)W+k_X+1} - r_{i+(X-1)W+1} \geq r_i + \frac{X+1}{\lceil\frac{T}{W}\rceil+1} - r_{i+(X-1)W+1}$, as $p_{i+(X-1)W+k_X}$ is acknowledged at $r_i + \frac{X+1}{\lceil\frac{T}{W}\rceil+1}$. Since $r_{i+(X-1)W+1} \leq a_{i+(X-1)W} + \frac{1}{\lceil\frac{T}{W}\rceil+1} = r_i + \frac{X}{\lceil\frac{T}{W}\rceil+1}$,

$$\frac{C_{\mathtt{WAIT}(\frac{1}{\lceil\frac{T}{W}\rceil+1})}(\tau_X)}{C_{OPT}(\tau_X)} \leq \frac{1 + r_i + \frac{X+1}{\lceil\frac{T}{W}\rceil+1} - r_{i+(X-1)W+1}}{r_i + \frac{X+1}{\lceil\frac{T}{W}\rceil+1} - r_{i+(X-1)W+1}} \leq \frac{1+\frac{1}{\lceil\frac{T}{W}\rceil}}{\frac{1}{\lceil\frac{T}{W}\rceil}} \leq \lceil\frac{T}{W}\rceil + 2. \tag{10}$$

Equations (7), (8), (9) and (10) altogether complete the proof. □

Acknowledgements

This work is supported by the Ministry of Education, Culture, Sports, Science and Technology, Grant-in-Aid for Young Scientists (B), 17700054, 2006.

References

1. Dooly, D.R., Goldman, S.A., Scott, S.D.: On-line analysis of the TCP acknowledgment delay problem. Journal of the ACM 48(2), 243–273 (2001)
2. Borodin, A., El-Yaniv, R.: Online Computation and Competitive Analysis. Cambridge University Press, Cambridge (1998)
3. Karlin, A.R., Kenyon, C., Randall, D.: Dynamic TCP acknowledgment and other stories about e/(e-1). Algorithmica 36(3), 209–224 (2003)
4. Noga, J., Seiden, S., Woeginger, G.J.: A faster off-line algorithm for the TCP acknowledgement problem. Information Processing Letters 81, 71–73 (2002)
5. Albers, S., Bals, H.: Dynamic TCP acknowledgment: Penalizing long delays. In: Proceedings of 14th ACM-SIAM Symposium on Discrete Algorithms, pp. 47–55 (2003)
6. Frederiksen, J.S., Larsen, K.S., Noga, J., Uthaisombut, P.: Dynamic TCP acknowledgement in the LogP model. Journal of Algorithms 48, 407–428 (2003)

Author Index

Lecture Notes in Computer Science

For information about Vols. 1–4536

please contact your bookseller or Springer

Vol. 4660: S. Džeroski, J. Todoroski (Eds.), Computational Discovery of Scientific Knowledge. X, 327 pages. 2007. (Sublibrary LNAI).

Vol. 4651: F. Azevedo, P. Barahona, F. Fages, F. Rossi (Eds.), Recent Advances in Constraints. VIII, 185 pages. 2007. (Sublibrary LNAI).

Vol. 4647: R. Martin, M. Sabin, J. Winkler (Eds.), Mathematics of Surfaces XII. IX, 509 pages. 2007.

Vol. 4632: R. Alhajj, H. Gao, X. Li, J. Li, O.R. Zaïane (Eds.), Advanced Data Mining and Applications. XV, 634 pages. 2007. (Sublibrary LNAI).

Vol. 4628: L.N. de Castro, F.J. Von Zuben, H. Knidel (Eds.), Artificial Immune Systems. XII, 438 pages. 2007.

Vol. 4619: F. Dehne, J.-R. Sack, N. Zeh (Eds.), Algorithms and Data Structures. XVI, 662 pages. 2007.

Vol. 4617: V. Torra, Y. Narukawa, Y. Yoshida (Eds.), Modeling Decisions for Artificial Intelligence. XII, 502 pages. 2007. (Sublibrary LNAI).

Vol. 4616: A. Dress, Y. Xu, B. Zhu (Eds.), Combinatorial Optimization and Application. XI, 390 pages. 2007.

Vol. 4613: F.P. Preparata, Q. Fang (Eds.), Frontiers in Algorithmics. XI, 348 pages. 2007.

Vol. 4612: I. Miguel, W. Ruml (Eds.), Abstraction, Reformulation, and Approximation. XI, 418 pages. 2007. (Sublibrary LNAI).

Vol. 4611: J. Indulska, J. Ma, L.T. Yang, T. Ungerer, J. Cao (Eds.), Ubiquitous Intelligence and Computing. XXIII, 1257 pages. 2007.

Vol. 4610: B. Xiao, L.T. Yang, J. Ma, C. Muller-Schloer, Y. Hua (Eds.), Autonomic and Trusted Computing. XVIII, 571 pages. 2007.

Vol. 4609: E. Ernst (Ed.), ECOOP 2007 — Object-Oriented Programming. XIII, 625 pages. 2007.

Vol. 4608: H.W. Schmidt, I. Crnkovic, G.T. Heineman, J.A. Stafford (Eds.), Component-Based Software Engineering. XII, 283 pages. 2007.

Vol. 4607: L. Baresi, P. Fraternali, G.-J. Houben (Eds.), Web Engineering. XVI, 576 pages. 2007.

Vol. 4606: A. Pras, M. van Sinderen (Eds.), Dependable and Adaptable Networks and Services. XIV, 149 pages. 2007.

Vol. 4605: D. Papadias, D. Zhang, G. Kollios (Eds.), Advances in Spatial and Temporal Databases. X, 479 pages. 2007.

Vol. 4604: U. Priss, S. Polovina, R. Hill (Eds.), Conceptual Structures: Knowledge Architectures for Smart Applications. XII, 514 pages. 2007. (Sublibrary LNAI).

Vol. 4603: F. Pfenning (Ed.), Automated Deduction – CADE-21. XII, 522 pages. 2007. (Sublibrary LNAI).

Vol. 4602: S. Barker, G.-J. Ahn (Eds.), Data and Applications Security XXI. X, 291 pages. 2007.

Vol. 4600: H. Comon-Lundh, C. Kirchner, H. Kirchner (Eds.), Rewriting, Computation and Proof. XVI, 273 pages. 2007.

Vol. 4599: S. Vassiliadis, M. Berekovic, T.D. Hämäläinen (Eds.), Embedded Computer Systems: Architectures, Modeling, and Simulation. XVIII, 466 pages. 2007.

Vol. 4598: G. Lin (Ed.), Computing and Combinatorics. XII, 570 pages. 2007.

Vol. 4597: P. Perner (Ed.), Advances in Data Mining. XI, 353 pages. 2007. (Sublibrary LNAI).

Vol. 4596: L. Arge, C. Cachin, T. Jurdziński, A. Tarlecki (Eds.), Automata, Languages and Programming. XVII, 953 pages. 2007.

Vol. 4595: D. Bošnački, S. Edelkamp (Eds.), Model Checking Software. X, 285 pages. 2007.

Vol. 4594: R. Bellazzi, A. Abu-Hanna, J. Hunter (Eds.), Artificial Intelligence in Medicine. XVI, 509 pages. 2007. (Sublibrary LNAI).

Vol. 4592: Z. Kedad, N. Lammari, E. Métais, F. Meziane, Y. Rezgui (Eds.), Natural Language Processing and Information Systems. XIV, 442 pages. 2007.

Vol. 4591: J. Davies, J. Gibbons (Eds.), Integrated Formal Methods. IX, 660 pages. 2007.

Vol. 4590: W. Damm, H. Hermanns (Eds.), Computer Aided Verification. XV, 562 pages. 2007.

Vol. 4589: J. Münch, P. Abrahamsson (Eds.), Product-Focused Software Process Improvement. XII, 414 pages. 2007.

Vol. 4588: T. Harju, J. Karhumäki, A. Lepistö (Eds.), Developments in Language Theory. XI, 423 pages. 2007.

Vol. 4587: R. Cooper, J. Kennedy (Eds.), Data Management. XIII, 259 pages. 2007.

Vol. 4586: J. Pieprzyk, H. Ghodosi, E. Dawson (Eds.), Information Security and Privacy. XIV, 476 pages. 2007.

Vol. 4585: M. Kryszkiewicz, J.F. Peters, H. Rybinski, A. Skowron (Eds.), Rough Sets and Intelligent Systems Paradigms. XIX, 836 pages. 2007. (Sublibrary LNAI).

Vol. 4584: N. Karssemeijer, B. Lelieveldt (Eds.), Information Processing in Medical Imaging. XX, 777 pages. 2007.

Vol. 4583: S.R. Della Rocca (Ed.), Typed Lambda Calculi and Applications. X, 397 pages. 2007.

Vol. 4582: J. Lopez, P. Samarati, J.L. Ferrer (Eds.), Public Key Infrastructure. XI, 375 pages. 2007.

Vol. 4581: A. Petrenko, M. Veanes, J. Tretmans, W. Grieskamp (Eds.), Testing of Software and Communicating Systems. XII, 379 pages. 2007.

Vol. 4580: B. Ma, K. Zhang (Eds.), Combinatorial Pattern Matching. XII, 366 pages. 2007.

Vol. 4579: B. M. Hämmerli, R. Sommer (Eds.), Detection of Intrusions and Malware, and Vulnerability Assessment. X, 251 pages. 2007.

Vol. 4578: F. Masulli, S. Mitra, G. Pasi (Eds.), Applications of Fuzzy Sets Theory. XVIII, 693 pages. 2007. (Sublibrary LNAI).

Vol. 4577: N. Sebe, Y. Liu, Y.-t. Zhuang (Eds.), Multimedia Content Analysis and Mining. XIII, 513 pages. 2007.

Vol. 4576: D. Leivant, R. de Queiroz (Eds.), Logic, Language, Information and Computation. X, 363 pages. 2007.

Vol. 4575: T. Takagi, T. Okamoto, E. Okamoto, T. Okamoto (Eds.), Pairing-Based Cryptography – Pairing 2007. XI, 408 pages. 2007.

Vol. 4574: J. Derrick, J. Vain (Eds.), Formal Techniques for Networked and Distributed Systems – FORTE 2007. XI, 375 pages. 2007.

Vol. 4573: M. Kauers, M. Kerber, R. Miner, W. Windsteiger (Eds.), Towards Mechanized Mathematical Assistants. XIII, 407 pages. 2007. (Sublibrary LNAI).

Vol. 4572: F. Stajano, C. Meadows, S. Capkun, T. Moore (Eds.), Security and Privacy in Ad-hoc and Sensor Networks. X, 247 pages. 2007.

Vol. 4571: P. Perner (Ed.), Machine Learning and Data Mining in Pattern Recognition. XIV, 913 pages. 2007. (Sublibrary LNAI).

Vol. 4570: H.G. Okuno, M. Ali (Eds.), New Trends in Applied Artificial Intelligence. XXI, 1194 pages. 2007. (Sublibrary LNAI).

Vol. 4569: A. Butz, B. Fisher, A. Krüger, P. Olivier, S. Owada (Eds.), Smart Graphics. IX, 237 pages. 2007.

Vol. 4566: M.J. Dainoff (Ed.), Ergonomics and Health Aspects of Work with Computers. XVIII, 390 pages. 2007.

Vol. 4565: D.D. Schmorrow, L.M. Reeves (Eds.), Foundations of Augmented Cognition. XIX, 450 pages. 2007. (Sublibrary LNAI).

Vol. 4564: D. Schuler (Ed.), Online Communities and Social Computing. XVII, 520 pages. 2007.

Vol. 4563: R. Shumaker (Ed.), Virtual Reality. XXII, 762 pages. 2007.

Vol. 4562: D. Harris (Ed.), Engineering Psychology and Cognitive Ergonomics. XXIII, 879 pages. 2007. (Sublibrary LNAI).

Vol. 4561: V.G. Duffy (Ed.), Digital Human Modeling. XXIII, 1068 pages. 2007.

Vol. 4560: N. Aykin (Ed.), Usability and Internationalization, Part II. XVIII, 576 pages. 2007.

Vol. 4559: N. Aykin (Ed.), Usability and Internationalization, Part I. XVIII, 661 pages. 2007.

Vol. 4558: M.J. Smith, G. Salvendy (Eds.), Human Interface and the Management of Information, Part II. XXIII, 1162 pages. 2007.

Vol. 4557: M.J. Smith, G. Salvendy (Eds.), Human Interface and the Management of Information, Part I. XXII, 1030 pages. 2007.

Vol. 4556: C. Stephanidis (Ed.), Universal Access in Human-Computer Interaction, Part III. XXII, 1020 pages. 2007.

Vol. 4555: C. Stephanidis (Ed.), Universal Access in Human-Computer Interaction, Part II. XXII, 1066 pages. 2007.

Vol. 4554: C. Stephanidis (Ed.), Universal Acess in Human Computer Interaction, Part I. XXII, 1054 pages. 2007.

Vol. 4553: J.A. Jacko (Ed.), Human-Computer Interaction, Part IV. XXIV, 1225 pages. 2007.

Vol. 4552: J.A. Jacko (Ed.), Human-Computer Interaction, Part III. XXI, 1038 pages. 2007.

Vol. 4551: J.A. Jacko (Ed.), Human-Computer Interaction, Part II. XXIII, 1253 pages. 2007.

Vol. 4550: J.A. Jacko (Ed.), Human-Computer Interaction, Part I. XXIII, 1240 pages. 2007.

Vol. 4549: J. Aspnes, C. Scheideler, A. Arora, S. Madden (Eds.), Distributed Computing in Sensor Systems. XIII, 417 pages. 2007.

Vol. 4548: N. Olivetti (Ed.), Automated Reasoning with Analytic Tableaux and Related Methods. X, 245 pages. 2007. (Sublibrary LNAI).

Vol. 4547: C. Carlet, B. Sunar (Eds.), Arithmetic of Finite Fields. XI, 355 pages. 2007.

Vol. 4546: J. Kleijn, A. Yakovlev (Eds.), Petri Nets and Other Models of Concurrency – ICATPN 2007. XI, 515 pages. 2007.

Vol. 4545: H. Anai, K. Horimoto, T. Kutsia (Eds.), Algebraic Biology. XIII, 379 pages. 2007.

Vol. 4544: S. Cohen-Boulakia, V. Tannen (Eds.), Data Integration in the Life Sciences. XI, 282 pages. 2007. (Sublibrary LNBI).

Vol. 4543: A.K. Bandara, M. Burgess (Eds.), Inter-Domain Management. XII, 237 pages. 2007.

Vol. 4542: P. Sawyer, B. Paech, P. Heymans (Eds.), Requirements Engineering: Foundation for Software Quality. IX, 384 pages. 2007.

Vol. 4541: T. Okadome, T. Yamazaki, M. Makhtari (Eds.), Pervasive Computing for Quality of Life Enhancement. IX, 248 pages. 2007.

Vol. 4539: N.H. Bshouty, C. Gentile (Eds.), Learning Theory. XII, 634 pages. 2007. (Sublibrary LNAI).

Vol. 4538: F. Escolano, M. Vento (Eds.), Graph-Based Representations in Pattern Recognition. XII, 416 pages. 2007.

Vol. 4537: K.C.-C. Chang, W. Wang, L. Chen, C.A. Ellis, C.-H. Hsu, A.C. Tsoi, H. Wang (Eds.), Advances in Web and Network Technologies, and Information Management. XXIII, 707 pages. 2007.